D1652822

Handbook of solid-state lasers

© Woodhead Publishing Limited, 2013

Related titles:

Semiconductor lasers: Fundamentals and applications (ISBN 978-0-85709-121-5)

Lasers for medical applications: Diagnostics, therapy and surgery (ISBN 978-0-85709-237-3)

Laser spectroscopy for sensing: Fundamentals, techniques and applications (ISBN 978-0-85709-273-1)

Details of these books and a complete list of titles from Woodhead Publishing can be obtained by:

- visiting our web site at www.woodheadpublishing.com
- contacting Customer Services (e-mail: sales@woodheadpublishing.com; fax: +44 (0) 1223 832819; tel.: +44 (0) 1223 499140 ext. 130; address: Woodhead Publishing Limited, 80, High Street, Sawston, Cambridge CB22 3HJ, UK)
- in North America, contacting our US office (e-mail: usmarketing@woodheadpublishing.com; tel.: (215) 928 9112; address: Woodhead Publishing, 1518 Walnut Street, Suite 1100, Philadelphia, PA 19102-3406, USA

If you would like e-versions of our content, please visit our online platform: www.woodheadpublishingonline.com. Please recommend it to your librarian so that everyone in your institution can benefit from the wealth of content on the site.

We are always happy to receive suggestions for new books from potential editors. To enquire about contributing to our Electronic and Optical Materials series, please send your name, contact address and details of the topic/s you are interested in to laura.pugh@woodheadpublishing.com. We look forward to hearing from you.

The Woodhead team responsible for publishing this book:
Commissioning Editor: Laura Pugh
Publications Co-ordinator: Adam Davies
Project Editor: Cathryn Freear
Editorial and Production Manager: Mary Campbell
Production Editor: Richard Fairclough
Cover Designer: Terry Callanan

© Woodhead Publishing Limited, 2013

Woodhead Publishing Series in Electronic and Optical Materials: Number 35

Handbook of solid-state lasers

Materials, systems and applications

Edited by

B. Denker and E. Shklovsky

WP
WOODHEAD PUBLISHING

Oxford Cambridge Philadelphia New Delhi

© Woodhead Publishing Limited, 2013

Published by Woodhead Publishing Limited,
80 High Street, Sawston, Cambridge CB22 3HJ, UK
www.woodheadpublishing.com
www.woodheadpublishingonline.com

Woodhead Publishing, 1518 Walnut Street, Suite 1100, Philadelphia,
PA 19102-3406, USA

Woodhead Publishing India Private Limited, G-2, Vardaan House, 7/28 Ansari Road,
Daryaganj, New Delhi – 110002, India
www.woodheadpublishingindia.com

First published 2013, Woodhead Publishing Limited
© Woodhead Publishing Limited, 2013. Note: the publisher has made every effort to ensure that permission for copyright material has been obtained by authors wishing to use such material. The authors and the publisher will be glad to hear from any copyright holder it has not been possible to contact.
The authors have asserted their moral rights.

This book contains information obtained from authentic and highly regarded sources. Reprinted material is quoted with permission, and sources are indicated. Reasonable efforts have been made to publish reliable data and information, but the authors and the publisher cannot assume responsibility for the validity of all materials. Neither the authors nor the publisher, nor anyone else associated with this publication, shall be liable for any loss, damage or liability directly or indirectly caused or alleged to be caused by this book.

Neither this book nor any part may be reproduced or transmitted in any form or by any means, electronic or mechanical, including photocopying, microfilming and recording, or by any information storage or retrieval system, without permission in writing from Woodhead Publishing Limited.

The consent of Woodhead Publishing Limited does not extend to copying for general distribution, for promotion, for creating new works, or for resale. Specific permission must be obtained in writing from Woodhead Publishing Limited for such copying.

Trademark notice: Product or corporate names may be trademarks or registered trademarks, and are used only for identification and explanation, without intent to infringe.

British Library Cataloguing in Publication Data
A catalogue record for this book is available from the British Library.

Library of Congress Control Number: 2012955451

ISBN 978-0-85709-272-4 (print)
ISBN 978-0-85709-750-7 (online)
ISSN 2050-1501 Woodhead Publishing Series in Electronic and Optical Materials (print)
ISSN 2050-151X Woodhead Publishing Series in Electronic and Optical Materials (online)

The publisher's policy is to use permanent paper from mills that operate a sustainable forestry policy, and which has been manufactured from pulp which is processed using acid-free and elemental chlorine-free practices. Furthermore, the publisher ensures that the text paper and cover board used have met acceptable environmental accreditation standards.

Typeset by Replika Press Pvt Ltd, India
Printed and bound in the UK by MPG Books Group

© Woodhead Publishing Limited, 2013

Contents

Contributor contact details *xiii*
Woodhead Publishing Series in Electronic and Optical Materials *xix*
Foreword *xxiii*
Preface *xxvii*

Part I Solid-state laser materials 1

1 Oxide laser crystals doped with rare earth and transition metal ions 3
K. Petermann, University of Hamburg, Germany

1.1 Introduction 3
1.2 Laser-active ions 4
1.3 Host lattices 10
1.4 Laser medium geometry 12
1.5 Rare earth-doped sesquioxides 15
1.6 Mode-locked sesquioxide lasers 23
1.7 Future trends 23
1.8 References 24

2 Fluoride laser crystals 28
R. Moncorgé, A. Braud, P. Camy and J. L. Doualan, University of Caen, France

2.1 Introduction 28
2.2 Crystal growth, structural, optical and thermo-mechanical properties of the most important fluoride crystals 32
2.3 Pr^{3+} doped crystals for RGB video-projection and quantum information experiments 38
2.4 Yb^{3+} doped fluorides for ultra-short and high-power laser chains 43
2.5 Undoped crystals for nonlinear optics and ultra-short pulse lasers 48
2.6 References 49

© Woodhead Publishing Limited, 2013

3 Oxide laser ceramics 54
V. B. KRAVCHENKO and Y. L. KOPYLOV, V. A. Kotel'nikov Institute of Radioengineering and Electronics, Russian Academy of Sciences, Russia

3.1 Introduction 54
3.2 Ceramics preparation 56
3.3 Physical properties of oxide laser ceramics 65
3.4 Solid-state lasers using oxide ceramic elements 71
3.5 Conclusion 74
3.6 Acknowledgements 75
3.7 References 75

4 Fluoride laser ceramics 82
P. P. FEDOROV, A. M. Prokhorov General Physics Institute, Russian Academy of Sciences, Russia

4.1 Introduction 82
4.2 Fluoride powders: chemistry problems and relevant technology processes 84
4.3 Fluoride ceramics as optical medium 87
4.4 Development of the fluoride laser ceramics synthesis protocol 90
4.5 Microstructure, spectral luminescence and lasing properties 92
4.6 $CaF_2:Yb^{3+}$ system 100
4.7 Prospective compositions for fluoride laser ceramics 101
4.8 Conclusion 104
4.9 Acknowledgments 104
4.10 Note to reader 104
4.11 References 105

5 Neodymium, erbium and ytterbium laser glasses 110
V. I. ARBUZOV, Research and Technological Institute of Optical Material Science, Russia and N. V. NIKONOROV, Saint-Petersburg National Research University of Information Technologies, Mechanics and Optics (ITMO), Russia

5.1 Introduction 110
5.2 The history of laser glasses 111
5.3 Commercial laser glasses 116
5.4 Modern neodymium and erbium laser glasses 122
5.5 Ytterbium glasses 127
5.6 Future trends in glass-based laser materials 129
5.7 References 132

© Woodhead Publishing Limited, 2013

6 Nonlinear crystals for solid-state lasers 139
V. Pasiskevicius, Royal Institute of Technology, Sweden

6.1 Introduction 139
6.2 Second-order frequency conversion 140
6.3 Nonlinear crystal development 152
6.4 Nonlinear crystals: current status and future trends 156
6.5 Sources of further information and advice 161
6.6 References 161

Part II Solid-state laser systems and their applications 169

7 Principles of solid-state lasers 171
N. N. Il'ichev, A. M. Prokhorov General Physics Institute, Russian Academy of Sciences, Russia

7.1 Introduction 171
7.2 Amplification of radiation 172
7.3 Optical amplifiers 176
7.4 Laser resonators 179
7.5 Model of laser operation 181
7.6 Conclusion 189
7.7 References 190

8 Powering solid-state lasers 193
C. R. Hardy, Kigre, Inc., USA

8.1 Introduction 193
8.2 Safety 195
8.3 Flashlamp pumping 195
8.4 Laser diode pumping 212
8.5 Control features 223
8.6 Conclusion 224
8.7 References 225

9 Operation regimes for solid-state lasers 227
R. Paschotta, RP Photonics Consulting GmbH, Germany

9.1 Introduction 227
9.2 Continuous-wave operation 228
9.3 Pulsed pumping of solid-state lasers 234
9.4 Q-switching 236
9.5 Mode locking 242
9.6 Chirped-pulse amplification 248
9.7 Regenerative amplification 251
9.8 References 253

© Woodhead Publishing Limited, 2013

10 Neodymium-doped yttrium aluminum garnet (Nd:YAG) and neodymium-doped yttrium orthovanadate (Nd:YVO_4) 256
A. AGNESI and F. PIRZIO, University of Pavia, Italy

10.1 Introduction 256
10.2 Oscillators for neodymium lasers 257
10.3 Power/energy limitations and oscillator scaling concepts 267
10.4 Power scaling with master oscillator/power amplifier (MOPA) architectures 273
10.5 Future trends 277
10.6 Sources of further information and advice 279
10.7 References 279

11 System sizing issues with diode-pumped quasi-three-level materials 283
A. JOLLY, Commissariat à l'Energie Atomique, Centre d'Etudes Scientifiques et Techniques d'Aquitaine, France

11.1 Introduction 283
11.2 Ytterbium-doped materials and bulk operating conditions 284
11.3 Overview of Yb-based systems pump architectures and modes of operation 293
11.4 YAG–KGW–KYW-based laser systems for nanosecond and sub-picosecond pulse generation 304
11.5 Conclusion and future trends 318
11.6 References 320

12 Neodymium doped lithium yttrium fluoride (Nd:$YLiF_4$) lasers 323
N. U. WETTER, Centro de Lasers e Aplicações – IPEN/SP-CNEN, Brazil

12.1 Introduction 323
12.2 Pumping methods of Nd:YLF lasers 325
12.3 Alternative laser transitions 335
12.4 Future trends 336
12.5 References 337

13 Erbium (Er) glass lasers 341
B. I. DENKER, B. I. GALAGAN and S. E. SVERCHKOV, A. M. Prokhorov General Physics Institute, Russian Academy of Sciences, Russia

13.1 Introduction 341
13.2 Flashlamp pumped erbium (Er) glass lasers 343
13.3 Laser diode (LD) pumped erbium (Er) glass lasers 345

© Woodhead Publishing Limited, 2013

13.4 Means of Q-switching for erbium (Er) glass lasers 348
13.5 Applications of erbium (Er) glass lasers 351
13.6 Crystal lasers emitting at about 1.5 microns: advantages and drawbacks 352
13.7 References 354

14 Microchip lasers 359
J. J. ZAYHOWSKI, Massachusetts Institute of Technology, USA

14.1 Introduction 359
14.2 Microchip lasers: a broadly applicable concept 362
14.3 Transverse mode definition 364
14.4 Spectral properties 367
14.5 Polarization control 374
14.6 Pulsed operation 374
14.7 Nonlinear frequency conversion 387
14.8 Microchip amplifiers 390
14.9 Future trends 392
14.10 Sources of further information and advice 393
14.11 References 393

15 Fiber lasers 403
B. SAMSON, Nufern, USA and L. DONG, Clemson University, USA

15.1 Introduction and history 403
15.2 Principle of fiber lasers 406
15.3 High power continuous wave (CW) fiber lasers 420
15.4 Pulsed fiber lasers 429
15.5 Ultrafast fiber lasers 437
15.6 Continuous wave (CW) and pulsed fiber lasers at alternative wavelengths 444
15.7 Emerging fiber technologies for fiber lasers 448
15.8 Conclusion and future trends 455
15.9 References 455

16 Mid-infrared optical parametric oscillators 463
M. HENRIKSSON, Swedish Defence Research Agency, Sweden

16.1 Introduction 463
16.2 Nonlinear optics and optical parametric devices 465
16.3 Nonlinear optical materials for the infrared region 472
16.4 Tuneable single frequency optical parametric oscillators (OPOs) for spectroscopy 474
16.5 High power and high energy nanosecond pulselength systems 476

© Woodhead Publishing Limited, 2013

© Woodhead Publishing Limited, 2013

© Woodhead Publishing Limited, 2013

Contributor contact details

(* = main contact)

Editors

Boris Denker*
A. M. Prokhorov General Physics Institute
Russian Academy of Sciences
38 Vavilova Str.
Moscow 119991
Russia

E-mail: denker@lst.gpi.ru; bid44@list.ru

Eugene Shklovsky
Optech Incorporated
Canada

E-mail: esh_42@yahoo.ca

Chapter 1

Klaus Petermann
Institute of Laser-Physics
University of Hamburg
Luruper Chaussee 149
22761 Hamburg
Germany

E-mail: klaus.petermann@physnet.uni-hamburg.de

Chapter 2

Richard Moncorgé,* A. Braud, P. Camy and J. L. Doualan
Centre de Recherche sur les Ions, les Matériaux et la Photonique (CIMAP)
UMR 6252 CEA-CNRS-ENSICaen
University of Caen
6 boulevard Maréchal Juin
14050 Caen cedex
France

E-mail: richard.moncorge@ensicaen.fr

Chapter 3

V. B. Kravchenko* and Y. L. Kopylov
V. A. Kotel'nikov Institute of Radioengineering and Electronics
Russian Academy of Sciences, Fryazino Branch
1 Vvedenskogo Sq.
Fryazino
Moscow 141190
Russia

E-mail: vbk219@ire216.msk.su

© Woodhead Publishing Limited, 2013

Chapter 4

Pavel P. Fedorov
Laser Materials and Technology Research Center
A. M. Prokhorov General Physics Institute
Russian Academy of Sciences
38 Vavilova Str.
Moscow 119991
Russia

E-mail: ppf@lst.gpi.ru

Chapter 5

Valerii I. Arbuzov*
Research and Technological Institute of Optical Material Science
All-Russian Scientific Center 'S. I. Vavilov State Optical Institute'
36/1 Babushkin Street
Saint Petersburg 192171
Russia

E-mail: arbuzov@goi.ru

N. V. Nikonorov
Saint-Petersburg National Research University of Information Technologies, Mechanics and Optics (ITMO)
49 Kronverkskiy Avenue
Saint Petersburg 197101
Russia

E-mail: nikonorov@oi.ifmo.ru

Chapter 6

Valdas Pasiskevicius
Department of Applied Physics
Royal Institute of Technology (KTH)
Roslagstullsbacken 21
10691 Stockholm
Sweden

E-mail: vp@laserphysics.kth.se

Chapter 7

Nikolay N. Il'ichev
A. M. Prokhorov General Physics Institute
Russian Academy of Sciences
38 Vavilova Str.
Moscow 119991
Russia

E-mail: ilichev@kapella.gpi.ru

Chapter 8

Christopher R. Hardy
Kigre, Inc.
100 Marshland Road
Hilton Head Island
SC 29926
USA

E-mail: hardy1@hargray.com

Chapter 9

Rüdiger Paschotta
RP Photonics Consulting GmbH
Waldstr. 17
78073 Bad Dürrheim
Germany

E-mail: Paschotta@rp-photonics.com

© Woodhead Publishing Limited, 2013

Chapter 10

Antonio Agnesi* and Federico Pirzio
Department of Industrial and Information Engineering
University of Pavia
Via Ferrata 1
27100 Pavia
Italy

E-mail: agnesi@unipv.it; federico.pirzio@unipv.it

Chapter 11

Alain Jolly
Commissariat à l'Energie Atomique
Centre d'Etudes Scientifiques et Techniques d'Aquitaine
Chemin des sablières
33114 Le Barp
France

E-mail: alain.jolly@cea.fr; alain.jolly@alphanov.com

Chapter 12

Niklaus Ursus Wetter
Centro de Lasers e Aplicações (IPEN/SP-CNEN)
R. Prof. Lineu Prestes 2242, Cid. Universitária
05508-000 São Paulo-SP
Brazil

E-mail: nuwetter@ipen.br

Chapter 13

Boris I. Denker, Boris I. Galagan and Sergei E. Sverchkov*
A. M. Prokhorov General Physics Institute
Russian Academy of Sciences
38 Vavilova Str.
Moscow 11999
Russia

E-mail: denker@lst.gpi.ru; bid44@list.ru; glasser@lst.gpi.ru

Chapter 14

John J. Zayhowski
Lincoln Laboratory
Massachusetts Institute of Technology
244 Wood Street
Lexington, MA 02420-9108
USA

E-mail: zayhowski@ll.mit.edu

Chapter 15

Bryce Samson*
Nufern
7 Airport Park Road
East Granby, CT 06026
USA

E-mail: BSamson@nufern.com

Liang Dong
Holcombe Department of Electrical and Computer Engineering
Clemson University
Clemson, SC 29634
USA

E-mail: dong4@clemson.edu

© Woodhead Publishing Limited, 2013

Chapter 16

Markus Henriksson
The Swedish Defence Research Agency (FOI)
Box 1165
581 11 Linköping
Sweden

E-mail: markus.henriksson@foi.se

Chapter 17

Helen Pask* and Jim Piper
MQ Photonics
Department of Physics and Astronomy
Macquarie University
NSW 2109
Australia

E-mail: tzuk@soreq.gov.il

Chapter 18

Darren Rand,* J. Hybl and T. Y. Fan
Lincoln Laboratory
Massachusetts Institute of Technology
244 Wood Street
Lexington, MA 02420-9108
USA

E-mail: drand@ll.mit.edu

Chapter 19

Celio Pasquini
Instituto de Química
Universidade Estadual de Campinas (UNICAMP)
Caixa Postal 6154
13084-971 Compinas SP
Brazil

E-mail: pasquini@iqm.unicamp.br

Chapter 20

David G. Kochiev*
A. M. Prokhorov General Physics Institute
Russian Academy of Sciences
38 Vavilova Str.
Moscow 119991
Russia

E-mail: dkochiev@gmail.com

Alexei V. Lukashev
Stemedica Cell Technologies
5375 Mira Sorento Place
Suite 100
San Diego, CA 92121
USA

E-mail: alukashev@stemedica.com

Ivan A. Shcherbakov and Sergei K. Vartapetov
A. M. Prokhorov General Physics Institute
Russian Academy of Sciences
38 Vavilova Str.
Moscow 119991
Russia

E-mail: director@gpi.ru; svart@pic.troitsk.ru

© Woodhead Publishing Limited, 2013

Chapter 21

Yehoshua Kalisky
Nuclear Research Center
R&D Division
P.O. Box 9001
Beer-Sheva
84190 Israel

E-mail: kalisky@netvision.net.il

Chapter 22

Aladar Czitrovszky
Wigner Research Centre for Physics
Institute for Solid State Physics and Optics
29–33 Konkoly-Thege M. Street
Budapest XII
Hungary

E-mail: czitrovszky.aladar@wigner.mta.hu

© Woodhead Publishing Limited, 2013

Woodhead Publishing Series in Electronic and Optical Materials

1 **Circuit analysis**
J. E. Whitehouse

2 **Signal processing in electronic communications: For engineers and mathematicians**
M. J. Chapman, D. P. Goodall and N. C. Steele

3 **Pattern recognition and image processing**
D. Luo

4 **Digital filters and signal processing in electronic engineering: Theory, applications, architecture, code**
S. M. Bozic and R. J. Chance

5 **Cable engineering for local area networks**
B. J. Elliott

6 **Designing a structured cabling system to ISO 11801: Cross-referenced to European CENELEC and American Standards**
Second edition
B. J. Elliott

7 **Microscopy techniques for materials science**
A. Clarke and C. Eberhardt

8 **Materials for energy conversion devices**
Edited by C. C. Sorrell, J. Nowotny and S. Sugihara

9 **Digital image processing: Mathematical and computational methods**
Second edition
J. M. Blackledge

10 **Nanolithography and patterning techniques in microelectronics**
Edited by D. Bucknall

© Woodhead Publishing Limited, 2013

11 **Digital signal processing: Mathematical and computational methods, software development and applications**
Second edition
J. M. Blackledge

12 **Handbook of advanced dielectric, piezoelectric and ferroelectric materials: Synthesis, properties and applications**
Edited by Z.-G. Ye

13 **Materials for fuel cells**
Edited by M. Gasik

14 **Solid-state hydrogen storage: Materials and chemistry**
Edited by G. Walker

15 **Laser cooling of solids**
S. V. Petrushkin and V. V. Samartsev

16 **Polymer electrolytes: Fundamentals and applications**
Edited by C. A. C. Sequeira and D. A. F. Santos

17 **Advanced piezoelectric materials: Science and technology**
Edited by K. Uchino

18 **Optical switches: Materials and design**
Edited by S. J. Chua and B. Li

19 **Advanced adhesives in electronics: Materials, properties and applications**
Edited by M. O. Alam and C. Bailey

20 **Thin film growth: Physics, materials science and applications**
Edited by Z. Cao

21 **Electromigration in thin films and electronic devices: Materials and reliability**
Edited by C.-U. Kim

22 ***In situ* characterization of thin film growth**
Edited by G. Koster and G. Rijnders

23 **Silicon-germanium (SiGe) nanostructures: Production, properties and applications in electronics**
Edited by Y. Shiraki and N. Usami

24 **High-temperature superconductors**
Edited by X. G. Qiu

25 **Introduction to the physics of nanoelectronics**
S. G. Tan and M. B. A. Jalil

© Woodhead Publishing Limited, 2013

26 **Printed films: Materials science and applications in sensors, electronics and photonics**
Edited by M. Prudenziati and J. Hormadaly

27 **Laser growth and processing of photonic devices**
Edited by N. A. Vainos

28 **Quantum optics with semiconductor nanostructures**
Edited by F. Jahnke

29 **Ultrasonic transducers: Materials and design for sensors, actuators and medical applications**
Edited by K. Nakamura

30 **Waste electrical and electronic equipment (WEEE) handbook**
Edited by V. Goodship and A. Stevels

31 **Applications of ATILA FEM software to smart materials: Case studies in designing devices**
Edited by K. Uchino and J.-C. Debus

32 **MEMS for automotive and aerospace applications**
Edited by M. Kraft and N. M. White

33 **Semiconductor lasers: Fundamentals and applications**
Edited by A. Baranov and E. Tournie

34 **Handbook of terahertz technology for imaging, sensing and communications**
Edited by D. Saeedkia

35 **Handbook of solid-state lasers: Materials, systems and applications**
Edited by B. Denker and E. Shklovsky

36 **Organic light-emitting diodes: Materials, devices and applications**
Edited by A. Buckley

37 **Lasers for medical applications: Diagnostics, therapy and surgery**
Edited by H. Jelínková

38 **Semiconductor gas sensors**
Edited by R. Jaaniso and O. K. Tan

39 **Handbook of organic materials for optical and optoelectronic devices: Properties and applications**
Edited by O. Ostroverkhova

© Woodhead Publishing Limited, 2013

40 **Metallic films for electronic, optical and magnetic applications: Structure, processing and properties**
Edited by K. Barmak and K. Coffey

41 **Handbook of laser welding technologies**
Edited by S. Katayama

42 **Nanolithography: The art of fabricating nanoelectronics, nanophotonics and nanobiology devices and systems**
Edited by M. Feldman

43 **Laser spectroscopy for sensing: Fundamentals, techniques and applications**
Edited by M. Baudelet

44 **Chalcogenide glasses: Preparation, properties and applications**
Edited by J.-L. Adam and X. Zhang

45 **Handbook of MEMS for wireless and mobile applications**
Edited by D. Uttamchandani

46 **Subsea optics and imaging**
Edited by J. Watson and O. Zielinski

47 **Carbon nanotubes and graphene for photonic applications**
Edited by S. Yamashita, Y. Saito and J. H. Choi

48 **Optical biomimetics: Materials and applications**
Edited by M. Large

49 **Optical thin films and coatings**
Edited by Angela Piegari and François Flory

50 **Computer design of diffractive optics**
Edited by V. A. Soifer

© Woodhead Publishing Limited, 2013

Foreword

The theoretical basis of stimulated emission of electromagnetic radiation was presented by Albert Einstein in the early 20th century. However, almost half a century passed until the practical realisation of the idea. The first innovation was in the microwave frequency range – the maser (microwave amplification of stimulated emission of radiation) was invented.

The solution, that is stimulated emission of radiation as a way to obtain amplification and oscillation at microwave frequencies, had independently been proposed by Townes in 1951, Weber in 1953, and Prokhorov and Basov in 1954. The first Nobel Prize in the field went to Townes, Prokhorov and Basov in 1964. In 1956 Blombergen proposed a successful and versatile maser pumping scheme. The demonstration of microwave amplification by stimulated emission of radiation or maser action by Gordon, Zeiger and Towns in 1954, and Scovi, Feher and Seidel in 1957, and subsequent research and development efforts verified the potential of this principle towards the realisation of highly monochromatic oscillators and amplifiers with very low noise.

The first proposal to extend the principle from microwaves to optical frequencies was made in 1958 by Shawlow and Towns. Their papers assembled and extrapolated knowledge from the field of microwave masers and optical spectroscopy. They suggested that an open resonator such as a Fabry–Perot interferometer would select a few radiation modes with extremely high efficiency, leaving an enormous number of remaining modes unamplified. They then went on to predict the unique properties of laser light such as high coherence, low divergence, narrow line-width and low noise. Laser action was first observed by Maiman in 1960. It was obtained from ruby (sapphire crystal doped with Cr^{3+}) at room temperature, optically pumped by a flashlamp similar to those used in photography, giving light pulses in the millisecond range. This laser is shown in the photo taken of the original exhibited at the '50 ans du Laser' Conference in Paris in 2010 at the Palace de Louvre (see page xxiv).

The first laser was one where the active material was a solid (single crystal). The significance of this type of solution has been increasing in recent years

© Woodhead Publishing Limited, 2013

The original laser used by Maiman to observe laser action in 1960 (source A. Czitrovsky).

in a growing number of scientific, technological and medical applications. This is why this handbook will be a useful guide for both professionals and students of the field.

This discovery triggered an avalanche of laser work in laboratories everywhere and soon led to the demonstration of laser action in other active materials, in other solids including optical fibres, in gases, in semiconductors, and in liquids. Other Nobel Prizes followed the first one, like that in the semiconductor field of Z. Alferov and numerous further ones in applications. In time, all the specific laser parameters went into the direction of extremes, short pulses down to attoseconds, extremely large intensities (and powers) and narrow beams both in space and in frequency line-widths.

The laser found a large number of applications in the second half of the 20th century. The rapid progress in laser physics resulted in breakthroughs in nonlinear optics, light-matter interaction, quantum optics, metrology, technological applications and medicine.

Not only was the first laser made of solid-state materials, but many of the later, most versatile lasers covering a wide range of continuous wave power or pulse peak power, pulse duration and energy, wavelength, divergence and coherence are solid-state lasers, so their applications are very wide. Fine mechanics as well as theoretical physics have played crucial roles in the development of solid-state lasers, materials science and optics and

© Woodhead Publishing Limited, 2013

inversely, the new developments have led to new discoveries in these fields and beyond.

This book provides a description of laser theory, the characterisation of the main laser parameters, the principles of the operation of lasers, a classification of laser materials (Nd: YAG, Nd: YVO_4, YLF, Er:glass, etc.) and a review of the different type of lasers (microchip lasers, fibre lasers, Raman-lasers, cryogenic lasers, YLF-lasers, LIBS-lasers, etc.). The final section is devoted to solid-state laser applications in different technological processes (medicine, including surgery and ophthalmology, defence science and environmental protection).

I hope that this book will be widely used and will stimulate new, groundbreaking ideas both in laser development and applications.

Professor Norbert Kroo
Hungarian Academy of Sciences

© Woodhead Publishing Limited, 2013

Preface

More than 50 years ago the first solid-state laser (SSL) was made by Theodore Maiman. Since then SSLs have found diverse applications in different areas of industry and human activity: technological processes, scientific research, medicine and entertainment to name but a few. SSLs being broadly used, are still a fruitful area for researchers aiming to improve laser performance and develop novel SSLs with unique properties. That is why the regular appearance of books covering different aspects of SSLs is very important.

This handbook on SSLs will be interesting both for laser professionals and those who are impressed by their remarkable achievements in the science and technology. It comprises a collection of papers on different aspects of SSLs written by experts in their fields. We take this opportunity to express our gratitude to all authors for their valuable contributions to this book.

We regret that not all the topics of SSLs are contained in this book, due to the limited time frame of the project as well as the restricted volume of the book. Nevertheless, this handbook, in our opinion, gives an overview of modern state-of-the-art of laser materials, the most important SSLs and their applications. We hope that it will capture the interest of the reader who takes it from a bookstand and looks through the contents.

B. Denker
E. Shklovsky

Part I

Solid-state laser materials

© Woodhead Publishing Limited, 2013

1 Oxide laser crystals doped with rare earth and transition metal ions

K. PETERMANN, University of Hamburg, Germany

DOI: 10.1533/9780857097507.1.3

Abstract: At the beginning of this chapter the most prominent transition metal- and rare earth-doped oxide lasers and their emission wavelengths are introduced. After a short section about the fabrication of the laser crystals, some aspects concerning the geometry of the active medium for high-power lasers are presented. Finally, the spectroscopy as well as the laser results of rare earth-doped sesquioxides are reported in more detail. The chapter ends with some expected trends for the future.

Key words: high-power solid-state lasers, transition metal- and rare earth-doped oxides, growth of laser crystals, spectroscopy of rare earth-doped sesquioxides, sesquioxide lasers.

1.1 Introduction

Since the renaissance of solid-state lasers in the mid-1980s numerous types of continuous and pulsed lasers on the basis of transition metal (TM) and rare earth (RE) ions have been developed. Many new laser wavelengths have been realised, but a few wavelength gaps still exist, for example in the yellow and blue/near UV spectral range. Especially, UV lasers are challenging, because high-energy photons very often create colour centres in the active medium, resulting in high laser losses. Furthermore, compact and efficient pump sources are not yet available, except (In,Ga)N-diode lasers. So, further research is necessary to close these wavelength gaps.

Also, many new host materials like oxides, fluorides, and glasses have been investigated in the past. However, the most successful family of host lattices for high-power lasers are still the garnets. With Nd- and Yb-doped YAG ($Y_3Al_5O_{12}$) multi-kilowatt lasers are available nowadays. But also the vanadates (YVO_4 and $GdVO_4$) are very important laser materials due to their high absorption and emission cross-sections. A new class of hosts are the sesquioxides (Sc_2O_3, Y_2O_3, and Lu_2O_3), which exhibit high thermal conductivity and can be doped with high concentrations of RE ions. Thus, these laser materials are predestined for high-power solid-state lasers, although the growth of single crystalline material is quite complicated.

This chapter on TM- and RE-doped oxides is structured as follows. In the next two sections the most important laser-active TM and RE ions are

© Woodhead Publishing Limited, 2013

introduced as well as the commonly used host crystals and their growth techniques. In Section 1.4 a short introduction to thin-disk lasers is given, and in Section 1.5 the spectroscopic properties of RE-doped sesquioxides are presented as well as the results of the cw-laser experiments. With a brief section about pulsed laser systems and a short outlook into future developments, this chapter will close.

1.2 Laser-active ions

The basic properties of a solid-state laser are dominated by the interaction of the laser-active ion and the host lattice. This interaction is quite strong in the case of transition metal ions (TM ions) due to the non-shielded 3d-electrons, which couple easily with the phonons of the surrounding oxygen ligands resulting in broad 3d–3d absorption and emission bands. In contrast, the 4f-electrons of the rare earth ions (RE ions) are shielded by the electrons in the 5s- and 5p-orbitals and thus the 4f–4f transitions are narrow and only weakly influenced by the crystal field provided by the ligands. Consequently, the transition energies or laser wavelengths are rather independent of the host lattice. In Fig. 1.1 the basic difference between the energy level scheme of RE ions with weak electron–phonon coupling and the vibronic energy levels of the TM ions with strong electron–phonon coupling is demonstrated.

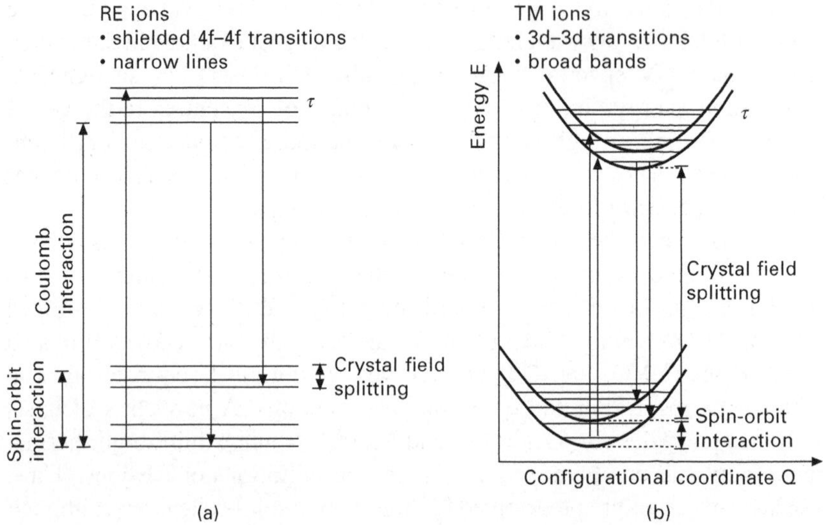

1.1 (a) Weak electron–phonon coupling of RE ions; (b) relatively strong electron–phonon coupling of TM ions described by the configurational coordinate model.

© Woodhead Publishing Limited, 2013

1.2.1 Transition metal ions

Beside Ti^{3+} the most important TM laser ion is chromium with its different valencies, i.e. Cr^{2+}, Cr^{3+}, and Cr^{4+}. Cr^{3+}:Al_2O_3 was the very first solid-state laser, developed by T. H. Maiman in 1960 (Maiman 1960), and Cr^{3+} doped into garnets like $Gd_3Sc_2Ga_3O_{12}$ (GSGG) was the first tunable TM ion laser in the deep red spectral range (Struve *et al.*, 1983; Huber and Petermann 1985). The tuning range of all Cr^{3+} lasers is limited to about 100 nm, but the great advantage is that they can be pumped by diode lasers.

Due to the broader tuning range extending from about 680 nm to 1100 nm the Ti:Al_2O_3 laser has replaced the Cr^{3+} lasers and can be regarded as the most important tunable laser nowadays (Moulton 1985). Ti:sapphire lasers are typically pumped by frequency doubled Nd^{3+} lasers at 532 nm, but pumping by (Ga, In)N diode lasers will be possible in future, when diodes with sufficient cw-power are available. Presently the output power of green diode lasers is limited to about 50 mW (Avramescu *et al.* 2010). Tests with GaN diodes at 452 nm wavelength have already been performed, but the efficiency and stability were very low due to the creation of unknown parasitic losses (Roth *et al.* 2009).

For the near- to mid-infrared wavelength range between 1.9 μm and 3.4 μm Cr^{2+} is a suitable laser ion, if it is doped into various zinc-chalcogenides like ZnS or ZnSe (DeLoach *et al.* 1996). Presently, Cr^{2+}:ZnSe is the most efficient TM laser system with more than 1 W output power and a slope efficiency of 73%, which is very close to the quantum limit of 77%. These excellent data are due to the tetrahedral coordination of the Cr^{2+} ion resulting in a huge emission cross-section of more than 10^{-18} cm^2. Because of the wide tuning range of 1100 nm, Cr^{2+}:ZnSe may be regarded as the 'Ti:sapphire laser of the infrared' (Sorokina 2004). However, a Cr^{2+}:ZnO laser has not yet been developed because of the quite large ionic radius of the Cr^{2+} ion, which apparently prevents the diffusion of the ion into the host lattice. Here, other preparation techniques like layer growth by pulsed laser deposition (PLD) may be successful in future.

One interesting aspect of the Cr^{2+}-doped chalcogenides is the fact that they could be pumped electrically, since they are semiconductors (Fedorov *et al.* 2007). Recently, the first electrically pumped Cr^{2+}:ZnS waveguide laser was demonstrated (Vlasenko *et al.* 2009) and it can be expected that such laser structures will be realised in future also with ZnO as host lattice.

Another tunable laser ion in the near infrared (NIR) is tetrahedrally coordinated Cr^{4+}. In the past it was doped into a large variety of oxides, that is garnets, silicates, and germanates (Kück 2001). The best laser performances have been obtained with Cr^{4+}:$Y_3Al_5O_{12}$ (YAG) and Cr^{4+}:Mg_2SiO_4 (forsterite). Nearly 2 W cw output power at 42% slope efficiency and a tuning range from 1340 nm to 1570 nm for YAG (Sennaroglu *et al.* 1995) and 1.5 W

© Woodhead Publishing Limited, 2013

output power at 34% slope efficiency for forsterite have been measured (Zhavoronkov *et al.* 1997). However, the efficiency as well as the tuning range is limited in many oxides by excited state absorption (ESA) within the Cr^{4+} ion (Kück 2001).

More exotic laser ions are Fe^{2+}, Ni^{2+}, Co^{2+}, and Mn^{5+}, which have been lasing mostly at low temperatures with quite low efficiencies and thus are more of academic value. Other ions like Mn^{3+} (octahedral coordination) or the $3d^1$ ions V^{4+}, Cr^{5+}, and Mn^{6+} (tetrahedral coordination) have been investigated, with the result that they exhibit an extremely low quantum efficiency due to non-radiative transitions or suffer from strong ESA at the emission wavelengths preventing laser action. A summary of all TM ions with their various valencies is given in Table 1.1. Roman lettering denotes octahedral coordination in the host lattice and italic lettering tetrahedral coordination. Ions which have shown laser action are distinguished by bold characters (Kück 2001).

Another interesting, new laser ion for the NIR spectral range is bismuth, which was investigated in various glasses (Bufetov and Dianov 2009). However, the structure of the Bi-centre has not yet been identified. Nevertheless, high power cw-, Q-switched, and mode-locked lasing has been achieved with Bi-doped fibre lasers (Bufetov *et al.* 2010) though not yet with crystalline laser materials.

1.2.2 Rare earth ions

The largest class of laser ions are the rare earth ions (RE). Hundreds of lasers have been realised with Nd^{3+} as active ion, most of them in oxide crystals. The Nd ion exhibits various groups of NIR laser transitions around 900 nm, 1060 nm, and 1300 nm wavelength. The strongest and mostly used laser transition $^4F_{3/2} \rightarrow {}^4I_{11/2}$ is emitted at wavelengths between 1050 nm and 1100 nm depending on the host lattice. Typically, Nd lasers are pumped around 806 nm by diode lasers, which is a great advantage for the construction of compact and reliable laser modules.

For high-power lasers Yb^{3+} is a more suitable ion than Nd^{3+}, because it exhibits only two manifolds, i.e. the $^2F_{7/2}$ ground state and the $^2F_{5/2}$ excited state. Thus, ESA of the pump and laser wavelength as well as cross-relaxation, which are the major loss mechanisms, are absent. Furthermore, due to the small Stokes shift between the pump and laser transition of about 500 cm^{-1} the heat generation is comparatively small, resulting in highly efficient laser systems at high average power. As a disadvantage it has to be noted that Yb lasers operate in a quasi-three-level scheme with temperature-dependent reabsorption at the laser wavelength. This leads, in contrast to four-level systems like Nd lasers, to increased temperature sensitivity and higher laser thresholds.

© Woodhead Publishing Limited, 2013

Table 1.1 Overview of TM ions with different valencies and coordinations. Roman lettering denotes octahedral coordination, italic lettering denotes tetrahedral coordination, and bold characters indicate laser oscillation

Ion	$3d^1$	$3d^2$	$3d^3$	$3d^4$	$3d^5$	$3d^6$	$3d^7$	$3d^8$	$3d^9$
Ti	$\mathbf{Ti^{3+}}$	$\mathrm{Ti^{2+}}$							
V	$\mathrm{V^{4+}}$, $\mathit{V^{4+}}$	$\mathrm{V^{3+}}$, $\mathit{V^{3+}}$	$\mathbf{V^{2+}}$						
Cr	$\mathit{Cr^{5+}}$	$\mathrm{Cr^{4+}}$, $\boldsymbol{\mathit{Cr^{4+}}}$	$\mathbf{Cr^{3+}}$	$\mathrm{Cr^{2+}}$, $\boldsymbol{\mathit{Cr^{2+}}}$					
Mn	$\mathit{Mn^{6+}}$	$\mathrm{Mn^{5+}}$, $\boldsymbol{\mathit{Mn^{5+}}}$	$\mathrm{Mn^{4+}}$	$\mathrm{Mn^{3+}}$, $\mathit{Mn^{3+}}$	$\mathrm{Mn^{2+}}$, $\mathit{Mn^{2+}}$				
Fe		$\mathit{Fe^{6+}}$			$\mathrm{Fe^{3+}}$, $\mathit{Fe^{3+}}$	$\mathrm{Fe^{2+}}$, $\mathit{Fe^{2+}}$			
Co						$\mathrm{Co^{3+}}$, $\mathit{Co^{3+}}$	$\mathbf{Co^{2+}}$, $\mathit{Co^{2+}}$		
Ni							$\mathrm{Ni^{3+}}$, $\mathit{Ni^{3+}}$	$\mathbf{Ni^{2+}}$, $\mathit{Ni^{2+}}$	
Cu								$\mathrm{Cu^{3+}}$	$\mathrm{Cu^{2+}}$, $\mathit{Cu^{2+}}$

© Woodhead Publishing Limited, 2013

Another important RE ion especially for the NIR spectral range is Er^{3+}. Its emission wavelength around 1.5 μm is widely used in optical fibre communication systems and also for eye-safe laser applications. Er^{3+} can be pumped to some extent by energy transfer in Yb/Er codoped crystals or directly by diode lasers at 975 nm. Pump wavelengths between 1.45 μm and 1.5 μm allow in-band pumping of the $^4I_{13/2}$ upper laser level. Since high-power diode lasers are now available in this wavelength range, very efficient lasers with extremely low quantum defect can be realised as with Yb^{3+}. However, Er^{3+} suffers from low absorption and emission cross-sections, which cannot be compensated by high doping concentrations due to increased up-conversion loss processes. An upper limit for the Er concentration is about 1% depending on the host lattice.

Further well-known NIR laser ions are Tm^{3+} and Ho^{3+} emitting in the 2 μm spectral range, which are useful for medical applications and remote sensing. Pumping of Tm^{3+} is normally done with AlGaAs diode lasers at about 790 nm into the $^3F_4$ excited state. If high Tm concentrations are doped into the laser crystal, an efficient Tm^{3+}–Tm^{3+} cross-relaxation process occurs, resulting in two excitations in the $^3H_4$ upper laser level (two-for-one pumping scheme, see Fig. 1.2). By this down-conversion process the thermal load of the gain material is largely reduced.

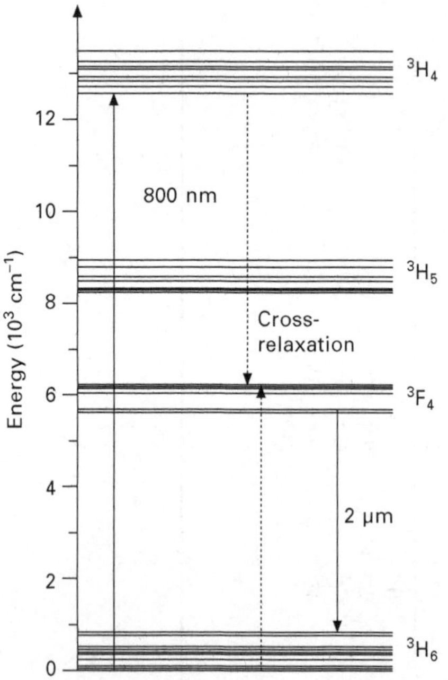

1.2 Energy-level scheme of Tm^{3+} with the very efficient cross-relaxation process being useful for pumping the NIR laser.

© Woodhead Publishing Limited, 2013

When codoping with Ho^{3+} the nearly resonant $Tm^{3+} \rightarrow Ho^{3+}$ energy transfer can be used for excitation of the Ho ions. Similar to Er lasers, both Tm^{3+} and Ho^{3+} can also be pumped directly in-band into the upper laser level, which reduces further the thermal losses. However, the required diodes at these long wavelengths are not yet well developed and high output powers without large wavelength shift are not yet available.

A very attractive ion for the generation of visible laser radiation is Pr^{3+}. The optical transitions starting from the $^3P_0$ and thermally populated $^3P_1/^1I_6$ and $^3P_2$ levels have cross-sections on the order of 10^{-19} cm^2, which are comparable to those of Nd^{3+}. Host lattices for Pr^{3+} should feature low effective phonon energies in order to avoid non-radiative $^3P_0 \rightarrow {}^1D_2$ transitions depopulating the upper laser level. For this reason most hosts for Pr^{3+} are fluorides, but also several oxides like phosphates, tungstates, and aluminates have been lasing in the orange (~600 nm) and red (~640 nm) spectral range. Till now, the green transition $^3P_1 \rightarrow {}^3H_5$ at around 523 nm with an emission cross-section about one order of magnitude less than the strong orange and red transitions has been realised only in halogenides, for example in $YLiF_4$ or $LiLuF_4$ (Richter *et al.* 2007).

In the past the development of Pr lasers was hampered by the availability of suitable pumping sources in the blue spectral region. Nowadays, frequency-doubled, optically pumped semiconductor lasers (OPS) emitting at 480 nm and (Ga,In)N diode lasers emitting around 444 nm allow direct pumping of the upper laser level, leading to an improved laser performance. Meanwhile, cw Pr^{3+}:$YLiF_4$ lasers have been operated diode-pumped with nearly 1 W output power (Gün *et al.* 2011) and OPS-pumped in the multi-watt regime (Ostroumov *et al.* 2007). Slope efficiencies up to 63.6% have been obtained in Pr^{3+}:$YLiF_4$ and of 45% in Pr^{3+}:$YAlO_3$ (Fibrich *et al.* 2009) as well as nearly 50% in diode pumped Pr^{3+}, Mg^{2+}:$SrAl_{12}O_{19}$ (Fechner *et al.* 2011). These data demonstrate clearly the high potential of Pr lasers.

1.2.3 Interconfigurational transitions of rare earth ions

4f $\rightarrow$ 5d Transitions of the RE ions are also useful for tunable solid-state lasers, because these transitions are electric-dipole allowed and thus provide high absorption and emission cross-sections. The $4f^{n-1}\,5d^1$ energy levels of all RE ions are in the range between 50,000 cm^{-1} and 160,000 cm^{-1} as can be found in the classic textbook of Dieke (Dieke 1968). The most prominent ion with 5d $\rightarrow$ 4f emission is Ce^{3+}. Laser oscillation has been realised around 300 nm in several fluorides like $YLiF_4$, $LiLuF_4$, $LiCaAlF_6$, $LiSrAlF_6$, and LaF_3 with efficiencies up to 50% (Alderighi *et al.* 2006). However, the main difficulty with respect to laser operation is still the lack of efficient, compact, and reliable excitation sources as well as solarisation effects of the active media by the UV radiation.

© Woodhead Publishing Limited, 2013

Table 1.2 Overview of RE ions and the corresponding laser wavelengths

Laser ion	Laser wavelengths	Remarks
Nd^{3+}	0.9 μm, 1.06 μm, 1.3 μm	NIR RE lasers
Yb^{3+}	1.0–1.1 μm	
Tm^{3+}	2 μm	
Ho^{3+}	2 μm	
Er^{3+}	1.6 μm, 3 μm	
Pr^{3+}	0.52 μm, 0.60 μm	Visible RE lasers
	0.64 μm, 0.72 μm	
Er^{3+}	0.55 μm	
Ce^{3+}	0.3 μm	UV RE lasers

In oxides no Ce laser is known. Here, Ce:YAG with its unusual emission in the yellow-green spectral range was investigated in detail and strong ESA into the conduction band was found to prevent laser oscillation (Hamilton *et al.* 1989). Whether oxides with reduced or even absent ESA exist is not yet clear and is presently under investigation.

An overview of the laser wavelengths of the most important RE ions is given in Table 1.2.

1.3 Host lattices

In order to develop a highly efficient solid-state laser material, several fundamental parameters are of great importance. So, the active medium has to be fabricated on a large scale and with high optical quality, that is parasitic absorptions and scattering centres should be absent. A high mechanical and chemical stability is desirable and for high-power lasers a high heat conductivity and a high damage threshold are essential. A low temperature dependence of the refractive index and a small second-order refractive index are advantageous for achieving a Gaussian beam profile and preventing self-focusing, respectively. Taking into account all these demands in parallel is nearly impossible. For example, sapphire or even diamond exhibit the highest heat conductivities among the dielectrics, but they cannot be doped with RE ions due to the very large ionic size in comparison to the available lattice sites. Other important factors are the crystal structure and the thermal expansion coefficients as well as the local site symmetry for the active ion, which influences the optical properties of the laser medium considerably.

For the preparation of large-scale laser materials various growth techniques have been developed in the past. Traditionally, laser-active crystals are grown from the melt by the Czochralski technique. A well-known example is Nd- and Yb-doped YAG ($Y_3Al_5O_{12}$), which belongs to the large family of garnet crystals and exhibits a cubic crystal structure. YAG and LuAG

© Woodhead Publishing Limited, 2013

($Lu_3Al_5O_{12}$) are routinely grown at around 1950°C from iridium crucibles with dimensions of up to 80 mm in diameter and 200 mm in length. From these boules rods, slabs, and thin disks are prepared for the corresponding types of solid-state lasers. Figure 1.3 shows as an example an Yb:YAG crystal grown with a flat interface for the preparation of homogeneous thin disks for high-power Yb lasers.

The perovskites $YAlO_3$ and $LuAlO_3$ are also commercially grown by the Czochralski technique. Like the garnets they have high heat conductivity and can be doped with RE and TM ions. However, the crystal structure is orthorhombic, which very often causes twinning during crystal growth. Other disadvantages are the non-isotropic thermal expansion coefficient and the tendency to colour centre formation, which makes the perovskites less favourable for high-power lasers.

Also the vanadates YVO_4, $GdVO_4$, and $LuVO_4$ are important laser materials due to their excellent heat conductivities as well as high absorption and emission cross-sections. However, the Czochralski growth is more difficult because of the quite low viscosity of the melt and the formation of colour centres (Kochurikhin *et al.* 1995).

For KYW, KGW, and KLuW from the tungstate family with the general formula $KRE(WO_4)_2$ and RE = Y, Gd, and Lu, the top seeded solution growth (TSSG) technique has to be applied because of a tetragonal/monoclinic phase transition in these crystals slightly below the melting point (Solé *et al.* 1996). $K_2W_2O_7$ is typically used as a solvent since it has a quite low melting point of 619°C and does not introduce any foreign impurities. By slow cooling of the saturated solvent, crystals of centimetre dimensions with high optical quality are obtained.

Another technique for growing oxides from the melt is the heat exchanger

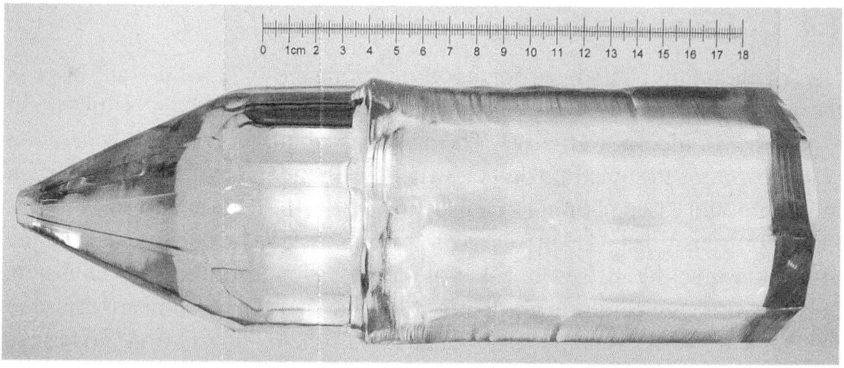

1.3 Czochralski-grown Yb:YAG crystal 80 mm in diameter and 200 mm in length; the left part is grown with convex interface and the right part with flat interface (courtesy of FEE, Idar-Oberstein, Germany).

© Woodhead Publishing Limited, 2013

method (HEM). Large Ti:sapphire boules of 208 mm diameter and 15 kg weight (Joyce and Schmid 2010) as well as undoped boules of 34 cm diameter and 65 kg weight are fabricated commercially by this method. With this technique a helium gas flow is cooling a seed crystal at the bottom of a molybdenum crucible containing the Al_2O_3 melt. With increasing gas flow the sapphire crystal grows nearly stress-free into the melt until it reaches the crucible wall at the end of the growth run. From these boules of high optical quality, i.e. low parasitic absorption by Ti^{3+}/T^{i4+} ion pairs, amplifier crystals of 175 mm diameter are fabricated for ultra-short pulse laser systems in the petawatt power range.

The same technique can also be used for the growth of sesquioxides like Sc_2O_3 (scandia), Y_2O_3 (yttria), and Lu_2O_3 (lutetia) (Peters *et al.* 2010). Due to the extremely high melting temperature of these oxides ($T_m \approx 2450°C$), rhenium crucibles ($T_m = 3180°C$) have to be used in combination with a slightly reducing growth atmosphere. The experimental setup that has been used for several years at the University of Hamburg is depicted in Fig. 1.4. Under optimal growth conditions colourless boules about 35 mm in diameter and 35 mm in height containing typically three to five monocrystalline blocks are obtained.

The sesquioxides may also be grown by the micro-pulling-down technique (Fukabori *et al.* 2011). With this method shaped rods of millimetre dimensions are pulled out of the nozzle at the bottom of a rhenium crucible. Furthermore, the laser heated pedestal growth (LHPG) technique can be used, which yields thin crystal fibres growing from the molten top of a polycrystalline feed rod. However, cracks are very often observed due to the strong thermal gradients (Laversenne *et al.* 2001). Small crystals have also been grown hydrothermally in an autoclave under 2 kbar pressure (MacMillen *et al.* 2011).

1.4 Laser medium geometry

One of the main challenges for good performance of a solid-state laser is the removal of the waste heat from the active medium. This is important especially for high-power lasers. The main contribution to the heat generation is in most cases the quantum defect, which is the energetic difference between the pump and laser photons being deposited in the gain medium via non-radiative transitions. An increased temperature leads to an increased line broadening and thus to lower peak cross-sections for absorption and emission. Furthermore, temperature gradients result in thermal lensing effects and mechanical stress, causing low beam quality of the laser radiation. To reduce the thermal load one can try to minimise the quantum defect by pumping the laser as close as possible to the laser wavelength. This was successfully demonstrated with Nd:YAG, where several laser transitions were pumped at 885 nm (Pavel *et al.* 2006) or even at 938 nm (Sangla *et al.* 2009) with

© Woodhead Publishing Limited, 2013

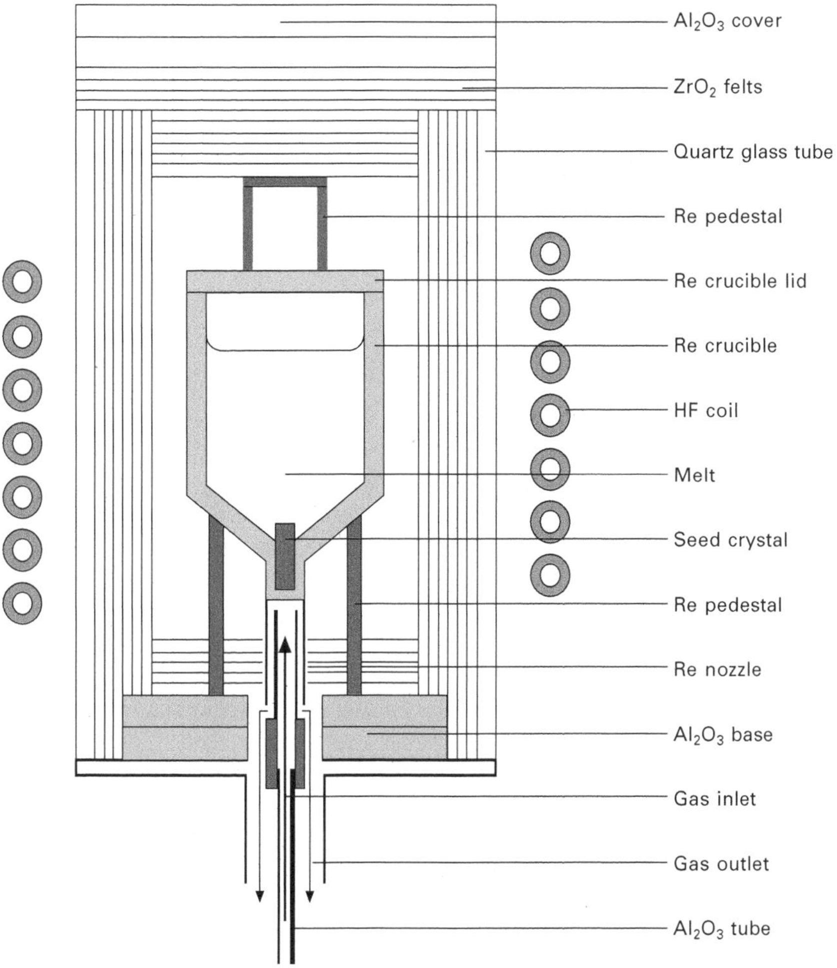

1.4 High-temperature HEM setup being used at the University of Hamburg for growing sesquioxide crystals from the melt.

laser emission at 1064 nm. According to the Nd-level scheme in Fig. 1.7 below, these lasers are in principle quasi-three-level systems like the Yb, Tm, Ho, and Er NIR lasers if they are in-band pumped.

The thermal effects can be further reduced if the host lattice exhibits high heat conductivity as in the case of YAG or the sesquioxides and by choosing a proper geometry of the gain medium. A large ratio between the cooled surface and the pumped volume is advantageous, which is given for the rod geometry and even more pronounced for the thin-disk laser (Giesen *et al.* 1994). Here, the active medium has the shape of a disk, which according to Fig. 1.5 on one side is highly reflective-coated for the pump and laser

© Woodhead Publishing Limited, 2013

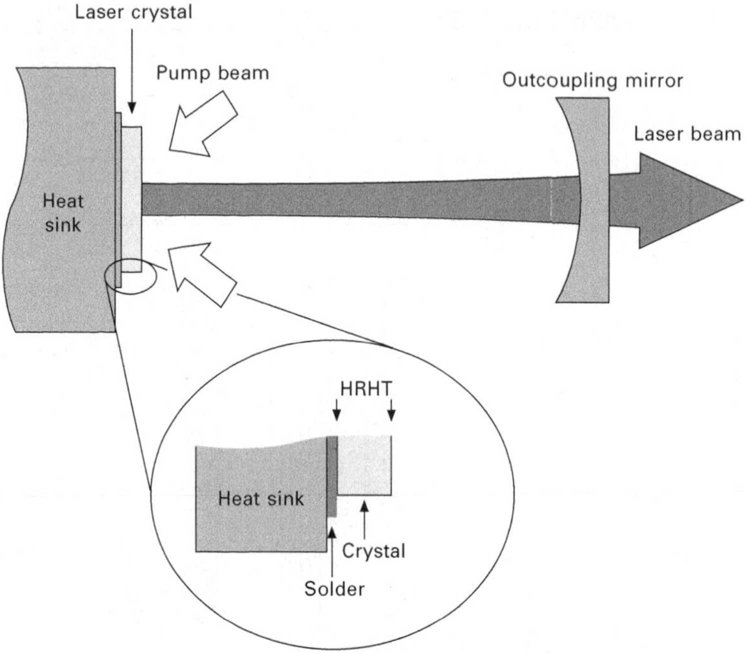

1.5 Schematic of a thin-disk laser; the HR/AR-coated disk is soldered on a copper heat sink and together with the outcoupling mirror forms the laser resonator.

wavelength and mounted on a water-cooled heat sink, whereas the opposite side is AR-coated. When pumping the disk, the temperature gradient is built up perpendicular to the disk surface and almost no radial gradients appear. As a consequence no thermal lensing occurs and thus a high beam quality is obtained. To achieve a high absorption of the pump light despite the short interaction length with the gain medium, the disk is placed in a special pump module, which allows for 16–32 passes of the pump light through the disk. The output power from a thin-disk laser can be scaled easily by increasing the pump spot diameter and the laser mode diameter on the disk. In this way output powers exceeding 5 kW from one Yb:YAG disk have been demonstrated with pump spot diameters of more than 10 mm (Giesen and Speiser 2007). Even better results can be expected with Yb-doped LuAG and especially with sesquioxides due to their higher heat conductivity.

The following sections will focus now on RE spectroscopy and the present status of Nd- and high-power Yb-sesquioxide lasers as well as the development of Tm- and Ho-doped Lu_2O_3 lasers for medical purposes. Furthermore, wavelength tuning of Er-doped sesquioxide lasers for LIDAR applications will be reported.

© Woodhead Publishing Limited, 2013

1.5 Rare earth-doped sesquioxides

1.5.1 Crystal structure and thermo-physical properties

The sesquioxides Sc_2O_3, Y_2O_3, and Lu_2O_3 as well as their solid solutions belong to the cubic Bixbyite structure (Pauling and Shappell 1930). The heat conductivity of all sesquioxides exceeds that of YAG by about 50% as can be seen in Fig. 1.6 and in Table 1.3 (Peters *et al.* 2002). Especially for

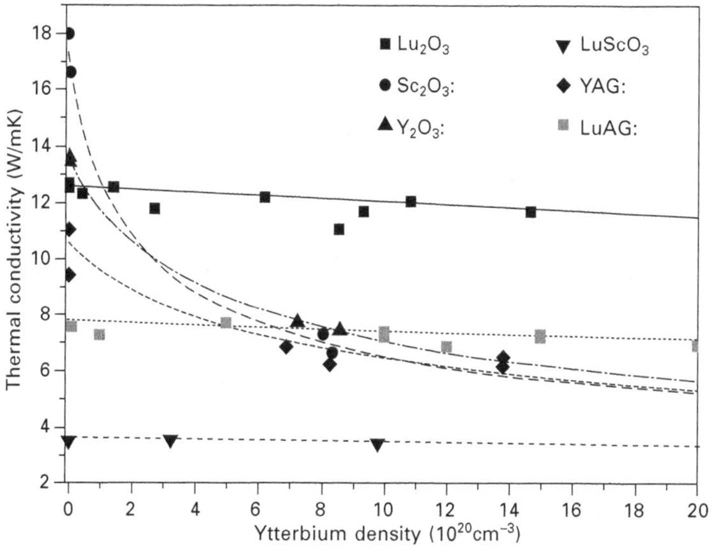

1.6 Thermal conductivity of sesquioxides and garnets as a function of the Yb-doping concentration.

Table 1.3 Material properties of the sesquioxides in comparison to YAG

	Sc_2O_3	Y_2O_3	Lu_2O_3	YAG
Melting point (°C)	2430	2430	2450	1940
Structure	cubic	cubic	cubic	cubic
Space group	$Ia\,3/T_h^7$	$Ia\,3/T_h^7$	$Ia\,3/T_h^7$	$Ia\,3d/O_h^{10}$
Site symmetry	C_2, C_{3i}	C_2, C_{3i}	C_2, C_{3i}	D_2, C_{3i}, S_4
Density of cation sites (10^{20} cm^{-3})	335.5	268.7	285.2	138
Lattice constant (Å)	9.844	10.603	10.391	12.00
Coordination numbers	6	6	6	8, 6, 4
Effective ionic radius of the host cation (Å)	0.75	0.90	0.86	1.02
Thermal conductivity at 30°C (W/mK), undoped	16.5	13.6	12.5	11.0
Doped with 3% Yb (W/mK)	6.6	7.7	11.0	6.8
Transparency range (μm)	0.22–8	0.23–8	0.23–8	0.18–6
Maximum phonon energy (cm^{-1})	672	597	618	857
Moh's hardness	6.5	6.5	6–6.5	8.5

© Woodhead Publishing Limited, 2013

Yb doping, but also for Tm and Er doping, Lu_2O_3 is the most favourable sesquioxide host due to its distribution coefficient close to 1 and the fact that the very high heat conductivity of 12.6 W/mK of undoped Lu_2O_3 is only slightly reduced by about 10% if for example the crystal is doped with high Yb concentrations (Peters *et al.* 2011a). In the case of YAG this reduction amounts to nearly a factor of two because of the large mass difference between ytterbium and yttrium. Thus, Yb-doped $Lu_3Al_5O_{12}$ (LuAG) is more suitable than Yb:YAG, but Yb:Lu_2O_3 turns out to be the most promising active medium for high-power continuous wave and pulsed lasers.

1.5.2 Spectroscopy and laser action of rare earth-doped sesquioxides

Nd^{3+}-doped sesquioxides

The Nd^{3+} ion offers a large number of emission lines in the NIR spectral range, starting from the $^4F_{3/2}$ upper laser level and ending in the $^4I_{13/2}$, $^4I_{11/2}$, and $^4I_{9/2}$ manifolds. These groups of emission lines are shown in Fig. 1.7 for all three sesquioxides together with the corresponding level scheme of Nd:YAG for comparison. The strongest and most commonly used laser

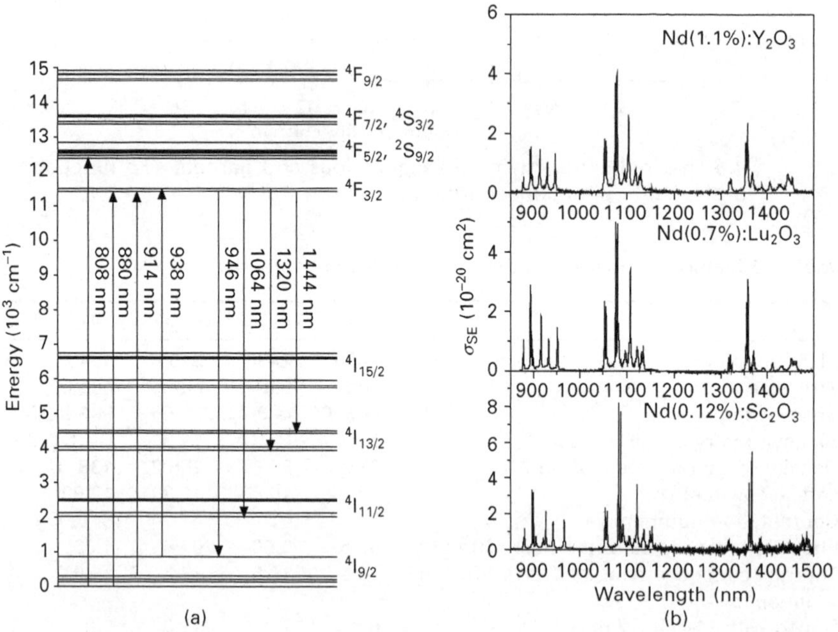

1.7 (a) Energy-level scheme of Nd^{3+}:YAG with the well-known laser transitions; (b) room-temperature emission spectra of Nd-doped sesquioxides (Fornasiero 1999).

© Woodhead Publishing Limited, 2013

transition $^4F_{3/2} \rightarrow {}^4I_{11/2}$ is observed between 1050 nm and 1150 nm with maximum emission cross-sections between 4×10^{-20} cm^2 and 8×10^{-20} cm^2, which is a factor 4–8 less than in YAG (Fornasiero 1999). The absorption cross-sections around 800 nm, which are the typical pumping wavelengths of Nd lasers, are also about 5×10^{-20} cm^2. The measured radiative lifetime of the $^4F_{3/2}$ multiplet is, with 400 µs in Nd:Y_2O_3 and Nd:Lu_2O_3, very similar to that of Nd:YAG, but a little shorter in Nd:Sc_2O_3 (300 µs). Non-radiative transitions are less probable in all sesquioxides due to the quite low phonon energies of about 600 cm^{-1}.

Nd lasers have been realised successfully in all three sesquioxides (Stone and Burrus 1978) as well as in the mixed oxides $YScO_3$ (Bagdasarov *et al.* 1975) and very recently in $Lu_{2-x}Sc_xO_3$ (Reichert *et al.* 2012). Slope efficiencies up to 58% have been achieved so far under Ti:sapphire pumping (Fornasiero 1999). The longest wavelengths for all Nd transitions, being 20–40 nm longer than in Nd:YAG, are obtained with Nd:Sc_2O_3 (Fornasiero *et al.* 1999). The reason is that Sc_2O_3 provides the largest crystal field splitting due to the small Sc-lattice sites being substituted by the large Nd ions.

Yb^{3+}-doped sesquioxides

The absorption and emission spectra of Yb^{3+} are less structured than those of Nd^{3+} because of the simple level scheme with the ground state $^2F_{7/2}$ and the only excited state $^2F_{5/2}$. The quite broad spectra of Yb:Lu_2O_3 displayed in Fig. 1.8 indicate a rather strong electron–phonon coupling of the Yb^{3+} ion, which is useful for short pulse generation and for tuning of the Yb^{3+} laser emission. The absorption cross-section of the 0-phonon transition at 976 nm is as high as 3.4×10^{-21} cm^2, which is higher by a factor of four than in Yb-doped YAG and indicates that this is the most effective wavelength for pumping (also in view of a low quantum defect). However, the line width is only 2.5 nm (FWHM), making it necessary to use wavelength-stabilised pump diodes.

The reabsorption-free fluorescence spectrum consists – besides the zero-phonon line – of two emission bands at 1032 nm and 1079 nm caused by the Stark splitting of the Yb ground state. The calculated cross-sections are 13×10^{-21} cm^2 and 4×10^{-21} cm^2, respectively, compared to 19×10^{-21} cm^2 and 3×10^{-21} cm^2 in Yb:YAG. Also the fluorescence lifetimes are comparable, that is 820 µs in Yb:Lu_2O_3 and 954 µs in Yb:YAG (Peters *et al.* 2008). Similar spectroscopic data have been measured in Yb:Sc_2O_3, Yb:Y_2O_3, and several mixed crystals (Peters *et al.* 2010). It is worth noting that the Yb^{3+} lifetimes of all sesquioxides and of many other oxides may be quenched at doping concentrations above about 1.5×10^{21} cm^{-3} by energy transfer to impurities or even divalent ytterbium ions (Fagundes-Peters *et al.* 2007).

Comparable laser performances have been achieved with Yb:Lu_2O_3,

© Woodhead Publishing Limited, 2013

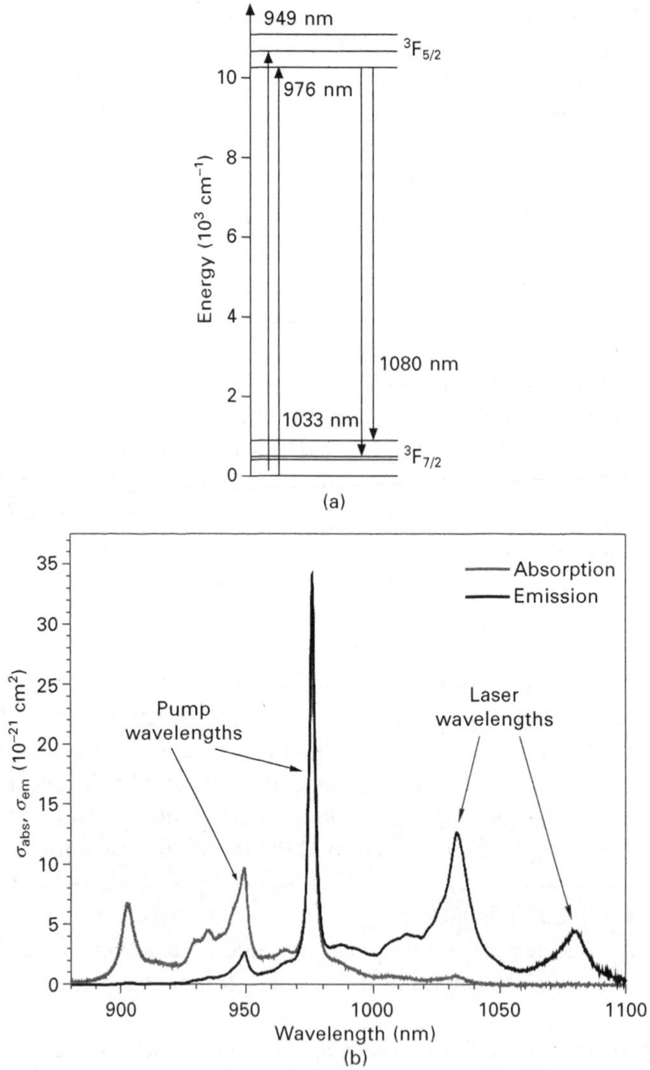

1.8 (a) Energy-level scheme of Yb^{3+}; (b) room-temperature absorption and emission spectra of Yb^{3+}:Lu_2O_3; the emission spectrum is corrected for reabsorption losses by the reciprocity method and the Fürchtbauer–Ladenburg equation.

Yb:Sc_2O_3, and Yb:$LuScO_3$, whereas Yb:Y_2O_3 seems to suffer from its hexagonal → cubic phase transition at about 2330°C, causing a large number of scattering centres (Zinkevich 2007). Nevertheless, Ti:sapphire- and diode-pumped laser tests have been performed in a nearly concentric linear resonator (Petermann *et al.* 2002) and slope efficiencies up to 61% for different outcoupling mirrors have been obtained.

© Woodhead Publishing Limited, 2013

For the high-power experiments, according to Fig. 1.5, thin $Yb:Lu_2O_3$ and $Yb:Sc_2O_3$ disks with diameters between 5 mm and 6.5 mm and thicknesses between 150 μm and 400 μm have been soldered on the copper heat sink. At Yb concentrations between 2% and 5% with respect to the cation sites more than 95% of the incident pump power was absorbed using a pump module with 24 light passes. As pump source a volume Bragg-grating (VBG) stabilised diode laser with more than 400 W output power was used, which was designed to fit exactly the zero-phonon line absorption of the investigated sesquioxides. Due to the grating stabilisation only a very low wavelength shift was observed with increasing output power and temperature.

In Fig. 1.9 the input/output characteristics of the 2% $Yb:Lu_2O_3$ laser in a linear resonator setup is displayed with optimised disk thickness and output coupler. The curves for the two different pump spot sizes clearly demonstrate the power scaling ability of the thin-disk design, keeping the pump power density constant and thus limiting the risk of catastrophic disk damage. More than 300 W output power (142 W in fundamental mode) at 1034 nm wavelength have been measured at a slope efficiency of 85%, which corresponds to an optical-to-optical efficiency of 73% (Peters *et al.* 2011b). Decreasing the pump spot size to 1.9 mm diameter results in an even higher slope efficiency of 88%; indeed, the theoretical limit due to the quantum defect is 94%.

Also $Yb:Sc_2O_3$ and the mixed oxide $Yb:LuScO_3$ showed slope efficiencies above 80% and output powers in excess of 250 W. In Table 1.4 the optimised laser parameters and the corresponding cw-output data are summarised. These

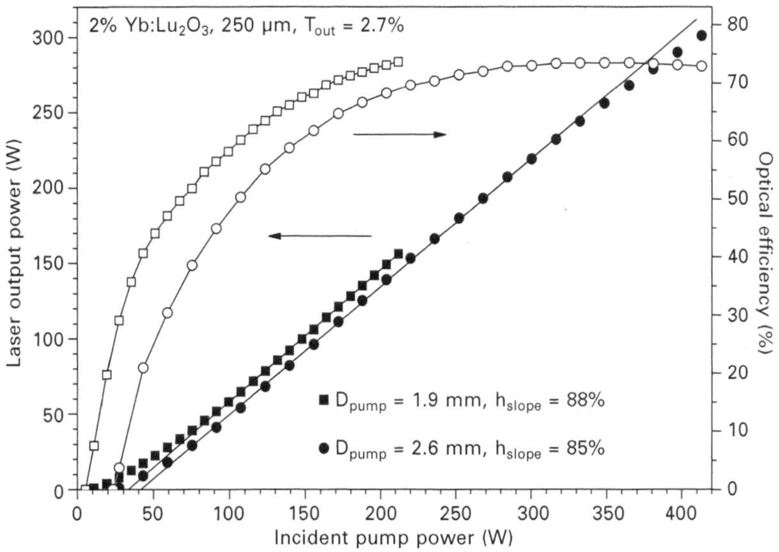

1.9 Laser characteristics of a 250 μm $Yb^{3+}:Lu_2O_3$ disk for two different pump spot diameters and an optimum outcoupling rate of 2.7%.

© Woodhead Publishing Limited, 2013

Table 1.4 Thin-disk laser data of Yb-doped sesquioxides

	$LuScO_3$	Sc_2O_3	Lu_2O_3
Laser wavelength (nm)	1041	1042	1034
Outcoupling transmission (%)	1.2	1.2	2.7
Disk thickness (μm)	200	200	250
Doping concentration (%)	3	2.4	2
Pump spot diameter (mm)	4	4	2.6
Pump power (W)	365	380	413
Output power (W)	250	264	301
Optical-to-optical efficiency (%)	69	70	73
Slope efficiency (%)	81	80	85
Pump power density (kW/cm^2)	2.9	3.0	7.8

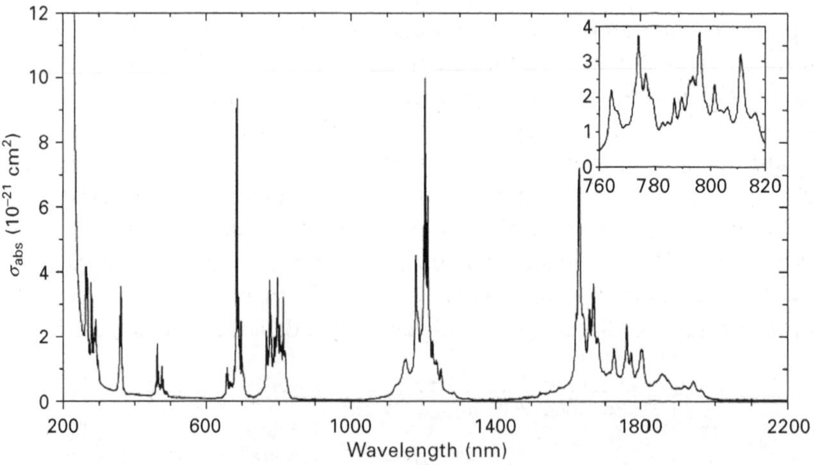

1.10 Room-temperature absorption spectrum of Tm^{3+}:Lu_2O_3; the inset shows the absorption in the 800 nm region.

data clearly demonstrate the high potential of the Yb-doped sesquioxides for high-power solid-state lasers.

Additionally, it should be mentioned that due to the broad gain spectrum of the Yb ion all sesquioxide lasers can be tuned continuously with a birefringent filter from about 1000 nm to more than 1100 nm at power levels between 10 W and 20 W (Peters *et al.* 2007).

Tm- and Ho-doped sesquioxides

The absorption spectra of Tm^{3+} look very similar in all sesquioxides. The room temperature spectrum of a Tm:Lu_2O_3 crystal is shown in Fig. 1.10. The inset indicates that the Tm ion can be pumped efficiently in a quite broad spectral region around 800 nm with no special requirements for the pump

© Woodhead Publishing Limited, 2013

source such as wavelength or temperature stabilisation. The emission spectrum in the 2 μm wavelength region is displayed in Fig. 1.11. The calculated emission cross-sections are three times higher than those of Tm:YAG or Tm:$YLiF_4$. The fluorescence lifetime of the $^3F_4$ excited state measured with a 1% doped sample was determined to be 3.8 ms (reabsorption-free measurement by the pinhole method (Kühn *et al.* 2007)). The broad gain spectrum makes the Tm:Lu_2O_3 laser tunable and a promising candidate for the generation of short pulses.

In the same spectral range as Tm:Lu_2O_3 also Ho^{3+}-doped Lu_2O_3 is emitting. In the fluorescence spectrum two prominent peaks are located at 2124 nm and 2134 nm with cross-sections of 4.5×10^{-21} cm^2 and 2.3×10^{-21} cm^2, respectively, where free-running laser action can be expected. The highest absorption cross-sections of 11.7×10^{-21} cm^2 and 10.2×10^{-21} cm^2 are found at 1928 nm and 1940 nm, respectively, being suitable for in-band pumping of the Ho laser (Koopmann *et al.* 2011a).

For the laser experiments a very compact setup was designed with a 1.0% Tm-doped HEM-grown rod 2.5 mm in diameter and 15 mm in length and with a 110 W diode laser 796 mm wavelength as pump source. With an output coupler of T_{oc} = 3.8% as well as T_{oc} = 7%, slope efficiencies of 42% with respect to the incident power (46% for absorbed power) and a maximum output power of 41 W at 2065 nm wavelength have been achieved

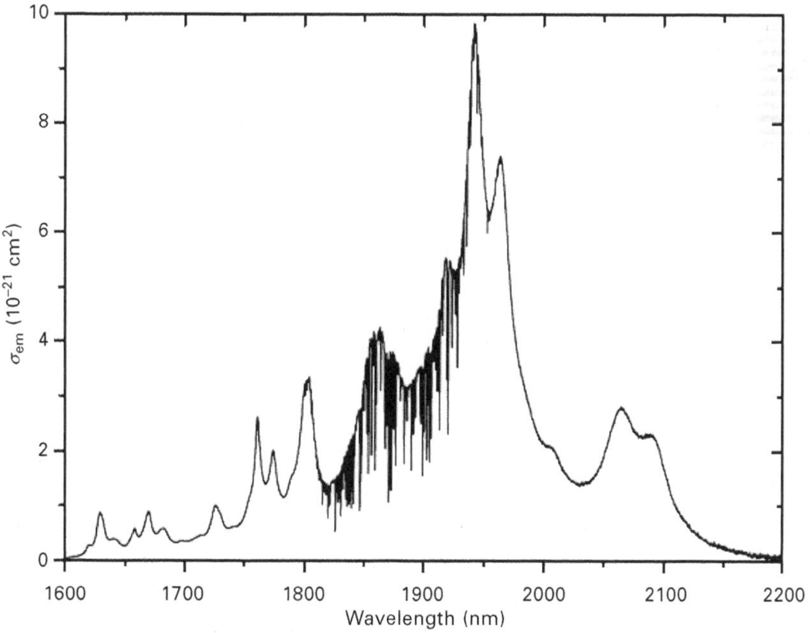

1.11 Reabsorption-free emission spectrum of Tm^{3+}:Lu_2O_3 in the 2 μm range; the noise around 1900 nm is due to water vapour absorption.

© Woodhead Publishing Limited, 2013

(Koopmann *et al.* 2011b). At higher outcoupling rates the laser wavelength switched to 1965 nm as expected from the higher gain at that wavelength. With $Tm:Sc_2O_3$ very similar laser results have been achieved under similar experimental conditions (Koopmann *et al.* 2011c).

Even longer wavelengths may be realised with Ho-doped lutetia (Koopmann *et al.* 2011a). This laser material can be pumped very effectively by a Tm-fibre laser as was done in the past with Ho:YAG (Shen *et al.* 2004) or in a more compact design with GaSb-based high-power laser diodes (Lamrini *et al.* 2011). In the case of pumping the $Ho:Lu_2O_3$ laser with an amplified Tm-fibre laser a slope efficiency of 76% and an output power of 25 W at 40 W absorbed pump power were obtained recently (Koopmann 2012).

Er-doped sesquioxides

For LIDAR applications, for example for detection of carbon dioxide or methane in the atmosphere, special absorption wavelengths in the NIR spectral range have to be addressed. The specific CO_2 absorption wavelengths at 1579 nm and 1603 nm can be provided by Er^{3+}-doped mixed garnets, vanadates as well as sesquioxide lasers. This is demonstrated by the absorption and emission spectra of $Er^{3+}:Lu_2O_3$ in Fig. 1.12. The vertical lines of the CO_2

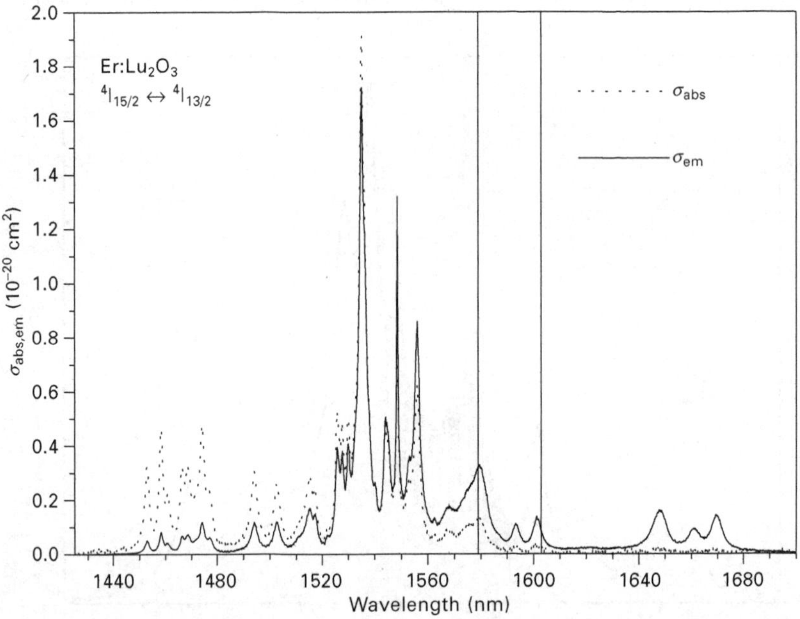

1.12 Room-temperature absorption and emission spectra of $Er^{3+}:Lu_2O_3$ in the NIR spectral range; the vertical lines around 1600 nm indicate specific CO_2 absorption lines suitable for LIDAR measurements.

© Woodhead Publishing Limited, 2013

absorption coincide quite well with two emission bands of the Er ion (Brandt *et al.* 2010). Fortunately, no excited state absorption can be detected in this wavelength range. The large absorption peak at 1535 nm with an absorption cross-section of 1.9×10^{-20} cm^2 is highly suitable for in-band pumping of the $Er:Lu_2O_3$ laser. Also with $Er^{3+}:Sc_2O_3$ the 1579.3 nm absorption line can nearly be reached (Fechner *et al.* 2008).

Since Er lasers in the eye-safe 1.6 μm wavelength region suffer from strong parasitic up-conversion processes, Er concentrations of less than 0.5% are mandatory in all oxides. The desired wavelength of 1579.3 nm has been realised nearly with a 19 mm long $Er(0.3\%):Lu_2O_3$ laser rod in-band pumped at 1536 nm with an Er-fibre laser (Brandt *et al.* 2010). An output power of 1.3 W at nearly 8 W incident pump power was achieved. The laser wavelength was 1580 nm and the maximum slope efficiency 32%. However, by using an $Er:(Lu,Y)_2O_3$ solid-solution crystal as active medium with an optimised Lu/Y ratio, that is by compositional tuning, the CO_2 absorption wavelength of 1579.3 nm could be reached exactly.

1.6 Mode-locked sesquioxide lasers

Yb- and Tm-doped sesquioxides are also useful for short-pulse generation due to the broad emission bandwidths of these ions. The techniques that have been used successfully are passive mode-locking with semiconductor saturable absorber mirrors (SESAM) or with single-walled carbon nanotubes as saturable absorbers. Furthermore, with Kerr-lens mode-locking the shortest pulses in the sub-100 fs range have been achieved (Tokurakawa *et al.* 2011).

If short pulses with high average power are required from a single oscillator, SESAM mode-locked thin-disk lasers are presently the best choice. With $Yb:Lu_2O_3$ as gain medium 141 W average power of 738 fs pulses have been demonstrated recently (Baer *et al.* 2010). Pulses as short as 74 fs have been obtained by Kerr-lens mode-locking with the mixed sesquioxide $LuScO_3$ providing an even broader gain bandwidth than the parent single oxides (Schmidt *et al.* 2010). Power scaling with this laser material has been performed recently in a thin-disk laser configuration resulting in 235 fs pulses and an average power of 23 W (Saraceno *et al.* 2011).

Also with Tm-doped Lu_2O_3 ultrashort pulses of 175 fs have been created in the 2 μm wavelength range with carbon nanotubes as saturable absorbers (Schmidt *et al.* 2012). According to the extremely large emission bandwidth of the Tm ion even shorter pulse widths are expected in future laser experiments.

1.7 Future trends

For efficient heat removal in high-power thin-disk lasers, the disk thickness should be as small as possible and thus the Yb concentration as high as

© Woodhead Publishing Limited, 2013

possible. However, in several oxides a very efficient non-linear loss process was observed at high Yb concentrations and high inversion densities, decreasing the laser efficiency drastically (Fredrich-Thornton 2010). This problem is not yet fully understood and should be solved in future.

Another point is that tunable lasers in the green to yellow wavelength range are still missing, closing the gap between the fundamental and frequency-doubled Ti:sapphire laser. Here, Ce^{3+}-doped lasers based on oxide host lattices may become available in future. Till now, only Ce-doped fluorides exist emitting in the 300 nm spectral range. Fixed wavelengths in the green spectral region can be realised in principle by Pr^{3+}, Ho^{3+}, and Er^{3+}, but only in fluorides and not in oxides. Maybe suitable oxide hosts will be found also for these ions.

Furthermore, a powerful tunable laser in the blue/near-UV spectral range is urgently needed for various applications in chemistry, biology, microscopy, and all types of spectroscopy. Again, Ce^{3+} could be a promising candidate, but the question of a suitable host lattice is still open.

1.8 References

Alderighi D, Toci G, Vannini H, Parisi D, Bigotta S, Tonelli M (2006), 'High efficiency solid state lasers based on Ce:$LiCaAlF_6$ crystals', *Appl. Phys. B*, 83, 51–54.

Avramescu A, Lermer T, Müller J, Eichler C, Bruederl G (2010), 'True green laser diodes at 524 nm continuous wave output power on c-plane GaN', *Appl. Phys. Express*, 3, 061003 1–3

Baer C R E, Kränkel C, Saraceno C J, Heckl O H, Golling M, Peters R, Petermann, K, Südmeyer T, Huber G, Keller U (2010), 'Femtosecond thin disk laser with 141 W of average power', *Opt. Lett.*, 35, 2302–2304.

Bagdasarov K S, Kaminskii A A, Kevorkov A M, Li L, Prokhorov A M, Tevosyan T A, Sarkisov S E (1975), 'Investigation of the stimulated emission of cubic crystals of $YScO_3$ with Nd^{3+} ions', *Sov. Phys. Dokl.*, 20, 681–683.

Brandt C, Tolstik N A, Kuleshov N V, Petermann K, Huber G (2010), 'Inband pumped Er:Lu_2O_3 and Er,Yb:YVO_4 lasers near 1.6 μm for CO_2 LIDAR', in *Advanced Solid-State Photonics*, OSA Technical Digest, paper AMB 15.

Bufetov I A and Dianov E M (2009), 'Bi-doped fibre lasers', *Laser Phys. Lett.*, 6, 487–504.

Bufetov I A, Melkumov M A, Khopin V F, Firstov S V, Shubin A V, Medvedkov O I, Guryanov A N, Dianov E M (2010), 'Efficient Bi-doped fibre lasers and amplifiers for the spectral region 1300–1500 nm', *Proc. SPIE* 7580, 758014.

DeLoach L D, Page R H, Wilke G D, Payne S A, Krupke W F (1996), 'Transition metal-doped zinc chalcogenides: spectroscopy and laser demonstration of a new class of gain media', *IEEE J. Quantum Electron.*, 32, 885–895.

Dieke G H (1968), *Spectra and Energy Levels of Rare Earth Ions in Crystals*, Baltimore, MD, John Wiley & Sons.

Fagundes-Peters D, Martynyuk N, Lünstedt K, Peters V, Petermann K, Huber G, Basun S, Laguta V, Hofstaetter A (2007), 'High quantum efficiency YbAG-crystals', *J. Lum.*, 125, 238–247.

© Woodhead Publishing Limited, 2013

Fechner M, Peters R, Kahn A, Petermann K, Heumann E, Huber G (2008), 'Efficient in-band-pumped Er:Sc_2O_3-laser at 1.58 μm', in *Conference on Lasers and Electro-Optics*, OSA Technical Digest, paper CTuAA 3.

Fechner M, Reichert F, Hansen N-O, Petermann K, Huber G (2011), 'Crystal growth, spectroscopy, and diode pumped laser performance of Pr,Mg:$SrAl_{12}O_{19}$', *Appl. Phys. B*, 102, 731–735.

Fedorov V V, Gallian A, Moskalev I, Mirov S B (2007), 'En route to electrically pumped broadly tunable middle infrared lasers based on transition metal doped II–VI semiconductors', *J. Lum.*, 125, 184–195.

Fibrich M, Jelínková, Šulc, Nejezchleb K (2009), 'Visible cw laser emission of GaN-diode pumped Pr:YAlO3 crystal', *Appl. Phys. B*, 97, 363–367.

Fornasiero L (1999), 'Nd^{3+}- and Tm^{3+}-dotierte Sesquioxide', PhD dissertation, University of Hamburg, Shaker Verlag, Aachen, Germany (in German).

Fornasiero L, Mix E, Peters V, Heumann E, Petermann K, Huber G (1999), 'Efficient laser operation of Nd:Sc_2O_3 at 966 nm, 1082 nm, and 1486 nm', *Advanced Solid-State Lasers,* OSA Trends in Optics and Photonics, 26, 249–252.

Fredrich-Thornton S T (2010), 'Nonlinear losses in single crystalline and ceramic Yb:YAG thin-disk lasers', PhD dissertation, University of Hamburg, Shaker Verlag, Aachen, Germany (in English).

Fukabori A, Chani V, Kamada K, Yanagida T, Yokota Y, Moretti F, Kawaguchi N, Yoshikawa A (2011), 'Growth of Y_2O_3, Sc_2O_3, and Lu_2O_3 crystals by the micro-pulling-down method and their optical and scintillation characteristics', *J. Cryst. Growth*, 318, 823–827.

Giesen A and Speiser J (2007), 'Fifteen years of work on thin-disk lasers: Results and scaling laws', *IEEE J. Sel. Top. Quantum Electron.*, 13, 598–609.

Giesen A, Hügel H, Voss A, Wittig K, Brauch U, Opower H (1994), 'Scalable concept for diode-pumped high-power solid-state lasers', *Appl. Phys. B*, 58, 365–372.

Gün T, Metz P, Huber G (2011), 'Power scaling of laser diode pumped Pr^{3+}:$LiYF_4$ cw lasers: efficient laser operation at 522.6 nm, 545.9 nm, 607.2 nm, and 639.5 nm', *Opt. Lett.*, 36, 1002–1004.

Hamilton D S, Gayen S K, Pogatshnik G J, Ghen R D, Miniscalco W J (1989), 'Optical-absorption and photoionization measurements from the excited states of Ce^{3+}:$Y_3Al_5O_{12}$', *Phys. Rev. B*, 39, 8807–8815.

Huber G and Petermann K (1985), 'Laser action in Cr-doped garnets and tungstates', in *Springer Series in Optical Sciences*, 47, Springer Verlag, 11–19.

Joyce D and Schmid F (2010), 'Progress in the growth of large scale Ti:sapphire crystals by the heat exchanger method (HEM) for petawatt class lasers', *J. Cryst. Growth*, 312, 1138–1141.

Kochurikhin V V, Shimamura K, Fukuda T (1995), 'Czochralski growth of gadolinium vanadate single crystals', *J. Cryst. Growth*, 151, 393–395.

Koopmann P (2012), 'Thulium- and holmium-doped sesquioxides for 2 μm lasers', PhD dissertation, University of Hamburg, Shaker Verlag, Aachen, Germany (in English).

Koopmann P, Lamrini S, Scholle K, Schäfer M, Fuhrberg P, Huber G (2011a), 'Multi-watt laser operation and laser parameters of Ho-doped Lu_2O_3 at 2.12 μm', *Opt. Mater. Express*, 1, 1447–1456.

Koopmann P, Lamrini S, Scholle K, Fuhrberg P, Petermann K, Huber G (2011b), 'Efficient diode-pumped laser operation of Tm:Lu_2O_3 around 2 μm', *Opt. Lett.*, 36, 948–950.

Koopmann P, Lamrini S, Scholle K, Fuhrberg P, Petermann K, Huber G (2011c), 'Long

© Woodhead Publishing Limited, 2013

wavelength laser operation of Tm:Sc_2O_3 at 2116 nm and beyond', *Advanced Solid-State Photonics Conference,* OSA Technical Digest, paper ATuA 5.

Kück S (2001), 'Laser-related spectroscopy of ion-doped crystals for tunable solid-state lasers', *Appl. Phys. B*, 72, 515–562.

Kühn H, Fredrich-Thornton S, Kränkel C, Peters R, Petermann K (2007), 'Model for the calculation of radiation trapping and description of the pinhole method', *Opt. Lett.*, 32, 1908–1910.

Lamrini S, Koopmann P, Schäfer M, Scholle K, Fuhrberg P (2011), 'Efficient high-power Ho:YAG laser directly in-band pumped by a GaSb-based laser diode stack at 1.9 μm', *Appl. Phys.*, 1, 1447–1456.

Laversenne L, Guyot Y, Goutaudier C, Cohen-Adad M T, Boulon G (2001), 'Optimization of spectroscopic properties of Yb^{3+}-doped refractory sesquioxides: cubic Y_2O_3, Lu_2O_3 and monoclinic Gd_2O_3', *Opt. Mater.*, 16, 475–483.

MacMillen C, Thompson D, Tritt T, Kolis J (2011), 'Hydrothermal single-crystal growth of Lu_2O_3 and lanthanide-doped Lu_2O_3', *Cryst. Growth Des.*, 11, 4386–4391.

Maiman T H (1960), 'Stimulated optical radiation in ruby', *Nature*, 187, 493–494.

Moulton P F (1985), 'Spectroscopic and laser characteristics of Ti:Al_2O_3', *J. Opt. Soc. Am. B*, 3, 125–133.

Ostroumov V G, Seelert W R, Hunziker L E, Ihli C, Richter A, Heumann E, Huber G (2007), 'UV generation by intracavity frequency doubling of an OPS-pumped Pr:YLF laser with 500 mW of cw power at 360 nm', *Photonics West* 6451, Technical Program and Proceedings of SPIE, 6451–02.

Pauling L and Shappell M D (1930), 'The crystal structure of bixbyite and the C-modification of the sesquioxides', *Z. Kristallographie*, 75, 128–142.

Pavel N, Lupei V, Saikawa J, Taira T, Kan H (2006), 'Neodymium concentration dependence of 0.94-, 1.06-, and 1.34-μm laser emission and of heating effects under 809- and 885-nm diode laser pumping of Nd:YAG', *Appl. Phys. B*, 82, 599–605.

Petermann K, Fornasiero L, Mix E, Peters V (2002), 'High melting sesquioxides: crystal growth, spectroscopy, and laser experiments', *Opt. Mater.*, 19, 67–71.

Peters V, Bolz A, Petermann K, Huber G (2002), 'Growth of high-melting sesquioxides by the heat exchanger method', *J. Cryst. Growth*, 237–239, 879–883.

Peters R, Kränkel C, Petermann K, Huber G (2007), 'Broadly tunable high-power Yb:Lu_2O_3 thin disk laser with 80% slope efficiency', *Opt. Express*, 15, 7075–7082.

Peters R, Kränkel C, Petermann K, Huber G (2008), 'Crystal growth by the heat exchanger method, spectroscopic characterisation, and laser operation of high-purity Yb:Lu_2O_3', *J. Cryst. Growth*, 310, 1934–1938.

Peters R, Petermann K, Huber G (2010), 'Growth technology and laser properties of Yb-doped sesquioxides', in Capper P and Rudolph P, *Crystal Growth Technology*, Weinheim, Germany, Wiley-VCH Verlag.

Peters R, Kränkel C, Fredrich-Thornton S T, Beil K, Petermann K, Huber G, Heckl O H, Baer C R E, Saraceno C J, Südmeyer T, Keller U (2011a), 'Thermal analysis and efficient high power continuous-wave and mode-locked thin disk laser operation of Yb-doped sesquioxides', *Appl. Phys. B*, 102, 509–514.

Peters R, Kränkel C, Beil K, Petermann K, Huber G, Heckl O H, Baer C R E, Saraceno C J, Südmeyer T, Keller U (2011b), 'Yb-doped sesquioxide thin disk laser exceeding 300 W of output power in continuous-wave operation', *Conference on Lasers and Electro-Optics (CLEO)*, San Jose, CA, 16 May 2010.

Reichert F, Fechner M, Koopmann P, Brandt C, Petermann K, Huber G (2012), 'Spectroscopy and laser operation of Nd-doped mixed sesquioxides $(Lu_{1-x}Sc_x)_2O_3$', *Appl. Phys. B*, 108, 475–478.

© Woodhead Publishing Limited, 2013

Richter A, Heumann E, Huber G (2007), 'Power scaling of semiconductor laser pumped praseodym-lasers', *Opt. Express*, 15, 5172–5178.

Roth P W, Maclean A J, Burns D, Kemp A J (2009), 'Directly diode-laser-pumped Ti:sapphire laser', *Opt. Lett.*, 34, 3334–3336.

Sangla D, Balembois F, Georges P (2009), 'Nd:YAG laser diode-pumped directly into the emitting level at 938 nm', *Opt. Express*, 17, 10091–10097.

Saraceno C J, Heckl O H, Baer C R E, Golling M, Südmeyer T, Beil K, Kränkel C, Petermann K, Huber G, Keller U (2011), 'SESAM's for high-power femtosecond modelocking: power scaling of an Yb:$LuScO_3$ thin disk laser to 23 W and 235 fs', *Opt. Express*, 19, 20288–20300.

Schmidt A, Petrov V, Griebner U, Peters R, Petermann K, Huber G (2010), 'Diode-pumped mode-locked Yb:$LuScO_3$ single crystal laser with 74 fs pulse duration', *Opt. Lett.*, 35, 511–513.

Schmidt A, Petrov V, Griebner U, Choi S Y, Yeom D I, Rotermund F, Koopmann P, Petermann K, Huber G, Fuhrberg P (2012), 'Femtosecond mode-locked thulium-doped Lu_2O_3 laser around 2 μm', *Advanced Solid-State Photonics Conference,* OSA Technical Digest, paper AM2A.5.

Sennaroglu A, Pollock C R, Nathel H (1995), 'Efficient continuous-wave chromium-doped YAG laser', *J. Opt. Soc. B*, 12, 930–937

Shen D Y, Abdolvand A, Cooper L J, Clarkson W A (2004), 'Efficient Ho:YAG laser pumped by a cladding pumped tunable Tm:silica-fibre laser', *Appl. Phys. B*, 79, 559–561.

Solé R, Nikolov V, Ruiz X, Gavaldà J, Solans X, Aguiló M, Diaz F (1996), 'Growth of β-$KGd_{1-x}Nd_x(WO_4)_2$ single crystals in $K_2W_2O_7$ solvents', *J. Cryst. Growth*, 169, 600–603.

Sorokina I T (2004), 'Crystalline lasers' in *Springer Topics in Applied Physics*, 89, Springer Verlag, 255–349.

Stone J and Burrus C A (1978), 'Nd:Y_2O_3 single-crystal fibre laser: Room temperature cw operation at 1.07- and 1.35-μm wavelength', *J. Appl. Phys.*, 49, 2281–2287.

Struve B, Huber G, Laptev V V, Shcherbakov I A, Zharikov E V (1983), 'Tunable room-temperature cw laser action in Cr^{3+}:GdScGa-garnet', *Appl. Phys. B*, 30, 117–120.

Tokurakawa M, Shirakawa A, Ueda K, Peters R, Fredrich-Thornton S T, Petermann K, Huber G (2011), 'Ultrashort pulse generation from diode pumped mode-locked Yb^{3+}:sesquioxide single crystal laser', *Opt. Express*, 19, 2904–2909.

Vlasenko N A, Oleksenko P F, Mukhlyo M A, Veligura L I, Denisova Z L (2009), 'Stimulated emission of Cr^{2+} ions in ZnS:Cr thin-film electroluminescent structures', *Semicond. Phys., Quantum Electron. Optoelectron.*, 12, 362–365.

Zhavoronkov N, Avtukh A, Mikhailov V (1997), 'Chromium-doped forsterite laser with 1.1 W of continuous-wave output power at room temperature', *Appl. Opt.*, 36, 8601–8605.

Zinkevich M (2007), 'Thermodynamics of rare earth sesquioxides', *Prog. Mater. Sci.*, 52, 597–647.

© Woodhead Publishing Limited, 2013

2
Fluoride laser crystals

R. MONCORGÉ, A. BRAUD, P. CAMY and J. L. DOUALAN, University of Caen, France

DOI: 10.1533/9780857097507.1.28

Abstract: Fluoride crystals have been studied for a long time and many review papers have been produced in the past to give the state of the art in this field. So, concerning this past work, the present chapter only describes the materials which gave rise to the most interesting laser results or which could be reconsidered in the near future because of specific applications or because of new forthcoming technologies. The chapter first concentrates on general considerations of the crystal growth, structural, optical and thermo-mechanical properties of fluoride crystals, then discusses the investigation and development of some rare-earth doped and undoped fluoride crystals for up-to-date laser applications such as XPW pulse cleaning, RGB colour display and high-power laser chains.

Key words: fluoride, crystal growth, rare-earth, spectroscopy, solid-state laser.

2.1 Introduction

Interest in fluoride laser crystals dates back to the early 1960s with the beginning of solid-state lasers, and many review papers have been published [1–6]. Fluorites like CaF_2 or SrF_2 doped with divalent as well as trivalent lanthanides and actinides such as Sm^{2+} [7], Tm^{2+} [8], Dy^{2+} [9] and U^{3+} [10] were indeed among the first crystalline and transparent ceramic host laser materials which operated in the CW regime and were diode-pumped. This interest is due to a number of advantages which make them either complementary or even preferable to oxide materials, their usual competitors.

Fluoride materials, first, have generally much higher bandgaps (up to about 12 eV) and lower phonon frequencies (down to about 400 cm^{-1}) than the usual oxides. This means wider transparency ranges which extend on both the short and long wavelength sides and which allow a number of applications such as UV-photolithography and nonlinear frequency conversion in the near-UV, in the case of undoped materials, or laser emission, when doped with appropriate laser-active ions, either in the near-UV or in the mid-infrared. Wide bandgap and low birefringence fluorides like CaF_2 or BaF_2 have to be used when operating UV molecular and excimer lasers like F_2 and ArF lasers around 156 and 193 nm to avoid any damage to the optics. Fluorides with high nonlinear susceptibilities and high transmittance in the near-UV

© Woodhead Publishing Limited, 2013

are also actively investigated for nonlinear frequency conversion of visible to UV laser radiations and for some nonlinear wave-mixing processes used to improve the temporal contrast of intense laser pulses (see Section 2.5). Last but not least, wide bandgap materials such as fluorides are often preferred to avoid detrimental optical losses due to excited-state and multi-photon absorptions within the energy levels of the laser-active ions up into the conduction band, which is especially important for laser materials operating at the shortest wavelengths but also for laser materials, such as Yb^{3+} doped laser amplifiers (see Section 2.4), operating at high power levels.

Low phonon frequencies are recommended to reduce multi-phonon non-radiative relaxations and to produce laser transitions between adjacent energy levels with a maximum quantum efficiency. This is called the 'energy gap law' according to which the non-radiative relaxation probability between two adjacent energy levels becomes negligible when the considered energy gap exceeds about five times the highest phonon energy of the material. This is the case, for instance, for the RGB (red, green, blue) laser transitions observed from the $^3P_0$ energy level of Pr^{3+} ions in most of the fluorides (see Section 2.3).

Another advantage of fluorides lies in their low refractive index and negative thermo-optic coefficient. A low refractive index is favourable to achieve a high stimulated emission cross-section and a low laser threshold. A negative thermo-optic coefficient dn/dT is useful to compensate for thermal lensing effects resulting from positive thermal expansion and photo-elastic coefficients which appear in most laser materials at high pump and laser levels. It often maintains the stability of the laser resonators without using additional compensating optics.

According to these requirements (see Table 2.1), colquiirite and scheelite crystals like Li(Ca or Sr)AlF_6 and $LiLuF_4$ doped with Ce^{3+} ions, for instance, led to tunable laser operation between about 280 and 340 nm with laser slope efficiencies up to 55% [11, 12]. Some authors even reported laser operation of Nd^{3+} in LaF_3 and $LiYF_4$ around 172 and 260 nm, respectively [13, 14]. On the other hand, $LiYF_4$, BaY_2F_8, KY_3F_{10} and CaF_2 crystals doped with Tm^{3+} or Er^{3+} ions led to tunable laser operation around 1.9 μm [15] and 2.8 μm [16, 17], respectively, with laser slope efficiencies up to 50%. Room-temperature laser operation was even achieved on a 4.3 μm laser transition of a Dy^{3+} doped $LiYF_4$ crystal [18].

A number of laser materials based on fluoride crystals doped with transition metal ions of the iron group such as Cr^{3+}, V^{2+}, Ni^{2+} and Co^{2+} were also successfully operated in the past in the near- and the mid-infrared thanks to the phonon properties of fluorides. However, electron–phonon coupling and non-radiative relaxations as well as excited-state absorptions in the emitting levels of the laser systems [19–23] often remain too strong and very few of them could be operated at room temperature. As a matter of fact, only

© Woodhead Publishing Limited, 2013

Table 2.1 Structural properties of most important fluoride crystals and comparison with $Y_3Al_5O_{12}$

Matrices	CaF_2 [39, 40]	$LiYF_4$ [43]	$LiLuF_4$	KY_3F_{10} [47, 49, 50, 52]	BaY_2F_8 [43]	KYF_4 [53]	$Li(Ca,Sr)AlF_6$ [59, 60]	$Y_3Al_5O_{12}$
Structure	Cubic	Tetragonal	Tetragonal	Cubic	Monoclinic	Trigonal	Trigonal	Cubic
Space group (Schönflies)	Fm3m (O_h^5)	I4$_1$/a (C_{4h}^6)	I4$_1$/a (C_{4h}^6)	Fm3m (O_h^5)	C2/m (C_{2h}^3)	P3$_1$ (C_3^2)	$P\bar{3}_{1c}$ (D_{3d}^2)	Ia3d (O_h^{10})
Lattice constants (Å)	$a = 5.46$	$a = 5.16$, $c = 10.85$	$a = 5.123$, $c = 10.52$	$a = 11.54$	$a = 6.97$, $b = 10.46$, $c = 4.26$, $\beta = 99°7'$	$a = 14.06$, $c = 10.103$	$a = 5$, $c = 9.65$ (Ca), $a = 5.07$, $c = 10.19$ (Sr)	$a = 12$
Cell volume (Å^3)	$a^3 = 162.77$	$a^2c = 288.88$	$a^2c = 276.10$	$a^3 = 1532.80$	$abc\sin(\beta) = 303.57$	$\sqrt{3}\,a^2c/2 = 1729.62$	208.73 (Ca), 229.51 (Sr)	1729
Nb formula/unit cell	4	4	4	8	2	18	2	8
Site symmetry occupied by active ion	Ca^{2+} (O_h, C_{4v}, C_{3v}) [40]	Y^{3+}(S_4)	Y^{3+}(S_4)	Y^{3+}(C_{4v}) [52]	Y^{3+}(C_2)	–	Ca^{2+}, Sr^{2+}, Al^{3+} (O_h)	Y^{3+}(D_2)
RE^{3+} site density (10^{22} cm^{-3})	2.45	1.39	1.39	1.56	1.30	1.04	0.8	1.4

© Woodhead Publishing Limited, 2013

some Cr^{3+} doped colquiirites Cr:Li(Ca, Sr)(Al or Ga)F_6 and Co^{2+}:MgF_2 really gave rise to efficient room-temperature laser operation. The former could be tuned around 800 nm and operated CW as well as mode-locked [24, 25]. Optically pumped with low-power diode lasers, they were and remain the most compact diode-pumped and short-pulse solid-state laser systems ever achieved in this wavelength domain. Co^{2+}:MgF_2 was also operated at room temperature within a wide spectral range around 1.9 μm [26, 27], but it could be operated CW and mode-locked only at cryogenic temperatures [28, 29].

Today, however, except for Nd^{3+}:$LiYF_4$, which is often used, after frequency-doubling, as a pump source for Ti:sapphire laser amplifiers, and for Tm^{3+}:$LiYF_4$ which is used as a tunable laser source for LIDAR applications or as a pump source for 2 μm Ho lasers [30, 31], most of these laser systems have been abandoned or wait for new technology breakthroughs.

This is the case, for instance, for the Ce^{3+} doped UV laser materials mentioned above. These systems indeed were operated by pumping either with the frequency-quadrupled radiation of Q-switched Nd:YAG lasers at 266 nm, or with the frequency-doubled radiation of Cu-vapour lasers at 289 nm or even with another Ce-based laser system like Ce:$LiCaAlF_6$ around 290 nm [32], which was rather complicated and too expensive to allow any further development. These laser materials, however, remain quite interesting and might be used again in the near future once UV semiconductor diode lasers will be available at the watt level at the required pump wavelengths between about 250 and 300 nm. Indeed, thanks to their broadband tunability (30 nm around 320 nm in the case of Ce:$LiLuF_4$, for example [12]) compact and efficient diode-pumped sub-picosecond UV laser sources and laser amplifiers could be produced very simply at low cost.

This was the case of the Yb^{3+} doped laser materials, which we have not mentioned so far, with the emergence of high-power semiconductor diode lasers operating around 980 nm. This seems also to be the case for Pr^{3+} doped materials which have been known for a long time for their multicolour laser transitions and have found renewed interest in recent years thanks to the availability of blue semiconductor laser diodes around 445 nm.

So, to complete the reviews which have already been made in the past on fluoride laser crystals [2–6], the rest of the presentation will first focus on general considerations on their crystal growth and structural, optical and thermo-mechanical properties, then on the investigation and the development of some rare-earth doped and undoped fluoride crystals for up-to-date laser applications such as XPW (Cross-Polarized Wave) pulse cleaning, RGB colour display and high-power laser chains.

© Woodhead Publishing Limited, 2013

2.2 Crystal growth, structural, optical and thermo-mechanical properties of the most important fluoride crystals

Many fluoride crystals have been investigated in the past. However, only a few of them can be easily grown to a reasonable size (a few cm^3) with high enough optical quality and with sufficiently good thermo-mechanical properties to be prepared and used as laser media.

There is first *CaF_2 and its isotypes SrF_2 and BaF_2*. These fluorites occupy a particular place in the series since they are the only crystals which can be grown to a very large size (CaF_2 crystals up to 60 cm diameter have been grown in the past by using the Bridgman technique) while keeping a high optical quality [33], and they are the only ones with a thermal conductivity exceeding that of standard oxides. In fact, as reported in Table 2.2, pure CaF_2 has a thermal conductivity comparable to that of YAG ($Y_3Al_5O_{12}$), around 9.7 and 10.7 $Wm^{-1}K^{-1}$, respectively. This is very important for high-power laser applications. With a hardness of 4 mohs, fluorites remain rather smooth and fragile compared to oxides, so special care is needed to handle them. In particular, because of their high thermal expansion coefficient, they are characterized by a low thermal shock parameter. This means that the warming and cooling of laser elements should be made progressively. Moreover, one should avoid squeezing these laser elements in too rigid sample holders (indium foil is recommended). Once these requirements are respected, there is no problem in warming up and cooling down the crystals as much as necessary. Another advantage of fluorites is the possibility of growing them in the form of thin films by using either MBE (molecular beam epitaxy) or LPE (liquid phase epitaxy) techniques. Rare-earth doped CaF_2 layers of several tens of micrometres thus could be grown with an excellent quality onto pure CaF_2 substrates [34]. Moreover, when CaF_2 is doped with rare-earth ions, its refractive index can be increased by several percent [35]. This is the reason why efforts are presently being made to prepare such layers for waveguide and thin-disk laser applications.

The *scheelites $LiYF_4$ (LYF) and $LiLuF_4$ (LLF)* come immediately after the fluorites in terms of maturity and applications. Both crystals can be grown by using standard techniques (Bridgman or Czochralski) with an excellent quality but their size usually remains in the order of a few centimetres in diameter by a few centimetres long, which is largely enough for medium-power laser systems. However (see Fig. 2.1), while $LiLuF_4$ grows congruently, $LiYF_4$ does not. This means that it is often easier to get excellent optical quality $LiLuF_4$ crystals. This distinction was found to be especially critical when the crystals were doped with Ce^{3+} ions for UV laser applications. Laser results were indeed very different between Ce:LLF for which laser efficiency could reach up to 55%, and Ce:YLF for which laser efficiency was reduced to a few

© Woodhead Publishing Limited, 2013

Table 2.2 Optical and thermo-mechanical properties of typical fluoride crystals at room temperature (300 K)

Matrices	CaF_2 [39]	$LiYF_4$	$LiLuF_4$	KY_3F_{10}	BaY_2F_8	KYF_4	$Li(Ca,Sr)AlF_6$ [59, 60]	$Y_3Al_5O_{12}$ [88]
Initial melt composition	CaF_2	52%LiF–48%YF_3	50%LiF–50%LuF_3	25%KY–75%YF_3	33%BaF_2–67%YF_3	57%KY–43%YF_3	LiF–CaF_2–AlF_3	
Melting temperature (°C)	1420	810 [43]	860 [45]	990 [41], 1030 [51]	987–995 [43], 960 [44]	Slightly above 800 [55]	810 (Ca), 766 (Sr)	1940
Density (g/cm^3)	3.181	3.99 [42]	6.186 [42]	4.28	4.97–5.80 [44]	3.49 [61]	2.99 (Ca), 3.45 (Sr)	4.56
Max phonon frequency (cm^{-1})	495	460	460	495	420	–	580	
Transparency range (μm)	0.13–12	0.12–7.5	–	0.2–8.5	0.2–9.5 [61]	–	>0.115 (Ca), >0.12 (Sr)	0.25–5
Refractive indices	~1.42 at λ = 1 μm	n_o = 1.453, n_e = 1.475at λ = 639.5 nm [61]	n_o = 1.468, n_e = 1.494 at λ = 632.8 nm [61]	n = 1.4856 at 0.6 μm [44]	n_x = 1.5142, n_y = 1.5232, n_z = 1.5353, λ = 632.8 nm [44]	1.42	~1.39 (Ca), ~1.4 (Sr)	1.82 at 0.8 μm
dn/dT (10^{-6} K^{-1})	–11.5 [59]	–4.6 (E//a), –6.6 (E//c) [88]	–3.6 (E//a), –6.0 (E//c) [88]	–	–	–	–4.6 (E//c), –4.2 (E//a) (Ca)	7.3
Young modulus (GPa)	80–110	85 [60]	–	–	–	–	109 (Sr), 96 (Ca)	280
Hardness (moh)	4 158 knoop	4–5 [44]	3.5–4.5 [61]	4.5 [44]	4–5 [44]	3 [61]	–	8.5
Water [58] durability D_w at 50°C (mg/(cm^2day))	0.08	0.03	–	–	–	–	0.24 (Ca), 2.7 (Sr)	–

(Continued)

© Woodhead Publishing Limited, 2013

Table 2.2 Continued

Matrices	CaF_2 [39]	$LiYF_4$	$LiLuF_4$	KY_3F_{10}	BaY_2F_8	KYF_4	$Li(Ca,Sr)AlF_6$ [59, 60]	$Y_3Al_5O_{12}$ [88]
Thermal conductivity ($W\ m^{-1}\ K^{-1}$)	9.7, 6 (1% Yb) [91]	6 [42], 5.3 (*a*-axis), 7.2 (*c*-axis) [88]	6 [42], 5 (*a*-axis), 6.3 (*c*-axis) [88]	3 [49]	6 [42], 3.5 [88]	–	3.3 (Sr), ~4.6 to 5.1 (Ca)	11.2, 8.6 (2% Yb)
Thermal diffusivity (m^2/s) [87]	–	17–23	15–20	–	12	–	1.08 (Sr), ~1.7 (Ca)	41
Thermal expansion coefficient ($10^{-6}\ K^{-1}$)	18.9 [60]	14.31 along *a*-axis, 10.05 along *c*-axis [88]	13.6 along *a*-axis, 10.8 along *c*-axis [88]	14.5 [42]	17 (E//a), 18.7 (E//b), 19.4 (E//c) [44]	–	–10 along *c*-axis, +25 along *a*-axis (Sr)	7.7–8
Thermal shock parameter R'_T ($W/m^{1/2}$)	0.78 [60]	1.12 [60]	–	–	–	–	0.8 along *c*-axis, 0.42 along *a*-axis (Sr), 0.53 (Ca)	5.2 [60]

© Woodhead Publishing Limited, 2013

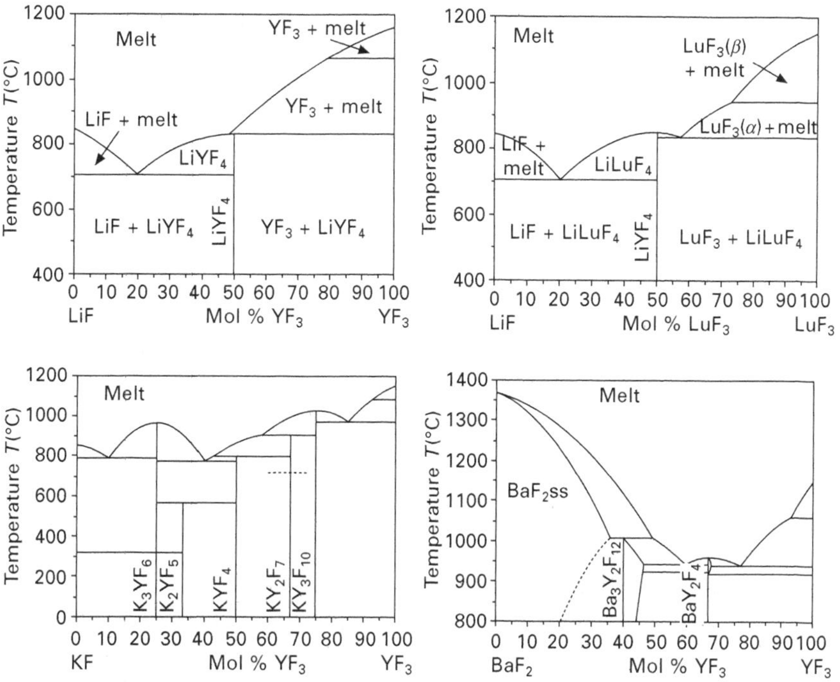

2.1 Phase diagrams of $LiYF_4$ [41], $LiLuF_4$ [42], KY_3F_{10} and KYF_4 [43, 48] and BaY_2F_8 [46] after Ref. 44.

percent. This was partly explained by the different bandgap energies of the materials (11 eV in LLF and 10 eV in YLF) but also by their crystal growth which probably favours the appearance of F^- vacancies, thus of electron traps and colour centres, in the case of YLF [36]. Both crystals are uniaxial crystals. Some care thus is needed in the orientation of the laser elements depending on the spectroscopic properties of the considered systems.

Scheelites, like fluorites, are very stable crystals with excellent water durability (important for water-cooled laser rods). They also have large, although lower than other fluorides, thermal expansion coefficients, but these coefficients are anisotropic, which has to be taken into account in high-power laser systems. Like fluorites, these large expansion coefficients are somewhat compensated, in the presence of thermal lensing effects, by their negative thermo-optical coefficients. We remind readers here, for the sake of clarity, that the thermal lens focal length f_T which appears in laser materials at high pump power is expressed as:

$$f_T = \frac{2\pi K_c \omega_p^2}{\eta_h P_{abs}}[dn/dT + (n-1)(1+\nu)a_T + n^3 a_T C_{r,\phi}]^{-1}$$

where ω_p, η_h and P_{abs} stand for the pump waist radius, the fractional thermal

© Woodhead Publishing Limited, 2013

load and the absorbed pump power, respectively, and K_c, dn/dT, a_T and $C_{r,\phi}$ represent the thermal conductivity, the thermo-optical, the thermal expansion and the photo-elastic coefficients of the material; n is its refractive index and v the Poisson ratio.

For instance, thanks to this compensating effect, it has been proved several times that thermal lensing effects were much more reduced in Nd:YLF than in Nd:YAG [37, 38].

Third in the series is the *cubic system* KY_3F_{10}. This crystal grows congruently and its thermo-optical and thermo-mechanical properties are comparable to those of the scheelites. It is more difficult, however, to get really good optical quality crystals. They often appear as milky. Good quality crystals are really obtained by using absolutely moisture-free raw materials and a very clean growing procedure. From the structural point of view, the unit cell is made of two structural units ($[KY_3F_8]^{2+}$ and $[KY_3F_{12}]^{2-}$) which alternate along the three crystallographic axes (see Fig. 2.2) and the trivalent rare-earth ions substitute for Y^{3+} ions in sites of C_{4v} symmetry. They are surrounded by eight fluorine ions forming square-based antiprisms. Because of the existence of two types of arrangements of fluorine ions, cubes and cubo-octahedra, and the distortion introduced by the rare-earth dopants, the interatomic distances slightly change from one site to the other. From the spectroscopic point of view, this partially disordered structure generally results in broader absorption and emission lines, thus in reduced cross-sections compared to that found in the case of scheelites.

The most popular fluoride, which then comes in the series, is the *monoclinic crystal* BaY_2F_8 (BYF). This is a biaxial crystal where the main symmetry axis is the b crystallographic axis (see Fig. 2.2) which is perpendicular to the crystallographic axes a and b (which form an angle $\beta = 99.7°$) and to the

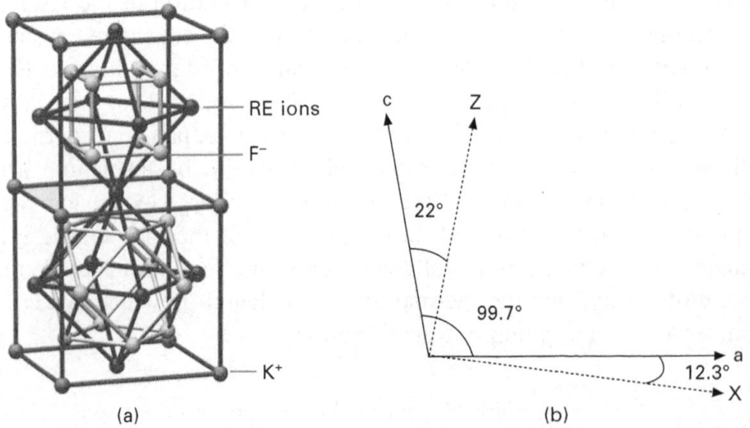

2.2 (a) Structure of KY_3F_{10} [49]; (b) respective positions of the (a, c) crystallographic and (X, Z) optical axes for BaY_2F_8.

© Woodhead Publishing Limited, 2013

optical axes X and Z. BaY_2F_8 also melts congruently but is not as easy to grow as the previous fluoride crystals. This is probably due to the large size of the Ba^{2+} ion and to its particular phase diagram (see Fig. 2.1). Nevertheless, good quality crystals with reasonable size can be grown by using either the Bridgman or the Czochralski technique. Due to its monoclinic properties, more efforts are necessary to cut and orient the crystals, then to record the absorption and emission spectra of the rare-earth doped crystals. Indeed, the spectra need to be recorded not only with light beams polarized along the three optical axes X, Y = b and Z but also (in case of magnetic dipole allowed optical transitions) for light propagating along these three optical directions. These spectra usually present sharp and strongly polarized lines with high absorption and emission cross-sections. BaY_2F_8 has a higher refractive index but its thermo-optical and mechanical properties are comparable to those of the other fluorides.

The fifth kind of fluoride crystal is *KYF_4*. KYF_4 melts non-congruently, which makes crystal growth rather critical. Good optical quality crystals, however, can be grown. Like KY_3F_{10}, this crystal is a superstructure derived for fluorite. As a matter of fact, it has been sometimes considered as a multisite host material [54, 55] and sometimes as a disordered one [53, 56, 57]. It results in broad absorption and emission lines. This implies reduced absorption and emission cross-sections but also complementary wavelength domains which can be implemented for some applications which necessitate more flexibility for the pump wavelength or laser operation at a specific wavelength. This is the reason why it has been developed and is still being investigated (see Section 2.4), despite less favourable thermo-optical and mechanical properties.

The last of this non-exhaustive list of fluoride laser materials are the *colquiirites $LiSrAlF_6$ and $LiCaAlF_6$* and, to a lesser extent, $LiSrGaF_6$. As indicated above, these host crystals have been mainly investigated with Ce^{3+} and Cr^{3+} dopants. Ce^{3+} ions substitute for Li^+ (Ca^{2+}), which was often accommodated, for charge compensation, by codoping with monovalent Na^+ ions. Very few works have been produced on other trivalent rare-earth dopants. The reason resides in the fact that rare-earth ions in these materials occupy cubic environments and that the oscillator strengths of the optical transitions are usually very weak. Ce^{3+} was more particularly investigated because the involved optical transitions were not f–f forbidden transitions, as in the case of the other rare-earth ions, but strong and broad electric-dipole allowed f–d transitions. These host materials were also very successfully developed with Cr^{3+} dopants because of their broadly tunable and ultra-short pulse laser operation. In that case, Cr^{3+} substitute for the Al^{3+} or Ga^{3+} ions and there is no need for a charge compensator. Very good optical quality and large-sized single crystals could be grown (see [3] for pictures), for UV-photolithography, because of their exceptionally high transmittance up

© Woodhead Publishing Limited, 2013

to about 120 nm, and for the above-mentioned laser applications. However, nowadays, the enthusiasm for this type of material has been somewhat stopped, maybe in part because of their fragility (sensitivity to moisture, low hardness) and their poor thermo-mechanical properties (low thermal conductivity, low thermal shock parameters).

Of course, many other fluoride laser hosts have been investigated in the past and have shown interesting properties, for instance LaF_3, MgF_2, $KZnF_3$, $KMgF_3$, $SrAlF_5$, $BaMgF_4$, $BaMnF_4$, K_2YF_5, $BaLiF_3$, etc. However, either due to complicated crystal growth procedures or because of spectroscopic properties of rare-earth or transition metal ions of limited interest, none of them presently really gives rise to any application.

2.3 Pr^{3+} doped crystals for RGB video-projection and quantum information experiments

Many efforts have been dedicated in the past to the investigation of compact solid-state lasers operating in the visible range. There was even for a while a conference entitled 'Compact blue-green solid-state lasers' in parallel to the other well-known one 'Advanced solid-state lasers'. For that purpose, many rare-earth doped materials and excitation/emission processes were considered [62]. The most important ones were based on fluoride crystals and glasses doped with Pr^{3+}, Er^{3+} and Tm^{3+} ions, eventually co-doped with Yb^{3+} sensitizer ions, and most of them were operated via some up-conversion pumping schemes (excited-state absorption, up-conversion energy transfer, photo-avalanche). However, despite all these efforts, none of these systems has been sufficiently efficient in terms of wall-plug laser efficiency to really justify any industrial development. The situation has considerably changed in the last few years with the emergence and the development of semiconductor diodes based on gallium nitride (GaN) emitting a UV-blue laser light around 445 nm which perfectly fits the $^3P_2$ absorption band of Pr^{3+} [63] (see Fig. 2.3).

This was immediately followed by the demonstration of efficient diode-pumped laser operation of a number of Pr^{3+}-doped fluoride crystals [44, 45, 64–66] at red, orange, green and blue wavelengths (which had already been demonstrated in the past, but with much lower efficiencies, by using an argon ion laser as a pump source [67]) (Figs 2.3 and 2.4). These results now pave the way for the development of new RGB (red, green, blue) solid-state laser sources for the new generation of high-definition laser TVs and for giant or miniature video-projection systems. The emission lines offered by the Pr^{3+} ion, at least in the red and the green, indeed match very well what is needed in terms of brightness, contrast and colour gamut for these applications. They also offer the possibility of building a compact and stable orange solid-state laser source which can be used for quantum information experiments [68] instead of using cumbersome dye lasers [69].

© Woodhead Publishing Limited, 2013

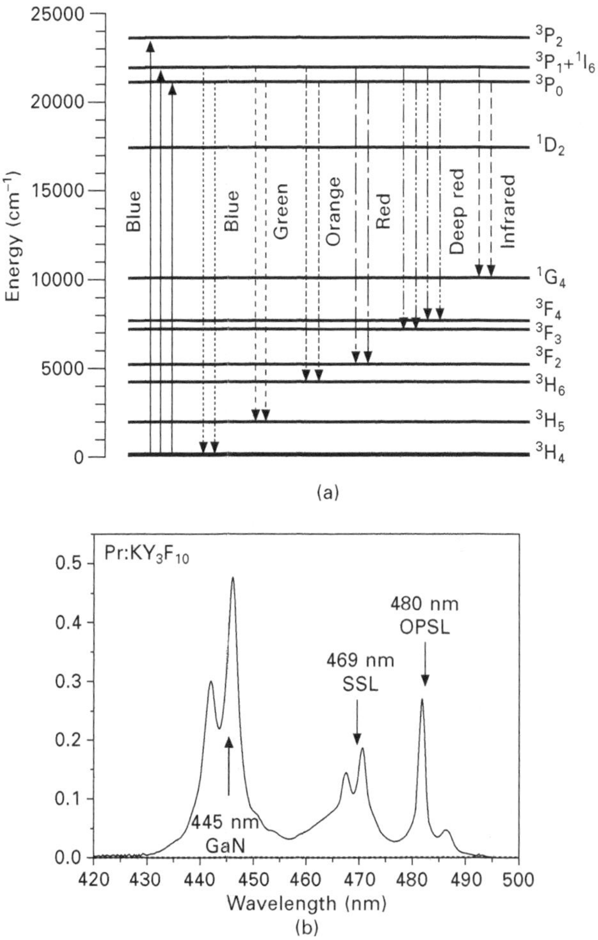

2.3 (a) Energy level diagram and pump and emission transitions of Pr^{3+}-doped fluorides; (b) blue absorption spectrum and main excitation wavelengths and pump sources for Pr:KY_3F_{10}.

However, before reaching these objectives, a number of improvements still need to be realized and new systems need to be discovered. Indeed, first, the commercially available blue diodes do not exceed about 1 watt and their beam quality is very poor. This is the reason why two other alternative solid-state laser pump sources have been considered in the past few years, i.e. intracavity frequency-doubled infrared solid-state lasers based on optically pumped semiconductors (OPSL) and Nd-doped laser crystals (NDLC) operating as three-level laser systems.

Efficient lasing of Pr:$LiYF_4$ and Pr:$LiLuF_4$ has thus been reported in the green (523 nm), orange (607 nm) and red (640 nm) by pumping the crystals

© Woodhead Publishing Limited, 2013

with an intracavity frequency-doubled OPSL delivering up to 1.6 W at 479.5 nm [66], which perfectly fits the $^3P_0$ absorption peak of Pr^{3+} in $LiYF_4$ and $LiLuF_4$. This solution has the advantage of being easily scalable, whenever high-power devices really find an interest in the marketplace. However, there are also several drawbacks. This is not a really compact solution, OPSLs are only commercially available on demand and remain very expensive and they involve an absorption line around 480 nm which is usually intense but also very sharp, which means a reduced wavelength flexibility for pumping a particular system or switching for another one. The second solution which is based, for instance, on a standard Nd:YAG laser crystal pumped by a fibre-coupled laser diode and associated with a LBO (LiB_3O_5) nonlinear crystal

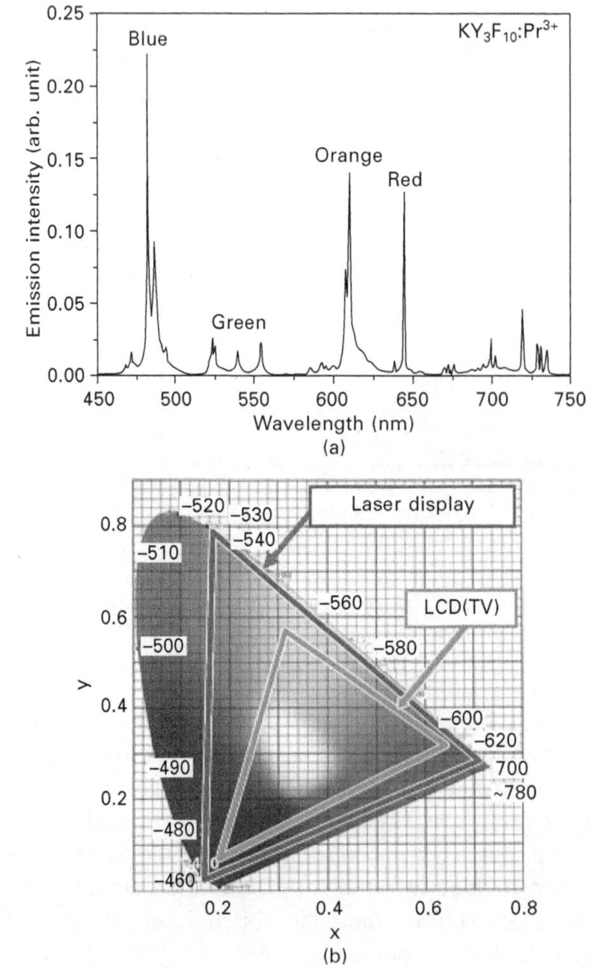

2.4 (a) Blue–red emission spectrum of Pr^{3+}:KY_3F_{10}; (b) colour triangle for laser display and (LCD) color TV.

© Woodhead Publishing Limited, 2013

placed inside the cavity, thus might be more favourable. In this case, the Nd:YAG laser operates on a three-level laser transition around 938 nm and gives rise, after frequency doubling, to a laser emission around 469 nm. This wavelength matches another absorption peak of Pr^{3+} which is usually less intense than the others located around 445 and 480 nm but which gives more flexibility thanks to a slightly larger bandwidth. Such a system has already been implemented to produce a blue laser radiation of nearly 1.4 W as well as several hundreds of mW at 471 and 473 nm, thanks to intracavity frequency doubling and frequency summing of the other well-known three-level laser transition at 946 nm [70, 71]. Again, using such a pump source, efficient lasing of Pr:$LiYF_4$ and Pr:$LiLuF_4$ could be obtained [72]. Lasing was also obtained at several wavelengths with Pr:KY_3F_{10}, and red and orange laser emissions at 642.3 and 605.5 nm were also obtained for the first time with Pr:KYF_4 [73]. This last result is particularly interesting, because it has been demonstrated that the 605.5 nm orange laser emission could be tuned up to 605.86 nm by slightly tilting the laser crystal. It thus proves that a laser operating with Pr:KYF_4 in this orange spectral domain is a highly suitable source for quantum information experiments.

Namely, a second challenge in this research field is to grow better quality single crystals like Pr:KYF_4, which is not so easy, and to extend the fabrication of these materials to thin films to form waveguides, thus to realize more compact laser systems which can be used for miniature RGB laser applications. From this point of view, several ways can be explored: (1) pulsed laser deposition (PLD) and (2) liquid phase epitaxy (LPE), followed by rib-waveguide tracing using RIE (reactive ion etching) or short-pulse laser writing [74]. At the moment, only LPE has led to significant results [34, 75–78] and it is probably this solution which will be adopted in the future. In each case, however, suitable fluxes, solvents and substrates have to be found.

In the case of Pr:CaF_2, the most straightforward substrate is pure CaF_2, which is easy to buy or to grow, and the most adequate solvent is $CaCl_2$ [34]. For $LiYF_4$ (or $LiLuF_4$), the substrate is an undoped crystal plate and the flux (and solvent) consists of a charge made of about 25% YF_3 and 75% LiF (see Fig. 2.1) to lower the melt temperature down to about 750°C.

A third objective is to explore new materials offering more intense emission lines around 525 and 605.5 nm. Indeed, Pr:$LiYF_4$ and Pr:$LiLuF_4$ are the only materials which offer emission lines with satisfactory emission cross-sections around 525 nm and Pr:KYF_4 is the only crystal with an emission line peaking around 605.5 nm. Pr:CaF_2 codoped with other ions (Y^{3+} for instance) and mixed compounds could offer better solutions. Concerning the orange emission, there is another problem that should be kept in mind. Indeed, depending on the considered host material, such emission can be subject to reabsorption because of the existence of weak but non-negligible

© Woodhead Publishing Limited, 2013

absorption lines (assigned to $^3H_4 \rightarrow {}^1D_2$ optical transitions within the Pr^{3+} ions) peaking around 580 nm and extending up to about 610 nm. From this point of view, as can be seen in Fig. 2.5, lasing at 605.86 nm, because of

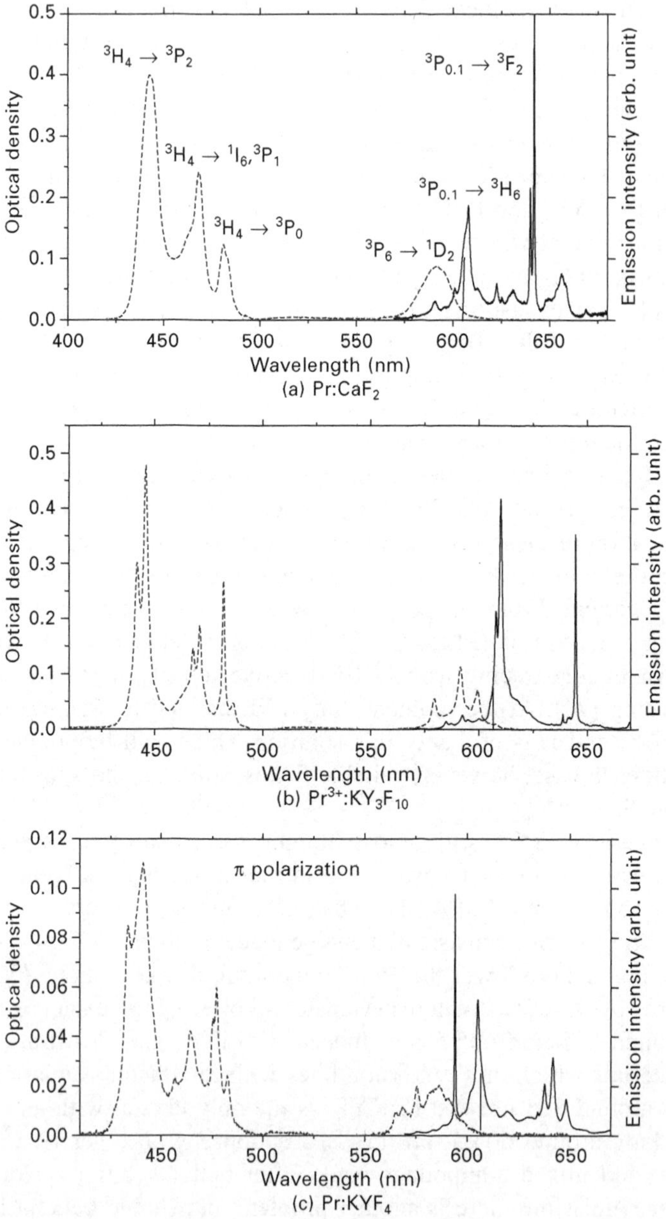

2.5 Room-temperature absorption and emission spectra of some Pr^{3+}-doped fluoride crystals in the RGB spectral domain.

© Woodhead Publishing Limited, 2013

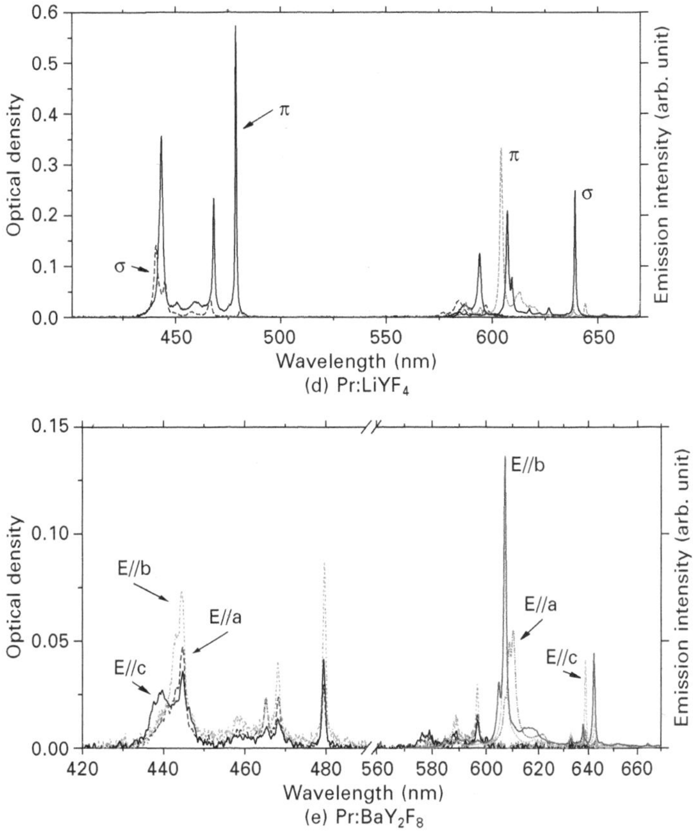

2.5 Continued

reduced reabsorption, might be more efficient in $Pr:LiYF_4$ or $Pr:BaY_2F_8$ than in $Pr:KYF_4$, and the optimum would be found in π polarization for $Pr:LiYF_4$ and with E//b or E//c for $Pr:BaY_2F_8$.

2.4 Yb^{3+} doped fluorides for ultra-short and high-power laser chains

Thanks to several technological breakthroughs – high-power laser diodes around 980 nm, for instance – and also to specific advantages such as reduced thermal loadings, zero excited-state absorption losses, larger energy storage capabilities and better laser efficiencies, Yb^{3+} doped crystals and glasses prepared in the form of bulk materials, thin-disk and fibres have become the most popular laser media for the development of short-pulse as well as high-energy and high-power laser systems. As a matter of fact, many efforts have been dedicated during recent decades to the development of mature

© Woodhead Publishing Limited, 2013

materials such as YAG and some (fluoro)phosphate glasses, and to the search for and the implementation of new ones which could combine (1) a relatively broad absorption band (for improved laser diode wavelength flexibility), (2) wide laser wavelength tunability (which is essential for the production of ultra-short pulses), (3) reasonably high absorption and emission cross-sections (for low laser thresholds and high gains), (4) long emission lifetime (for improved energy storage), and last but not least (5) reasonably good thermo-optical and thermo-mechanical properties. In that search, however, the distinction has to be made between the materials supposed to be used in laser oscillators and as laser amplifiers. Indeed, while points (1), (2) and (3) are essential for laser oscillators, it is clear that points (2), (4) and (5) are more important for laser amplifiers. Added to the fact that all considered materials need to be grown with an excellent optical quality and that they need to be grown in large sizes for laser amplifiers, several systems can be chosen and exploited for laser oscillators and low to medium power laser applications, while only a few of them are presently considered for laser amplifiers and high-power laser chains.

As a matter of fact Yb^{3+}-doped fluoride laser crystals like $Yb:CaF_2$ and $Yb:LiYF_4$ have recently proved to be very attractive for both laser oscillators and high-power laser chains. This is due to a number of specific advantages which were only (re)discovered over the past few years and which are detailed extensively, in the case of $Yb:CaF_2$ and its isotypes $Yb:SrF_2$ and $Yb:BaF_2$, in recent references and review articles [79–82].

Without going into precise details, the main advantages of a system like $Yb:CaF_2$ over an oxide like Yb:YAG are the following: (1) a broad and reasonably intense absorption peak around 980 nm (see Fig. 2.6) allowing the use of high-power fibre-coupled laser diodes and diode stacks now commercially available at that wavelength, (2) a broad, smooth and reasonably intense emission band around 1035 nm, allowing the production and the amplification of ultra-short laser pulses (down to about 100 fs at the moment) while avoiding parasitic losses due to ASE (amplified spontaneous emission), (3) a large energy storage capability, with an emission lifetime of about 2.3 ms (while most of the Yb-doped oxides have an emission lifetime below about 1 ms), (4) an exceptionally high (for a fluoride) thermal conductivity (as mentioned previously) and reasonably good mechanical properties, and, again, (5) a well-mastered crystal growth allowing the preparation of laser elements from thin films, for thin-disk laser applications, to bulk crystals of very large size (diameter of several tens of centimetres) for high-power laser chains. As a matter of fact, the laser properties of these Yb-doped fluorites have been also extensively investigated at cryogenic temperatures and other specific advantages have been demonstrated, compared to other systems like Yb:YAG and $Yb:LiYF_4$ [83–85], i.e. (1) larger absorption and emission cross-sections while keeping about the same absorption and emission

© Woodhead Publishing Limited, 2013

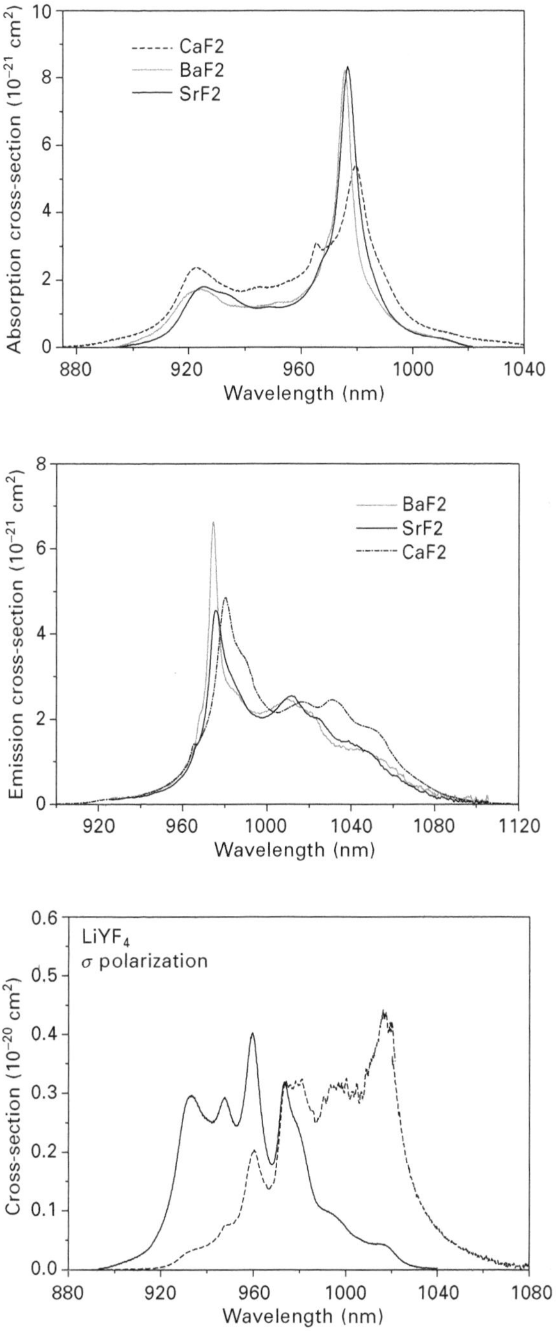

2.6 Room-temperature absorption and emission spectra of the main Yb^{3+}-doped fluoride crystals.

© Woodhead Publishing Limited, 2013

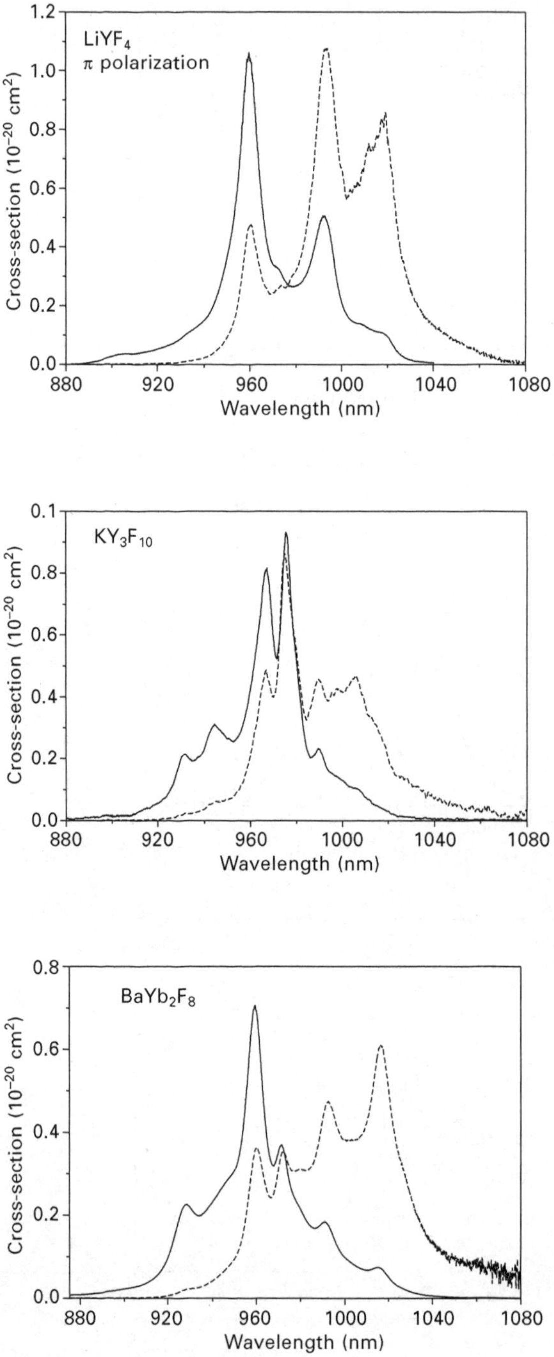

2.6 Continued

© Woodhead Publishing Limited, 2013

bandwidths, which is important for laser amplifiers, and (2) the possibility of operating the systems at very short laser wavelengths [86], close to the so-called 'zero-line' located around 980 nm, thus with very reduced thermal losses (small quantum defect) which is, again, very attractive for high-power laser applications. It has been shown, however, that lowering the temperature of these Yb-doped fluorite crystals at such cryogenic temperatures did not necessarily increase their thermal conductivity, as it does in pure materials [83, 87, 88]. For instance, the thermal conductivity of 3%Yb:CaF_2 would remain constant and equal to about 4.2 W/(m.K) down to 120 K and decrease down to 3 W/(m.K) at 77 K, while it would hyperbolically increase up to 40 and 100 W/(m.K) at the same temperatures for pure CaF_2. In fact, while the influence of the Yb-doping concentration on the room temperature value of the thermal conductivity of the materials has been extensively investigated [89, 90], almost no data exist concerning the influence of this Yb-doping concentration on the thermal behaviour of this thermal conductivity at low temperatures. According to Popov *et al.* [91, 92], the behaviour observed in heavily doped CaF_2 crystals would originate from the Yb_6F_{37} clusters which are formed in these materials [40] and which are responsible for their particular luminescence and laser properties [82]. The formation of these clusters would lead to some structural disorder and the observed glass-like thermal behaviour of the thermal conductivity. This should not be observed with perfectly ordered and single-site systems like Yb:YAG, Yb:FAP [93] or Yb:$LiYF_4$, but it remains to be proved.

To conclude, it is clear that Yb:CaF_2 is a very promising material which already surpasses most of the existing oxides and could enter in the construction of many future laser systems. Other Yb-doped fluorides, like Yb:$LiYF_4$, Yb:BaY_2F_8, and to a lesser extent Yb:KY_3F_{10} [96], however, should be kept in mind and might be worth exploiting. Their absorption and emission bands are more structured than in the case of fluorites, but the bands remain wide and the emission lifetimes remain long (2.1 ms for Yb:$LiYF_4$, 2.02 ms for $BaYb_2F_8$ and 1.77 ms for Yb:KY_3F_{10}) and their emission cross-sections are about three times larger than that of Yb:CaF_2. Indeed, a cryogenically cooled Yb:$LiYF_4$ laser with 224 W linearly polarized output power and a slope efficiency of 68% has already been demonstrated at a laser wavelength of 995 nm by pumping with diodes at 960 nm, thus with a very small quantum defect of about 3.6% [85]. Yb:BaY_2F_8 might be even more interesting. As can be seen in Fig. 2.6, its emission bandwidth is even wider, which is attractive for femtosecond pulse generation, and it presents several peaks around 993 and 1016 nm which, combined with an absorption around 960 nm, could give rise to an even smaller quantum defect. Moreover, as in the case of $LiYF_4$ (with $LiYbF_4$), the dopant concentration can be increased up to fully concentrated samples, i.e. $BaYb_2F_8$, while keeping the same spectroscopic features and the same emission lifetime, something which can be exploited in the future for compact and thin-disk lasers.

© Woodhead Publishing Limited, 2013

At the end, it is worth mentioning a final advantage of Yb-doped fluorides compared to Yb-doped oxides which is not often considered but which might be important at high power densities. It is related to the existence of intense charge-transfer bands which occur in Yb-doped oxides, but not in Yb-doped fluorides, above about 250 nm. These strongly electric-dipole allowed absorption bands have been proved to be responsible for non-negligible pump-induced refractive index changes down to the near-infrared [94, 95] and could be responsible, at high pump power densities, to laser damage caused by non-negligible multi-photon absorptions.

2.5 Undoped crystals for nonlinear optics and ultra-short pulse lasers

Some fluoride crystals have also been investigated over the past few years because of their high transmittance in the near-UV spectral range and their non-negligible second and/or third-order nonlinear susceptibility.

Only fluorides have a high enough transmittance to allow for second-harmonic frequency conversion at wavelengths shorter than about 200 nm. The borate CLBO ($CsLiB_6O_{10}$) transmits up to about 180 nm but it is very hygroscopic, thus too difficult to handle. Consequently, at the moment, efforts are preferentially dedicated to fluoro-borates like KBBF ($KBe_2BO_3F_2$) [97, 98] and CBF ($Ca_5(BO_3)_3F$) [99]. KBBF crystals were investigated, for instance, for the fourth and fifth harmonics of Ti:sapphire at 197 and 156 nm. However, their achieved conversion efficiencies were very small and only significant around 197 nm, i.e. 0.64% compared to 17% at the same wavelength of 197 nm in BBO. Concerning CBF, works are still in progress to improve the quality of the crystals. At the moment, it is considered as a promising competitor to LBO (LiB_3O_5).

Ferroelectric domain phenomena and periodic domain patterning were also investigated in as-grown orthorhombic $BaMgF_4$ single crystals [100–102] to produce a visible-to-UV second-harmonic frequency converter based on quasi phase-matching (QPM).

On the other hand, cubic fluorites like BaF_2 are presently being experimented with for their third-order nonlinearities to improve the temporal contrast of short and intense laser pulses [103, 104]. This consists in the creation of a four-wave mixing process governed by the anisotropy of the real part of the nonlinear susceptibility of the material that generates what is called a crossed-polarized wave (XPW), thus with an orthogonal polarization. By focusing an ultra-short laser pulse on such a nonlinear crystal placed between crossed polarizers, it is possible to reduce the part of the pedestal surrounding the pulse and which results from amplified spontaneous emission and other spectral distortions to improve its temporal contrast. As the efficiency of such a process is proportional to the product of the

© Woodhead Publishing Limited, 2013

nonlinear susceptibility by some anisotropy parameter, BaF_2 compared to other well-known materials like YVO_4, CaF_2 or LiF, appears as a good compromise, its only real competitor at the moment being diamond (but of course with specific drawbacks). This good compromise is also due to its cubic symmetry which is preferable to tetragonal (to avoid parasitic signals coming from residual linear birefringence), to its very reduced chromatic dispersion and, because of its wide bandgap transparency, to negligible multiphoton absorption processes. Thus, thanks to all these properties, BaF_2 could help to increase pulse contrasts by several orders of magnitude with the achievement of improved spectral purities. Moreover, by working on its orientation, it was possible to achieve XPW efficiencies up to about 30%, which is excellent. So, now, the challenge consists in growing crystals with even better optical quality, no refractive index variations and no voids, evaluating other fluorites like CaF_2 more carefully, and searching for some codopants which may increase the anisotropy of the crystal while preserving about the same nonlinear susceptibility.

2.6 References

1. F. Träger (ed.), *Springer Handbook of Lasers and Optics*, Chapter 11.2, 'Solid-state lasers', pp. 614–694 (2007)
2. P.P. Fedorov and V.V. Osiko, 'Crystal growth of fluorides', in *Bulk Crystal Growth of Electronic, Optical and Optoelectronic Materials*, ed. P. Capper. Wiley Series in Materials for Electronic and Optoelectronic Applications, John Wiley & Sons, pp. 339–356 (2005)
3. R. Moncorgé, 'Current topics of rare-earth lasers', in *Spectroscopic Properties of Rare-earth in Optical Materials*, eds G. Liu and B. Jacquier. Springer Series in Material Sciences no. 83, Chapter 6, pp. 320–378 (2004)
4. R. Moncorgé and K. Shimamura, *Mater. Sci. Technol.* **41**(3), 24–32 (2004)
5. J.L. Doualan and R. Moncorgé, *Ann. Chim. Sci. Mat.* **28**, 5–20 (2003)
6. R. Moncorgé, 'Spectroscopy of broad-band UV-emitting materials based on trivalent rare-earth ions', in *Ultraviolet Spectroscopy and UV Lasers*, eds P. Misra and M.A. Dubinskii. Marcel Dekker, Chapter 9 (2002)
7. P.P. Sorokin and M.J. Stevenson, *IBM Journal*, 56–58 (1961)
8. R.C. Duncan Jr. and Z. J. Kiss, *Appl. Phys. Lett.* **3**, 23–24 (1963)
9. S.E. Hatch, W.F. Parsons and R.J. Weagley, *Appl. Phys. Lett.* **5**, 153–154 (1964)
10. R.J. Keyes and T.M. Quist, *Appl. Phys. Lett.* **4**, 50–51 (1964)
11. C.D. Marshall, J.E. Speth, S.A. Payne, W.K. Krupke, G.J. Quarles, V. Castillo and B.H.T. Chai, *J. Opt. Soc. Am.* **B 11**, 2054 (1994)
12. A.J.S. McGonigle, D.W. Coutts, S. Girard and R. Moncorgé, *Opt. Comm.* **193**, 233 (2001) and references therein
13. R.W. Waynant and P.H. Klein, *Appl. Phys. Lett.* **46**, 14 (1985)
14. M.A. Dubinskii and A.C. Cefalas, *J. Opt. Soc. Am.* **B9**, 1148 (1992)
15. P. Camy, J.L. Doualan, S. Renard, A. Braud, V. Ménard and R. Moncorgé, *Opt. Comm.* **236**, 395–402 (2004) and references therein
16. C. Wyss, W. Luthy, H.P. Weber, P. Rogin and J. Hulliger, *Opt. Comm.* **139**, 215 (1997)

© Woodhead Publishing Limited, 2013

17. C. Labbe, J.L. Doualan, P. Camy, R. Moncorgé and M. Thuau, *Opt. Comm.* **209**, 193 (2002)
18. N.B. Barnes and R.E. Allen, *IEEE J. Quant. Electr.* **27**, 277 (1991)
19. H.W.H. Lee, S.A. Payne and L.L. Chase, *Phys. Rev.* **B 39**, 8907–8914 (1989)
20. R. Moncorgé and T. Benyattou, *Phys. Rev.* **B 37**(16) 9177 (1988)
21. R. Moncorgé and T. Benyattou, *Phys. Rev.* **B 37**(16) 9186 (1988)
22. S.A. Payne, L.L. Chase and G.D. Wilke, *Phys. Rev.* **B 37**, 998 (1988)
23. H. Manaa and R. Moncorgé, *Opt. Quant. Elect.* **22**, 219 (1990)
24. S.A. Payne, L.L. Chase, H.W. Newkirk, L.K. Smith and W.F. Krupke, *IEEE J. Quant. Electron.* **24**, 2243–2252 (1988)
25. J.M. Hopkins, G.J. Valentine, W. Sibbett, J. Aus der Hau, F. Morier-Genoud, U. Keller and A. Valster, *Opt. Comm.* **154**, 54–58 (1998)
26. D.M. Rines, P.F. Moulton, D. Welford and G.A. Rines, *Opt. Lett.* **19**, 628–630 (1994)
27. H. Manaa, Y. Guyot and R. Moncorgé, *Phys. Rev.* **B 48**, 3633 (1993)
28. P.F. Moulton, *IEEE J. Quant. Electron.* **QE 21**, 1582–1595 (1985) and references therein
29. B. C. Johnson, P. F. Moulton and A. Mooradian, *Opt. Lett.* **9**(4) 116–118 (1984)
30. P.A. Budni, M.L. Lemons, J.R. Mosto and E.P. Chicklis, *IEEE J. Sel. Top. Quant Electron.* **6**(4) 629–635 (2000)
31. M. Schellhorn, A. Hirth and C. Kieleck, *Opt. Lett.* **28**(20) 1933–1935 (2003)
32. P. Rambaldi, R. Moncorgé, J.P. Wolf, C. Pédrini and J.Y. Gesland, *Opt. Comm.* **146** (1998) 163
33. T. Yonezawa, K. Matsuo, H. Tamada and Y. Kawamoto, *J. Cryst. Growth* **236**, 281 (2002)
34. A. Pena, P. Camy, A. Benayad, J.L. Doualan, C. Maurel, M. Olivier, V. Nazabal and R. Moncorgé, *Opt. Mater.* **33**, 1616–1620 (2011)
35. T. Balaji, G. Lifante, E. Daran, R. Legros and G. Lacoste, *Thin Solid Films* **339**, 187–193 (1999)
36. V.V. Semashko, M.A. Dubinskii, R.Yu. Abdulsabirov, S.L. Korableva *et al.*, SPIE Proc. XI Int. Feofilov Symp. (Kazan, Russia), vol. 4766 (2002)
37. C. Pfistner, R. Weber, H.P. Weber, S. Merazzi and R. Gruber, *IEEE J. Quant. Elec.* **30**(7), 1605–1615 (1994)
38. M. Pollnau, P.J. Hardman, M.A. Kern, W.A. Clarkson and D.C. Hanna, *Phys. Rev.* **B 58**(24), 16076–16092 (1998)
39. J.L. Doualan, P. Camy, R. Moncorgé, E. Daran, M. Couchaud and B. Ferrand, J. *Fluorine Chem.* **128**, 459–464 (2007)
40. C.R.A. Catlow, A.V. Chadwick and J. Corish, *J. Solid State Chem.* **48**, 65–76 (1983)
41. R.E. Thoma, C.F. Weaver, H.A. Friedman, H. Insley and L.A. Harris, *J. Phys. Chem.* **65** (1961) 1096
42. I.R. Harris, H. Safi, N.A. Smith, M. Altunbas, B. Cockayne and J.G. Plant, *J. Mater. Sci.* **18** (1983) 1235
43. A. Braud, S. Girard, J.L. Doualan, M. Thuau, R. Moncorgé and A.M. Tkachuk, *Phys. Rev.* **B 61**(8), 5280–5292 (2000)
44. A. Richter, PhD Dissertation, Hamburg University (2008)
45. F. Cornacchia, A. Richter, E. Heumann, G. Huber, D. Parisi and M. Tonelli, *Opt. Expr.* **15**(3), 992–1002 (2007)
46. B.P. Sobolev and N.L. Tkachenko, *J. Less-Com. Met.* **85**, 155 (1982)

© Woodhead Publishing Limited, 2013

47. A. Braud, P.Y. Tigreat, J.L. Doualan and R. Moncorgé, *Appl. Phys.* **B 72**, 909 (2001)
48. R.Y. Abdulsabirov, M.A. Dubinski, B.N. Kazanov, N.I. Silkin and S.I. Yagudin, *Sov. Phys. – Crystallography* **34** (1987)
49. S.L. Chamberlain and L.R. Coruccini, *Phys. Rev.* **B 71**, 024434 (2005)
50. T.M. Pollak, R.C. Folweiler, E.P. Chicklis, J.W. Baer, A. Linz and D. Gabbe, *Chem. Abs.* **93**(568), 127 (1980)
51. B.H.T. Chai, J. Lefaucheur and A. Pham, *SPIE* 1863, 131 (1993)
52. M. Ito, S. Hraiech, C. Goutaudier, K. Lebbou and G. Boulon, *J. Cryst. Growth* **310**, 140–144 (2008)
53. B.Y. Le Fur, N.M. Khaidukov and S. Aleonard, *Acta Cryst.* C**48**, 978–982 (1992)
54. T.H. Allik, L.D. Merkle, R.A. Utano, B.H.T. Chai, J.L.V. Lefaucheur, H. Voss and G.J. Dixon, *J. Opt. Soc. Am.* **B 10**(4), 633–637 (1993)
55. X.X. Zhang, P. Hong, M. Bass and B.H.T. Chai, *Appl. Phys. Lett.* **66**(8), 926–928 (1995)
56. J. Sytsma, S.J. Kroes, G. Blasse and N.M. Khaidukov, *J. Phys. Condens. Matter* **3**, 8959–8966 (1991)
57. E. Sani, A. Toncelli, M. Tonelli and F. Traverso, *J. Phys. Condens. Matter* **16**, 241–252 (2004)
58. T. Danger, T. Sandrock, E. Heumann, G. Huber and B. Chai, *Appl. Phys.* **B 57**, 239–241 (1993)
59. S.A. Payne, L.K. Smith, R.J. Beach, B.H.T. Chai, J.H. Tassano, L.D. DeLoach, W.L. Kway, R.W. Solarz and W.F. Krupke, *Appl. Opt.* **33**(24), 5526–5536 (1994)
60. B.W. Woods, S.A. Payne, J.E. Marion, R.S. Hughes and L.E. Davis, *J. Opt. Soc. Am.* **B 8**(5), 970–977 (1991)
61. M.J. Weber, *Handbook of Optical Materials*, CRC Press (2003)
62. R. Moncorgé, L.D. Merkle and B. Zandi, *MRS Bull.* **21–25** (Sept. 1999)
63. S. Khiari, M. Velazquez, R. Moncorgé, J.L. Doualan, P. Camy, A. Ferrier and M. Diaf, *J. Alloy. Comp.* **451**, 128–131 (2008)
64. A. Richter, E. Heumann, E. Osiac, G. Huber, W. Seelert and A. Diening, *Opt. Lett.* **29**(22), 2638–2640 (2004)
65. P. Camy, J.L. Doualan, R. Moncorgé, J. Bengoechea and U. Weichmann, *Opt. Lett.* **32**(11), 1462–1464 (2007)
66. A. Richter, E. Heumann, G. Huber, V. Ostroumov and W. Seelert, *Opt. Expr.* **15**(8), 5172–5178 (2007)
67. T. Sandrock, T. Danger, E. Heumann, G. Huber and B.H.T. Chai, *Appl. Phys.* **B 58**, 149–151 (1994)
68. J. Klein, F. Beil and T. Halfmann, *Phys. Rev. Lett.* **99**(11), 113003 (1999)
69. Tekhnoscan Corp., http://www.tekhnoscan.ru/english/DYE-SF-077.htm
70. B. Xu, P. Camy, J.L. Doualan, A. Braud, Z.P. Cai, A. Brenier and R. Moncorgé, *Laser Phys. Lett.* **9**(4), 295 (2012)
71. B. Xu, P. Camy, J.L. Doualan, Z.P. Cai, F. Balembois and R. Moncorgé, *J. Opt. Soc. Am.* **B 29**(3), 346 (2012)
72. B. Xu, P. Camy, J.L. Doualan, Z. Cai and R. Moncorgé, *Opt. Expr.* **19**(2), 1191–1197 (2011)
73. B. Xu, F. Starecki, P. Camy, J.L. Doualan, Z.P. Cai, A. Braud, R. Moncorgé, D. Paboeuf, P. Goldner and F. Bretenaker, submitted to *Opt. Exeter*.
74. D. Beckmann, D. Esser and J. Gottmann, *Appl. Phys.* **B 104** (3), 619 (2011)

© Woodhead Publishing Limited, 2013

75. P. Rogin and J. Hulliger, *J. Cryst. Growth* **179**, 551 (1997)
76. L. Douysset-Bloch, B. Ferrand, M. Couchaud, L. Fulbert, M.F. Joubert, G. Chadeyron and B. Jacquier, *J. Alloy. Comp.* **275–277**, 67–71 (1998)
77. B. Ferrand, B. Chambaz and M. Couchaud, *Opt. Mater.* **11**, 101–114 (1999)
78. S. Renard, P. Camy, J.L. Doualan, R. Moncorgé, M. Couchaud and B. Ferrand, *Opt. Mater.* **28**, 1289–1291 (2006)
79. V. Petit, J.L. Doualan, P. Camy, V. Menard and R. Moncorgé, *Appl. Phys.* **B 78**, 681–684 (2004)
80. M. Siebold, S. Bock, U. Schramm, B. Xu, J.L. Doualan, P. Camy and R. Moncorgé, *Appl. Phys.* **B 97**(2), 327–338 (2009)
81. F. Druon, S. Ricaud, D.N. Papadopoulos, A. Pellegrina, P. Camy, J.L. Doualan, R. Moncorgé, V. Cardinali, B. Le Garrec, A. Courjaud, E. Mottay and P. Georges, *Opt. Mat. Expr.* **1**, 489–502 (2011)
82. R. Moncorgé, P. Camy, J.L. Doualan, A. Braud, J. Margerie, L.P. Ramirez, A. Jullien, F. Druon, S. Ricaud, D.N. Papadopoulos and P. Georges, J. Lumin. 133, 276 (2013)
83. D.C. Brown, IEEE J. Sel. Topics Quant. Electron. **11**(3) 587–599 (2005), and D.C. Brown, R.L. Cone, Y. Sun and R.W. Equall, *IEEE J. Sel. Topics Quant. Electron.* **11**(3) 604–612 (2005)
84. T.Y. Fan, D.J. Ripin, R.L. Aggarwal, J.R. Ochoa, B. Chann, M. Tilleman and J. Spitzberg, *IEEE Sel. Topics Quant. Electron.* **3**, 448–459 (2007)
85. L.E. Zapata, D.J. Ripin and T.S. Fan, *Opt. Lett.* **35**(11), 1854–1856 (2010)
86. S. Ricaud, D.N. Papadopoulos, A. Pellegrina, F. Balembois, P. Georges, A. Courjaud, P. Camy, J.L. Doualan, R. Moncorgé and F. Druon, *Opt. Lett.* **36**, 1602–1604 (2011)
87. G.A. Slack, *Phys. Rev.* **122**(5), 1451–1464 (1961)
88. R.L. Aggarwal, D.J. Ripin, J.R. Ochoa and T.Y. Fan, *J. Appl. Phys.* **98**, 103514 (2005)
89. R. Gaumé, B. Viana and D. Vivien, *Appl. Phys. Lett.* **83**(7), 1355–1357 (1983)
90. J. Petit, B. Viana, P. Goldner, J.P. Roger and D. Fournier, *J. Appl. Phys.* **108**, 123108 (2010)
91. P. Popov, P. Fedorov, S. Kuznetsov, V. Konyushkin, V. Osiko and T. Basiev, 'Thermal conductivity of single crystals of $Ca_{1-x}Yb_xF_{2+x}$ solid solution', *Dokl. Phys.* **53**(4), 198–200 (2008)
92. P. Popov, P. Fedorov, V. Konyushkin, A.N. Nakladov, S. Kuznetsov, V. Osiko and T. Basiev, *Dokl. Phys.* **53**(4), 413–415 (2008)
93. L.D. DeLoach, S.A. Payne, L.L. Chase, L.K. Smith, W.L. Kway and W.F. Krupke, *IEEE J. Quant. Electron.* **29**(4), 1179–1191 (1993)
94. R. Moncorgé, O.N. Ereymekin, J.L. Doualan and O.L. Antipov, *Opt. Comm.* **281**, 2526–2530 (2008)
95. R. Soulard, R. Moncorgé, A. Zinoviev, K. Petermann, O. Antipov and A. Brignon, *Opt. Expr.* **18**(11), 1173–1180 (2010)
96. M. Ito, G. Boulon, A. Bensalah, Y. Guyot, C. Goutaudier and H. Sato, *J. Opt. Soc. Am.* **B 24**(12), 3023–3033 (2007)
97. B. Wu, D. Tang, N. Ye and C. Chen, *Opt. Mater.* **5**(1,2), 105–109 (1996)
98. T. Kanai, T. Kanda, T. Sekikawa, S. Watanabe, T. Togashi, C. Chen, C. Zhang, Z. Xu and J. Wang, *J. Opt. Soc. Am.* **B 21**(2), 370–375 (2004)
99. K. Xu, P. Loiseau, G. Aka and J. Lejay, *Cryst. Growth Des.* **9**(5), 2235–2239 (2009)

© Woodhead Publishing Limited, 2013

100. H.R. Zheng, K. Shimamura, E.A.G Villora, S. Takekawa, K. Kitamura, G.R. Li and Q.R. Yin, *Phys. Stat. Sol. – Rapid Res. Lett.* **2**(3), 123–125 (2008)
101. H. Zeng K. Shimamura, E.A.G. Villora, S. Takekawa and K. Kitamura, *J. Mater. Res.* **22**(4), 1072–1076 (2007)
102. H. Zeng, K. Shimamura, C. Kannan, E.A.G. Villora, S. Takekawa and K. Kitamura, *Appl. Phys. A – Mater. Sci. Proc.* **85**(2), 173–176 (2006)
103. A. Jullien, X. Chen, A. Ricci, J.P. Rousseau, R. Lopez-Martens, L.P. Ramirez, D. Papadopoulos, A. Pellegrina, F. Druon and P. Georges, *Appl. Phys.* **B 102**, 769 (2011) and references therein
104. L.P. Ramirez, D.N. Papadopoulos, A. Pellegrina, P. Georges, F. Druon, P. Monot, A. Ricci, A. Jullien, X. Chen, J.P. Rousseau and R. Lopez-Martens, *Opt. Expr.* **19**, 93–98 (2011) and references therein

© Woodhead Publishing Limited, 2013

3
Oxide laser ceramics

V. B. KRAVCHENKO and Y. L. KOPYLOV,
V. A. Kotel'nikov Institute of Radioengineering and
Electronics, Russian Academy of Sciences, Russia

DOI: 10.1533/9780857097507.1.54

Abstract: Three main groups of oxide laser ceramics include garnets, rare earth oxides and perovskites. Their fabrication processes and main physical properties are reviewed. Laser applications of oxide ceramics are considered. Fabrication trends and possible fields of future applications are summarized.

Key words: laser ceramics, yttrium aluminium garnet, rare earth oxides, nanopowders, nanopowders compaction, ceramics sintering, ceramic lasers.

3.1 Introduction

Three main groups of materials are used to fabricate active elements of solid-state laser (SSL): crystals, glasses and ceramics. They all include dopants (mainly, ions of rare earths or transition elements) and have their own advantages and drawbacks in laser applications.

Laser crystals with ordered structures usually have narrow luminescence lines of rare earth activators, and correspondingly they have high emission cross-sections, high mechanical and thermal stability, and high heat conductivity and fracture toughness compared to laser glasses. On the other hand, laser crystals are usually grown from melt at high temperatures with the application of expensive crucibles, which makes them expensive to produce, the crystal growing times are long, the size of crystals is restricted, and there are usually restrictions in doping level, there are problems with optical uniformity and there is little possibility of adjusting their properties. Laser glasses can be fabricated as samples with large dimensions, excellent optical uniformity, lower price, high doping level, and reduced costs compared with laser crystals and it is possible to change their physical properties in a broad range, but they have disordered structure, lower emission cross-sections of activators and lower heat conductivity, thermal shock resistance and chemical stability. Fortunately enough, polycrystalline oxide laser ceramics combine all the physical advantages of the corresponding crystals with much simpler fabrication technology similar to the glass fabrication in cost, making it possible to produce large, optically uniform samples and to change the doping level and the physical properties to a considerable extent. The technology enables the fabrication of various complicated laser

© Woodhead Publishing Limited, 2013

structures, like waveguides and systems with heat dissipation, so efforts to obtain laser quality oxide ceramics have been underway for many years. The materials investigated until recently have cubic symmetry, so there are no additional scattering losses induced by the refraction coefficient difference in adjacent grains. Anisotropic crystals drew some attention recently, and the first successful results on such laser ceramics will be discussed at the end of the chapter.

Reviews of oxide laser ceramics fabrication history can be found in some papers (e.g., Ikesue and Aung, 2006, 2008; Lu *et al.*, 2002a; Kaminskii, 2007; Garanin *et al.*, 2010, 2011; Ikesue *et al.*, 2006; Taira, 2007a,b; Sanghera *et al.*, 2011a). The first cold-pressed Y_2O_3/ThO_2 (Yttralox) Nd-doped laser ceramics were reported by Anderson (1970, 1972), and the laser action of improved Yttralox:Nd (2.5–10 mol% ThO_2) ceramics was reported a little later (Greskovich and Chernoch, 1973, 1974). The efficiency of laser generation in ceramics was around 0.3% with single-shot flashlamp pumping, and the attenuation losses at 1.064 μm were three times higher (at least 0.02 cm^{-1}) than for the commercial silicate ED-2 Nd glass. No sufficient improvement was obtained for those ceramics for many years.

Great attention was paid to the preparation of transparent ceramics of one of the best materials for solid-state lasers – yttrium aluminium garnet, $Y_3Al_5O_{12}$ (YAG). De With (1984) prepared YAG ceramics with almost 100% relative density using powders obtained by spray-drying of sulphate solutions with subsequent calcining and adding SiO_2 and MgO as sintering aids.

M. Sekita and coworkers (Sekita *et al.*, 1990, 1991) fabricated transparent YAG ceramics using the urea precipitation method. It proved possible to reduce non-active losses in the ceramics from 2.7 cm^{-1} (Sekita *et al.*, 1990) to 0.2 cm^{-1} (Sekita *et al.*, 1991) but still the Nd lasing threshold was too high in comparison with Nd:YAG single crystals to obtain laser generation.

The first real breakthrough in YAG laser ceramics fabrication was made by A. Ikesue and co-workers (Ikesue *et al.*, 1995a,b) who prepared the ceramics by mixing oxide powders with subsequent spray-drying, uniaxial pressing of compacts followed by cold isostatic pressing (CIP) and reactive sintering of the compacts above 1700°C in a vacuum. Nd laser action with satisfactory results was obtained on these ceramics for the first time (Ikesue *et al.*, 1995b). This solid-state reaction method was later developed and improved by many teams and is one of the most popular methods of oxide laser ceramics fabrication at present. It will be described in more detail in the next section.

The second breakthrough in YAG laser ceramics fabrication was made by Konoshima Co., Japan (Yanagitani *et al.*, 1998a,b) where a non-reactive sintering method was proposed. Starting YAG nanopowders were prepared by calcination of precursors obtained by chemical precipitation from chlorides

© Woodhead Publishing Limited, 2013

solutions with the addition of ammonium sulphate and with ammonium hydrocarbonate as a precipitant. Compacts were prepared by slip casting and sintered at temperatures above 1700°C in a vacuum. Details of non-reactive sintering fabrication procedures are described in the next section. Highly transparent ceramics of YAG and rare earth oxides with different dopants and substitutions and non-active losses smaller than in the corresponding crystals were prepared by this method and were used for many solid-state lasers.

These important achievements, especially after the first successful laser experiments (e.g., Ikesue *et al.*, 1995b, Taira *et al.*, 1998, 1999; Lu *et al.*, 2000a,b,c, 2002a,b,c,d, and references therein), stimulated very active research in fabrication, investigation and application of oxide laser ceramics of three main groups, which are garnets, mainly YAG with different dopants and some substitution of matrix ions, then yttrium and rare earth oxides and during recent years some perovskites. More than 400 publications on oxide laser ceramics have appeared since the first information in 1995. More than 60 reports were presented in each of seven annual Laser Ceramics Symposia held in different countries starting from 2005. Certainly it is not possible to include all this material in our short review so we shall discuss further the main features of oxide laser ceramics preparation methods, some results of physical properties investigations and some data on laser experiments with oxide laser ceramics. References are given mainly on the review articles and on some publications with data included in our review.

3.2 Ceramics preparation

3.2.1 Two methods of ceramics fabrication

Two main methods of oxide laser ceramics fabrication have been developed during the last 20 years. These methods are both well known in classic ceramics processing as the one-step and two-step processes. In the one-step process (which is also called reactive sintering) the simplest oxides are mixed in the necessary stoichiometric ratio and compacted and then synthesis of chemical composition and sintering takes place within the single heating procedure. In the two-step process (non-reactive sintering) the synthesis of powders of the desired chemical composition is made at the first step and then powder compaction and compact sintering are made at the second one. In classic ceramics processing, two-step methods are considered to be preferable due to the well-controlled sintering process and the very small volume chemical composition deviation during standard ceramics technology procedures such as milling, mixing, spray-drying, slip and tape casting. The most significant successes in laser ceramics fabrication were achieved in Konoshima Chemical Corp. by Yagi *et al.* (2007a) in a two-step process. On

© Woodhead Publishing Limited, 2013

the other hand, the first laser ceramics and the first lasing were obtained by Ikesue *et al.* (1995a) by a one-step ceramics process. General diagrams of one-step and two-step processes are shown in Figs 3.1 and 3.2 respectively. There is one common feature in both these technologies despite the large difference. Particles in powders before compaction and sintering must be of sub-micrometre size. Really, the driving force of sintering is Laplace pressure which is increased with decreasing diameter of the powder particles. Therefore particles in powder must have small dimensions to reach the high sintering ability of the powder, but small particles tend to agglomerate. Therefore a compromise between the two tendencies should be reached. Adding to these two characteristics the ability of particles for homogeneous arrangement (Krell *et al.*, 2009) determines the optimal characteristics of particles in powders before compaction. This optimal dimension in oxide powders for laser ceramics is 200–300 nm.

3.2.2 Powder production

As shown above, for both one- and two-step process morphology, the dimensions and agglomeration of the initial powders are very important. Generally for the one-step process commercial powders are available with the appropriate chemical purity, morphology and agglomeration. For the two-step

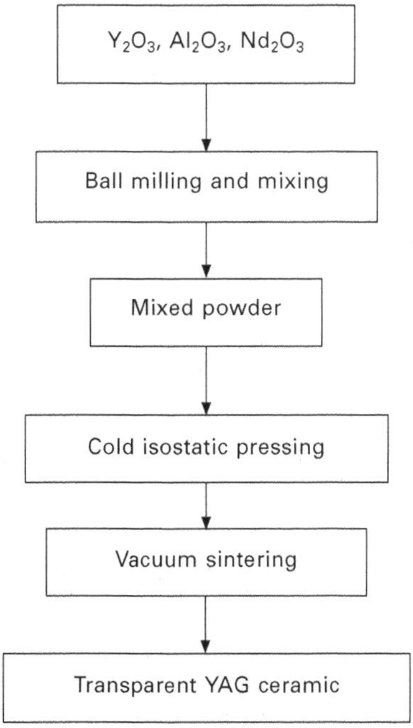

3.1 Diagram of one-step ceramics process.

© Woodhead Publishing Limited, 2013

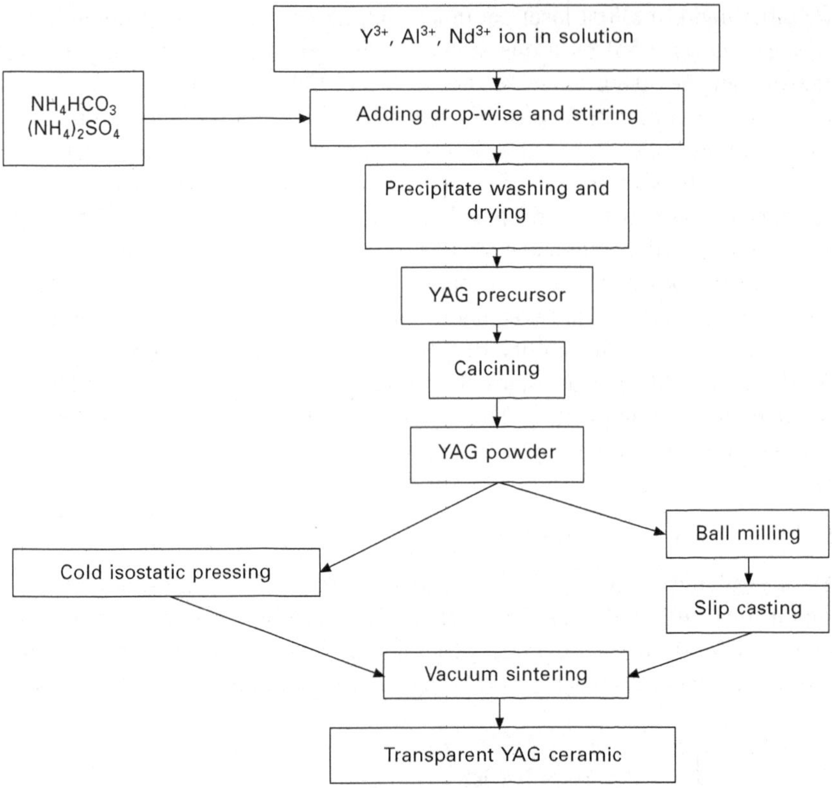

3.2 Diagram of two-step ceramics process.

process commercial powders with the desired compositions and properties are usually not available, at least for laser ceramics, and the fabrication of powders of the desired chemical composition and morphology is the key problem in this case. However, as shown by Sanghera *et al.* (2011a,b) and Soules (2007), the optical quality of laser ceramics is in each case better if specially produced initial powders are used.

One of the most effective methods of such powder production was proposed by Yanagitani *et al.* (1998a,b) for the fabrication of YAG:Nd laser ceramics. This is a method of chemical co-precipitation from aqueous solutions. The method can be used for the production of different oxide compositions and simple oxides like Y_2O_3, Lu_2O_3, etc., doped by Yb, or Gd, or Nd and other cations which are necessary for one- and two-step processes of ceramics fabrication. In the case of YAG:Nd ceramics the precipitation procedure is as follows. The aqueous solutions of Al, Nd and Y salts are used as starting materials. The mixture of these solutions is used in ratios of Al, Nd and Y in accordance with the stoichiometry of $(Nd_x,Y_{1-x})_3Al_5O_{12}$ garnet. This mixture is added drop-wise into a water solution of ammonia hydrogen carbonate

© Woodhead Publishing Limited, 2013

(NH_4HCO_3) as a precipitant under constant mild stirring. This titration process is called reverse strike (RS). The normal strike (NS) process when the precipitant is added to a mixture of salt solutions can also be used. After titration the precipitate is separated from the mother solution and washed with water and alcohol several times to remove any byproducts. The precipitate is heated in an oven at temperatures of 30–120°C during 10–24 hours after washing to obtain the precursor. The precursor is a powder with particles of usually 30–200 nm dimensions which consists of hydrates ($Y(OH)_3$, $Nd(OH)_3$, $Al(OH)_3$) and carbonates ($Y(CO_3)_3$, $Nd(CO_3)_3$, $Al(CO_3)_3$), and possibly double salts. This precursor is calcined at temperatures of 1000–1250°C to produce a powder with particles of $(Nd_xY_{1-x})_3Al_5O_{12}$ composition having dimensions 60 to 200 nm. The particle dimensions, their composition and the degree of agglomeration depend on the solution temperatures, their concentration, the type of starting salts (nitrates, sulphates, chlorides, etc.), the molar ratio of cations and precipitant, the volume of the reactor, and the rate of titration.

The main problem of powder production in a co-precipitation process is particle agglomeration, especially for particles of complicated chemical composition, for example $(Nd_xY_{1-x})_3Al_5O_{12}$. In Fig. 3.3, 3D agglomerates on YAG:Nd powder are shown. To prevent this agglomeration, the addition of ammonia sulphate in precipitant solution was proposed (references in Yagi *et al.*, 2007a). This additive acts as a dispersant during precipitation and retains this dispersive ability during the oxide synthesis process at temperatures up to 1100°C, as was shown by Li *et al.* (2000). The addition of $Mg(NO_3)_2$ into nitrate solutions of Y, Nd and Al for the precipitation of YAG:Nd precursor was proposed by Liu *et al.* (2010). This additive also

3.3 3-D agglomerates of YAG powder. This picture was created in the Center of Nanomaterials and Nanotechnologies of Tomsk Polytechnical University.

© Woodhead Publishing Limited, 2013

decreases the quantity of 3D agglomerates in the powder as shown in Fig. 3.4. Good results for YAG:Nd powder as shown in Fig. 3.5 were achieved by a combination of the presence of ammonia sulphate in high concentration in the precipitant solution, Y chloride in cation solutions, and high temperature (up to 1250°C) of precursor calcination (Stevenson *et al.* 2011). Only 2D agglomerates which include two to three initial particles are seen here. The powder of complex oxides like $(Nd_xY_{1-x})_3Al_5O_{12}$ can be obtained free of any

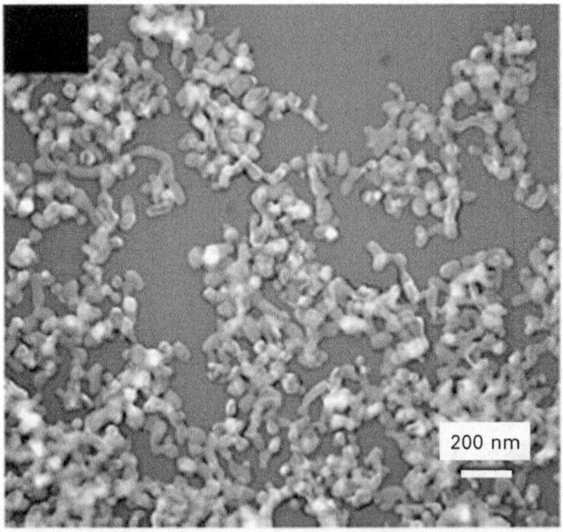

3.4 YAG powder obtained by co-precipitation with addition of MgO (Liu *et al.*, 2010).

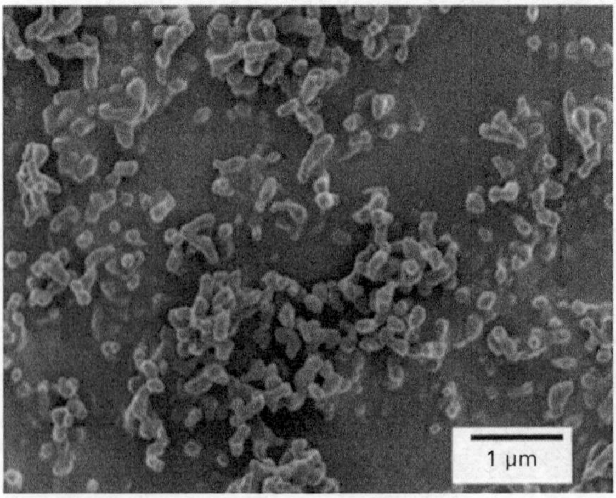

3.5 YAG powder obtained by co-precipitation with YCl_3 (Stevenson *et al.*, 2011).

© Woodhead Publishing Limited, 2013

particle agglomeration by hydrothermal or solvothermal methods (Li *et al.*, 2004). A SEM image of such powder is shown in Fig. 3.6 (Korjinsky and Kopylov, 2008). There are no agglomerates here, and particle dimensions are close to optimal.

The method of Y_2O_3 spherical particle preparation by homogeneous precipitation was first proposed by Verlinden (1987). The decomposition of urea under heating a mixture of urea and yttrium nitrate solutions was used in this method. Amorphous precipitate spherical particles with the chemical formula $Y(CO_3)OH.H_2O$ produced by this process maintained their sphericity when calcined to polycrystalline Y_2O_3. In the work of Qin *et al.* (2011) powder with spherical Y_2O_3 particles was successfully used to produce the excellent $(Nd_xY_{1-x})_3Al_5O_{12}$ ceramics in one step: the solid-state reactive sintering method. A similar method of homogeneous precipitation was proposed by Li *et al.* (2008) for the preparation of solid solution Y_2O_3–Gd_2O_3 spherical particles. The SEM image of these particles is shown in Fig. 3.7. It is possible to see that the powder is highly monodisperse and free of any agglomerates and particle dimensions are close to optimal.

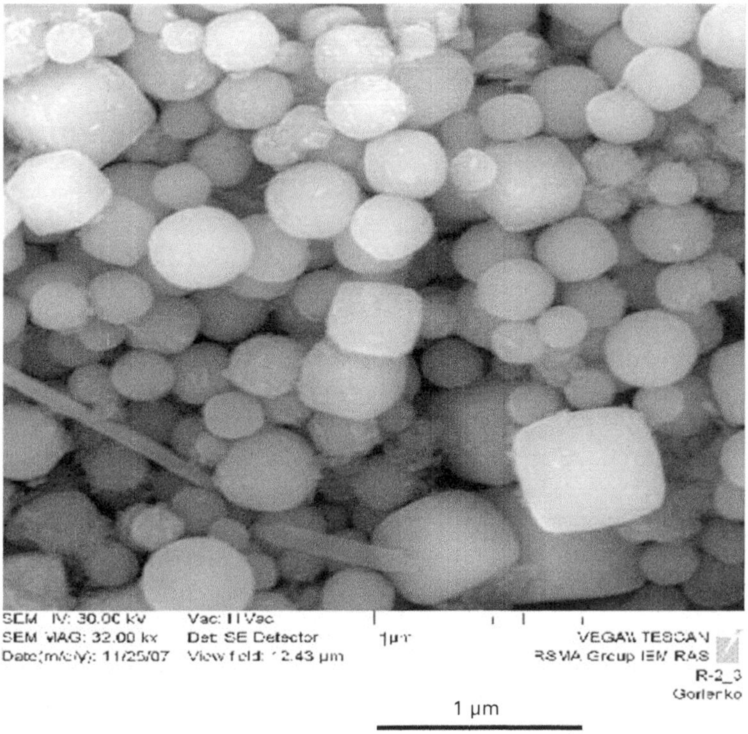

3.6 YAG nanocrystals obtained by solvothermal method (Korjinsky and Kopylov, 2008).

© Woodhead Publishing Limited, 2013

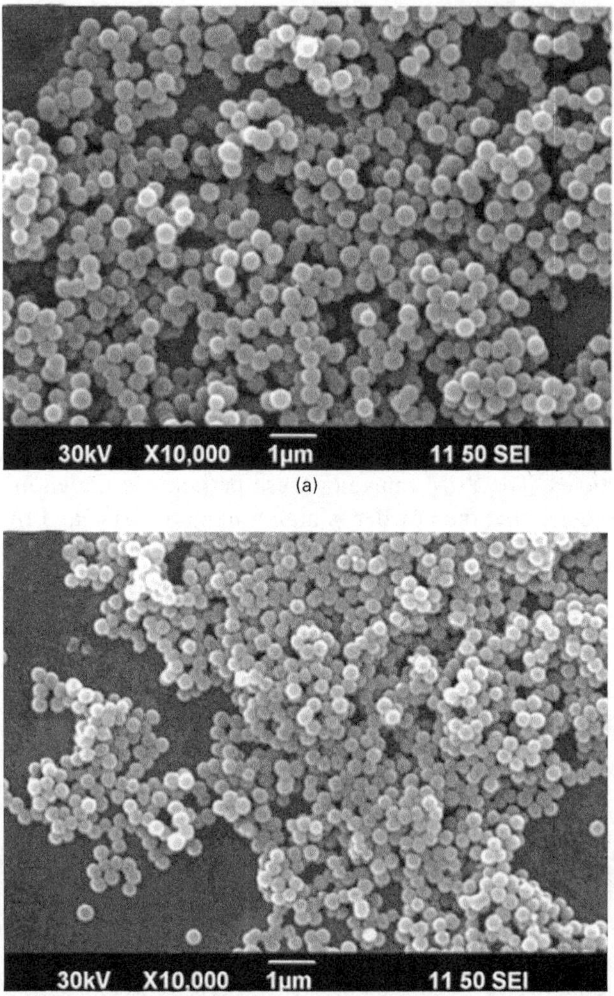

3.7 Y_2O_3 powder obtained by homogeneous precipitation (Qin *et al.*, 2011): (a) precursor, (b) Y_2O_3 after calcination at 800°C for 2 hours.

3.2.3 Powder compaction, slip casting and tape casting

The most efficient compaction method for the one-step process is to dry-press the initial powders. Powders of starting oxides are carefully mixed in an appropriate ball mill, dried and pressed in two steps. The first step is uniaxial pressing at low (15–20 MPa) pressures. The second step is cold isostatic pressing (CIP) at higher (200–300 MPa) pressures. To improve the uniformity of compact density the initial powders are sometimes granulated

© Woodhead Publishing Limited, 2013

after mixing by spray-drying. This technology was first introduced by Ikesue *et al.* (1995a). More often the granulation is produced by sieving through a 200-mesh screen. Any possible organic residue is removed by heating the compact to 600–1000°C. The usual compact density after dry pressing is 50–53%.

For a one-step process slip casting is possible, but it is rather difficult to obtain highly concentrated slurry (and consequently high density of the compact) due to the ability of some oxides to hydrolyse in water suspensions. On the other hand, tape casting is acceptable for the one-step process because organic solvents are used for the initial oxide mixing. A mixture of ethanol and xylene in a weight ratio of 1:1 is usually applied in the tape-casting process as a solvent. The solvent and powder weight ratio is about 1:5. The amount of binder usually is about 6–8 wt% to powders, and the total amount of plasticizers is about 5–7 wt%. The more preferable binder is polyvinyl butyral (PVB); possible plasticizers are butyl benzyl phthalate (BBP) and polyalkylene glycol (PAG) in weight ratio 1:1 to each other. All these organics are carefully mixed with powders, de-aired in a vacuum and then cast in a tape-casting machine to obtain the single layer tape 0.1–0.5 mm thick. The tape is then cut into pieces, stacked and laminated at a pressure of 10–20 MPa (better in isostatic conditions) and a temperature of 110–120°C. After burning of organics the compact is sintered in vacuum. This technology gives a unique possibility of producing active laser media with the desired distribution of dopant which it is not possible to obtain in glasses and is difficult in single crystals. High optical quality ceramics YAG:Er (Kupp *et al.*, 2010) and YAG:Yb (Tang *et al.*, 2012) were obtained using this technology.

For the two-step process all methods of compaction mentioned above are suitable but the most preferable one is slip casting. Laser ceramics with the best optical quality were obtained just by slip casting and the two-step process. The slip casting method consists of water-based slurry preparation and its casting into a suction porous mould or in a special mould for casting at high external pressure. It is necessary to increase as much as possible the powder content in the slurry and to maintain at the same time the lowest slurry viscosity to obtain high density of the compact. The mobility of particles in a slurry is determined generally by the particle separation distance: the greater the separation, the lower the slurry viscosity and the higher the particle mobility. These interparticle distances are controlled by the presence in the slurry of polyelectrolytes which dissociate in water, and their anionic part is absorbed on the particle surface, that provides electrostatic and spatial separation of the particles. As shown in Bagayev *et al.* (2011) the effect of such separation for a fixed powder content is greater for bigger particle dimensions and smaller molecular weight of polyelectrolytes. Optimal slurry parameters are controlled by changing the powder particle dimensions, the

© Woodhead Publishing Limited, 2013

molecular weight and the amount of polyelectrolyte. When the slip 'ages' during casting, changes in the interparticle forces may cause a change in the particle packing, and the cast's density near the mould's surface and near the upper surface of the cast may be different. That means there is a limit of the cast thickness with uniform density. For powders with nanoparticles this limit in thickness is about 10 mm. To increase this limit colloidal slip casting at high pressure was developed (Kopylov *et al.*, 2009). Additional external pressure is applied in this method to the upper surface of slurry that increases the uniform thickness limit up to several centimetres. The compact density after slip casting can be as high as 60–63% at the optimal conditions of the process and particle dimensions are about 200 nm.

3.2.4 Sintering

The sintering temperature of laser ceramics compacts is usually about 200–300°C below the melting point of the corresponding composition. The choice of sintering temperature is determined by the growth of ceramics grains because there is a temperature point where the rate of grain growth increases sharply. For example, this point for YAG ceramics is 1750°C. In many investigations the sintering temperature of YAG ceramics is really close to this remarkable point. In most investigations the sintering environment is a vacuum that facilitates the closing of remaining pores. Very similar results were obtained within a hydrogen environment due to the very high mobility of hydrogen atoms.

For the two-step process of ceramics fabrication the sintering is a simple densification of material and growth of ceramics grains. For the one-step process the sintering is accompanied by some chemical reactions with one or several crystal phases forming. For YAG ceramics, for example, five or six crystal phases can be observed during the sintering process. These phases are transformed one into another during heating. The final formation of the YAG phase is completed at a temperature of about 1500°C. In the case when powder particles have a higher reactive ability this temperature can be lower, 1300–1400°C. For laser ceramics the sintering is not a simple densification. There is no sense in talking about densification when the volume of the residual pores is less than 10^{-2}%, which is usual for laser ceramics. In contrast to ordinary ceramics including optical ceramics, for laser ceramics the supertask is the removal of residual pores. It was shown in Bagayev *et al.* (2011) that some pores cannot be removed by regular sintering procedures if the pore diameter is bigger than the mean diameter of the initial particles in the compact. These pores can be removed by a high-temperature isostatic pressing (HIP) process. In this process a sample of ceramic, for example YAG, is heated up to 1750°C inside a chamber filled by argon gas at a pressure of 200 MPa. High external pressure and high

© Woodhead Publishing Limited, 2013

temperature facilitate fast and complete removal of residual pores. Moreover, HIP may be considered as a very effective supplement to ordinary sintering in a vacuum, see, e.g., Lee *et al.* (2009). It is necessary to produce some reoxidation after sintering in a vacuum environment by heating ceramics in an oxygen-containing atmosphere. Usually annealing in air at temperatures of 1300–1450°C is sufficient to complete the reoxidation.

3.3 Physical properties of oxide laser ceramics

We shall consider here mainly those physical properties of the ceramics which are important for their applications in solid-state lasers.

3.3.1 Thermal conductivity and mechanical properties

Many properties of polycrystalline oxide laser ceramics are close to or practically identical to the properties of corresponding single crystals. This refers to thermal conductivity as well. YAG ceramics were investigated in detail using normal thermal conductivity measurement methods and heat phonons (sub-terahertz frequency) propagation (e.g., Bisson *et al.*, 2007; Gaume *et al.*, 2003; Huie *et al.*, 2006; Kaminskii, 2007; Kaminskii *et al.*, 2011; Lu *et al.*, 2002a; Sato and Taira, 2006; Taira, 2007a,b; Yagi *et al.*, 2007b). Thermal conductivity values obtained for room temperature were close to the values for single crystal YAG, correspondingly 10.5 and 10.7 W/m.K (Yagi *et al.*, 2007b), and the difference remains small when cooling from room temperature down to about 150 K. However, differences among poly- and single crystal samples become apparent below 100 K. At 25 K, thermal conductivity of the single crystal was largest, at about 8 W/cm.K. On the other hand, the thermal conductivity of polycrystalline ceramics was about seven times lower than that of the single crystal at low temperature. The maximum thermal conductivity of ceramic samples, which had 3, 4, and 7.5 mm grain size, was 0.64, 0.99, and 1.1 W/cm.K, at a temperature of 46, 39, and 37 K, respectively. In Yagi *et al.* (2007b) the difference was explained by greater phonon scattering at the grain surfaces in the ceramic at low temperature as compared to crystals. On the other hand, it was found during experiments with heat phonon propagation that in good-quality YAG laser ceramics at mean size of grains around 10 μm, the intergrain boundaries (IBs) are well stabilized, and the mean thicknesses of IBs are less than the lattice constant of the grain material. The samples can therefore exhibit relatively high optical and thermophysical properties (Kaminskii *et al.*, 2011). This question is important for powerful lasers operating at liquid nitrogen temperature, so additional investigations are needed.

Many measurements of thermal conductivity were made for yttrium and rare earth oxide crystals and laser ceramics (references as for YAG).

© Woodhead Publishing Limited, 2013

The crystals have excellent thermal conductivities (13.6 W/m.K for Y_2O_3, 12.5 W/m.K for Lu_2O_3, and 16.5 W/m.K for Sc_2O_3, all greater than that of around 11 W/m.K for YAG) (references in Zhang *et al.*, 2011a, Sanghera *et al.*, 2011b). The addition of dopants usually reduces the heat conductivity of the ceramics, including YAG (e.g., Jacinto *et al.*, 2008). For instance the thermal conductivity of oxide ceramics doped with 10 mol% Yb_2O_3 decreased in most cases rather substantially, to around 9 W/m.K for Sc_2O_3 and 9 W/m.K for Y_2O_3 (according to Aggarwal *et al.*, 2010; Sanghera *et al.*, 2011a) or even to 6.3 W/m.K for Y_2O_3 and 4.2 W/m.K for Sc_2O_3 according to Bourdet *et al.* (2010). The thermal conductivity of YAG with 10 at% Yb is around 7 W/m.K (Aggarwal *et al.*, 2010; Sanghera *et al.*, 2011a). It is necessary to note that different methods and different authors give slightly differing thermal conductivity values. The only oxide which does not change its thermal conductivity at high Yb doping levels is Lu_2O_3, with its greatest thermal conductivity being around 12.5 W/m.K at room temperature according to Aggarwal *et al.* (2010) or 10.5 W/m.K according to Bourdet *et al.* (2010).

The mechanical properties of oxide laser ceramics are very important because they restrict input and correspondingly the output power of solid-state lasers. Comparisons of YAG ceramics properties with data for single crystals have been made in some publications (e.g., Huie *et al.*, 2006; Kaminskii *et al.*, 2003, 2005; Kaminskii, 2007; Mezeix and Green, 2006; Yagi *et al.*, 2007b; Feldman *et al.*, 2011). The mechanical behaviour of the polycrystalline YAG was very similar to that of the single-crystal material, as were the elastic constants of polycrystalline YAG. The polycrystalline material showed slight advantages in hardness and fracture toughness (Mezeix and Green, 2006). The Young's modulus of YAG ceramics was found to be 308 GPa at room temperature and decreased to 264 GPa at 1400°C, and Poisson's ratio was around 0.23 in this temperature range (Yagi *et al.*, 2007b). The fracture strength and fracture toughness of the ceramics was found to be higher than for the crystal, but these values differ rather strongly – from three times difference in fracture toughness according to Kaminskii *et al.* (2003, 2005) to around 50% above the crystal value according to Huie *et al.* (2006) and Sanghera *et al.* (2011a). These discrepancies may be partly connected to the different grain size of YAG ceramics and to different grinding of the samples tested. For laser applications, real tests of YAG crystal and ceramic rod samples, using both four-point flexure tests and experiments in real laser heads under increasing pump power until the rods were broken, were performed by Feldman *et al.* (2011). Approximately 50% higher tensile strength was fixed for ceramic rods. But they gave worse results as compared to single-crystalline rods after strengthening by treatment in hot phosphoric acid. The procedure increased fracture resistivity two times for ceramic rods and five times for single-crystalline ones. This may signify also that some

© Woodhead Publishing Limited, 2013

changes in the strengthening procedures should be done probably for the two materials.

Y_2O_3 ceramics also have a much greater microhardness (around 10 GPa) and fracture toughness (at around 2.5 $MPa.m^{1/2}$) as compared to single crystals. These values are approximately 60% lower than for YAG ceramics (Kaminskii *et al.*, 2005), but it is necessary to note that the thermal shock figure of merit, which is very important for solid-state lasers with a high mean output power and correspondingly with high thermal gradients inside the laser active element, is higher for Y_2O_3 crystal, 1.4×10^{-20} m^2/W, than for YAG crystal, around 1×10^{-20} m^2/W (Sanghera *et al.*, 2011a). Additional investigations are needed to compare these two groups of laser ceramics.

3.3.2 Optical properties

The main problem in the fabrication of oxide laser ceramics is the necessity to obtain high optical transmission and low losses, especially at laser wavelength. The main sources of losses in the ceramics are pores and secondary phases (e.g., Boulesteix *et al.*, 2010; Ikesue and Kamata, 1996; Ikesue *et al.*, 1997, Ikesue and Yoshida, 1999; Sanghera *et al.*, 2011a,b,c,d; Zhang *et al.*, 2011b). As discussed earlier, the intergrain borders in the ceramics practically do not influence the absorption losses (Kaminskii *et al.*, 2011). The purity of the starting compounds is important. If extra-pure starting compounds are not available, additional purification using chemical precipitation methods can considerably improve optical transmittance of ceramics (e.g. Aggarwal *et al.*, 2010; Sanghera *et al.*, 2011a,b,c,d; Gaume *et al.*, 2010; Fig. 3.8). The shape of the starting oxide nanoparticles, the degree of particle agglomeration and the compaction procedures are of extreme importance, as they mainly influence pore concentration in the ceramics after the sintering process, as was discussed in Section 3.2.

Figure 3.9 shows the dependence of calculated scattering losses at 1 micron wavelength and the concentration of pores in YAG ceramics (Zhang *et al.*, 2011b). One can see that it is necessary for pore concentration to be around 1 ppm in order to have losses similar to commercial YAG crystals. The same conclusion, even with lower pore concentration, was previously obtained (Boulesteix *et al.*, 2010), using direct comparison of laser efficiencies of YAG crystal and YAG ceramics with different pore concentrations that were estimated using a combination of scanning electron microscopy and confocal laser scanning microscopy.

Pore concentration depends on ceramics technology. It was shown recently (Stevenson *et al.*, 2011), that it is possible to decrease the number of pores using colloidal processing of YAG nanopowders (e.g., tape casting), which leads to lower sintering times and temperatures as well as to decreased grain size at the same density as compared to dry-pressed samples. Samples with

© Woodhead Publishing Limited, 2013

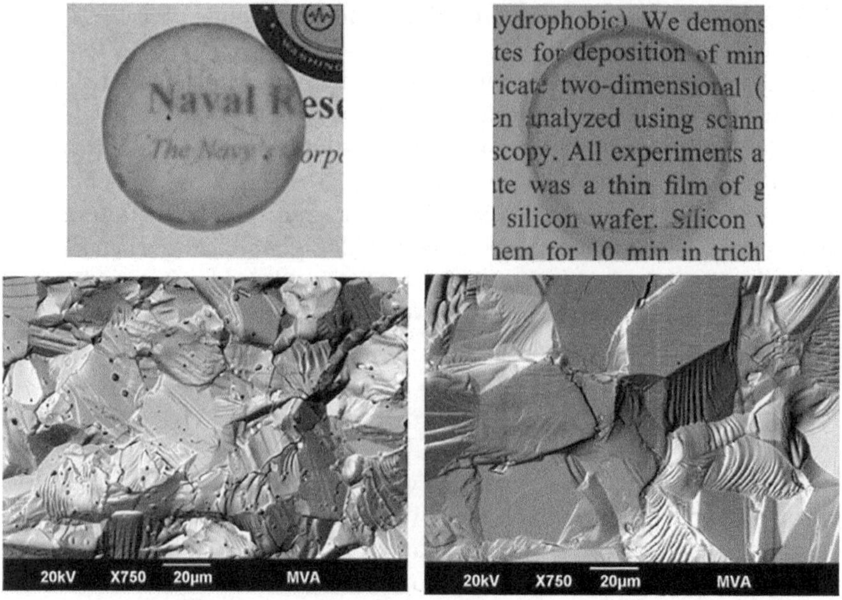

(a) (b)

3.8 Visible photograph and scanning electron microscope image of 10% Yb^{3+}:Lu_2O_3 transparent ceramic samples made by hot pressing and HIP using (a) commercial Lu_2O_3 and Yb_2O_3 powders that were mixed and heat treated, and (b) NRL co-precipitated powder. Samples are 25 mm diameter by 3 mm thick (Sanghera *et al.*, 2011d).

higher silica content (a sintering aid) densify at lower temperatures and times than lower silica content samples. Lower silica content samples may achieve full density by sintering at relatively high temperatures, but the more densifying sintering trajectory leads to smaller average grain sizes than for higher silica content samples sintered at lower temperatures.

One important property of oxide laser ceramics is the rather high value of thermooptic distortions in active elements under a high pumping level as compared to so-called athermal phosphate glasses (Alexeev *et al.*, 1980). Measurements of ceramics refraction index change d*n*/d*T* with temperature (Bourdet *et al.*, 2010; Cardinali *et al.*, 2011) gave at 270 K values around 8–9 ppm/K for Y_2O_3 and Lu_2O_3 ceramics, as well as for YAG ceramics with different Yb doping. A highly improbable value of –21 ppm/K was indicated for Sc_2O_3 Yb-doped ceramics. These values together with measured thermal expansion coefficients of the ceramics (around 6 ppm/K), also responsible for final distortions, show that thermooptic distortions in oxide ceramics will be much greater than for phosphate laser glasses where great negative values of d*n*/d*T* are normal (Alexeev *et al.*, 1980). A value of –6 ppm/K was given for LHG-8 Nd Hoya phosphate laser glass used in high-energy

© Woodhead Publishing Limited, 2013

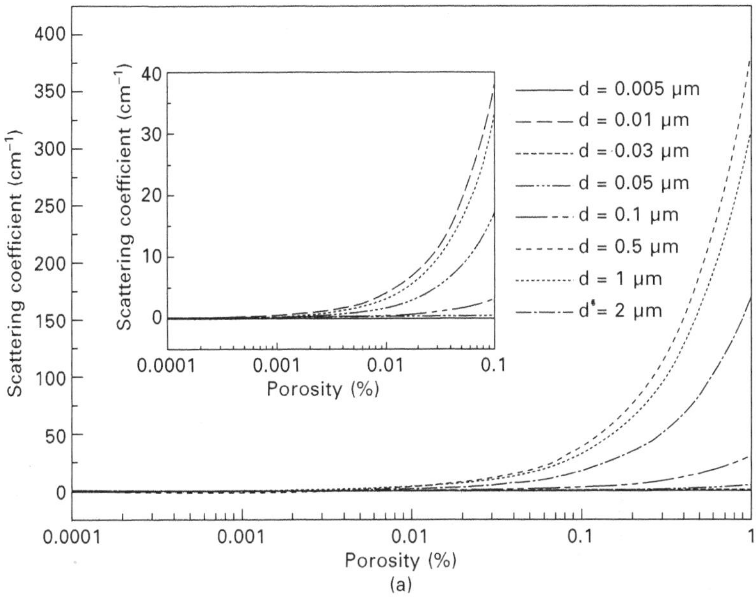

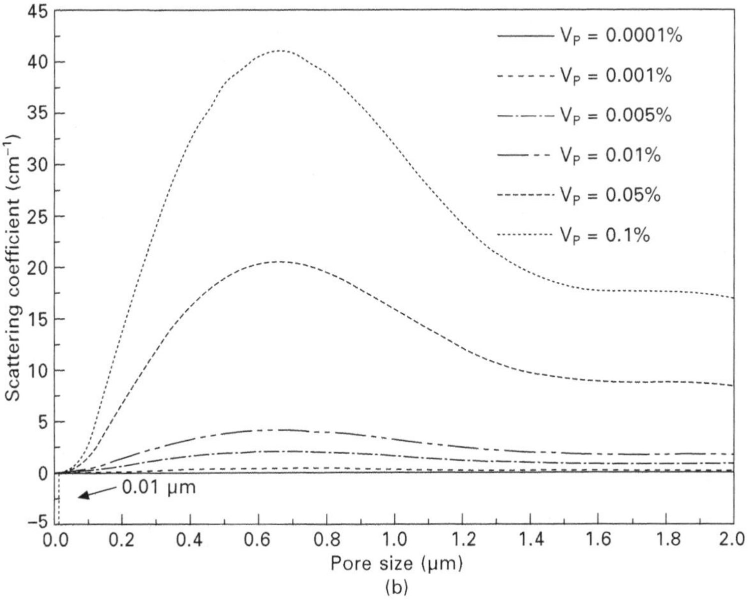

3.9 Simulated scattering coefficients as a function of (a) porosity for different pore sizes, and (b) pore size for different porosities, where d = 0.005–2 µm, porosity V_p = 0.0001%–1%, incident wavelength λ = 1000 nm. Inset: simulated scattering coefficients as a function of porosity for different pore sizes, where porosity V_p = 0.0001%–0.1% (Zhang *et al.*, 2011b).

© Woodhead Publishing Limited, 2013

laser systems for plasma investigations; this value is usual for athermal laser glasses giving the lowest thermooptic distortions. More detailed studies of thermal lenses appearing in ceramic rod or slab laser active elements are necessary to obtain precise values of beam distortions, like those done for YAG ceramics in Jacinto *et al.* (2008), and to develop some countermeasures similar to changes in the Nd doping profile along the rod active element length which greatly improve temperature gradient distribution and beam quality at longitudial pumping by a semiconductor laser (Sanghera *et al.*, 2011a). It is necessary to note that the thermooptic behaviour of polycrystalline ceramics is more complicated than that of single crystals and depends not only on the physical properties of the material but also on the ratio of grain size to active element length (e.g., Khandokhin *et al.*, 2011, and references therein)

3.3.3 Spectroscopic properties of oxide laser ceramics

Oxide laser ceramics doped with lasing trivalent Nd, Yb, Er, Ho and Tm ions were obtained both for YAG and for yttrium and rare-earth oxide compositions. Absorption and luminescence spectra of all lasing ions are practically identical to those in the corresponding single crystals (Kaminskii, 1990, 1996, 2007, and references therein). Lasing wavelengths will be indicated in Section 3.4. The main difference expected between single crystals and ceramics is connected with the difference in their structure. One can expect different concentration and correspondingly different cross-relaxation behaviour for activators in the grains and on the grain borders. Until now there has been no definite conclusion about activator distribution and their segregation in laser ceramics. This problem was discussed in detail for YAG ceramics in some papers (e.g., Merkle *et al.*, 2006; Garanin *et al.*, 2011 and references therein). It is probable that Nd segregation in YAG ceramics depends on the material fabrication method and the grain size. It was found that grain boundaries in fine-grained ceramics prepared by non-reactive sintering have a reduced spatial extension, and their presence has no influence on the fluorescence properties of Nd^{3+} ions located in the surroundings of the grain boundaries (Jacinto *et al.*, 2008). This is different from the case of large grain ceramics in which confocal fluorescence imaging experiments have revealed a relevant Nd^{3+} fluorescence quenching at the grain boundaries (Ramirez *et al.*, 2008).

Non-uniformity of activator distribution is shown to depend upon the difference in ionic radii of activator and matrix (Chani *et al.*, 2010): for YAG it is great for Ce and Nd ions and practically absent for Yb ions. This correlates with distribution coefficients of corresponding ions during YAG crystal melt growth which are small for Ce and Nd and close to unity for Yb.

© Woodhead Publishing Limited, 2013

Garnets and rare earth oxide crystals composing laser ceramics are usually well-ordered and have rather narrow absorption and luminescence lines, even at partial substitution of matrix ions. This is good for high emission cross-sections, giving lower thresholds and higher efficiency in SSL. But there are applications when broader luminescence lines are desirable, for instance to obtain short laser pulses in the pico- and femtosecond range. Such a situation was observed in doped oxides with disordered perovskite structure. Two lasing ceramics of this type were obtained recently (Zhao *et al.*, 2011). For instance, Nd^{3+} in well-known electrooptical cubic lead–lanthanum zirconate–titanate (PLZT) ceramics has a luminescence line centred at 1064.4 nm, with FWHM of 35.6 nm, which is about 22.7 times broader than that obtained in well-known Nd^{3+}:YVO_4 laser crystals (Kurokawa *et al.*, 2011; Zhao *et al.*, 2011). Some laser experiments began with this ceramic recently. Another perovskite-type laser ceramic was based on Nd-doped solid solution of barium–zirconium–magnesium tantalate (BZMT) developed by Murata Co., Japan, and laser pulses of 1.4 ps were obtained using these ceramics (Kurokawa *et al.*, 2011).

3.4 Solid-state lasers using oxide ceramic elements

Some excellent reviews of the application of oxide laser ceramics in SSL have appeared (e.g., Garanin *et al.*, 2011; Ikesue *et al.*, 2006a; Ikesue and Aung 2006, 2008; Kaminskii, 2007; Sanghera, 2011a,b,c,d; Taira, 2007a,b, 2009, 2011). Different types of lasers were developed and their parameters are improving continuously. We shall review here three different types of SSL with ceramic AE: continuous-work (CW), repetition-rate and short-pulse ones.

3.4.1 CW SSL with ceramic elements

The most impressive results were obtained using a CW SSL with laser diode (LD) pumping of Nd-doped ceramics. Figure 3.10 (based on Kaminskii, 2007) illustrates the impressive progress in this field for several years after the first successful Nd-YAG laser ceramics experiment (Ikesue *et al.*, 1995b). The references mainly show results obtained with Nd:YAG ceramics produced by Konoshima Co., Japan. The rods, 8×203 mm in size, gave an output CW power above 1 kW in 2002 (Lu *et al.*, 2002b). The large-sized plates, $10 \times 10 \times 2$ cm, with excellent optical quality fabricated by Konoshima Co. were used in a superpowerful 67 kW CW heat capacity laser in the Lawrence Livermore Laboratory, USA (Yamamoto *et al.*, 2008; Sanghera *et al.*, 2011a, Soules, 2007). Recently, Northrop Grumman Co., USA, and Textron, USA, managed to obtain a CW output power above 100 kW, using different

© Woodhead Publishing Limited, 2013

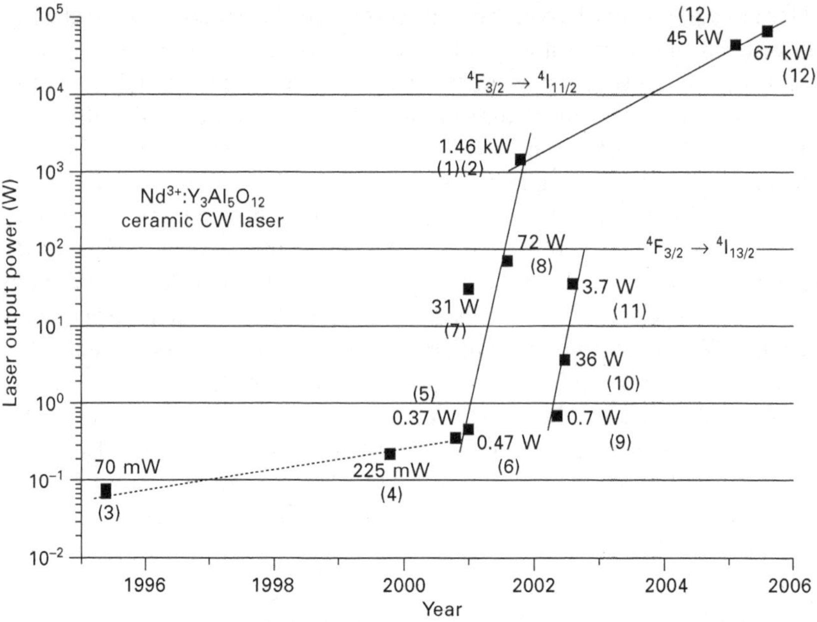

3.10 Progress in the development of LD-pumped $Y_3Al_5O_{12}:Nd^{3+}$ ceramic CW lasers emitting at wavelengths of two $^4F_{3/2} \rightarrow {}^4I_{11/2}$ and $^4F_{3/2} \rightarrow {}^4I_{3/2}$ generation channels at 300 K with end (longitudinal, ref. 3,4,5,6,9,10) and side (or transverse, ref. 1,2,7,8,11,12) pumping geometries. The CW laser output power is given in mW, W, kW (based on Kaminskii, 2007). References in the figure 1 = Lu *et al.* (2001b), 2 = Lu *et al.* (2002b), 3 = Ikesue *et al.* (1995b), 4 = Taira *et al.* (1998), 5 = Lu *et al.* (2000a), 6 = Lu *et al.* (2000b), 7 = Lu *et al.* (2000c), 8 = Lu *et al.* (2001a), 9 = Lu *et al.* (2002c), 10 = Kaminskii *et al.* (2003), 11 = Lu *et al.* (2002d), 12 = Sanghera *et al.* (2011a).

Nd:YAG ceramics slabs, with lengths up to 350 mm. This was end-pumped for Northrop Grumman and used 'thinzag' geometry with zig-zag beam propagation for Textron (Marmo *et al.*, 2009 Quarles, 2009; McNaught *et al.*, 2009 and references in Sanghera *et al.*, 2011a). Plug efficiency around 20% was obtained in the Northrop Grumman laser with excellent beam quality. It is likely that this is not the upper limit for ceramic CW SSL, because the coherent combination of high-brightness, high-power modular lasers into a single beam with excellent optical quality, which these designs successfully developed, suggests further increase of SSL CW output power is possible. It is necessary to note that the ceramic AE size can be made very big, up to 1×1 m^2 plates (Ueda, 2007, 2010).

Other laboratories are also working to increase the output power of CW ceramic lasers, e.g., Bo *et al.* (2011), where 2440 W output was obtained using a $93 \times 30 \times 3$ mm slab with scattering coefficient at 1064 nm less

© Woodhead Publishing Limited, 2013

than 0.004 cm^{-1}, and an optical-to-optical conversion efficiency of 36.5% was obtained.

Considerable progress was achieved in flash-lamp pumped Nd:YAG ceramic lasers, and their efficiency was increased using double doping Cr–Nd with energy transfer from Cr^{3+} to Nd^{3+} ions (Yagi *et al.*, 2006).

YAG ceramics 1 mm thick doped with 20 at% ytterbium ions were used for a thin-disk laser with slope efficiency of 52% and later with 61% (Dong *et al.*, 2007, and references in Sanghera *et al.*, 2011a). Heavy-doped Yb:YAG ceramic is more suitable for a thin-disk laser than a single crystal with the same Yb^{3+}-ion lasants. Output power of 6.5 kW from a thin-disk laser based on a 200 μm thick 9%Yb:YAG ceramic active medium with a 1 mm thick undoped YAG cap to mitigate thermal loading was reported, and several similar thin-disk lasers were combined to generate a total of >25 kW output (references in Sanghera *et al.*, 2011a).

3.4.2 Repetition-rate ceramic lasers

There are two main research directions here. The first one is the development of repetition-rate microchip lasers Slow but steady progress in output parameters has been made here. A composite, all-ceramic, high-peak power Nd:YAG/Cr^{4+}:YAG monolithic micro-laser with multiple-beam output was developed for engine ignition (Pavel *et al.*, 2011). Nd:YAG/Cr^{4+}:YAG monolithic cavity was pumped by three independent lines. At a 5 Hz repetition rate, each line delivered laser pulses with ~2.4 mJ energy and 2.8 MW peak power. The M2 factor of a laser beam was 3.7, and stable air breakdowns were realized. The increase of pump repetition rate up to 100 Hz improved the laser pulse energy by 6% and required ~6% increase of the pump pulse energy. Such lasers are proposed to improve the fuel combustion process in engine ignition, with expected large-scale applications (Sakai *et al.*, 2008, Tsunekane and Taira, 2006, 2007, Tsunekane *et al.*, 2010, Taira, 2007b, 2009, 2010, 2011).

Another research trend is connected with extremely high power repetition-rate lasers for inertial confinement fusion in future thermonuclear plants. The estimates show the necessary output energy should be around 1–2 MJ in approximately 1 ns pulse with 10–20 Hz repetition rate, which corresponds to mean output power of 15–40 MW (e.g., Byer, 2007; Ueda, 2007, 2010). It can be realized using laser ceramics. Similar requirements refer to systems for laser plasma accelerators (Ueda, 2011). A further approach may be taken with reference to the fabrication of non-cubic laser ceramics. For instance, fluorapatite non-cubic laser ceramics with Nd and Yb doping were obtained using particle orientation in an external magnetic field, and laser action was demonstrated in these samples (Akiyama *et al.*, 2010, 2011; Akiyama and Taira, 2011; Taira, 2011).

© Woodhead Publishing Limited, 2013

3.4.3 Short-pulse SSL

Considerable progress was demonstrated by ultrashort pulse ceramic lasers, as 10 ps pulses were obtained in Nd:$Y_3ScAl_4O_{12}$ (YSAG) ceramics. The slope efficiency was 44.4% with 610 mW output power (Saikawa *et al.*, 2004; Ikesue *et al.*, 2006). Pulses of 68 fs and even 53 fs with ~1 W average power output were obtained in the laser, where a 1.5 mm thick Yb:Y_2O_3 ceramic containing 2.5 at% Yb was placed behind the Yb:Sc_2O_3 ceramic plate (Tokurakawa *et al.*, 2009). Some new possibilities of obtaining short pulses that are connected with the application of disordered perovskite ceramics with broad luminescence lines were discussed in Section 3.3.3.

The main attention in laser experiments has been paid to oxide laser ceramics doped with Nd and Yb, but some data have also been reported for such ceramic lasers with Er, Ho and Tm ions.

3.5 Conclusion

Existing technologies enable the fabrication of oxide laser ceramics with plates of size 10 × 10 × 2 cm and slabs 1 × 5 × 35 cm with absorption losses less than 100 ppm/cm at lasing wavelength and optical uniformity better than $\lambda/30$ at 100 cm^2 area; this is better than for the usual oxide laser crystals. Laser AE preforms with areas of around 1 × 1 m^2 can be produced using ceramics technology development. Most physical properties of oxide ceramics are equal to those of the corresponding single crystals but the mechanical properties of ceramics are better than those of the crystals. The possibility of both increasing the size and decreasing the costs of oxide laser ceramics exists. The highest output power above 100 kW for a CW SSL laser was achieved using YAG ceramic active elements. Femtosecond lasers are realized using different doped oxide ceramics, and microchip Nd:YAG lasers with an impulse output power above 6 MW at 532 nm were designed with prospects of large-scale applications in combustion engines. Currently in progress are investigations into the production of repetition-rate ceramic lasers with a mean output power of up to 30–40 MW, for use in thermonuclear power stations.

There are further avenues of research that will improve oxide laser ceramics technology, such as the development of anisotropic material ceramics and increasing the size and quality of laser ceramic samples. Further useful solid-state laser designs and a transition to industrial production, along with the development of laser-assisted thermonuclear power plants to enable an unrestricted energy supply, will prove to be paramount (Akiyama *et al.*, 2010, 2011; Akiyama and Taira, 2011; Boulon, 2012; Kartner *et al.*, 2010; Sanghera *et al.*, 2011a,b,c,d; Taira, 2011; Ueda, 2007, 2010, 2011).

© Woodhead Publishing Limited, 2013

3.6 Acknowledgements

This work has been supported by the Program of the Presidium of the Russian Academy of Sciences, grants RFBR 10-02-00844-a, 10-02-00705a, 11-02-90465-Ukr-f-a and 11-02-12128-ofi-m. The authors thank Academician RAS S. N. Bagayev and Corresponding Member RAS A. A. Kaminskii for useful discussions Prof. G. Boulon and Prof. K. Ueda for permission to use illustrations and to Dr. E. S. Dvilis from Tomsk Polytechnical University for help in investigations of powders by SEM.

3.7 References

Aggarwal I, Sanghera J, Kim R, Villalobos G, Baker C, Frantz J, Shaw B, Sadowski B, Miklos F, Hunt M, and Kung F (2010), 'High efficiency lasing using 10% Yb^{3+} doped Lu_2O_3 ceramics', in *6th Laser Ceramics Symposium: International Symposium on Transparent Ceramics for Photonic Applications*, Münster, Germany, 6–8 December 2010. Technical Digest 5.

Akiyama J, and Taira T (2011), 'Fabrication of rare-earth patterned laser ceramics by use of gradient magnetic field', in *Advances in Optical Materials*, OSA Technical Digest (CD), Optical Society of America, paper AIWA3.

Akiyama J, Sato Y, and Taira T (2010), 'Laser ceramics with rare-earth-doped anisotropic materials', *Opt. Lett.*, 35(21), 3598–3600.

Akiyama J, Sato Y, and Taira T (2011), 'Laser demonstration of diode-pumped Nd^{3+}-doped fluorapatite anisotropic ceramics', *Appl. Phys. Express*, 4(2), 022703.

Alexeev N E, Gapontsev V P, Kravchenko V B, Rudnitskii Yu P, and Zhabotinskii M E. (1980), *Phosphate Laser Glasses*, Moscow, Nauka.

Anderson RC (1970), 'Transparent yttria-based ceramics and method of producing same', US Patent 3545987, pat. 08.12.1970.

Anderson RC (1972), 'Transparent zirconia-, hafnia- and thoria-rare earth ceramics', US Patent 3640887, pat. 08.02.1972.

Bagayev S, Kaminskii A, Kopylov Y and Kravchenko V (2011), 'Problems of YAG nanopowders compaction for laser ceramics', *Opt. Mater.*, 33, 702–705.

Bisson J-F, Yagi H, Yanagitani T, Kaminski A, Barabanenkov Y N, and Ueda K (2007), 'Influence of the grain boundaries on the heat transfer in laser ceramics', *Opt. Rev.*, 14(1), 1–13.

Bo Y, Liu W, Chen Y, Jiang B, Xu J, Li J, Xu Y, Pan Y, Xu J, Guo Y, Yang F, Peng Q, Cui D, and Xu Z (2011), '2440 W QCW diode-pumped Nd:YAG ceramic slab laser', in *7th Laser Ceramics Symposium, International Symposium on Laser Ceramics and Photonic Appplications*, Singapore, 14–17 November 2011. Abstracts, 46.

Boulesteix R, Maître I A, Baumard J-F, Rabinovitch I A, and Reynaud F (2010), 'Light scattering by pores in transparent Nd:YAG ceramics for lasers: correlations between microstructure and optical properties', *Opt. Express*, 18(14), 14992–15002.

Boulon G (2012), 'Fifty years of advances in solid-state laser materials', *Opt. Mater.*, 34(3), 499–512.

Bourdet G, Cardinali V, Marmois E, and Le Garrec B (2010), 'Determination of the thermo-optic coefficient d*n*/d*T* and the thermal conductivity of ytterbium doped sesquioxides ceramics (Sc_2O_3, Y_2O_3, Lu_2O_3) at cryogenic temperature', in *6th Laser*

© Woodhead Publishing Limited, 2013

Ceramics Symposium: International Symposium on Transparent Ceramics for Photonic Applications, Münster, Germany, 6–8 December 2010. Technical Digest 14.

Byer R L (2007), 'Progress in engineering ceramics for advanced solid state lasers', *3rd Laser Ceramics Symposium: International Symposium on Transparent Ceramics for Photonic Applications*, Paris, CNRS, 8–10 October, 2007, P. IO-G-1.

Cardinali V, Marmois E, Le Garrec B, and Bourdet G (2011), 'Determination of the thermo-optic coefficient dn/dT of ytterbium doped ceramics (Sc_2O_3, Y_2O_3, Lu_2O_3, YAG), crystals (YAG, CaF_2) and neodymium doped phosphate glass at cryogenic temperature', *Opt. Mater.*, doi: 10.1016/j.optmat.2011.05.035.

Chani V I, Boulon G, Zhao V, Yanagida T, and Yoshikawa A (2010), 'Correlation between segregation of rare earth dopants in melt crystal growth and ceramic processing for optical applications', *Jap. J. Appl. Phys.*, 49(7), 0756011–0756016.

De With G and van Dijk H J A (1984), 'Translucent Y3Al5O12 ceramics', *Mater. Res. Bull.*, 19, 1669–1674.

Dong J, Shirakawa A, Ueda K, Yagi H, Yanagitani T, and Kaminskii A A (2007), 'Laser-diode pumped heavy-doped Yb:YAG ceramic lasers', *Opt. Lett.*, 32, 1890–1892

Feldman R, Golan Y, Burshtein Z, Jackel S, Moshe I, Meir A, Lumer Y, and Shimony Y, (2011), 'Strengthening of poly-crystalline (ceramic) Nd:YAG elements for high-power laser applications', *Opt. Mater.*, 33, 695–701.

Garanin S G, Dmitriuk A V, Zhlin A A, Mikhailov M D, and Rukavishnikov N N (2010), 'Laser ceramics. 1. Preparation methods', *J. Opt. Technol.*, 77(6), 52–68.

Garanin S G, Dmitriuk A V, Zhlin A A, Mikhailov M D, and Rukavishnikov N N (2011), 'Laser ceramics. 2. Spectroscopic and laser properties', *J. Opt. Technol.*, 78(6), 60–70.

Gaume R, Viana B, and Vivien D (2003), 'A simple model for the prediction of thermal conductivity in pure and doped insulating crystals', *Appl. Phys. Lett.*, 83, 1355.

Gaume R, He Y, Markosyan A, and Byer R L (2010), 'Characterization of optical losses in transparent YAG ceramics', *6th Laser Ceramics Symposium: International Symposium on Transparent Ceramics for Photonic Applications*, Münster, Germany, 6–8 December 2010. Technical Digest 5.

Greskovich C, and Chernoch J P (1973), 'Polycrystalline ceramic lasers', *J. Appl. Phys.*, 44, 4599–4606.

Greskovich C, and Chernoch J P (1974), 'Improved polycrystalline ceramic lasers', *J. Appl. Phys.*, 45, 4495–4502.

Huie J C, Gentilman R, Stefanik T, and Rockosi D (2006), 'Recent advances in onshore produced ceramic laser gain materials', in *Laser Source and System Technology for Defense and Security II*, edited by Gary L. Wood and Mark A. Dubinskii. Proc. SPIE, 6216, 62160L.

Ikesue A, and Aung Y L (2006), 'Synthesis and performance of advanced ceramic lasers', *J. Am. Ceram. Soc.*, 89(6), 1936–1944.

Ikesue A, and Aung Y L (2008), 'Ceramic laser materials', *Nature Photonics*, 2(12), 721–727.

Ikesue A, and Kamata K (1996), 'Microstructure and optical properties of hot isostatically pressed Nd:YAG ceramics', *J. Am. Ceram. Soc.*, 79, 1927–1933.

Ikesue A, and Yoshida K (1999), 'Influence of pore volume on laser performance of Nd:YAG ceramics', *J. Mater. Sci.*, 34, 1189–1195.

Ikesue A, Furusata I, and Kamata K (1995a), 'Fabrication of polycrystalline, transparent YAG ceramics by a solid-state reaction method', *J. Am. Ceram. Soc.*, 78, 225–228.

Ikesue A, Kinoshita T, Kamata K, and Yoshida K (1995b), 'Fabrication and optical

© Woodhead Publishing Limited, 2013

properties of high-performance polycrystalline Nd:YAG ceramics for solid-state lasers', *J. Am. Ceram. Soc.*, 78(4), 1033–1040.

Ikesue A, Yoshida K, Yamamoto T, and Yamaga I (1997), 'Optical scattering centers in polycrystalline Nd:YAG laser', *J. Am. Ceram. Soc.*, 80(6), 1517–1522.

Ikesue A, Aung Y L, Taira T, Kamimura T, Yoshida K, and Messing G (2006), 'Progress in ceramic lasers', *Annu. Rev. Mater. Res.*, 36(1), 397–429.

Jacinto C, Benayas A, Catunda T, García-Solé J, Kaminskii A A, and Jaque D (2008), 'Microstructuration induced differences in the thermo-optical and luminescence properties of Nd:YAG fine grain ceramics and crystals', *J. Chem. Phys.*, 129, 104705.

Kaminskii A A (1990), *Laser Crystals: Their Physics and Properties*, Springer, Berlin.

Kaminskii A A (1996), *Crystalline Lasers: Physical Processes and Operating Schemes*, CRC Press, Boca Raton, FL.

Kaminskii A A (2007), 'Laser crystals and ceramics: recent advances', *Laser and Photon. Rev.*, Wiley-VCH Verlag, Weinheim, 1(2), 93–177.

Kaminskii A A, Akchurin M Sh, Alshits V I, Ueda K, Takaichi T, Lu J, Uematsu T, Musha M, Shirakawa A, Gabler V, Eichler H J, Yagi H, Yanagitani T, Bagayev S N, Fernandez J, and Balda R (2003), 'New data on the physical properties of $Y_3Al_5O_{12}$-based nanocrystalline laser ceramics', *Crystalogr. Rep.*, 48, 515–519.

Kaminskii A A, Akchurin M Sh, Gainutdinov R V, Takaichi K, Shirakava A, Yagi H, Yanagitani T, and Ueda K (2005), 'Microhardness and fracture toughness of Y_2O_3- and $Y_3Al_5O_{12}$-based nanocrystalline laser ceramics', *Crystalogr. Rep.*, 50, 869–873.

Kaminskii A A, Taranov A V, and Khazanov E N (2011), 'Structural study of oxide ceramics using the phonon spectroscopy', *J. Commun. Technol. Electron*, 56(10), 1234–1241.

Kartner F, Pollnau M, Ueda K-I, and Van Driel H (2010), 'Lasers: the next 50 years', *J. Opt. Soc. Amer. B*, 27(11), LF1–LF2.

Khandokhin P A, Ievlev I V, Lebedeva Yu S, Mukhin I B, Palashov O V, and Khazanov E A (2011), 'Polarisation dynamics of a Nd:YAG ceramic laser', *Quantum Electronics*, 41(2), 103–109.

Kopylov Y, Kravchenko V, Bagayev S, Shemet V, Komarov A, Karban O, and Kaminskii A (2009), 'Development of Nd: $Y_3Al_5O_{12}$ laser ceramics by high-pressure colloidal slip-casting (HPCSC) method', *Opt. Mater.*, 31, 707–710.

Korjinsky M, and Kopylov Y (2008), 'The new hydrothermal synthesis of yttrium aluminum garnet powders', *II Russian-French Seminar 'Nanotechnology, Energy, Plasma, Lasers' (NEPL-2008), Abstracts*. Tomsk Polytechnic University, Tomsk, 46–47.

Krell A, Hutzler T, and Klimke J (2009), 'Transmission physics and consequences for materials selection, manufacturing, and applications', *J. Eur. Ceram. Soc.*, 29, 207–221.

Kupp E R, Messing G L, Anderson J M, Gopalan V, Dumm J Q, Kraizinger C, Ter-Gabrielyan N, Merkle L D, Dubinskii M, Simonaitis-Castillo V K, and Quarles G J (2010), 'Co-casting and optical characteristics of transparent segmented composite Er:YAG laser ceramics', *J. Mater. Res.*, 25(3), 476–483.

Kurokawa H, Shirakawa A, Tokurakawa M, Ueda K, Kuretake S, Tanaka N, Kintaka Y, Kageyama K, Takagi H, and Kaminskii A A (2011), 'Broadband-gain Nd^{3+}-doped Ba(Zr, Mg, Ta)O_3 ceramic lasers for ultrashort pulse generation', *Opt. Mater.*, 33(5), 667–669.

Lee S-H, Kupp E R, Stevenson A J, Anderson J M, Messing G L, Li X, Dickey E C, Dumm J Q, Simonaitis-Castillo V K, and Quarles G J (2009), 'Hot isostatic pressing of transparent Nd:YAG ceramics', *J. Am. Ceram. Soc.*, 92(7), 1456–1463.

© Woodhead Publishing Limited, 2013

Li J, Ikegami T, Lee J, and Mori T (2000), 'Low-temperature fabrication of transparent yttrium aluminum garnet (YAG) ceramics without additives', *J. Am. Ceram. Soc.*, 83(4), 961–963.

Li J, Li X, Sun X, Ikegami T and Ishigaki T (2008), 'Uniform colloidal spheres for $(Y_{1-x}Gd_x)_2O_3$ (x = 0–1): Formation mechanism, compositional impacts, and physicochemical properties of the oxides', *Chem. Mater.*, ACS Publications, 20, 2274–2281.

Li X, Liu H, Wang J, Cui H, and Han F (2004), 'Production of nanosized YAG powders with spherical morphology and nonaggregation via a solvothermal method', *J. Am. Ceram. Soc.*, 87(12), 2288–2290.

Liu W, Zhang W, Li J, Kou H, Zhang D, and Pan Y (2010), 'Synthesis of Nd:YAG powder leading to transparent ceramics. Effect of MgO dopant', *J. Eur. Ceram. Soc.*, 31, 653–657.

Lu J, Prabhu M, Song J, Li C, Xu J, Ueda K, Kaminskii A A, Yagi H, and Yanagitani T (2000a), 'Optical properties of highly Nd^{3+}-doped $Y_3Al_5O_{12}$ ceramics', *Appl. Phys. B*, 71, 469–473.

Lu J, Prabhu M, Xu J, Ueda K, Yagi H, Yanagitani T, and Kaminskii A A. (2000b), 'Highly efficient 2% Nd:yttrium aluminum garnet ceramic laser', *Appl. Phys. Lett.*, 77, 3707–3709.

Lu J, Song J, Prabhu M, Xu J, Ueda K, Yagi H, Yanagitani T, and Kudryashov A (2000c), 'High-power Nd:$Y_3Al_5O_{12}$ ceramic laser', *Jpn. J. Appl. Phys.*, 39 (Part 2, No. 10B), L1048–L1050.

Lu J, Murai T, Takaichi K, Uematsu T, Misawa K, Probhu M, Xu J, Ueda K, Yagi H, Yanagitani T, Kaminskii A A, and Kudryashov A (2001a), '72 W Nd:$Y_3Al_5O_{12}$ ceramic laser', *Appl. Phys. Lett.*, 78, 3586–3588.

Lu J, Prabhu M, Ueda K, Yagi H, Yanagitani T, Kudryashov A, and Kaminskii A A (2001b), 'Potential of ceramic YAG lasers', *Laser Physics*, 11(10), 1053–1057.

Lu J, Lu J, Murai T, Takaichi K, Uematsu K, Ueda K, Yagi H, Yanagitani T, Akiyama Y, and Kaminskii A A (2002a), 'Development of Nd:YAG ceramic lasers', in *OSA Topics, vol. 68, Advanced Solid-State Lasers*, edited by M. E. Fermann and L. R. Marinelly, 507–517.

Lu J, Ueda K, Yagi H, Yanagitani T, Akiyama Y, and Kaminskii A A (2002b), 'Neodymium doped yttrium aluminum garnet ($Y_3Al_5O_{12}$) nanocrystalline ceramics – a new generation of solid state laser and optical materials', *J. Alloys Compounds*, 341, 220–225.

Lu J, Lu J, Shirakawa A, Ueda K, Yagi H, Yanagitani T, Gabler V, Eichler H J, and Kaminskii A A (2002c), 'New highly efficient 1.3 mcm CW generation in the $^4F_{3/2}$ $^4I1_{3/2}$ channel of the nanocrystalline Nd^{3+}:$Y_3Al_5O_{12}$ ceramic laser under diode pumping', *Phys. Stat. Sol. (a)*, 189(2), R11–R13.

Lu J, Lu J, Murai T, Takaichi K, Uematsu T, Xu J, Ueda K, Yagi H, Yanagitani T, and Kaminskii A A (2002d), '36-W diode-pumped continuous-wave 1319-nm Nd:YAG ceramic laser', *Opt. Lett.*, 27(13), 1120–1122.

Marmo J, Injeyan H, Komine H, McNaught S, Machan J, and Sollee J (2009), 'Joint high power solid state laser program advancement at Northrop Grumman', *Proc. SPIE*, 7195, 719507-1–719507-6.

McNaught S J, Komine H, Weiss S B, Simpson R, Johnson A M F, Machan J, Asman C P, Weber M, Jones G C, Valley M M, Jankevics A, Burchman D, McClellan M, Sollee J, Marmo J, and Injeyan H (2009), '100 kW coherently combined slab MOPAs', in *Conference on Quantum Electronics and Laser Science Conference on Lasers and Electro-Optics*, CLEO/QELS, paper CThA1.

© Woodhead Publishing Limited, 2013

Merkle L D, Dubinskii M, Schepler K L, and Hegde S M (2006), 'Concentration quenching in fine-grained ceramic Nd:YAG', *Opt. Express*, 14(9), 3893–3905.

Mezeix L, and Green D J (2006), 'Comparison of the mechanical properties of single crystal and polycrystalline yttrium aluminum garnet', *Int. J. Appl. Ceram. Technol.*, 3(2), 166–176.

Pavel N, Tsunekane M, and Taira T (2011), 'Composite, all-ceramics, high-peak power Nd:YAG/Cr^{4+}:YAG monolithic micro-laser with multiple-beam output for engine ignition', *Opt. Express*, 19(10), 9378–9384.

Qin X, Yang H, Zhou G, Lou D, Zhang J, Wang S, and Ma J (2011), 'Synthesis of submicron-sized spherical Y_2O_3 powder for transparent YAG ceramics', *Mater. Res. Bull.*, 46, 170–174.

Quarles G J (2009), 'Studies of the manufacturing, fabrication and applications of next-generation oxide polycrystalline laser gain materials', *5th Laser Ceramics Symposium: International Symposium on Transparent Ceramics for Photonic Applications*, Bilbao, Spain, 9–11 December, 2009, P. I-5.

Ramirez M O, Wisdom J, Li H, Aung Y L, Stitt J, Messing G L, Dierolf V Liu Z, Ikesue A, Byer R L, and Gopalan V, (2008), 'Three-dimensional grain boundary spectroscopy in transparent high power ceramic laser materials', *Opt. Express*, 16(9), 5965–5973.

Saikawa J, Sato Y, Taira T, and Ikesue A (2004), 'Absorption, emission spectrum properties, and efficient laser performances of Yb:$Y_3ScAl_4O_{12}$ ceramics', *Appl. Phys. Lett.*, 85(11), 1898–1900.

Sakai H, Kan H, and Taira T (2008), '>1 MW peak power single-mode high-brightness passively Q-switched Nd^{3+}:YAG microchip laser', *Opt. Express*, 16(24), 19891–19899.

Sanghera J, Shaw B, Kim W, Villalobos G, Baker C, Frantz J, Hunt M, Sadowski B, and Aggarwal I (2011a), 'Ceramic laser materials', *Proc. SPIE*, 7912, 79121Q-1–15.

Sanghera J, Kim W, Villalobos G, Baker C, Frantz J, Shaw B , Bayya Sh, Sadowski B, Hunt M, and Aggarwal I (2011b), 'Transparent ceramics for high power solid state lasers', *Proc. SPIE*, 8039, 803903-1–8.

Sanghera J, Bayya Sh, Villalobos G, Kim W, Frantz J, Shaw B, Sadowski B, Miklos R, Baker C, Hunt M, Aggarwal I, Kung F, Reicher D, Peplinski S, Ogloza A, Langston P, Lamar C, Varmette P, Dubinskiy M, and DeSandre L (2011c), 'Transparent ceramics for high-energy laser systems', *Opt. Mater.*, 33, 511–518.

Sanghera J, Kim W, Baker C, Villalobos G, Frantz J, Shaw B, Lutz A, Sadowski B, Miklos R, Hunt M, Kung F, and Aggarwal I (2011d), 'Laser oscillation in hot pressed 10% Yb^{3+}:Lu_2O_3 ceramic', *Opt. Mater.*, 33, 670–674.

Sato Y, and Taira T (2006), 'The studies of thermal conductivity in $GdVO_4$, YVO_4, and $Y_3Al_5O_{12}$ measured by quasione-dimensional flash method', *Opt. Express*, 14(22), 10528–10536.

Sekita M, Haneda H, Yanagitani T, and Shirasaki S (1990), 'Induced emission cross section of Nd:$Y_3Al_5O_{12}$ ceramics', *J. Appl. Phys.*, 67, 453–458.

Sekita M, Haneda H, Shirasaki S, and Yanagitani T (1991), 'Optical spectra of undoped and rare-earth- (= Pr, Nd, Eu and Er) doped transparent ceramic $Y_3Al_5O_{12}$', *J. Appl. Phys.*, 68, 3709–3718.

Soules T (2007), 'Transparent laser ceramics at Lawrence Livermore National Laboratory (LLNL)', *3rd Laser Ceramics Symposium: International Symposium on Transparent Ceramics for Photonic Applications*, Paris, CNRS, 8–10 October, 2007, P. O-G-2.

Stevenson A J, Li X, Martinez M A, Anderson J M, Suchy D L, Kupp E L, Dickey E C, Mueller K T, and Messing G L (2011), 'Effect of SiO_2 on densification and

© Woodhead Publishing Limited, 2013

microstructure development in nd:yag transparent ceramics', *J. Am. Ceram. Soc.*, 94(5), 1380–1387.

Taira T (2007a), 'RE^{3+}-ion-doped YAG ceramic lasers', *IEEE J. Sel. Top. Quantum Electron.*, 13(3), 798–809.

Taira T (2007b), 'Ceramic YAG lasers', *C. R. Phys.*, 8(2), 138–152.

Taira T (2009), 'Micro solid-state photonics – review', *Rev. Laser Eng.*, 37, 227–234.

Taira T (2010), 'High brightness microchip laser and engine ignition', *Rev. Laser Eng.*, 38, 576.

Taira T (2011), 'Domain-controlled laser ceramics toward giant micro-photonics [invited]', *Optical Materials Express*, 1(5), 1040–1050.

Taira T, Ikesue A, and Yoshida K (1998), 'Diode pumped Nd:YAG ceramics lasers', in *Advanced Solid State Lasers*, edited by W. Bosenberg and M. Fejer, Vol. 19 of OSA Trends in Optics and Photonics Series (Optical Society of America), paper CS4.

Taira T, Ikesue A, and Yoshida K (1999), 'Performance of highly Nd^{3+}-doped YAG ceramic microchip laser', in *Proc. Conf. Lasers Electro-Opt.* (1999), paper CTak39, 136–137.

Tang F, Cao Y, Huagang L, Guo W, and Wang W (2012), 'Fabrication and laser behavior of composite Yb:YAG ceramic', *J. Am. Ceram. Soc.*, 95(1), 56–69.

Tokurakawa M, Shirakawa A, Ueda K, Yagi H, Noriyuki M, Yanagitani T, and Kaminskii A A (2009), 'Diode-pumped ultrashort-pulse generation based on $Yb^{3+}:Sc_2O_3$ and $Yb^{3+}:Y_2O_3$ ceramic multi-gain-media oscillator', *Opt. Express*, 17(5), 3353–3361.

Tsunekane M, and Taira T (2006), '300 W continuous-wave operation of a diode edge-pumped, hybrid composite Yb:YAG microchip laser', *Opt. Lett.*, 31(13), 2003–2005.

Tsunekane M, and Taira T (2007), 'High-power operation of diode edge-pumped, composite all-ceramic $Yb:Y_3Al_5O_{12}$ microchip laser', *Appl. Phys. Lett.*, 90(12), 121101.

Tsunekane M, Inohara T, Ando A, Kido N, Kanehara K, and Taira T (2010), 'High peak power, passively Q-switched microlaser for ignition of engines', *IEEE J. Quant. Electron.*, 46(2), 277–284.

Ueda K (2007), 'Ceramics for ultra high power lasers', *3rd Laser Ceramics Symposium: International Symposium on Transparent Ceramics for Photonic Applications*, Paris, CNRS, 8–10, October 2007, P. IO-C-1.

Ueda K (2010), 'Towards ultra high intensity lasers. Application of ceramic lasers', *6th Laser Ceramics Symposium: International Symposium on Transparent Ceramics for Photonic Applications*, Münster, Germany, 6–8 December 2010. Technical Digest 44.

Ueda K (2011), 'High power ceramic lasers for laser plasma accelerators: Report of 2nd ICFA/ICUIL Joint Workshop at LBNL', *7th Laser Ceramics Symposium: International Symposium on Transparent Ceramics for Photonic Applications*, Singapore, 14–17 November, 2011, Abstracts, 2.

Verlinden M (1987), 'Preparation of monodispersed yttrium hydroxycarbonate spheres by homogeneous precipitation', Thesis, Ames Lab., IA (USA), OSTI ID: 6427782; DE87013071.

Yagi H, Yanagitani T, and Ueda K (2006), '$Nd^{3+}:Y_3Al_5O_{12}$ laser ceramics: Flashlamp pumped laser operation with a UV cut filter', *J. Alloys Compounds*, 421, 195–199.

Yagi H, Yanagitani T, Takachi K, Yeda K, and Kaminskii A (2007a), 'Characterization and performance of highly transparent $Nd^{3+}:Y_3Al_5O_{12}$ laser ceramics', *Opt. Mater.*, 29, 1258–1262.

Yagi H, Yanagitani T, Numazawa T, and Ueda K (2007b), 'The physical properties of

© Woodhead Publishing Limited, 2013

transparent $Y_3Al_5O_{12}$. Elastic modulus at high temperature and thermal conductivity at low temperature', *Ceram. Int.*, 33, 711–714.

Yamamoto R, Bhachu B S, Cutter K P, Fochs S N, Letts S A, Parks C W, Rotter M D, and Soules T F (2008), 'The use of large transparent ceramics in a high powered, diode pumped solid state laser', in *Advanced Solid-State Photonics*, OSA Technical Digest Series (CD), Optical Society of America, paper WC5.

Yanagitani T, Yagi H, and Yamasaki Y (1998a), 'Production of fine powder of yttrium aluminum garnet', Japanese patent 10-101411.

Yanagitani T, Yagi H, and Ichikawa M (1998b), 'Production of yttrium-aluminum-garnet fine powder', Japanese patent 10-101333.

Zhang H, Yang Q, Lu S, and Shi Z (2011a), 'Structural and spectroscopic characterization of Yb^{3+} doped Lu_2O_3 transparent ceramics', *Opt. Mater.*, 34(6), 969–972.

Zhang W, Lup T, Wei N, Wang Y, Ma B, Li F, Lu Z, and Qi J (2011b), 'Assessment of light scattering by pores in Nd:YAG transparent ceramics', *J. Alloys Compounds*, 520(6), 36–41.

Zhao H, Sun X, Zhang J W, Zou Y K, Li K K, Wang Y, Jiang H, Huang P-L, and Chen X (2011), 'Lasing action and optical amplification in Nd^{3+} doped electrooptic lanthanum lead zirconate titanate ceramics', *Opt. Express*, 19(4), 2965–2971.

© Woodhead Publishing Limited, 2013

4 Fluoride laser ceramics

P. P. FEDOROV, A. M. Prokhorov General Physics Institute,
Russian Academy of Sciences, Russia

DOI: 10.1533/9780857097507.1.82

Abstract: Current progress in the research and development of fluoride laser ceramics seriously lags behind the achievements attained in the area of oxide-based ceramic materials. Despite the fact that the first polycrystalline CaF_2:Dy laser sample had been obtained in 1964, the next report of lasing on LiF ceramics with color centers appeared only in 2007. However, despite this delay in studying the materials, an advanced technique of preparation of fluoride laser ceramics by hot-forming is currently available. This has permitted the creation of lasing with $CaF_2:R^{3+}$ (R = Yb, Nd, Tm, Er), $SrF_2:R^{3+}$ (R = Nd, Pr), and $Ca_{0.65}Sr_{0.30}Yb_{0.05}F_{2.05}$ polycrystalline ceramic materials. Spectral and other physical properties of these ceramics are almost identical to those of single crystals with the same chemical composition, whereas the mechanical quality and opportunity to shape fluoride ceramics to manufacture laser elements exceed similar properties of single crystals. The latter factors open a very broad and attractive prospect for the use of fluoride laser ceramics in the construction of various laser systems.

Key words: fluorides of lithium, calcium, strontium, barium, rare earth metals; solid solutions with fluorite structure, pyrohydrolysis, optical ceramics, microhardness, fracture toughness, laser materials, color centers, optical losses.

4.1 Introduction

Originally, the word 'ceramics' was applied to thermally treated mixtures of clay, water and some additional supplements. Currently this term is used for polycrystalline inorganic materials obtained by the thermal treatment of precursor batches in order to assure mechanical integrity and property continuity of the manufactured item. Ceramics can contain an amorphous component; however, one would use a more specific name such as 'glass ceramics' only in the case of the domination of non-crystalline components over crystalline ones in the described sample.

Optical ceramic materials, and in particular fluoride optical ceramics, have been known for several decades (Vydrik *et al.*, 1980). Since the time of their discovery in the middle of the twentieth century, many researchers from various laboratories have worked on the application of these polycrystalline transparent optical materials in various areas. Both American and Soviet techniques for preparing ceramic magnesium, calcium and barium fluorides

© Woodhead Publishing Limited, 2013

(US trademark IRTRAN, USSR trademark KO) for use in infrared (IR) optics were developed in the 1960s and 1970s (Carnall *et al.*, 1966; Swinehart and Packer, 1978; Volynets, 1973a,b). Bulk fluoride ceramic samples were produced by a hot-pressing technique via sintering powders of the corresponding substances under pressure. The first noteworthy preparations of optical ceramics for use as laser materials were carried out specifically with metal fluorides by the US company Eastman Kodak (Hatch *et al.*, 1964; Carnall *et al.*, 1966, 1969). This included the first hot-pressed polycrystalline $CaF_2:Dy^{2+}$ laser-active elements. It is worth noting that these laser materials should not have optical losses at the lasing wavelength of more than 10^{-3} cm^{-1}, while gradients of the refractive index should not exceed 10^{-5}–10^{-6} cm^{-1}. In order to achieve such a high quality product, the earliest technology for manufacturing laser materials included the preparation of starting components from their melts, followed by their pulverization and pressing. Complex procedures of powder surface treatment and ceramics translucence were also utilized. These fluoride samples exhibited the ability to maintain sustained lasing, but the power of the oscillation beams was much lower than those produced with single-crystal laser elements. Similar results were obtained with the use of oxide laser ceramic materials in the earlier papers (Greskovich and Chernoch, 1973, 1974).

A new phase of laser ceramics development began at the end of the twentieth century, when the high quality neodymium-doped yttrium–aluminum garnet ceramic materials were prepared (Ikesue *et al.*, 1995; Lu *et al.*, 2002; Sanghera *et al.*, 2011). In contrast with preceding experiments of the 1970s, the lasing properties of these products were comparable or superior to single crystals. These results renewed interest in the study of fluoride laser ceramics, including their starting materials and preparation. Various ideas concerning the manufacture of fluoride laser ceramics, as well as raw materials for their preparation, were expressed by Grass (Grass and Stark, 2005), Ishizawa (Ishizawa, 2005), and Mortier (Mortier *et al.*, 2007; Aubry *et al.*, 2009). However, preparation of laser fluoride ceramics has been successfully carried out by different research groups from the A. M. Prokhorov General Physics Institute (Moscow, Russia), the State Optical Institute (St Petersburg, Russia) and the INCROM Company (St Petersburg, Russia), cooperating under the auspices of Russian Federation state contracts (Fedorov *et al.*, 2006, 2009; Basiev *et al.*, 2007a,b,c, 2008a,b; Alimov *et al.*, 2008).

Compared to single crystals and glasses, ceramic materials possess many advantages. First of all, ceramic optical products have an essential degree of freedom in choosing the shape and the size of manufactured specimens. In particular, use of ceramics allows manufacturing of large-sized samples. Secondly, one can produce ceramic materials with high and uniform (or otherwise specifically designed) distribution of dopants, which is not easily achievable (or even possible) in the case of single crystals. Thirdly, ceramic

© Woodhead Publishing Limited, 2013

preparations have substantial tenacity and durability, whereas crystal and glass optical materials are notoriously fragile. Single crystals become more prone to cracking at the cleavage planes with an increase in their chemical purity and perfection of their structure, while ceramics do not have this tendency.

For example, the growth of single crystals of fluorite-type $M_{1-x}R_xF_{2+x}$ (M = alkali earth metal, R = rare earth element) solid solutions from the melt is hampered by instability of the crystallization front. As a result, the obtained crystals have a honeycomb substructure and are optically inhomogeneous. There are some areas of chemical composition where one cannot grow single crystals of laser quality from the melt by directed crystallization, or such growth becomes a very difficult procedure despite the existence of the corresponding thermodynamically stable solid phase (Fedorov and Osiko, 2005; Kouznetsov *et al.*, 2007). Thus fluoride ceramics technology becomes a very appealing alternative option.

4.2 Fluoride powders: chemistry problems and relevant technology processes

Elimination of pores and achieving practically 100% density is a necessary (but insufficient) requirement for laser optic materials. It is a serious problem for oxide ceramic materials. However, physical and chemical properties of fluorides (binary and complex as well) are different from those of the corresponding oxides, and, therefore, it is easier to achieve the above target by using fluoride systems. Unfortunately, there is an additional obstacle in the case of fluoride ceramics: in order to obtain highly optically transparent fluoride ceramics, several very complex chemical problems have to be overcome.

One such difficulty stems from the fact that fluorides are prone to pyrohydrolysis, i.e. reaction with water and/or water vapor at elevated temperatures (Fedorov and Osiko, 2005; Sobolev, 2000), leading to the formation of the corresponding oxides and oxofluorides. These products are not isomorphous to the initial fluorides and, being accumulated in the initial material, usually form the second phase as finely dispersed inclusions. This results in the irreversible destruction of optical homogeneity in pyrohydrolyzed materials. Therefore, one must take all necessary precautions to prevent hydrolysis (pyrohydrolysis) at all stages of the manufacturing process. The well-developed surfaces of fluoride particles create additional obstacles in overcoming pyrohydrolysis because these surfaces adsorb various molecules, including water. Fedorov *et al.* (2008) estimate that fluorides precipitated from aqueous solutions retain up to 15 adsorbed monolayers of water at their surface. These water molecules, which are strongly bound to the particle surface, are not easily removed. This situation becomes even more

© Woodhead Publishing Limited, 2013

intractable in the case of nanoparticles with their vast surface area making their pyrohydrolysis much easier than that of macroparticles (Kouznetzov *et al.*, 2006b).

One may minimize the effect of hydrolysis by using various active fluorinating atmospheres to convert the formed oxides and/or oxofluorides back to the desired fluorides (Fedorov and Osiko, 2005). Gaseous carbon tetrafluoride, CF_4, is the reagent most frequently used for this purpose. This or similar fluorinating technologies have been developed to improve the optical quality of fluoride ceramics by conversion of oxide impurities to fluorides, but combining batch fluorination and ceramics manufacturing is very technical. It requires a hermetically sealed vacuum furnace and complete absence of air in the system.

In order to avoid the above complications, researchers developed different protocols to synthesize fluoride ceramics. One simpler process is hot pressing. This method turns polycrystalline powders into a monolithic body via particle agglomeration (Fig. 4.1). Usually, this process is performed at $T = (0.5–0.8)*T_{melting}$ and up to 300 MPa load under $10^{-2}–10^{-3}$ torr vacuum. This technique also has its shortcomings, including reducing conditions which may lead to the appearance of color centers or transition of ions-activators to their lower valence states. Such samples usually exhibit yellow, gray or even black coloration, and they may also have inclusions. Their treatment in a fluorinating atmosphere can result in discoloration and, therefore, remedies this problem.

The quality of the starting materials is one of the most important factors affecting the synthesis of the optical ceramics. Transparency data for different

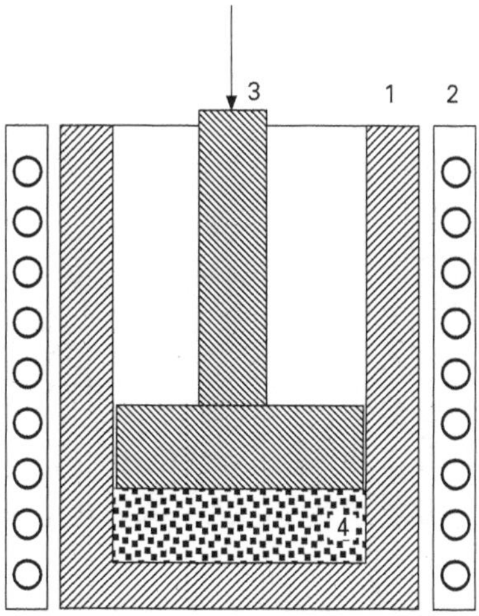

4.1 Preparation of ceramics by hot pressing with agglomeration of powder particles. 1 cylindrical chamber, 2 heating jacket, 3 piston, 4 sample.

© Woodhead Publishing Limited, 2013

fluorite ceramics samples (1–10) prepared from different quality batches are presented in Fig. 4.2(a). A proper choice of the initial ingredients allowed the creation of products of the same quality, as in Carnall *et al.* (1966) (Fig. 4.2(b)).

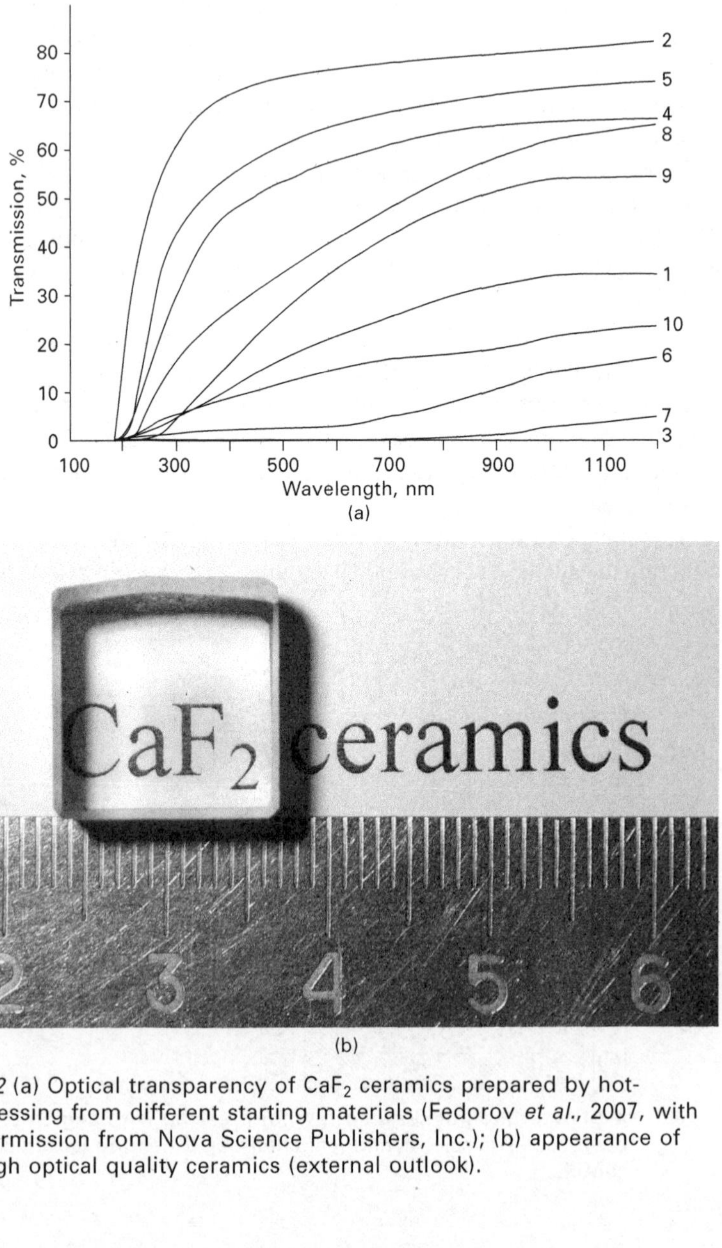

4.2 (a) Optical transparency of CaF_2 ceramics prepared by hot-pressing from different starting materials (Fedorov *et al.*, 2007, with permission from Nova Science Publishers, Inc.); (b) appearance of high optical quality ceramics (external outlook).

© Woodhead Publishing Limited, 2013

4.3 Fluoride ceramics as optical medium

Samples of CaF_2 optical ceramics prepared by hot-pressing were studied by Akchurin *et al.* (2006), Popov *et al.* (2007), Basiev *et al.* (2008a,b), Palashov *et al.* (2007, 2009) and Khazanov *et al.* (2009) as the model object. For comparison, Akchurin *et al.* (2006), Popov *et al.* (2007), Basiev *et al.* (2008a,b), Palashov *et al.* (2007, 2009) and Khazanov *et al.* (2009) have also used standard CaF_2 single crystals from the State Optical Institute, St Petersburg, Russia, and a CaF_2 natural optical ceramics specimen (Suranskoe deposit, Ural Mountains; average crystalline domain size about 37 nm). The latter ceramic sample was a unique natural object formed approximately one billion years ago.

UV/visible and IR absorption spectra of the synthesized CaF_2 ceramics (curves 1 and 2) as well as CaF_2 single crystal (curve 3), are presented in Fig. 4.3.

The transmission of the ceramics samples is somewhat inferior to that of the single crystal in the near-UV part of the spectrum; however, at visible and IR wavelengths, the corresponding transmission values coincide or almost coincide.

Two different methods were used to measure small optical losses in synthesized CaF_2 ceramics under monochromatic laser radiation. One method was specifically designed to estimate the losses in optical fibers and was based on measuring the evolving heat from the weakly absorbing medium when a laser beam passed through it (Plotnichenko *et al.*, 1981). This method was used by Basiev *et al.* (2008a,b). The second method of thermally induced depolarization of laser radiation was suggested by Palashov *et al.* (2007, 2009). Both types of measurements were carried out at 1.06 micron wavelength. The first method gave a loss value of 9×10^{-3} cm^{-1}, whereas the second method produced a value of $(1.1–1.5) \times 10^{-3}$ cm^{-1}. Both numbers indicate a high transparency of the prepared ceramic samples suitable for laser materials. Low values of the optical losses in synthesized CaF_2 ceramics were determined by the specific structure of the boundaries between crystalline grains.

The study of transport of sub-THz heat acoustic phonons in CaF_2 single crystals and ceramics showed the length of their free path at 3.82 K to be equal to 1.2×10^{-1}, 6.6×10^{-2} and 1.4×10^{-2} cm respectively for single-crystal, synthesized and natural ceramic samples (Khazanov *et al.*, 2009). The actual thickness of the above boundary layers between crystalline domains in natural ceramics has been estimated to be about 1.5 ± 0.3 nm. At the same time, actual samples of the synthesized CaF_2 ceramics exhibited properties with zero or almost no boundary layer.

A comparative study of the thermal conductivity of CaF_2 single-crystal, synthesized and natural ceramic samples has shown that all have practically

© Woodhead Publishing Limited, 2013

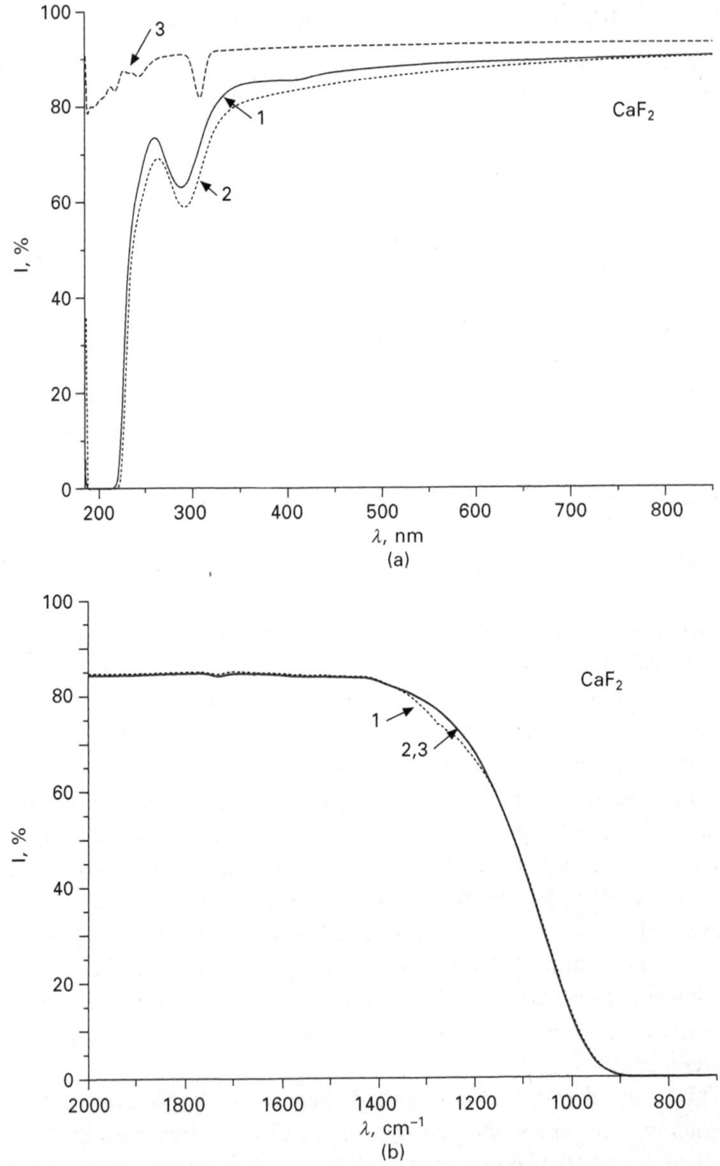

4.3 (a, b) Optical transparency of synthesized CaF_2 ceramics (thickness 8.8 mm) (1, 2) and CaF_2 single crystal (thickness 8.3 mm) (3) (Fedorov *et al.*, 2007, with permission from Nova Science Publishers, Inc.).

the same properties over the 50–300 *K* temperature interval (Popov *et al.*, 2007) (Fig. 4.4).

Carnall *et al.* (1966) found that the refraction and optical dispersion

© Woodhead Publishing Limited, 2013

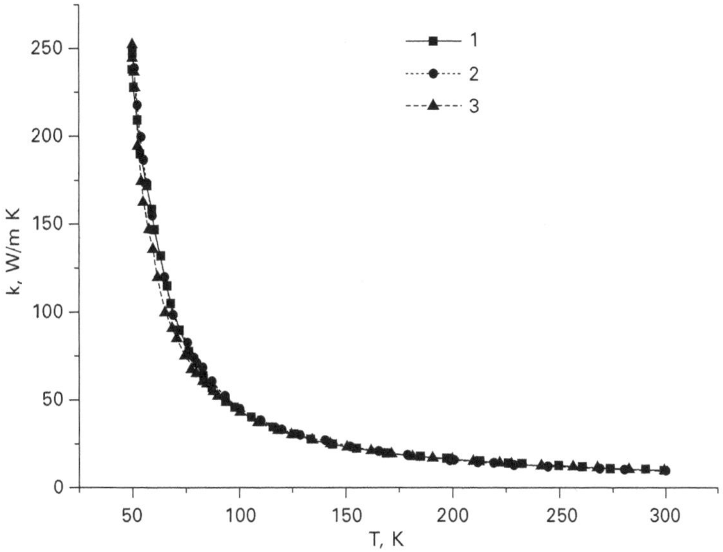

4.4 Thermal conductivity of CaF_2 samples: natural sample from Suranskoe deposit, Ural Mountains (1), single crystal (2) and synthesized optical ceramics (3) (Popov *et al.*, 2007, with permission from Pleiades Publishing Ltd).

Table 4.1 Microhardness (H) and fracture toughness (K_{1C}) of CaF_2 single crystal and optical ceramic samples

Material type	H (GPa)	K_{1C} (MPa·m$^{1/2}$)
Single crystal	2.00 ± 0.07	1.5 ± 0.15
Natural ceramic sample	2.25 ± 0.05	6.3 ± 0.60
Synthesized ceramics sample	2.60 ± 0.10	4.7 ± 0.3

coefficients of CaF_2 single-crystal and synthesized ceramic specimens are practically the same. This was confirmed by Fedorov and Lapin for the pure CaF_2 as well as doped CaF_2 single crystals and ceramics samples (Fedorov *et al.*, 2009).

In order to characterize mechanical properties of the prepared CaF_2 ceramic samples, their microhardness and fracture toughness were studied by Akchurin (Akchurin *et al.*, 2006; Fedorov *et al.*, 2007; Basiev *et al.*, 2008a,b) (Table 4.1). These samples were subjected to mechanical load, followed by measuring the linear size of the formed radial cracks (C), which appeared near the site of load application, as a function of the load on the indenter. Calculations of the coefficient of stress intensity (fracture toughness) K_{1C} exhibited the following correlation:

$$K_{1C} = 0.016(E/H)^{1/2}P/C^{3/2}$$

© Woodhead Publishing Limited, 2013

where H is the microhardness, P is the load, and E is Young's modulus (Table 4.1). K_{1C} values for the synthesized and natural CaF_2 ceramic specimens were more than three and four times higher than K_{1C} values for single crystals.

Thus, fluoride ceramics appear to have sufficiently good optical properties for use in laser development, while their mechanical properties absolutely exceed those of single crystals.

4.4 Development of the fluoride laser ceramics synthesis protocol

Laser materials usually need doping with activator admixtures (unless they work exclusively utilizing color centers). Therefore, several authors (Bensalah *et al.*, 2006; Kouznetzov *et al.*, 2006a, 2007; Fedorov *et al.*, 2007) have attempted the synthesis of two-component rare-earth activated batches that provided activated ceramics of the appropriate quality.

Preliminary experiments (Fedorov *et al.*, 2007) showed that mixing powders of calcium and erbium fluorides does not lead to products which have an acceptable level of composition homogeneity, due to incomplete diffusion of components. Product samples were optically inhomogeneous and lasing was not obtained. This highlights the obvious obstacle with samples of complex chemical composition and batches with multiple precursor components. Kouznetsov *et al.* (2006a, 2007), Fedorov *et al.* (2007) and Bensalah *et al.* (2006) simultaneously attempted similar approaches to address the latter problem by implementing a co-precipitation from aqueous solutions via slow dropwise addition of starting nitrate solutions to aqueous hydrofluoric acid, e.g.:

$$0.95Ca(NO_3)_2 + 0.05Yb(NO_3)_3 + 2.05HF$$
$$\rightarrow Ca_{0.95}Yb_{0.05}F_{2.05}\downarrow + 2.05HNO_3$$

Whereas target single-phase $M_{1-x}R_xF_{2+x}$ solid solutions have been easily precipitated by the addition of HF to the aqueous systems containing calcium and/or strontium ions along with lanthanide ions, similarly treated barium–lanthanide fluoride samples appeared to be a mixture of BaF_2 and $Ba_4R_3F_{17}$ phases instead (Kouznetsov *et al.*, 2010). The latter circumstances prohibited the synthesis of optically homogeneous ceramics due to the fluctuation of the chemical composition within the sample. In contrast, $Ca_{1-x}R_xF_{2+x}$ batches – after necessary treatment – produced transparent ceramics (Fig. 4.5). Some of these $Ca_{1-x}R_xF_{2+x}$ specimens, e.g., CaF_2:Yb ceramics (Fig. 4.5(a), (b)), in accordance with their UV spectra, accommodate significant Yb^{2+} concentrations.

However, ceramics prepared from batches initially co-precipitated from aqueous solutions do not lase. The strongly adsorbed water layer on the

© Woodhead Publishing Limited, 2013

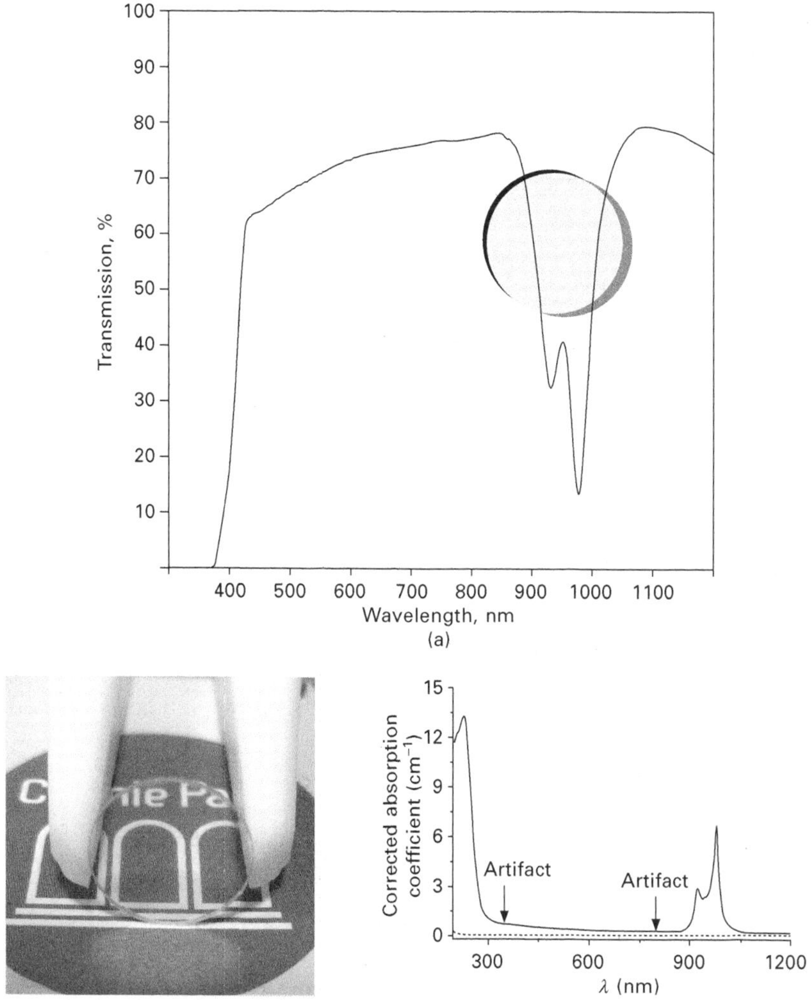

4.5 CaF_2:Yb ceramics prepared by hot-pressing according to (a) Fedorov *et al.* (2007), with permission from Nova Science Publishers, Inc. (Basiev *et al.*, 2006) and (b) Lyberis *et al.* (2011), reprinted from *Journal of the European Ceramic Society*, Lyberis A, Patriarche G, Gredin P, Vivien D, and Mortier M, 'Origin of light scattering in ytterbium doped calcium fluoride transparent ceramic for high power lasers', vol. 31, pp. 1619–1630, copyright 2011 with permission from Elsevier.

surface of the batch particles could be a reason for the above effect. This layer can not be removed under a fluorinating atmosphere or under translucence treatment. Lanthanide fluorides undergo hydrolysis much more easily than

© Woodhead Publishing Limited, 2013

alkali earth metal fluorides (Fedorov and Osiko, 2005; Kouznetzov *et al.*, 2006b), generating oxygen-containing phases between particles and thus causing the light beam scattering. The excessive content of ytterbium at the inter-grain boundaries creates an additional problem of refractive index variations (Lyberis *et al.*, 2011).

The deformation of heated crystals under pressure (hot-forming process) for the synthesis of ceramics mentioned above led to successful avoidance of product contamination with oxygen via pyrohydrolysis (Fig. 4.6). Whereas Hatch *et al.* (1964) and Carnall *et al.* (1966, 1969) hot-pressed preliminary pulverized solids, Basiev *et al.* (2007a,b,c, 2008a,b), Fedorov *et al.* (2009) and Alimov *et al.* (2008) used the single crystals as starting materials for hot-pressing protocols. Indeed, parameters of the above hot-forming and hot-pressing processes are different due to the different physical mechanisms of single-crystal deformation and consolidation of powder particles in the course of the formation of polycrystalline optic materials.

4.5 Microstructure, spectral luminescence and lasing properties

4.5.1 Microstructure

The microstructure of ceramics samples depends on the synthesis method used (Figs 4.7 and 4.8).

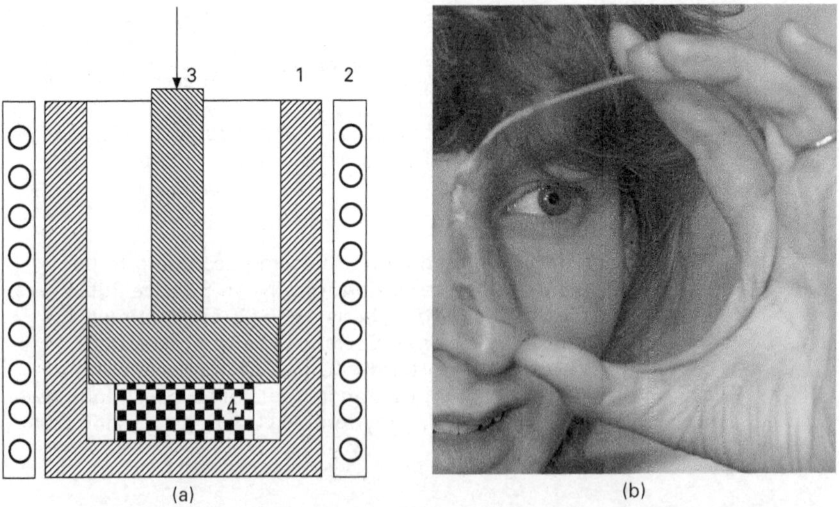

4.6 (a) Preparation of ceramics by hot-forming via crystal deformation: 1 cylindrical chamber, 2 heating jacket, 3 piston, 4 sample; and (b) CaF_2:Yb^{3+} laser ceramics prepared by this method (with permission of E. V. Chernova).

© Woodhead Publishing Limited, 2013

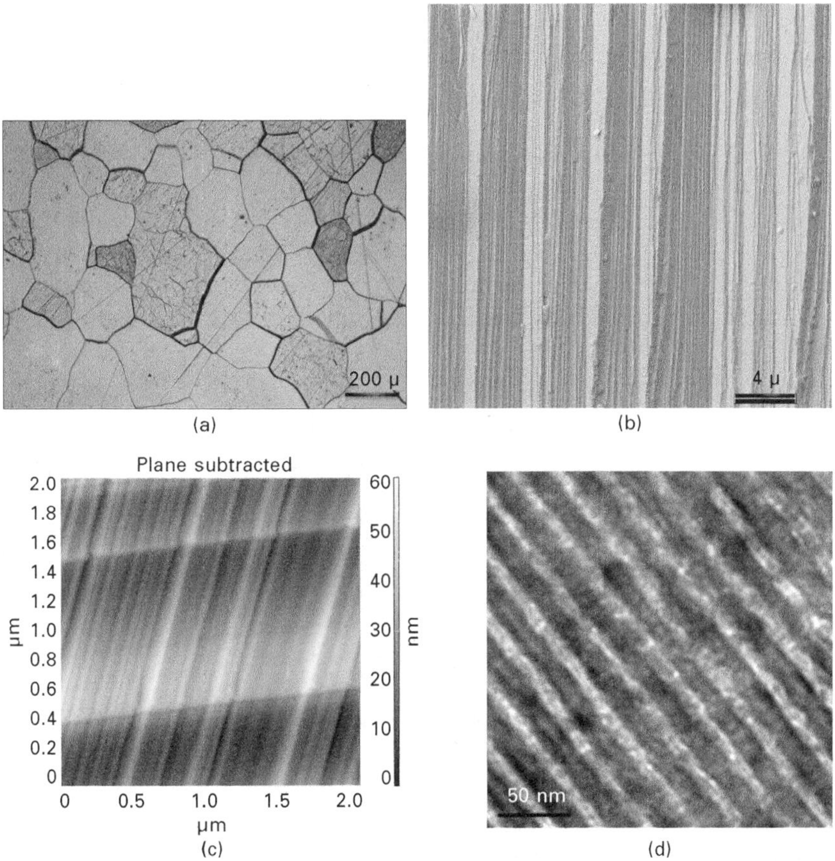

4.7 Microstructure of hot-pressed CaF_2 ceramics by (a) optical microscopy, (b) electron microscopy (shearing cut), (c) atomic force microscopy (shearing cut), and (d) high resolution electron microscopy (Akchurin *et al.*, 2011, with permission from Pleiades Publishing Ltd).

Hot-pressing fluoride powders leads to a homogeneous microstructure with relatively large (80–150 nm) domains. This has been observed for pure CaF_2 as well as for doped ceramics, such as CaF_2:Yb. The layered fine nanostructure within these domains has been detected by electron microscopy and atomic force microscopy (AFM): the fine layer width for CaF_2 is *ca.* 25 nm (Akchurin *et al.*, 2011). The most probable guess is that this nanostructure stems from crystal twinning within the grains of the sample.

When they are deformed at higher temperatures, single-crystal samples have a different microstructure. For low degrees of deformation, one can observe a fractal crack-type structure of the borders of crystalline domains (Fig. 4.8(a)). This structure is most likely not stable and should relax with sample aging.

© Woodhead Publishing Limited, 2013

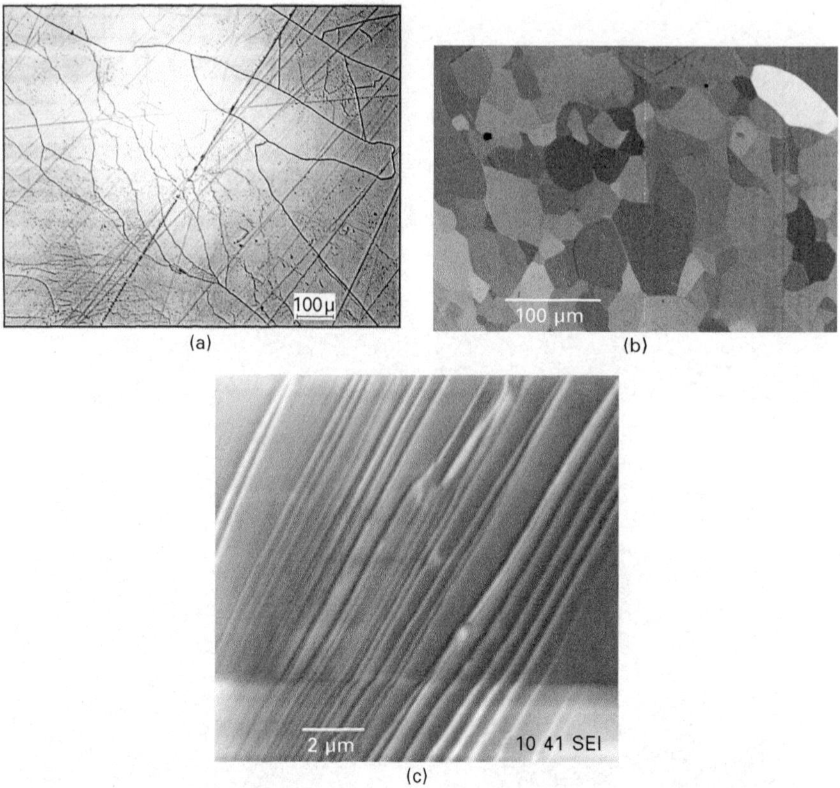

4.8 Microstructure of samples of hot-formed (a) $Ca_{0.65}Sr_{0.30}Yb_{0.05}F_{2.05}$ (Fedorov *et al.*, 2009) and (b), (c) CaF_2:Tm (Bol'shchikov *et al.* 2011, with permission from Quantum Electronics).

When the degree of deformation is increased, the sample microstructure becomes inhomogeneous and contains relatively large areas (*ca.* 1 mm) that have a single-crystal-type structure along with grain-type structured areas, typical of classic ceramics. Only at high degrees of deformation does the structure of the deformed single crystal turn into a completely homogeneous typical ceramic structure. One can observe that the layered nanostructure of the ceramics domains of the deformed single crystals is very similar to the nanostructure of the hot-pressed powders (Akchurin *et al.*, 2011) (Fig. 4.8(b), (c)).

The above observation can be illustrated with a typical correlation between the average size of BaF_2 crystalline domains and the degree of deformation of the single crystal (Fig. 4.9); only highly deformed samples correspond to the 'completely' ceramic state of the studied material.

© Woodhead Publishing Limited, 2013

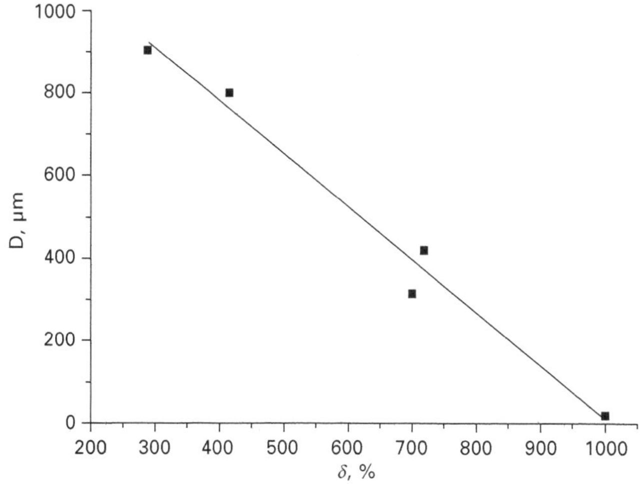

4.9 Average size of crystalline domains *vs.* deformation degree of BaF_2 single crystal (δ) at 800°C; $\delta = (h_0 - h)/h \times 100\%$, where h_0 = height of the initial sample and h = height of the formed plate.

4.5.2 Spectral luminescence and lasing properties

Lithium fluoride crystals with aggregate color centers are among the most well-known laser materials for a smoothly tunable generation of pico- and femto-second pulses in the near-IR region. These crystals are ideal for application in tunable lasers and lasers with mode synchronization, for multiple reasons. They have a broad amplification band in the 1–1.3 micron range and a high cross-section of luminescence transition (*ca.* $10^{-17}cm^2$). They also have a broad absorption band for pumping by lasers that are irradiating near 1 micron, and high thermal conductivity.

Since laser diodes with a radiation wavelength of 960–980 nm are now achievable, it is possible to use lithium fluoride crystals to create efficient and compact solid-state tunable lasers. However, these crystals have some serious drawbacks, such as a low mechanical strength that restricts the limiting power of lasing radiation. The creation of lithium fluoride optical ceramics with color centers can eliminate this obstacle.

Such LiF laser materials with the F_2^- color centers were produced by the irradiation of pre-synthesized LiF optical ceramics. Comparison of lasing of LiF–F_2^- ceramics and single crystals of similar optical density was carried out (Basiev *et al.*, 2007a,b).

Absorption, luminescence and lasing spectra of LiF–F_2^- ceramics are shown in Fig. 4.10. Figure 4.11 shows the dependence of the average lasing output power of ceramic and single-crystal LiF–F_2^- samples from the absorbed average power (Basiev *et al.*, 2007b). Ceramic samples were characterized by lower lasing thresholds than single crystals. The maximum

© Woodhead Publishing Limited, 2013

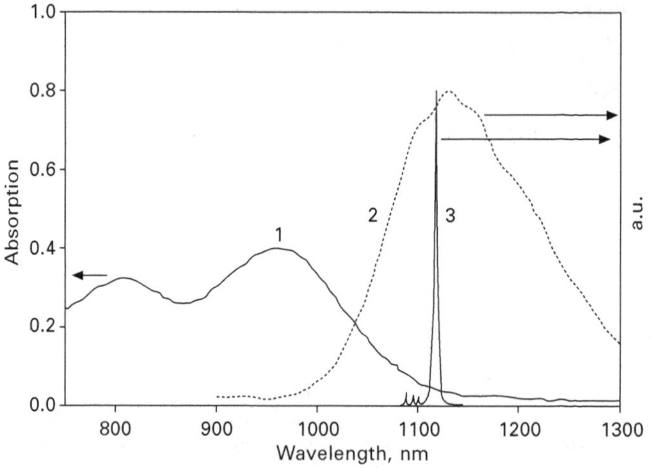

4.10 Absorption (1), fluorescence (2) and oscillation (3) spectra of LiF ceramics with F_2^- color centers (Basiev *et al.*, 2007b).

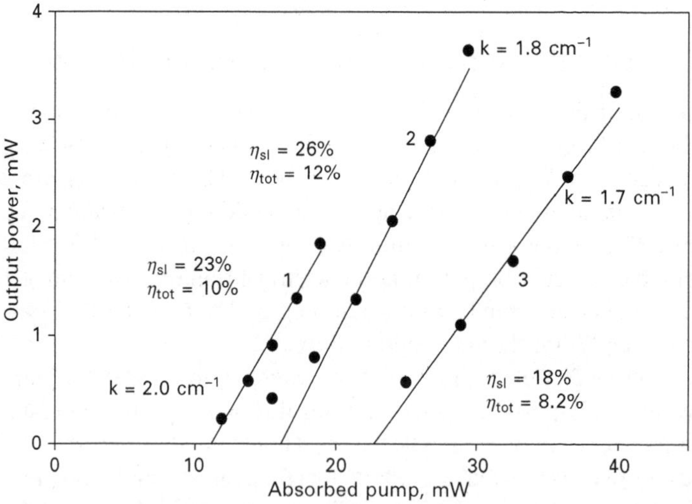

4.11 Average output power *vs.* absorbed average power for ceramic and single-crystal samples of LiF:F_2^- under laser diode pumping measured under similar conditions (laser diode pumping, λ = 967 nm, τ = 1 ms, f = 5Hz): 1 = 1 mm ceramics, 2 = 4 mm ceramics, 3 = 7 mm single crystal (Basiev *et al.*, 2007a,b); k = absorption coefficient.

slope lasing efficiency of 26% has been observed for a 4 mm thick ceramic sample (*vs.* 18% for the single crystal sample). Unfortunately, there are no data available to compare mechanical properties of ceramic and single-crystal LiF–F_2^- lasers.

© Woodhead Publishing Limited, 2013

The absorption and luminescence spectra of fluoride ceramics and single crystals doped with rare-earth ions (Er, Yb, Tm) are essentially identical (Fedorov *et al.*, 2007; Bol'shchikov *et al.*, 2011). As an example the typical Tm absorption and emission spectra are presented in Fig. 4.12. The set of optical centers, lifetimes of the excited states, absorption coefficients,

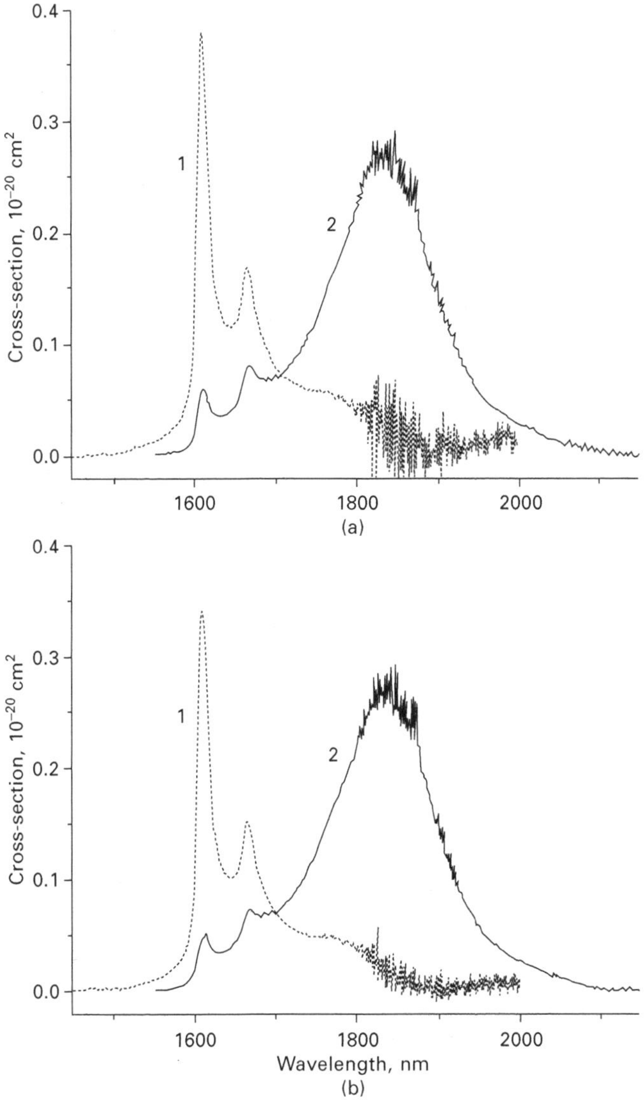

4.12 Absorption (1) and emission (2) cross-section spectra of CaF_2:Tm^{3+} single crystals (a) and ceramics (b) ($^3H_6 \leftrightarrow {}^3F_4$ transition, $T = 300$ K) (Bol'shchikov *et al.*, 2011, with permission from Quantum Electronics).

© Woodhead Publishing Limited, 2013

wavelengths of absorption bands and luminescence lines are very close for the samples of the same chemical composition. Excellent quality of rare earth-doped fluoride ceramic specimens, prepared by hot-forming of crystals, allows the achievement of lasing parameters that are almost equivalent to the properties of the single crystals of the same chemical composition (Fig. 4.13). Lasing properties of $CaF_2:Yb^{3+}$ single crystals and ceramics were studied by Doroshenko (Fedorov *et al.*, 2009) with the use of laser diode pumping (100 micrometer fiber radiation output with the impulse set at 2 ms duration of impulse at 10 GHz repetition frequency). The LD pump

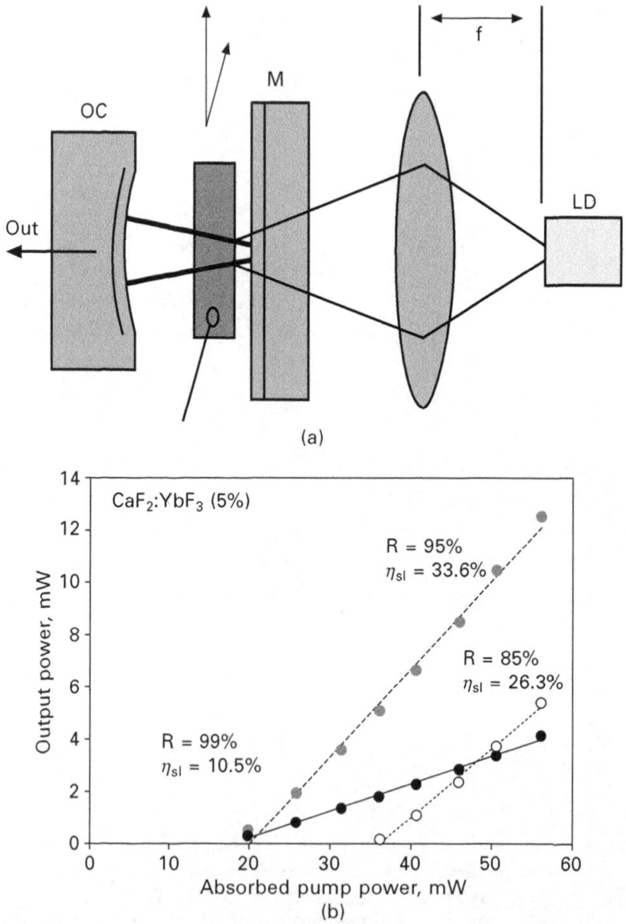

4.13 Lasing of $CaF_2:Yb^{3+}$ (5%) samples (Fedorov *et al.*, 2009): (a) optical scheme for studying the lasing properties (OC output coupler, M dichroic mirror, LD laser diode); output power vs. absorbed pump power for (b) single crystal and (c) ceramics for various reflection coefficients of the output mirror.

© Woodhead Publishing Limited, 2013

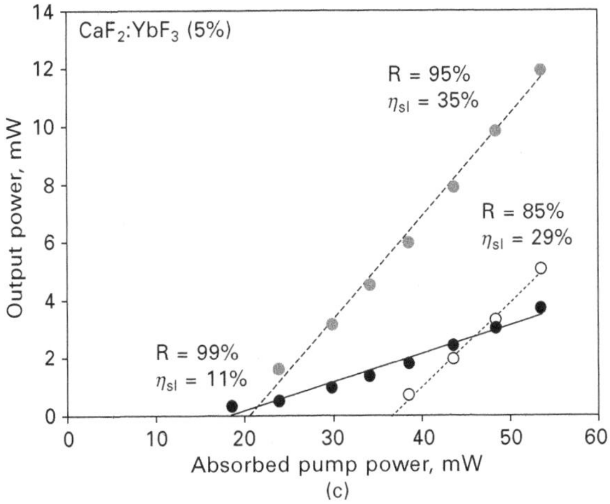

4.13 Continued

wavelength (λ = 967 nm at room temperature) was thermally adjusted to coincide with the absorption maximum of Yb^{3+} ions in the CaF_2 matrix (972 nm). LD radiation was focused by the single lens (focus distance f = 5 mm). A hemispherical resonator for measuring lasing properties was formed by a flat dichroic mirror (maximum transparency at the pump wavelength of 972 nm and maximum reflection within 1020–1080 nm spectral range suitable for lasing) and a spherical output mirror (50 mm radius) (Fig. 4.13(a)).

The lasing parameters (output power, lasing threshold and slope efficiency) of $Ca_{0.95}Yb_{0.05}F_{2.05}$ single-crystal (Fig. 4.13(b)) and ceramic (Fig. 4.13(c)) samples for the different reflection coefficients of the output mirror R are very similar. The highest slope efficiency coefficient of 35% has been obtained for $Ca_{0.95}Yb_{0.05}F_{2.05}$ ceramics at R = 95%.

Lasing has been observed for $CaF_2:R^{3+}$, where R = Yb, Nd, Tm, Er (Alimov *et al.*, 2008; Fedorov *et al.*, 2009; Bol'shchikov *et al.*, 2011; M.E. Doroshenko, private communication), $SrF_2:R^{3+}$, where R = Nd (Basiev *et al.*, 2010), Pr (Basiev *et al.*, 2011), and $Ca_{0.65}Sr_{0.30}Yb_{0.05}F_{2.05}$ (Basiev *et al.*, 2008a,b) ceramic samples. This variety of doping rare-earth ions has allowed laser oscillations in the near-IR–visible spectral range. However, most efforts have been concentrated on $CaF_2:Yb^{3+}$ ceramics. Ytterbium ion, compared to the widely utilized neodymium ions, have numerous advantages as ions-activators, when they are used in diode-pumped laser materials (Krupke, 2000). High optical quality of the above laser ceramics has provided high oscillation slope efficiencies: up to 45% in pulsed operation mode, and up to 30% in CW mode under laser diode pumping (Basiev and Doroshenko, 2011).

© Woodhead Publishing Limited, 2013

4.6 $CaF_2:Yb^{3+}$ system

Whereas laser applications (especially in the powerful laser systems) of $CaF_2:Yb^{3+}$ single crystals have been actively studied recently (Lucca *et al.*, 2004; Ito *et al.*, 2004; Kaczmzrek *et al.*, 2005; Bensalah *et al.*, 2007; Petit *et al.*, 2007, 2008; Hraiech *et al.*, 2007, 2010; Jouini *et al.*, 2008; Siebold *et al.*, 2009; Druon *et al.*, 2011; Ricaud *et al.*, 2011; Pena *et al.*, 2011), one of the first spectroscopic investigations of ytterbium admixtures in CaF_2 crystals was done as early as in 1969 (Voron'ko *et al.*, 1969).

The absorption bands in a $Ca_{1-x}Yb_xF_{2+x}$ solid solution spectrum, which associates the couples of Yb^{3+} ions or combination of several complex optical centers (Voron'ko *et al.*, 1969; Petit *et al.*, 2007), can be attributed to the spectral properties of the Yb_6F_{37} cluster (Fig. 4.14), for these bands stay unchanged in a very wide concentration interval of ytterbium.

The preparation of $Ca_{1-x}Yb_xF_{2+x}$ ceramics and single crystals under reducing conditions (e.g., by contacting the graphite crucible surface) is accompanied with the partial conversion of Yb^{3+} to Yb^{2+} ions and the subsequent appearance of the corresponding absorption bands in the UV segment of the absorption spectrum (Kaczmzrek *et al.*, 2005; Nicoara *et al.*, 2008) and possible thermal non-radiative losses under lasing (Druon *et al.*, 2011).

It is very important to keep in mind that according to the existing

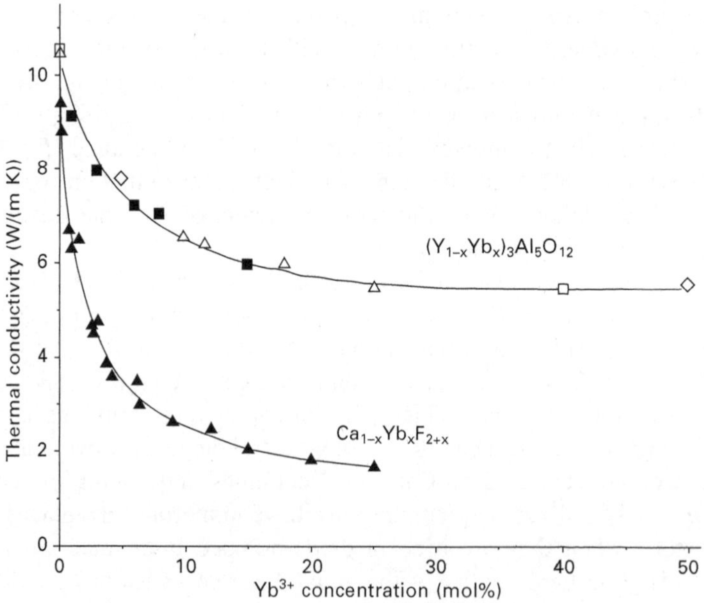

4.14 Thermal conductivity of $(Y_{1-x}Yb_x)Al_5O_{12}$ (Sanghera *et al.*, 2011) and $Ca_{1-x}Yb_xF_{2+x}$ (Popov *et al.*, 2008; P. A. Popov, private communication) solid solutions at 300 K.

© Woodhead Publishing Limited, 2013

CaF_2–YbF_3 phase diagram, $Ca_{1-x}Yb_xF_{2+x}$ solid solution exists under non-equilibrium conditions at room temperature (Sobolev and Fedorov, 1978; Fedorov, 2010). Usually, transformations in these solid solutions using a few mol% of lanthanide trifluoride are extremely slow, meaning there were no detectable changes in the samples over several decades. This may be related to the very low values of the cations' diffusion coefficients and relatively high activation energy barriers for the formation of the centers for crystallization of the new phase. These factors, however, can cause certain problems for ceramic materials designated for use under high energy flow conditions. Phase transformation also can be initiated at the boundaries between twinning domains. One can describe the border between twinning domains in a fluorite-type crystal lattice as a two-layer cation pack instead of the regular three-layered matrix, corresponding to the closest packing of cations in a fluorite-type crystal lattice. These borders can play the role of the nucleation centers for the formation of the new tysonite-related phase (low temperature orthorhombic modification of β-YbF_3), but this suggestion requires further study.

Earlier, alleged equal thermal conductivity of both laser ceramics and pure fluorite was considered as an important advantage of ceramic materials (Ito *et al.*, 2004; Jouini *et al.*, 2008; Siebold *et al.*, 2009). However, actual measurements of this parameter disproved this assumption and demonstrated that the introduction of a heterovalent admixture to a fluorite structure leads to the dramatic decrease of the thermal conductivity, especially at lower temperature. The reason for such a substantial decrease of thermal conductivity in the case of samples with heterovalent isomorphic doping *vs.* specimens with isovalent substitutions is that the scattering of acoustic phonons occurs at defect clusters for materials with heterovalent doping instead of the single atoms in the isovalent-substituted samples (Fig. 4.14).

Ceramics with a relatively high concentration of trifluoride dope exhibit a glass-type temperature dependency on their thermal conductivity (Popov *et al.*, 2008). A comparison of the thermal conductivity of $Ca_{1-x}Yb_xF_{2+x}$ solid solution and ytterbium-doped Al–Y garnet (Sanghera *et al.*, 2011) (Fig. 4.15) shows that fluoride ceramics are inferior to oxide materials. A decrease of ytterbium content in $Ca_{1-x}Yb_xF_{2+x}$ leads to the increase of its thermal conductivity, but relatively insufficient improvement of this property significantly limits prospective use of CaF_2:Yb^{3+} materials, including ceramics, for powerful lasers.

Properties of various CaF_2:Yb ceramics are presented in Table 4.2.

4.7 Prospective compositions for fluoride laser ceramics

Crystal symmetry plays a decisive role in the preparation of laser ceramics. Both fluoride and oxide laser ceramics have cubic crystal lattices. Oxides have

© Woodhead Publishing Limited, 2013

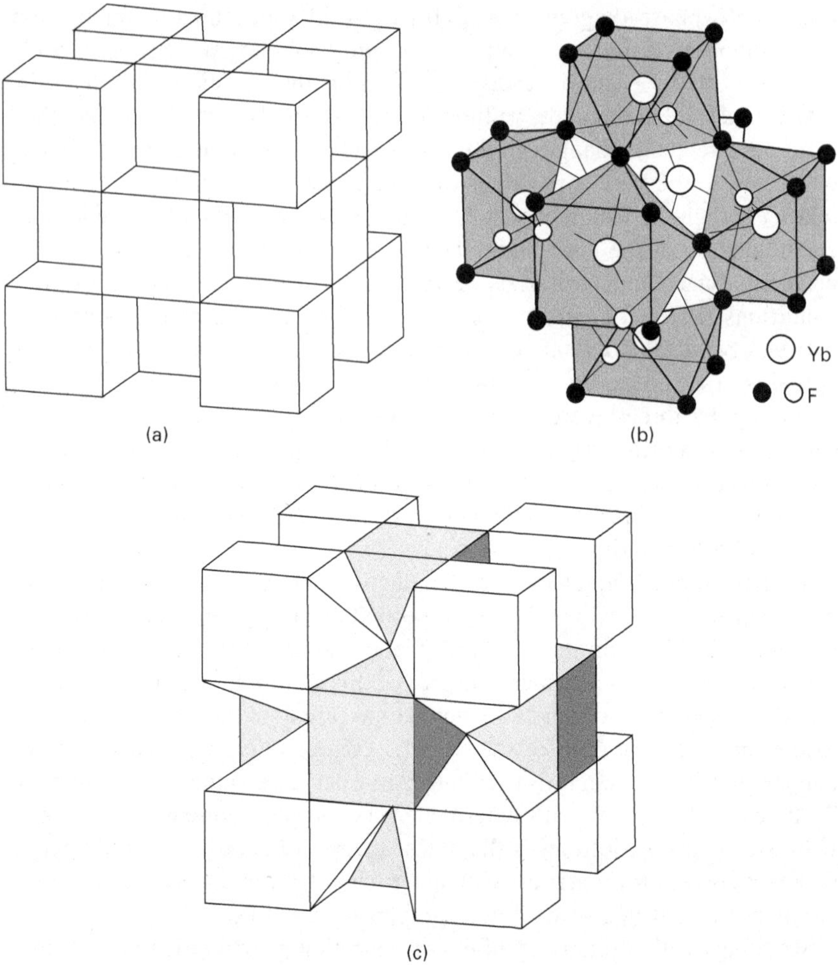

4.15 (a) CaF_2 crystal lattice unit, (b) Yb_6F_{37} cluster, and (c) its positioning in fluorite crystal lattice (Greis and Haschke, 1982; Bevan *et al.*, 1980, 1988; Kazanskii *et al.*, 2005).

a garnet (*Ia*3*d* space symmetry group, or SSG) or yttrium oxide (*I*a3 SSG) structure, whereas fluorides are represented by CaF_2, SrF_2 or their combinations (fluorite-type solid solutions) and lithium fluoride (all fluorides belong to *Fm*3*m* SSG). These cubic crystals are very likely to form boundaries between crystalline domains that correspond to polysynthetic twinning and, therefore, produce low optical losses. From this point of view, cubic KY_3F_{10} (*Fm*3*m* SSG) could be a very good candidate for the same laser application.

This may also be true for a CaF_2:Yb,Na solid solution (Hraiech *et al.*, 2007). Fluorite, double-doped with sodium and ytterbium, does not produce good quality single crystals as per the phase diagram for the NaF–CaF_2–YbF_3

© Woodhead Publishing Limited, 2013

Table 4.2 Properties of $CaF_2:Yb^{3+}$ laser ceramics

Property	Symbol, dimension	$Ca_{0.97}Yb_{0.03}F_{2.03}$	$Ca_{0.95}Yb_{0.05}F_{2.05}$
System		Cubic	Cubic
Space group		*Fm*-3*m*	*Fm*-3*m*
Lattice constant	*a*, Å	5.466	5.469
Density	ρ, g/cm^3	3.48	3.48
Yb^{3+} concentration	cm^{-3}	0.73×10^{21}	1.22×10^{21}
Melting temperature	*T*, °C	1417	1417
Thermal conductivity	*k*, W/m.K		
at 300 K		4.5	3.5
at 50 K		2.5	1.5
Refraction coefficient	n_{589nm}	1.4412	1.4468
	n_{lasing}	1.4366	1.4422
Lifetime of the excited state	τ, ms	2.2	2.1
Pumping wavelength	$\lambda_{pumping}$, nm	967	967
Lasing wavelength	λ_{lasing}, nm	1025; 1040	1025; 1040
Efficiency	%	48	35

system (Fedorov and Rappo, 2008), thus ceramic CaF_2:Yb,Na looks like a very promising alternative.

Compounds with non-cubic crystal lattices do not form laser-quality ceramic materials. Because of birefringence and light scattering at the incoherent boundaries between crystalline domains, their samples exhibit essential optical losses. One could produce acceptable values of light transmittance for Y_2O_3-based ceramics, but a similar result was impossible to achieve for Al_2O_3-based ceramics. The same is true for MgF_2 (tetragonal system, P42/mnm SSG), YF_3 (orthorhombic system, Pnma SSG), $LiRF_4$, and the R – rare earth elements (tetragonal system, $I4_1/a$ SSG), and BaR_2F_8 (monoclinic system, C2/m SSG) fluorides, too.

As previously stated, CaF_2-based ceramics exhibit twinning at the nano-level and, therefore, generate low optical losses because of perfectly well-structured boundaries between the crystalline domains. Polysynthetic twinning is known for the second type of phase transitions (or close to the second type), when lower-temperature modifications are slightly distorted in comparison with more symmetric higher-temperature modifications. This type of twinning with lowering crystal lattice symmetry is especially common for the phase transitions of the higher-temperature cubic phases. Ferroelectrics (Zheludev, 1971) are an example of this phenomenon. In particular, A_2BRF_6 elpasolites (A, B- alkaline metals with $r_A > r_B$, R – rare earth elements) undergo multiple phase transitions at different temperatures (Flerov *et al.*,

© Woodhead Publishing Limited, 2013

1998), and Rb_2NaAlF_6 transparent ceramics were synthesized by Ahrens *et al.* (2007) by cold-pressing of an elpasolite precursor produced by the sol-gel method. Another example of the same kind is $NaMgF_3$ (Ocafrain *et al.*,1996; Yoshiasa *et al.*, 2003). This compound has an orthorhombic crystal lattice at room temperature, but at 700°C it undergoes phase transition and forms high-temperature cubic modification, before its melting point of 1020°C. Preliminary studies have shown an opportunity to synthesize transparent twinned polycrystalline samples at room temperature (Fedorov *et al.*, 2009).

Ordered fluorite-type phases in MF_2–RF_3 systems (Fedorov *et al.*, 1974; Sobolev and Fedorov, 1978; Sobolev, 2000) also have slightly distorted CaF_2 structures and, therefore, may be considered as prospective materials for transparent laser ceramics. For example, optically homogeneous crystal domains of trigonally distorted fluorite-type $Ca_8R_5F_{31}$ have been observed in polished cross-sections of two-phase samples in CaF_2–RF_3 systems by Fedorov *et al.* (1974).

However, one can make highly transparent ceramics from non-cubic crystalline substances only if the crystalline domains will be less than the wavelength of the used electromagnetic radiation (Basiev *et al.*, 2008a,b).

4.8 Conclusion

The presented data show that modern fluoride ceramics technology provides materials with high optical transparency and homogeneity. It allows the manufacture of fluoride laser ceramics with properties similar to the properties of corresponding single crystals. Thus fluoride laser ceramics are a very prospective and easily available material for solid-state lasers of various types and different designations.

4.9 Acknowledgments

The author wishes to express his personal thanks to V. V. Osiko, M. A. Doroshenko, E. A. Garibin and I. A. Mironov for fruitful discussions, and to A. I. Popov, R. Simoneaux and E. V. Chernova for the preparation of this manuscript.

4.10 Note to the reader

While this manuscript was under preparation, Akiyama *et al.* (2011) suggested a novel technique for the synthesis of anisotropic laser ceramics. Their paper opens a very broad perspective to the synthesis of the textured fluoride laser ceramic materials, e.g., with LaF_3 structure. Also Akchurin *et al.* (2013) further developed the hot-pressing technique for inorganic

© Woodhead Publishing Limited, 2013

fluoride materials and reported sustained lasing with the use of hot-pressed CaF_2:Yb powder.

4.11 References

Ahrens M, Schuchke K, Redmer S, and Kemnitz E (2007), 'Transparent ceramics from sol-gel derived elpasolites by cold pressing', *Solid State Sciences*, 9, 833–837.

Akchurin M Sh, Gainutdinov R V, Smolyanskii P L, and Fedorov P P (2006), 'Anomalously high fracture toughness of polycrystalline optical fluorite from the Suran Deposit (South Urals)', *Doklady Physics*, 51(1), 10–12.

Akchurin M Sh, Gainutdinov R V, Garibin E A, Golovin Yu I, Demidenko A A, Dukel'skii K V, Kuznetsov S V, Mironov I A, Osiko V V, Smirnov A N, Tabachkova N Yu, Tyurin A I, Fedorov P P, and Shindyapin V V (2011), 'Nanostructure of optical fluoride ceramics', *Inorganic Materials: Applied Research*, 2(2), 7–13.

Akchurin M Sh, Basiev T T, Demidenko A A, Doroshenko M E, Fedorov P P, Garibin E A, Gusev P E, Kuznetsov S V, Krutov M A, Mironov I A, Osiko V V, and Popov P A (2013), 'CaF_2:Yb laser ceramics', *Optical Materials*, in press (DOI 10.1016/j.optmat.2012.09.035).

Akiyama J, Sato Y, and Taira T (2011), 'Laser demonstration of diode-pumped Nd^{3+}-doped fluorapatite anisotropic ceramics', *Applied Physics Express*, 4, 022703.

Alimov O K, Basiev T T, Doroshenko M E, Fedorov P P, Konyushkin V A, Kouznetsov S V, Nakladov A N, Osiko V V, and Shlyakhova O V (2008), 'Spectroscopic and laser properties of Yb^{3+} ions in BaF_2–SrF_2–CaF_2 single crystals and nanoceramics', *15th International Conference on Luminescence and Optical Spectroscopy of Codence Matter*, Lyon, July 2008, 147.

Aubry P, Bensalah A, Gredin P, Patriarche G, Vivien D, and Mortier M (2009), *Optical Materials*, 31, 750–753.

Basiev T T, and Doroshenko M E (2011), *7-th Laser Ceramics Symposium*, Singapore, 14–17 November 2011. Abstracts, 67.

Basiev T T, Bolyasnikova L S, Demidenko V A, Fedorov P P, Kouznetsov S V, Mironov I A, Ovsyannikova O P, Osiko V V, and Akchurin M Sh (2006), 'Fluoride nanoceramics', *National Russian Conference on Crystal Growth 12*, Moscow, 23–27 October, 2006. Abstract, 433 (in Russian).

Basiev T T, Doroshenko M E, Fedorov P P, Konyushkin V A, Kouznetsov S V, and Osiko V V (2007a), 'LiF:F_2^- color center ceramic laser material', *3rd Laser Ceramics Symposium: International Symposium on Transparent Ceramics for Photonic Applications*, Paris, 8–10, 2007 October Abstract O-L-1.

Basiev T T, Doroshenko M E, Konyushkin V A, Osiko V V, Ivanov L I, and Simakov S V (2007b), 'Lasing in diode-pumped fluoride nanostructure F_2^-:LiF color centers ceramics', *Quantum Electronics*, 37(11), 989–990.

Basiev T T, Doroshenko M E, Kouznetsov S V, Konyushkin V A, Osiko V V, and Fedorov P P (2007c), *Ceramic laser microstructured material with twinned nanostructure and method of manufacture*. RU Patent no. 2,358,045.

Basiev T T, Doroshenko M E, Fedorov P P, Konyushkin V A, Kouznetsov S V, Osiko V V, and Akchurin M Sh (2008a), 'Efficient laser based on CaF_2–SrF_2–YbF_3 nanoceramics', *Optical Letters*, 33(5), 521–523.

Basiev T T, Doroshenko M E, Konyushkin V A, Osiko V V, Fedorov P P, Demidenko V A, Dukelîskii K V, Mironov I A, and Smirnov A N (2008b), 'Fluoride optical nanoceramics', *Russian Chemical Bulletin*, International Edition, 57(5), 877–886.

© Woodhead Publishing Limited, 2013

Basiev T T, Doroshenko M E, Konyushkin V A, and Osiko V V (2010), '$SrF_2:Nd^{3+}$ laser fluoride ceramics', *Optics Letters*, 35(23), 4009–4011.

Basiev T T, Konyushkin V A, Konyushkin D V, Doroshenko M E, Huber G, Reishert F, Hansen N-O, and Fechner M (2011), 'First ceramic laser in the visible spectral range', *Optical Materials Express*, 1(8), 1511–1514.

Bensalah A, Mortier M, Patriarche G, Gredin P, and Vivien D (2006), *Journal of Solid State Chemistry*, 179, 2636–2644.

Bensalah A, Ito M, Guyot Y, Goutaudier C, Jouini A, Brenier A, Sato H, Fukuda T, and Boulon G (2007), 'Spectroscopic properties and quenching processes of Yb^{3+} in fluoride single crystals for laser applications', *Journal of Luminescence*, 122–123, 444–446.

Bevan D J M, Greis O, and Strahle J (1980), 'A new structural principle in anion-excess fluorite-related superlattices', *Acta Crystallographica*, 36(6), 889–890.

Bevan D J M, Ness S E, and Taylor M R (1988), 'On the crystal chemistry of Ca_2YbF_7 and other closely-related structures with cuboctahedral anion clusters', *European Journal of Solid State Inorganic Chemistry*, 25(5–6), 527–534.

Bol'shchikov F A, Garibin E A, Gusev P E, Demidenko A A, Kruglova M V, Krutov M A, Lyapin A A, Mironov I A, Osiko V V, Reiterov V M, Ryabochkina P A, Sakharov N V, Smirnov A N, Ushakov S N, and Fedorov P P (2011), 'Nanostructured Tm:CaF_2 ceramics: potential gain media for two-micron lasers', *Quantum Electronics*, 41(3), 193–197.

Carnall E, Hatch S E, and Parsons W E (1966), *Materials Science Research*, 3, 165.

Carnall E, Hatch S E, Parsons W F, and Weagley R J (1969), *Hot-pressed polycrystalline laser materials*, US Patent 3,453,215.

Druon F, Ricaud S, Papadopoulos D N, Pellegrina A, Camy P, Doualan J L, Moncorge R, Courjaud A, Mottay E, and Georges P (2011), 'On Yb:CaF_2 and Yb:SrF_2: review of spectroscopic and thermal properties and their impact on femtosecond and high power laser performance', *Optical Materials Express*, 1(3), 489–502.

Fedorov P P (2010), 'Third law of thermodynamics as applied to phase diagrams', *Russian Journal of Inorganic Chemistry*, 55(11), 1722–1739.

Fedorov P P, and Osiko V V (2005), 'Crystal growth of fluorides', in: *Bulk Crystal Growth of Electronic, Optical and Optoelectronic Materials*, ed. P. Capper. Wiley Series in Materials for Electronic and Optoelectronic Applications. John Wiley & Sons, 339–356.

Fedorov P P, and Rappo A V (2008), '$NaF–CaF_2–YbF_3$ phase diagram', *Russian Journal of Inorganic Chemistry*, 53(7), 1126–1129.

Fedorov P P, Izotova O E, Alexandrov V B, and Sobolev B P (1974), 'New phases with fluorite-derived structure in CaF_2–(Y, Ln)F_3 systems', *Journal of Solid State Chemistry*, 9(4), 368–374.

Fedorov P P, Osiko V V, Basiev T T, *et al.* (2006), Report No. 4 (2006), 212 pp, Russian state contract no. 02.435.11.2011, 'Development of technology for nanoceramics with low optical losses on the basis of metal fluorides doped with rare-earth elements' (in Russian).

Fedorov P P, Osiko V V, Basiev T T, Orlovskii Yu V, Dykel'skii K V, Mironov I A, Demidenko V A, and Smirnov A N (2007), 'Optical fluoride and oxysulphide ceramics: preparation and characterization', in: *Developments in Ceramic Materials Research*, Nova Science Publishers, 53–95.

Fedorov P P, Kuznetsov S V, Voronov V V, Yarotskaya I V, and Arbenina V V (2008), 'Soft chemical synthesis of $NaYF_4$ nanopowders', *Russian Journal of Inorganic Chemistry*, 53(11), 1681–1685.

© Woodhead Publishing Limited, 2013

Fedorov P P, Osiko V V, Basiev T T, *et al.* (2009), Report No. 4 (2009), 72 pp, Russian state contract no. 02.513.12.3029, 'Nanoceramics on the base of compounds with high optical transmission in the mid-IR range for coherent and incoherent light sources' (in Russian).

Flerov I N, Gorev M V, Aleksandrov K S, Tressaud A, Grannec J, and Couzi M (1998), 'Phase transitions in elpasolites (ordered perovskites)', *Materials Science and Engineering*, R24(3), 81–151.

Grass R N, and Stark W J (2005), 'Flame synthesis of calcium-, barium fluoride nanoparticles and sodium chloride', *Chemical Communications*, 1767–1769.

Greis O, and Haschke J M (1982), in: *Rare Earth Fluorides. Handbook on the Physics and Chemistry of Rare Earths*, ed. K A Gscheidner and L Eyring. Elsevier, Amsterdam, New York, Oxford, 5(45), 387–460.

Greskovich C, and Chernoch J P (1973), 'Polycrystalline ceramic lasers', *Journal of Applied Physics*, 44(10), 4599–4606.

Greskovich C, and Chernoch J P (1974), 'Improved polycrystalline ceramic lasers', *Journal of Applied Physics*, 45(10), 4495–4502.

Hatch S E, Parsons W F, and Weagley R J (1964), 'Hot-pressed polycrystalline $CaF_2:Dy^{2+}$ laser', *Applied Physics Letters*, 5(8), 153–154.

Hraiech S, Jouini A, Kim K J, Guyot Y, Goutaudier C, Yoshikawa A, Trabelsi-Ayadi M, and Boulon G (2007), 'Breakage of Yb^{3+} pairs by Na^+ in Yb^{3+}-doped CaF_2 laser host', *Annales de Physique*, 32(2–3), 59–61.

Hraiech S, Jouini A, Kim K J, Guyot Y, Yoshikawa A, and Boulon G (2010), 'Role of monovalent alkali ions in the Yb^{3+} centers of CaF_2 laser crystals', *Radiation Measurements*, 45, 323–327.

Ikesue A, Kinoshita T, Kamata K, and Iosida K (1995), *Journal of the American Ceramics Society*, 78, 1033.

Ishizawa H (2005), *Proceedings of the 13th International Workshop on Sol-Gel Sciences Technology*, Los Angeles, CA, 289.

Ito M, Goutaudier Ch, Guyot Y, Lebbou Kh, Fukuda T, and Boulon G (2004), 'Crystal growth, Yb^{3+} spectroscopy, concentration quenching analysis and potentiality of laser emission in $Ca_{1-x}Yb_xF_{2+x}$', *Journal of Physics: Condensed Matter*, 16, 1501–1521.

Jouini A, Brenier A, Guyot Y, Boulon G, Sato H, Yoshikawa A, Fukuda K, and Fukuda T (2008), 'Spectroscopic and laser properties of the near-infrared tunable laser material Yb^{3+}-doped CaF_2 crystal', *Crystal Growth and Design*, 8(3), 808–811.

Kaczmzrek S M, Tsuboi T, Ito M, Boulon G, and Leniec G (2005), 'Optical study of Ym^{3+}/Yb^{2+} conversion in CaF_2 crystals', *Journal of Physics: Condensed Matter*, 17(25), 3771–3786.

Kazanskii S A, Ryskin A I, Nikiforov A E, Zaharov A Yu, Ougrumov M Yu, and Shakurov G S (2005), 'EPR spectra and crystal field of hexamer rare-earth clusters in fluorites', *Physical Review B*, 72, 014127.

Khazanov E N, Taranov A V, Fedorov P P, Kuznetsov S V, Basiev T T, Mironov I A, Smirnov A N, Dukel'skii K V, and Garibin E A (2009), 'A Study of the transport of thermal acoustic phonons in CaF_2 single crystals and ceramics within the subterahertz frequency range', *Doklady Physics*, 54(1), 14–17.

Kouznetsov S V, Basiev T T, Voronov V V, Lavristchev S V, Osiko V V, Tkatchenko E A, Fedorov P P, and Yarotzkay I V (2006a), 'Preparation of nanoparticles of $M_{1-x}R_xF_{2+x}$ from water solutions', *Proceedings of ISIF-2006, Second International Siberian Workshop INTERSIBFLUORINE – 'Advanced Inorganic Fluorides'*, Tomsk, 11–16 June 2006, 124–127.

© Woodhead Publishing Limited, 2013

Kouznetsov S V, Osiko V V, Tkatchenko E A, and Fedorov P P (2006b), 'Inorganic nanofluorides and related nanocomposites', *Russian Chemical Reviews*, 75(12), 1065–1082.

Kouznetsov S V, Yarotzkay I V, Fedorov P P, Voronov V V, Lavristchev S V, Basiev T T, and Osiko V V (2007), 'Preparation of nanopowdered $M_{1-x}R_xF_{2+x}$ (M = Ca, Sr, Ba; R = Ce, Nd, Er, Yb) solid solution', *Russian Journal of Inorganic Chemistry*, 52, 315–320.

Kouznetsov S V, Fedorov P P, Voronov V V, Samarina K S, Ermakov R P, and Osiko V V (2010), 'Synthesis of $Ba_4R_3F_{17}$ (R stands for rare-earth elements) powders and transparent compacts on their base', *Russian Journal of Inorganic Chemistry*, 55(4), 484–493.

Krupke W F (2000), 'Ytterbium solid-state lasers. The first decade', *IEEE Journal of Quantum Electronics*, 6(6), 1287–1296.

Lu J, Ueda K, Yagi H, Yanagitani T, Akiyama Y, and Kaminskii A (2002), 'Doped and undoped yttrium aluminium garnet ($Y_3Al_5O_2$) nano-crystalline ceramics – a new generation of solid state laser and optical methods', *Journal of Alloys and Compounds*, 341, 220–225.

Lucca A, Dbourg G, Jacquemet M, Druon F, Balembois F, and Georges P (2004), 'High-power diode-pumped Yb^{3+}:CaF_2 femtosecond laser', *Optics Letters*, 29(23), 2767–2769.

Lyberis A, Patriarche G, Gredin P, Vivien D, and Mortier M (2011), 'Origin of light scattering in ytterbium doped calcium fluoride transparent ceramic for high power lasers', *Journal of the European Ceramic Society*, 31, 1619–1630.

Mortier M, Bensalah A, Dantelle G, Patriarche G, and Vivien D (2007), 'Rare-earth doped oxyfluoride glass-ceramics and fluoride ceramics: Synthesis and optical properties', *Optical Materials*, 29, 1263–1270.

Nicoara I, Pecingina-Garjoaba N, and Bunoiu O (2008), 'Concentration distribution of Yb^{2+} and Yb^{3+} ions in YbF_3:CaF_2 crystals', *Journal of Crystal Growth*, 310(7–9), 1476–1481.

Ocafrain A, Chaminade J P, Viraphong O, Cavagnat R, Couzi M, and Pouchard M (1996), 'Growth by the heat exchanger method and characterization of neighborite, $NaMgF_3$', *Journal of Crystal Growth*, 166, 414–418.

Palashov O V, Khazanov E A, Mukhin I B, Mironov I A, Smirnov A N, Dukel'skii K V, Fedorov P P, Osiko V V, and Basiev T T (2007), 'Comparison of the optical parameters of a CaF_2 single crystal and optical ceramics', *Quantum Electronics*, 37(1), 27–28.

Palashov O V, Khazanov E A, Mukhin I B, Smirnov A N, Mironov I A, Dukel'skii K V, Garibin E A, Fedorov P P, Kuznetsov S V, Osiko V V, Basiev T T, and Gainutdinov R V (2009), 'Optical absorption in CaF_2 nanoceramics', *Quantum Electronics*, 39(10), 943–947.

Pena A, Camy P, Benayad A, Doualan J-L, Maurel C, Olivier M, Nazabal V, and Moncorge R (2011), 'Yb:CaF_2 grown by liquid phase epitaxy', *Optical Materials*, 33, 1616–1620.

Petit V, Camy P, Doualan J-L, and Moncorge R (2007), 'Refined analysis of the luminescent centers in the Yb^{3+}:CaF_2 laser crystal', *Journal of Luminescence*, 122–123, 5–7.

Petit V, Voretti P, Camy P, Doualan J-L, and Moncorge R (2008), 'Active waveguides produced in Yb^{3+}:CaF_2 by H^+ implantation for laser applications', *Journal of Alloys and Compounds*, 451, 68–70.

Plotnichenko V G, Sysoev V K, and Firsov I G (1981), 'Analysis of absorption coefficient calorimeteric measurements for high-transparent solid materials', *Zhurnal Tekhnicheskoi Fiziki*, 51(B.9), 1903–1908 (in Russian).

© Woodhead Publishing Limited, 2013

Popov P A, Dykel'skii K V, Mironov I A, Demidenko V A, Smirnov A N, Smolyanskii P L, Fedorov P P, Osiko V V, and Basiev T T (2007), 'Thermal conductivity of CaF_2 optical ceramics', *Doklady Physics*, 52, 7–9.

Popov P A, Fedorov P P, Kuznetsov S V, Konyushkin V A, Osiko V V, and Basiev T T (2008), 'Thermal conductivity of single crystals of $Ca_{1-x}Yb_xF_{2+x}$ solid solutions', *Doklady Physics*, 53(4), 198–200.

Ricaud S, Papadopoulos D N, Pellegrina A, Balembois F, Georges P, Courjaud A, Camy P, Doualan J L, Moncorge R, and Druon F (2011), 'High-power diode-pumped cryogenically cooled Yb:CaF_2 laser with extremely low quantum defect', *Optics Letters*, 36(9), 1602–1604.

Sanghera J, Shaw B, Kim W, Villalobos G, Baker C, Frantz J, Hunt M, Sadowski B, and Aggarwal I (2011), 'Ceramic laser materials', *Proceedings of the SPIE*, 7912, 79121Q–1.

Siebold M, Bock S, Schramm U, Xu B, Doualan J L, Camy P, and Moncorge R (2009), 'Yb:CaF_2 – a new old laser crystal', *Applied Physics B*, 97, 327–338.

Sobolev B P (2000), *The Rare Earth Trifluorides*, p. 1, 'The high-temperature chemistry of the rare earth trifluorides', Institute D'Estudis Catalans, Barcelona, 520 pp.

Sobolev B P, and Fedorov P P (1978), 'Phase diagrams of the CaF_2–(Y,Ln)F_3 systems. I. Experimental', *Journal of Less-Common Metals*, 60, 33–46.

Swinehart C F, and Packer H (1978), US Patent 4,089,937.

Volynets F K (1973a), 'Preparation methods, structure and physico-chemical properties of optical ceramics', *Optiko-mekhanicheskaya prom-st [Optical Mechanical Industry]*, 9, 48–51 (in Russian).

Volynets F K (1973b), 'Optical properties and application of optical ceramics', *Optiko-mekhanicheskaya prom-st [Optical Mechanical Industry]*, 10, 47–51 (in Russian).

Voron'ko Yu K, Osiko V V, and Shcherbakov I A (1969), 'Optical centers and the interaction of ions Yb^{3+} in cubic fluorite crystals', *Soviet Physics* JETP, 29(1), 86.

Vydrik G A, Solov'eva T V, and Kharitonov F Ya (1980), *Prozrachnaya Keramica* [*Transparent Ceramics*], Moscow: Energy (in Russian).

Yoshiasa A, Sakamoto D, Okudera H, Ohkawa M, and Ota K (2003), 'Phase relation of $Na_{1-x}K_xMgF_3$ ($0 \leq x \leq 1$) perovskite-type solid solutions', *Materials Research Bulletin*, 38, 421–427.

Zheludev I S (1971), *Physics of Crystalline Dielectrics*, vol. 2, *Electrical Properties*, Plenum Press.

© Woodhead Publishing Limited, 2013

5
Neodymium, erbium and ytterbium laser glasses

V. I. ARBUZOV, Research and Technological Institute of Optical Material Science, Russia and N. V. NIKONOROV, Saint-Petersburg National Research University of Information Technologies, Mechanics and Optics (ITMO), Russia

DOI: 10.1533/9780857097507.1.110

Abstract: This chapter is devoted to a description of laser glasses. It gives an insight into the story of laser glasses and familiarizes readers both with the classification of laser neodymium and erbium glasses, and with their basic properties. The necessity of introducing sensitizing additions (ytterbium or chromium and ytterbium simultaneously) into erbium glasses is substantiated. Requirements for neodymium phosphate glasses for large-sized active elements of powerful high-energy amplifying facilities destined for investigations in the field of laser-controlled thermonuclear synthesis are characterized. Modern erbium and ytterbium–erbium media are described.

Key words: glasses doped with rare earth ions, absorption and luminescence spectra, classification and properties of neodymium laser glasses, properties of erbium glasses and glass-ceramics, gain and loss spectra of erbium media.

5.1 Introduction

The silicate and phosphate neodymium glasses as well as the phosphate erbium glasses are among the classic laser materials that have been thoroughly investigated. Snitzer (1961) was the first to observe laser action in glasses doped with neodymium. Later on, Snitzer and Woodcock (1965) reported on Yb–Er glass lasers and a number of companies and institutes took part in laser glass development, including the S. I. Vavilov State Optical Institute, the Institute for Radio Engineering and Electronics of the USSR Academy of Sciences (USSR AS), the P. N. Lebedev Physical Institute of the USSR AS, the A. M. Prokhorov General Physics Institute of the USSR AS, and the Otto-Schott-Institute for Glass Chemistry, as well as such companies as Schott Glass, Hoya (this corporation no longer exists but its glasses are still used in lasers), Kigre, Corning, and Chinese Institutions (Chinese Atom Optics Co. Ltd, the Shanghai Institute of Optics and Fine Mechanics, and the Chinese Academy of Sciences).

Neodymium glasses represent the widest class of laser glasses that are produced in different countries of the world and that find a wide application

© Woodhead Publishing Limited, 2013

in various laser or amplifier systems. In particular, rod active elements with diameters from a few to more than 100 mm, and lengths up to 1000 mm, can be produced from these glasses (Alekseev *et al.*, 1980; Bayanov *et al.*, 1984; Lunter *et al.*, 1991). Besides, neodymium glasses can be used for manufacturing large-sized disk active elements for multichannel high-peak power and high-energy amplifying facilities (Arbuzov *et al.*, 2002a, 2002b, 2003; Brown, 1981; Campbell *et al.*, 1999, Campbell and Suratwala, 2000; Dmitriev *et al.*, 2006; Jiang *et al.*, 1998; Martin *et al.*, 1981; Sirazetdinov *et al.*, 2006). For this reason, the interest shown in neodymium laser glasses is clear.

As to erbium glasses, because of the small intensity of absorption bands of Er^{3+} ions (Auzel *et al.*, 1975), additions of Yb^{3+} ions are often introduced into the composition of erbium glasses that act as sensitizers of erbium luminescence (Alekseev *et al.*, 1980, 2003; Fang Yong-Zheng *et al.*, 2007; Galant *et al.*, 1976a; Gapontsev *et al.*, 1978; Kalinin *et al.*, 1974; Karlsson *et al.*, 2002; Maksimova *et al.*, 1991; Snitzer and Woodcock, 1965). Attempts were made (Byshevskaya-Konopko *et al.*, 2001; Denker *et al.*, 1992; Izyneev and Sadovskii, 1997; Sverchkov *et al.*, 1992; Vorob'ev *et al.*, 1987) to elaborate erbium glasses for miniature lasers with a high repetition rate of pumping pulses. Ytterbium–erbium glasses became widely used as the active media in lasers and optical amplifiers. This is associated with the fact that the lasing wavelength of erbium ions (1.5 μm), firstly, is optimal for information transfer along fiber-optic communication lines and, secondly, lies in an eye-safe region. Unlike erbium ions with their low intensity absorption, ytterbium ions have an intense absorption band in the region of 1 μm, and this allows one to use them for pumping powerful semiconductor laser diodes.

5.2 The history of laser glasses

The creation of lasers on the basis of ruby crystals has caused a boom in the study of doped materials and, in fact, launched an era of solid-state lasers. At the same time, there is information in Alekseev *et al.* (1980) about attempts to obtain generation of stimulated emission in different spectral regions using not only crystals but also glasses doped with rare-earth ions such as the following:

- Gd^{3+} (0.3125 μm) (Gandy and Ginther, 1962a; Ostrovskaya and Ostrovskii, 1963; Karapetyan and Reishakhrit, 1967)
- Tb^{3+} (0.542 μm) (Feofilov, 1961; Peterson and Bridenbaugh, 1962; Vargin and Karapetyan, 1964; Karapetyan and Reishakhrit, 1967)
- Yb^{3+} (1.015 μm) (Etzel *et al.*, 1962; Vargin and Karapetyan, 1964; Snitzer, 1966; Young, 1969
- Pr^{3+} (1.047 μm) (Karapetyan, 1963; Karapetyan and Reishakhrit, 1967)

© Woodhead Publishing Limited, 2013

- Nd^{3+} (1.05–1.06 μm) (Snitzer, 1961; Vargin and Karapetyan, 1964; Pearson *et al.*, 1964; Young, 1963, 1969; Snitzer, 1966; Karapetyan and Reishakhrit, 1967
- Nd^{3+} (0.918 μm) (Mauer, 1963a; Pearson *et al.*, 1964)
- Nd^{3+} (1.37–1.401 μm) (Mauer, 1963b; Pearson *et al.*, 1964)
- Er^{3+} (1.55 μm) (Vargin and Karapetyan, 1964; Gandy *et al.*, 1965; Snitzer, 1966; Karapetyan and Reishakhrit, 1967; Young, 1969)
- Ho^{3+} (2.046 μm) (Gandy and Ginther, 1962b; Karapetyan and Reishakhrit, 1967; Young, 1969)
- Tm^{3+} (1.85 μm) (Gandy *et al.*, 1967; Young, 1969).

In the last 25 years, some new prospective types of doped glasses have been objects of investigation, for instance:

- Fluoride glasses:
  - Er^{3+} (Reisfeld *et al.*, 1982)
  - Nd^{3+} and Ho^{3+} (Reisfeld *et al.*, 1986)
  - Ho^{3+} and Er^{3+} (Eyal *et al.*, 1987)
  - Nd^{3+} (Amaranath *et al.*, 1991)
  - Nd^{3+} (Fernandez *et al.*, 1994)
  - Er^{3+} (Javorniczky *et al.*, 1995)
  - Pr^{3+} (Seeber *et al.*, 1995)
  - Nd^{3+}, Yb^{3+}, Pr^{3+} (Kwasny *et al.*, 2000)
  - Er^{3+}, Er^{3+}–Pr^{3+} (Golding *et al.*, 2000)
  - Tm^{3+} (Doulan *et al.*, 2003)
  - Pr^{3+} (Olivier *et al.*, 2011)
- Fluoride phosphate (Philipps *et al.*, 2001: Er^{3+}:Yb^{3+}) glasses
- Fluoro-aluminate (Ehrt, 2003: Nd^{3+}, Yb^{3+}, Pr^{3+}, Ho^{3+}, Er^{3+}, Tm^{3+}) glasses
- Bi-doped silica glasses (Dianov, 2008)
- New phosphate (Nd^{3+}) glasses (Galagan *et al.*, 2009a)
- Germanate (Tm^{3+}) glasses (Fusari *et al.*, 2009)
- Telluride (Ho^{3+}) glasses (Milanese *et al.*, 2011).

Before we proceed to familiarize ourselves with practical glasses, it makes sense to briefly describe the absorption and luminescence spectra of neodymium, ytterbium and erbium ions.

5.2.1 Nd^{3+}

The ground and radiating levels of Nd^{3+} ions are $^4I_{9/2}$ and $^4F_{3/2}$, respectively (Fig. 5.1) (Diecke and Crosswhite, 1963). Within the transparency region of glasses, Nd^{3+} ions have a number of absorption bands, which differ from each other in their structure and intensity. For example, Fig. 5.2 shows the measured absorption spectrum of the Russian phosphate laser glass GLS21.

© Woodhead Publishing Limited, 2013

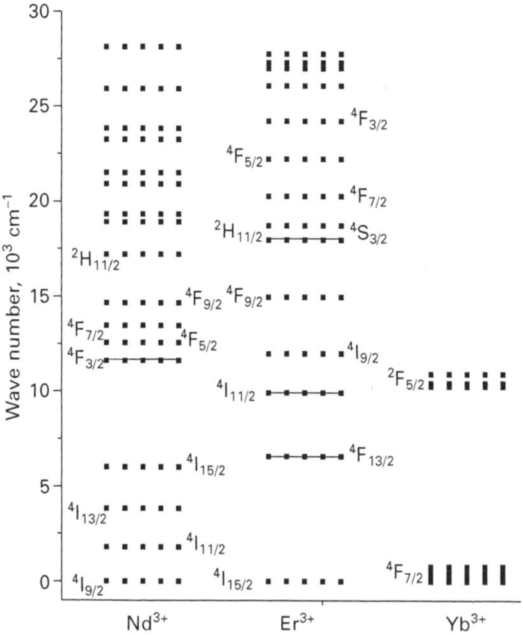

5.1 Diagram of energy levels of Nd^{3+}, Er^{3+} and Yb^{3+} ions (from the data of Diecke and Crosswhite, 1963).

A similar spectrum in silicate glasses looks qualitatively just as in phosphate glasses, but it is characterized by a greater diffusivity of bands than the spectrum shown in Fig. 5.2.

The most intensive bands in the absorption spectrum are bands with maxima at 581 (transitions $^4I_{9/2} \rightarrow {}^2G_{7/2}$, $^4G_{5/2}$), 794 ($^4I_{9/2} \rightarrow {}^4F_{5/2}$), 735 ($^4I_{9/2} \rightarrow {}^4F_{7/2}$), 515 ($^4I_{9/2} \rightarrow {}^2G_{9/2}$) and 365 ($^4I_{9/2} \rightarrow {}^2D_{3/2}$) nm. Pumping just in these absorption bands provides an inverse population of radiating level $^4F_{3/2}$. Absorption bands of neodymium ions in UV spectrum areas (when the wavenumber exceeds 384 nm) are superimposed on rising absorption of light by the glass matrix. It is clear that pumping in this region of the spectrum can lead basically to heating active elements.

Luminescence is observed at excitation within all these absorption bands of Nd^{3+} ions . Its spectrum consists of four bands caused by electron transitions $^4F_{3/2} \rightarrow {}^4I_{9/2}$ (a resonant band with a maximum in the region of 900 nm), $^4F_{3/2} \rightarrow {}^4I_{11/2}$ (1060 nm), $^4F_{3/2} \rightarrow {}^4I_{13/2}$ (1300 nm) and $^4F_{3/2} \rightarrow {}^4I_{15/2}$ (1800 nm). As this takes place, about 90% of the total numbers of luminescence light quanta are emitted in the bands 900 and 1060 nm. The band at 1060 nm is more intensive in all glass-forming matrices (Fig. 5.3). So, depending on the glass structure, 45–48% of the total number of emitted quanta in silicate and germanate glasses are concentrated in these bands, and up to 50–53% in phosphate and borate glasses (Brachkovskaya *et al.*, 1976).

© Woodhead Publishing Limited, 2013

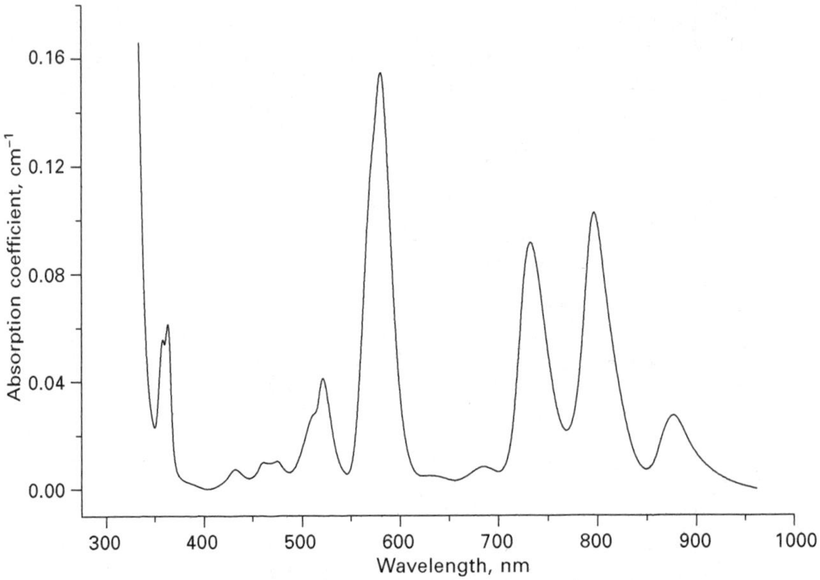

5.2 Absorption spectrum of a phosphate laser glass GLS21.

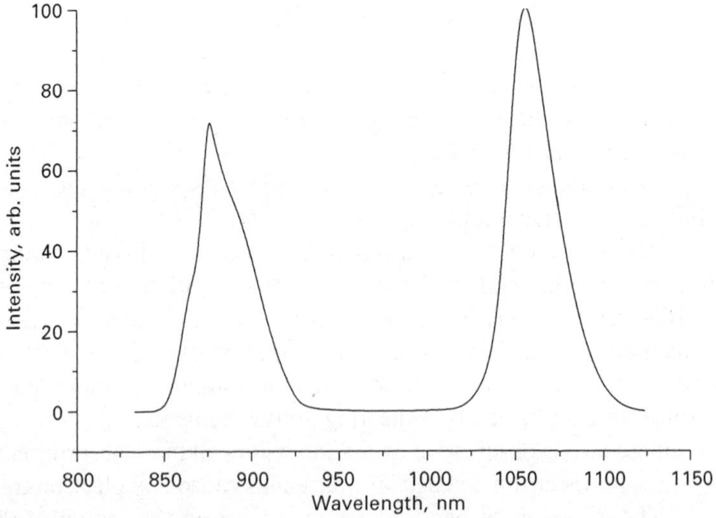

5.3 Luminescence spectrum of a phosphate laser glass GLS21.

5.2.2 Yb^{3+}

Absorption and emission of radiation by Yb^{3+} ions are caused by the presence of only two split energy states: the ground level $^2F_{7/2}$ and the excited level $^2F_{5/2}$ (Fig. 5.1). The ground level is split by a matrix crystal field on four,

© Woodhead Publishing Limited, 2013

and the excited level on threes sublevels. The distances between the next sublevels of each manifold are of the order of several hundred cm^{-1} (Diecke and Crosswhite, 1963). There is a resonant band in absorption and luminescence spectra of Yb^{3+} ions that lies in the region of 975 nm in phosphate glasses. This accounts for reabsorption of emitted radiation. The degree of absorption and luminescence spectra overlap increases with temperature growth, giving rise to the growth of the reabsorbed luminescence light part.

5.2.3 Er^{3+}

The only intensive band with a maximum at 1540 nm is observed in Er^{3+} luminescence spectra in the majority of oxide glass matrices. It is caused by resonant electron transitions $^4I_{13/2} \rightarrow {}^4I_{15/2}$. However, in the cases of fluorberyllate and telluride glasses, it is possible to register also luminescence bands corresponding to electron transitions $^4S_{3/2} \rightarrow {}^4I_{15/2}$ (550 nm), $^4I_{11/2} \rightarrow {}^4I_{15/2}$ (990 nm) and $^2P_{3/2} \rightarrow {}^4I_{15/2}$ (320 nm). The reason for such a depleting erbium luminescence spectrum in glasses is a strong ion–phonon interaction caused by small values of energy gaps between the majorities of neighboring excited levels (Fig. 5.1). As to absorption spectra, some bands with rather weak intensity in the visible and near-IR regions of the spectrum are presented. The most intensive absorption bands are those with maxima at 522 ($^4I_{15/2} \rightarrow {}^2H_{11/2}$) and 378 ($^4I_{15/2} \rightarrow {}^4G_{11/2}$) nm; however, even their oscillator forces are rather small as they make up only 5.7×10^{-6} and 12×10^{-6}, respectively (Alekseev *et al.*, 1980). Other bands rank below them in intensities approximately by an order of magnitude. For this reason, to provide inverse density of population of level $^4I_{13/2}$ at lamp pumping is very difficult. However, the situation can be improved by introducing some sensitizers into the composition of erbium glasses. Yb^{3+} ions can play such a role, for example. The width of the absorption band of ytterbium ions is of the order of 100 nm that exceeds by several times the bandwidths of erbium ions. At a high concentration of ytterbium, the pumping radiation is almost completely absorbed in a range from approximately 870 to 1100 nm.

As appears from the scheme in Fig. 5.1, the energy gap between radiating ($^2F_{5/2}$) and one of the sublevels of the the ground manifold ($^2F_{7/2}$) of Yb^{3+} ions almost precisely corresponds to the gap between levels $^4I_{15/2}$ and $^4I_{11/2}$ of Er^{3+} ions, providing effective transfer of excitation energy from ytterbium (energy donor, D) ions to erbium (energy acceptor, A) ions. The energy transfer efficiency, W(DA), strongly depends on the concentration of co-activator ions and on the type of glass matrix, showing the highest values in phosphate and borate glasses. Optimal concentrations of Yb^{3+} and Er^{3+} ions in the two-activated phosphate glasses depend on many factors and can be varied up to 4×10^{21} and 1.3×10^{20} cm^{-3}, respectively (Alekseev *et al.*, 1980). Under such a ratio of co-activator concentrations, the quantum yield

© Woodhead Publishing Limited, 2013

of energy transfer can reach 100%. Erbium concentration in silicate glasses should be twice as large as that in phosphate glasses to maintain limiting values of the quantum yield of energy transfer.

Efficiency of generation in erbium glasses can go down due to excited state absorption of pumping and emitted radiation by the excited Er^{3+} ions. In other words, induced electron transitions in absorption from the meta-stable level $^4I_{13/2}$ to levels $^4F_{9/2}$ (8900 cm^{-1}) and $^4I_{9/2}$ (6100 cm^{-1}) are possible. The excited state absorption in phosphate glasses is expressed slightly less effectively than in silicate glasses. There is another reason for low efficiency of ytterbium–erbium silicate laser glasses, namely the reverse energy transfer from erbium ions back to ytterbium ions because of the rather long lifetime of Er^{3+} ions at the $^4I_{11/2}$ level in silicate glasses (tens of microseconds compared to 2–3 μs in phosphate glasses). Exactly for this reason, all the commercially available ytterbium–erbium laser glasses are phosphate-based.

It should be noted that laser media doped with Er^{3+} ions are characterized by a three-level generation scheme. This complicates creating an inverse population of the upper laser level and leads to an increase in laser threshold.

5.3 Commercial laser glasses

5.3.1 Advantages and disadvantages of different types of laser glasses

The most simple and natural way of making commercial laser glasses at the initial stage of development of laser material technology was from experience of optical glass production (Alekseev *et al.*, 1980). In the early 1960s optical glasses were predominantly silicate, and the technology of their industrial production has been fulfilled to perfection. Silicate glasses are easy and cheap to manufacture, and have good mechanical durability and chemical stability. The basic physical and chemical properties of silicate glasses, as well as their relation to composition, were well known, allowing one to optimize the compositions of laser silicate glasses.

Attempts to develop practical neodymium laser glasses on the basis of borate and germanate systems failed. Germanate glasses have no advantages over silicate glasses in spectral and lasing properties, being, at the same time, more expensive. This excluded the possibility of developing practical compositions of germanate laser glasses. Borate glasses can show large values of the stimulated emission cross-section; however, their quantum yield and luminescence lifetime are very small because of strong suppression of Nd^{3+} luminescence due to ion–phonon interaction.

As to phosphate glasses, they were considered as non-technological for a long time, their area of practical application being very narrow; therefore nobody included them on a prospective basis for laser glass development.

© Woodhead Publishing Limited, 2013

At the same time, laboratory investigations showed that the phosphate basis allowed one to replace P_2O_5 by oxides and fluorides of other glass-forming elements, giving the opportunity of improving many physical and chemical properties of phosphate glasses and reducing their crystallization ability. Fluoro-beryllate glasses proved to be close in their properties to phosphate glasses; however, the raised aggressiveness, rather low chemical durability and volatility of fluorine-containing melts, as well as the low damage threshold, served as a brake in working out practical compositions (Galaktionova *et al.*, 1974). Another disadvantage of fluoro-beryllate glasses was that they were ecologically more dangerous than phosphate glasses, as not only beryllium itself but all its chemical compounds are toxic. Besides, it was difficult to get highly homogeneous blanks of active elements from these glasses. For these reasons, fluorine-containing glasses were replaced later by more stable and non-toxic fluoro-zirconate glasses. As for fluoro-aluminate and fluoride phosphate laser glasses, they are melted and used up till now. In particular, fluoro-aluminate glasses doped with Er^{3+} ions are the basis for broadband lasers and amplifiers for the 1.5 μm region (Ehrt, 2003). Further, Yb^{3+}-doped fluoroaluminate glasses (Ehrt, 2003) and fluoride phosphate glasses (Hein *et al.*, 2004) allow one to develop diode pumped high-peak-power femtosecond lasers of the petawatt level.

Despite the aforesaid comments about phosphate glasses, catalogs of laser glasses which included not only silicate but also phosphate glasses doped with different types of laser ions were developed by scientists in Russia, France, the USA, Japan, China and Germany.

The most widespread laser glasses are neodymium laser glasses. The values of the nonlinear index of refraction in phosphate glasses are rather small (no more than 1.3×10^{-13} cm^2/V^2), the luminescence band in the region of 1.06 μm is narrow (no more than 23 nm), and values of the stimulated emission cross-section are high enough (up to 4.7×10^{-20} cm^2) to bring glasses of this system to be among the most important practical laser glasses (USSR Branch Standard OST3-30-77 ‘Optical glass GLS Technical conditions’, Russian Catalog ‘Commercial laser glasses’; US Schott Glass Technologies Catalog ‘Laser Glass’; US Catalog ‘Laser Glasses from Schott’; Japanese Catalog ‘Laser Glass from Hoya’; USA Kigre, Inc.; Chinese Atom Optics Co. Ltd).

As for the lack of phosphate glasses, they are more expensive and mechanically and thermally less strong than silicate glasses and their moisture resistance is less than that of silicate glasses. At the same time, the mechanical and thermal resistance of phosphate glasses could be essentially raised by special methods of surface processing.

Detailed studies of functionally important properties of laser glasses have allowed their merits and demerits in comparison with laser crystals to be revealed (Kaminsky, 1975). Advantages of laser glasses are as follows:

© Woodhead Publishing Limited, 2013

- High optical homogeneity, isotropy of properties and small losses by inactive absorption of radiation at the lasing wavelength
- The possibility of getting castings of great volume at rather low cost
- High manufacturability at melting and machining; possibility of production of elements of almost any sizes and types
- The possibility of a variation of important functional properties over wide limits by a change in glass composition
- Broad bands of absorption of activator ions (because of non-homogeneous broadening of their energy levels) that raise the efficiency of usage of pumping lamps in the creation of inverse density of population in laser ions
- The possibility of introducing large concentrations of activator ions, or sensitizer ions (Nd^{3+}, Yb^{3+}, Cr^{3+}) into glasses doped with Er^{3+} or Yb^{3+}, or protector additives that bring down efficiency of solarization (coloring glass under the action of radiation of pumping lamps).

The lack of laser glasses in comparison with laser crystals is caused by:

- Smaller (at least by an order of magnitude) heat conductivity
- Smaller values of such characteristics as hardness (from three to eight times), elasticity module (from several times to 1–2 orders) and mechanical rigidity (within an order)
- The large width of luminescence bands, and smaller values of the emitted radiation cross-sections of rare-earth ions leading to smaller amplification factors and higher thresholds of generation.

One of the major parameters of any laser material is the quantum yield of luminescence. As a rule, it falls with growth of activator concentration. The basis of concentration quenching of luminescence is the migration of excitation energy on activator ions (Dexter, 1953; Foerster, 1949). If laser material contains foreign quenching impurities, their action radii increase strongly with activator concentration due to energy migration. However, in the case of neodymium laser materials, concentration quenching of luminescence can take place even in the absence of foreign quenchers, proceeding in accordance with the cross-relaxation scheme. The nearest unexcited activator ion will act as a quencher for the excited activator ion of the same nature. In the case of Nd^{3+}, the quenching cross-relaxation scheme will look as follows:

$$Nd^{3+}(^4F_{3/2}) + Nd^{3+}(^4I_{9/2}) = 2\ Nd^{3+}(^4I_{15/2}, {}^4I_{13/2}) \qquad [5.1]$$

where $^4F_{3/2}$ is the radiating, $^4I_{9/2}$ is the ground, and $^4I_{15/2}$ and $^4I_{13/2}$ are the upper levels of the multiplet ^4I of Nd^{3+} ions.

The first phosphate glasses with high neodymium concentrations and a low effect of concentration quenching of Nd luminescence were developed by Voronko *et al.* (1976) and Batygov *et al.* (1976).

© Woodhead Publishing Limited, 2013

The IR luminescence quenching is also possible due to ion-phonon interaction (Alekseev *et al.*, 1980; Basiev *et al.*, 1975, Ermolaev *et al.*, 1977; Heber, 1976; Layne, *et al.*, 1975; Rieseberg and Weber, 1976; Tolstoj, 1970). The basis for the ion–phonon interaction (multi-phonon relaxation) is exchange of energy of electron transitions in emission for vibration excitation of the matrix basic structural groupings (phonons). There are two major critical parameters of the ion–phonon interaction: first, the value of the energy gap, ΔE, between the excited and the nearest lower levels of the center giving energy; and second, the process order, P, which is equal to the number of phonons with energy $\hbar\omega$ (the high-frequency boundary of the matrix vibration spectrum is usually understood under $\hbar\omega$), which need to be excited simultaneously to satisfy the energy conservation law:

$$\Delta E = P * \hbar\omega. \quad [5.2]$$

The probability of multi-phonon relaxation is close to zero for processes with participation of more than five phonons (Heber, 1976). At $P \approx 4$, radiating decay of the excited state starts to compete with multi-phonon non-radiating relaxation of excitation. In a general view, the dependence of the multi-phonon relaxation rate W on the order of relaxation P and on the energy gap ΔE is expressed by a simple empirical relation:

$$W = A * \varepsilon^P = A * \varepsilon^{\Delta E/\hbar\omega} \quad [5.3]$$

where A and ε are the constants characterizing the force of ion–phonon interaction in the given material, and $\varepsilon << 1$.

It follows from [5.3] that the ion–phonon interaction in glasses will have an impact on the luminescence efficiency of rare-earth ions to a greater extent than in crystals as the high-frequency boundary of the phonon spectrum of glasses is shifted strongly towards the bigger frequencies in comparison with those in crystals. Really, the high-frequency boundary of the phonon spectrum lies at 1150 cm^{-1} in silicate, at 1250 cm^{-1} in phosphate and at 1500 cm^{-1} in borate glasses (Alekseev *et al.*, 1980), and only at 700 cm^{-1} in one of the most important laser crystals, $Y_3Al_5O_{12}$ (Rieseberg and Weber, 1976). Because of effective ion-phonon interaction, a set of radiating levels of rare-earth ions in glasses is more depleted in comparison with the above set in crystals: in glasses radiating transitions are observed almost always from the lowermost levels of the excited multiplets. A consequence of this is that the spectrum of luminescence of one or another rare-earth activator in glasses contains a smaller number of bands than in crystals.

If the rates of radiating and non-radiating decay of the excited state are close to each other at $P \approx 4$, then the average value of the energy gap $\Delta E = P * \hbar\omega$ (at which the ion–phonon interaction should already adversely affect luminescence intensity) will be close to 6000 cm^{-1} for borate, 5000 cm^{-1} for phosphate and 4600 cm^{-1} for silicate glasses. As the value of the energy

© Woodhead Publishing Limited, 2013

gap $\Delta E(^4F_{3/2} - {}^4I_{15/2}) \approx 5500\ cm^{-1}$ for Nd^{3+} ions, therefore it becomes clear that, at least in borate glasses, non–radiating ion-phonon interaction proves to be rather effective. For this reason, the quantum yield of neodymium luminescence in borate glasses can differ considerably from 100%.

When investigating processes of non-radiating decay of excited states of rare-earth ions in glasses, it is necessary to consider vibrations of $(OH)^-$-groups, which always are available in all glasses. As $(OH)^-$-groups possess higher (~3000 cm^{-1}) frequencies of vibrations (Alekseev *et al.*, 1980) than those of the basic structural groupings of glass, then the non-radiative relaxation of excited rare-earth ions with excitation of $(OH)^-$-group vibrations has lower order, i.e. it appears to be very effective. Rather strong quenching of Nd^{3+}, Yb^{3+} and Er^{3+} luminescence in glasses is explained precisely by this phenomenon (Alekseev *et al.*, 1975; Bondarenko *et al.*, 1975; Galant *et al.*, 1976b; Kovalyova *et al.*, 1975). So, an increase in relative quantum yield, q, and luminescence lifetime, τ, of Nd^{3+}, Er^{3+} and Yb^{3+} ions is observed on reduction of the water content in a glass (control over the water content is carried out on the value of the absorption of $(OH)^-$-groups at wavelength 3.45 or 2.86 microns), and the ratio q/τ remains constant for all activator concentrations (Bondarenko *et al.*, 1975). It should be noted that changes in q and τ values are not accompanied by a reorganization of both absorption and luminescence spectra. This testifies that probabilities, A, of radiating transitions are remaining invariable in the course of dehydration or saturation glasses with water, whereas probabilities, W, of non-radiating transitions are changing. $(OH)^-$ groups are able also to decrease luminescence efficiency of Er^{3+} ions at 1.5 μm as they have some absorption at this wavelength caused by their overtone vibrations.

Results of work described by Arbuzov (1980, 1982, Arbuzov *et al.*, 1976) allow one to judge a water role in concentration quenching of rare-earth ion IR luminescence. Measurements of absolute quantum yield of Yb^{3+} ion luminescence in phosphate glasses both dehydrated and saturated with water were carried out with variation of activator concentration over wide limits (Table 5.1). As the table shows, ytterbium quantum yield falls with growth of its concentration; however, its fall in dehydrated glasses begins at an activator concentration of about $6.5 \times 10^{20}\ cm^{-3}$ whereas in glasses saturated with water the value of q is already small enough (63%) even at a small ytterbium concentration ($0.5 \times 10^{20}\ cm^{-3}$) and shows a tendency to sharp reduction with growth of ytterbium concentration.

5.3.2 Classification and main properties of commercial laser glasses

Now we will discuss the commercial laser glasses which are industrially produced in different countries. The Russian catalog of neodymium laser

© Woodhead Publishing Limited, 2013

Table 5.1 Concentration dependence of quantum yield of Yb^{3+} luminescence in phosphate glasses dehydrated and saturated with water

$[Yb^{3+}]$, 10^{20} cm^{-3}	0.5	0.8	1.8	2.9	3.9	6.5	11.2	20.7
q in dehydrated glasses, %	100	100	100	100	100	99	87	63
q in water-containing glasses, %	63	48	28	20	15	11	7	2

Source: Arbuzov (1980, 1982).

Table 5.2 Neodymium concentrations in commercial laser glasses

Silicate glasses		Phosphate glasses	
Glass mark	$[Nd^{3+}]$, 10^{20} cm^{-3}	Glass mark	$[Nd^{3+}]$, 10^{20} cm^{-3}
GLS1	1.9	GLS21*	1.4
GLS2	1.9	GLS22*	2.0
GLS3	4.6	GLS23*	3.6
GLS5	0.97	GLS24*	5.7
GLS6	1.96	GLS25	2.33
GLS7	3.05	GLS26	3.3
GLS9	4.56	GLS27	12.7
GLS10	2.34	GLS32	2.0
GLS14	4.56	KGSS0178/125	12
ED-2	2.83	KGSS0180	0.5–5.0
LG-670	≤3.18	Q100*	10.7
LG-680	0.91, 1.82, 3.18	LG-700	2.82, 5.64
LSG-91H	3.0	LG-750*	≤4
		LG-760*	≤7.5
		HAP-4	3.2
		LHG-5	3.2
		LHG-8*	3.1
		LHG-80	3.1

* Athermal glasses.

glasses includes glasses of types GLS (Russian Catalog 'Commercial laser glasses', 1982) and KGSS (Table 5.2). Depending on their marks, concentration of neodymium ions in the Russian catalog glasses lies in the range from 0.97×10^{20} to 12.7×10^{20} cm^{-3} (Table 5.2). In glasses of the KGSS 0180 type, neodymium concentration can vary from 0.5×10^{20} (KGSS 0180/5) to 5×10^{20} (KGSS 0180/50) cm^{-3}. The catalog of the American firm Schott Glass Technologies (US Catalog 'Laser Glass', US Catalog 'Laser Glasses from Schott') is presented by silicate glasses LG-670 and LG-680 and phosphate glasses LG-700, LG-750 and LG-760. Depending on customer requirements, neodymium concentration in them can vary up to 7.5×10^{20} cm^{-3}. The list of neodymium laser glasses which were manufactured by the Japanese company Hoya Corporation (Japan Catalog 'Laser Glass from Hoya') included silicate glass LSG-91H and phosphate glasses HAP-4, LHG-5, LHG-8 and LHG-80. Kigre, Inc. (Kigre, Inc., 2009; Kigre 'QX-type glass') produces

© Woodhead Publishing Limited, 2013

both phosphate (QX/Nd, Q-98, Q-100) and silicate (Q-246) glasses doped with neodymium. Two types of Nd^{3+} doped phosphate glasses – N21 and N31 – are produced by the Chinese Atom Optics Co. Ltd that is supported by the Shanghai Institute of Optics and Fine Mechanics, and by the Chinese Academy of Sciences. The glass N31 has virtually the same properties as Hoya's glass LHG-8.

The nomenclature of commercial laser glasses is not confined only by neodymium glasses. The second most widespread laser activator is erbium. Snitzer and Woodcock (1965) were the first to obtain generation at the wavelength of 1536 nm using Yb^{3+}–Er^{3+} glasses. Erbium laser materials represent a great interest for the purposes of distance measurement, because radiation at the wavelength of 1540 nm is safe for human eyes as opposed to the radiation of neodymium lasers (1.06 μm). Scientists from the Institute of Radio Engineering and Electronics of the USSR Academy of Sciences developed a composition of the first commercial ytterbium–erbium glasses (Alekseev *et al.*, 1980). Their glasses have received designation LGS-E. The concentration of Yb^{3+} ions in these glasses was equal to 1.5×10^{21} cm^{-3}, and Er^{3+} concentration could be varied from 3×10^{19} to 9×10^{19} cm^{-3}. Scientists from the S. I. Vavilov State Optical Institute have created erbium glasses with two and with three activators. The first glass (KGSS 0135/15) contains Yb^{3+} (2.1×10^{21} cm^{-3}) and Er^{3+} (2×10^{19} cm^{-3}) ions simultaneously, and the second glass (KGSS 0135/153) – also Cr^{3+} (3×10^{19} cm^{-3}) ions. The ytterbium and chromium are introduced into these glasses as sensitizers of Er^{3+} ion luminescence. Erbium glass LEG-30 was produced by the Hoya Corporation (Japan Catalog 'Laser Glass from Hoya'). Kigre, Inc. (Kigre, Inc., 2009; Kigre 'QX-type glass') produces also erbium (QE-7 and QE-7s, QX/Er), and ytterbium (QX/Yb) laser glasses as well as glasses doped with holmium, thulium, dysprosium, europium and praseodymium. Glasses of the QX-type are characterized by heightened chemical and thermal durability; glasses Q-98 and Q-100 (highly concentrated) belong to the athermal glasses. Phosphate glasses co-doped with ytterbium and erbium as well as simply erbium glasses EP6, WM4 and Cr14 are produced by the Chinese Atom Optics Co. Ltd.

Data on optical constants, n_D and v_D, on refractive indices at the lasing wavelength 1.06 (1.054) μm, as well as on non-linear refraction indices, n_2, are presented in Table 5.3 for the majority of the above laser glasses.

5.4 Modern neodymium and erbium laser glasses

5.4.1 Neodymium glasses

From the beginning of the twenty-first century, phosphate glasses LHG-8 and LG-750 (Campbell *et al.*, 1999; Campbell and Suratwala, 2000), Russian

© Woodhead Publishing Limited, 2013

Table 5.3 Optical properties of commercial laser glasses

Glass mark	n_D	ν_D	n_{1060} (n_{1054})	n_2, 10^{-13} esu
GLS1	1.534	57.9	1.521	1.40
GLS2	1.529	56.9	1.518	1.43
GLS3	1.528	57.0	1.518	1.42
GLS5	1.545	52.4	1.534	1.71
GLS6	1.551	52.2	1.538	1.75
GLS7	1.554	51.2	1.542	1.82
GLS8	1.560	50.6	1.548	1.89
GLS9	1.527	57.8	1.516	1.40
GLS10	1.546	56.3	1.535	1.54
GLS14	1.537	56.9	1.524	1.47
LG-680	1.570	57.4	1.560	1.60
LSG-91H	1.561	56.6	1.550	1.58
GLS21–GLS24	1.594	59.3	1.582	1.65
GLS25	1.578	65.3	1.568	1.64
GLS26	1.574	65.7	1.564	1.35
GLS32	1.594	56.9	1.587	1.32
LG-700	1.515	67.5	1.504	1.10
LG-750	1.526	68.2	1.516	1.01
LG-760	1.519	69.2	1.508	1.08
LHG-80	1.543	64.7	(1.533)	1.02
HAP-4	1.543	64.6	(1.533)	1.25
LHG-8	1.530	66.5	(1.520)	1.13
LHG-5	1.541	63.5	(1.531)	1.28
LEG-30 (Er^{3+})	1.542	65.4	(1.527)	1.22
N21	1.576	65.3	1.565	1.30
N31	1.536	66.2	1.528	1.1

glass KGSS 0180/35 (Arbuzov *et al.*, 2002a, 2002b, 2003), Kigre's high-energy neodymium athermal glasses as well as Chinese neodymium glass N31 (Atom Optics Co. Ltd) are used in the largest high-peak-power and high-energy laser facilities for investigations in the field of laser-controlled thermonuclear synthesis (LCTS).

Glasses LHG-8 and LG-750 are manufactured by continuous glass melting technology (Campbell *et al.*, 1999), and the glass KGSS 0180/35 by classical two-stage technology (Arbuzov *et al.*, 2003). A huge number of large-sized active elements is required for equipment of such laser facilities. Compositions and the basic properties of these glasses are presented in Table 5.3. Despite small differences in concentration and the nomenclature of components, glasses of the above marks (Table 5.4) are close to each other in basic properties (lasing wavelength, $\lambda_{gen;}$ half-width of luminescence band, $\Delta\lambda_{eff}$; cross-section of stimulated emission, σ; refractive index, n_D; non-linear refractive index, n_2) that, in principle, allow one to replace one glass with another in laser facilities.

Glasses for the LCTS have to meet very rigid requirements on a wide variety of parameters. One part of the functionally important properties of

© Woodhead Publishing Limited, 2013

Table 5.4 Compositions and properties of glasses for LCTS

Component, wt%; Property	KGSS 0180/35 (RTIOMS, Russia)	LHG-8 (Hoya, Japan)	LG-750 (Schott, USA)
P_2O_5	57–63	56–60	55–60
SiO_2	1.8–2.5	–	–
B_2O_3	2.5–3.4	–	–
Al_2O_3	6–10	8–12	8–12
K_2O	10–13	13–17	13–17
BaO	10–13	10–15	10–15
n_D	1.530	1.523	1.526
n_2, 10^{-13}esu	1.02	1.12	1.01
σ, 10^{-20} cm^2	3.6	3.6	3.9
λ_{gen}, nm	1053.5	1053.5	1053.5
$\Delta\lambda_{eff}$, nm	25.5	26.5	25.4
α, 10^{-7} K^{-1}	116	127	134
T_g, °C	460	485	460
ρ, g/cm^3	2.83	2.83	2.59

Table 5.5 Changes in luminescence lifetime of Nd^{3+} ions in a KGSS 0180/35 glass in the course (hours) of its dehydration

τ, hours	0	2	4	5	6	7	8	9	10	11	12	13	14	15	16
τ, μs	130	180	220	235	250	260	270	278	285	290	294	297	300	300	300

Source: Arbuzov *et al.* (2003).

glasses is defined by their composition, and another depends on the technology of their synthesis. Quantum yield of Nd^{3+} ion luminescence, beam resistance (damage threshold), coefficient of inactive absorption of radiation at the lasing wavelength, bubbles and optical homogeneity belong to the second group of properties. The formulation of these requirements is as follows.

1. The luminescence quantum yield should not be less than 0.70 (corresponding values of the luminescence lifetime $\tau \geq 280$ μs). Because of water presence in glasses it is an uneasy problem at Nd^{3+} ion concentrations from 3.5×10^{20} to 4.2×10^{20} cm^{-3}. Excitation energy migration that takes place at the above neodymium concentrations gives rise to an increase in the action radius of the remaining hydroxyl groups in the glass matrix. To ensure so high a value of the luminescence quantum yield, it is required to lower the concentration of hydroxyl groups in the course of glass dehydration by more than an order of magnitude and to finish it at the level for which the natural absorption coefficient in the region of the maximum of hydroxyl group absorption, a_{nat} (3.55 μm) $\leq$ 4.6 cm^{-1} (decimal absorption coefficient, a_{dec} (3.55 μm) $\leq$ 2 cm^{-1}). Prolonged operation of the glass melt dehydration is required to achieve this objective. As an example, Table 5.5 demonstrates values of neodymium luminescence lifetime at various stages of dehydration of the glass KGSS 0180/35 melt. As can be seen from this table, it takes 14

© Woodhead Publishing Limited, 2013

hours of dehydration to raise τ from 130 to 300 μs (quantum yield increases from 0.32 to 0.73, respectively, and a (3.55 μm) diminishes from 20 to 1.5 cm^{-1}). The τ value remains at the reached level for a further 2 hours of dehydration. Invariability of τ values at this dehydration stage allows one to finish this operation and to pass on to another one.

2. Beam resistance (damage threshold) should not be below 20 J/cm^2 at laser pulse duration 3.0–3.5 ns.

3. Glass should not contain striae and inclusions of metallic Pt that reduce the glass damage threshold by more than one order of magnitude. It is very difficult to meet this requirement as glasses are melted in platinum crucibles to get glasses with high optical homogeneity. At the same time, the surfaces of Pt crucibles are destroyed under the action of aggressive high-temperature glass melt.

4. The natural coefficient of inactive absorption at the lasing wavelength should not exceed 0.0015 cm^{-1}. It should be noted that the value of this coefficient grows with concentrations both of Nd^{3+} and of micro-impurities of Cu^{2+} and Fe^{2+} ions (Arbuzov *et al.*, 2008).

5. For suppression of parasitic super-luminescence in a disk active element, its side surfaces should be pasted over with plates of the glass completely absorbing amplified radiation and having a refractive index n_c which exceeds that of the active element glass, n_a (Arbuzov *et al.*, 2002a). These plates are made usually from copper-containing glass and pasted to an active element by means of special glue with the value of the refractive index n_g lying between corresponding values of n_a and n_c (Arbuzov *et al.*, 2002a). The decimal absorption coefficient of copper-containing glass at the lasing wavelength and the thickness of plates are chosen in such a way that optical density is close to 2. Radiation quanta that either extend along the big surfaces of an active element towards its side edges or are reflected or scattered by the active element metal frame can give rise to an undesirable dump of the population inverse density of activator ions. This means that the output signal in the direction of radiation to be amplified could decrease due to this process. However, in the presence of such plates, this undesirable process in the active element is prevented because the above quanta are absorbed by the cladding plates. As a result, the signal amplification coefficient in an active element with cladding plates appears to be 1.25 times bigger than that in an active element without cladding.

5.4.2 Highly concentrated metaphosphate ytterbium–erbium glasses

As has been mentioned more than once, ytterbium–erbium glasses are widely used as the active media in lasers and optical amplifiers. This is associated with the fact that the lasing wavelength of the erbium ion (1.5 μm), firstly, is

© Woodhead Publishing Limited, 2013

optimal for information transfer along fiber-optic communication lines and, secondly, lies in an eye-safe spectral range (Desurvire, 1994). The ytterbium ions are also introduced into the glass to increase the pumping efficiency, since they play the role of sensitizers for the erbium ions (Alekseev *et al.*, 1980; Desurvire, 1994). It is desirable that ytterbium concentration in laser glasses with two activator types should be as high as possible to ensure the most effective pumping erbium luminescence. For this reason, attempts were made to develop highly concentrated ytterbium–erbium glasses. For example, borate (Vorob'ev *et al.*, 1987) and phosphate (Maksimova *et al.*, 1991; Denker *et al.*, 1992; Sverchkov *et al.*, 1992) glasses doped with up to 7×10^{21} and with 4.2×10^{21} Yb^{3+}/cm^{-3}, respectively, were developed and tested.

Typical ytterbium-ion concentrations, for example, in commercial ytterbium–erbium laser phosphate glasses are $(19–21) \times 10^{20}$ cm^{-3} (Lunter and Fyodorov, 1994; Schott Glass Technologies; Kigre, Inc.). Such ytterbium-ion concentrations make it possible to carry out efficient pumping and excitation transfer for erbium-ion concentrations all the way to 1×10^{20} cm^{-3} (Alekseev *et al.*, 1980). A further increase of the erbium-ion concentration reduces the efficiency of non-radiating energy transfer from ytterbium to erbium, and this, along with other factors, for example up-conversion and concentration quenching, reduces the laser efficiency. An additional point to emphasize is that laser media doped with Er^{3+} ions are characterized by a three-level generation scheme. This complicates creating an inverse population of the upper laser level and leads to an increase in laser threshold.

Increasing the ytterbium concentration can reduce the lasing threshold, increase the limiting erbium-ion concentration, and improve the lasing properties of lasers that operate in the Q-switched regime (Georgiou *et al.*, 2001).

Today the design and fabrication of novel laser glasses doped with high concentration of activators (for example, with concentration of ytterbium ions more than 20×10^{20} cm^{-3}) presents a topical issue. The increase of concentration of activator can be achieved by design of special glasses, where the rare-earth ions can play a role not only as activators but also in the glass-forming system. For example, some novel metaphosphate glasses doped with erbium and ytterbium have been developed in joint cooperation between the University of ITMO and the Research and Technological Institute of Optical Materials Science (RTIOMS, St Petersburg) (Aseev *et al.*, 2003; Nikonorov *et al.*, 2003). Ytterbium concentration achieves saturation value of $N_{Yb} = 52.4 \times 10^{20}$ cm^{-3} in these glasses. The next example illustrates features of the highly concentrated laser metaphosphate ytterbium–erbium glasses.

Figure 5.4 illustrates the gain coefficient at 1535 nm for erbium transition of $^4I_{13/2} \rightarrow {}^4I_{15/2}$ as a function of pump power at 980 nm for different concentrations of ytterbium and fixed concentration of erbium ($N_{Er} = 0.29 \times$

© Woodhead Publishing Limited, 2013

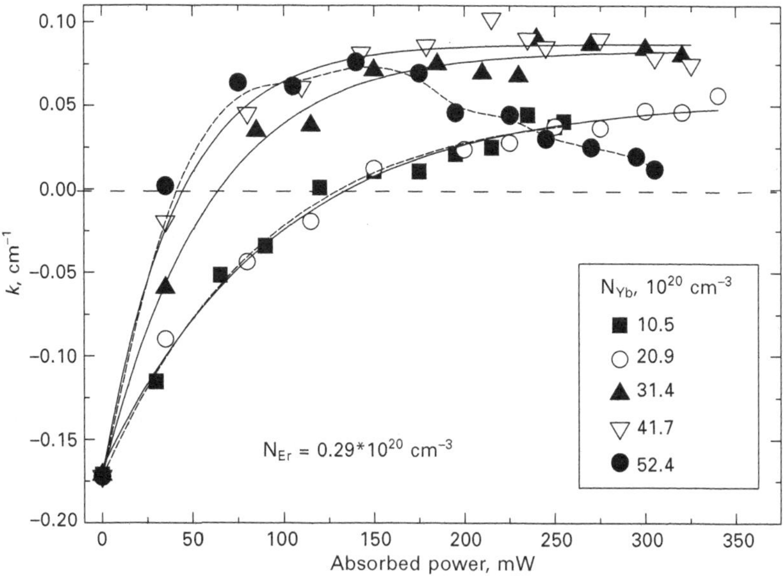

5.4 Gain at 1535 nm *vs.* pump power for different ytterbium concentrations.

10^{20} cm^{-3}). It can be seen from Fig. 5.4 that for ytterbium concentrations of 10.5×10^{20} and 20.9×10^{20} cm^{-3}, typical of commercial ytterbium–erbium lasers, the probe signal begins to increase ($k > 0$) when the absorbed pump power is more than 150 mW and reaches its maximum value of 0.05 cm^{-1} at pump powers of 300–350 mW.

Increasing the ytterbium concentration ($N_{Yb} > 20.9 \times 10^{20}$ cm^{-3}) reduces the gain threshold. Thus, for the concentration $N_{Yb} = 41.7 \times 10^{20}$ cm^{-3}, the gain becomes $k > 0$ at a pump power of 50 mW and reaches a maximum value of $k = 0.08$–0.1 cm^{-1} at a pump power of 200 mW. A further increase of pump power to 300 mW causes virtually no increase of the gain. At the maximum ytterbium concentration ($N_{Yb}= 52.4 \times 10^{20}$ cm^{-3}), gain ($k > 0$) is already observed at an absorbed pump power of 35 mW. The maximum gain for this limiting concentration reaches a value of 0.07 cm^{-1} at a pump power of 70 mW. However, at powers above 150 mW, there is a decrease of the gain. Thus, at a pump power of 300 mW, the gain decreases to 0.01 cm^{-1}.

5.5 Ytterbium glasses

Despite the fact that Yb^{3+} ions are usually used in glasses doped with Er^{3+} ions as their sensitizers, they can play the role of an activator in glasses as well. For this reason, in the last 15–20 years, ytterbium crystals and glasses

© Woodhead Publishing Limited, 2013

have drawn increasing attention from researchers as an independent class of laser materials (Mix *et al.*, 1995; Griebner *et al.*, 1996; Jiang S. *et al.*, 1997; Honninger C. *et al.*, 1998, 1999; Pascotta *et al.*, 2000; Jiang C. *et al.*, 2000; Karlsson *et al.*, 2002; Dai *et al.*, 2002; Ehrt, 2003; Hein *et al.*, 2004; Galagan *et al.*, 2009b). The fact is that ytterbium glasses have a number of advantages. First of all, like neodymium glasses, they allow one to generate radiation in the same spectral region of 1 μm. In the case of ytterbium glasses, high doping levels are possible without a reduction of luminescence radiation lifetime. A simple energy level scheme of Yb^{3+} ions causes small quantum defect, low heat generation in active elements as well as lack of excited state absorption, up-conversion, and concentration quenching due to the scheme of cross-relaxation that is rather efficient in concentrated neodymium glasses. Further, due to the broad luminescence band of Yb^{3+} ions, ytterbium glasses are very attractive for the generation of ultrashort laser pulses at the femtosecond level with tuning laser wavelength in a rather wide spectral range. On the other hand, a broad absorption band of ytterbium ions and their long luminescence lifetime (up to 2 ms) are favorable for efficient laser (Ti:sapphire or Nd:YAG) or diode (InGaAs) pumping.

To the disadvantages of ytterbium lasers belong the low emission cross-sections (10^{-21}–10^{-20} cm^2) and the quasi-three-level scheme of generation that does not exclude ground state absorption at the laser wavelength (Honninger *et al.*, 1998). The thermal population of the ground manifold sublevels leads to an increase in the laser threshold. All of this requires a high intensity of pumping and cooling laser to increase its efficiency (Honninger *et al.*, 1999; Dai *et al.*, 2002).

As for the ytterbium glass types, the QX/Yb phosphate glass (Mix *et al.*, 1995; Griebner *et al.*, 1996; Jiang S. *et al.*, 1997; Honninger *et al.*, 1998, 1999), niobium phosphate glasses (Jiang C. *et al.*, 2000), aluminum boron-phosphate glasses (Dai *et al.*, 2002; Galagan *et al.*, 2009b), fluoride phosphate glass (Honninger *et al.*, 1999; Hein *et al.*, 2004), and a Q-246/Yb silicate glass (Honninger *et al.*, 1998) were investigated to obtain generation of ultrashort pulses. All the glasses under study were well dehydrated, and the ytterbium concentration was varied up to 2×10^{21} cm^{-3}.

The tuning was shown to stretch from 1025 to 1065 nm in QX/Yb phosphate glasses, from 1030 to 1082 nm in Q-246/Yb silicate glasses, from 1055 to 1069 nm in a fluoride phosphate glass (Honninger *et al.*, 1998), and from 1008 to 1080 nm in a new aluminum boron-phosphate glass (Galagan *et al.*, 2009b). A minimal duration of generated pulses was observed in phosphate (58 fs) and in silicate (61 fs) glasses QX/Yb and Q-246/Yb, respectively (Honninger *et al.*, 1998). The slope efficiency of generation may reach 49% (Jiang S. *et al.*, 1997).

Thus, glasses doped with ytterbium are promising laser materials for development of diode or laser pumped femtosecond lasers.

© Woodhead Publishing Limited, 2013

5.6 Future trends in glass-based laser materials

Currently, glass-ceramic materials are of great interest as laser media. These glass-ceramics combine the best properties of crystals (high emission cross-section, quantum yield of luminescence, mechanical and thermal strength, etc.) and glasses (possibilities of pressing and molding, spattering, pulling optical fibers, and carrying out ion exchange to fabricate waveguide structures). One of the major drawbacks of glass-ceramic materials is high light scattering occurring at the boundary of the crystalline phase and the glass phase. That is why current research in the development of optical glass-crystalline materials is aimed at decreasing light scattering by means of formation of nanosized (5–30 nm) crystals or nanoparticles in the glass matrix. Only the nanoscale nature of the crystalline phase can significantly reduce the light scattering in heterophase composites (where the extinction coefficient can reach less than 0.01 cm^{-1}), increase its transparency and classify these materials as optical ones. Today the novel transparent glass-ceramics doped by rare-earth ions can compete in their spectral and luminescent properties with traditional laser materials (crystals and glasses), operating at 1.5 μm. In this section we focus on novel transparent oxy-fluoride nanoglass-ceramics doped by erbium.

The story of oxy-fluoride glass-ceramics production technology starts in the 1970s. Auzel *et al.* (1975) made an attempt to synthesize oxy-fluoride glasses containing rare-earth ions. This attempt resulted in production of non-transparent glass-ceramic materials containing microcrystals with a diameter of about 10 μm. The efficiency of luminescence revealed by these media was in several times larger than that of the etalon luminophore LaF_3:Yb:Er. Later on, in 1993, there was published the first paper devoted to synthesis of transparent glass-ceramics containing the cubic fluoride phase, activated by erbium and ytterbium ions (Wang and Ohsaki,1993). For the first time, materials were produced which combine all the advantages of the glass-like aluminosilicate matrix and the optical features of the low-phonon fluoride crystals.

Recently, the transparent fluorine-containing glass-ceramics matrices, containing the rare-earth ions, included in the fluoride-like nanocrystalline phases (10–40 nm), are drawing attention due to the series of spectroscopy advantages. It is obvious that from the viewpoint of laser active media development, the optimal materials are characterized by the low-frequency phonon spectrum and by the low content of the OH-groups, because, in this case, one can reduce the excitation energy losses due to the multi-phonon quenching process. For a long time, there was a common opinion that only the fluorine-containing materials (like fluoride glasses and crystals) were optimal for the problem solution. However, recently the synthesis of glass-ceramics materials like oxy-fluoride silicate glasses has become the priority direction

© Woodhead Publishing Limited, 2013

of studies (Tikhomirov *et al.*, 2004; Dantelle *et al.*, 2006; Beggiora *et al.*, 2003; Mortier *et al.*, 2001; Kolobkova *et al.*, 2007, 2010; Aseev *et al.*, 2012). Such composite materials combine the optical parameters of the low-phonon fluoride crystals and the good mechanical and chemical features of the silicate glasses. It was also revealed that some of the oxy-fluoride glass-like materials have a feature of forming the fluoride nanocrystals doped with the rare-earth ions, during the process of heat treatment of the raw primary glass. Hence, such materials combine all the positive features of the fluoride nanocrystals, which control the optical properties of the rare-earth ions, with that of oxide glasses, such as easy production technology and excellent macroscopic features like chemical and mechanical strength and high optical quality.

The possibilities of design and fabrication of transparent oxy-fluoride nanoglass-ceramics for laser applications have been demonstrated for the first time by the research group of St Petersburg University of ITMO (Aseev *et al.*, 2009, 2010). Glasses of the system SiO_2–Al_2O_3/–CdF_2–PbF_2–ZnF_2–YF_3–LnF_3, where Ln = La, Pr, Dy, Nd, Tb, Eu, Er, Yb, Sm, Tm, Ho, have been synthesized. It was shown that due to spontaneous crystallization in a glass, the new crystalline phases are growing – the yttrium oxy-fluoride of lead, the lanthanide oxy-fluoride of lead and the yttrium–lanthanide oxy-fluoride of lead. The size of the crystalline phase is 15–40 nm, thus providing transparency of nanoglass-ceramics in the visible and near-IR spectral ranges and putting it into the class of optical materials. The next examples illustrate features of the laser oxy-fluoride nanoglass-ceramics doped with Er.

Figure 5.5 illustrates the luminescence spectra of erbium ions of a novel

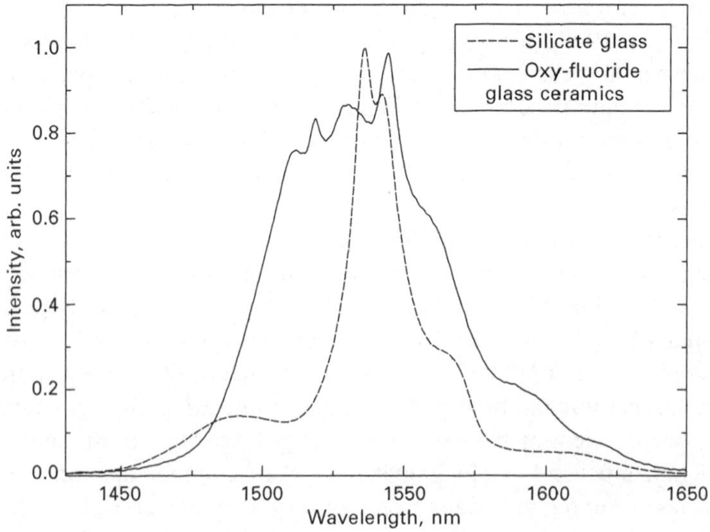

5.5 Luminescence spectra of erbium ions in a silicate glass (dotted curve) and oxy-fluoride nanoglass ceramics (solid curve).

© Woodhead Publishing Limited, 2013

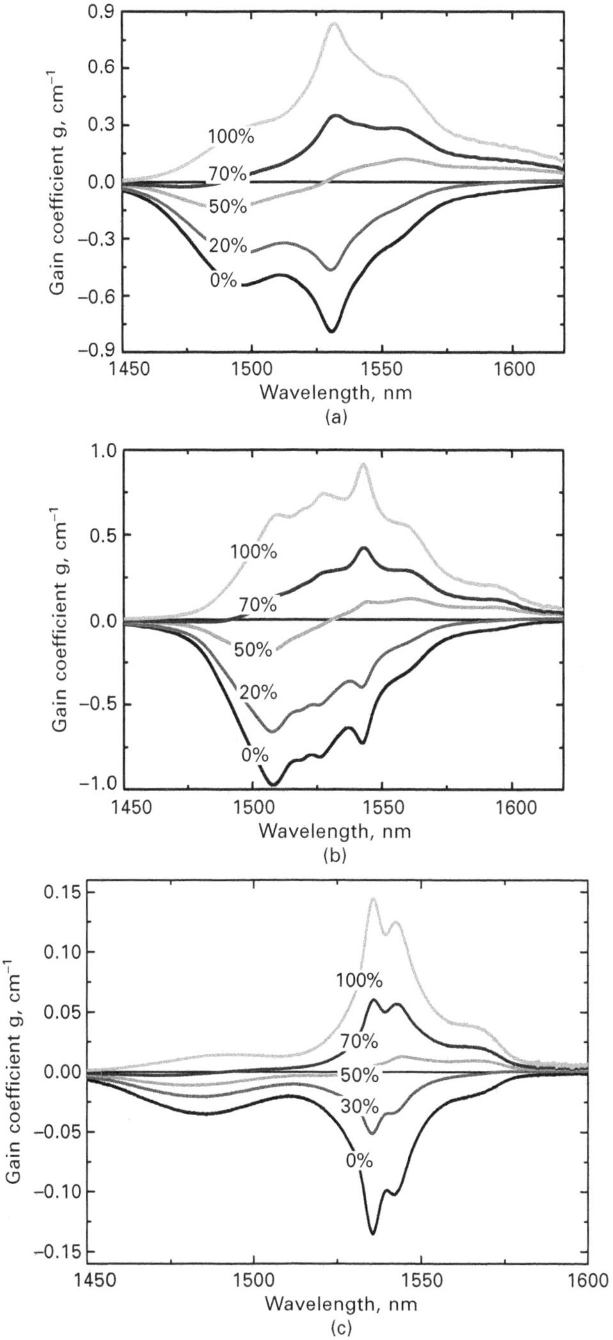

5.6 Gain/loss spectra of (a) initial glass, (b) oxy-fluoride nanoglass ceramics, and (c) commercial silicate glass, doped with erbium, for various pumping ratios N_2/N_{Er}. Pump wavelength 980 nm.

© Woodhead Publishing Limited, 2013

oxy-fluoride nanoglass-ceramics made in St Petersburg University of ITMO in comparison with the commercially available silicate glass (KGSS-134). One can see that the spectrum half-width is equal to $\Delta\lambda = 66$ nm for the nanoglass-ceramics, while that for the silicate glass it is equal to $\Delta\lambda = 20$ nm. One has also to note that the new nanoglass-ceramics reveal the high (>80%) quantum yield of luminescence for the transition $^4I_{13/2} \rightarrow {}^4I_{15/2}$.

Figure 5.6 illustrates the gain/loss spectra for the new nanoglass-ceramics in comparison with the starting glass (before treatment) and with the commercially available silicate glass (KGSS-134). One can see that the transfer from the starting glass to the nanoglass-ceramics by heat treatment results in increase of the amplification range from 48 to 64 nm. This is accompanied by the increase of the maximal amplification gain for the same pumping level. For instance, for the starting glass for pumping level 70% the gain is equal to $g = 0.35$ cm^{-1}, while for the glass-ceramics it is equal to $g = 0.42$ cm^{-1}. One can also see that the amplification spectrum for the commercial silicate glass is much worse than that of both the untreated fluorine-containing glass and the nanoglass-ceramics on its basis.

Hence, the novel transparent oxy-fluoride nanoglass-ceramics doped by erbium reveal the good spectral and luminescent properties, which are no worse than that of the well-known commercial glasses, and exceed them for such parameters as the amplification spectrum width and gain coefficient. Thus, the novel transparent oxy-fluoride nanoglass-ceramics are very promising candidates for 1.5 μm lasers and amplifiers.

5.7 References

Alekseev N. E., Gapontsev V. P., Gromov A. K. *et al.* (1975) 'Impact of hydroxyl groups on luminescent properties of phosphate glasses doped with rare earth ions', *Izvestiya AN SSSR, Seriya Neorganicheskie Materialy (Bulletin of USSR Academy of Sciences, Inorganic Materials Series)*, 11(2), 323–327 (in Russian).

Alekseev N. E., Gapontsev V. P., Zhabotinsky M. E. *et al.* (1980) *Laser Phosphate Glasses*, Moscow, Nauka.

Alekseev N. E., Byshevskaya-Konopko L. O., Vorob'ev I. L. *et al.* (2003) 'Continuous wave lasing at 1.54 μm in a flashlamp-pumped ytterbium– erbium-doped glass', *Quantum Electronics*, 33(12), 1062–1064.

Amaranath G., Buddhudu S., Bryant F. J. *et al.* (1991) 'Spectroscopic properties of Nd^{3+}-doped heavy metal fluoride glasses', *Journal of Luminescence*, 47, 255–269.

Arbuzov V. I. (1980) 'Measurement of absolute quantum yield of resonance luminescence of Yb^{3+} in glass by the modulation method', *Zhurnal Prikladnoi Spectroskopii (Journal of Applied Spectroscopy)*, 33(6), 1030–1035 (in Russian).

Arbuzov V. I. (1982) *Inhomogeneous structure of spectra and efficiency of rare earth ions luminescence in glass*, PhD Dissertation. Leningrad, S. I. Vavilov State Optical Institute.

Arbuzov V. I., Brachkovskaya N. B., Zhmyreva I. A. *et al.* (1976) 'Absolute luminescence quantum yield of glasses doped with neodymium', *Kvantovaya Elektronika (Quantum Electronics)*, 3(9), 2005–2013 (in Russian).

© Woodhead Publishing Limited, 2013

Arbuzov V. I., Vakhmyanin K. P., Volynkin V. M. *et al.* (2002a) 'Absorbing claddings for big-sized disc active elements of neodymium phosphate glass KGSS 0180/35 for laser amplifiers', *Opticheskii Zhurnal (Journal of Optical Technology)*, 69(1), 16–20 (in Russian).

Arbuzov V. I., Charukhev A. V., Fyodorov Yu. K. *et al.* (2002b) 'Neodymium phosphate glasses for high-energy and high-pick-power lasers', *Glasstechnische Berichte – Glass Science and Technology*, 75C2, 209–214.

Arbuzov V. I., Volynkin V. M., Lunter S. G. *et al.* (2003) 'Big-sized disc active elements of neodymium phosphate glass for high-peak-power high-energy lasers', *Opticheskii Zhurnal (Journal of Optical Technology)*, 70(5), 68–78 (in Russian).

Arbuzov V. I., Gusev P. E., Dukel'sky K. V. *et al.* (2008) 'Problem of inactive absorption of radiation at the generation wave length of neodymium phosphate glasses for active elements high-power high-energy lasers', *Proceedings of International Conference 'X Kharitonovskie Tematicheskie Yauchnye Chteniya' ('X Khariton's Thematic Scientific Readings') 'High-power Lasers and Investigations on Physics of High-Energy-Densities'*, 69–74 (in Russian).

Aseev V. A., Nikonorov N. V., Przhevusky A. K. *et al.* (2003) 'Measuring the gain/loss spectrum in high-concentration ytterbium-erbium-doped laser glasses', *Journal of Optical Technology*, 70, 778.

Aseev V. A., Golubkov V. V., Klementeva A. V. *et al.* (2009), 'Spectral luminescence properties of transparent lead fluoride nanoglassceramics doped with erbium ions', *Optics and Spectroscopy*, 106(5), 691–696.

Aseev V., Kolobkova E., Korchagin E. *et al.* (2010) 'Rare-earth doped transparent nano-glassceramics for photonic applications', *Proceedings of Xth International Conference on Fiber Optics and Photonics*, Guwahati, India, p. 526.

Aseev V. A., Golubkov V. V., Kolobkova E. V., Nikonorov N. V. (2012) 'Lead oxyfluoride lanthanides in glass-like matrix', *Glass Physics and Chemistry*, 38(1), 11–18.

Atom Optics Co. Ltd., Shanghai, P.R. China.

Auzel F., Pecile D., Morin D. (1975) 'Er^{3+} doped ultra-transparent oxy-fluoride glass-ceramics for application in the 1.54 μm telecommunication window', *Journal of the Electrochemical Society*, 122, 101–108.

Basiev T. T., Mamedov T. G., Shcherbakov I. A. (1975) 'Investigation of non-radiative relaxation of the metastable state Nd^{3+} $^4F_{3/2}$ in silicate glass', *Kvantovaya Elektronika (Quantum Electronics)*, 2(6), 1269–1277 (in Russian).

Batygov S. Kh., Voronko Yu. K., Denker B. I. *et al.*, (1976) 'Physicochemical, spectral, luminescent, and stimulated emission properties of phosphate glasses with high neodymium concentrations', *Soviet Journal of Quantum Electronics*, 6(10), 1220–1222.

Bayanov V. I., Bordachev E. G., Kryzhanovsky V. I. *et al.* (1984) 'Rod amplifiers on neodymium phosphate glass with a diameter of 60 mm with high amplifier coefficient', *Kvantovaya Elektronika (Quantum Electronics)*, 11(2), 310–316 (in Russian).

Beggiora M., Reaney I. M., Seddon A. B. *et al.* (2003) 'Phase evolution in oxy-fluoride glass ceramics', *Journal of Non-Crystalline Solids*, 326–327, 476–483.

Bondarenko E. G., Galant E. I., Lunter S. G. *et al.* (1975) 'Impact of water in glass on quenching luminescence of rare earth activator', *Optiko-mekhanicheskay Promyshlennost*, 6, 42–44 (in Russian).

Brachkovskaya N. B., Grubin A. A., Lunter S. G. *et al.* (1976) 'Intensities of optical transitions in absorption and luminescence spectra of neodymium in glasses', *Kvantovaya Elektronika (Quantum Electronics)*, 3(5), 998–1005 (in Russian).

© Woodhead Publishing Limited, 2013

Branch Standard OST 3 - 30 - 77, 'Optical glass GLS. Technical conditions', USSR, 1977 (in Russian).

Brown D. C. (1981) *High-peak-power Nd: Glass Laser System*, Berlin, Heidelberg and New York, Springer.

Byshevskaya-Konopko L. O., Vorob'ev I. L., Izyneev A. A. *et al.* (2001) 'Optimisation of the pumping parameters of a repetitively pulsed erbium laser', *Quantum Electronics*, 31(10), 861–863.

Campbell J. H., Suratwala T. I. (2000) 'Nd-doped phosphate glasses for high-energy/ high-peak-power lasers', *Journal of Non-Crystalline Solids*, 263–264, 318–331.

Campbell J. H., McLean M. J., Hawley-Fedder R. *et al.* (1999) 'Development of continuous glass melting for production of Nd-doped phosphate glasses for the NIF and LMJ laser systems', *SPIE*, 3492, 778–786.

Dai S., Sugiyama A., Hu L. *et al.* (2002) 'The spectrum and laser properties of ytterbium doped phosphate glass at low temperature', *Journal of Non-Crystalline Solids*, 311(2), 138–144.

Dantelle G., Mortier M., Patriarche G., Vivien D. (2006) 'Comparison between nanocrystals in glass-ceramics and bulk single crystals Er^{3+}-doped PbF_2', *Journal of Solid State Chemistry*, 179, 2003–2006.

Denker B., Maximova G., Osiko V. *et al.* (1992) 'Lasing tests of novel erbium laser glasses' , *Russian Journal of Quantum Electronics*, 19(9), 1063–1067.

Desurvire E. (1994) *Erbium-doped fiber Amplifiers*, New York, Wiley.

Dexter D. (1953) 'Theory of sensitized luminescence in solids', *Journal of Chemical Physics*, 21(5), 836–850.

Dianov E. M. (2008) 'Bi-doped optical fibers: a new active medium for NIR lasers and amplifiers', in *Optical Components and Materials V*, edited by M. J. F. Digonnet, S. Jiang, J. W. Glesener and J. C. Dries, *Proceedings of SPIE*, 6890, 68900H-1.

Diecke G. H., Crosswhite H. M. (1963) 'The spectra of the doubly and triply ionized rare earths', *Applied Optics*, 2(7), 675–686.

Dmitriev D. I., Arbuzov V. I., Dukelsky K. V. *et al.* (2006) 'Testing of KGSS 0180 laser glass for platinum micro-inclusions', *Proceedings of SPIE*, 6594, 202–209.

Doulan J. L., Girard S., Haquin H. *et al.* (2003) 'Spectroscopic properties and laser emission of Tm doped ZBLAN glass at 1.8 μm', *Optical Materials*, 24, 563–574.

Ehrt D. (2003) 'Fluoroaluminate glasses for lasers and amplifiers', *Current Opinion in Solid State and Material Science*, 7, 135–141.

Ermolaev V. L., Bodunov E. N., Sveshnikova E. B., Shakhverdov T. L. (1977) *Non-radiative Transfer of Electron Excitation Energy* Moscow, Nauka.

Etzel H. W., Gandy H. W., Ginther R. J. (1962) 'Stimulated emission of infrared radiation from ytterbium-activated silica glass', *Applied Optics*, 1, 534–536.

Eyal M., Reisfeld R., Jorgensen C. K., Bendow B. (1987) 'Laser properties of holmium and erbium in thorium-, zinc- and yttrium-based fluoride glass', *Chemical Physics Letters*, 139(5), 395–400.

Fang Yong-Zheng, Jin Ming-Lin, Wen Lei *et al.* (2007) 'Er^{3+} and Yb^{3+} codoped phosphate laser glass for high power flashlamp pumping', *Chinese Physics Letters*, 24(5), 1283–1286.

Feofilov P. P. (1961) 'About spectra and luminescence kinetics of single crystals CaF_2–Tb', *Optika I Spektroskopiya (Optics and Spectroscopy)*, 10, 142–144 (in Russian).

Fernandez J., Balda R., Arriandiga M. A. (1994) 'Spectroscopic and laser properties of Nd^{3+} in fluoride glasses', *Optical Materials*, 4, 91–97.

Foerster T. (1949) 'Experimentalle und theoretische Untersuchung des zwischenmolekularen

© Woodhead Publishing Limited, 2013

Uebergangs von Electronenaufregungsenergie', *Zeitschrift für Naturforschung*, 4A(5), 321–327.

Fusari F., Lagatsky A., Jose G. *et al.* (2009) 'Continuous-wave laser operation of a bulk Tm:germanate glass laser around 2 μm with 40% internal slope efficiency', *3rd International Conference on Middle Infrared Coherent Sources*, Trouville, France, Book of Papers, PO5.

Galagan B. I., Glushchenko I. N., Denker B. I. *et al.* (2009a) 'New high-strength neodymium phosphate laser glass', *Kvantovaya Elektronika (Quantum Electronics)*, 39(12), 1117–1120 (in Russian).

Galagan B. I., Glushchenko I. N., Denker B. I. *et al.* (2009b) 'New ytterbium-phosphate glass for diode-pumped lasers', *Kvantovaya Elektronika (Quantum Electronics)*, 39(10), 891–894 (in Russian).

Galaktionova N. M., Garkavi G. A., Zubkova V. S. *et al.* (1974) 'Continuous action laser on neodymium glass', *Optika i Spektroskopiya (Optics and Spectroscopy)*, 37(1), 162–165 (in Russian).

Galant E. I., Kalinin V. M., Lunter S. G. (1976a) 'On generation of glasses with ytterbium and erbium at laser pumping', *Quantovaya Elektronika (Quantum Electronic)*, 3(10), 2187–2196 (in Russian).

Galant E. I., Lunter S. G., Mironov A. N., Fyodorov Yu. K. (1976b) 'Impact of deuteriation on luminescent and optical properties of doped glasses', *Fizika i Khimiya Stekla (Glass Physics and Chemistry)*, 2(4), 351–352 (in Russian).

Gandy H. W., Ginther R. J. (1962a) 'Stimulated emission of ultraviolet radiation from gadolinium-activated glass', *Applied Physics Letters*, 1, 25–27.

Gandy H. W., Ginther R. J. (1962b) 'Stimulated emission from holmium activated silicate glass', *Proceedings of IRE*, 50(10), 2113–2114.

Gandy H. W., Ginther R. J., Weller J. F. (1965) 'Laser oscillations in erbium activated silicate glass', *Applied Optics*, 16(3), 266–267.

Gandy H. W., Ginther R. J., Weller J. F. (1967) 'Stimulated emission of Tm radiation in silicate glass', *Journal of Applied Physics*, 38, 3030–3031.

Gapontsev V. P., Zhabotinsky M. E., Izyneev A. A. *et al.* (1978) 'Effective transformation of induced radiation 1.054–1.54 mkm', *Pis'ma v Zhurnal Experimental'noi I Teoreticheskoi Fiziki (Letters to the Journal of Experimental and Theoretical Physics)*, 18(7), 428–431 (in Russian).

Georgiou E., Musset O., Boquillon J.-P., Denker B., Sverchkov S. E. (2001) '50mJ/30ns FTIR Q-switched diode pumped Er:Yb glass 1.54 μm laser', *Optical Communications*, 198, 147–153.

Golding P. S., Jackson S. D., King T. A., Pollnau M. (2000) 'Energy transfer process in E^{3+}-doped and Er^{3+}, Pr^{3+}-codoped ZBLAN glasses', *Physical Review B*, 62(2), 856–864.

Griebner U., Koch R., Schonnagel H. *et al.* (1996) 'Laser performance of a new ytterbium doped phosphate laser glass', *OSA TOPS on Advanced Solid-State Lasers*, 1, 26–29.

Heber J. (1976) 'Fluoreszenzlebensdauern und Mehrphononenprozesse in wasserhaltigen Salzen von Eu^{3+} und Tb^{3+}', *Phyzik kondensierten Materie*, 6(6), 381–402.

Hein J., Podleska S., Siebold M. *et al.* (2004) 'Diode-pumped chirped pulse amplification to the joule level', *Applied Physics B*, 79, 419–422.

Honninger C., Morier-Genoud F., Moser M. *et al.* (1998) 'Efficient and tunable diode-pumped femtosecond Yb:glass lasers', *Optics Letters*, 23(2), 126–128.

Honninger C., Paschotta R., Graf M. *et al.* (1999) 'Ultrafast ytterbium-doped bulk lasers and laser amplifiers', *Applied Physics B*, 69, 3–17.

© Woodhead Publishing Limited, 2013

Izyneev A. A., Sadovskii P. I. (1997) 'New highly efficient LGS-KhM erbium-doped glass for uncooled miniature lasers with a high pulse repetition rate',*Quantum Electronics*, 27(9), 771–775.

Japanese Catalog 'Laser Glass from Hoya', Hoya Corporation, Tokyo.

Javorniczky J. S., Newman P. J., MacFarlane D. R. *et al.* (1995) 'High erbium content heavy metal fluoride glasses', *Journal of Non-Crystalline Solids*, 184, 249–253.

Jiang C., Luo T., Myers M. *et al.* (1998) 'Phosphate glasses for high average power lasers', *Rare Earth Doped Devices II*, *SPIE Proceedings*, 3280, Photonic West.

Jiang C., Liu H., Zeng Q. *et al.* (2000) 'Yb:phosphate laser glass with high emission cross-section', *Journal of Physics and Chemistry of Solids*, 61, 1217–1223.

Jiang S., Myers M. J., Rhonehouse D. L. *et al.* (1997) 'Ytterbium doped phosphate laser glasses', *SPIE Proceedings*, 2986, 10–15.

Kalinin V. M., Mak A. A., Prilezhaev D. S., Fromzel V. A. (1974) 'On special features of generation of optical quantum generators on glasses doped with Yb^{3+} and Er^{3+} at laser pumping', *Zhurnal Tekhnicheskoi Fiziki (Journal of Technical Physics)*, 44(6), 1328–1331 (in Russian).

Kaminsky A. A. (1975) *Laser Crystals*, Moscow, Nauka.

Karapetyan G. O. (1963) 'Luminescence of glasses with rare-earth activators', *Izvestiya AN SSSR, Seriya Physicheskaya (Bulletin of USSR Academy of Sciences, Physical Series)*, 27, 799–802 (in Russian).

Karapetyan G. O., Reishakhrit A. L. (1967) 'Luminescent glasses as materials for optical quantum generators', *Izvestiya AN SSSR, Seriya Neorganicheskie Materialy (Bulletin of USSR Academy of Sciences, Inorganic Material Series)*, 3, 216–259 (in Russian).

Karlsson G., Laurell F., Tellefsen J. *et al.* (2002) 'Development and characterization of Yb–Er laser glass for high average power laser diode pumping', *Applied Physics B*, 75, 1–6.

Kigre, Inc. (2009) Catalog 'Rare Earth Doped Laser Glass', Hilton Head, USA.

Kigre 'QX-type glass', Web page 'Innovations in Solid-State Laser Technology', Hilton Head, USA.

Kolobkova E. V., Melekhin V. G., Penigin A. N. (2007) 'Optical glass-ceramics based on fluorine-containing silicate glasses doped with rare-earth ions', *Glass Physics and Chemistry*, 33(1), 8–13.

Kolobkova E. V., Tagil'tseva N. O., Lesnikov P. A. (2010) 'Specific features of the formation of oxyfluoride glass-ceramics of the SiO_2–PbF_2–CdF_2–ZnF_2–Al_2O_3–Er(Eu,Yb)F_3 system', *Glass Physics and Chemistry*, 36(3), 317–324.

Kovalyova I. V., Kolobkov V. P., Tatarintsev B. V., Yakhkind A. K. (1975) 'Regularities of rare earth activators luminescence quenching in telluride glasses', *Zhurnal Prikladnoj Spectroskopii (Journal of Applied Spectroscopy)*, 23(6), 1021–1025 (in Russian).

Kwasny M., Mierczyk Z., Stepien R., Jedrzejewski K. (2000) 'Nd^{3+}-, Er^{3+}- and Pr^{3+}-doped fluoride glasses for laser applications', *Journal of Alloys and Compounds*, 300–301, 341–347.

Layne C. B., Lowdermilk W. H., Weber M. J. (1975) 'Nonradiative relaxation of rare-earth ions in silicate glass', *IEEE Journal of Quantum Electronics*, 11(9), 798–799.

Lunter S. G., Fyodorov Yu. K. (1994) 'Development of erbium laser glasses', *Proceedings of Fiber Symposium on Light Materials, Laser Technology Material for Optical Telecommications*, 2, 327–333.

Lunter S. G., Mak A. A., Starikov A. D. *et al.* (1991) 'Six-beam laser facility on neodymium glass 'Progress' with a power of 3 TWt', *XIV International Conference on Coherent and Non-linear Optics –91*. Leningrad, S. I. Vavilov State Optical Institute.

© Woodhead Publishing Limited, 2013

Maksimova G. V., Sverchkov S. E., Sverchkov Yu. E. (1991) 'Lasing tests on new ytterbium–erbium laser glass pumped by neodymium lasers', *Soviet Journal of Quantum Electronics*, 21(12), 1324–1325.

Martin W. E., Trenholme J. B., Yarema S. M., Hurley C. A. (1981) 'Solid–state disk amplifier for fusion-laser system', *IEEE Journal of Quantum Electronics*, 17(9), 1744–1754.

Mauer P. B. (1963a) 'Operation of a Nd^{3+}-glass optical maser at 9180 Å', *Applied Optics*, 2, 87–88.

Mauer P. B. (1963b) 'Laser action in neodymium-doped glass at 1.37 microns', *Applied Optics*, 3, 153–154.

Milanese D., Lousteau J., Gomes L. *et al.* (2011) 'Ho-doped tellurite glasses for emission in the mid infrared wavelength region', *CLEO/Europe_EQEC 2011, Munich, Germany*, CE. P. 6 WED.

Mix E., Heumann E., Huber G. *et al.* (1995) Efficient cw-laser operation of Yb-doped fluoride phosphate glass at room temperature', *Journal of the Optical Society of America*, 24, 339–342.

Mortier M., Goldner P., Chateau C., Genotelle M. (2001) 'Erbium doped glass–ceramics: concentration effect on crystal structure and energy transfer between active ions', *Journal of Alloys and Compounds*, 323–324, 245–249.

Nikonorov N. V., Przhevuskii A. K., Chukharev A. V. (2003) 'Characterization of non-linear upconversion quenching in Er-doped glasses: modeling and experiment', *Journal of Non-Crystalline Solids*, 324, 92.

Olivier M., Pirasteh P., Doulan J.-L. *et al.* (2011) 'Pr^{3+}-doped ZBLA fluoride glasses for visible laser emission', *Optical Materials*, 33, 980–984.

Ostrovskaya G. V., Ostrovskii Yu. K. (1963) 'Determination of oscillator forces of absorption band of gadolinium ions', *Optika I Spectroskopiya (Optics and Spectroscopy)*, 14, 161–163 (in Russian).

Pascotta R., Aus der Au J., Spuehler G.J. *et al.* (2000) Diode-pumped passively mode-locked laser with high average power', *Applied Physics B* 70, 25–31.

Pearson A. D., Porto S. P. S., Northover W. R. (1964) 'Laser oscillations at 0.918, 1.057 and 1.401 microns in Nd^{3+}-doped borate glass', *Journal of Applied Physics*, 35, 1704–1706.

Peterson G. E., Bridenbaugh P. M. (1962) 'Direct observation of rise fluorescence in Tb^{3+} following a short burst of ultraviolet excitation', *Journal of the Optical Society of America*, 52(9), 1079–1080.

Philipps J. F., Topfer T., Ebendorf-Heideprim H. *et al.* (2001) 'Spectroscopic and lasing properties of Er^{3+}:Yb^{3+}-doped fluoride phosphate glasses', *Applied Physics B*, 72, 399–405.

Reisfeld R., Katz G., Spector N. *et al.* (1982) 'Optical transition probabilities of Er^{3+} in fluoride glasses', *Journal of Solid State Chemistry*, 41, 253–261.

Reisfeld R., Eyal M., Jorgensen C. K. (1986) 'Comparison of laser properties of rare earths in oxide and fluoride glasses, *Journal of Less-Common Metals*, 126, 187–194.

Rieseberg L. A., Weber M. J. (1976) 'Relaxation phenomena in rare-earth luminescence', *Progress in Optics*, 14, 89–159.

Russian Catalog 'Commercial laser glasses' (1982), Moscow, Dom Optiki (in Russian).

Schott Glass Technologies, 'IOG-1 laser glasses'.

Seeber W., Downing E. A., Hesselink L. *et al.* (1995) 'Pr^{3+}-doped fluoride glasses', *Journal of Non-Crystalline Solids*, 189, 218–226.

© Woodhead Publishing Limited, 2013

Sirazetdinov V. S., Arbuzov V. I., Dmitriev D. I. *et al.* (2006) 'Resistance of KGSS 0180 neodymium glass to laser-induced damage under different irradiation conditions', *Proceedings of SPIE*, 6610, 216–224.

Snitzer E. (1961) 'Optical maser action of Nd in barium crown glass', *Physical Review Letters*, 7(12), 444–446.

Snitzer E. (1966) 'Glass lasers', *Applied Optics*, 5(10), 1487–1499.

Snitzer E., Woodcock R. (1965) 'Yb^{3+}–Er^{3+} glass laser', *Applied Physics Letters*, 6(3), 45–46.

Sverchkov Yu., Denker B., Maximova G. *et al.* (1992) 'Lasing parameters of GPI erbium glasses', in *Solid State Lasers III*, edited by G. J. Quarles, *Proceedings of SPIE*, 1627, 37–41.

Tikhomirov V. K., Rodriguez V. D., Mendez-Ramos J. *et al.* (2004) 'Comparative spectroscopy of $(ErF_3)(PbF_2)$ alloys and Er^{3+}-doped oxyfluoriode glass-ceramics', *Optical Materials*, 27, 543–547.

Tolstoj M. N. (1970) 'Non-radiative energy transfer between rare earth ions in crystals and glasses', in: *Spektroskopiya Kristallov (Spectroscopy of Crystals)*, Moscow, Nauka (in Russian).

US Catalog 'Laser Glass', Schott Glass Technologies, Inc., Duryea, PA, USA.

US Catalog 'Laser Glasses from Schott'.

Vargin V. V., Karapetyan G. O. (1964) 'Luminescence of glasses doped with rare earth elements', *Optiko-mekhanicheskaya Promyshlennost (Optical and Mechanical Industry)*, 2, 2–6 (in Russian).

Vorob'ev I., Gapontsev V., Gromov A. *et al.* (1987) 'New active media for 1.5 μm Er minilasers', *Laser Optics Conference*, Leningrad, Book of Abstracts, 242.

Voronko Yu. K., Denker B. I., Zlenko A. A. *et al.* (1976) 'Spectral and lasing properties of Li–Nd-phosphate glass', *Optics Communications*, 18, 88–89.

Wang Y. H., Ohsaki J. (1993) 'New transparent vitroceramics colored with Er^{3+} and Yb^{3+} for efficient frequency up-conversion', *Journal of Physics Letters*, 63(24), 3268–3270.

Young C. G. (1963) 'Continuous glass laser', *Applied Physics Letters*, 2, 151–152.

Young C. G. (1969) 'Glass lasers', *Proceedings of IEEE*, 57(7), 1267–1282.

© Woodhead Publishing Limited, 2013

6

Nonlinear crystals for solid-state lasers

V. PASISKEVICIUS, Royal Institute of Technology, Sweden

DOI: 10.1533/9780857097507.1.139

Abstract: This chapter aims first to guide the reader through the basic concepts which are encountered in the field of nonlinear optical materials and frequency conversion. The second aim is to show the historical developments in the area of nonlinear crystals and to expose the current status in this field. Due to remarkable progress in the development of quasi-phase-matched nonlinear media with associated numerous advantages for applications, the development focus is steadily shifting from synthesis of ever newer nonlinear crystals and towards demonstrations of structured or engineered media, as well as towards further improvement of the quality of the crystals used for fabricating such functional structures.

Key words: nonlinear optical materials, frequency conversion, quasi-phase matching, optical parametric oscillators, optical parametric amplifiers, mid-infrared nonlinear optical materials.

6.1 Introduction

This chapter will focus on second-order nonlinear media suitable for frequency conversion by either upconversion or downconversion processes using laser sources described in other chapters of this book. The need for and the utility of nonlinear optical frequency conversion is rather obvious due to the fact that solid state lasers can generate output with rather narrow spectral lines and within limited spectral ranges. Tunability, achievable in the second-order nonlinear optical processes, on the other hand, is in principle determined only by the transparency range of the nonlinear crystal if quasi-phase-matched (QPM) engineered nonlinear crystals are used, as will be shown in Section 6.2. For birefringence phase-matching (BPM) there are more restrictions on tunability, namely, the dispersion of the indices of refraction, as well as dependence of the effective nonlinear coefficients on propagation direction. Moreover, nonlinear frequency conversion is subjected to much lower thermal loads as compared to those typically experienced by active laser materials. The thermal load in nonlinear crystals is associated with small parasitic linear and nonlinear absorption, present to a certain degree in all crystals. Reducing those small absorption losses is a very important and actively pursued avenue for crystal quality improvement.

In this chapter we will first introduce the reader to the basic concepts of frequency conversion in second-order nonlinear media (Section 6.2), then

© Woodhead Publishing Limited, 2013

very briefly describe the remarkable progress in development of nonlinear crystals over the last 50 years (Section 6.3). The current status of the development of nonlinear materials will be discussed in Section 6.4 focusing primarily on the 'success stories', i.e. nonlinear crystals which found their application niches. Through this historic expose, we hope, the logic and the directions of future progress in the area of nonlinear crystals will become apparent.

6.2 Second-order frequency conversion

6.2.1 Basic concepts

Optical nonlinear response can be understood phenomenologically as a nonlinear response of radiating dipoles constituted by valence electrons and ions in the material driven by external electric fields. This response is quantified by a macroscopic polarization (dipole moment per unit volume). The macroscopic polarization $\boldsymbol{P}$ enters into the Maxwell equations through the constitutive relation defining the dielectric displacement vector $\boldsymbol{D}$:

$$\boldsymbol{D} = \varepsilon_0 \boldsymbol{E} + \boldsymbol{P} \qquad [6.1]$$

where $\boldsymbol{E}$ is the external electric field and ε_0 is the dielectric permittivity of vacuum. The polarization then becomes a source term in wave equations derived from the Maxwell equation. Standard derivations can be found in many textbooks, e.g. Butcher and Cotter (1990) where the mks system of units is adopted.

It is common to make an assumption that the macroscopic polarization as a function of time can be expanded in a power series with respect to the external electric field:

$$P(t) = \varepsilon_0 \left[\begin{array}{l} \int_{-\infty}^{\infty} \rho_1(\tau) E(t-\tau) \mathrm{d}\tau + \int_{-\infty}^{\infty} \mathrm{d}\tau' \\ \int_{-\infty}^{\infty} \rho_2(\tau, \tau') E(t-\tau) E(t-\tau')\, \mathrm{d}\tau + \ldots \end{array} \right] \qquad [6.2]$$

where ρ_n are the medium response functions of order n. These response functions are real and satisfy the causality condition, i.e. they are equal to a zero negative time argument. Equation 6.2 includes memory effects of the medium, and thus dispersion, and therefore is suitable not only for monochromatic fields but also for ultrashort and/or spectrally broadband pulses. Here we are interested in the second-order nonlinear response and therefore limit the expansion to the term ρ_2. It is common to use an alternative, Fourier-transformed form of Eq. 6.2 where polarization and external fields are functions of cyclic frequency, ω:

© Woodhead Publishing Limited, 2013

$$P(\omega) = \varepsilon_0(\chi^{(1)}(-\omega;\omega)E(\omega) + \chi^{(2)}(-\omega;\omega',\omega'')E(\omega')E(\omega'') + \ldots \quad [6.3]$$

where $\chi^{(1)}$, $\chi^{(2)}$ are the linear and the second-order susceptibilities of the medium.

The susceptibilities are related to the response functions, ρ_1 and ρ_2, via one-dimensional and two-dimensional Fourier transforms, respectively. As per convention, the frequency arguments in the susceptibility tensor notation reveal the photon energy conservation condition, i.e. the sum of the frequency arguments should be equal to zero. For instance, the second-order term in Eq. 6.3 would correspond to the sum-frequency generation process, $\omega = \omega' + \omega''$. The expansion in Eq. 6.3 simplifies even further if dispersion in the medium can be neglected, in which case the susceptibilities become simple constants. The expansions in Eq. 6.2 and Eq. 6.3 are valid, obviously, only if the series converge, i.e. if the nonlinear response can be treated as perturbation. This often is the case for an electric field that is much weaker than the ionization threshold and at frequencies far away from material electronic and ionic resonances.

According to Neumann's principle, the symmetry group of the physical property described by a tensor such as second-order susceptibility must include all symmetry elements of the point group of the crystal. Brief inspection of the second-order term in Eq. 6.3 immediately tells us that in order to have non-zero second-order response the medium should lack inversion symmetry. There are 21 crystal symmetry classes where the inversion symmetry element is missing. These classes are called acentric or non-centrosymmetric. Symmetry elements in the crystal point symmetry group also contribute by making some of the tensor elements equal to zero. The second-order polarization can be defined from Eq. 6.3 and assuming monochromatic electric fields can be written as follows:

$$(P^{(2)}_{\omega_3})_i = \varepsilon_0 \sum_{j,k}\sum_{\omega} K(-\omega_3;\omega_1,\omega_2)\chi^{(2)}_{\mathrm{ijk}}(-\omega_3;\omega_1,\omega_2)(E_{\omega_1})_j(E_{\omega_2})_k \quad [6.4]$$

where the indices *i, j, k* denote Cartesian components of the electric fields and the polarization, and the summation over frequencies denotes summation over all distinct frequency sets ω_1, ω_2 in the field spectrum satisfying the energy conservation condition $\omega_3 = \omega_1 + \omega_2$. The coefficient $K(-\omega_3;\omega_1,\omega_2)$ takes into account the fact that for some processes some frequencies can be equal, and the terms are indistinguishable. The rule for calculating $K(-\omega_3;\omega_1,\omega_2)$ can be found in Butcher and Cotter (1990), and the values for some second-order processes are given in Table 6.1.

The susceptibility tensors always possess intrinsic permutation symmetry, which, in the context of Eq. 6.4 means that exchanging Cartesian index and corresponding frequency pairs $(j,\omega_1) \leftrightarrow (k,\omega_2)$ would not change the value of the susceptibility, $\chi^{(2)}_{ijk}(-\omega_3;\omega_1,\omega_2) = \chi^{(2)}_{ikj}(-\omega_3;\omega_2,\omega_1)$. Kleiman symmetry

© Woodhead Publishing Limited, 2013

Table 6.1 Second-order nonlinear processes

Process	Energy conservation	Susceptibility	$K(-\omega_3; \omega_1, \omega_2)$
Second harmonic generation (SHG)	$2\omega = \omega + \omega$	$\chi^{(2)}_{ijk}(-2\omega; \omega, \omega)$	1/2
Sum-frequency generation (SFG)	$\omega_3 = \omega_1 + \omega_2$	$\chi^{(2)}_{ijk}(-\omega_3; \omega_1, \omega_2)$	1
Difference frequency generation (DFG)	$\omega_3 = \omega_1 - \omega_2$	$\chi^{(2)}_{ijk}(-\omega_3; \omega_1, -\omega_2)$	1
Optical rectification	$0 = \omega - \omega$	$\chi^{(2)}_{ijk}(0; \omega, -\omega)$	1/2
Pockels effect	$\omega = \omega - 0$	$\chi^{(2)}_{ijk}(-\omega; 0, \omega)$	2

Note: For electrooptic effect and optical rectification, the zero frequency in practice means that the frequency is much lower than the optical frequency. Care should be taken in evaluating susceptibilities for these processes due to the large lattice contribution to the nonlinear response (Flytzanis, 1969; Yablonowitch *et al.*, 1972; Boyd *et al.*, 1971a).

Table 6.2 Correspondence between the indices in second-order susceptibility $\chi^{(2)}_{ijk}$ and the nonlinear coefficient d_{im}

m	1	2	3	4	5	6
j, k	xx	yy	zz	$zy\ yz$	$zx\ xz$	$xy\ yx$

goes even further by assuming that in the absence of absorption, the frequency dispersion of the nonlinear susceptibility can be neglected and therefore one can cyclically permute Cartesian indices without permuting frequencies at the same time. Such a condition allows one to reduce the maximum number of independent tensor elements from 27 to 18. Due to this reduction, it is very common in the context of second-order nonlinear interactions to use a matrix of nonlinear coefficients, d, instead of second-order susceptibility. The d-matrix is defined as follows:

$$\chi^{(2)}_{ijk}(-\omega_3; \omega_1, \omega_2) = 2d_{im}(-\omega_3; \omega_1, \omega_2) \qquad [6.5]$$

where the index $m = 1, 2, ..., 6$ corresponds to the two possible permutation of the last pair of Cartesian coordinates. Table 6.2 shows this correspondence. Table 6.3 lists point-symmetry groups corresponding to the acentric crystals together with the number of independent nonlinear coefficients in the d_{im} matrix and examples of nonlinear crystals corresponding to these groups. For a more visual depiction of the nonlinear coefficient matrices the reader is referred to one of the textbooks in nonlinear optics, e.g. Yariv (1989). The second-order susceptibility tensors, like all other tensors describing physical properties, are always given in an orthogonal Cartesian coordinate

© Woodhead Publishing Limited, 2013

Table 6.3 Crystal symmetry classes, corresponding symmetry groups, and number of independent coefficients without and with assumption of Kleiman permutation symmetry (KPS). The last column lists the nonlinear crystals which can either be obtained from commercial sources or whose further development might result in commercial applications

Symmetry class, point symmetry group	Anisotropy	Maximum number of d_{im}	Number of independent d_{im} with KPS	Nonlinear crystals
Triclinic	Biaxial			
1		18	18	
Monoclinic	Biaxial			
2		8	4: 14 = 25 = 36, 16 = 21, 22, 23 = 34	BiBO
m		10	6: 11, 12 = 26, 13 = 35, 15 = 31, 24 = 32, 33	$BaGa_4Se_7$, DAST*
Orthorhombic	Biaxial			
222		3	1: 14 = 25 = 36	
mm^2		5	3: 15 = 31, 24 = 32, 33	KTP, KTA, RTP, RTA, $KNbO_3$, $AgGaGeSe_2$, LBO, $BaGa_4S_7$, $MgBaF_4$
Tetragonal	Uniaxial			
$\bar{4}$		4	2: 14 = 25 = 36, 15 = −24 = 31 = −32	
4		4	2: 15 = 24 = 31 = 32, 33	
$\bar{4}2m$		2	1: 14 = 25 = 36	KDP, CLBO, $ZnGeP_2$, $CdSiP_2$, $AgGaS_2$, $AgGaSe_2$
4 2 2		1	0	
4 m m		3	2: 15 = 24 = 31 = 32, 33	
Trigonal	Uniaxial			
3		6	4: 12 = 26 = −11, 15 = 24 = 31 = 32, 16 = 21 = −22, 33	
3 2		2	1: 12 = 26 = −11	α-Quartz, Te, YAB

© Woodhead Publishing Limited, 2013

Table 6.3 Continued

Symmetry class, point symmetry group	Anisotropy	Maximum number of d_{im}	Number of independent d_{im} with KPS	Nonlinear crystals
3 m		4	3: 15 = 24 = 31 = 32, 16 = 21 = −22, 33	$LiNbO_3$, $LiTaO_3$, BBO
Hexagonal				
$\bar{6}$		2	2: 12 = 26 = −11, 16 = 21 = −22	
$\bar{6}$2m		1	1: 16 = 21 = −22	GaSe
6		4	2: 15 = 24 = 31 = 32, 33	$LiIO_3$
6 2 2		1	0	
6 m m		3	2: 15 = 24 = 31 = 32, 33	CdSe, CdS, GaN
Cubic	Isotropic			
2 3		1	1: 14 = 25 = 36	
4 3 2		0	0	
$\bar{4}$3m		1	1: 14 = 25 = 36	GaAs, GaP, InP, InAs, InSb, GaSb, ZnTe, CdTe, GaN

*DAST = 4-*N,N*-dimethylamino-4′-*N*′-methyl-stilbazolium tosylate.

system, which has to be referenced to the crystallographic axes of a particular crystal. The convention which is used for this purpose can be found in the IEEE standard for piezoelectric crystals (IEEE Std 176-1949), which was subsequently confirmed in 1987 (IEEE/ANSI Std 176-1987). In 1992, a small revision of IEEE/ANSI Standard 176 was proposed to better specify conventions for nonlinear crystals from the orthorhombic class (Roberts and Meeker, 1992).

6.2.2 Phase-matching considerations

The induced polarization plays the role of a source in coupled-wave equations describing three-wave mixing (TWM) processes listed in Table 6.1. Our goal in this subsection is to define the conditions for an efficient frequency conversion process and design figure of merit (FOM) which could be used for comparison of different nonlinear materials.

© Woodhead Publishing Limited, 2013

Consider a TWM process satisfying the energy conservation condition:

$$\omega_1 = \omega_2 + \omega_3 \qquad [6.6]$$

This condition is satisfied by all the processes listed in Table 6.1. For instance, it can correspond to a sum-frequency generation (SFG) where two photons at frequencies ω_2 and ω_3 combine to produce a single photon at frequency ω_1. The reverse process of photon splitting is also possible, where the highest-energy photon at frequency ω_1 is split into two lower-energy photons at frequencies ω_2 and ω_3. The splitting or downconversion process can be 'stimulated' if two waves, say those at frequencies ω_1 and ω_3, are simultaneously present and overlapping in space. This is the difference frequency generation (DFG) process, the result of which is reduction of the number of photons at ω_1, the pump wave, and an increase by an equal number of photons in the wave at frequency ω_3, which is called the signal wave, and at frequency ω_2, the newly generated wave, which is called the idler wave. In this way the signal wave is coherently amplified. The processes are schematically illustrated in Fig. 6.1. The parametric gain is used in optical parametric amplifiers. In optical parametric generators (OPG), on the other hand, the DFG process is seeded by quantum noise, so only a single pump wave is present at the input of a crystal. An optical parametric oscillator (OPO) is essentially an OPG process realized inside a resonant cavity. Due to the quantum optical nature of the initiation of the OPG process, classical coupled-wave equations fail to describe the physics relevant at low pump powers. For quantum optical treatment of the parametric downconversion process the reader should refer to the pioneering papers on that subject (Louisell *et al.*, 1961; Mollow and Glauber, 1967).

Strict conservation of the photon numbers in the nonlinear interaction is expressed by Manley–Rowe relations, which, using the definition of Eq. 6.6, read:

$$-\frac{\mathrm{d}}{\mathrm{d}z}\frac{I_1}{\omega_1} = \frac{\mathrm{d}}{\mathrm{d}z}\frac{I_2}{\omega_2} = \frac{\mathrm{d}}{\mathrm{d}z}\frac{I_3}{\omega_3} \qquad [6.7]$$

where I_j, $j = 1, 2, 3$, are the intensities of the wave at frequencies ω_j and z is the coordinate along the wave propagation direction.

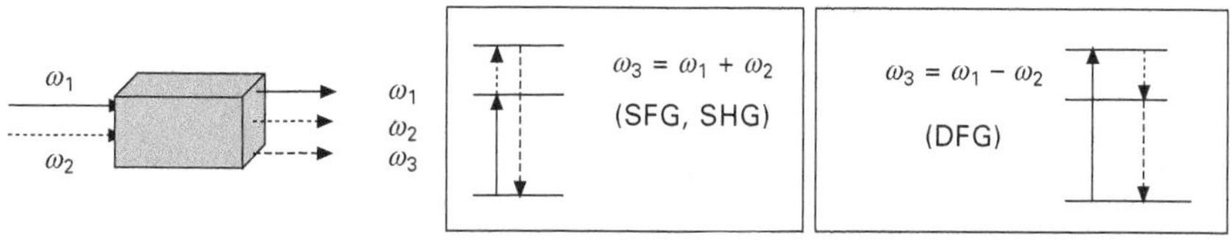

6.1 Schematic illustration of three-wave mixing processes in second-order nonlinear medium. Second-harmonic generation (SHG) is a degenerate case of SFM, where the waves ω_1 and ω_2 are spectrally and spatially degenerate.

© Woodhead Publishing Limited, 2013

In TWM as in all scattering processes the efficiency crucially depends on momentum conservation, i.e. the wave-vector mismatch:

$$\Delta \boldsymbol{k} = \boldsymbol{k}_1 - \boldsymbol{k}_2 - \boldsymbol{k}_3 \qquad [6.8]$$

The wave-vector magnitude by definition is $k_j = n_j\omega_j/c$ (n_j is the index of refraction), and the direction is perpendicular to the dielectric displacement vector $\boldsymbol{D}$.

If the wave-vector mismatch $\Delta\boldsymbol{k} \neq 0$ and the interaction length becomes longer than the characteristic coherence length:

$$L_{\text{coh}} = \pi/|\Delta \boldsymbol{k}| \qquad [6.9]$$

then the power flow in the TWM process reverses direction, e.g. resulting in back-conversion of the signal and the idler waves to the pump. Therefore high efficiency of the TWM process requires phase-matching, $\Delta\boldsymbol{k} = 0$. Consider a TWM process (DFG, SFG, SHG) where two waves at ω_2 and ω_3 are much stronger than the third one at ω_1 and assume that depletion is not important. If a nonlinear crystal of length L contains a one-dimensional structure of nonlinear coefficient distributed along the x-axis coinciding with the direction of wave-vectors, $d(x) = d_{ij}g(x)$, where $g(x)$ can take values of ± 1 inside the crystal and is zero elsewhere, then the generated wave will be:

$$E_1(L) = -i\,\frac{\omega_1}{n_1 c}\,d_{jm}E_2E_3\int_{-\infty}^{\infty} g(x)\exp(i\Delta kx)\mathrm{d}x \qquad [6.10]$$

It is clear from Eq. 6.10 that the field magnitude at the sum frequency will be proportional to the Fourier transform of the spatial distribution of the nonlinear coefficient. For a homogeneous crystal, i.e. $g(x) = 1$, for $-L/2 \leq (x) \leq L/2$, the generated intensity will be:

$$I_1(L) = \frac{2\omega_1^2}{n_1 n_2 n_3 c^3 \varepsilon_0}\,d_{jm}^2 I_2 I_3 L^2\ \mathrm{sinc}^2(\Delta kL/2) \qquad [6.11]$$

There are two main methods for achieving phase matching: (1) by exploiting birefringence which is characteristic for all acentric nonlinear crystals except those in the cubic crystallographic class (Table 6.3), and (2) by employing quasi-phase-matching (QPM) as was first proposed in Armstrong *et al.* (1962).

A simple case of birefringence phase-matching (BPM) in a positive uniaxial crystal (where the difference between the extraordinary and ordinary indices of refraction $n^{\text{e}} - n^{\text{o}} > 0$) with normal group velocity dispersion is illustrated in Fig. 6.2(a). In general in uniaxial crystals two types of phase-matching are determined, type-I and type-II. Type-I employs ω_2 and ω_3 (cf. Eq. 6.6) of the same polarization, so the processes oo-e or ee-o are possible. By convention, the type-II processes employ orthogonally polarized waves at

© Woodhead Publishing Limited, 2013

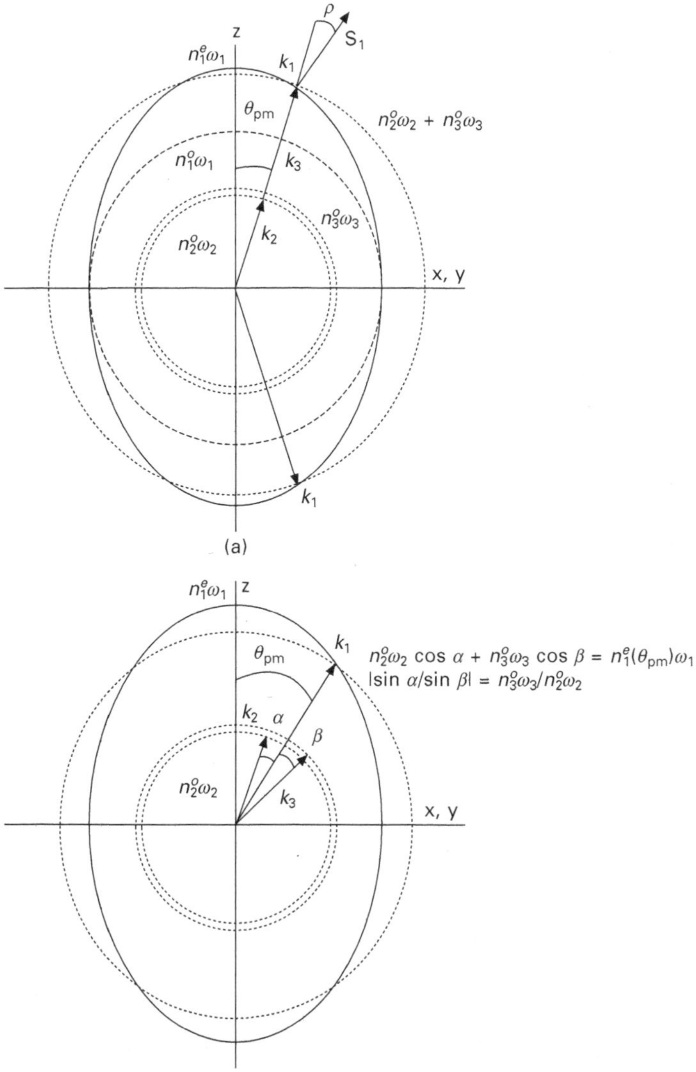

6.2 (a) Collinear (scalar) type-I oo-e BPM in positive uniaxial crystal. (b) Noncollinear (vector) type-I BPM in the same material.

ω_2 and ω_3, therefore there are four possible combinations of polarizations. Figure 6.2(a) illustrates collinear or scalar type-I oo-e BPM. The $\Delta \boldsymbol{k} = 0$ condition will be satisfied at the polar angles θ_{pm}, where the extraordinary index ellipsoid $n_1^e\omega_1$ crosses the circumference with a radius of $n_2^o\omega_2 + n_3^o\omega_3$. The two phase-matching directions located symmetrically with respect to the x–y plane as illustrated in Fig. 6.2(a) in 3D geometry would correspond to

© Woodhead Publishing Limited, 2013

two symmetrically located phase-matching loci. Unless the waves propagate along the indicatrix axes, the wave-vector and the Poynting vector directions of the extraordinary-polarized wave ($\mathbf{S}_1$ on Fig. 6.2(a)) will be spatially diverging with the walk-off angle:

$$\rho = -\frac{1}{n(\theta)}\frac{\mathrm{d}n(\theta)}{\mathrm{d}\theta} \qquad [6.12]$$

This is the case of critical phase-matching where the spatial walk-off limits the nonlinear interaction length and introduces beam astigmatism. On the other hand, the walk-off is essential for achieving high beam quality in large-Fresnel number OPO ring resonators with image rotation (Armstrong and Smith, 2003). At some specific wavelengths the phase-matching direction is parallel to one of the indicatrix axes for biaxial crystals or is propagating perpendicular to the optical axis in uniaxial crystals (in the x–y plane in Fig. 6.2). This geometry is called noncritical phase-matching and is usually the preferred geometry of interaction. Substantial effort has been devoted in developing solid-solutions for some crystal systems, where the noncritical phase matching wavelength could be tuned by changing the crystal composition (Badikov *et al.*, 2009). The BPM can be realized in noncollinear or so-called vectorial configuration and is shown in Fig. 6.2(b). The equations given in the figure allow calculation of the noncollinear angle α (or β) and the phase matching polar angle, θ_{pm}.

In biaxial crystals the situation is substantially more complicated and analytical expressions exist only for BPM when the interaction takes place in principal planes. For more detailed reference the reader can consult nonlinear optics handbooks, e.g. Sutherland (1996). The formulae for calculating d_{eff} for different symmetry classes can be found in Sutherland (1996) and Dimitriev *et al.* (2010).

6.2.3 Quasi-phase-matching second-order interactions

QPM, in the simplest and most useful embodiment, is realized by periodically inverting the sign of the nonlinear coefficient for every coherence length L_{coh}. This is equivalent to periodically adding a phase-shift π into the TWM interaction, thus preventing the reversal of the energy flow direction. The momentum conservation condition for the QPM case is

$$\Delta k' = \boldsymbol{k}_1 - \boldsymbol{k}_2 - \boldsymbol{k}_3 - \boldsymbol{K}_g = 0 \qquad [6.13]$$

where the additional QPM structure vector,

$$\boldsymbol{K}_g = \pi/L_{\mathrm{coh}} = 2\pi/\Lambda \qquad [6.14]$$

is a design parameter which assures momentum conservation. Different possibilities in the simplest one-dimensional QPM structure are shown

© Woodhead Publishing Limited, 2013

in Fig. 6.3. For the simplest case of collinear interaction the QPM period required for achieving the wavevector mismatch compensation can be directly calculated from Eq. 6.13:

$$\Lambda = \frac{2\pi c}{n_1\omega_1 - n_2\omega_2 - n_3\omega_3} \quad [6.15]$$

QPM gives substantially more freedom in designing the interaction geometry, including counter-propagating interactions (Canalias *et al.*, 2005; Canalias and Pasiskevicius, 2007) which are virtually impossible to realize by using BPM. Other crucial advantages of the QPM nonlinear frequency conversion are: (a) the possibility of realizing noncritical interaction throughout the transparency range of nonlinear material; (b) the possibility of employing the highest nonlinearity, and if that happens to be a diagonal element of the second-order susceptibility, it cannot be exploited in BPM; (c) the possibility of realizing simultaneous QPM interactions using several different nonlinear coefficients (Pasiskevicius *et al.*, 2002); (d) the possibility of designing the

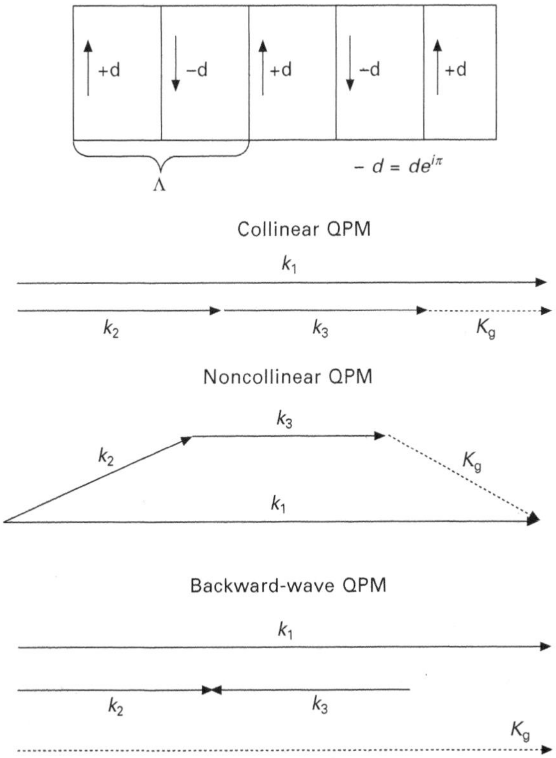

6.3 Sketch of a 1D QPM structure with inverted nonlinear coefficients and illustration of the types of QPM which can be realized in this simple case.

© Woodhead Publishing Limited, 2013

interaction bandwidth (Arbore *et al.*, 1997); and (e) the possibility of achieving efficient frequency conversion in optically isotropic materials such as glass (Kashyap *et al.*, 1994) or III–V semiconductors (Lallier *et al.*, 1998).

The effective nonlinear coefficient of a QPM structure can be determined by Fourier expansion of the spatial modulation function $g(x)$:

$$g(x) = \sum_{p=-\infty}^{\infty} G_p \exp(-iK_p x) \qquad [6.16]$$

and inserting it into Eq. 6.10 (Fejer *et al.*, 1992). For a one-dimensional periodic modulation with a period of Λ and a duty cycle $D = l/\Lambda$, the Fourier coefficients are

$$G_p = \frac{2}{\pi p} \sin(\pi D p) \qquad [6.17]$$

and the reciprocal lattice vector is

$$K_p = \frac{2\pi p}{\Lambda} \qquad [6.18]$$

The index p signifies the order of the quasi-phase-matching. For a given TWM process with frequencies of the interacting waves varying only slightly around their central frequencies ω_2, ω_3 it is possible to find a single reciprocal vector K_p which would maximize the output at ω_1. Namely, the structure should be designed to satisfy the QPM condition $\Delta k = K_p$ preferably using the first-order QPM ($p = 1$) as this also maximizes the Fourier coefficient in Eq. 6.17. The SFM intensity at the end of the QPM structure can be expressed as:

$$I_1(L) = \frac{2\omega_1^2}{n_1 n_2 n_3 c^3 \varepsilon_0} d_{jm}^2 \left(\frac{2}{\pi p}\right)^2 \sin^2(\pi p D) I_2 I_3 L^2 \mathrm{sinc}^2((\Delta k - K_p)L/2) \qquad [6.19]$$

It is evident that the QPM interaction is formally analogous to the birefringence phase matched process (Eq. 6.19), though with the nonlinear coefficient reduced by a factor of $(2/(\pi p))\sin(\pi p D)$. The benefits offered by having an additional design parameter, K_p, in most cases far outweigh the drawback of the reduced nonlinear response for QPM structures. Figure 6.4(a) shows SHG power generated in non-phase-matched QPM and in a homogeneous crystal where it is assumed that a perfect BPM can be achieved with the same nonlinear coefficient d_{jm}. The effects of reduced effective nonlinear coefficient in the QPM case are clearly seen. In reality, however, the QPM enables exploiting the largest nonlinear coefficient inaccessible for BPM. Such a case is shown in Fig. 6.4(b), where the noncritically birefringence

© Woodhead Publishing Limited, 2013

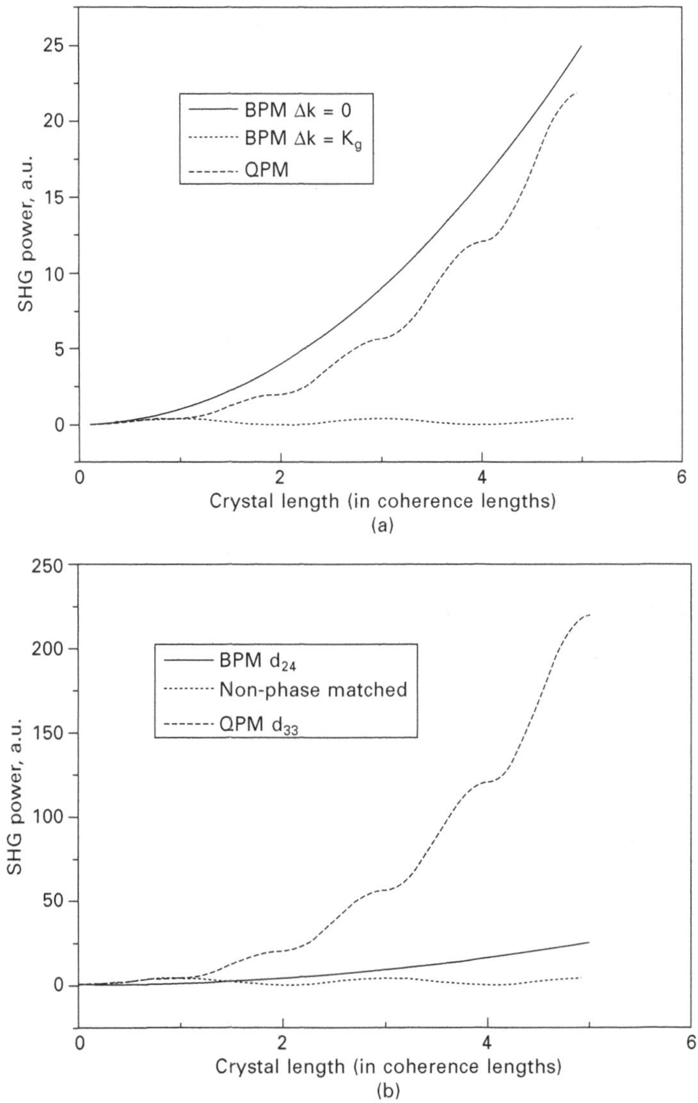

6.4 Power of second-harmonic as a function of crystal length for birefringence phase-matched, quasi-phase-matched and non-phase-matched cases (a) when all processes employ the same nonlinear coefficient, and (b) in a more frequent case when QPM enables utilization of the highest diagonal nonlinear coefficients such as in PPKTP, PPLN, etc.

phase-matched SHG in $KTiOPO_4$ (KTP) for the coefficient d_{24} is compared with the QPM in periodically poled $KTiOPO_4$ (PPKTP), where the largest d_{33} coefficient can be used.

© Woodhead Publishing Limited, 2013

Due to the fact that the momentum conservation condition is altered by changing the beam incidence angles, or the wave frequency or crystal temperature, one can determine the acceptance bandwidth with respect to each of these parameters. An example of the effect of temperature change of 8 mm-long PPKTP designed for frequency doubling 1064 nm laser output is shown in Fig. 6.5. The acceptance bandwidth with respect to any parameter ψ can be calculated (Fejer *et al.*, 1992):

$$\delta\psi = \frac{5.57}{L}\left|\frac{\partial \Delta k'}{\partial \psi}\right|^{-1} \qquad [6.20]$$

where L is the crystal length and $\Delta k'$ is given by Eq. 6.13.

Finally, it is useful to define a measure which would allow comparison of these crystals in terms of performance as second-order frequency converters. From Eq. 6.11 and Eq. 6.19 a figure of merit (FOM) determining the strength of nonlinear coupling can be defined:

$$\text{FOM} = d_{\text{eff}}^2/n^3 \qquad [6.21]$$

This definition, although somewhat arbitrary, is useful for comparing materials operating in a similar spectral range.

6.3 Nonlinear crystal development

Spatial and temporal coherence of laser radiation enables generation of high peak power and high intensity beams suitable for realizing efficient

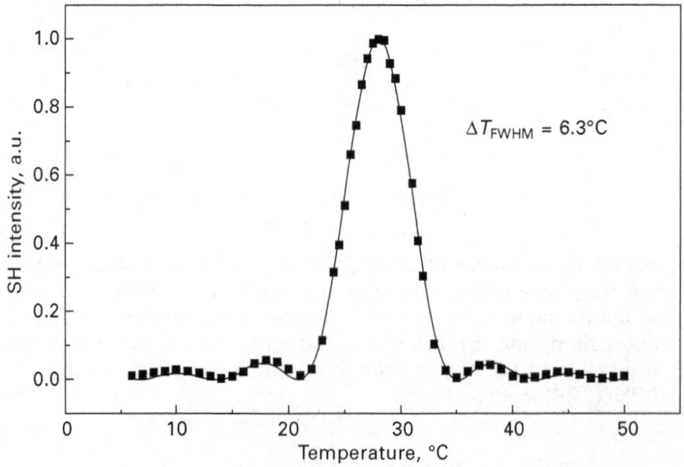

6.5 Temperature acceptance bandwidth in 8-mm–long PPKTP for frequency doubling of 1064 nm laser radiation. Solid line = theory (Eq. 6.24), squares = experimental data.

© Woodhead Publishing Limited, 2013

nonlinear optical interactions. Indeed, just a year after the invention of the laser, Franken *et al.* demonstrated second harmonic generation (SHG) in quartz (Franken *et al.*, 1961). This experiment marked the birth of the field of nonlinear optics. By employing second-order parametric upconversion and downconversion processes, wavelength ranges unreachable by laser sources became available.

The low efficiency issue of the first experiment had been solved very quickly. In the 1962 papers by P. A. Giordmaine, and by P. D. Maker *et al.*, published in the same issue of *Physical Review Letters* (Giordmaine, 1962; Maker *et al.*, 1962), the authors pointed out the need for phase matching and demonstrated birefringence phase-matching in KH_2PO_4 (KDP). The paper by Maker contains the first demonstration of what has become known as the Maker fringe technique used to this day for characterizing second-order nonlinear coefficients in new nonlinear materials. The seminal paper by J. A. Armstrong and coworkers, published in the same year (Armstrong *et al.*, 1962), provided a very detailed theoretical framework for treatment of nonlinear interactions of electromagnetic waves in dielectric media, including QPM structures. There, accurate solutions of the coupled-wave equations for TWM in the plane-wave approximation were obtained. G. D. Boyd and D. A. Kleinman extended the theory to the interactions of more realistic focused Gaussian beams (Boyd and Kleinman, 1968).

Over the past 50 years a large number of second-order nonlinear crystals have been synthesized and demonstrated (Dimitriev *et al.*, 2010). The prolific bibliography of nonlinear crystals is somewhat misleading, however, as the majority of synthesized crystals have never reached beyond the initial characterization stage. From the list of birefringence phase-matched nonlinear crystals developed before the 1980s there remain only the two oldest compounds, KDP (Mitsui and Furuichi, 1953) and crystalline quartz, which remain of practical importance today due to their excellent optical transmission in the ultraviolet and the possibility of growing these crystals to unprecedented large sizes, suitable for frequency conversion of laser beams employed in inertial fusion experiments (Sasaki and Yokotani, 1990). It should be stressed that in the development of nonlinear materials the economic considerations have very strong weight.

Significant breakthroughs were achieved in developing novel birefringence phase-matched nonlinear crystals in the 1980s and 1990s by successful synthesis and large crystal growth of materials from the borate family, most notably of β-BaB_2O_4 (BBO) (Chen *et al.*, 1984; Jiang *et al.*, 1986), LiB_3O_5 (LBO) (Chen *et al.*, 1989), $CsLiB_6O_{10}$ (CLBO) (Mori *et al.*, 1995) and BiB_3O_6 (BiBO) (Becker *et al.*, 1999). These crystals are characterized by good transparency in the ultraviolet below 300 nm, high optical damage threshold, relatively high nonlinearities as compared to KDP, and the possibility of birefringence phase-matching from ultraviolet to near-infrared. Due to such

© Woodhead Publishing Limited, 2013

advantageous properties, these materials are broadly used today in a large variety of commercial devices.

The QPM techniques make the second-order nonlinear interactions engineerable and independent of refractive index dispersion or birefringence. The initial article on the QPM technique (Armstrong *et al.*, 1962) contained several proposals of how this could be achieved. The simplest embodiment was QPM by periodic total internal reflection of the interacting waves. Historically this was the first method to be demonstrated experimentally. In 1966 Boyd and Patel employed total internal reflection in ZnSe and GaAs plates to demonstrate enhanced second-harmonic generation of the recently invented CO_2 laser (Boyd and Patel, 1966). This apparently simple QPM method suffers from a couple of drawbacks, one of them being surface scattering losses at each reflection, the other the so-called Goos–Hänchen shift at the crystal–air boundary which spatially separates the interacting beams after several reflections (Kogelnik and Weber, 1974). Although this QPM method was used later on a few occasions for frequency conversion in the mid-infrared spectral region (Komine *et al.*, 1998; Haidar, 2006), the above-mentioned drawbacks and the emergence of better alternatives make this method less relevant.

A substantially more versatile QPM technique employs spatially structured second-order nonlinearity in dielectrics. Initially, the spatial structuring was attempted by stacking 180°-rotated thin plates having lengths equal to or an odd multiple of the coherence length. Such a QPM method has been demonstrated in nominally non-birefringent semiconductors, CdTe (Pitch *et al.*, 1976) and GaAs (Szilagyi *et al.*, 1976; Thompson *et al.*, 1976), as well as in crystalline quartz, $LiNbO_3$ (Okada *et al.*, 1976) and LiB_3O_5 (LBO) (Mao *et al.*, 1992). Diffusion-bonding of thin plates was later demonstrated in GaP (Okuno *et al.*, 1997) and GaAs (Lallier *et al.*, 1998) as a way to reduce Fresnel reflectivity losses inherent in multi-stack designs. Due to practical difficulties involved in handling nonlinear crystal plates thinner than about 100 μm, the diffusion bonding technique might only be viable for the long-wavelength mid-infrared and far-infrared spectral ranges.

The current trend in second-order nonlinear optical materials is characterized by a shift of the focus from the synthesis of new compounds and towards the development of nonlinear structures with spatially engineered nonlinearity as well as improvement of the material quality of the crystals where such structuring can be accomplished. This trend established itself starting from the 1990s when reliable and well-controlled techniques for structuring of second-order nonlinearity were developed in oxide ferroelectrics, $LiNbO_3$, $LiTaO_3$ and $KTiOPO_4$ (KTP) isomorphs, synthesized in the 1980s (Zumsteg *et al.*, 1976; Bierlein and Vanherzeele, 1989), and later in semiconductor GaAs. The first three are ferroelectric crystals where the sign of the second-order nonlinear coefficient is reversed by inverting the polarity of spontaneous

© Woodhead Publishing Limited, 2013

polarization. The ferroelectric properties of $LiNbO_3$ and $LiTaO_3$ had been well known before the invention of lasers (Matthias and Remeika, 1949). For a long time $LiNbO_3$ was considered a 'frozen ferroelectric' due to the fact that the controlled reversal of spontaneous polarization was difficult to achieve. Improvements in the crystal quality by using Czochralski growth allowed observation and identification of the ferroelectric domain structure (Nassau *et al.*, 1966). An important breakthrough happened when it was demonstrated that $LiNbO_3$ crystals with periodically inverted ferroelectric domains could be produced directly during Czochralski growth by keeping the growing crystal at an angle with respect to the temperature field (Feng *et al.*, 1980). A similar procedure was later demonstrated for $LiTaO_3$ (Wang *et al.*, 1986).

The real breakthrough came when the periodic poling of ferroelectric crystals was demonstrated. This technique allows accomplishing the design and structuring of second-order nonlinearity by using standard lithography techniques. Therefore this technique offers high adaptability to specific requirements of an application. Prior to the optical applications, the periodic structuring of ferroelectrics by using an interdigital finger electrode pattern had been demonstrated in the field of ultrasonic transducers (Nakamura and Shimizu, 1983). Periodic poling using diffusion and thermal treatment was later studied for the fabrication of nonlinear optical waveguides in $LiNbO_3$ and KTP (Webjörn *et al.*, 1989, Lim *et al.*, 1989; Van der Poel *et al.*, 1990). In 1993 Yamada and coworkers (Yamada *et al.*, 1993) achieved a breakthrough in small-period electric-field poling in $LiNbO_3$ waveguides which were intended for blue light generation in the first-order QPM process. This immediately opened up opportunities for using periodically poled $LiNbO_3$ (PPLN) and PPKTP waveguides for frequency conversion of low-power laser diodes.

Periodically poled nonlinear materials started receiving substantially larger interest when the structuring technology was demonstrated for bulk crystals. Owing to engineerability of the QPM structures as well as large nonlinearities they became potential replacements for BPM crystals. Periodic electric-field poling of Czochralski-grown bulk congruent $LiNbO_3$ (CLN) was reported in 1994 (Webjörn *et al.*, 1994). The thickness of the structures was initially limited to 0.5 mm due to a large coercive field in CLN (~21 kV/mm). A better understanding of the ferroelectric domain kinetics during polarization inversion was required to achieve periodicities below 20 μm in 0.5 mm-thick PPLN. Miller and coworkers demonstrated a substantial improvement in poling PPLN structures with sub-10 μm periodicities (Miller *et al.*, 1996). A backswitching method was investigated in CLN in order to reduce the QPM periodicity even further (Batchko *et al.*, 1999).

Approximately at the same time as in CLN, the electric-field poling of the bulk hydrothermally grown KTP was demonstrated, producing QPM structures for green light generation (Chen and Risk, 1994). The coercive field in KTP

© Woodhead Publishing Limited, 2013

is an order of magnitude lower than in CLN, so it is substantially easier to fabricate QPM structures thicker than 0.5 mm. These initial demonstrations used hydrothermally grown KTP wafers. Hydrothermal growth results in good-quality homogeneous KTP crystals but this growth process is very slow and has to proceed at high pressures, so, eventually, commercial crystals were predominantly grown by a flux method. This growth method produces crystals with large ionic conductivity, making standard poling methods difficult to apply. Eventually, the periodic poling techniques were developed for this material, including a Rb-exchange-assisted electric field poling (Karlsson and Laurell, 1997) and a low-temperature poling (Rosenman *et al.*, 1998).

Over the years other ferroelectric crystals have been explored for their potential as QPM nonlinear media. For instance, $BaTiO_3$, a well-known photorefractive crystal, was one of the first materials to be poled (Miller and Savage, 1959), but it took 40 years before an optical parametric oscillator was demonstrated in periodically poled $BaTiO_3$ structure (Setzler *et al.*, 1999). Another photorefractive crystal, $Sr_{0.6}Ba_{0.4}Nb_2O_6$ (SBN), was used to demonstrate quasi-phase matched SHG (Horowitz *et al.*, 1993; Zhu *et al.*, 1997). Great hopes for blue light generation were associated with $KNbO_3$, a crystal with a complicated ferroelectric domain structure. So far, however, it has proved to be too prone to self-restructuring during nonlinear interaction (Meyn *et al.*, 1999; Hirohashi *et al.*, 2004; Hirohashi and Pasiskevicius, 2005). A substantial effort has been devoted to developing self-frequency-doubling laser crystals, the most successful being $YAl_3(BO_3)_4$ (YAB), which can be doped with rare-earth active ions (Filimonov *et al.*, 1974; Wang *et al.*, 1999). YAB is isostructural to quartz and therefore is an interesting crystal for further development of QPM structures. Indeed, QPM has been observed in Yb:YAB due to natural piezoelastic twinning (Dekker and Dawes, 2006).

6.4 Nonlinear crystals: current status and future trends

From Eqs 6.19 and 6.11 it is apparent that TWM efficiency is proportional to the frequencies of the generated waves. However, at lower frequencies this dependence can be well compensated by higher nonlinearity. From the definition of FOM, Eq. 6.21, it might appear that there is a strong inverse dependence on the index of refraction. However, as noted by Miller (1964), the ratio of the second-order susceptibility and the cube of the first-order susceptibility is a quantity (called Miller's delta) which changes surprisingly little over many material classes. For deeper insights the reader is referred to Jackson *et al.* (1997) and references therein. Considering that the Miller's delta, δ_{jm}, is a slowly changing quantity and expressing the second-order nonlinear coefficient through it, the FOM then becomes very strongly directly dependent on the index of refraction:

© Woodhead Publishing Limited, 2013

$$\text{FOM} \approx \delta_{jm}^2 (n^2 - 1)^6 / n^3 \approx \delta_{jm}^2 n^9 \quad [6.22]$$

If we next use the approximate dependence (Moss, 1985) between the index of refraction and the electronic bandgap, $n \propto E_g^{-1/4}$, then

$$\text{FOM} \approx \delta_{jm}^2 / E_g^{9/4} \quad [6.23]$$

and therefore we should expect a strong increase in the efficiency of TWM in nonlinear crystals with smaller bandgaps. The increased nonlinearity in mid-infrared materials more than compensates the decrease of the coupling due to the decreasing frequency of the generated waves. This dependence of the FOM on the bandgap for a number of crystals used in the mid-infrared is investigated in detail in a recent paper (Petrov, 2011). Clearly, the FOM is not the only consideration in choosing a nonlinear crystal. Other properties can prevent it from being adopted, for instance the presence of parasitic effects such as two-photon absorption and/or absorption by native defects or inadvertent impurity doping during growth.

In Table 6.4 we list the most widely used nonlinear crystals, the vast majority of which are commercially available. The crystals in the table are ranked in the order of increasing FOM. In accordance with Eq. 6.23, the crystals

Table 6.4 Transparency range, maximum effective nonlinear coefficient and figure of merit for second-order frequency conversion in most commonly used and commercially available crystals

Crystal	Transparency, μm	Maximum* d_{eff}, pm/V	FOM $(pm/V)^2$	QPM demonstrated
DAST	0.6–6.7	290	4840	no
GaP	0.6–25	117	458	yes
$CdSiP_2$	0.56–6.8	90	277	no
$ZnGeP_2$	1.5–12	75.4	182	no
GaSe	0.65–18	54	166	no
GaAs	0.876–13	60	100	yes
$LiNbO_3$	0.33–5.5	15.9	24	yes
KTP	0.35–3.6	10	17	yes
RTP	0.35–3.6	10	17	yes
KTA	0.35–4	10	17	yes
$AgGaS_2$	0.5–13	13	12.2	no
$LiTaO_3$	0.28–5.5	8.2	7.3	yes
BiBO	0.286–2.5	2.8	1.34	no
BBO	0.185–2.6	2.2	1.11	no
GaN	0.365–13.6	3.4	0.95	yes
CLBO	0.18–2.75	0.74	0.18	no
LBO	0.16–2.6	0.85	0.18	no
KDP	0.17–1.7	0.43	0.057	no
Quartz	0.14–2.5	0.35	0.045	yes
$MgBaF_4$	0.17– 10	0.06	0.0013	yes

*For the materials where QPM has been demonstrated the maximum effective nonlinear coefficient is given as $d_{\text{eff max}} = 2d_{\text{jm max}}/\pi$.

© Woodhead Publishing Limited, 2013

at the top are those developed for far-infrared and mid-infrared frequency conversion. In that spectral area semiconductors are widely used, and DAST, the only organic crystal, is here an exception. DAST was initially intended for electrooptic modulation of lasers at telecommunications wavelengths (Pan *et al.*, 1996), but its large electrooptic coefficients also meant that the nonlinear coefficients for DFG in the far-infrared are also very large and could be exploited for THz generation (Tang *et al.*, 2011).

The mid-infrared spectral range, somewhat arbitrarily defined as that stretching from 2 μm to 13 μm, is the range where active research and synthesis of new materials is happening at the moment, especially in the II–IV–V_2 group of semiconductors such as $ZnGeP_2$ (ZGP) (Ray *et al.*, 1969; Boyd *et al.*, 1971b) and $CdSiP_2$ (CSP) (Schunemann *et al.*, 2008). Such crystals have very high nonlinearities and can be grown to large sizes and therefore have certain advantages over QPM structures. ZGP crystal, which has been known since the 1970s, has a relatively narrow bandgap and thus must be pumped at wavelengths above 2 μm. The recent arrival, CSP crystal, first grown by Shunemann and coworkers at BAE systems (Schunemann *et al.*, 2008), has a relatively large bandgap and therefore can be pumped by well-established and high-power lasers at 1 μm (Petrov *et al.*, 2009).

Of the far-infrared and mid-infrared crystals at the top of Table 6.4 there are only two materials which have been employed for making QPM structures. QPM structures fabricated by diffusion bonding of GaP plates have been used for THz generation using the DFG process (Tomita *et al.*, 2006). Due to the fact that the coherence length for the DFG process into THz is rather long, of a few hundred microns in both GaP and GaAs, it is actually beneficial to employ the diffusion bonding process instead of lithographical patterning and overgrowth processes. Such processes, however, are necessary for the frequency converters in the mid-infrared spectral region. Much progress has occurred over the past decade in periodic structures fabricated in GaAs. Improving on the earlier technique of stacking and bonding of GaAs plates having opposite orientations of the $\chi^{(2)}$ sign (Lallier *et al.*, 1998), two slightly different methods were developed, in which epitaxial growth over an orientationally patterned GaAs (OP-GaAs) template was exploited (Eyres *et al.*, 2001; Faye *et al.*, 2007). The key to both methods was the availability of fast epitaxial regrowth using hydride vapour phase epitaxy (HVPE) over the GaAs template. This template can be produced either by molecular beam epitaxy growth of a thin Ge layer followed by an inverted layer of GaAs with subsequent steps of lithography and chemical etching enabling very precise control over the periodicity and the duty cycle; or alternatively by bonding of two GaAs plates with opposite orientations and subsequent steps of mechanical polishing, lithography and etching. Both methods produce very high-quality structures and the HVPE GaAs layer

© Woodhead Publishing Limited, 2013

shows very low absorption losses in the near-infrared: lower, in fact, than that in the original GaAs substrate. OP-GaAs bulk structures with thickness exceeding 1 mm can be produced by these methods.

In terms of FOM and nonlinearity of the materials used to generate mid-infrared radiation by parametric downconversion, the exception is the semiconductor crystal $AgGaS_2$. This material, although inferior in terms of nonlinearity and thermal conductivity with respect to ZGP and CSP, is included in Table 6.4 due to the fact that it is rather broadly commercially available.

Further down the list in Table 6.4 there are oxide ferroelectric materials primarily employed for QPM interactions. These structures can be designed for generating frequency-converted output anywhere from 350 nm using periodically poled MgO-doped stoichiometric $LiTaO_3$ (PPSLT) and down to 5 μm by employing periodically poled $LiNbO_3$ (PPLN). Quick progress in the structuring of the most promising QPM materials, congruent $LiNbO_3$ (CLN) and KTP, made it evident that these crystals should be further improved in order to reduce the effects of photorefraction and blue- and green-light induced infrared absorption (BLIIRA, GRIIRA) (Wang *et al.*, 2004, Hirohashi *et al.*, 2007). Previous research, which focused on CLN as a potential material for photorefractive components, revealed that doping with about 5% MgO substantially reduces photorefraction in this crystal (Zhong *et al.*, 1980). MgO:$LiNbO_3$ was successfully poled and it was also shown that the coercive field in this material is reduced down to about 4.5 kV/cm (Kuroda *et al.*, 1996). This fact allowed fabrication of MgO:PPLN structures having thickness of 5 mm, therefore allowing handling of substantial optical powers (Ishizuki and Taira, 2005). A threshold in MgO doping concentration for reduction of the photorefraction as well as reduction of GRIIRA was indeed confirmed and attributed to the reduction of native defects, Nb in Li, when Mg is incorporated on Li sites (Furukawa *et al.*, 2001). Further reduction in coercive field in $LiNbO_3$ and $LiTaO_3$ was achieved by increasing [Li] concentration in the crystals to the extent of making them close to stoichiometric. Three methods were demonstrated to achieve this goal successfully: a double crucible Czochralski growth (Kitamura *et al.*, 1992), Li vapour-phase equilibration (Katz *et al.*, 2004) and top-seeded solution growth (Polgár *et al.*, 1997). Strong photorefraction (Kitamura *et al.*, 1997), however, prevented this material from wider applications in frequency conversion. Near-stoichiometric $LiNbO_3$ (SLN) and near-stoichiometric $LiTaO_3$ (SLT), additionally doped with MgO in order to reduce photorefraction, became new materials for QPM structure fabrication using electric-field poling and capable of handling substantial powers in the visible spectral range. Recently, a successful periodic poling of 5-mm-thick congruent $LiTaO_3$ heavily (7 at%) doped with MgO has been achieved at elevated temperatures, making this material an interesting candidate for high-power applications (Ishizuki and

© Woodhead Publishing Limited, 2013

Taira, 2010). Efforts to further reduce absorption due to inadvertent doping during growth are continuing (Schwesyg *et al.*, 2010a, 2010b).

Material research efforts were required to improve performance of PPKTP as well. This material is susceptible to colour centre formation and a concomitant induced absorption when irradiated with high-peak power blue or green laser beams (Wang *et al.*, 2004; Hirohashi *et al.*, 2007). Previous material research in mixed $Rb_xK_{1-x}TiOPO_4$ crystals showed that for $x \sim 0.01–0.02$ the ionic conductivity mediated by K^+ vacancies is reduced by orders of magnitude and the material still retains similar ferroelectric properties as KTP (Thomas *et al.*, 1994; Kriegel *et al.*, 2001; Jiang *et al.*, 2002). Fabrication of the QPM structures revealed superior properties of this solid solution in terms of ferroelectric domain control as well as in terms of optical performance (Wang *et al.*, 2007). A drastically reduced ionic conductivity in $Rb_xK_{1-x}TiOPO_4$ allowed fabrication of 5-mm-thick periodically poled crystals for high energy frequency conversion (Zukauskas *et al.*, 2011). Figure 6.6 shows selectively etched surfaces of the 5-mm-thick PPKTP crystal revealing high-quality structure of the nonlinear coefficient d_{33}. Another approach for increasing the aperture of the PPKTP is by template growth where the seed crystal already contains a periodic structure. Such template growth has been recently accomplished by employing special composition flux, which permitted crystal growth below the Curie temperature (Peña *et al.*, 2011).

Further down Table 6.4 we encounter large bandgap materials with predictably low FOM. CLBO in the borate family is an example of an application-driven crystal. This crystal was developed with the view of the optical lithography roadmap which required shifting the wavelength of the optical lithography sources down to 193 nm. CLBO is promising due to the

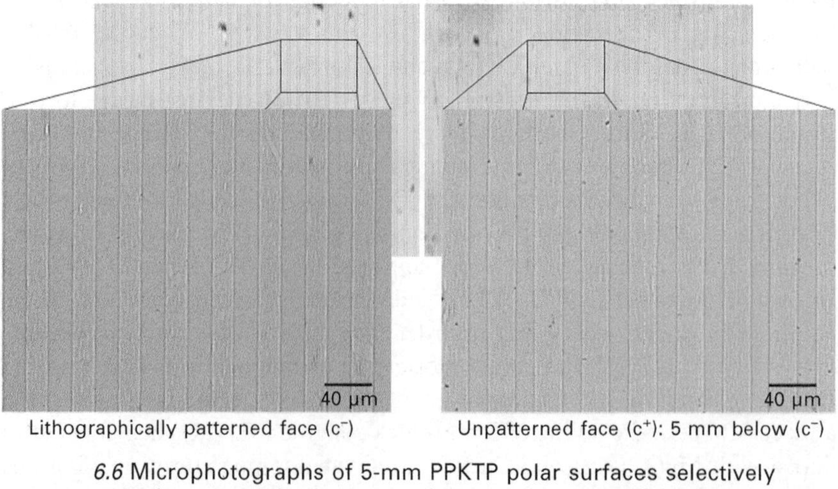

6.6 Microphotographs of 5-mm PPKTP polar surfaces selectively etched to reveal the structure of the nonlinear coefficient d_{33}.

© Woodhead Publishing Limited, 2013

possibility of phase matching below 200 nm, a substantially lower Poynting vector walkoff as compared to BBO and a potential for growth of large crystal sizes (Mori *et al.*, 1995, Merriam *et al.*, 2007). There are also choices of QPM media in the ultraviolet such as periodically poled $MgBaF_4$ (Buchter *et al.*, 2001), a crystalline quartz, which has been structured by applying mechanical stress (Harada *et al.*, 2004), and orientationally patterned GaN (Chowdhury *et al.*, 2003).

Probably the most important lesson that the nonlinear optical materials community had to learn over the past two decades was the economic one. Due to a long development and technology refinement time with associated expense, one needs a very strong strategic application with a long-term horizon in order to motivate an effort in material development. The flexibility in wavelength and interaction design provided by QPM techniques will probably become even more appealing in the future.

6.5 Sources of further information and advice

For a quick introduction into the topics of nonlinear optics it is worth checking out the most recent, surprisingly concise, textbook by Geoffrey New (New, 2011), where the most important aspects are very well accentuated. More detailed descriptions of the calculation methods and collections of relevant formulae as well as material property tables can be found in Sutherland (1996). An even more comprehensive collection of the published material parameter data can be found in the continuously updated nonlinear materials handbook (Dimitriev *et al.*, 2010). The most recent updates on the properties of mid-infrared materials including the technology of OP-GaAs can be found in Petrov (2011).

6.6 References

Arbore M. M., Marco O. and Fejer M. M. (1997), 'Pulse compression during second-harmonic generation in aperiodic quasi-phase-matching gratings', *Opt. Lett.* 22, 865–867.

Armstrong D. J. and Smith A. V. (2003), 'Design and laboratory characterization of a highly efficient all solid state 200 mJ UV light source for ozone dial measurements', *SPIE Proc.* 4893, 105–120.

Armstrong J. A., Bloembergen N., Ducuing J. and Pershan, P. S. (1962), 'Interactions between light waves and nonlinear dielectrics', *Phys. Rev.* 127, 1918–1939.

Badikov V., Mitin K., Noack F., Panyutin V., Petrov V., Seryogin A. and Shevyrdyaeva G. (2009), 'Orthorhombic nonlinear crystals of $Ag_xGa_xGe_{1-x}Se_2$ for the mid-infrared spectral range', *Opt. Mater.* 31, 590–597.

Batchko R. G., Shur V. Y., Fejer M. M. and Boyd R. L (1999), 'Backswitch poling in lithium niobate for high-fidelity domain patterning and efficient blue light generation', *Appl. Phys. Lett.* 75, 1673–1675.

© Woodhead Publishing Limited, 2013

Becker P., Liebertz J. and Bohaty L. (1999), 'Top-seeded growth of bismuth triborate, BiB_3O_6', *J. Cryst. Growth*, 203, 149–155.
Bierlein J. D. and Vanherzeele H. (1989), 'Potassium titanyl phosphate: properties and new applications', *J. Opt. Soc. Am. B*, 6, 622–633.
Boyd G. D. and Kleinman D. A. (1968), 'Parametric interaction of focused Gaussian light beams', *Phys. Rev.* 39, 3597–3639.
Boyd G. D. and Patel C. K. N. (1966), 'Enhancement of optical second-harmonic generation (SHG) by reflection phase matching in ZnSe and GaAs', *Appl. Phys. Lett.* 8, 313–315.
Boyd G. D., Bridges T. J., Pollack M. A. and Turner E. H. (1971a), 'Microwave nonlinear susceptibilities due to electronic and ionic anharmonicities in acentric crystals', *Phys. Rev. Lett.* 26, 387–390.
Boyd G. D., Burhler E. and Storz F. G. (1971b), 'Linear and nonlinear optical properties of $ZnGeP_2$ and CdSe', *Appl. Phys. Lett.* 18, 301–304.
Buchter S. C., Fan T. Y., Liberman V., Zayhowsky J. J., Rothschild M., Mason E. J., Cassanho A., Jenssen H. P. and Burnett J. H. (2001), 'Periodically poled $BaMgF_4$ for ultraviolet frequency generation', *Opt. Lett.* 26, 1693–1695.
Butcher P. N. and Cotter D. (1990), *The Elements of Nonlinear Optics*, Cambridge University Press.
Canalias C. and Pasiskevicius V. (2007), 'Mirrorless optical parametric oscillator' *Nature Photonics*, 1, 459–462.
Canalias C., Pasiskevicius V., Fokine M. and Laurell F. (2005), 'Backward quasi-phase matched second harmonic generation in submicrometer periodically poled $KTiOPO_4$', *Appl. Phys. Lett.* 86, 181105.
Chen C., Wu B. and Jiang A. (1984), 'The optical properties and growth of new ultraviolet frequency multiplication monocrystal β-BaB_2O_4', *Scientia Sinica* 28, 598.
Chen C., Wu Y., Jiang A., Wu B., Li R. and Li S. (1989), 'New nonlinear crystal: LiB_3O_5', *J. Opt. Soc. Am.* B 6, 616–321.
Chen Q. and Risk W. P. (1994), 'Periodic poling of $KTiOPO_4$ using an applied electric field', *Electron. Lett.* 30, 1516–1517.
Chowdhury A., Ng H. M., Bhardwaj M. and Weimann N. G. (2003), 'Second-harmonic generation in periodically poled GaN', *Appl. Phys. Lett.* 83, 1077–1079.
Dekker P. and Dawes J. (2006), 'Twining and "natural quasi-phase-matching" in Yb:YAB', *Appl. Phys. B*, 83, 267–271.
Dimitriev V. G., Gurzadyan G. G. and Nikogosyan, D. N. (2010), *Handbook of Nonlinear Optical Crystals*, Berlin and Heidelberg, Springer-Verlag.
Eyres L. A., Tourreau P. J., Pinguet T. J., Ebert C. B., Harris J. S., Fejer M. M., Becouarn L., Gérard B. and Lallier E., (2001), 'All-epitaxial fabrication of thick, orientation-patterned GaAs films for nonlinear optical frequency conversion', *Appl. Phys. Lett.* 79, 904–906.
Faye D., Lallier E., Grisard A. and Gérard B. (2007), 'Thick low-loss orientation-patterned gallium arsenide (OP-GaAs) samples for mid-infrared laser sources', *Proc. SPIE* 6740, 67400I.
Fejer M. M., Magel G. A., Jundt D. H. and Byer R. L. (1992), 'Quasi-phase-matched second harmonic generation: tuning and tolerances', *IEEE J. Quantum Electron.* 28, 2631–2654.
Feng D., Ming N.-B., Hong J.-F., Yang Y.-S., Zhu J.-S., Yang Z. and Wang Y.-N. (1980), 'Enhancement of second-harmonic generation in $LiNbO_3$ crystals with periodic laminar ferroelectric domain structure', *Appl. Phys. Lett.* 37, 607–609.

© Woodhead Publishing Limited, 2013

Filimonov A. A., Leonyuk N. I., Meissner L. B., Timchenko T. I. and Rez I. S. (1974), 'Nonlinear optical properties of crystals with yttrium–aluminium borate (YAB) structure', *Kristall und Technik* 9, 63–66.

Flytzanis C. (1969), 'Electro-optic coefficients in III-V compounds', *Phys. Rev. Lett.* 23, 1336–1339.

Franken P. A., Hill A. E., Peters C. W. and Weinreich, G. (1961), 'Generation of optical harmonics', *Phys. Rev. Lett.* 7, 118–119.

Furukawa Y., Kitamura K., Alexandrovski A., Route R. K., Fejer M. M. and Foulon G. (2001), 'Green-induced infrared absorption in MgO doped $LiNbO_3$', *Appl. Phys. Lett.* 78, 1970–1972.

Giordmaine J. A. (1962), 'Mixing of light beams in crystals', *Phys. Rev. Lett.* 8, 19–20.

Haidar R. (2006), 'Fractional quasi-phase-matching by Fresnel birefringence', *Appl. Phys. Lett.* 88, 211102.

Harada M., Muramatsu K., Iwasaki Y., Kurimura S. and Taira T. (2004), 'Periodic twinning in crystal quartz for optical quasi-phase matched secondary harmonic conversion', *J. Mater. Res.* 19, 969–972.

Hirohashi J. and Pasiskevicius V. (2005), 'Second-harmonic blue generation using periodic 90° domain structures in $KNbO_3$', *Appl. Phys. B* 81, 761–763.

Hirohashi J., Yamada K., Kamio H. and Shichijyo S. (2004), 'Embryonic nucleation method for fabrication of uniform periodically poled structures in potassium niobate for wavelength conversion devices', *Jpn. J. Appl. Phys.* 43, 559–566.

Hirohashi J., Pasiskevicius V., Wang S. and Laurell F. (2007), 'Picosecond blue light-induced absorption in single-domain and periodically poled ferroelectrics', *J. Appl. Phys.* 101, 033105.

Horowitz M., Bekker A. and Fischer B. (1993), 'Broadband second-harmonic generation in $Sr_xBa_{1-x}Nb_2O_6$ by spread spectrum phase matching with controllable domain gratings', *Appl. Phys. Lett.* 62, 2619–2621.

IEEE Std 176–1949 (1949), 'Standards on piezoelectric crystals, 1949', *Proc. IRE* 37, 1378–1395.

IEEE/ANSI Std 176–1987 (1987), 'IEEE standards on piezoelectricity'.

Ishizuki H. and Taira T. (2005), 'High-energy quasi-phase-matched optical parametric oscillation in a periodically poled MgO:$LiNbO_3$ device with a 5mm × 5mm aperture', *Opt. Lett.* 30, 2918–2920.

Ishizuki H. and Taira T. (2010), 'High energy quasi-phase matched optical parametric oscillation using Mg-doped congruent $LiTaO_3$ crystal', *Opt. Express* 18, 253–258.

Jackson A. G., Ohmer M. C. and LeClair S. R. (1997), 'Relationships of the second order nonlinear optical coefficient to energy gap in inorganic non-centrosymmetric crystals', *Infrared Phys. Technol.* 38, 233–244.

Jiang A., Cheng F., Lin Q., Cheng Z. and Zheng Y. (1986), 'Flux growth of large single crystals of low temperature phase barium metaborate', *J. Cryst. Growth* 79, 963–969.

Jiang Q., Thomas P. A., Hutton K. B. and Ward R. C. C. (2002), 'Rb-doped potassium titanyl phosphate for periodic ferroelectric domain inversion', *J. Appl. Phys.* 92, 2717–2723.

Karlsson H. and Laurell F. (1997), 'Electric field poling of flux grown $KTiOPO_4$', *Appl. Phys. Lett.* 71, 3474–3476.

Kashyap R., Veldhuis G. J., Rogers D. C. and McKee P. F. (1994), 'Phase-matched second-harmonic generation by periodic poling of fused silica', *Appl. Phys. Lett.* 64, 1332–1334.

© Woodhead Publishing Limited, 2013

Katz M., Route R. K., Hum D. S., Parameswaran K. R., Miller G. D. and Fejer M. M. (2004), 'Vapor-transport equilibrated near-stoichiometric lithium tantalate for frequency-conversion applications', *Opt. Lett.* 29, 1775–1777.

Kitamura K., Yamamoto J. K., Iyi N., Kimura S. and Hayashi T. (1992), 'Stoichiometric $LiNbO_3$ single crystal growth by double crucible Czochralski method using automatic powder supply system', *J. Cryst. Growth* 116, 327–332.

Kitamura K., Furukawa Y., Ji Y., Zgonik M., Medrano C., Montemezzani G. and Günter P. (1997), 'Photorefractive effect in $LiNbO_3$ crystals enhanced by stoichiometry control', *J. Appl. Phys.* 82, 1006–1009.

Kogelnik H. and Weber H. P. (1974), 'Rays, stored energy and power flow in dielectric waveguides', *J. Opt. Soc. Am.* 64, 174–185.

Komine H., Long W. H., Jr., Tully J. W. and Stappaerts E. A. (1998), 'Quasi-phase-matched second-harmonic generation by use of a total-internal-reflection phase shift in gallium arsenide and zinc selenide plates', *Opt. Lett.* 23, 661–663.

Kriegel R., Wellendorf R. and Kaps C. (2001), 'Variation of lattice parameters in the mixed crystal system $K_{1-x}Rb_xTiOPO_4$ and inherent stress situation in ion exchanged single crystals', *Mater. Res. Bull.* 36, 245–252.

Kuroda A., Kurimura S. and Uesu Y. (1996), 'Domain inversion in ferroelectric MgO:$LiNbO_3$ by applying electric field', *Appl. Phys. Lett.* 69, 1565–1567.

Lallier E., Brevignon M. and Lehoux J. (1998), 'Efficient second-harmonic generation of a CO_2 laser with quasi-phase-matched GaAs crystal', *Opt. Lett.* 23, 1511–1513.

Lim E. J., Fejer M. M., Byer R. L. and Kozlovsky W. J. (1989), 'Blue light generation by frequency doubling in periodically poled lithium niobate channel waveguides', *Electron. Lett.* 25, 731–732.

Louisell W. H., Yariv A. and Siegman A. E. (1961), 'Quantum fluctuations and noise in parametric processes. I', *Phys. Rev.* 124, 1646–1654.

Maker P. D., Terhune R. W., Nicenoff M. and Savage, C. M. (1962), 'Effects of dispersion and focusing on the production of optical harmonics', *Phys. Rev. Lett.* 8, 21–22.

Mao H., Fu F., Wu B. and Chen C. (1992), 'Noncritical quasiphase-matched second harmonic generation in LiB_3O_5 crystal at room temperature', *Appl. Phys. Lett.* 61, 1148–1150.

Matthias B. T. and Remeika J. P. (1949), 'Ferroelectricity in ilmenite structure', *Phys. Rev.* 76, 1886–1887.

Merriam J., Bethune D. S., Hoffnagle J. A., Hinsberg W. D., Jefferson C. M., Jacob J. J. and Litvin T. (2007), 'A solid-state 193 nm laser with high spatial coherence for sub-40 nm interferometric immersion lithography', *Proc. SPIE* 6520, 65202Z.

Meyn J.-P., Klein M. E., Woll D., Wallenstein R. and Rytz D. (1999), 'Periodically poled potassium niobate for second-harmonic generation at 463 nm', *Opt. Lett.* 24, 1154–1156.

Miller G. D., Batchko R. G., Fejer M. M. and Byer R. L. (1996), 'Visible quasi-phasematched harmonic generation by electric-field-poled lithium niobate', *Proc. SPIE* 2700, 34–45.

Miller R. C. (1964), 'Optical second harmonic generation in piezoelectric crystals', *Appl. Phys. Lett.* 5, 17–19.

Miller R. C. and Savage A. (1959), 'Direct observation of antiparallel domains during polarization reversal in single-crystal barium titanate', *Phys. Rev. Lett.* 2, 294–296.

Mitsui T. and Furuichi J. (1953), 'Domain structure of Rochelle Salt and KH_2PO_4', *Phys. Rev.* 90, 193–202.

Mollow B. R. and Glauber R. J. (1967), 'Quantum theory of parametric amplification. I', *Phys. Rev.* 160, 1076–1096.

© Woodhead Publishing Limited, 2013

Mori Y., Kuroda I., Nakajima S., Sasaki T. and Nakai S. (1995), 'New nonlinear optical crystal: Cesium lithium borate', *Appl. Phys. Lett.* 67, 1818–1820.
Moss T. S. (1985), 'Relations between the refractive index and energy gap of semiconductors', *Phys. Stat. Sol.* B 131, 415–417.
Nakamura K. and Shimizu H. (1983), 'Poling of ferroelectric crystals by using interdigital electrodes and its application to bulk-wave transducers', *1993 IEEE Ultrasonics Symp.* 527–530.
Nassau K., Levinstein H. J. and Loiacono G. M. (1966), 'Ferroelectric lithium niobate. 1. Growth, domain structure, dislocations and etching', *J. Chem. Phys. Sol.* 27, 989–996.
New G. (2011), *Introduction to Nonlinear Optics*, Cambridge University Press.
Okada M., Takizawa K. and Ieri S. (1976), 'Second harmonic generation by periodic laminar structure of nonlinear optical crystal', *Opt. Commun.* 18, 331–334.
Okuno Y., Uomi K., Aoki M. and Tsuchiya T. (1997), 'Direct wafer bonding of III–V compound semiconductors for free-material and free-orientation integration', *IEEE J. Quantum Electron.* 33, 959–969.
Pan F, Knöpfle G., Bosshard C., Follonier S., Spreiter R., Wong M. S. and Günter P. (1996), 'Electro-optic properties of the organic salt 4-*N*,*N*-dimethylamino-4′-*N*′-methyl-stilbazolium tosylate', *Appl. Phys. Lett.* 69, 13–15.
Pasiskevicius V., Holmgren S., Wang S. and Laurell F. (2002), 'Simultaneous second harmonic generation with two orthogonal polarization states in periodically poled KTP', *Opt. Lett.* 27, 1628–1630.
Peña A., Ménaert B., Boulanger B., Laurell F., Canalias C., Pasiskevicius V., Segonds P., Debray J. and Pairis S. (2011), 'Template-growth of periodically domain-structured $KTiOPO_4$', *Opt Mat. Express* 1, 185–191.
Petrov V. (2011), 'Parametric down-conversion devices: The coverage of the mid-infrared spectral range by solid-state laser sources', *Opt. Mater.* doi: 10.1016/j.optmat.2011.03.042.
Petrov V., Schunemann P. G., Zawilski K. T. and Pollak T. M. (2009), 'Noncritical singly resonant optical parametric oscillator operation near 6.2 μm based on a $CdSiP_2$ crystal pumped at 1064 nm', *Opt. Lett.* 34, 2399–2401.
Pitch M. S., Cantrell C. D. and Sze R. C. (1976), 'Infrared second-harmonic generation in nonbirefringent cadmium telluride', *J. Appl. Phys.* 47, 3514–3517.
Polgár K., Péter Á., Kovács L., Corradi G. and Szaller Zs. (1997), 'Growth of stoichiometric $LiNbO_3$ single crystals by top seeded solution growth method', *J. Cryst. Growth* 177, 211–216.
Ray B., Payne A. J. and Burrell G. J. (1969), 'Preparation and some physical properties of $ZnGeP_2$', *Phys. Stat. Sol.* B 35, 197–204.
Roberts D. A. and Meeker T. R. (1992), 'Proposed revision for the polar orthorhombic class in IEEE/ANSI Std 176', *IEEE Trans. Ultrasonics, Ferroelectrics and Frequency Control* 39, 165–166.
Rosenman G., Skliar A., Eger D., Oron M. and Katz M. (1998), 'Low temperature periodic electrical poling of flux-grown $KTiOPO_4$ and isomorphic crystals', *Appl. Phys. Lett.* 73, 3650–3652.
Sasaki T. and Yokotani A. (1990), 'Growth of large KDP crystals for laser fusion experiments', *J. Cryst. Growth* 99, 820–826.
Schunemann P. G., Zawilski K. T., Pollak T. M., Zelmon D. E., Fernelius N. C. and Hopkins F. K. (2008), 'New nonlinear optical crystal for mid-IR OPOs: $CdSiP_2$', Advanced Solid State Photonics, 28 January 2008, Nara, Japan, Post-Deadline Paper MG6.

© Woodhead Publishing Limited, 2013

Schwesyg J. R., Kajiyama M. C. C., Falk M., Jundt D. H., Buse K. and Fejer M. M. (2010a), 'Light absorption in undoped congruent and magnesium-doped lithium niobate crystals in the visible wavelength range', *Appl. Phys* B 100, 109–115.

Schwesyg J. R., Philips C. R., Ioakeimidi K., Kajiyama M. C. C., Falk M., Jundt D. H., Buse K. and Fejer M. M. (2010b), 'Suppression of mid-infrared light absorption in undoped congruent lithium niobate crystals', *Opt. Lett.* 35, 1070–1072.

Setzler S. D., Schunemann P. G., Pollak T. M., Pomeranz L. A., Missey M. J. and Zelmon D. E. (1999), 'Periodically poled barium titanate as a new nonlinear optical material', Trends in Optics and Photonics (OSA), *Advanced Solid State Lasers*, 26, 676–680.

Sutherland R. L. (1996), *Handbook of Nonlinear Optics*, Marcel Dekker, New York.

Szilagyi A., Hordvik A. and Schlossberg H. (1976), 'A quasi-phase-matching technique for efficient optical mixing and frequency doubling', *J. Appl. Phys.* 47, 2025–2032.

Tang M., Minamide H., Wang Y., Notake T., Ohno S. and Ito H. (2011), 'Tunable terahertz-wave generation from DAST crystal pumped by a monolithic dual-wavelength laser', *Opt. Express* 19, 779–786.

Thomas P. A., Duhlev R. and Teat S. J. (1994), 'A comparative structural study of a flux-grown crystal of $K_{0.86}Rb_{0.14}TiOPO_4$ and an ion-exchanged crystal of $K_{0.84}Rb_{0.16}TiOPO_4$', *Acta Cryst.* B 50, 538–543.

Thompson D. E., McMullen J. D. and Andersson D. B. (1976), 'Second-harmonic generation in GaAs "stack of plates" using high-power CO_2 laser radiation', *Appl. Phys. Lett.* 29, 113–115.

Tomita I., Suzuki H., Ito H., Takenouchi H., Ajito K., Rungsawang, R. and Ueno Y. (2006), 'Terahertz-wave generation from quasi-phase-matched GaP for 1.55 μm pumping', *Appl. Phys. Lett.* 88, 071118.

Van der Poel C. J., Bierlein J. D., Brown J. B. and Colak S. (1990), 'Efficient type I blue second-harmonic generation in periodically segmented $KTiOPO_4$ waveguides', *Appl. Phys. Lett.* 57, 2074–2076.

Wang P., Dawes J. M., Dekker P., Knowles D. S., Piper J. A. and Lu B. (1999), 'Growth and evaluation of ytterbium-doped yttrium aluminum borate as a potential self-doubling laser crystal', *J. Opt. Soc. Am.* B 16, 63–69.

Wang S., Pasiskevicius V. and Laurell F. (2004), 'Dynamics of green-light induced infrared absorption in KTP and PPKTP', *J. Appl. Phys.* 96, 2023–2028.

Wang S., Pasiskevicius V. and Laurell F. (2007), 'High efficiency, periodically poled Rb-doped $KTiOPO_4$ using in-situ monitoring', *Opt. Mater.* 30, 594–599.

Wang W. S., Zhou Q., Geng Z. H. and Feng D. (1986), 'Study of $LiTaO_3$ crystals grown with a modulated structure. I. Second harmonic generation in $LiTaO_3$ crystals with periodic laminar ferroelectric domains', *J. Cryst. Growth* 79, 706–709.

Webjörn J., Laurell F. and Arvidsson G. (1989), 'Fabrication of periodically domain-inverted channel waveguides in lithium niobate for second harmonic generation', *IEEE J. Lightwave Technol.* 7, 1597–1600.

Webjörn J., Pruneri V., Russel P. St J., Barr J. R. M. and Hanna D. C. (1994), 'Quasi-phase-matched blue light generation in bulk lithium niobate, electrically poled via periodic electrodes', *Electron. Lett.* 30, 894–895.

Yablonowitch E., Flytzanis C. and Bloembergen N. (1972), 'Anisotropic interference of three-wave and double two-wave frequency mixing in GaAs', *Phys. Rev. Lett.* 29, 865–868.

Yamada M., Nada N., Saitoh M. and Watanabe K. (1993), 'First-order quasi-phase matched $LiNbO_3$ waveguide periodically poled by applying an external field for efficient blue second-harmonic generation', *Appl. Phys. Lett.* 62, 435–437.

© Woodhead Publishing Limited, 2013

Yariv A. (1989), *Quantum Electronics,* third edition. John Wiley, Singapore, 380–384.

Zhong G.-G., Jian J. and Wu Z.-K. (1980), in: *Proceedings of the Eleventh International Quantum Electronics Conference, IEEE Cat. No. 80 CH 1561-0* (June 1980), p. 631.

Zhu Y. Y., Fu J. S., Xiao R. F. and Wong G. K. L. (1997), 'Second harmonic generation in periodically domain-inverted $Sr_{0.6}Ba_{0.4}Nb_2O_6$ crystal plate', *Appl. Phys. Lett.* 70, 1793–1795.

Zukauskas A., Thilmann N., Pasiskevicius V., Laurell F. and Canalias C. (2011), '5 mm thick periodically poled Rb-doped KTP for high energy optical parametric frequency conversion', *Opt. Mat. Express* 1, 201–206.

Zumsteg F. C., Bierlein J. D. and Grier T. E. (1976), '$K_xRb_{1-x}TiOPO_4$ a new nonlinear optical material', *J. Appl. Phys.* 47, 4980.

© Woodhead Publishing Limited, 2013

Part II

Solid-state laser systems and their applications

© Woodhead Publishing Limited, 2013

7
Principles of solid-state lasers

N. N. IL'ICHEV, A. M. Prokhorov General Physics Institute,
Russian Academy of Sciences, Russia

DOI: 10.1533/9780857097507.2.171

Abstract: The main goal of this chapter is to give an introduction to the principles of solid-state lasers. A short history of laser inventions is presented. The main schemes of active center levels used for obtaining amplification are listed. Equations that describe amplification of continuous wave radiation and short pulses under optical pumping are presented. The role of the laser resonator in the formation of the spectral and spatial characteristics of laser radiation is briefly discussed. Laser operation is described on the basis of a laser model. Within the framework of the model, equations that relate to output power, pump power and energy characteristics are derived. The limitations that are due to the laser model assumptions are discussed.

Key words: laser principles, light amplification, cw operation, active Q-switch, passive Q-switch, model of a laser, solid-state lasers.

7.1 Introduction

The laser (abbreviation of *Light Amplification by Stimulated Emission of Radiation*) is a light source. The peculiarities of light emitted by a laser are low divergence and high coherence. In any laser one type of energy, for example light, chemical energy, electrical energy and so on, is converted into the energy of light radiation.

Any laser consists of the two main elements: the amplifying (active) medium and the resonator. An external source of power or energy is necessary to make an active medium. The active material stores the energy of the external source in the form of so-called electronic or vibronic excitation of some centers (molecules, atoms or ions) that can absorb and emit light. This energy is extracted from an active material in a form of light radiation, into the resonator as a result of induced or stimulated transitions. Two parallel mirrors form the typical laser resonator. Energy is stored in the resonator in the form of light radiation. Radiation from the laser resonator is emitted into space through a semitransparent mirror. The combination of the resonator and amplifying material gives rise to low divergence and high coherence (narrow spectral width) of output laser radiation. The low divergence of the radiation means that the light propagates as a beam without substantially spreading sideways over long distance.

© Woodhead Publishing Limited, 2013

The state of an active material which enables it to amplify radiation is formed by the excitation of bounded electrons of some centers into the upper excited states, so that the number of the centers at the excited states is higher than that at some lower states. This state of the material is called the state with inverse population. In this state an active material can amplify light at some wavelengths when it passes through the material.

Each type of laser has its own amplifying material and a unique set of wavelengths it can generate. To date, the range of wavelengths spans from X-rays up to the far infrared part of the spectrum.

The first laser (Maiman, 1960) was a solid-state laser. The active material of this laser was a ruby crystal. The laser wavelength was 694.3 nm. The predecessor of the laser was the maser (abbreviation of *Microwave Amplification by Stimulated Emission of Radiation*). It was put into operation by C.H. Townes in 1955 (Gordon *et al.*, 1955). In 1964 C.H. Townes (in the USA), and N. G. Basov and A. M. Prokhorov (in the former USSR) were awarded the Nobel Prize 'for fundamental work in the field of quantum electronics, which has led to the construction of oscillators and amplifiers based on the maser–laser principle' (Nobelprize.org).

It is possible to classify lasers in various ways. If an active laser material is selected for classification there are solid-state, semiconductor, liquid, gaseous, and free electron lasers. Fiber lasers are considered as a separate laser class in this chapter. Concerning the operation mode, there are pulsed and continuous wave (cw) lasers. The principles of laser action are common for all types of lasers. In the following sections, attention will be paid to the energy characteristics of the solid-state laser, i.e. to its output power at cw operation, and its pulse energy at pulsed operation.

7.2 Amplification of radiation

As already mentioned, a part of the laser is active material. Energy that is stored in it is extracted by means of amplification of light. An amplifying medium is the source of energy that the laser radiation is extracted from. The mirrors of a laser resonator create positive feedback when radiation bounces back and forth inside the cavity. Amplification is necessary to compensate for the losses of radiation due to absorption or other reasons. Feedback means that some part of the radiation returns back into the amplifier.

In a solid-state laser an amplifying material is a crystal, a glass or ceramic doped with some impurity centers. For example, the ruby mentioned above is a sapphire crystal doped with Cr^{3+} ions. Doping centers in a solid-state material have some system of electronic or vibronic energy levels. Electron transitions between the levels lead to absorption or emission of light and in some conditions to amplification if certain conditions are fulfilled. Let us consider the two main schemes of energy levels in lasers. These schemes

© Woodhead Publishing Limited, 2013

in their most important parts were suggested by A. M. Prokhorov and N. G. Basov (Basov and Prokhorov, 1955) to get amplification in the microwave wavelength region. N. Bloembergen (Bloembergen, 1956) proposed using electronic states of doped centers in solid-state materials for a practical realization of the three-level scheme in masers.

Figure 7.1(a) shows one of these schemes, called the three-level laser scheme. Level 1 is a ground level, and level 2 is the upper laser level. Laser transition takes place between levels 1 and 2. Level 3 is used to absorb optical radiation from some other optical source via 1 to 3 transition. This radiation is called pump radiation. The lifetime of level 3 is usually short because of the high probability of non-radiative transition from level 3 to level 2.

The other important scheme is presented in Fig. 7.1(b). Laser radiation in this case is emitted when transition between levels 2 and 4 takes place. In contrast to the scheme in Fig. 7.1(a), lower level 4 has some energy and its population that is due to temperature (Boltzmann factor) may be much lower than the ground state. Usually in a solid-state material, the non-radiative probability of 4 and 1 transition is high compared to the probabilities of all other transitions. Pump radiation is absorbed at the 1 and 3 transition. This scheme is called the four-level laser scheme.

Both schemes have a feature that is peculiar to solid-state lasers: fast non-radiative decay of excitation from level 3 to level 2 (Fig. 7.1(a) and (b)) and from level 4 to level 1 (Fig. 7.1(b)) is possible due to the fact that doped centers are embedded in solid-state materials and this decay is due to multiphonon transitions. The ruby laser is an example of a laser with a three-level scheme; a neodymium laser is an example of a laser with a four-level scheme. The characteristics of laser crystals can be found, e.g., in Kaminskii (1981).

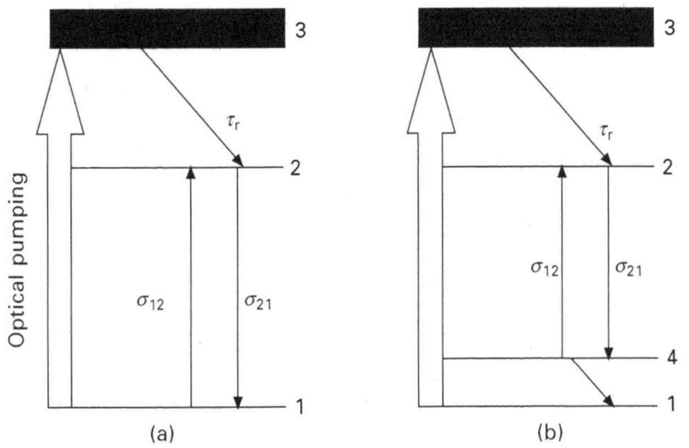

7.1 Scheme of laser levels: (a) three-level and (b) four-level schemes.

© Woodhead Publishing Limited, 2013

Let us consider propagation of a parallel beam of light through the material that contains dope centers. If frequency of light is in resonance with 1 to 2 transition (Fig. 7.1(a)) then the light will be absorbed by the centers. The dependence of light intensity on a path in the material is described by the well-known equation:

$$\frac{dI}{dz} = -\gamma I$$

where z is the coordinate along the direction of light propagation, and γ is the light absorption coefficient. The solution of this equation gives the Bouguer–Lambert–Beer law (P. Bouguer, 1729; J. H. Lambert, 1760; A. Beer, 1852; see e.g. Prokhorov (1984), p. 60):

$$I(z) = I(0)\exp(-\gamma z)$$

where $I(z)$ is the intensity of light at the point with z-coordinate, and $I(0)$ is the intensity at $z = 0$. The absorption coefficient for resonant radiation is proportional to the concentration of absorbing centers (the Beer law is established for solutions), i.e. $\gamma = \sigma_{12}N_0$, where σ_{12} is the so-called absorption cross-section of a dope center level at 1 to 2 transition, and N_0 is the number of absorbing centers per unit volume of material.

The Bouguer–Lambert–Beer law is only valid for low light intensity, when all absorbing centers are considered to be at the ground level. Generally, it is necessary to take into account that when a dope center absorbs photons it becomes excited and moves to level 2, and the number of absorbing centers changes. There are several possible ways for the excited center to return back to ground level. The emission of photons with the energy characteristic for transition from 2 to 1 is called spontaneous emission. The opposite process to absorption is called stimulated emission. In this case, an excited center moves from the upper to the ground level emitting a new photon under the action of external photons and this new photon has the same characteristics as the external one. The concept of stimulated emission was extended (generalized) by A. Einstein (Einstein, 1916) in the case of the quantum description of light interaction with matter.

The introduction of dope centers to solid-state materials makes it necessary to mention a further method of excited decay, which is due to multi-phonon transitions when only phonons and not photons are emitted.

Let us denote the total concentration of the centers as N_0 and their concentration at levels 1 and 2 (Fig. 7.1(a)) as N_1 and N_2, respectively. Values N_1 and N_2 will be referred to below as populations of level 1 and level 2. Conservation of the total number of centers takes place: $N_0 = N_1 + N_2$. Let us assume that there is radiation with intensity I and frequency ω corresponding to transition from 1 to 2 at some point of the material. And at the same point there is other additional radiation with intensity I_p and

© Woodhead Publishing Limited, 2013

frequency that is in resonance to level 1 to 3 transition. The last radiation we will call pump radiation. The rate of change of ground level population N_1 at this point under the action of radiation with intensity I (the dimensions of I are quanta/s cm^2) due to absorption from level 1 to level 2 is $-\sigma_{12}N_1I$. The rate of change of the level 1 population due to stimulated emission from level 2 under the action of radiation with intensity I is $\sigma_{21}N_2I$, where σ_{21} is the cross-section of stimulated transitions from level 2 to level 1. The population of the ground level is changed due to spontaneous emission and non-radiative decay processes from level 2. The rate of this change is N_2/τ, where τ is the lifetime of the excited center at level 2. There is also the rate of change of level 1 population due to the absorption of pump radiation through 1 to 3 transition $-\sigma_{13}N_1I_p$, and the assumption that all centers that are excited to level 3 rapidly transit from level 3 to level 2. The sum of all these rates gives the equation for the rate of ground level population change:

$$\frac{dN_1}{dt} = -\sigma_{12}N_1I + \sigma_{21}N_2I + \frac{N_2}{\tau} - I_p\sigma_{13}N_1 \qquad [7.1]$$

The additional equation is the conservation condition for the total number of centers $N_0 = N_1 + N_2$. Equation [7.1] is called the rate equation (Statz and deMars, 1960). Intensities of pump radiation I_p and radiation I have different wavelengths, and their propagation directions are determined by the specific construction of a laser.

Cross-sections σ_{12} and σ_{21} depend on the wavelength and the type of dope center. We consider these cross-sections as known values.

If stimulated transitions from upper level 2 (Fig. 7.1) are taken into account then the absorption coefficient γ will be:

$$\gamma = \sigma_{12}N_1 - \sigma_{21}N_2 = \sigma_{12}N_0 - (\sigma_{12} + \sigma_{21})N_2$$

Usually $N_2 << N_0$ and light is absorbed during its propagation in the material, i.e. $\gamma > 0$. But if:

$$N_2 > N_0 \frac{\sigma_{12}}{\sigma_{12} + \sigma_{21}}$$

then $\gamma < 0$ and this means that the intensity of radiation increases during its propagation and amplification takes place.

Let us assume in equation [7.1] that $I = 0$; then in a steady state ($dN_1/dt = 0$) the value $\gamma < 0$ if

$$I_p > \frac{hv}{\tau\sigma_{13}} \frac{\sigma_{12}}{\sigma_{21}}$$

i.e. at some intensity of pump radiation it is possible to obtain amplification. The third level plays an important role in the possibility of obtaining amplification (Fig. 7.1).

© Woodhead Publishing Limited, 2013

It is impossible to obtain amplification for a system consisting of only two levels. To show this, let us assume that $I_p = 0$ and there is only radiation I and level 1 and level 2 interact with radiation. In a steady state, $dN_1/dt = 0$ and from equation [7.1] it follows that

$$\gamma = \frac{\sigma_{12} N_0}{1 + (\sigma_{12} + \sigma_{21})\tau I}$$

and from the last expression it becomes clear that $\gamma > 0$ because $I \geq 0$; σ_{12}, $\sigma_{21} \geq 0$, $N_0 > 0$. This means that in a system of two levels for radiation with intensity I resonant to the transition 1 to 2, it is impossible to get amplification at this wavelength. Notice that when the populations of the ground and upper levels are at thermal equilibrium, the levels follow Boltzmann distribution $N_2/N_1 = \exp(-\Delta E/k_b T_K)$(degeneracy of levels is assumed to be $g_1 = g_2 = 1$ for simplicity), where ΔE is the energy difference between level 1 and level 2, k_b is Boltzmann's constant, and T_k is temperature; then it is impossible to have amplification because $N_2 < N_1$.

7.3 Optical amplifiers

Let us consider the characteristics of amplifiers. We will discuss continuous wave amplifiers and short pulse amplifiers.

7.3.1 Continuous wave amplifiers

In a continuous wave amplifier, the time intensity of the pump is constant, as is the amplified radiation. Let us denote the amplification coefficient as α and

$$\alpha = (\sigma_{12} + \sigma_{21})N_2 - \sigma_{12}N_0 \qquad [7.2]$$

The propagation of a parallel beam of light with rectangular transverse distribution with intensity I in a material with amplification and absorption is described by the known equation

$$\frac{dI}{dz} = \alpha I - \gamma_0 I \qquad [7.3]$$

The amplification coefficient α is described by [7.2] and γ_0 is the absorption coefficient of non-resonant losses that are not connected with the dope centers that are used to obtain amplification. Spectral and transverse characteristics of light are changing during its propagation in an amplifier. We will not take into account these changes and we will consider only the change of light intensity as a function of the coordinate.

It is possible to convert [7.3] by using equations [7.1] and [7.2] into

© Woodhead Publishing Limited, 2013

$$\frac{1}{I}\frac{dI}{dz} = \left(\frac{I_p}{I_{ps}}(\sigma_{21}N_0 - \gamma_0) - (\sigma_{12}N_0 + \gamma_0) - \gamma_0\frac{I}{I_s}\right)\left(1 + \frac{I_p}{I_{ps}} + \frac{I}{I_s}\right)^{-1} \quad [7.4]$$

where $I_{ps} = h\nu/\sigma_{13}\tau$ is the saturation intensity for pump radiation; $I_s = 1/(\sigma_{12} + \sigma_{21})\tau$ is the saturation intensity for radiation resonant to 1 to 2 transition in Fig. 7.1. It is assumed that the pump intensity does not depend on the coordinate. This is the case, for example, when the amplifying material is a cylinder and amplifying radiation propagates along the axis of the cylinder while the pump radiation propagates perpendicularly to the axis of the cylinder as shown in Fig. 7.2.

Let us discuss equation [7.4] when the intensity I is low. Low intensity means that it is much less than saturation intensity I_s: $I << I_s$. In this case one can see that for $I_p = 0$ equation [7.4] describes the propagation of radiation in the material with the total absorption coefficient $\gamma_c = \sigma_{12}N_0 + \gamma_0$ that is a sum of the absorption coefficient of the non-pumped active material and the non-resonant one. The value of the pump intensity can also take a special value, such as I_p^{th}, when the total absorption coefficient for low intensity radiation is equal to zero. This value can be found from equation [7.4]: $I_p^{th} = I_{ps}(\sigma_{12}N_0 + \gamma_0)/(\sigma_{21}N_0 - \gamma_0)$. If the pump intensity is $I_p > I_p^{th}$ then radiation with low intensity will rise with distance during its propagation in the material; if $I_p < I_p^{th}$ the light intensity will diminish with distance.

From equation [7.4] it follows that if $I_p > I_p^{th}$ then the increasing intensity of amplifying radiation with distance leads to the decreasing of amplification and at some point its value $I = I_{cw}$ will not change with distance. It is easy

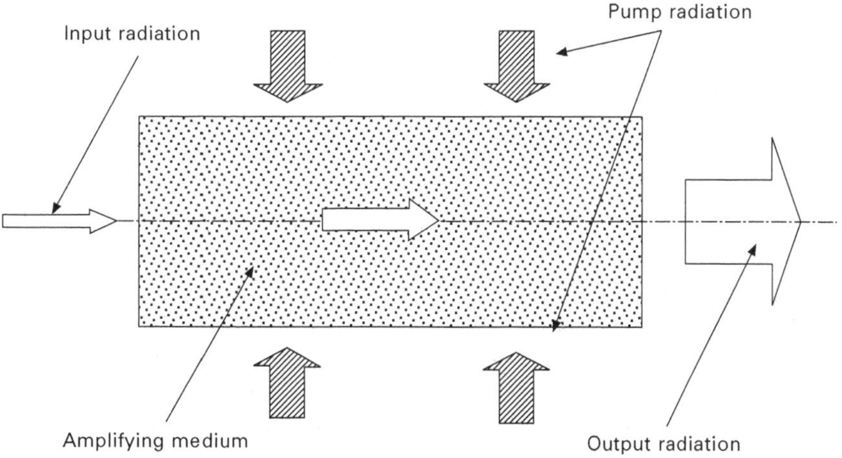

7.2 Propagation of radiation in amplifier. Pump radiation is shown.

© Woodhead Publishing Limited, 2013

to find this value I_{cw} from [7.4]. Let us denote $I_p = I_p^{th} n$, where n is excess of pump intensity over I_p^{th}. Then if $n > 1$ one can obtain

$$I_{cw} = I_s \left(1 + \frac{\sigma_{12} N_0}{\gamma_0}\right)(n - 1) \qquad [7.5]$$

Intensity I_{cw} does not depend on the value of intensity I_{in} at the amplifier input. The only condition is $I_{in} > 0$.

7.3.2 Short pulse amplifiers

Let us consider amplification of short light pulses with a duration that is much less than the lifetime of upper level 2 (Fig. 7.1). Here, it is possible to neglect the change of population due to pump and spontaneous transitions because the rates of these processes are much less than the rate caused by the stimulated transitions. The equation for the upper level population in this case is

$$\frac{dN_2}{dt} = [-(\sigma_{12} + \sigma_{21}) N_2 + \sigma_{12} N_0] I \qquad [7.6]$$

Equation [7.6] has a solution

$$(\sigma_{12} + \sigma_{21}) N_2(t) - \sigma_{12} N_0$$
$$= ((\sigma_{12} + \sigma_{21}) N_2(t = 0) - \sigma_{12} N_0) \exp\left(-(\sigma_{12} + \sigma_{21}) \int_0^t I(t')dt'\right) \qquad [7.7]$$

Time $t = 0$ in [7.7] corresponds to the instant of time before the light pulse comes at the given point of the amplifier material. Let us introduce a new variable $x(t)$ according to the next definition

$$x(t) = (\sigma_{12} + \sigma_{21}) \int_0^t I(t')dt' \qquad [7.8]$$

This dimensionless variable is the ratio of fluence inside the material $F(t) = \hbar\omega \int_0^t I(t')dt'$ to saturation fluence F_s for the transition between levels 1 and 2: $F_s = \hbar\omega/(\sigma_{12} + \sigma_{21})$, where $\hbar\omega$ is the energy quantum of radiation, $\hbar$ is Planck's constant, and ω is the frequency of radiation. In other words $x(t)$ is normalized fluence. If one is interested in only the amount of pulse energy after amplification at the given point z of the amplifier, then from [7.3], [7.6] and [7.7] one can obtain the equation that describes the change of pulse normalized fluence $u = x(t \to \infty)$ during its propagation through the amplifying material

© Woodhead Publishing Limited, 2013

$$\frac{du}{dz} = \alpha_0[1 - \exp(-u)] - \gamma_0 u \qquad [7.9]$$

where α_0 is the amplification coefficient at the time before the arrival of the light pulse. This equation is called the Avizonis–Grotbeck equation (Avizonis and Grotbeck, 1966). If $\gamma_0 = 0$, equation [7.9] has an analytical solution that connects normalized fluences at input u_{in} and output u_{out} of an amplifier:

$$u_{out} = \ln[1 + (\exp(u_{in}) - 1)\exp(\alpha_0 L)] \qquad [7.10]$$

where L is the length of an amplifier. This equation is called the Frantz–Nodvik equation (Frantz and Nodvik, 1963). It should be mentioned that in this paper, the change of pulse time shape during its propagation in an amplifier is also considered.

One can see from [7.9] that at low normalized fluence $u << 1$ amplification takes place only if condition $\alpha_0 > \gamma_0$ holds. As a pulse propagates in the amplifier with $\alpha_0 > \gamma_0$ its fluence rises with distance and there is some normalized fluence x_∞ when the pulse energy does not change with distance: $du/dz = 0$. As follows from [7.9] the value x_∞ is the root of the equation

$$x_\infty = \frac{\alpha_0}{\gamma_0}(1 - \exp(-x_\infty)) \qquad [7.11]$$

If the $\gamma_0 > \alpha_0$ equation [7.11] only has the solution $x_\infty = 0$, then the amplification coefficient is less than the absorption one and the pulse energy is diminishing on the path of light in the material.

7.4 Laser resonators

The wavelength of laser radiation is mainly given by the nature of the dope centers and material where the centers are embedded. For example, YAG doped with Nd^{3+} (YAG:Nd^{3+}) usually has a wavelength of 1064 nm; YAG:Er^{3+} has wavelength 2940 nm; see, e.g., Kaminskii (1981). The spectral width of laser radiation depends not only on the width of the dope center transition but on the laser resonator characteristics. Light from a laser is emitted as a highly directed beam, i.e. laser radiation is characterized by low divergence. The resonator's role in the formation of the spectral and spatial characteristics of laser radiation is briefly discussed below.

7.4.1 Spectral and spatial characteristics of laser radiation

As mentioned above, a laser resonator consists of two mirrors that are parallel each to other. This construction is known in optics as the Fabry–Perot interferometer. The idea of implementing this interferometer was independently

© Woodhead Publishing Limited, 2013

suggested in Prokhorov (1958), Dicke (1958) and Schawlow and Townes (1958). Radiation is spread between these mirrors in two directions at each point of the resonator. Usually one of the mirrors is semi-transparent so part of radiation can escape from the resonator. The second mirror has a reflection coefficient close to 100% at the laser wavelength. This resonator is called an open resonator in contrast to an ultrahigh-frequency resonator, which as a rule is some closed volume of space that is restricted by conductive walls. Introducing radiation into an empty resonator will cause decay due to losses in it. These losses can be due to radiation leakage through the semi-transparent mirror or to reflection, absorption and scattering at the resonator elements. If one places the amplifying material inside the resonator then it is possible in general to overcome these losses. As laser radiation goes back and forth in the resonator it acquires some special characteristics, such as a narrow spectral width and transverse spatial distribution.

7.4.2 Resonator modes

Electromagnetic field distribution in open empty resonators was calculated by A. G. Fox and T. Li (Fox and Li, 1960, 1961) using the Huygens–Fresnel principle: see Born and Wolf (2005). It was shown that a number of special transverse electromagnetic field distributions exist. These distributions reproduce themselves after a full pass through a resonator with almost exponential decay of amplitude with the number of passes. These distributions have their own frequencies and losses due to diffraction on the resonator mirror apertures.

Theoretical study of open resonators having curved mirrors have shown (Boyd and Gordon, 1961; Boyd and Kogelnik, 1962; Kogelnik and Li, 1966) that for some configurations, depending on resonator length and mirror curvatures, these field distributions are expressed by Hermite–Gauss functions for rectangular mirrors and by Laguerre–Gauss functions for circular mirrors. These are usually denoted as TEM_{mnq} and TEM_{plq}, accordingly. These functions play the role of eigenfunctions in an empty resonator. Any field in an empty resonator can be represented as a superposition of these functions. Indices *m, n* or *p, l* are integers and refer to the transverse field distributions. Its numerical value is not high and for most practical cases it is restricted by the first several integers. Index *q* is an integer and describes the longitude field distribution in a resonator. Its value is about several orders of magnitude as a rule.

The divergence of laser radiation is defined by the modes that are excited in a laser resonator and may be close to the diffraction limit.

© Woodhead Publishing Limited, 2013

7.5 Model of laser operation

Let us derive the equations that describe the characteristics of a laser at different modes of operation. Output power and the laser's pulse energy are the main subjects of discussion.

The first laser equation is equation [7.1] for population of laser levels. It is necessary to add an equation for the radiation inside the laser resonator. It is possible to use the number of quanta in a laser mode or radiation intensity in a laser resonator; see, e.g., Siegman (1986), Svelto (1998 and Koechner (2006). Intensity is used below.

7.5.1 Laser equations

The typical scheme of a laser resonator with an active element and mirrors is presented in Fig. 7.3. Let us assume that the laser resonator has an output mirror with reflection coefficient R. We substitute this resonator with one that has the same geometry but both its mirrors have 100% reflectivity and we assume also that there is an effective absorption coefficient γ_R in the active element with $\gamma_R = (1/2l_a) \ln (1/R)$, where l_a is the length of the active element. We also assume that all other losses in the laser resonator are uniformly distributed along the active element with some effective absorption coefficient.

It is possible to prove that in a steady state, the intensity of waves traveling in opposite directions in a laser resonator formed by two mirrors with reflectivity 100% with uniformly distributed losses does not depend on the coordinate along the resonator. We neglect the interference of electromagnetic fields traveling in opposite directions with the same frequencies, which produce standing waves inside the resonator. We assume that transverse intensity distribution in the resonator is rectangular.

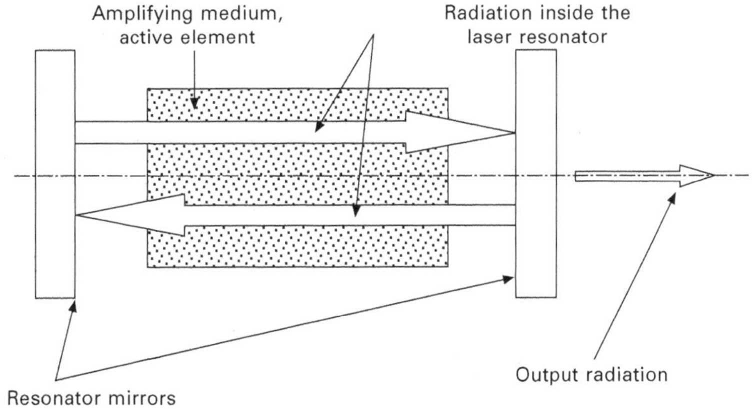

7.3 Laser resonator with active element.

© Woodhead Publishing Limited, 2013

Let us introduce the concept of losses per round trip of the laser resonator and denote the losses that are due to the semi-transparent mirror as $\delta_R = 2l_a\gamma_R = \ln(1/R)$, and the losses that are due to linear absorption γ_0 in the active medium as $\delta_0 = 2l_a\gamma_0$. We also assume that diffraction losses are included in δ_0. We denote the total resonator losses δ_c and $\delta_c = \delta_R + \delta_0 = 2l_a\,\gamma_c$, where $\gamma_c = \gamma_R + \gamma_0$ is the total effective absorption coefficient. From equation [7.1] for the population of the upper laser level, one can get

$$\frac{dN_2}{dt} = (-(\sigma_{12} + \sigma_{21})N_2 + \sigma_{12}N_0)\,I - \frac{N_2}{\tau} + \sigma_{13}(N_0 - N_2)I_p \quad [7.12]$$

where I is intensity inside a laser resonator. I is the sum of intensities of radiation propagating in opposite directions at all frequencies taking part in lasing; I_p is the pump intensity. We consider pump intensity to be uniform in volume of the amplifying material.

It is possible to obtain the equation for intensity inside the laser resonator from the following arguments. Let us denote the photon round-trip time inside the resonator as T. The number of photons that are extracted from the amplifying material during time T due to amplification is $\Delta n_{p+} = Al_a[(\sigma_{12} + \sigma_{21})N_2 - \sigma_{12}N_0]IT$, where A is the cross-sectional area that is filled by radiation in the resonator. The number of photons that are lost during the same time in the resonator is $\Delta n_{p-} = Al_a\gamma_c IT$. The number of photons that are stored in the resonator is $n_p = 0.5AIT$. The change over time T of this number is

$$\Delta n_p = A\left(\frac{dI}{dt}\frac{T}{2}\right)T$$

and equation $\Delta n_p = \Delta n_{p+} - \Delta n_{p-}$ holds. From here we can deduce the equation for intensity inside a resonator:

$$T\frac{dI}{dt} = 2l_a((\sigma_{12} + \sigma_{21})N_2 - \sigma_{12}N_0)\,I - \delta_c I \quad [7.13]$$

Expressions [7.12] and [7.13] are called the rate equations and are usually used to describe the time dependence of intensity inside a laser resonator. These equations do not take into account spontaneous emission. The equations have been implemented (Statz and deMars, 1960) to address the time dynamics of output power for masers.

7.5.2 Continuous wave operation

In a steady state, the left parts of [7.12] and [7.13] are equal to zero and we have the equation that binds the intensity I of laser radiation inside a laser resonator and pump intensity for continuous wave operation:

© Woodhead Publishing Limited, 2013

$$\frac{I_p}{I_{ps}}(\sigma_{21}N_0 - \gamma_c) - (\sigma_{12}N_0 + \gamma_c) - \gamma_c\frac{I}{I_s} = 0 \qquad [7.14]$$

Values I_s and I_{ps} have the same meaning as in [7.4]. Let us introduce the concept of threshold pump intensity I_p^{th} that follows from [7.14] at $I = 0$:

$$\frac{I_p^{th}}{I_{ps}} = \frac{\sigma_{12}N_0 + \gamma_c}{\sigma_{21}N_0 - \gamma_c}$$

There is no lasing if $I_p < I_p^{th}$ as follows from [7.14]. Let us express pump intensity as $I_p = I_p^{th}n$, where n is excess over threshold pump intensity. Finally, we have the equation for intensity inside the laser resonator for $n \geq 1$:

$$I = I_s\left(1 + \frac{\delta_{12}}{\delta_c}\right)(n - 1) \qquad [7.15]$$

where $\delta_{12} = 2l_a\sigma_{12}N_0$. If $n < 1$ then in the model frames $I = 0$. It is interesting to notice that [7.15] in its appearance coincides with [7.5].

Output power is allocated to the effective absorption coefficient γ_R introduced above. If the energy of the quanta of laser radiation is $\hbar\omega$ then it follows from [7.15] for output power P_{out} of the laser, see e.g. Koechner (2006), that:

$$P_{out} = AI_s\hbar\omega\frac{1}{2}\ln(1/R)\left(1 + \frac{\delta_{12}}{\delta_c}\right)(n - 1) \qquad [7.16]$$

The pump intensity and pump power are proportional to each other, so $n = P_p/P_p^{th}$ where P_p is the pump power and P_p^{th} is its threshold value ($P_p^{th} \sim I_p^{th}$ for given geometry of pumping). It follows from [7.16] that if $P_p > P_p^{th}$ ($n > 1$) then output power is a linear function of pump power, whereas if $P_p \leq P_p^{th}$ then $P_{out} = 0$.

7.5.3 Free-running operation

It is possible to raise the pump power if pulsed pumping is used, for example in a flashlamp. The first laser put into operation was inside a pulse-pumped flashlamp (Maiman, 1960). The output energy is characteristic of the laser radiation in this case. One can derive the output energy by using equations [7.12] and [7.13], but we will obtain the output energy with the help of simpler equations and arguments.

When there is a steady state, the amplification coefficient is equal to the absorption coefficient due to the saturation of amplification in the active material as follows from [7.13]. After the pump action begins there is a time interval before amplification in the active element reaches the threshold

© Woodhead Publishing Limited, 2013

value, when laser generation starts. Let us ignore this time and assume that during the pump action steady state takes place.

Let us denote the pump pulse energy as E_p, and the cross-section areas of radiation and the active element as A and A_a, accordingly. Energy $\eta(A/A_a)\ E_p$ is passed through the upper laser level at the region of the active element filled with the laser radiation during the time of pump action at the frequency of laser radiation; η is the coefficient depending on the geometry of pumping. Let us denote as E_1 the energy that is necessary for the creation of the threshold population N_2^{th} of the upper laser level at the region of the active element filled by radiation $E_1 = \hbar\omega A l_a N_2^{th}$, where $\hbar\omega$, A and l_a are defined above. From [7.13], one can obtain the threshold population of the upper laser level:

$$N_2^{th} = \frac{1}{\sigma_{12} + \sigma_{21}}\left(\frac{\delta_c}{2l_a} + \sigma_{12}N_0\right)$$

Energy is released due to the decay of the upper laser level after the pump action is finished. Spontaneous emission also takes place at the same time as generation. While a tiny part of this is emitted into the laser modes, the emission is spread in all directions. It is therefore possible to consider all this energy as lost, as well as the energy due to non-radiative decay in the upper laser level during the time of generation. This energy is

$$E_2 = Al_a\hbar\omega\frac{N_2^{th}}{\tau}T_p$$

where T_p is the time duration of the laser pulse and we assume that it is equal to the pump pulse duration time. The last assumption is valid if the pump energy is not too close to the threshold. The total energy that is due to decay of the upper laser level population is

$$E_{sp} = E_1 + E_2 = \hbar\omega Al_aN_2^{th}\left(1 + \frac{T_p}{\tau}\right)$$

Radiation in the laser resonator is dissipated due to absorption and this energy is

$$E_\gamma = \hbar\omega Al_a\gamma_c\int_0^{T_p} I(t)\mathrm{d}t = AF_{in}\frac{\delta_c}{2}$$

where

$$F_{in} = \hbar\omega\int_0^{T_p} I(t)\mathrm{d}t$$

is fluence inside the resonator. As energy is conserved, there should be $\eta(A/$

© Woodhead Publishing Limited, 2013

A_a) $E_p = E_{sp} + E_\gamma$. Output energy E_{out} is dissipated via effective absorption linked to the output mirror γ_R: $E_{out} = A\gamma_R l_a F_{in}$. Finally, we can obtain the expression for output energy as

$$E_{out} = E_p \eta \frac{A}{A_a}\frac{\ln(1/R)}{\delta_c} - \frac{1}{2}\ln(1/R)AF_s\left(1+\frac{\delta_{12}}{\delta_c}\right)\left(1+\frac{T_p}{\tau}\right) \qquad [7.17]$$

where $\delta_{12} = 2l_a\sigma_{12}N_0$. Expression [7.17] gets over into [7.16] in the limit $T_p >> \tau$. This is approximate because of the assumptions that were taken, but it seems to be useful especially for free-running lasers with a multimode transverse distribution. For a four-level system [7.17] is given in Avanesov *et al.* (1980).

7.5.4 Active and passive Q-switch operation

In the two cases considered above, the amplification coefficient during generation is equal to the threshold coefficient. This circumstance restricts laser output power by the power of the pump source. If high power is necessary there is one more method that allows high power output to be reached. This method is known as Q-switching (Hellwarth, 1961). It works by the intentional reduction of the resonator's quality factor (Q-factor) by introducing high losses in the resonator. Before laser pulse generation starts, the amplification coefficient is less than the total absorption coefficient, so no lasing takes place. If there is a rapid change to the losses so that the new absorption coefficient becomes less than the amplification coefficient, intensity in the resonator begins to rise from spontaneous background and short (typically 1–100 ns), powerful (up to several MW) pulses are generated. Such a laser pulse was called a giant pulse, although presently this term is out of use. Examples of active devices that control resonator losses are a Pockels's cell, an acousto-optic modulator, and a rotating prism. The method of forced modulation of resonator loss is called active Q-switching. The Q-switch mode of operation was described in Prokhorov (1963) with the help of the rate equations.

We will not dwell on a detailed description of the active Q-switched mode of operation. The main equations are [7.6] and [7.13]. Let us take into account that the threshold amplification coefficient is $\alpha_{th} = \gamma_c = \delta_c/2l_a$ and α_0 is the amplification coefficient in the laser material at the moment ($t = 0$) when the resonator losses drop. Let us also introduce excess over threshold n as $n = \alpha_0/\alpha_{th}$. From [7.6] and [7.13], taking into account that $x_\infty = (\sigma_{12} + \sigma_{21})\int_0^\infty I(t)\mathrm{d}t$, we obtain a transcendental equation connecting excess over threshold and normalized fluence in a laser resonator:

© Woodhead Publishing Limited, 2013

$$x_\infty = (1 - \exp(-x_\infty))n \qquad [7.18]$$

Equation [7.18] is the same as the well-known equation (Prokhorov, 1963) connecting initial, threshold and final inversion densities in other notation.

From [7.6] and [7.13], we find intensity I_{max} in the resonator when it reaches a maximum (Prokhorov, 1963):

$$I_{max} = \frac{\delta_c}{T(\sigma_{12} + \sigma_{21})}(n - 1 - \ln(n)) \qquad [7.19]$$

Let us define pulse duration τ_p as $\tau_p = \int_0^\infty I(t)\mathrm{d}t/I_{max}$. From [7.19] and the definition of x_∞, we obtain (see Koechner, 2006)

$$\tau_p = \left(\frac{T}{\delta_c}\right)\frac{x_\infty}{n - 1 - \ln(n)} \qquad [7.20]$$

Laser pulse output energy E_{out} can be found, as was already done for steady-state operation:

$$E_{out} = \frac{1}{2}\ln(1/R)\,AF_s x_\infty \qquad [7.21]$$

It follows from [7.18] that fluence inside a laser resonator at Q-switch depends only on the excess over the threshold n and saturation fluence that is specific for the given laser material: $F_{in} = x_\infty F_s$. For some materials, this saturation fluence has typically high values. For example, for phosphate glasses doped with Er^{3+} at wavelength 1540 nm, the cross-section is $\sigma_{21} = 0.56 \times 10^{-20}$ to 0.7×10^{-20} cm^2 (Alekseev *et al.*, 1980), depending on glass composition, so saturation fluence is $E_s \approx 10$ J/cm^2 ($\sigma_{12} \approx \sigma_{21}$) and for $n = 2$ we have from [7.18] $x_\infty \approx 1.6$ and $F_{in} \approx 16$ J/cm^2. The last value is high enough to produce optical damage of optical elements in the laser resonator. So for the laser materials with high saturation fluence it is preferable to avoid the use of high values of n to avoid damage of intra-resonator optical elements by laser radiation. Equation [7.18] in this notation is in Basiev *et al.* (1982).

Transmission of some materials becomes higher at higher intensities. If an optical element of such a material is placed into a resonator then the laser radiation itself begins to 'control' transmission of this element. This mode of operation is called passive Q-switching. An example of passive Q-switching or a saturable absorber is a YAG crystal doped with Cr^{4+} (Zharikov *et al.*, 1986).

Historically, one of the first successful passive Q-switchings of a ruby laser was reported in Masters *et al.* (1963) wherein dye film-covered glass substrate was used as a passive Q-switch. Dye solution as a reversible bleaching absorber was reported in Kafalas *et al.* (1964) and Soffer (1964).

© Woodhead Publishing Limited, 2013

Colored glass, compounds of selenium and cadmium sulfide, have also been used as passive Q-switches for ruby lasers (Bret and Gires, 1964).

The scheme of a saturable absorber's levels is presented in Fig. 7.4. Laser radiation is in resonance with level 1 to 2 transition. Transition between levels 2 and 3 is considered fast. If the lifetime of level 3 is large in comparison with the resulting laser pulse duration, for example for YAG:Cr^{4+} it is 2 μs (Kuck *et al.*, 1995), then passive Q-switching can be described as being similar to active Q-switching. In this case, almost all absorbing centers are at level 3 during Q-switch laser pulse generation. The efficiency of this type of passive Q-switch depends on the relation between cross-sections of the saturable absorber and the active laser centers at the wavelength of laser radiation. If the condition $\sigma_q >> \sigma_{12} + \sigma_{21}$ (where σ_q is the absorption cross-section of a saturable absorber) is fulfilled then it is possible to obtain an effective Q-switch. Practically, the characteristics of output laser radiation (pulse duration, output energy) in this mode of operation are close to those when active Q-switching is used.

If the absorbing centers level 3 lifetime is too short (for example, 1 ns or less), then the time dynamics of laser generation become complicated. For example, in some conditions Q-switching becomes impossible and this element acts as a simple absorbing filter. Effective Q-switch operation is also possible if $\tau_q\sigma_q >> \tau(\sigma_{12} + \sigma_{21})$, where τ_q is the recovery time of the population of level 1 (Fig. 7.4): see, e.g., Siegman (1986). The concept of the second threshold (New and O'Hare, 1978) was introduced to describe the characteristics of lasers with these passive Q-switches. It is necessary to mention that there is one more important mode of operation of the laser with a passive Q-switch element that has a short recovery time. It is called mode locking (Mocker and Collins, 1965; DeMaria *et al.*, 1966). This is when laser radiation is emitted in the form of a train of short (up to several femtoseconds) pulses; it will not be discussed here.

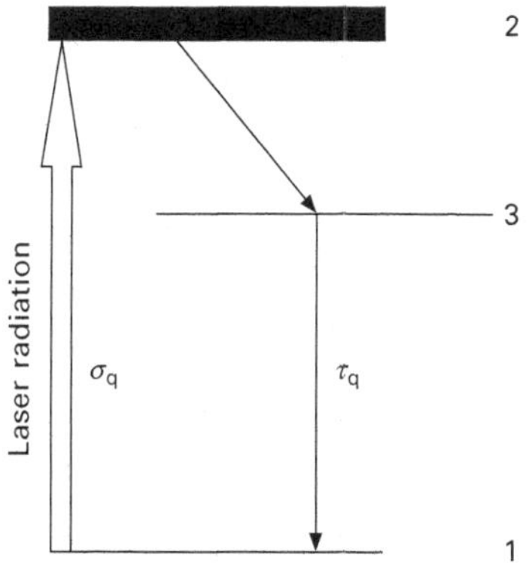

7.4 Scheme of levels of saturable absorber.

© Woodhead Publishing Limited, 2013

As mentioned above, in some conditions passive Q-switching can be considered as a variant of active Q-switching. The only difference is that it is necessary to take into account losses of energy to make a passive Q-switch transparent for laser radiation. We bring here final results referring to original papers (Mikaelyan *et al.*, 1965, 1967; Szabo and Stein, 1965). For x_∞ in the passive Q-switch mode of operation, instead of [7.18] we have equation

$$x_\infty + \frac{1}{\beta}(n-1)(1-\exp(-\beta x_\infty)) = (1-\exp(-x_\infty))n \qquad [7.22]$$

where $n = (2\ln(1/T_0) + \delta_c)/\delta_c$; T_0 is transmission of a saturable absorber at low laser intensity; and $\beta = \sigma_q/(\sigma_{12} + \sigma_{21})$ where σ_q is the absorption cross-section of a saturable absorber's centers. It is also assumed that the cross-sectional areas of radiation in the active material and the passive Q-switch are equal to each other and the recovery time of the saturable absorber is more than the duration of the light pulse. The output energy can be found using [7.21]. If $\beta >> 1$ we can neglect the second term in the left part of equation [7.22] which results in equation [7.18]. The pulse maximum intensity in a resonator and the pulse time duration in this case can be found from [7.19] and [7.20]. It is possible to show that equation [7.22] has the only root $x_\infty = 0$ if $\beta \leq n/(n-1)$, and for these values of β, Q-switch mode is impossible (Mikaelyan *et al.*, 1967).

It is necessary to note that the passive Q-switch operation has some peculiarities besides those that are named above. The spectral width of the laser radiation at passive Q-switching is narrower in comparison with active Q-switching (Sooy, 1965) due to a longer buildup time of radiation from spontaneous background to pulse maximum.

7.5.5 Model limitations

In our model we have made some assumptions that it is necessary to discuss. Pump intensity distribution inside the amplifying medium is not normally uniform and depends on a specific laser system. But practice shows that it is possible in many cases to use the mean in space values of pump intensity and obtain results with a level of accuracy that is acceptable for practical needs.

The intensity distribution of the laser radiation along a laser active element was considered to be uniform. For a steady state this statement follows from the used model when all inter-resonator losses are distributed uniformly along an active element as some efficient absorption coefficient and two 100% reflective resonators. This does not happen in reality, because most losses are concentrated at certain parts of the laser resonator, such as when the reflection coefficient of the output resonator mirror is less than 100%.

© Woodhead Publishing Limited, 2013

But this assumption is an important simplification to consider, as it makes the key points of laser physics much clearer. It is possible to take into account dependence on coordinate of intensity and amplification coefficient during generation. The paper by Rigrod (Rigrod, 1965) is an example of this consideration.

There is a standing wave for each mode that takes place during generation. This leads to an inhomogeneous distribution of amplification coefficient that is high at the nodes of the electric field and low at the antinodes due to a saturation of amplification. This in turn influences the spectrum of the laser radiation, as mentioned in Mak *et al.* (1968) and Siegman (1986). In our consideration this interference is not taken into account. The last assumption is applicable in the case of a wide generation spectrum when many longitude modes are taking part in the laser generation. This is normal for solid-state lasers.

The transverse distribution of the laser radiation in a laser resonator is not rectangular and saturation of amplification also distorts its transverse distribution. The Q-switched mode of operation is a special case of high saturation. The question of transverse distribution distortion due to saturation amplification has been discussed in many papers. We can refer to, for example, early papers by Fox and Li (1966) Statz and Tang (1965), Tang and Statz (1967), Bridges (1968), and McAllister *et al.* (1970). Approximate consideration of the Q-switched mode with TEM_{00} transverse distribution is presented in Il'ichev *et al.* (1997) and Bhardwaj *et al.* (2007). Transverse distribution $TEM_{n,m}$ $(m, n \geq 0)$ is also considered in Il'ichev *et al.* (2007).

7.6 Conclusion

This text presented a description of the principles of solid-state lasers at the main operating modes. A simple approximate model of a laser with uniformly distributed losses along the laser resonator was used for demonstrative purposes. We would like to underline the following points:

- Energy from an external source is stored in an active medium in the form of electronic excitation of dope centers. In order for amplification to take place, this energy is necessary. The amplification coefficient has to be large enough to overcome inter-resonator losses.
- Energy in the form of radiation is stored inside a laser resonator in a few selected modes of the resonator. The probability of radiation emission into these modes becomes much higher than in all others due to stimulated emission of radiation when laser generation takes place.
- Energy that is stored in a laser resonator is dissipated via inter-resonator losses.

Analytical expressions presented connect output power or energy characteristics

© Woodhead Publishing Limited, 2013

of laser radiation with the parameters of the amplifying medium, resonator, and pump power or energy.

References in this text reflect the historical development of understanding of solid-state laser principles and achievements in this area.

More than 50 years have passed since the first laser was put into operation and one would expect that all aspects of solid-state lasers have been researched. Nevertheless there are some problems that are still waiting for more attentive research. We single out those of them that are concerned with the influence of space gain saturation on output power, energy, divergence and spectrum characteristics of laser radiation.

The first problem is the influence of gain saturation on the transverse distribution of laser radiation inside a laser resonator. Gain saturation is a feature that is inherent to any laser mode of operation and especially to ones that are Q-switched. Further discussion of this problem can give us more precise evaluation of the mode volume and the radiation losses of a active resonator.

The second problem is that experiments also show that the spectrum of laser radiation for a dope center with a homogeneous broadened spectral line in a case with no special spectral mode discrimination is not a single longitude mode as could be expected. It is considered that existing standing waves of longitude modes lead to gain grating and this in turn is one of the factors responsible for the width of the laser radiation spectrum of solid-state lasers with a homogeneous broadened spectral line; see, e.g., the review by Mak *et al.* (1968) and Siegman (1986). More accurate consideration of the influence of the saturated gain grating on spectral, spatial and time characteristics of laser radiation is desirable.

These examples show that, despite the amount of time that has passed since the first laser was put into operation, our knowledge of solid-state lasers is still developing.

7.7 References

Alekseev N E, Gapontsev V P, Zhabotinskii M E, Kravchenko V B and Rudnitskii Yu P (1980), *Lasernye Fosfatnye Stekla* (*Laser Phosphate Glasses*), Moscow, Nauka.

Avanesov A G, Basov Yu G, Garmash V M, Denker B I, Il'ichev N N, Maksimova G V, Malyutin A A, Osiko V V, Pashinin P P, Prokhorov A M and Sychev V V (1980), 'High-efficiency pulse-periodic laser utilizing high concentration neodymium phosphate glass', *Sov J Quant Electr*, 10, 644–646.

Avizonis P V and Grotbeck R L (1966), 'Experimental and theoretical ruby amplifier dynamics', *J Appl Phys*, 37, 687–693.

Basiev G G, Denker B I, Il'ichev N N, Malyutin A A, Mirov S B, Osiko V V and Pashinin P P (1982), 'Passively Q-switched laser utilizing concentrated Li–Nd–La phosphate glass', *Sov J Quant Electr*, 12, 984–988.

Basov N G and Prokhorov A M (1955), 'About possible methods for obtaining active molecules for a molecular oscillator', *Sov Phys – JETP*, 1, 184–185.

© Woodhead Publishing Limited, 2013

Bhardwaj A, Agrawal L, Pal S and Kumar A (2007), 'Optimization of passively Q-switched Er:Yb:Cr:phosphate glass laser: theoretical analysis and experimental results', *Appl Phys*, B86, 293–301.
Bloembergen N (1956), 'Proposal for a new type solid state maser', *Phys Rev*, 104, 324–327.
Born M and Wolf E (2005), *Principles of Optics*, 7th edition, Cambridge University Press.
Boyd G D and Gordon J P (1961), 'Confocal multimode resonator for millimeter through optical wavelength masers', *Bell Sys Tech J*, 40, 489–508.
Boyd G D and Kogelnik H (1962), 'Generalized confocal resonator theory', *Bell Sys Tech J*, 41, 1347–1369.
Bret G and Gires F (1964), 'Giant-pulse laser and light amplifier using variable transmission coefficient glasses as light switches', *Appl Phys Lett*, 4, 175.
Bridges W B (1968), 'Gaussian beam distortion caused by saturable gain or loss', *IEEE J Quant Electr*, QE-4, 820–827.
DeMaria A J, Stetser D A and Heynau H (1966), 'Self mode-locking of lasers with saturable absorbers', *Appl Phys Lett*, 8, 174–176.
Dicke R H (1958), 'Molecular amplification and generation systems and methods', US Patent 2, 851, 652, 9 September 1958.
Einstein A (1916), 'Strahlungs-Emission und -Absorption nach der Quantentheorie', *Phys Ges*, 18, 318–323. Collection of studies, 3, Moscow, Nauka (1956), 386–392.
Fox A G and Li T (1960), 'Resonant modes in an optical maser', *Proc. IRE (Correspondence)*, 48, 1904–1905.
Fox A G and Li T (1961), 'Resonant modes in a maser interferometer', *Bell Sys Tech J*, 40, 453–488.
Fox A G and Li T (1966), 'Effect of gain saturation on the oscillating modes of optical masers', *IEEE J Quant Electr*, QE-2, 774–783.
Frantz L M and Nodvik J S (1963), 'Theory of pulse propagation in a laser amplifier', *J Appl Phys*, 34, 2346–2349.
Gordon J P, Zeiger H J and Townes C H (1955), 'The maser new type of microwave amplifier, frequency standard, and spectrometer', *Phys Rev*, 99, 1264–1274.
Hellwarth R W (1961), 'Control of fluorescent pulsations', in *Advances in Quantum Electronics*, ed. Singer J R, New York, Columbia University Press, pp. 334–341.
Il'ichev N N, Gulyamova E S and Pashinin P P (1997), 'Passive Q switching of a neodymium laser by a Cr^{4+}:YAG crystal switch', *Quant Electr*, 27, 972–977.
Il'ichev N N, Pashinin P P, Shapkin P V and Nasibov A S (2007), 'Passive Q-switching of an erbium-doped glass laser by using a Co^{2+}:ZnSe crystal', *Quant Electr*, 37, 974–980.
Kafalas P, Masters J I and Murray E M E (1964), 'Photosensitive liquid used as a nondestructive passive Q-switch in a ruby laser', *J Appl Phys*, 35, 2349.
Kaminskii A A (1981), *Laser Crystals*, New York, Springer-Verlag.
Koechner W (2006), *Solid State Laser Engineering*, 6th edition, New York, Springer Science+Media Inc.
Kogelnik H and Li T (1966), 'Laser beams and resonators', *Appl Opt*, 5, 1550–1567.
Kuck S, Petermann K, Pohlmann U and Huber G (1995), 'Near-infrared emission of Cr^{4+}-doped garnets', *Phys Rev B* 51, 17323–17331.
Maiman T H (1960), 'Stimulated optical radiation in ruby', *Nature*, 187, 493–494.
Mak A A, Anan'ev Yu A and Ermakov B A (1968), 'Solid state lasers', *Sov Phys Uspekhi*, 10, 419–452.

© Woodhead Publishing Limited, 2013

Masters J I, Ward J and Hartouni E (1963), 'Laser Q-spoiling using an exploding film', *Rev Sci Instr*, 34, 365–367.

McAllister G L, Mann M M and DeShazer L G (1970), 'Transverse-mode distortions in giant-pulse laser oscillators', *IEEE J Quant Electr*, QE-6, 44–48.

Mikaelyan A L, Antonyants V Ya, Dolgiy V A and Turkov Yu G (1965), 'About investigation of optical generator with passive Q-switch', *Radiotekhika i elektronika*, 10, 1350–1351.

Mikaelyan A L, Ter-Mikaelyan M L and Turkov Yu G (1967), *Opticheskiye Generatory na Tverdom Tele* (*Solid State Optical Generators*), Moscow, Soviet Radio.

Mocker H W and Collins R J (1965), 'Mode competition and self-locking effects in a Q-switched ruby laser', *Appl Phys Lett*, 7, 270–273.

New G H C and O'Hare T B (1978), 'A simple criterion for passive Q-switching of lasers', *Phys Lett*, 68A, 27–28.

Nobelprize.org, the official website of the Nobel Prize, 'The Nobel Prize in Physics 1964'. Available from http://www.nobelprize.org/nobel_prizes/physics/laureates/1964/

Prokhorov A M (1958), 'Molecular amplifier and generator for submillimeter waves', JETP (USSR), 34, 1658–1659; *Sov Phys JETP*, 7, 1140–1141.

Prokhorov A M (1963), 'Generation of optical amplifier at instantaneous switching of quality', *Radiotekhika i elektronika*, 8, 1073–1074.

Prokhorov A M, editor (1984), *Fizicheskii Enciklopedicheskii Slovar* (*Physical Encyclopedic Dictionary*), Moscow, Sovetskaya Enciklopediya.

Rigrod W W (1965), 'Saturation effects in high-gain lasers', *J Appl Phys*, 36, 2487–2490.

Schawlow A L and Townes C H (1958), 'Infrared and optical masers', *Phys Rev*, 29, 1940–1949.

Siegman A E (1986), *Lasers*, Sausalito, CA, University Science Books.

Soffer B H (1964), 'Giant pulse laser operation by a passive, reversibly bleachable absorber', *J Appl Phys*, 36, 2551.

Sooy R (1965), 'The natural selection of modes in a passive Q-switched laser', *Appl Phys Lett*, 7, 36–37.

Statz H and deMars G (1960), 'Transients and oscillation pulses in masers', in *Quantum Electronics*, ed. Towns C H, New York, Columbia University Press, pp. 530–537.

Statz H and Tang C L (1965), 'Problem of mode deformation in optical masers', *J Appl Phys*, 36, 1816–1819.

Svelto O (1998), *Principles of Lasers*, 4th edition, New York, Springer.

Szabo A and Stein R A (1965), 'Theory of laser giant pulsing by a saturable absorber', *J Appl Phys*, 36, 1562–1566.

Tang C L and Statz H (1967), 'Effect of intensity-dependent anomalous dispersion on the mode shapes of Fabry–Perot oscillators', *J Appl Phys*, 38, 886–887.

Zharikov E V, Zabaznov A M, Prokhorov A M, Shkadarevich A P and Shcherbakov I A (1986), 'Use of GSGG:Cr:Nd crystals with photochromic centers as active elements in solid lasers', *Sov J Quantum Electron*, 16, 1552–1554.

© Woodhead Publishing Limited, 2013

8
Powering solid-state lasers

C. R. HARDY, Kigre, Inc., USA

DOI: 10.1533/9780857097507.2.193

Abstract: This chapter discusses two types of laser power supply architectures for pumping solid-state laser materials. Designs for flashlamp pumped laser systems, including pulse-forming network (PFN) and energy storage unit (ESU) topologies, are discussed first. Advanced laser drivers for diode pumped solid-state (DPSS) laser systems are then presented. Design methods and calculations are given to ensure a safe, reliable, and efficient laser driver is chosen for the solid-state laser application.

Key words: pulse-forming network (PFN), energy storage unit (ESU), diode pumped solid-state (DPSS), laser driver, laser power supply.

8.1 Introduction

Lasers are often broken down into four general categories based on their lasing medium: gas tube discharge (chemical, excimer, gas and metal-vapor), liquid (dye), semiconductor (diode), and solid-state. Gas and dye lasers are not considered solid-state. Although semiconductor lasers have their own classification, and are also not considered solid-state, their driver electronics are identical to those of diode pumped solid state (DPSS) lasers. Therefore, in this section we elaborate on various architectures for driver electronics to power lamp-pumped and diode-pumped solid-state lasers.

The introduction of flashlamp pumped solid-state lasers in the early 1960s started a new branch in high energy power supply design (Koechner, 1976). Figure 8.1 shows a schematic of the first gigawatt (world record) ruby laser developed and manufactured by Lear Siegler Laser Systems Center (Myers, 1965). A primary power supply was used to drive the two parallel flashlamps while a secondary power supply was used to drive the flashlamp for the Q-switched element.

Figure 8.2 shows an actual ruby laser rod and uranium U^{6+} doped glass excited state absorber Q-switch cell used in this early world record laser system.

Crystalline and glass solid-state laser materials require high optical pump energies to lase, and gas discharge 'flashlamps' were developed to deliver the necessary electrical-to-optical conversion. These flashlamps provide the solid-state laser designer with a broadband radiation pump source that may be electrically optimized to overlap with the laser gain material's absorption

© Woodhead Publishing Limited, 2013

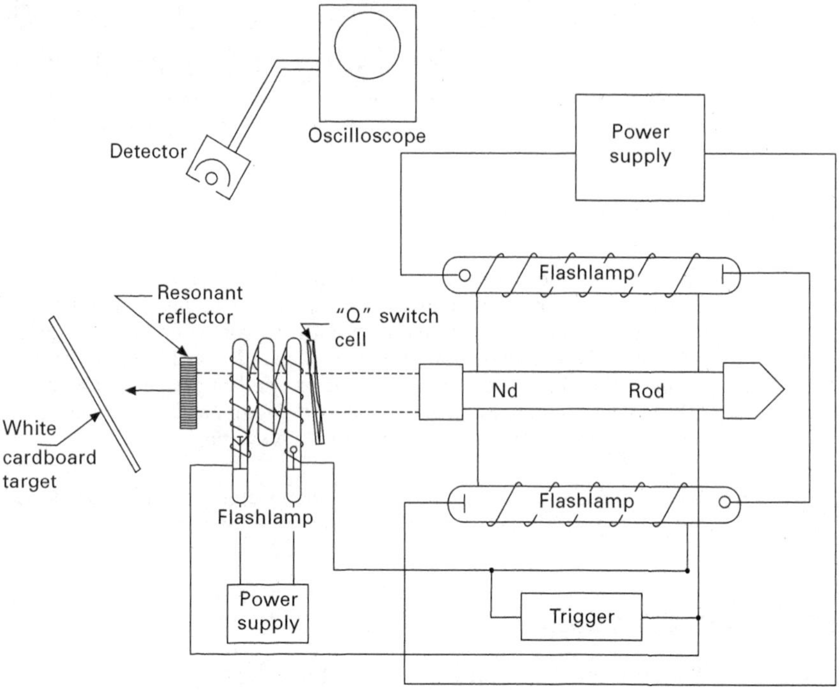

8.1 First gigawatt ruby laser.

8.2 Ruby laser rod and U6+ doped glass excited state absorber Q-switch.

spectra. For pulsed operation, the flashlamp's emission is optimized to match the laser's fluorescent lifetime.

However, controlling the hundreds or even thousands of amps (often at thousands of volts) required is not trivial. Capacitors, inductors, switches, connectors, and even wire must be properly rated for safe and reliable laser power supply operation. Protective circuitry should also be added to handle

© Woodhead Publishing Limited, 2013

current, voltage, and EMI transients. Most importantly, federal and local safety regulations need to be followed to prevent potential injury or fatality.

Diode pumped solid-state (DPSS) lasers, including fiber lasers, are now widely used in the industry. DPSS lasers replace the gas tube arclamp or flashlamp with a semiconductor diode laser as a pump source. Advantages of diode pumping (when compared to lamp-pumped systems) include higher efficiency, longer component lifetime, and lower maintenance requirements. DPSS systems have lower operating voltages and relatively less stored energy and are therefore much safer than flashlamp pumped systems. In some high peak power laser applications, the lamp pumped laser is still a popular option. The diode laser's maximum peak power is the same as its average power. A 1 cm pump diode laser array may provide ~100 watts per pulse of pump power. A flashlamp may produce orders of magnitude more peak power than a similarly priced diode array.

8.2 Safety

Laser power supply designers must understand and implement proper safety protocols for their product, application, and location. The designer should insure the device satisfies requirements for all areas where the equipment will be sold or potentially used, not just where it was developed. Regulations IEC 60825 internationally and ANSI Z136 in the United States outline requirements such as labeling, interlocks, and user safety for both the laser and laser-related equipment, including the power supply. In the United States, the Occupational Safety and Health Administration (OSHA) has standards and directives (instructions for compliance officers) related to laser hazards in the general laser industry that are specified by regulation 29CFR1910.

Class 3B and Class 4 lasers require interlocks that shut down a system whenever a safety condition is not met, such as when a room door or protective cover is opened. These interlocks are usually integrated into the laser power supply and therefore must be factored into the power supply design. An emergency stop switch must be in reach of the user and is typically located on the front panel or remote control section of the laser power supply. It is also good practice to include a key switch that can disable the laser and power supply when needed – during system maintenance, for example. A switch that allows the key to be removed in the OFF position is common as it allows the laser safety officer or similar designated responsible person to insure the laser and power supply will remain off when required.

8.3 Flashlamp pumping

A properly designed power supply will insure the flashlamp efficiently converts enough electrons to photons for reliable lasing in the solid-state

© Woodhead Publishing Limited, 2013

laser material. Xenon is typically used as the gas fill for flashlamps due to its relatively high input to output efficiency for solid-state laser material pumping. A graph of output efficiency versus wavelength for various fill gases is shown in Fig. 8.3.

Although a flashlamp's output is relatively broadband, specific regions of the spectrum can be emphasized by controlling current through the flashlamp (Buck *et al.*, 1963). At lower current densities, the spectral output is heavily weighted towards the visible and infrared end of the spectrum. As current densities increase, the lamp's spectral output shifts toward the blue and ultraviolet. Example emission spectra of a xenon flash lamp operating at a high current density (6.500 kA/cm^2 curve 1) and low current density (1000 kA/cm^2, curve 2) are shown in Fig. 8.4 (Penn State University, 2011).

Higher current density pulses often produce high energy short wavelength emission extending from the ultraviolet spectral region and dissipating through the visible and near infrared. This emission results from plasma energy that is

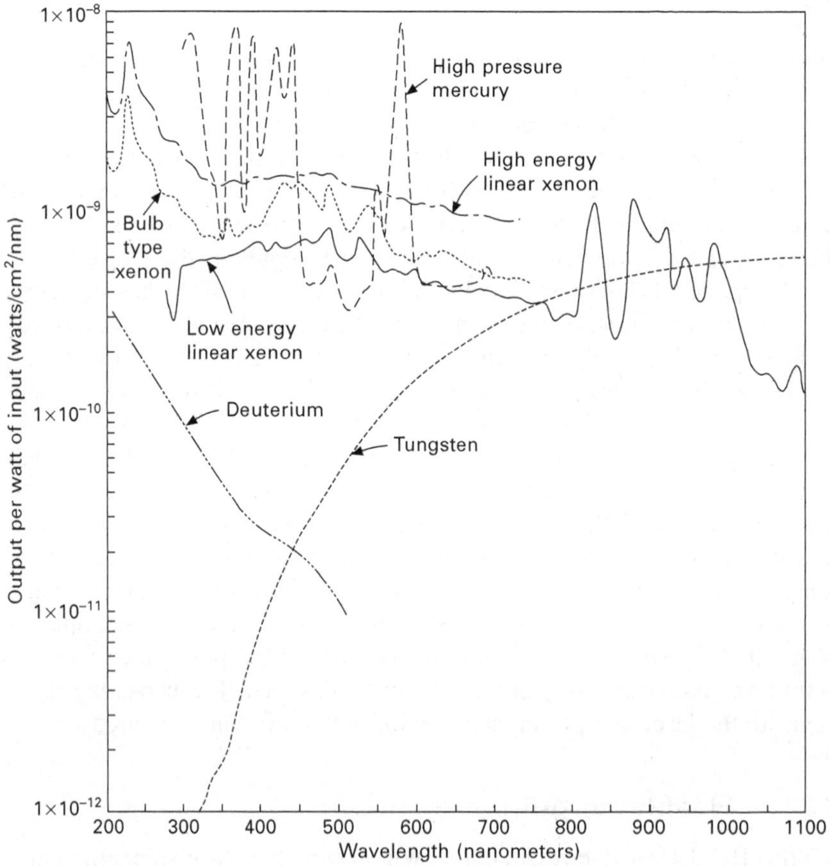

8.3 Fill gas comparison.

© Woodhead Publishing Limited, 2013

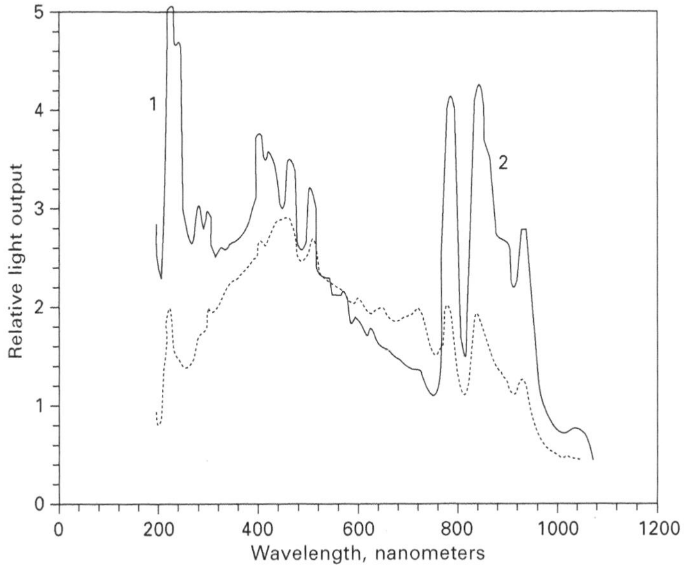

8.4 Emission spectra of a xenon flashlamp.

dominated by a strong 'white light' continuum, also described as broadband blackbody or *bremsstrahlung* radiation emission. Bremsstrahlung is from the German *bremsen*, to brake, and *Strahlung,* radiation, thus, 'braking radiation' or 'deceleration radiation'. This is essentially electromagnetic radiation that is produced by the acceleration and collision of charged particles (read electrons) with other charged particles such as atomic nuclei (Myers *et al.*, 2008).

The flashlamp's spectral output should match the laser's gain material as well as possible. Figure 8.5 shows a typical Nd:YAG absorption spectrum. At low power levels, the line radiation from krypton provides a good overlap. At higher power levels, xenon is preferred because, although its line structure is not perfectly matched to Nd:YAG, it is more efficient at converting electrical energy to blackbody radiation.

Flashlamp pumped power supplies may be operated either continuous (CW), modulated (quasi-CW), or pulsed with the proper choice depending on the laser application. The design of a flashlamp pumped solid-state laser includes optimization of key components such as the laser gain element, resonator optics, pump chamber, flashlamp and power supply. The laser's application dictates the targeted performance and operational envelope. Typical laser performance specifications include pulse repetition rate, peak power, average power, energy per pulse, power density, beam diameter, beam divergence, beam quality, electrical–optical efficiency, and wall plug efficiency.

© Woodhead Publishing Limited, 2013

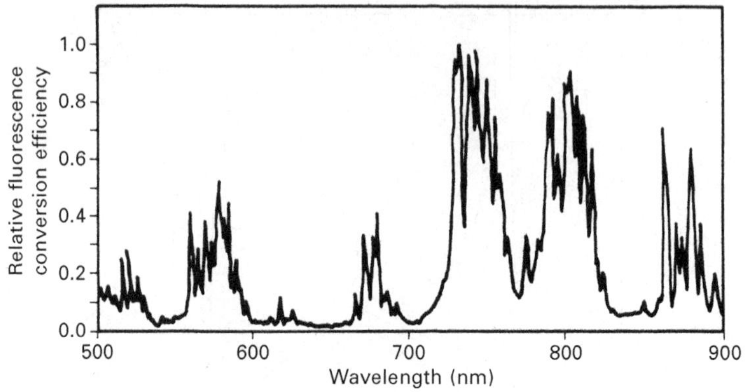

8.5 Nd:YAG absorption spectrum.

The required input energy (E_i) is a function of the laser's worst-case estimated efficiency and is defined as laser output energy (E_o) divided by resonator efficiency (*eff*) as shown in Equation 8.1:

$$E_i = \frac{E_o}{\text{eff}} \quad [8.1]$$

The laser application will determine the current pulsewidth requirement. Short, high energy pulses are typically best created with a pulse forming network (PFN) circuit, while pulses over 1 ms typically require an energy storage unit (ESU) approach. With this information, the proper energy storage, voltage, and current values can be determined.

8.3.1 Flashlamp ionization (triggering)

Flashlamps are gas discharge tubes that are usually filled with xenon and/or krypton. Four types of triggering are commonly used: external, series, simmer, and pseudo-simmer.

External trigger circuits

External triggering uses a step-up transformer to create a small arc (streamer) between the flashlamp's electrodes. To reduce the trigger energy required for a specific flashlamp, a thin, high temperature trigger wire is typically wrapped around the outside of the flashlamp. The trigger pulse is a damped oscillation from the transformer's secondary with peak amplitude voltage in the ± 10 kV to 30 kV range. The pulse duration is usually 200 ns per inch of arc-length. Since there is no switch holding back the PFN energy, current will begin to flow in the flashlamp as soon as the flashlamp is ionized by the fast, high voltage trigger pulse.

Specific trigger voltages depend on arc length, bore size, fill-pressure, electrode material and capacitor potential. The trigger pulse can also be applied to a metal bar or pump-cavity element as long as the conductive metal covers the entire distance between the flashlamp's electrodes and is close coupled to the flashlamp (typically under 6 millimeters). Because the transformer's secondary winding is outside the flashlamp's high current path, the transformer can be made relatively small. An external trigger circuit is shown in Fig. 8.6.

Series trigger circuits

Series triggering also uses a step-up transformer to create a small arc between the flashlamp's electrodes. However, unlike the external trigger circuit, the series design places the transformer 'in series with' the flashlamp. The series trigger transformer is larger and heavier than the external trigger design because the secondary winding must carry the full flashlamp current. The secondary winding also adds impedance to the PFN discharge circuit, which must be factored into the design.

However, by choosing the proper series transformer, it can replace the inductor element in the PFN circuit. Because the trigger pulse is applied directly to the flashlamp's electrode, the trigger voltage and energy can often be lower than required by the external trigger circuit. This results in less radiated electromagnetic interference (EMI) and more reliable flashlamp ionization. As with the external trigger circuit, current will begin to flow in the flashlamp as soon as the flashlamp is ionized by the trigger pulse from the transformer. A series trigger circuit is shown in Fig. 8.7.

Simmer trigger circuits

The simmer trigger circuit uses a separate DC current source (simmer power supply) to maintain a continuous DC current through the flashlamp. This small DC current (typically 100–500 mA) keeps the flashlamp ionized as soon as the external trigger pulse is applied. Because of the switch element,

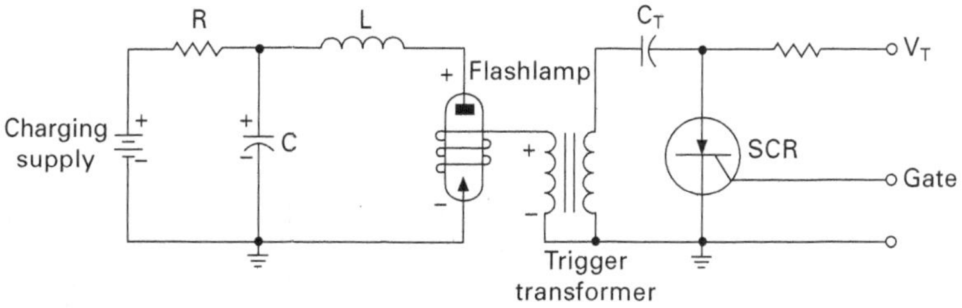

8.6 External trigger circuit.

© Woodhead Publishing Limited, 2013

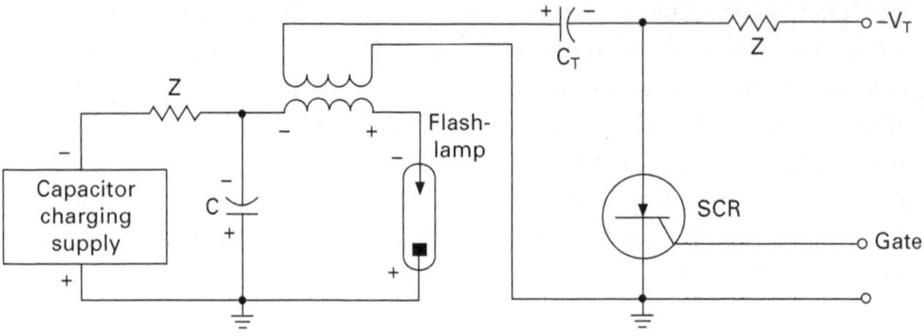

8.7 Series trigger circuit.

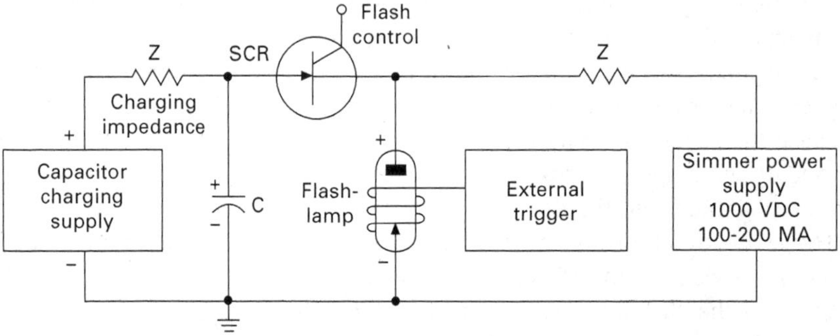

8.8 Simmer trigger circuit.

a silicon-controlled rectifier (SCR), for example, the PFN energy will not discharge into the flashlamp even though the lamp is ionized. Current from the PFN will flow through the flashlamp only when the separate flash control pulse is applied to the SCR switch.

Because the high voltage trigger pulse has to be applied only once, a simmer circuit is very advantageous for repetition rates above 1 hertz. However, simmering the lamp adds heat (often in the tens of watts) that must be removed from the flashlamp and laser pump cavity. A simmer trigger circuit is shown in Fig. 8.8.

Pseudo-simmer trigger circuits

Pseudo-simmer trigger circuits typically use a high power resistor and transistor switch instead of a separate simmer power supply. Although simple in its approach, the series simmer (pass) element can dissipate tens of watts. So this circuit often includes a secondary switch (not shown) in series with the resistor to enable the pseudo-simmer current just before the transistor, such as a metal oxide semiconductor field-effect transistor (MOSFET) or an insulated gate bipolar transistor (IGBT), to be turned on.

© Woodhead Publishing Limited, 2013

The time it takes for the flashlamp to start conducting current will depend on the lamp, PFN, and resistor value but is generally in the 1–100 μs range. A pseudo-simmer trigger circuit is shown in Fig. 8.9.

8.3.2 Pulsed forming network (PFN) design

A PFN design typically includes the AC-DC or DC-DC capacitor charging power supply, high voltage capacitor, air-core, wire-wound inductor, silicon controlled rectifier (SCR) switch, flashlamp, and related control circuitry. It is usually best to design the system by selecting the flashlamp, then calculating the proper PFN capacitor and inductor values to get the proper current pulse profile. A typical PFN circuit is shown in Fig. 8.10.

In most practical flashlamp circuits, the inductance, capacitance, and capacitor voltage are carefully chosen so that the energy is transferred to the flashlamp in a critically damped pulse. Critical damping is important to insure the most efficient transfer of energy from the capacitor to the flashlamp. For a flashlamp, a damping factor of 0.8 is considered to be optimal. A graph of a critically damped pulse with respect to normalized energy is shown in Fig. 8.11 (ILC Technology, 1983).

Damping factors over 0.8 are considered 'over-damped' and result in low peak current and power. A graph of over-damped ratios for normalized current versus normalized time is shown in Fig. 8.12.

Damping factors under 0.8 are 'under-damped' and result in high peak current, lower peak power, and lower efficiency energy transfer. Under-

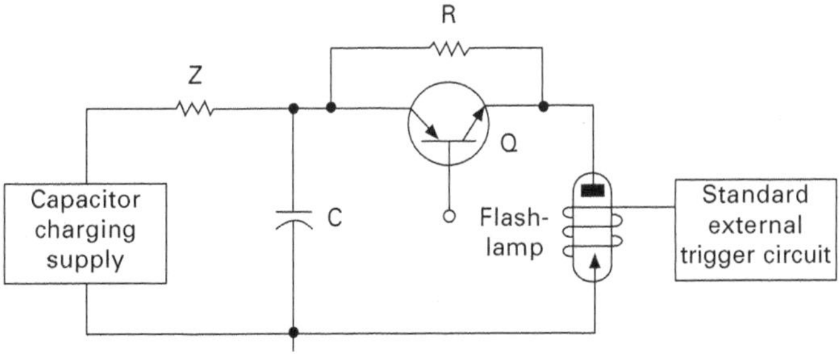

8.9 Pseudo-simmer trigger circuit.

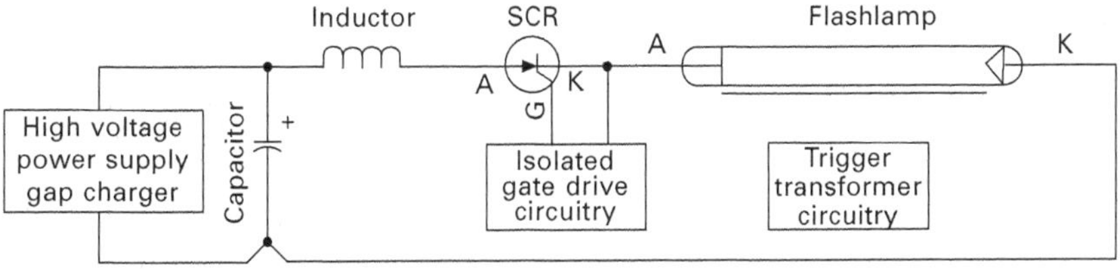

8.10 PFN circuit.

© Woodhead Publishing Limited, 2013

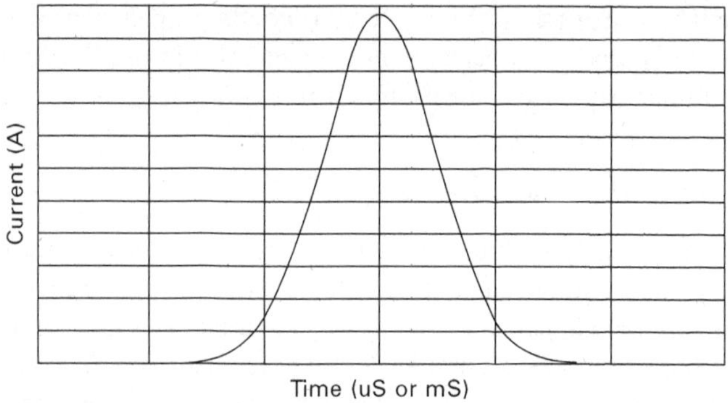

8.11 Critically damped pulse.

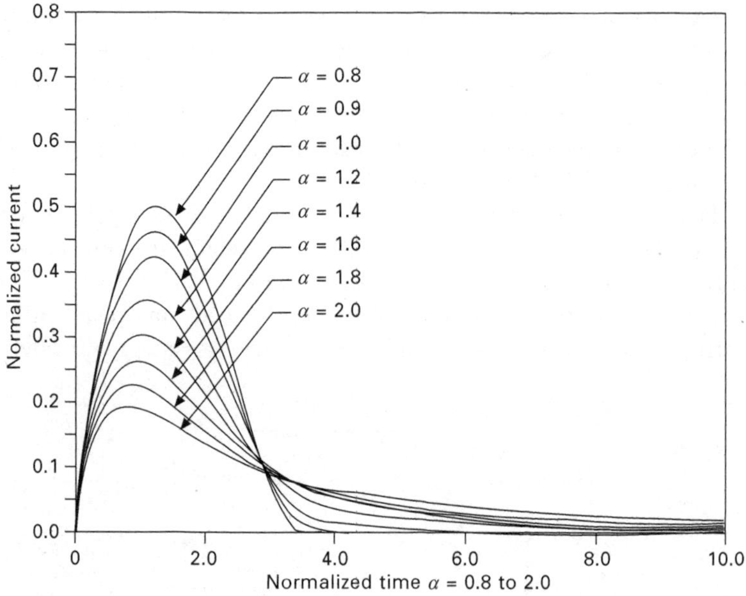

8.12 Over-damped pulse.

damped circuits also produce current reversal (ringing) which is detrimental to flashlamp lifetime. A graph of under-damped ratios for normalized current versus normalized time is shown in Fig. 8.13.

Designing a critically damped flashlamp circuit would be straightforward if it could be treated as a traditional RLC circuit. However, unlike a linear resistor, the flashlamp has dynamic impedance designated by K_o. The lamp impedance calculation is shown in Equation 8.2.

$$K_o = 1.28 * \left(\frac{\%Xe}{450} + \frac{\%Kr}{805}\right)^{1/5} * \frac{\ell}{d}\ \Omega\sqrt{A} \qquad [8.2]$$

© Woodhead Publishing Limited, 2013

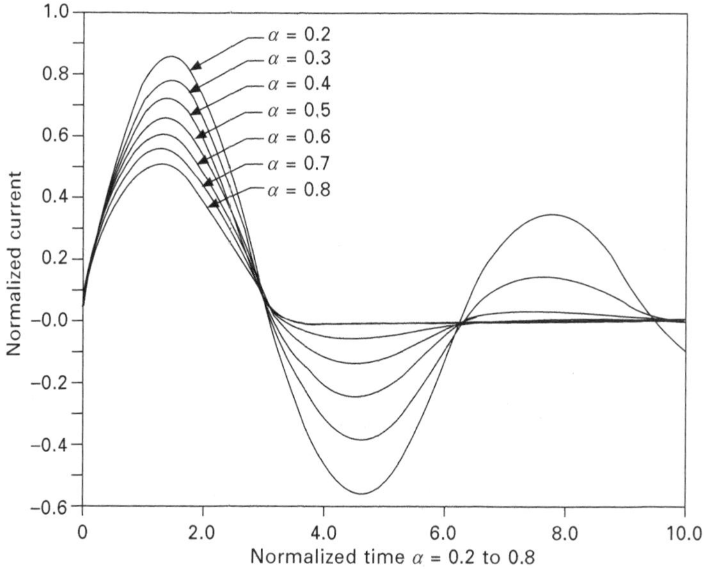

8.13 Under-damped pulse.

where fill pressure is in torr, ℓ = arc length in cm, and d = bore diameter in cm. Units are ohms*square-root of amps. The capacitance calculation is shown in Equation 8.3.

$$C = \left[\frac{2 * E_o * \alpha^4 * \left(\frac{t}{3}\right)^2}{K_o^4}\right]^{1/3} \quad [8.3]$$

where C = capacitance in farads; E_o = energy stored in capacitor in joules; α = unit-less damping parameter = 0.8 for critical damping; and t = desired current pulse-width at 10% points. The inductance calculation is shown in Equation 8.4.

$$L = \frac{\left(\frac{t}{3}\right)^2}{C} \quad [8.4]$$

where L = inductance in henries. The capacitor voltage calculation is shown in Equation 8.5.

$$V = \sqrt{\frac{2 * E}{C}} \quad [8.5]$$

where V = initial capacitor voltage in volts. The circuit impedance calculation is shown in Equation 8.6:

© Woodhead Publishing Limited, 2013

$$Z = \sqrt{\frac{L}{C}} \qquad [8.6]$$

where Z = circuit impedance in ohms. In reality, the exact capacitor and inductor values determined by calculation are rarely available off-the-shelf. So it is important to recalculate the actual damping factor using actual values for the capacitor and inductor. The actual damping factor calculation is shown in Equation 8.7, where α = the actual damping factor:

$$\alpha = \frac{K_o}{\sqrt{V_o * Z_o}} \qquad [8.7]$$

If necessary, lamp impedance can be adjusted to insure a critically damped current pulse is achieved. As seen in Equation 8.2, the lamp's K_o can be increased by increasing fill pressure. However, this will also affect the current pulse waveform. Therefore, actual circuit components should be selected to insure lamp impedance between 0.7 and 0.8. An anti-parallel shunt diode must be added directly across the PFN capacitor to eliminate the negative current swing of a potentially under-damped circuit.

The peak current delivered by the PFN to the flashlamp should be calculated to insure all of the circuit components are properly rated. The peak current calculation is shown in Equation 8.8:

$$I_{pk} = 0.94 * \left[\frac{(e^{-0.77 * \alpha}) * V_o}{Z_0} \right] \qquad [8.8]$$

where I_{pk} = peak current in amps. Flashlamps do not operate under standardized conditions and therefore cannot be given a specific lifetime rating. Instead, flashlamp lifetime is a function of bore size, arc length, input energy, and pulse width and is calculated in terms of total number of shots (flashes). The maximum input energy is referred to as explosion energy, Ex. The explosion energy calculation is shown in Equation 8.9:

$$Ex = 90 * d * \ell * \sqrt{t} \qquad [8.9]$$

where d = bore diameter in mm; ℓ = arc length in inches; and t = current pulse-width in milliseconds. The term explosion energy is used because this is the energy at which the envelope is likely to fracture. Flashlamp lifetime is a function of the ratio of the input energy to explosion energy. The flashlamp lifetime calculation is shown in Equation 8.10:

$$\text{Lifetime} = \left(\frac{E_{in}}{E_x} \right)^{-8.5} \qquad [8.10]$$

where E_{in} = input energy from capacitor, and E_x = explosion energy calculation.

© Woodhead Publishing Limited, 2013

In practice, actual lamp lifetimes may also be limited by electrode life due to sputtering of electrode material onto the flashlamp wall surface. In this case, light output drops gradually throughout the lamp's lifetime. These PFN equations can be modeled using various computer programs to calculate ideal values for the application. Computer modeling also allows the user to try 'what if' scenarios to potentially better optimize the laser system using off-the-shelf components.

Many types of PFN capacitor styles exist, including oil filled, metalized and discrete foil, film and paper dielectric. Companies such as CSI Technologies, Dearborn Electronics, and General Atomics Electronic Systems offer a wide capacitor selection and usually include helpful design guides and data-sheets on their websites. Metalized capacitors, shown in Fig. 8.14, are widely used in the pulsed laser industry.

Purchasing the proper PFN inductor can often prove tricky, so most companies manufacture their own. Thankfully the construction process is not too complex and simple loops of magnet wire around a plastic form will suffice. Magnet wire is available from companies such as MWS Wire Industries and is usually available through standard electronic product distributors.

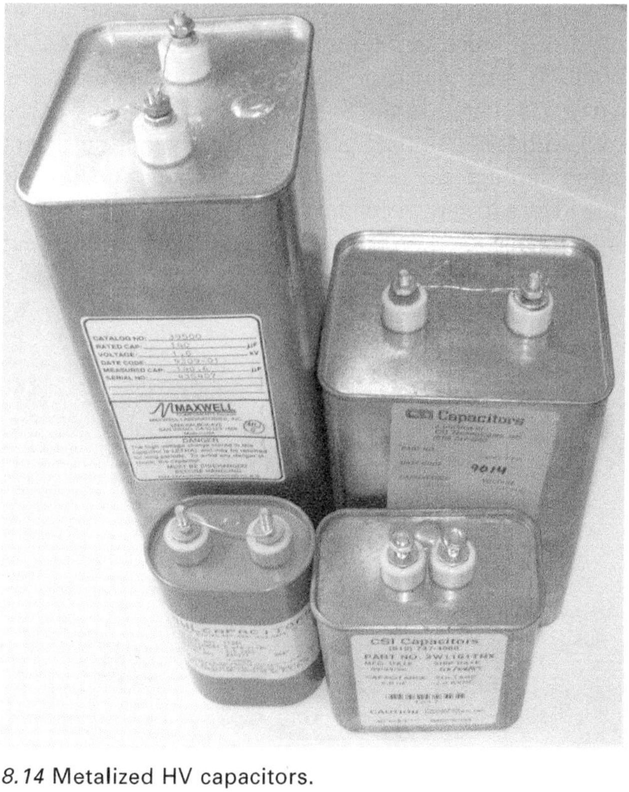

8.14 Metalized HV capacitors.

© Woodhead Publishing Limited, 2013

Polyimide-coated magnet wire is preferred for PFN applications. Transformer encapsulation material is generally available from companies such as Lord Corporation, Master-bond, and Solar Compounds Corporation.

A small, custom PFN is shown in Fig. 8.15. 20AWG magnet wire is wrapped around a 20 μF, 1 kV metal foil capacitor to provide 20 μH of inductance. This PFN assembly is part of a laser system used by StellarNet, Inc. in their PORTA-LIBS 2000 Laser Induced Breakdown Spectroscopy (LIBS) instrument. LIBS systems focus a high peak-power laser onto a small area at the surface of the specimen to create plasma. This permits real-time qualitative identification of trace elements in solids, gases, and liquids via optical detection of elemental emission spectra. The PORTA-LIBS 2000 instrument is shown in Fig. 8.16.

A similar PFN assembly was used by RCA Corporation (Burlington, MA) in their model AN/GVS-5 Near-Infrared (NIR) laser rangefinder. This unit uses a pulse time-of-flight (ToF) approach to determine accurate distance measurements to a remote target. A ToF laser rangefinder consists of a high peak power, pulsed laser transmitter, optical receiver, and range computer. Eye-safe and non-eye-safe laser wavelengths may be used depending on the application. The AN/GVS-5 PFN assembly can be seen in Fig. 8.17.

The high current SCR thyristor switch is available from manufacturers such as International Rectifier, Powerex, and Semikron. SCR thyristors conduct current only after they have been switched on by the gate terminal. Once conduction has started in the SCR, the device remains latched in the 'on' state, even without additional gate drive, as long as sufficient current continues to flow through the device's anode–cathode junction. In a PFN application, the SCR will stay in conduction as long as the current flowing to the lamp from the capacitor exceeds the SCR's latching current specification.

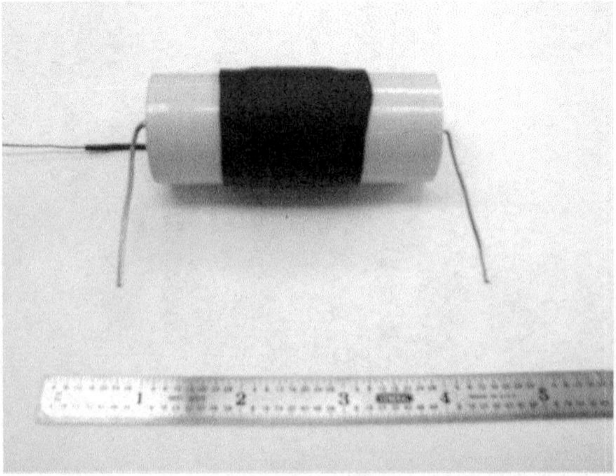

8.15 20 μF, 20 μH, 1 kV PFN assembly.

© Woodhead Publishing Limited, 2013

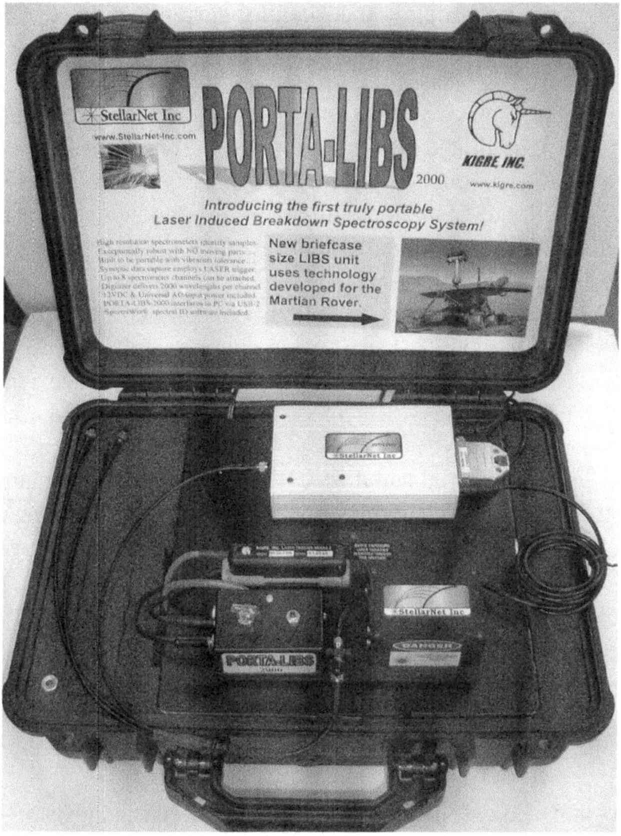

8.16 StellarNet PORTA-LIBS 2000 LIBS instrument.

8.17 RCA Corporation AN/GVS-5 NIR laser rangefinder.

© Woodhead Publishing Limited, 2013

As soon as current drops below the latching current rating, the SCR will stop conducting and will stay off until the next trigger pulse is received at its gate.

SCR latching currents are often under 1 amp, so it is important to turn off (quench) the high voltage power supply just before the trigger command is sent to the SCR. The quenching time is typically a few milliseconds to insure both the SCR and flashlamp come fully out of conduction. The power supply voltage and power rating are determined by the energy per pulse, the quench time, and the repetition rate.

SCRs for laser power supply applications typically come in three distinct package styles: stud mount, disc (hockey puck), and module. Modules are generally the easiest to use since their mounting surface tends to be electrically isolated from the SCR's anode and cathode terminals. Therefore, the module's heatsink does not require electrical isolation from the laser power supply enclosure. Stud and disc styles handle tremendous amounts of current, but their anode and cathode connections are the mounting surfaces, and are therefore not electrically isolated. Heatsinks for these types of devices must provide the necessary electrical hold-off for the application. Important specifications to consider when using an SCR include anode-to-cathode heat dissipation, maximum current, voltage, and d*v*/d*t* rating.

SCR gate drive circuits typically consist of a simple transformer to provide necessary high voltage isolation between the logic trigger pulse and the SCR's gate and cathode terminal. A typical SCR gate drive circuit is shown in Fig. 8.18.

Gate protection elements include an anti-parallel diode and resistor–capacitor (R–C) filter for transient absorption. The diode is often a high speed zener to clamp voltage spikes at the SCR's gate. The snubber circuit consists of a high current, anti-parallel diode and a series-connected resistor and capacitor with values selected for the necessary frequency response for the application.

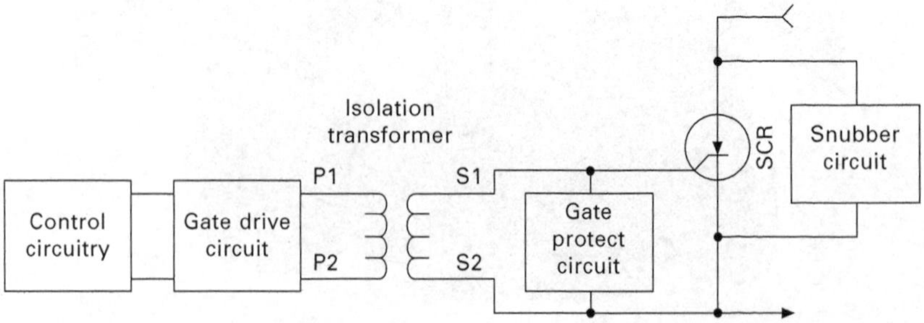

8.18 SCR gate circuit.

© Woodhead Publishing Limited, 2013

8.3.3 Multiple-section PFN design

Some applications require a square current profile instead of the standard Gaussian shape provided by a critically damped PFN. In this case, multiple PFN sections can be added together to create a 'mesh' circuit network. Three (or more) sections, with equal capacitor and inductor combinations, combined in a transmission line arrangement will result in a reasonably square current pulse profile. A three-mesh PFN developed for Fermi National Accelerator Laboratory in Batavia, IL, is shown in Fig. 8.19.

This PFN network was able to deliver 1700 V, 3200 A, 1.5 ms current pulses for a Nd:glass slab laser system. The custom multi-tapped inductor is located above the energy storage capacitors. The SCRs, current monitor circuits, discharge circuits, and fire control circuitry are located on the panel above the inductor. Each of these mesh PFN assemblies included optical transceivers to communicate with the user's fire control system. The laser system included four of these mesh PFN units, providing 32 kilowatts of average power and a large liquid cooling system for the Nd:glass slab and four large flashlamps. An oscilloscope trace of the 3200 A, 1.5 ms current pulse is shown in Fig. 8.20. The current scale is 500 A/division and the time scale is 500 μS/division. The current was measured using a Pearson Electronics model 5623, 0.001V/A, Hall-effect current probe. When using Hall-effect current monitors, it is important to use a probe with a sufficiently

8.19 Three mesh PFN.

© Woodhead Publishing Limited, 2013

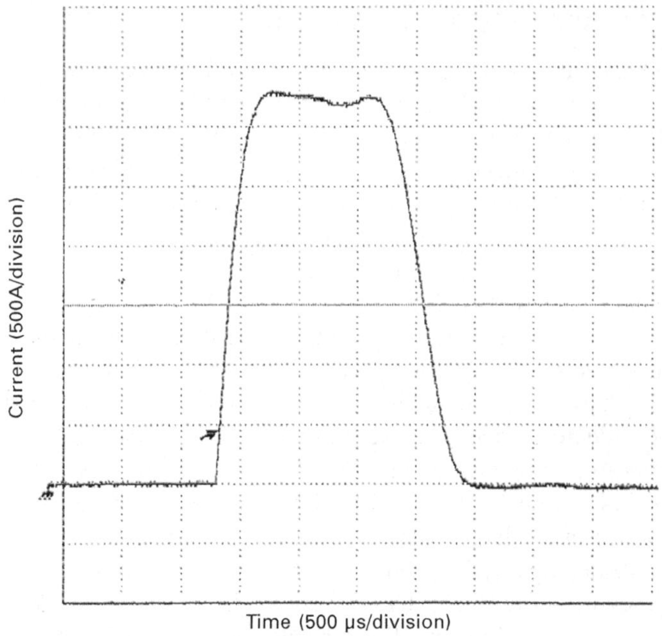

8.20 3200 A current pulse using a three-mesh PFN network.

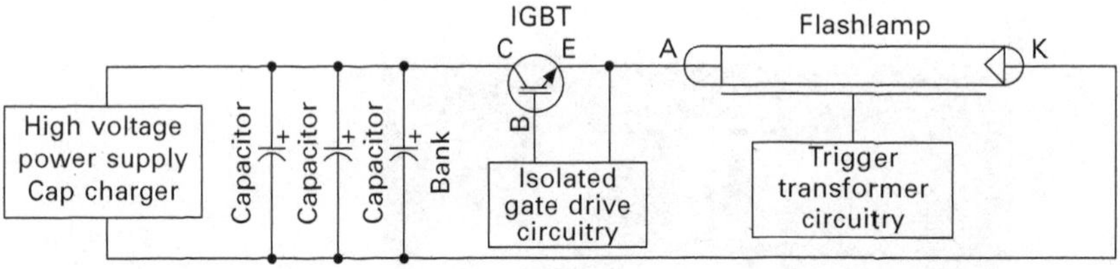

8.21 ESU circuit.

rated current–time product for the application, or waveform distortion due to probe saturation will result. DC probes are ideal for measuring long current pulses but are often limited by their relatively low peak current ratings.

8.3.4 Energy storage unit (ESU) design

An ESU design typically includes the AC-DC or DC-DC capacitor charging power supply, medium voltage energy storage capacitor bank, high current switch, flashlamp, and related control circuitry. It is usually best to design the system by selecting the flashlamp, then calculating the proper ESU capacitor bank values to get the proper current pulse profile. And unlike a fixed PFN design, the ESU circuit also allows for variable current pulse-widths. A typical ESU circuit is shown in Fig. 8.21.

© Woodhead Publishing Limited, 2013

ESU capacitor banks are typically a parallel or series–parallel combination of high energy density low equivalent series resistance (ESR) aluminum electrolytic capacitors. Companies such as Cornell Dubilier Electronics, Illinois Capacitor, TDK-Epcos, and Panasonic offer a wide capacitor selection and usually include helpful design guides and data-sheets on their websites. Laminated buss-bar assemblies can provide a reliable connection between energy storage capacitors. Due to their low impedance power path, buss-bars can reduce circuit losses and will help insure uniform current distribution. The high current IGBT thyristor switch is available from manufacturers such as International Rectifier, Microsemi, and Powerex.

Unlike SCRs, IGBTs have the ability to turn off via the gate control signal. This allows the power supply designer to create a specific current pulse shape without the added loss of an inductor element. The on-time of the IGBT's gate drive pulse will determine the flashlamp's current pulse-width. A single 600 A, 1400 V IGBT from Powerex, model CM600HA-28H, is shown in Fig. 8.22.

There are two important things to notice in this figure. First, the module's gate terminals are shorted to prevent damage during storage. Secondly, there is a large letter 'H' written on the top of the device. This letter is added by the manufacturer and corresponds to the measured on-resistance of the device. Specific on-resistance can vary from device to device, so it is extremely important to use modules that have the same letter when using them in parallel circuits for high current applications. This insures relatively

8.22 Powerex single IGBT module.

© Woodhead Publishing Limited, 2013

even current sharing through each IGBT in the parallel circuit. Modules with multiple IGBTs inside are also available.

High voltage, high current switches are widely used in ESU designs. Due to their relatively low gate charge, MOSFETs are often used in high frequency designs while IGBTs are typically preferred for low frequency (<20 kHz) applications. For a given package size, IGBTs generally have lower on-resistance and therefore lower drain-to-source voltage and are the choice for very high pulsed current applications.

IGBT gate drive circuits are required to properly get the IGBT into and out of conduction. Depending on the location of the IGBT in the ESU circuit, the gate drive circuit may need to be isolated. Small DC-DC converters are used to provide this isolation, but not all converters provide high isolation voltage, so it is important to refer to the manufacturer's datasheet to insure a sufficient isolation rating for your application. Gate drive circuits can also provide short circuit protection by monitoring the saturation voltage of the switching device. A typical IGBT gate drive circuit is shown in Fig. 8.23.

8.4 Laser diode pumping

Using semiconductor diode lasers instead of flashlamps as an optical pump source for solid-state lasers offers significant advantages such as higher optical efficiency and longer pump source lifetime. In the past decade, and particularly in the last three years, semiconductor laser diodes have made huge advances in power handling and reliability. Laser efficiency improvements have led system designers to expect ever higher efficiencies from their laser diode power supplies. Depending on the application, a properly designed DPSS laser can typically exceed 1 billion pulses. Therefore, the weak link in system design is now typically the driver (especially the energy storage and filter capacitors), not the flashlamp.

As is the case with a flashlamp-based power supply, a properly designed

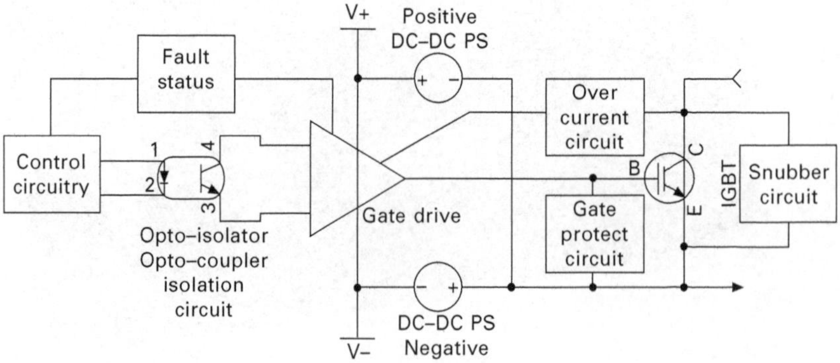

8.23 IGBT gate drive circuit.

© Woodhead Publishing Limited, 2013

DPSS laser driver is extremely important. The exact voltage and current control methods will greatly depend on the application. High average power systems typically require a switch-mode DC-DC converter with current feedback to the switching control circuitry. Low average power applications can often use a simpler linear current control circuit as long as the circuit's heat dissipation is acceptable. The decision whether to use an isolated or a non-isolated driver will also depend on the laser design and potential for external wiring faults, such as a short-circuit to chassis ground.

A typical DPSS driver will include input, soft-start, energy storage, current control, and laser diode protection circuits. An example of a 0 to 500 amp, 100 μs to 3 ms pulsed laser driver from Kigre, Inc. is shown in Fig. 8.24. Low profile energy storage capacitors (to handle the peak current discharge) are underneath the board.

8.4.1 Pump diode selection

Efficient optical pumping requires good spectral matching of the pump diode wavelength with the absorption spectrum of the laser material. Unfortunately, many laser materials have narrow absorption peaks, so careful attention is needed to insure the pump wavelength remains in the absorption band for all operating conditions. Table 8.1 lists the peak pump absorption wavelength and typical output wavelength for some of the most common laser materials.

It is important to consider the laser material's absorption spectra for the entire temperature range of one's application. The absorption spectrum should

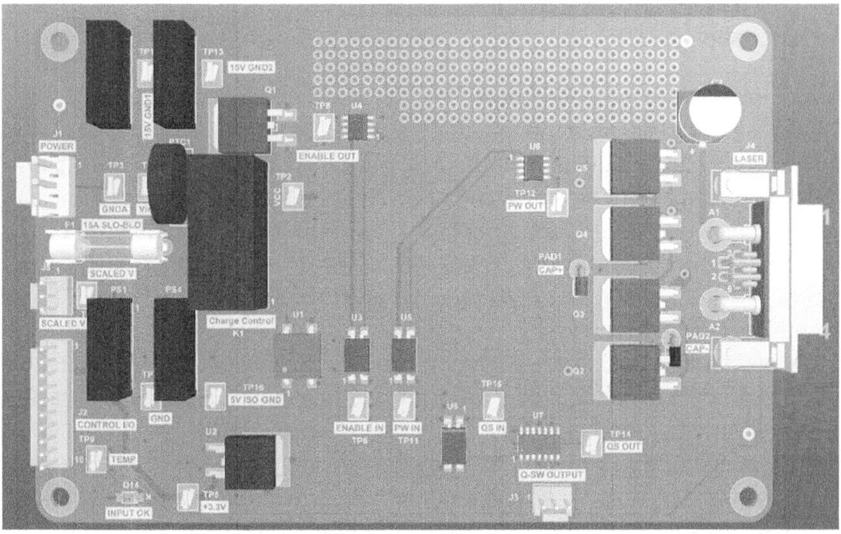

8.24 0–500 amp pulsed DPSS laser driver.

© Woodhead Publishing Limited, 2013

Table 8.1 Pump diode absorption wavelengths

Laser material	Er:YAG	Nd:YAG	Yb:YAG	Ho:Cr: Tm:YAG	Nd:glass (phosphate)	Er:glass (phosphate)
Peak absorption wavelength (25°C)	940 nm	808 nm	940 nm	781 nm	803 nm	975 nm
Output wavelength	2940 nm	1064 nm	1030 nm	2097 nm	1054 nm	1535 nm

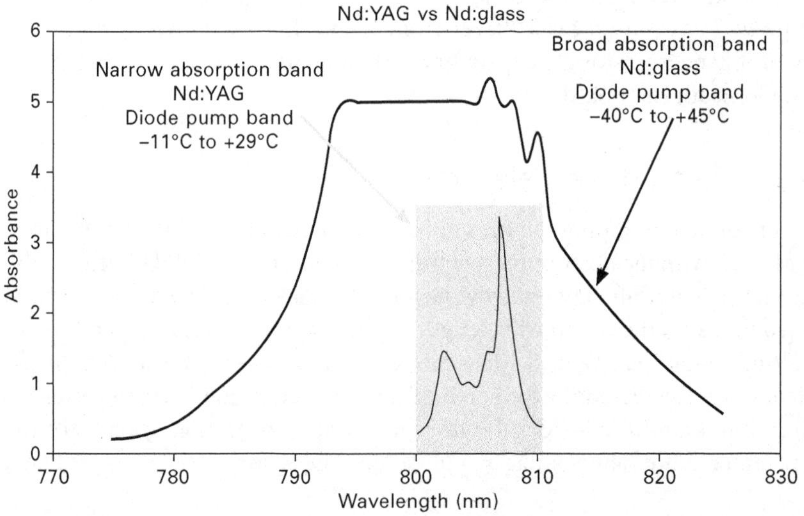

8.25 Effective diode drift range for Nd:YAG and Nd:glass pump bands.

be compared with the pump diode's center wavelength and wavelength shift specifications. The laser pump diode's wavelength will shorten at lower temperatures and lengthen at higher temperatures. Therefore, the best pump diode choice may not have its center wavelength at the laser material's peak absorption wavelength. Compromises must also be made with respect to standard pump diode wavelengths as most diode bars operate in the 780–860 nm or 940–980 nm wavelength regions with 808 nm, 940 nm, and 975 nm being the most common.

Wavelength shift due to temperature fluctuations can be especially troublesome in Nd:YAG and other crystalline host DPSS lasers as the width of their pump band tends to be very narrow. However, laser glass host gain materials provide relatively broad pump band widths, allowing them to function over wide temperature ranges without the need for diode thermal conditioning. A typical specification for diode wavelength drift with temperature is 0.25 nm/°C. Figure 8.25 illustrates the difference in pump

© Woodhead Publishing Limited, 2013

band width and how it affects the thermal stability of neodymium doped yttrium aluminum garnet (YAG) crystal and glass host lasers (Myers *et al.*, 2007).

A large number of individual diode emitters can be attached to a sub-mount and packaged into a diode bar array. Multiple diode bars can also be stacked vertically on the same sub-mount package as shown in Fig. 8.26. This approach can yield Quasi-CW (pulsed) output powers in the hundreds of watts.

Low power diode bars often use indium/tin (52% In, 48% Sn) solder due to its low melting point of 118°C. The low melting point allows the manufacturer to solder the laser diode emitters to the substrate without risking damage to the semiconductor material. However, high power diode bars are typically assembled using high temperature (hard) gold/tin (80% Au, 20% Sn) solder rather than lower temperature indium/tin solder. The higher melting point of gold/tin (281°C) allows the diode bar to operate at much higher temperatures, and therefore higher peak powers, than modules using indium/tin.

AuSn is a gold-based eutectic solder that does not need flux. In production, dry nitrogen is often used to displace oxygen during the gold/tin soldering process to insure oxidation does not occur. The diodes are mounted to a low coefficient of expansion substrate such as BeO ceramic.

Because the beam quality of high power diode bars can be poor, their low brightness is an issue. Therefore, collimation optics can be added to efficiently couple the diode bar's output to the laser material. Collimation of the diode bar's output to a fiber is commonly used in applications where the

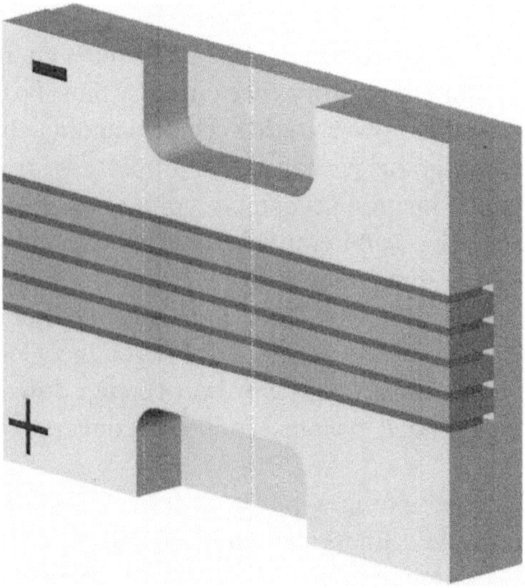

8.26 Diode bars stacked in vertical array.

© Woodhead Publishing Limited, 2013

pump source needs to be remotely located from the laser host material. A Volume bragg grating can also be used to narrow the pump diode's emission spectrum.

8.4.2 Input protection

Along with EMI transient protection, the input circuit should provide reverse voltage, over-voltage, and under-voltage protection.

Depending on the application, EMI protection can be handled by a combination of low, medium, and high frequency protection. Gas discharge tubes (GDTs) can handle low frequency, high energy transients, while transient voltage suppression (TVS) devices, such as metal oxide varistors (MOVs), can handle medium frequency, medium energy transients. Low-pass circuits of inductor elements, RF cores, and capacitors offer high frequency protection.

Under-voltage (UV), over-voltage (OV), and reverse supply protection are best handled with an IC controller such as the Linear Technology LTC4365. This device controls the gate voltages of a pair of back-to-back external N-channel MOSFETs to ensure the output stays within a safe operating range. Two precision comparators are used to monitor for over-voltage and under-voltage conditions. If the input supply rises above the OV threshold or falls below the UV threshold, the gate of the MOSFET is quickly turned off, disconnecting the output (load). An example input protection circuit is shown in Fig. 8.27.

8.4.3 Soft-start

Soft-start circuits are required to control the inrush of current from the power source to the driver's energy storage circuitry. The circuitry must be designed to protect both the driver and the power source. Understanding the power source's specifications and limitations is critical to a properly designed laser driver. Important specifications include acceptable voltage range, peak and average current and power ratings, and equivalent series resistance (ESR) values over the entire operating temperature range.

Thankfully, there are a variety of control ICs from manufacturers such as Linear Technology, Maxim Integrated Products, and Micrel, Inc. Control ICs are available for isolated, non-isolated, step-up (boost), step-down (buck), inverter (fly-back) and single-ended primary inductor converter (SEPIC) topologies.

Many DPSS laser applications are now powered from lithium batteries, such as the Energizer EL123AP (Energizer Holdings, Inc., 2011). This cylindrical lithium battery is nominally 3.0 volts and has a typical capacity of 1500 mAh down to 2.0 volts. It is important to note that below 2.0

© Woodhead Publishing Limited, 2013

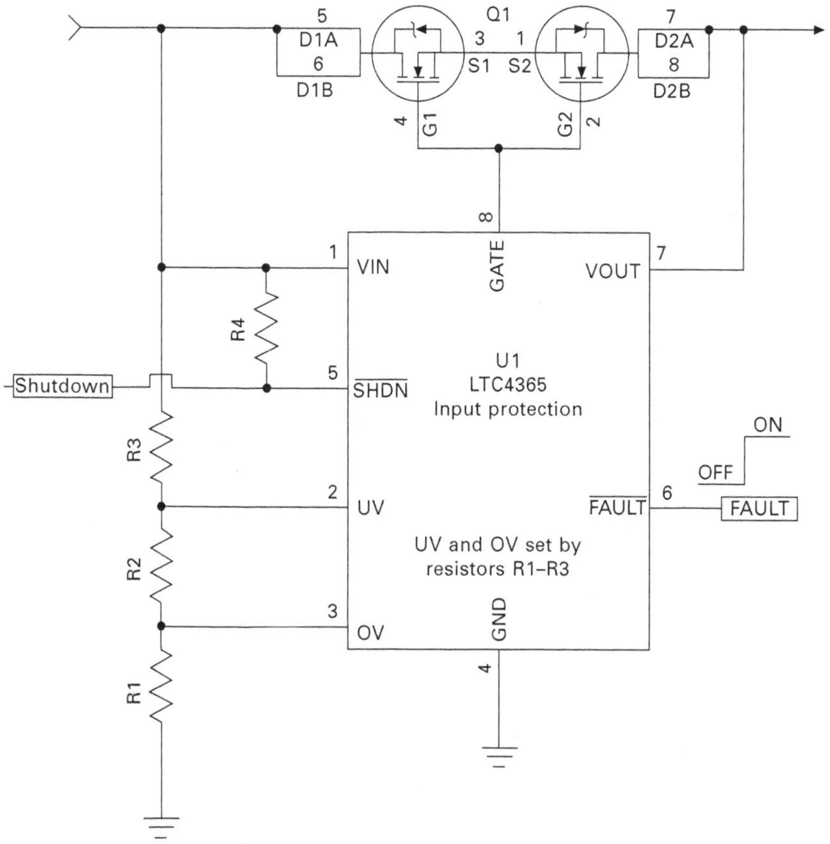

8.27 Input protection circuit.

volts, the usable energy and current quickly diminish. Manufacturers such as Energizer now often include simulated application test data on their battery datasheets. Pulsed DPSS laser applications are similar to photoflash applications, so battery requirement calculations for a particular application are straightforward.

8.4.4 Energy storage

Pulsed DPSS drivers often are required to deliver tens or even hundreds of amps to the pump diodes inside the laser. However, most power sources, including batteries, cannot directly handle this type of current surge. Therefore, DPSS drivers should include energy storage to handle the laser's peak current pulse requirements. The capacitors used must be low ESR types and are generally aluminum electrolytic, hybrid, or supercapacitor types.

The latest supercapacitors have extremely high energy density, but often

© Woodhead Publishing Limited, 2013

have too much internal resistance for DPSS driver applications. Relative to aluminum electrolytic capacitors, supercapacitors also have low voltage ratings. Series or series–parallel combinations of supercapacitors can be used for laser drive applications, but they must be properly balanced to avoid over-voltage conditions. A BEST-CAP supercapacitor from AVX is shown in Fig. 8.28.

A variety of control ICs are available to efficiently manage series-connected super-capacitors including the Linear Technology LTC3225, a two-cell, 150 mA supercapacitor charger. The Texas Instruments (TI) BQ33100 is a supercapacitor 'fuel gauge' that includes individual capacitor monitoring and voltage balancing for up to nine series-connected supercapacitors. A system partitioning diagram for the BQ33100 is shown in Fig. 8.29.

Voltage requirements for the energy storage section are a combination of the forward voltage of the laser diode at its operating current plus any voltage drops due to wiring, high current switch, and capacitor ESR resistances. This overall voltage drop, due to system losses, is typically referred to as 'compliance voltage' or 'overhead voltage'. Sufficient energy storage is required to insure the bank voltage remains above the compliance voltage.

8.4.5 Current control

The laser power supply designer has to determine the best current control method for the application. Contributing factors include space claim, compliance voltage, peak current, and heat dissipation requirements. Overall efficiency is also important, especially when operating from batteries.

8.4.6 Linear current control

Perhaps surprisingly, linear current control is often the best choice for a laser driver application. Linear control is non-switching and produces very low radiated EMI. It is also very immune to conducted EMI due to its inherent circuit impedance. Protecting the output against short-circuit faults is also

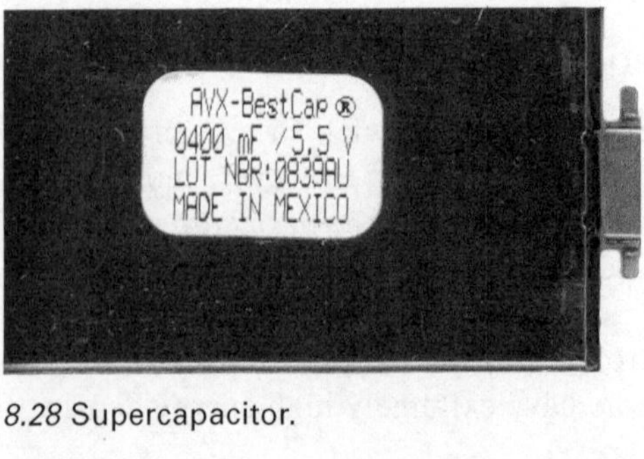

8.28 Supercapacitor.

© Woodhead Publishing Limited, 2013

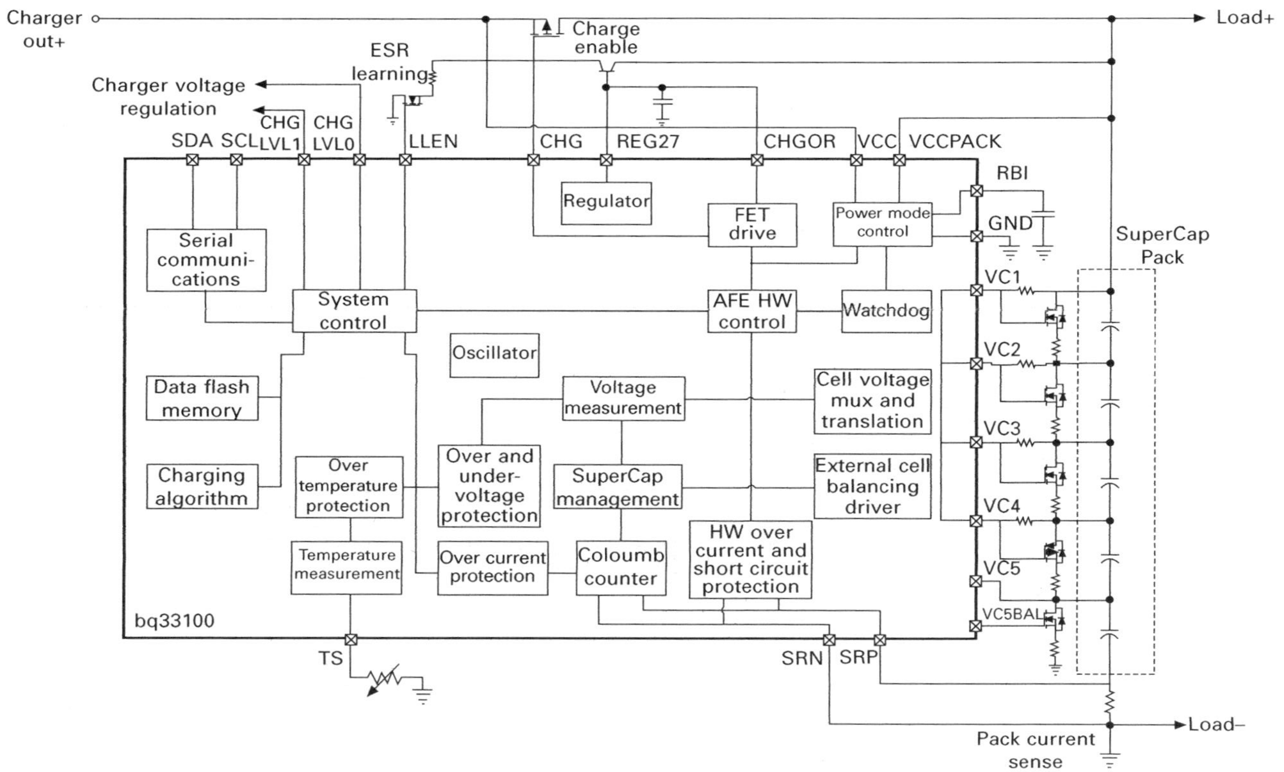

8.29 BQ33100 system partitioning diagram.

© Woodhead Publishing Limited, 2013

straightforward with a linear control approach and many devices such as the Texas Instruments (TI) OPA549 provide bullet-proof over-current protection. In addition, the OPA549 provides an accurate, user-selected current limit and the device senses the load indirectly, so no power resistor in series with the output current path is required. This allows the current limit to be adjusted from 0 A to 10 A with a simple potentiometer or controlled digitally with a digital-to-analog converter (DAC). Figure 8.30 demonstrates a basic 8 A (10 A peak) driver circuit using the OPA549.

Linear devices are also easily made parallel. An example laser drive circuit using two parallel Texas Instruments OPA548 surface mount devices is shown in Fig. 8.31. This circuit can provide 0–10 amps output with simple 0–5 volt user input.

Of course, the designer must be aware of the disadvantages of using linear current control, especially with respect to overall efficiency. One way to improve efficiency is to minimize the voltage drop from the input to the output. Adding an adjustable DC-DC converter to the front end of a linear current controller can maximize efficiency. A feedback circuit incorporating a programmable micro-controller or I^2C DAQ, such as the Maxim DS4432,

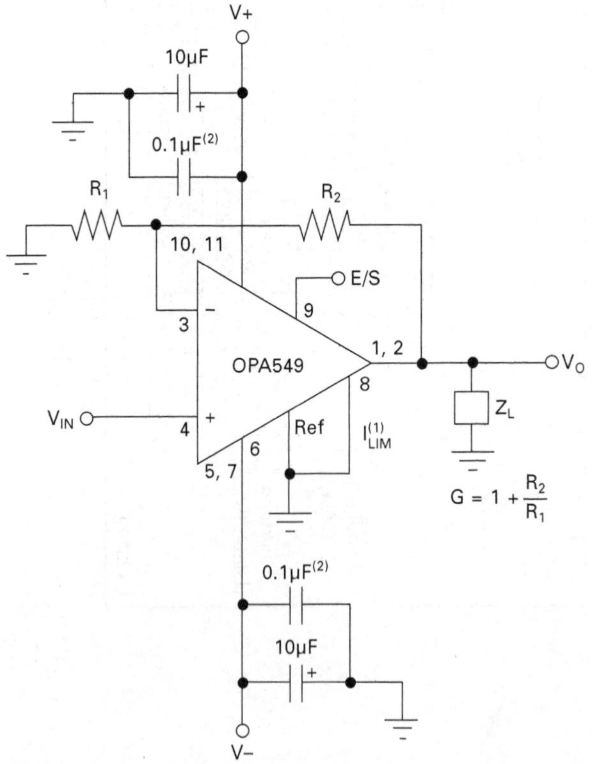

8.30 0–8 amp linear controller using an OPA549.

© Woodhead Publishing Limited, 2013

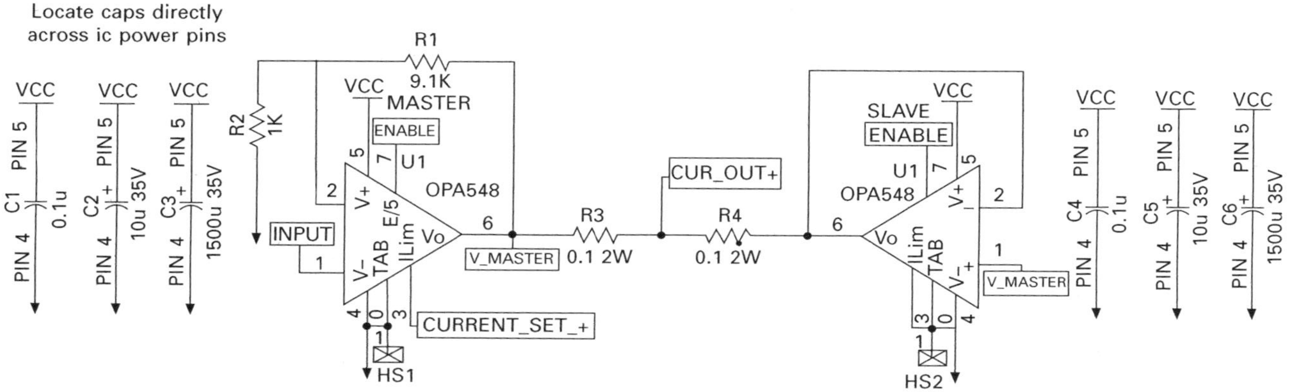

8.31 0–10 amp linear controller using parallel OPA548s

© Woodhead Publishing Limited, 2013

can be implemented. The feedback circuit should keep the linear driver's input voltage just above the minimum required compliance voltage value.

8.4.7 Switch-mode current control

A more common approach in DPSS laser power supply design is to use switch-mode control with feedback to regulate the laser driver's output current. Switch mode designs offer three main advantages when compared to linear regulator circuits. Their switching efficiency can be better; they can be smaller; and they can provide electrical isolation. Since less energy is lost in the transfer, smaller components and less thermal management are required. Current feedback can be used with boost, buck, fly-back, and SEPIC control topologies. Current sensing can be achieved with low value, ultra-stable power sense resistors or via non-contact devices such as Hall-effect current sensors.

Low average and/or peak current designs typically use a sense-resistor with a low resistive value in series with the output path. Current flowing through the precision resistor to the laser diode creates a voltage that can be measured by the feedback circuit. Series sense-resistor circuits are relatively small and low cost but can dissipate significant power at higher average power levels.

High current designs generally use a non-contact approach in feedback circuits to minimize power loss and to provide electrical isolation between the current-carrying conductor and the output signal. Hall-effect types can be either closed or open-loop style and should be sized for the peak current capability and saturation value. Higher repetition rate applications should consider a DC rated Hall-effect sensor to prevent degradation of the output signal due to sensor field saturation. However, Hall-effect sensors are relatively large when compared to series sense-resistor circuits. A clamp-on 70 A DC current probe from Tektronix, model A622, is shown in Fig. 8.32. This model is easy to use and connects directly to an oscilloscope. Its fast response time and 70 amps DC current rating make it ideal for most DPSS laser power supply current measurements.

8.32 Tektronix A622 DC current probe.

© Woodhead Publishing Limited, 2013

8.4.8 Laser diode protection

As with all semiconductor devices, laser diodes are very sensitive to over-voltage and over-current transients and proper protection is required. Laser driver designs should include a shorting relay, snubber circuit, EMI and transient protection. A sample output protection circuit is shown in Fig. 8.33.

8.5 Control features

Flashlamp and diode pumped laser controllers have many common features including various internal and external controls. Control signals are generally a mix of inputs and outputs and can be in either analog or digital form.

8.5.1 Inputs

The laser power supply designer should consider user inputs such as power on and off, laser standby and ready, and fire control settings (including repetition rate and burst). System inputs include feedback from the laser such as output pulse detection, temperature, laser energy and/or laser power. Other

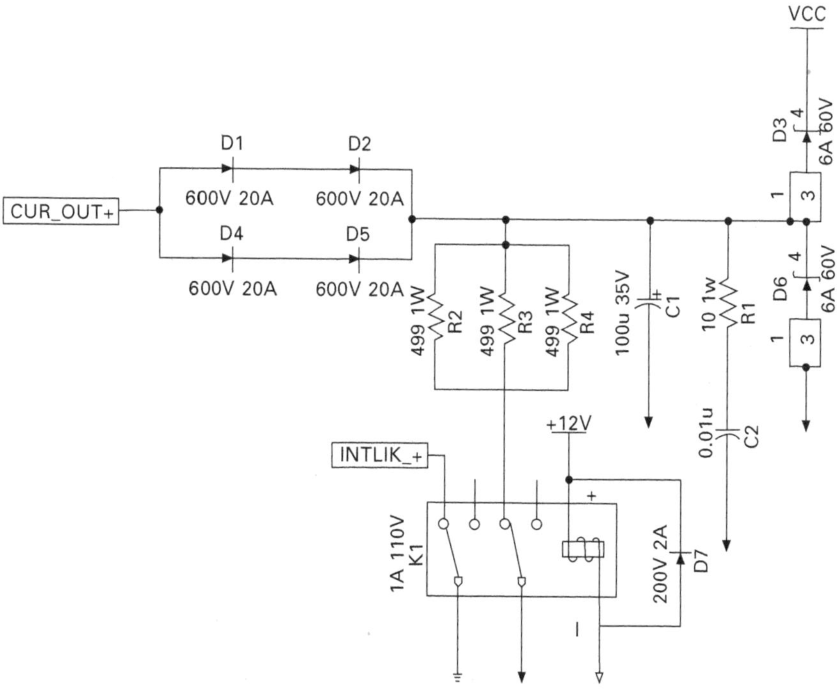

8.33 Laser diode protection.

© Woodhead Publishing Limited, 2013

system signals recognized by the controller as inputs may include coolant flow, pressure and temperature indication, current, voltage, or charge levels or system diagnostics such as board level DC-DC power supply status.

8.5.2 Outputs

In addition to controlling the flashlamp or pump diode current and energy, the laser power supply may also need to control hardware such as a Q-switch driver, cooling unit, or secondary laser amplifier system. Control of internal or external power supplies and/or energy storage devices is also common. Using a standard communication protocol, such as gigabit Ethernet or controller area network (CAN), allows microcontrollers, PLCs, and similar devices to communicate with each other without a host computer.

8.5.3 Software and hardware platform

Real-time software and hardware platforms are often necessary for laser systems that need reliability, timing precision and/or fast input/output (I/O) response times. Real-time operating systems from companies such as QNX Software Systems, LynuxWorks or Wind River Systems are readily available to the laser power supply designer. Development tools, such as National Instruments LabVIEW, that automatically integrate real-time operating system software are also available.

Systems that do not require real-time control can often use simple programmable logic controllers (PLC) or off-the-shelf data acquisition (DAQ) modules from companies such as Measurement Computing or National Instruments. These manufacturers also provide graphical programming languages that communicate directly with their DAQ modules for stand-alone or personal computer (PC) control via Ethernet, USB, IEEE-488 (GPIB) or PCI interface. A sample graphical user interface (GUI) from Kigre, Inc. is shown in Fig. 8.34.

8.6 Conclusion

As output energy and power density of DPSS lasers increase, they will continue to replace flashlamp-based laser systems. And because DPSS laser systems typically require lower operating voltages and energies, their power supplies are inherently safer than flashlamp-based supplies of the past.

Although laser system power supplies have been around for over 50 years, the latest designs and component materials are hardly mature. Powerful, low cost, off-the-shelf microcontrollers have pushed control and communication levels higher and higher. Semiconductor switch products are significantly improved and datasheets are readily available to the designer. Capacitor

© Woodhead Publishing Limited, 2013

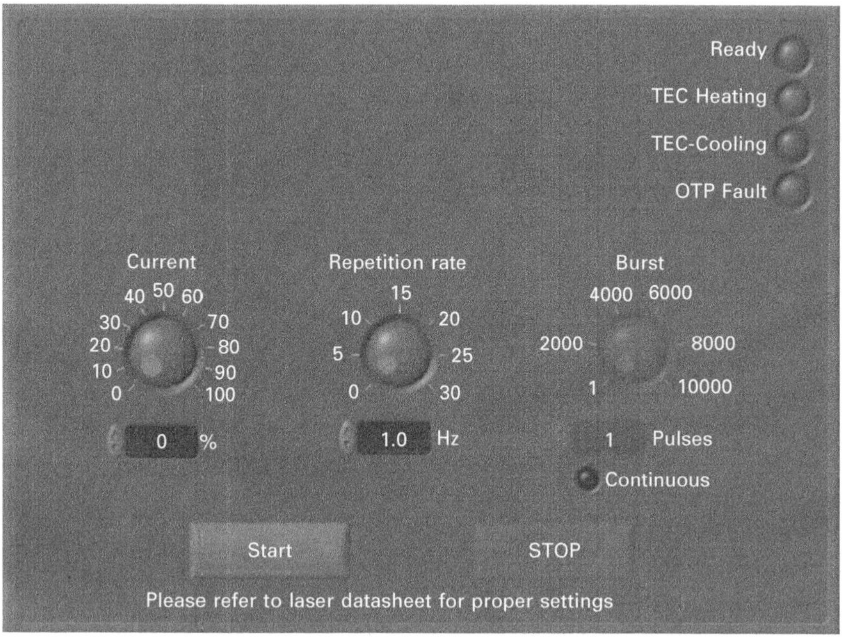

8.34 Graphical user interface (GUI).

and battery chemistries are evolving rapidly and the power supply designer must stay abreast of current technology to insure a competitive, efficient, and reliable product.

Due to the incredible advancement of laser applications, laser power supplies can be found in the medical, industrial, communication, environmental, research, and military fields. Wherever there are lasers you will find laser power supplies and the driver designs often vary as greatly as the laser application itself. From extremely small laser endoscopes to very large directed energy weapons, lasers continue to find a home in a variety of unique applications.

8.7 References

Buck, A., Erickson, R. & Barnes, F. (1963). Design and operation of xenon flashtubes. *Journal of Applied Physics*, 34, 2115–2116.

Energizer Holdings, Inc. (2011). *Energizer 123 Datasheet* [online]. Available at http://data.energizer.com/PDFs/123.pdf (accessed 11 July 2011).

ILC Technology (1983). *A Guide to Flashlamps for Pulsed Solid State Lasers*. Sunnyvale, CA: ILC Technology.

Koechner, W. (1976). *Solid-State Laser Engineering*. New York: Springer-Verlag.

Myers, J. D. (1965). *First GigaWatt (World Record) Ruby Laser*. Ann Arbor, MI: Lear Siegler Laser Systems Center.

© Woodhead Publishing Limited, 2013

Myers, M., Myers, J. D. & Guo, B. (2007). Practical internal combustion engine laser spark plug. *SPIE Optics & Photonics 2007* (pp. 3–4). San Diego, CA: SPIE.

Myers, M., Myers, J. D. & Myers, A. (2008). *Laser-induced Breakdown Spectroscopy (LIBS)*. Weinheim, Germany: Wiley-VCH.

Penn State University (2011). *Pulsed Light & PEF*. University Park, PA: Penn State University, Department of Architectural Engineering. Available at http://www.engr.psu.edu/iec/abe/control/pulsed_light.asp (accessed 15 July 2011).

© Woodhead Publishing Limited, 2013

9
Operation regimes for solid-state lasers

R. PASCHOTTA, RP Photonics Consulting GmbH, Germany

DOI: 10.1533/9780857097507.2.227

Abstract: A number of profoundly different laser operation modes can be used, which substantially add to the versatility of laser technology. Continuous-wave operation can be combined with different methods for spectral control and for influencing noise properties or beam quality. Q-switching is a method for generating intense nanosecond pulses, whereas ultrashort pulses with durations of picoseconds or femtoseconds are obtained with different variants of mode locking. Higher pulse energies are obtained with cavity dumping or regenerative amplification, possibly combined with chirped-pulse amplification. The focus of this chapter is on the realization of all those operation modes with solid-state bulk lasers.

Key words: solid-state lasers, continuous-wave operation, spectral control, wavelength tuning, beam quality, laser noise, Q-switching, nanosecond pulses, cavity dumping, mode locking, ultrashort pulses, picosecond pulses, femtosecond pulses, chirped-pulse amplification, regenerative amplification.

9.1 Introduction

The amazing versatility of laser technology results to a substantial degree from the possibility of exploiting a number of very different laser operation modes. Some of these are variations of continuous-wave operation, mostly focusing on spectral control, and partly on the noise properties or beam quality. Several other operation modes are used to generate short or ultrashort laser pulses in quite different parameter regimes. Q-switched lasers offer high pulse energies in nanosecond (or shorter) pulses; that technique has been successfully applied with many different laser crystals and glasses. Although neither active nor passive Q-switching involves particularly complicated physical effects, certain performance aspects are limited in non-obvious ways, as discussed in Section 9.3.3.

Ultrashort pulses can be generated with different techniques for mode locking. Here, the physical aspects involved are often more sophisticated, and a decent understanding of those is essential for fully realizing the performance potentials. Additional techniques such as cavity dumping and regenerative amplification, the latter possibly combined with chirped-pulse amplification, can be used to obtain higher pulse energies in still very short pulses.

This chapter is not intended to provide an overview on historical

© Woodhead Publishing Limited, 2013

developments or on the performance figures reached so far. It rather explains the basic principles and discusses a number of aspects which are relevant for understanding and realizing the performance potentials.

9.2 Continuous-wave operation

9.2.1 Basic aspects

Continuous-wave operation means that a laser is continuously emitting light. This requires a continuous supply of energy from a continuous-wave pump source. Usually, all parameters of the laser resonator such as output coupler transmission and additional optical power losses are constant during operation, but there may be adjustments of the resonator length or the pass band of an optical filter for spectral control during operation, as discussed below.

In this section, a number of aspects are explained which apply not only to continuous-wave lasers, but also to Q-switched lasers, mode-locked lasers, etc. In particular, many details of resonator design apply to such lasers in the same manner. However, there may be some additional requirements, such as appropriate mode areas in a Q-switch or a saturable absorber medium.

9.2.2 Resonator design for continuous-wave lasers

This text is not meant to be an introduction to laser resonator design – a complex topic which cannot be fully covered by a part of a book chapter. Instead, a number of important aspects of resonator designs specifically for solid-state lasers are discussed.

The first aspects to be determined are the beam radius (or area) and the path length of the laser beam in the crystal. The choice of values can depend on a number of important aspects:

- The laser crystal needs to be long enough to generate a sufficiently high laser gain.
- For an end-pumped laser, the crystal should also be long enough to allow for efficient pump absorption. On the other hand, too long a gain medium may impose too tight constraints on the pump beam quality. For a side-pumped laser, the laser beam diameter needs to be large enough to obtain efficient pump absorption within the volume of the laser beam.
- Too large a pumped volume can cause too high a pump threshold power. Essentially, it takes more pump power to achieve a certain laser gain over a larger transverse area.
- The pumped volume should not be too small, as this can lead to excessive thermal effects – in extreme cases, the laser crystal may be fractured due to thermally induced stress.

© Woodhead Publishing Limited, 2013

The decisions concerning beam radius and crystal length should be made carefully, as they have many consequences on the laser design. Changing them at a later stage of the development process may force one to change both the pump optics and the whole resonator design.

For the resonator design, the following key aspects have to be considered:

- A short resonator may be desirable to obtain a compact setup, or to obtain a large longitudinal mode spacing as required for single-frequency operation (see the following section). Still other cases require a small longitudinal mode spacing and thus a long resonator – for example, when mode hops during wavelength tuning should cause only small changes in optical frequency.
- A fundamental decision to be made for the resonator is the mode radius of the fundamental (Gaussian) resonator mode within the gain medium. If that mode radius is approximately equal to the radius of the pumped region, one may achieve operation on the fundamental transverse mode(s) only, with a high beam quality and a well-defined shape of the intensity profile of the generated laser beam. However, if the pumped area has to be large, it may be difficult to find a resonator with a correspondingly large mode area and still reasonable stability against thermal effects and misalignment. One may therefore be forced to design the resonator for a smaller mode area, accepting that multiple higher-order transverse modes will be excited.
- Thermal effects in solid-state laser gain media become relevant already at moderate continuous pump power levels of the order of 1 W. In particular, thermal lensing is often important: the temperature gradients within the gain medium lead to a focusing action on the laser beam, which can affect the mode radii everywhere inside and outside the resonator. In many cases, it is highly desirable to design a resonator such that the mode radii will not change substantially within some range of pump powers. The resonator should be operated well within one of its stability zones concerning the dioptric power of the thermal lens. These stability zones can have very different alignment sensitivity, but other aspects may force one to use the more sensitive zone. A particularly useful discussion of such issues has been given by Magni (1987).
- Most solid-state laser resonators have a linear topology, where the laser beam bounces back and forth between two end mirrors. However, a ring laser resonator is sometimes preferable, as it avoids the formation of a standing-wave pattern of the laser beam in the gain medium, if unidirectional operation can be enforced. Spatial hole burning effects are then avoided, which may otherwise make single-frequency operation difficult.

© Woodhead Publishing Limited, 2013

- It can be important for the spectral characteristics (see Section 9.2.3), for example, whether or not the laser crystal is placed at the end of a linear resonator.

A simple type of laser resonator (see Fig. 9.1) consists only of two mirrors around the laser crystal, with some air space on both sides of the crystal. Often, the output coupler mirror is flat, so that a collimated output beam is obtained; the other mirror is then concave, unless the thermal lens of the crystal already provides for sufficient focusing action.

It is often advantageous to use a resonator length of the order of the Rayleigh length of the intracavity beam. For example, a small mode radius of 100 μm in a low-power Nd:YAG laser leads to a Rayleigh length of ≈30 mm, and a resonator length of 20 mm is sufficient. For a high-power laser with a 1-mm mode radius, the Rayleigh length becomes ≈3 m, and the resonator then needs to be much longer. Otherwise, the beam would have a nearly constant beam radius within the whole resonator. This implies operation near the edge of a stability zone and thus very critical dependence on thermal lensing and alignment.

End-pumped lasers are often pumped directly through an end mirror, which can be transparent for the pump light but highly reflecting for the laser light. It is usually desirable to position the laser crystal close to that end mirror. The end mirror may even be realized as a dielectric coating directly on the laser crystal. Note that a zone-I resonator according to the notation of Magni (1987), which could have a lower alignment sensitivity, can then not be realized. Also, spatial hole burning may lead to an increased spectral width (see Section 9.2.3) when the laser crystal is close to an end mirror.

For not too high output powers, a microchip laser setup may be used, where both resonator mirrors are placed at the end faces of the laser crystal, and normally realized as dielectric coatings on the crystal. Flat end faces can be used, provided that the thermal lens at the intended pump power level has a suitable dioptric power to operate well within the resonator's stability zone. The pump power range which is appropriate in that respect is largely determined by the properties of the gain medium, and cannot be rescaled by modifying the crystal length.

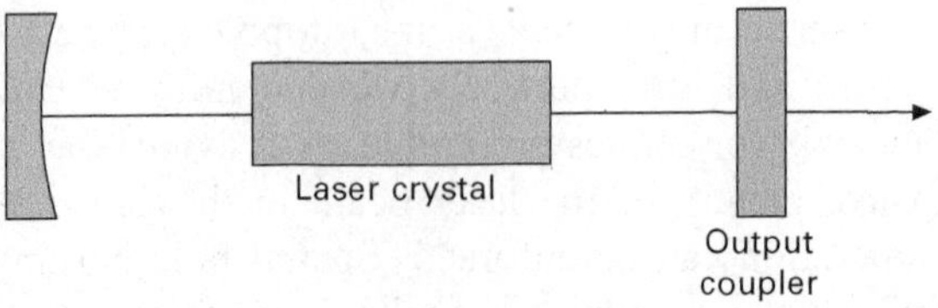

9.1 A simple laser resonator with two mirrors around the laser crystal.

© Woodhead Publishing Limited, 2013

Laser resonators with one or several additional folding mirrors are also often used. Such an approach often allows one to meet additional requirements. Some examples are briefly discussed here:

- A long resonator may be folded simply to limit the outer dimensions of the setup.
- In a long resonator with tight focus in the gain medium, one or several curved folding mirrors may be needed to avoid excessive increases of beam radius within the resonator, and to allow for operation well within a stability zone.
- A second beam focus may be obtained, which is suitable for an intracavity frequency doubler.
- If a laser crystal is operated under Brewster's angle, reflections on a curved folding mirror can be used for compensation of astigmatism.

9.2.3 Control of the emission spectrum

Intrinsically, many laser crystals and glasses exhibit homogeneous gain saturation. It is then relatively easy to obtain laser emission with a very small emission bandwidth, far below the gain bandwidth. One may even obtain single-frequency operation, i.e., operation only on a single longitudinal resonator mode. This is because the mode with highest gain will saturate the gain medium such that its own net gain per round trip is exactly zero in the steady state. All other modes will then have a negative net gain and thus 'die out' quickly.

An extrinsic effect leading to inhomogeneous saturation characteristics is spatial hole burning (Tang *et al.* 1963), as occurs whenever counterpropagating laser beams occur within the gain medium. These then create a standing-wave pattern, which is differently located for each resonator mode. Therefore, a single mode cannot fully saturate the gain for other modes, and can thus not prevent other modes from acquiring a positive net gain. As a result, the laser will often tend to oscillate on multiple modes. The covered optical frequency interval strongly depends on the circumstances:

- If the laser crystal is close to an end mirror, a substantial emission bandwidth will be required to more or less wipe out the standing-wave pattern, particularly when the laser crystal is short.
- If a thin laser crystal is at the centre of a resonator, lasing on two adjacent longitudinal modes can be sufficient to fully wipe out the interference patterns. This condition can lead to very narrowband laser emission (although not yet to single-frequency operation), even if the gain bandwidth is large and no additional bandpass filter is used.

Although a narrowband bandpass filter may enforce oscillation on few modes or a single mode, it becomes apparent that it becomes much easier to

© Woodhead Publishing Limited, 2013

enforce narrowband operation when spatial hole burning is minimized (e.g., with the laser crystal in the middle of the resonator) or eliminated (with a unidirectional ring laser design).

If single-frequency operation is required, this can often be made easier by certain measures:

- Spatial hole burning can be eliminated, usually with a unidirectional ring laser setup.
- The laser may be operated at a low power level, reducing thermal effects, and not too high above threshold in order to avoid strong gain saturation effects.
- A short resonator length can be used. This leads to a large longitudinal mode spacing and consequently to larger differences in modal gain.
- The choice of a gain medium with small gain bandwidth can further increase the difference in modal gain.
- Ideally, a compact and stable monolithic setup is used, largely eliminating thermal drifts (except for those in the laser crystal).

Once stable single-frequency operation is achieved (i.e., mode hops are largely suppressed), the emission bandwidth is often a tiny fraction of the longitudinal mode spacing, and many orders of magnitude smaller than the gain bandwidth. A fundamental limit arises from quantum noise. However, that fundamental limit, with operation at the Schawlow–Townes linewidth (Schawlow and Townes 1958), is hard to reach in practice – particularly with solid-state lasers, having a very low fundamental bandwidth limit. The emission bandwidth is nearly always limited by technical noise, for example by acoustic noise or (at lower frequencies) by thermal fluctuations. More information on the noise of solid-state lasers is available in Paschotta *et al.* (2007).

Apart from controlling the emission bandwidth, it is often desirable to tune the emission wavelength. The most common technique is to use a tuneable bandpass filter within the laser resonator. For example, many Ti:sapphire lasers contain a Lyot filter, which can be tuned by varying the orientation of its birefringent plates. Emission will then often still occur on multiple longitudinal modes, although usually with a bandwidth far below the transmission bandwidth of the filter.

If a very narrowband filter is used (for example, a high-finesse resonant filter), so that single-frequency operation is achieved, wavelength tuning will usually involve mode hops, i.e., jumps from one resonator mode to an adjacent mode, accompanied by a corresponding discrete change of optical frequency. Avoiding such mode hops is possible in principle (but difficult) by suitable tuning of the resonator length in synchronism with tuning the filter frequency.

© Woodhead Publishing Limited, 2013

9.2.4 Output power stabilization

Depending on the operation regime, the output power of a solid-state laser can exhibit different kinds of intensity noise:

- In single-frequency operation, the output power exhibits some low-frequency noise (depending on pump power variations, stability of the setup, thermal effects, etc.) and also some relaxation oscillations, but is normally very stable over short time scales, i.e., it has a very low level of high-frequency noise.
- For operation on multiple longitudinal modes, but only fundamental (Gaussian) transverse modes, there is mode beating, leading to strong power modulations at frequencies which are integer multiples of the round-trip frequency. At low noise frequencies (well below the round-trip frequency), the noise is usually higher than in single-mode operation, but it may be relatively low if a large number of resonator modes is involved. Few-mode operation often exhibits higher low-frequency noise.
- If a laser also emits on multiple transverse modes, additional strong frequency components can occur in the noise spectrum, as the higher-order transverse modes normally have optical frequencies between those of the longitudinal modes.

The single-frequency regime has the best potential for very low-intensity noise. However, operation on many longitudinal modes may also lead to reasonably low noise, at least at noise frequencies well below the round-trip frequency.

For minimizing the low-frequency noise, different measures can be used:

- The pump source should exhibit low-intensity noise. For diode-pumped lasers, this condition can be easily achieved when the pump diodes are operated with a well stabilized current. Note that pump noise is relevant below and particularly around the relaxation oscillation frequency of the solid-state laser.
- The resonator setup should be mechanically stable, avoiding excessive acoustic noise picked up from the surroundings. External noise sources such as pumps or ventilators from a cooling system should be acoustically decoupled from the laser setup as far as possible.
- The resonator design should be made such that its sensitivity to misalignment is minimized, as this also leads to a minimized reaction to acoustic noise.
- The output power can be further stabilized with an electronic feedback system (Fig. 9.2). Here, a photodiode is used to monitor the output power (using some parasitic reflection from the laser setup, for example). An electronic circuit uses that information to automatically readjust the pump

© Woodhead Publishing Limited, 2013

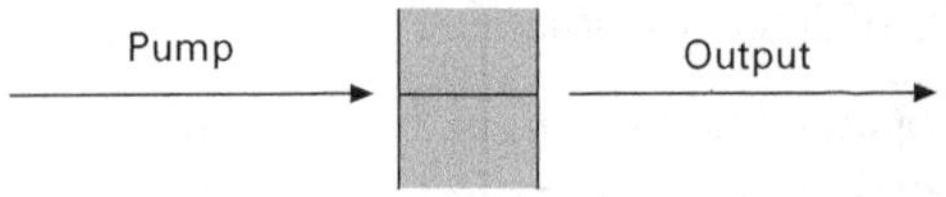

9.2 Diode-pumped solid-state laser with a feedback system for stabilization of the output power.

power. Well below the relaxation oscillation frequency, such a feedback system may reduce the intensity noise by several tens of decibels. By careful control of the phase relations, it is also possible to make a system very effective around the relaxation oscillation frequency, more or less suppressing the relaxation oscillation peak in the noise spectrum. See, for example, Harb *et al.* (1994).

9.2.5 Achievements

A few examples are given for technical achievements with continuous-wave lasers:

- Kilowatt output powers combined with diffraction-limited beam quality are best achieved with thin-disk lasers (Giesen and Speiser 2007). Even tens of kilowatts are possible with somewhat poorer beam quality.
- Multi-kilowatt output powers with reasonable beam quality are also possible with other technical approaches, such as the more traditional rod lasers (Bruesselbach and Sumida 2005).
- Simple and very compact microchip lasers (Zayhowski and Mooradian 1989) are suitable for moderate output powers with single-frequency output.
- Higher powers and narrower emission bandwidths in single-frequency operation are possible with monolithic nonplanar ring oscillators (Freitag *et al.* 1995).

9.3 Pulsed pumping of solid-state lasers

For a limited time, a substantially increased output power can be obtained from a solid-state laser if a short but intense pump pulse is used. Depending on various parameters, different operation regimes are possible; these are discussed in the following subsections.

© Woodhead Publishing Limited, 2013

9.3.1 Quasi-continuous-wave operation

If the pump pulse is relatively long, the laser may reach steady-state conditions within the pumping time. Apart from the turn-on dynamics at the beginning of the pump pulse (typically with some delayed start, followed by spiking and damped relaxation oscillations), the laser may then basically operate just as for ordinary continuous-wave operation, only possibly at a higher power level. During each pump cycle, the temperature of the gain medium rises steadily, but if the pump cycle is not too long, excessive heating is still avoided. The excess heat may be largely stored in the gain medium and released in the time between the pump cycles. Such lasers are sometimes called heat capacity lasers (Albrecht *et al.* 1998) as the heat capacity of the gain medium is exploited.

9.3.2 Gain switching

By appropriately adjusting the laser parameters, it is possible to achieve the emission of a relatively short pulse with a duration well below that of the pump pulse. Ideally, the pump power is applied just until the first spike of laser power is emitted (see Fig. 9.3). That spike may then extract most of the energy which has been stored in the gain medium, and no further spikes occur.

For intense pumping, the duration of the generated pulse (spike) can be well below the inverse relaxation oscillation frequency. It is essentially determined by the laser gain at the end of the pump pulse, the resonator losses, and the resonator length.

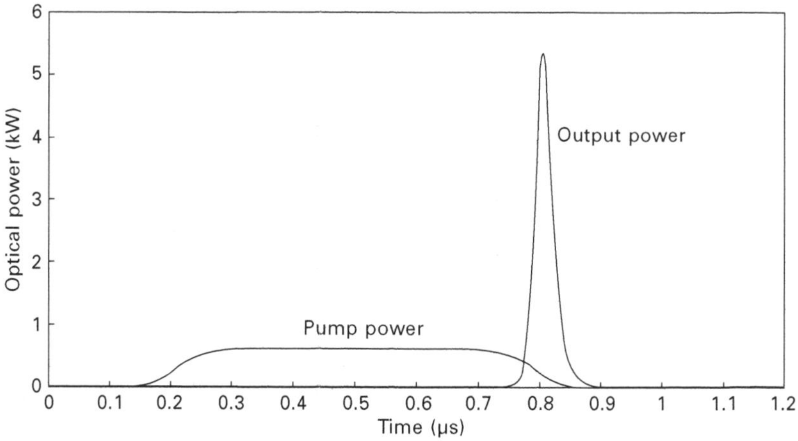

9.3 Simulated evolution of the output power of a gain-switched solid-state laser. The pump pulse ends approximately when the first spike is emitted, so that only that spike is obtained as the laser output.

© Woodhead Publishing Limited, 2013

The time required for build-up of the pulse limits the possible pumping time; pumping for a longer time would only lead to additional spikes. Increasing the pump power has a limited potential for increasing the output pulse energy, as it also increases the speed of pulse build-up and thus reduces the possible pump time. More effective measures may be to increase the resonator length and/or to increase the pumped area in the laser crystal, so that the pulse build-up will be slower.

Gain-switched operation of solid-state lasers is rarely used, as the possible performance, particularly in terms of pulse energy and pulse duration, is better with Q-switching (Section 9.3.3). That technique also leads to less critical constraints on the pump pulse parameters.

9.3.3 Achievements

A few examples of free-running lasers with pulsed pumping are as follows:

- Very compact lasers or lasers operating on low-gain transitions can generate high output powers for short pump pulses, where excessive heating is avoided. For example, this has been done with microchip lasers (Dascalu *et al.* 2002) and with 1123-nm Nd:YAG lasers (Li *et al.* 2010).
- Microchip lasers have been gain-switched to obtain pulses with durations below 1 ns (Zayhowski *et al.* 1989). Also, it is possible to obtain short pulses with megahertz repetition rates (Wang *et al.* 2007).
- Some lasers are gain-switched by pumping them with a Q-switched solid-state laser. An example is a tunable Cr^{4+}:forsterite laser (Agnesi *et al.* 1996).

9.4 Q-switching

9.4.1 Basic principle

Q-switching is usually the preferred technique for generating nanosecond pulses (sometimes even sub-nanosecond pulses) with solid-state lasers.

The basic principle of Q-switching (McClung and Hellwarth 1962) is that the resonator losses are kept at a high level during a pump phase, allowing one to store substantial energy in the gain medium without premature lasing, and then to suddenly switch the resonator losses to a lower value, so that a short pulse extracts a substantial fraction of the stored energy from the gain medium. Two different ways of switching the losses can be distinguished:

- Active Q-switching means that the losses are controlled with an active device, i.e., a modulator.

© Woodhead Publishing Limited, 2013

- Passive Q-switching is accomplished with a saturable absorber, which is switched by the light field itself.

The following subsections discuss these techniques in some detail.

9.4.2 Active Q-switching

The modulator for switching the resonator losses is in most cases either an acousto-optic modulator (AOM) or an electro-optic modulator (EOM), although other types of modulators are sometimes used. For example, a spinning laser mirror can periodically close the resonator's beam path.

The pulse generation can be understood by considering the following phases:

- During some pump phase, the modulator prevents lasing by introducing resonator losses which, combined with other losses, are higher than the laser gain (per round trip). Therefore, lasing is prevented. Energy extraction by amplified spontaneous emission (ASE) is usually quite weak in solid-state bulk lasers (in contrast to fibre lasers, exhibiting higher laser gain). In some cases, parasitic lasing limits the energy which can be stored in the gain medium.
- The useful duration of the pump phase is limited to a small multiple of the upper-state lifetime of the gain medium, since for longer pumping most of the additional pump energy would be lost by spontaneous emission.
- The pulse generation is triggered by quickly switching off the additional losses caused by the modulator. However, the generated pulse comes with some time delay, which is much larger than its duration. This is because lasing usually starts with spontaneous emission, which is amplified to a high level within dozens or even hundreds of resonator round trips. That time delay exhibits some random variations due to the randomness of spontaneous emission and possibly of other factors (such as the pump energy); this can lead to substantial timing jitter. Only in rare cases is some weak seed light from an external source or from some weak prelasing (Bollig *et al.* 1995) used for starting with better defined initial conditions.

Many lasers are periodically Q-switched, forming a regular pulse train. If the pulse period is not larger than the upper-state lifetime of the gain medium, continuous pumping is appropriate. For longer pulse periods, pulsed pumping is more efficient: the pump energy is provided just before a pulse is generated within less than one upper-state lifetime, so that only a minor part of the energy is lost via spontaneous emission.

Once the time-integrated intracavity power has reached the order of the

© Woodhead Publishing Limited, 2013

gain saturation energy, the laser gain is substantially saturated. When the laser gain reaches the resonator losses, so that the net gain is zero, the pulse maximum is reached. After that, the net gain becomes negative, so that the pulse power drops.

The generated pulse can extract most of the stored energy, provided that the initial laser gain is at least twice the resonator losses in the pulse generation phase. Otherwise, the pulse may end before all energy is extracted, so that the pulse energy is lower. Depending on the pulse period, some part of the energy may remain available for the next pulse, increasing its energy.

The time required for raising the pulse power from 1% to 100% of the peak value depends mainly on the initial net gain per round trip and the resonator round-trip time. Similarly, the time for getting from 100% power down to 1% again largely depends on the resonator losses and the resonator round-trip time. If the initial laser gain is approximately twice the resonator losses, so that the initial net gain has about the same magnitude as the final net loss, the pulse shape is approximately symmetric in time. For higher initial gain, the leading edge of the pulse may be shorter than the trailing one.

As a numerical example, consider an end-pumped Nd:YAG laser with a mode radius of 200 μm in the laser crystal. The laser performance is simulated with the software RP Fiber Power (which can be used in this case, although it has been developed for fibre devices). The results can be compared with simple analytical estimates. The laser is pumped with 4 W at 808 nm and a pump beam radius of 250 μm for a time of 250 μs (slightly longer than the upper-state lifetime of 230 μs). Without spontaneous emission and for complete pump absorption, the final stored energy, available at the laser wavelength, would be 4 W × 250 μs × (808 nm/1064 nm) = 0.76 mJ. (The ratio of pump and signal wavelength takes into account the quantum defect.) The simulated stored energy is only 0.41 mJ, since pump absorption is not complete and a substantial amount of energy is lost via spontaneous emission in the relatively long pump phase.

The round-trip (double-pass) gain for 0.41 mJ of stored energy can be estimated as that energy divided by the effective (double-pass) saturation energy of the Nd:YAG crystal of 419 μJ, and multiplied by 4.34 to get the gain in decibels. That results in 4.2 dB. The simulated gain is only 3.2 dB, essentially because some of the pump energy is stored at larger radii where the signal intensity is weak. For that moderate amount of gain, an acousto-optic modulator can easily introduce sufficient loss to prevent premature lasing.

The generated pulse (Fig. 9.4) has a duration (full width at half maximum) of ≈15 ns, i.e., about 12 times the assumed resonator round-trip time of 1.24 ns. The output coupler transmission was assumed to be 25%, corresponding to an attenuation of 1.25 dB, somewhat less than half the round-trip laser gain. As expected in this situation, the obtained pulse shape exhibits a slightly

© Woodhead Publishing Limited, 2013

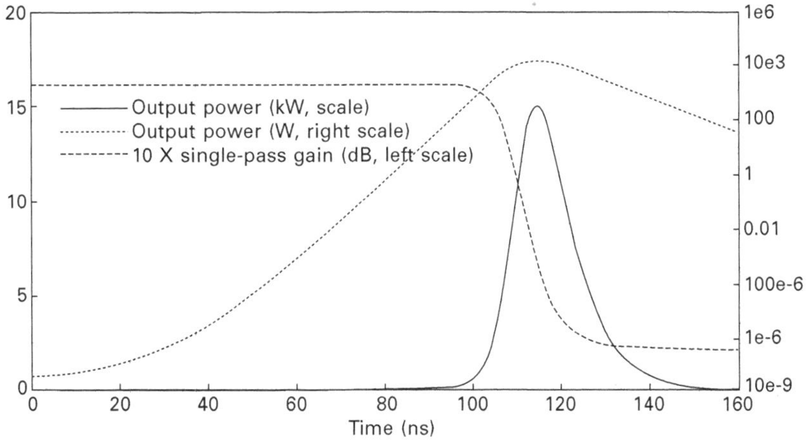

9.4 Simulated evolution of output power and single-pass gain of an actively Q-switched Nd:YAG laser.

longer trailing edge. The pulse duration is about three times the round-trip time divided by the resonator losses.

Figure 9.4 also shows the output power on a logarithmic scale (dotted curve). The rise of the power in the initial phase is exponential, apart from the first 40 nanoseconds, where the finite switching time of the Q-switch plays a role. The finite speed of the Q-switch only somewhat delays the pulse, but does not affect the power efficiency. Substantial performance losses would occur only when the Q-switch is not yet totally 'open' once the pulse power becomes substantial.

The gain (dashed curve) drops to ≈13% of its initial value, suggesting that 87% of the stored energy is extracted by the pulse. However, the simulation indicates that only 65% of the energy is extracted. The energy which remains stored after the pulse is mainly in a ring around the spatial pulse profile. Therefore, it can only weakly contribute to the gain, and the drop of gain is larger than the drop of stored energy. That discrepancy would become smaller if a pump beam with somewhat smaller width would be used.

The numerical comparisons demonstrate that reasonable estimates for key performance figures of a Q-switched laser can be obtained with very simple equations. For example, one can easily estimate the initial gain, the suitable output coupler transmission and the obtained pulse duration. For more precise calculations, taking into account spontaneous emission, spatial beam profiles, non-instantaneous switching, etc., numerical simulation software can be used.

The pulse repetition rate of an actively Q-switched laser can be easily controlled via the modulator driver. Typically, higher repetition rates lead to lower pulse energies for continuous pumping. At the same time, the

© Woodhead Publishing Limited, 2013

pulses become longer, as the initial laser gain becomes lower. For too high repetition rates, some pulses may be missing in the pulse train, so that the actual pulse train has a lower repetition rate. For rather low repetition rates, the pulse energy becomes high and the pulse duration short, but that effect diminishes once the pulse period exceeds the upper-state lifetime, as explained above.

The performance of a Q-switched laser can be limited by various issues. A typical one is related to physical limitations of laser resonators. For high pulse energies, a large laser mode area in the laser crystal is required, as otherwise the laser gain would become too high. While this reduces the beam quality requirements of the pump source, it also leads to a large Rayleigh length of the intracavity beam. As a consequence, a relatively long resonator should be used, as otherwise one would operate the laser close to a stability limit, where stable operation is hard to achieve. (It may not be a viable option to use a smaller beam focus somewhere in the resonator, because that would lead to excessive optical intensities, leading to damage or to breakdown in air.) Therefore, high-energy lasers tend to require long resonators, which lead to longer pulses.

Particularly long pulses result from the use of low-gain laser media. For example, thin-disk laser heads are suitable for very high output powers with good beam quality, but provide only a quite moderate gain. Therefore, the pulse duration is far above the resonator round-trip time. The latter cannot be made very short due to the resonator physics limitation discussed above. As a result, relatively long pulses (hundreds of nanoseconds) often have to be accepted.

9.4.3 Passive Q-switching

A saturable absorber is a simple and cheap substitute for an active modulator. Triggering of a pulse then occurs when the laser gain becomes so high that lasing begins despite the presence of the absorber. Once the optical power becomes high, the absorber is saturated, i.e., the losses which it introduces are reduced. Then, the net gain rises, and the optical power rises even faster. Finally, the laser gain is also saturated, and the power drops again.

Of course, a too high unsaturated loss would prevent lasing at all times. As far as that regime is avoided, higher unsaturated loss leads to a larger delay of pulse generation after turning on the pump source, and to a higher initial stored energy in the gain medium, and thus tentatively to a larger output pulse energy and a shorter pulse duration.

In principle, the extraction efficiency may be only slightly decreased by the losses in the saturable absorber, if two conditions are fulfilled: the saturation energy must be much smaller than that of the gain medium, and the absorber should exhibit only low non-saturable losses. The first condition

© Woodhead Publishing Limited, 2013

may be fulfilled by inserting the saturable absorber at a location where the laser beam is tightly focused; obviously, only a small volume of absorber medium then needs to be saturated, and this requires a low energy, whereas the unsaturated loss is not decreased by that measure. The second condition depends on material properties. Usual saturable absorber materials such as Cr^{4+}:YAG, as is often used for Q-switching Nd^{3+}:YAG lasers, are not ideal in that respect. Therefore, it is desirable to introduce not too high a saturable loss in order to limit the power losses due to non-saturable losses. This, however, often leads to lower energy stored in the gain medium and to a lower output pulse energy, compared with actively Q-switched operation. The pulse duration may nevertheless be short if the resonator can be made accordingly shorter – which, however, is not always the case, as the resonator length may not be limited by space constraints, but rather by resonator design issues, for example.

The absorber's recovery time does not have to be particularly short. Ideally, it is longer than the pulse duration, but still fast enough to provide high losses before the recovery of the gain leads to premature pulse emission.

Changes of the pump power of a passively Q-switched laser usually affect mostly the pulse repetition rate, but not substantially the pulse energy and duration. This is because the initial laser gain, where the pulse build-up starts, is essentially defined by the saturable loss. The pulse build-up is usually so fast that further pumping within that time has no substantial effect. The situation may change if the absorber does not exhibit complete recovery between subsequent pulses.

Extremely short pulse durations, sometimes well below one nanosecond, are possible with compact microchip laser setups (Spühler *et al.* 1999). The shortest pulses with durations even well below 0.1 ns have been obtained by Q-switching the high-gain medium Nd:YVO_4 with a semiconductor saturable absorber mirror (SESAM), which exhibits a negligible internal optical path length.

The optimization of passively Q-switched lasers has been discussed by Degnan (1995).

9.4.4 Extension with cavity dumping

For active Q-switching, the obtained pulse duration is usually at least several resonator round-trips, or more if the laser gain is low. This can be a problem particularly at high pulse repetition rates, where the stored energy is small, leading to a low gain. This limitation can be removed by combining Q-switching with cavity dumping. Here, the resonator has no ordinary output coupler; the pulse generation phase is effectively done with a 'closed' resonator, which does not allow energy to be extracted to the output. Once most of the stored energy has been transferred into the

© Woodhead Publishing Limited, 2013

circulating pulse, the energy is suddenly released with the cavity dumper, which is a fast optical switch. In that way, one can obtain a pulse duration down to one resonator round-trip time, independent of the initial gain. A disadvantage of the method is that very fast switching is necessary – much faster than for most ordinary Q-switched lasers, where the switching time can be much longer than the pulse duration. An electro-optic Q switch may then be required, as an acousto-optic device may be too slow.

9.4.5 Achievements

A few examples of achievements with Q-switched lasers are as follows:

- Pulse energies of 1 J or higher can be obtained by active Q-switching of diode-pumped lasers (Holder *et al.* 1992).
- Actively Q-switched high-gain high-power lasers allow for high average powers combined with short pulse durations (Du *et al.* 2003).
- A constant short pulse duration of 6 ns has been achieved by cavity dumping a Q-switched Nd:YAG laser at repetition rates up to 100 kHz (McDonagh *et al.* 2006).
- Passively Q-switched microchip lasers have been demonstrated to cover large parameter regions concerning pulse durations and pulse repetition rates (Spühler *et al.* 1999).

9.5 Mode locking

9.5.1 Basic aspects of mode locking

Mode locking (Hargrove *et al.* 1964) is a method for generating ultrashort laser pulses. Here, the pulse duration is usually far below the resonator round-trip time. Usually, a single ultrafast pulse is continuously circulating in the laser resonator. In the steady state, the pulse parameters may undergo changes within each resonator round trip due to effects like laser gain, losses, nonlinearities, chromatic dispersion, etc., but the pulse parameters are restored after each complete round trip. Each time the circulating pulse hits the output coupler mirror, an output pulse is emitted, having some fraction of the circulating pulse's energy. The obtained pulse repetition rate is the inverse resonator round-trip time.

For mode locking a laser, various methods can be used. Essential aspects are that the laser reliably develops a short circulating pulse within some time after switching on the pump source, and that this pulse remains stable over many resonator round trips. Mode-locking methods can be grouped into active and passive methods, as discussed in the following sections.

The term 'mode locking' refers to the fact that in the mode-locked state the light in different longitudinal modes of the laser resonator is locked in

© Woodhead Publishing Limited, 2013

phase, i.e., it has a rigid phase relationship. Breaking that phase relationship would correspond to a break-up of the pulse in the time domain. Most mode-locking mechanisms are actually more easily understood by considering the processes in the time domain.

In some cases (but rarely with solid-state bulk lasers), harmonic mode locking is implemented, where multiple pulses circulate in the resonator with a constant temporal spacing. In that way, very high pulse repetition rates can be achieved without using a very short laser resonator. A challenge, however, is to achieve a constant pulse spacing and stable pulse parameters. More or less complicated methods are used for that purpose, and not always with perfect success. In particular, harmonically mode-locked lasers often exhibit so-called supermode noise, leading to increased timing jitter.

9.5.2 Active mode locking

Active mode locking relies on an active modulator, which may be either a loss modulator or a phase modulator, driven with a high-frequency periodic signal. Pulse formation with a loss modulator is easily understood. If the loss modulator is close to one end of a linear laser resonator (or anywhere within a ring resonator), and the modulator frequency exactly corresponds to the resonator round-trip frequency, a short pulse passing the modulator at the 'correct' time will experience lower power losses than light coming at any other times. This mechanism favours the pulse against all other light in each round trip. After many round trips, only that pulse can circulate, because it saturates the laser gain such that its net gain is exactly zero, and all other light will have a negative net gain and will thus 'die out' in the long run.

In each round trip, the modulator tends to make the pulse shorter by attenuating its temporal wings more than the centre. That effect, however, becomes weaker and weaker as the pulse gets shorter. On the other hand, various effects such as the limited gain bandwidth, dispersion or nonlinearities may tend to increase the pulse duration. In the steady state, a balance of such effects determines the achieved pulse duration. Typically, one easily gets to the regime of tens of picoseconds, but not to much lower pulse durations, even with a strongly driven modulator. According to Kuizenga and Siegman (1970), the pulse duration depends only on the fourth root of the modulation depth, which therefore has only a weak influence.

Advantages of active mode locking are the reliable start of the mode-locking regime (although within many resonator round trips) and the possibility of having the pulse emissions synchronized with an electronic drive signal. However, the achievable pulse duration is not particularly short, and the costs of a modulator and the drive electronics add substantially to the overall cost. Also, precise matching of the modulator frequency and the resonator

© Woodhead Publishing Limited, 2013

round-trip frequency is required, often with an automatic feedback mechanism acting on the resonator length or the modulation frequency.

9.5.3 Passive mode locking

Similar to passive Q-switching, passive mode locking relies on a saturable absorber. Here, however, the recovery time of the absorber must be much shorter – typically, far below the resonator round-trip time. So-called fast absorbers have a recovery time even below the pulse duration, but in many cases a slow absorber is sufficient, with a recovery time which may be 10–20 times the desired pulse duration.

An essential advantage of a saturable absorber over an active modulator, apart from the typically lower cost, is that the loss modulation caused by the absorber becomes faster as the pulse duration gets shorter. (At least the loss saturation is always very fast; loss recovery may take some longer time, as discussed below.) Basically for that reason, passive mode locking allows one to achieve far lower pulse durations – in extreme cases, down to a few femtoseconds (Sutter *et al.* 1999, Morgner *et al.* 1999).

Although many different times of saturable absorbers have been used for mode locking of solid-state lasers, semiconductor saturable absorber mirrors (SESAMs) have been used most successfully (Keller *et al.* 1996) in a wide range of performance parameters, including average output powers well above 100 W (Baer *et al.* 2010) and pulse repetition rates of many tens of GHz or even above 100 GHz (Krainer *et al.* 2002). Key parameters of a SESAM are its operation wavelength range, the modulation depth (the magnitude of nonlinear reflectivity change), the saturation fluence and the recovery time. For extremely short pulses, the reflection bandwidth also needs to be optimized. All these parameters can be varied in substantial ranges through the choice of materials and also various design details. Particularly operation in extreme parameter ranges requires a careful choice of absorber parameters, based on a deep understanding of the various effects influencing the pulse formation and the interplay of such effects.

Instead of a true saturable absorber, an 'artificial saturable absorber' can be used, i.e., a mechanism which effectively provides saturable absorption. Perhaps the best known is Kerr lens mode locking (Spence *et al.* 1991, Salin *et al.* 1991), where nonlinear self-focusing of the circulating pulse leads to an improved overlap with the pumped region in the gain medium, favouring the pulse against a competing long background. That method provides a strong and fast saturable absorber, which is often used for generating the very shortest pulses with durations down to the order of 5 fs (Sutter *et al.* 1999, Morgner *et al.* 1999). Other techniques utilize a nonlinear mirror based on frequency conversion in a nonlinear crystal material (Stankov 1988) or nonlinear phase changes in an optical fibre (Mark *et al.* 1989).

© Woodhead Publishing Limited, 2013

It is important to note that the physical effects which are relevant for the pulse formation can be very different, depending on the circumstances. The two most important regimes for solid-state lasers are briefly discussed:

- For many picosecond lasers, the pulse shaping is done essentially by a SESAM, and the limit for the pulse duration is set by the gain bandwidth, or sometimes by chromatic dispersion. The Kerr nonlinearity is then often not very relevant. Typical passively mode-locked Nd:YAG lasers, for example, operate in that regime, generating pulses with durations of the order of 10 ps.
- If a broadband gain medium (e.g., a rare-earth-doped glass) is used and dispersive effects are minimized by dispersion compensation, the pulse duration soon gets into the sub-picosecond regime. In most cases, the effect of the Kerr nonlinearity of the gain medium (and possibly other optical components) then becomes strong. In that regime, it is not advantageous to minimize chromatic dispersion. Instead, it is better to have anomalous dispersion of a suitable magnitude, such that quasi-soliton pulses are formed in the laser resonator. This regime of soliton mode locking (Kärtner *et al.* 1996, Paschotta *et al.* 2001) leads to particularly well-shaped and stable pulses, if the laser is operated in a suitable regime concerning nonlinear phase shift per round trip, pulse duration and the gain bandwidth. The pulses are normally close to the bandwidth limit, i.e., they exhibit nearly no chirp.

Less frequently, chirped-pulse oscillators are used (Fernandez *et al.* 2004), where the laser resonator has a net chromatic dispersion in the normal dispersion (positive group delay dispersion) regime. Here, the circulating pulse develops a strong up-chirp and is relatively long; the extracted pulses are temporally compressed by applying a suitable amount of anomalous dispersion. The long intracavity pulse duration limits the strength of the Kerr effect and thus allows one to obtain stable mode locking in a regime of high pulse energy and large pulse bandwidth, as required for short compressed pulses.

Passively mode-locked lasers can suffer from various types of instabilities, some of which are briefly discussed in the following:

- Q-switching instabilities (Kärtner *et al.* 1995, Hönninger *et al.* 1999) are instabilities of the pulse energy, caused as a usually unwanted side-effect of the saturable absorber: an increase of the pulse energy leads to stronger saturation, thus to lower losses and an increased net gain, leading to an even more quickly rising pulse energy. Depending on a number of laser parameters, that type of instability can be fully suppressed, allowing a stable regime of continuous-wave mode locking. This is tentatively more difficult to achieve for very high pulse repetition rates.
- A totally different origin of quite similarly appearing instabilities is that the recovery time of a slow absorber may be too long (Paschotta *et al.*

© Woodhead Publishing Limited, 2013

2001). The maximum permissible recovery time can depend on various circumstances, such as the degree of absorber saturation and the gain bandwidth.

- Soliton mode-locked lasers can become unstable if the pulses experience too strong disturbances in each resonator round trip (for example, when the nonlinear phase change per round trip is too large) or when the gain bandwidth is too small.
- Even very weak parasitic reflections within a laser resonator can make the mode locking unstable, and it can also prohibit the reliable starting of the mode-locking regime.

A detailed understanding of many different physical mechanisms in the pulse formation mechanism is often vital for identifying the performance-limiting factors and finding a truly stable operation regime with the desired parameters. In this sense, mode-locked lasers, and particularly passively mode-locked lasers, are far more difficult to design than Q-switched lasers, for example.

9.5.4 Resonator design for mode-locked lasers

Compared with continuous-wave lasers, the resonator design of mode-locked lasers is subject to various additional constraints. Examples are briefly discussed:

- The resonator must be suitable for single transverse mode operation, as the excitation of higher-order modes would disturb the pulse generation process.
- As the resonator length determines the pulse repetition rate, the required length is often fixed or at least limited to some interval.
- A mode locker such as a modulator or a saturable absorber often has to be placed at or near one end of the resonator. In addition, particularly a saturable absorber needs to be operated with a suitable mode size, as this influences the saturation characteristics.
- Components required for dispersion control, such as dispersive mirrors or prism pairs, may have to be incorporated. Particularly for prism pairs, substantial space (i.e., a long resonator arm) may be required, and a collimated beam is needed in that part of the resonator.

Due to such constraints, it is often a highly non-trivial task to find a suitable resonator design. Note that a step-by-step approach cannot work, since most design changes intended to fulfil one more requirement would at the same time change many other properties. Carefully worked-out numerical design strategies are often essential.

Figure 9.5 shows the setup of a typical mode-locked low-power laser.

© Woodhead Publishing Limited, 2013

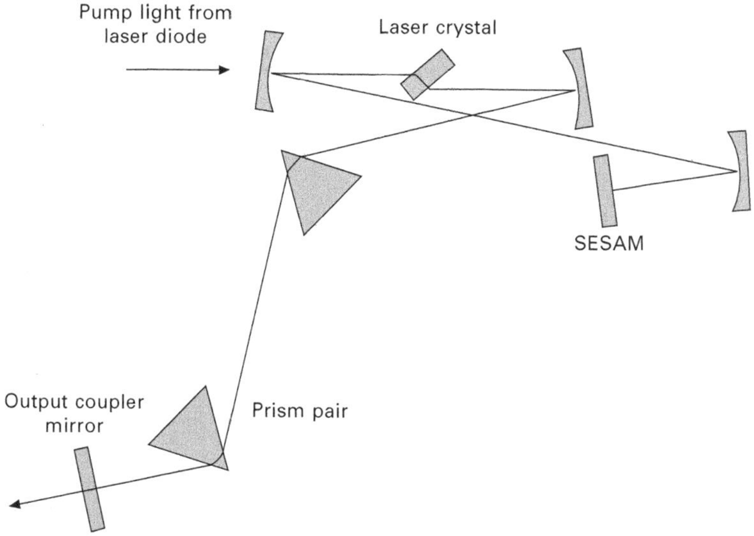

9.5 Setup of a typical mode-locked low-power laser.

The laser crystal is located between two focusing mirrors, through which the pump light can be injected. A prism pair in the resonator arm with the output coupler mirror serves for dispersion compensation, whereas the other resonator end is the SESAM. An appropriate mode size on the SESAM is obtained by using a focusing mirror next to it.

It is also often possible to use dispersive mirrors (e.g., with GTI or chirped-mirror designs) instead of a prism pair, or sometimes to work without any dispersion compensation. In some cases, the laser crystal is located close to an end mirror; there may then be substantial effects from spatial hole burning (see also Section 9.2.3), which may help to obtain a broader emission spectrum and thus shorter pulses, but can also make it more difficult to obtain bandwidth-limited pulses. Curved mirrors around the laser crystal or SESAM are often not required for high-power lasers, where the required mode areas on these devices are comparable with those throughout the laser resonator. These aspects show that many types of resonator designs can be used, although the one shown in Fig. 9.5 is quite common.

9.5.5 Combination with cavity dumping

The technique of mode locking can also be combined with cavity dumping (Gibson *et al.* 1996). Here, the pulses are amplified to higher and higher energies while the resonator losses are kept as low as possible (without any output coupling). When the maximum pulse energy is reached, much of the

© Woodhead Publishing Limited, 2013

energy of the pulse is extracted with a fast optical switch, which opens the resonator for a time somewhat below one round-trip time. Thereafter, the remaining pulse can be amplified again.

The optical switching is usually done with an electro-optic modulator combined with some polarization optics. The switching time needs to be shorter than one round-trip time.

This type of cavity dumping is suitable for obtaining higher-energy pulses, albeit at a reduced pulse repetition rate. The potential of this technique in terms of pulse energy is far below that of regenerative amplification (see Section 9.6.2). Also, the pulse duration is often increased, since the required optical switch introduces chromatic dispersion and particularly an optical nonlinearity. However, for high repetition rates this technique can be suitable and is simpler than realizing a regenerative amplifier.

9.5.6 Achievements

A few examples of achievements with mode-locked lasers are as follows:

- Pulse durations of the order of 5 fs have been achieved with passively mode-locked Ti:sapphire lasers (Sutter *et al.* 1999, Morgner *et al.* 1999).
- Passively mode-locked thin-disk lasers have been demonstrated with well above 100 W average output power and sub-picosecond pulse durations (Baer *et al.* 2010).
- Very compact passively mode-locked $Nd:YVO_4$ lasers have reached pulse repetition rates up to 160 GHz (Krainer *et al.* 2002).
- Mode-locked lasers with chirped pulses in a long laser resonator can generate particularly high pulse energies (Fernandez *et al.* 2004), even if the average output power is not particularly high.

9.6 Chirped-pulse amplification

9.6.1 Principle

The method of chirped-pulse amplification has been introduced by Strickland and Mourou (1985) in order to solve or at least mitigate the problem of high peak powers in amplifier systems. Whereas solid-state gain media can easily store millijoules or even joules of energy and transfer that energy to an amplified pulse within a very short time, the peak power and peak intensity of a pulse can then become so high that a number of detrimental nonlinear effects can occur. For example, large nonlinear phase shifts resulting from the Kerr effect can excessively broaden the pulse spectrum, or nonlinear self-focusing can lead to instant damage of the gain medium or other optical components.

© Woodhead Publishing Limited, 2013

The solution named chirped-pulse amplification involves the following steps (see Fig. 9.6):

- First, a large amount of chromatic dispersion is used to obtain strong temporal broadening (stretching) of the pulses to be amplified.
- Thereafter, the amplification is done. Due to the long pulse duration, the peak power remains moderate, avoiding a strong nonlinear effect.
- Finally, the pulse is compressed by using a large amount of chromatic dispersion, the sign of which is opposite to that in the pulse stretcher.

This method can solve the problem since a pulse compressor, in contrast to a laser amplifier, can be operated with a very large mode area, so that the peak intensity remains moderate despite an extremely high peak power.

The pulse compressor is often a set of diffraction gratings (Treacy 1969), which can provide a large amount of chromatic dispersion, resulting from wavelength-dependent beam path lengths, and a large operation bandwidth. Also, such gratings can be operated with large mode areas – in extreme cases with a beam diameter of several tens of centimetres, allowing for

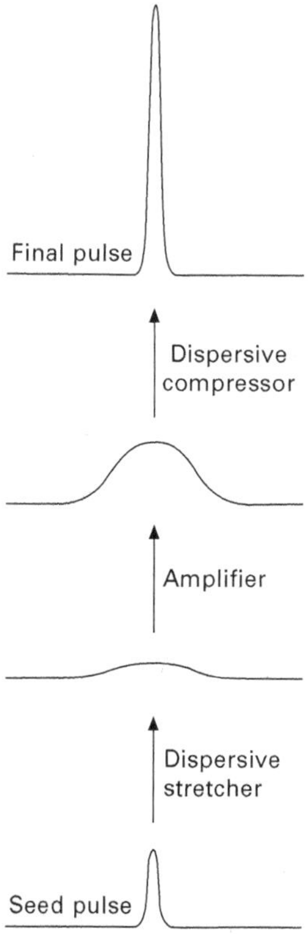

9.6 The principle of chirped-pulse amplification.

© Woodhead Publishing Limited, 2013

petawatt peak powers. Usually, such a compressor system contains either four diffraction gratings or two gratings (a grating pair) used in a double-pass configuration. For low stretched-pulse durations (below 1 ns) and small spectral widths, chirped volume Bragg gratings provide a simpler and more compact solution.

The possible amount of pulse stretching is actually not determined by the magnitude of the group delay dispersion, but rather by the spread of group delays within the available spectral width. That range of group delays is limited by the size of the compressor setup. Pulse stretching to a duration of several nanoseconds is feasible, but substantially longer stretched-pulse durations would require impractically large grating setups, which would also be very hard to align.

In the simplest case, the dispersion of the pulse stretcher and compressor should be just opposite to each other. Deviations from that balance are appropriate if significant dispersive or nonlinear effects occur in the amplifier, or when not bandwidth-limited seed pulses are used. In any case, the precision of matching the dispersion must be high, particularly when very short compressed pulses are needed, and when a very long stretched-pulse duration is realized.

Different kinds of laser amplifiers can be used, such as linear multipass amplifiers and regenerative amplifiers (Section 9.6.2), but also fibre amplifiers. A fibre amplifier would impose a peak power limit of a few megawatts due to self-focusing. This combined with a practical stretched-pulse duration of a few nanoseconds leads to a maximum pulse energy of the order of 10 mJ, not considering losses in the compressor (which can in principle be limited to a few percent). Using bulk laser crystals instead of fibres, one can effectively remove that self-focusing limitation, and much higher pulse energies become possible – even multiple joules. That combined with a compressed pulse duration far below 1 ps results in an output peak power of many terawatts or even in the petawatt regime (Perry *et al.* 1999).

9.6.2 Achievements

A few examples of achievements with chirped-pulse amplification lasers are as follows:

- Relatively compact Ti:sapphire amplifier systems can be made for ultrashort pulses with durations around 10 fs (Amani Eilanlou *et al.* 2008).
- Large laser/amplifier systems with multiple amplifier stages can generate pulses with hundreds of joules of energy and a few hundred femtoseconds duration, which lead to petawatt peak powers (Perry *et al.* 1999).

© Woodhead Publishing Limited, 2013

9.7 Regenerative amplification

9.7.1 Principle

A regenerative amplifier (Lowdermilk and Murray 1980) is similar to a mode-locked laser, as it also has an optical resonator in which an ultrashort pulse is circulated. There are important differences, however:

- The initial pulse is generated by an external seed laser, which is usually a mode-locked low-power laser. Therefore, special means for pulse formation within the resonator are not required.
- The laser gain is higher than the resonator losses, which are kept as low as possible (without output coupling), so that the pulse energy rises in each resonator round trip. Once the maximum pulse energy is reached, the pulse is ejected from the resonator, using a fast optical switch. (This can be called cavity dumping: see Section 9.5.5.)

Figure 9.7 shows a typical setup of a regenerative amplifier. A Pockels cell combined with a polarizer and a $\lambda/2$ waveplate serves as the fast optical switch, used both for injecting the seed pulse and extracting the amplified pulse. Additional polarizing optics separate the amplified pulses from the seed pulses.

Usually, repetitive amplification is done: after a pulse has been amplified, the energy in the gain medium is replenished by pumping, and another pulse can be amplified. The amplification phase usually takes much less time than the pumping phase.

For high pulse energies, excessive nonlinear effects can be avoided or mitigated with the method of chirped-pulse amplification, as explained in Section 9.5.6.

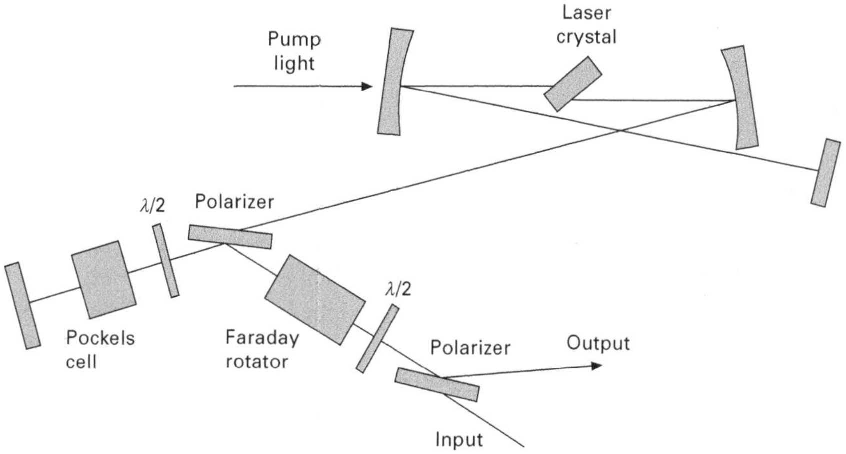

9.7 Typical setup of a regenerative amplifier.

© Woodhead Publishing Limited, 2013

The basic performance aspects of regenerative amplifiers can be compared with those for various other methods of pulse generation:

- The achievable pulse energy is in principle similar to that of a Q-switched laser, as much of the energy which is initially stored in the gain medium can be transferred to the amplified pulse. It is only that broadband gain media, as are required for amplifying femtosecond pulses, tend to exhibit a somewhat lower performance. For example, the frequently used Ti:sapphire gain medium has a low upper-state lifetime of ≈3 μs – to be compared with 230 μs for Nd:YAG, for example. This reduces the amount of energy which can be stored, at least with continuous pumping.
- The pulse energy can be substantially higher than for a cavity-dumped mode-locked laser, since the latter would not allow pumping the gain medium for a long time without energy extraction by a circulating pulse. A regenerative amplifier, other than a mode-locked laser, even allows for pulsed pumping – using a Q-switched pump source, for example.
- Still, the pulse energy is often more limited than for a linear multipass amplifier, since the latter can use a larger laser mode area and does not require an optical switch.
- Compared with a linear multipass amplifier, a regenerative amplifier can provide a much higher gain, since in principle an arbitrary number of resonator round trips can be used. This is controlled simply via the time of extracting the amplified pulse. Therefore, it is not necessary to use a high-gain laser material, although too low a gain may compromise the power efficiency if the resonator losses are not very small.
- The amplified pulse duration can be similar to the seed pulse duration from a mode-locked laser, but is often somewhat longer due to gain narrowing and possibly dispersive and nonlinear effects. Gain narrowing is often substantial due to the large overall gain. Sometimes, it is reduced by introducing a gain-flattening filter in the resonator, which, however, also somewhat reduces the power efficiency.

Typical femtosecond regenerative amplifiers are based on Ti:sapphire. With continuous pumping, the pulse energy is normally limited to the microjoule level by the short upper-state lifetime (see above), but the pulse repetition rate can easily be tens of kilohertz or even well above 100 kHz. With pulsed pumping, using a frequency-doubled Q-switched neodymium laser, pulse energies in the multiple-millijoule regime are easily possible, but then with lower repetition rates such as a few kilohertz. Some high-end systems, using a cryogenically cooled Ti:sapphire crystal and expensive pump sources, can be pushed into substantially higher performance regimes.

© Woodhead Publishing Limited, 2013

9.7.2 Achievements

The example setups named in Section 9.6.2 for chirped-pulse amplification also contain regenerative amplifiers. Some further examples are as follows:

- Regenerative amplifiers with a cryogenically cooled Ti:sapphire crystal can produce high output powers of tens of watts (Matsushima *et al.* 2006).
- Ytterbium-doped gain media, which can be easily diode-pumped, can also be used for high-power regenerative amplifiers. See, for example, Sayinc *et al.* (2009).

9.8 References

Agnesi A *et al.* (1996) 'All-solid-state gain-switched Cr^{4+}:forsterite laser', Opt. Commun. 127 (4–6), 273

Albrecht G *et al.* (1998), 'Solid state heat capacity disk laser', Laser and Particle Beams 16, 605–625

Amani Eilanlou A *et al.* (2008), 'Direct amplification of terawatt sub-10-fs pulses in a CPA system of Ti:sapphire laser', Opt. Express 16 (17), 13431–13438

Baer C R E *et al.* (2010), 'Femtosecond thin-disk laser with 141 W of average power', Opt. Lett. 35 (13), 2302–2304

Bollig C *et al.* (1995), 'Stable high repetition-rate single-frequency Q-switched operation by feedback suppression of relaxation oscillation', Opt. Lett. 20, 1383–1385

Bruesselbach H and Sumida D S (2005), 'A 2.65-kW Yb:YAG single-rod laser', IEEE J. Sel. Top. Quantum Electron. 11 (3), 600–603

Dascalu T *et al.* (2002), '100-W quasi-continuous-wave diode radially pumped microchip composite Yb:YAG laser', Opt. Lett. 27 (20), 1791–1793

Degnan J J (1995), 'Optimization of passively Q-switched lasers', IEEE J. Quantum Electron. 31 (11), 1890–1901

Du K *et al.* (2003), 'Electro-optically Q-switched Nd:YVO_4 slab laser with a high repetition rate and a short pulse width', Opt. Lett. 28 (2), 87–89

Fernandez A *et al.* (2004), 'Chirped-pulse oscillators: a route to high-power femtosecond pulses without external amplification', Opt. Lett. 29 (12), 1366–1368

Freitag I *et al.* (1995), 'Power scaling of diode-pumped monolithic Nd:YAG lasers to output powers of several watts', Opt. Commun. 115 (5–6), 511–515

Gibson G N *et al.* (1996), 'Electro-optically cavity-dumped ultrashort-pulse Ti:sapphire oscillator', Opt. Lett. 21 (14), 1055–1057

Giesen A and Speiser J (2007), 'Fifteen years of work on thin-disk lasers: results and scaling laws', IEEE J. Sel. Top. Quantum Electron. 13 (3), 598–609

Harb C C *et al.* (1994), 'Suppression of the intensity noise in a diode-pumped neodymium:YAG nonplanar ring laser', IEEE J. Quantum Electron. 30 (12), 2907–2913

Hargrove L E, Fork R L, and Pollack M A (1964), 'Locking of He–Ne laser modes induced by synchronous intracavity modulation', Appl. Phys. Lett. 5, 4–5

Holder L E *et al.* (1992), 'One joule per Q-switched pulse diode-pumped laser', IEEE J. Quantum Electron. 28 (4), 986–991

Hönninger C *et al.* (1999), 'Q-switching stability limits of cw passive mode locking', J. Opt. Soc. Am. B 16 (1), 46–56

© Woodhead Publishing Limited, 2013

Kärtner F X *et al.* (1995), 'Control of solid-state laser dynamics by semiconductor devices', Opt. Eng. 34, 2024–2036

Kärtner F X *et al.* (1996), 'Soliton modelocking with saturable absorbers', IEEE J. Sel. Top. Quantum Electron. 2, 540–556

Keller U *et al.* (1996), 'Semiconductor saturable absorber mirrors (SESAMs) for femtosecond to nanosecond pulse generation in solid-state lasers', IEEE J. Sel. Top. Quantum Electron. 2, 435–453

Krainer L *et al.* (2002), 'Compact Nd:YVO_4 lasers with pulse repetition rates up to 160 GHz', IEEE J. Quantum Electron. 38 (10), 1331–1338

Kuizenga D J and Siegman A E (1970), 'FM and AM mode locking of the homogeneous laser – Part I: Theory', IEEE J. Quantum Electron. 6, 694–708

Li C Y *et al.* (2010), 'QCW diode-side-pumped Nd:YAG ceramic laser with 247 W output power at 1123 nm', Appl. Phys. B 103 (2), 285–289

Lowdermilk W H and Murray J E (1980), 'The multipass amplifier: theory and numerical analysis', J. Appl. Phys. 51 (5), 2436–2444

Magni V (1987), 'Multielement stable resonators containing a variable lens', J. Opt. Soc. Am. A 4, 1962–1969

Mark J *et al.* (1989), 'Femtosecond pulse generation in a laser with a nonlinear external resonator', Opt. Lett. 14 (1), 48–50

Matsushima I *et al.* (2006), '10 kHz 40 W Ti:sapphire regenerative ring amplifier', Opt. Lett. 31 (13), 2066–2068

McClung F J and Hellwarth R W (1962), 'Giant optical pulsations from ruby', J. Appl. Phys. 33, 828–829

McDonagh L *et al.* (2006), '47 W, 6 ns constant pulse duration, high-repetition-rate cavity-dumped Q-switched TEM_{00} Nd:YVO_4 oscillator', Opt. Lett. 31 (22), 3303–3305

Morgner U *et al.* (1999), 'Sub-two cycle pulses from a Kerr-lens mode-locked Ti:sapphire laser', Opt. Lett. 24 (6), 411–413

Paschotta R *et al.* (2001), 'Passive mode locking with slow saturable absorbers', Appl. Phys. B 73 (7), 653–662

Paschotta R, Telle H R, and Keller U (2007), 'Noise of solid state lasers', in *Solid-State Lasers and Applications* (ed. A. Sennaroglu), CRC Press, Boca Raton, FL, Chapter 12, pp. 473–510

Perry M D *et al.* (1999), 'Petawatt laser pulses', Opt. Lett. 24 (3), 160–162

Salin F *et al.* (1991), 'Modelocking of Ti:sapphire lasers and self-focusing: a Gaussian approximation', Opt. Lett. 16 (21), 1674–1676

Sayinc H *et al.* (2009), 'Ultrafast high power Yb:KLuW regenerative amplifier', Opt. Express 17 (17), 15068–15071

Schawlow A L and Townes C H (1958), 'Infrared and optical masers', Phys. Rev. 112, 1940–1949

Spence D E, Kean P N, and Sibbett W (1991), '60-fsec pulse generation from a self-mode-locked Ti:sapphire laser', Opt. Lett. 16 (1), 42–44

Spühler G J *et al.* (1999), 'Experimentally confirmed design guidelines for passively Q-switched microchip lasers using semiconductor saturable absorbers', J. Opt. Soc. Am. B 16 (3), 376–388

Stankov K A (1988), 'A mirror with an intensity-dependent reflection coefficient', Appl. Phys. B 45, 191–195

Strickland D and Mourou G (1985), 'Compression of amplified chirped optical pulses', Opt. Commun. 56, 219–221

Sutter D H *et al.* (1999), 'Semiconductor saturable-absorber mirror-assisted Kerr lens

© Woodhead Publishing Limited, 2013

modelocked Ti:sapphire laser producing pulses in the two-cycle regime', Opt. Lett. 24 (9), 631–633

Tang C L *et al.* (1963), 'Spectral output and spiking behavior of solid-state lasers', J. Appl. Phys. 34, 2289–2295

Treacy E B (1969), 'Optical pulse compression with diffraction gratings', IEEE J. Quantum Electron. QE-5, 454–458

Wang Y *et al.* (2007), '1 MHz repetition rate single-frequency gain-switched Nd:YAG microchip laser', Laser Phys. Lett. 4 (8), 580–583

Zayhowski J J and Mooradian A (1989), 'Single-frequency microchip Nd lasers', Opt. Lett. 14 (1), 24–26

Zayhowski J J, Ochoa J, and Mooradian A (1989), 'Gain-switched pulsed operation of microchip lasers', Opt. Lett. 14 (23), 1318–1320

© Woodhead Publishing Limited, 2013

10 Neodymium-doped yttrium aluminum garnet (Nd:YAG) and neodymium-doped yttrium orthovanadate (Nd:YVO_4)

A. AGNESI and F. PIRZIO, University of Pavia, Italy

DOI: 10.1533/9780857097507.2.256

Abstract: In this chapter we review the history of neodymium-doped diode-pumped solid-state lasers (DPSSLs), since the first successful concepts exploited for industrial applications until the most recent state-of-the-art solutions. We also discuss the major issues related to power or energy scaling of DPSSLs and some representative examples of design solutions proposed to overcome such limitations. The most recent exciting developments that promise further power and brightness upscaling through the affirmation of new laser materials and laser architectures are also presented.

Key words: lasers, diode-pumped, neodymium, Q-switched, amplifiers.

10.1 Introduction

Neodymium solid-state lasers were quickly affirmed as industry standards after the invention of the ruby laser in 1960; in particular Nd:glass and Nd:YAG proved soon to be high-quality laser materials relatively easy to grow. The excellent thermo-mechanical and optical properties of Nd:YAG allowed a fast development of laser technology which has benefited many application fields. Q-switching and mode-locking techniques were readily invented to generate either high-peak power energetic pulses with nanosecond pulse width or very short picosecond and sub-picosecond pulses at the limit of the gain bandwidth. Harmonic and parametric conversion techniques were also exploited to further expand the range of applications of solid-state lasers, covering the optical spectrum from UV to mid-IR, with operating regimes ranging from continuous-wave (CW) to sub-picosecond pulses.

A significant boost to the success of solid-state lasers was given by the introduction of diode-pumping technology at the beginning of the 1990s, owing to increasing availability of affordable and powerful semiconductor lasers emitting at 800 nm, which eventually replaced short-lifetime and inefficient lamps as the pump means for neodymium lasers in many applications.

One of the first solid-state laser materials developed in the 1960s, Nd:YVO_4, was soon rediscovered and became a key element of low and medium-power diode-pumped solid-state lasers (DPSSLs), owing to its unique high emission

© Woodhead Publishing Limited, 2013

cross-section and large pump absorption coefficient at 808 nm: the fact that this material could not be grown with high-quality in large size was no longer relevant, since crystal volumes of only a few cubic millimeters were necessary for DPSSL technology.

Nowadays, Nd:YAG and Nd:YVO_4 lasers are widely employed in the industry for many applications, such as marking, drilling, cutting, soldering, trimming and scribing, but also as rangefinders, target illuminators and designators in military systems, as well as in medical applications such as skin treatments, ophthalmology and surgery. While the most precise micro-machining tasks are being accomplished by relatively low-power systems, very high-power multi-kilowatt cw lasers are employed for heavy tasks such as cutting in automotive industry, competing with CO_2 lasers and also with the rapidly emerging and extremely successful fiber lasers.

We review here the history of neodymium-doped DPSSLs, from the first successful concepts exploited for industrial applications until the most recent exciting developments that promise further boost in power and brightness increase through the affirmation of new laser materials and laser architectures.

10.2 Oscillators for neodymium lasers

The availability in the early 1990s of high-power diode bars was the technological turning point that allowed the development of new generations of high-brightness, compact and efficient Q-switched solid-state lasers. Since the optical power available from a single diode emitter is limited to only few watts, the output power of conventional diode lasers can be scaled up by increasing the number of emitters, realizing linear arrays. This results in the highly asymmetric, line-shaped emission of diode-bars. Consequently, lateral pumping is the natural choice to ease the coupling of high power diode radiation in an active medium, only requiring, in the simplest design, a cylindrical lens for diode fast-axis collimation. TEM_{00} transverse mode oscillation is required by most critical applications demanding the highest beam quality. However, mode-matching between resonator fundamental mode and the optically pumped active volume, crucial to get high efficiency in TEM_{00} mode operation, is much more easily obtained, relying on the longitudinal pumping concept. Unfortunately, this approach requires a proper reshaping of the diode emission (usually obtained by coupling the diode radiation into a multimode, depolarizing fiber, through complex and relatively expensive optical systems) and a proper resonator design to overcome, at least partially, thermally induced diffractive losses, thermal aberrations and, in the end, thermal-stress fracture. This is especially important in the case of active materials exhibiting relatively poor thermo-optical properties, such as Nd:YVO_4.

© Woodhead Publishing Limited, 2013

A representative example of an efficient diffraction-limited, multiwatt end-pump laser developed in the early 1990s is described in Nighan *et al.* (1995), where for the first time was reported a diode-bar pumped Nd:YVO_4 laser delivering more than 10 W cw power with high (exceeding 50%) optical conversion efficiency. The resonator was based on a Z-folded setup (Fig. 10.1) and when the oscillator was actively Q-switched with an acousto-optical modulator, it produced 8–20 ns long pulses with average power of 4 W up to 100 kHz repetition rate, in a circular beam with the divergence close to the diffraction limit. This cavity design allowed for efficient low-order transverse mode operation even in the presence of high-order non-parabolic thermally induced phase aberrations in the active medium. The key elements in the design of such lasers are as follows:

- A cavity fundamental mode intentionally kept smaller than the pump mode, in order to avoid overlap with the portion of the active medium where thermally induced aberrations are stronger
- A sufficiently low dopant concentration to distribute pump power absorption and thermal load along all the crystal length, still maintaining a reasonable overall pump absorption
- Ideally a flat-top pump profile, more often a super-Gaussian one (reasonably provided by a multi-mode pump delivery fiber), in order to confine the gain in a volume in which radial variation of refractive index is reduced or negligible.

While in the case of end-pumping a major effort should be spent in pump reshaping and cavity design, in the case of side-pumping the main issue is related to efficient filling of a large pumped volume with a low-order resonator

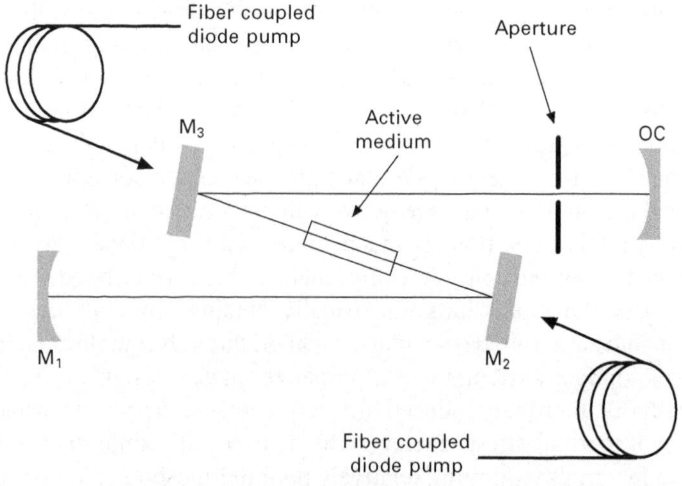

10.1 The Z-folded end-pumped laser setup.

© Woodhead Publishing Limited, 2013

mode, whose transverse dimension is usually smaller than the pumped volume cross-section. One possible solution, exploited earlier during DPSSL technology development, was the realization of a zig-zag beam path within a slab-shaped active medium, side-pumped by diode bars. This solution offers the inherent advantage of a good spatial overlap between the resonator mode and the high-gain layer near the pumped faces of a laterally pumped crystal slab. An example of such a solution is the coplanar pumped folded slab (CPFS) design (Richards and McInnes 1997) shown in Fig. 10.2. It consists of a Nd:YAG slab, side-pumped from one or both sides by cw diode arrays equipped with a simple fast-axis collimating fiber lens. The cavity beam is coupled and extracted from the active medium through Brewster-cut facets that transform a circular cavity mode in air into an elliptical mode inside the active medium, further increasing the active volume filling factor. With this design, employing 40 W cw pumping, more than 7 W in Q-switching at 10 kHz repetition rate with 5 ns long pulses were demonstrated in a nearly TEM_{00} beam.

Other hybrid solutions, trying to exploit the simplicity of pump power coupling typical of side-pumped systems, and the optimum overlap between pump and cavity mode offered by the end-pumping approach, were also investigated earlier. Such a design was called a tightly folded resonator (TFR) (Baer *et al.* 1992). As shown in Fig. 10.3, the path of the laser beam inside the active medium was arranged in such a way that bounces on the pumped face occur just in front of each diode-bar emitter, allowing a good overlap with the excited volume and high slope efficiency, close to 50%, in cw TEM_{00} operation. By Q-switching the laser by a Pockels cell, pulses as short as 5 ns were obtained in this configuration with average power of a few watts at 10 kHz maximum repetition rate.

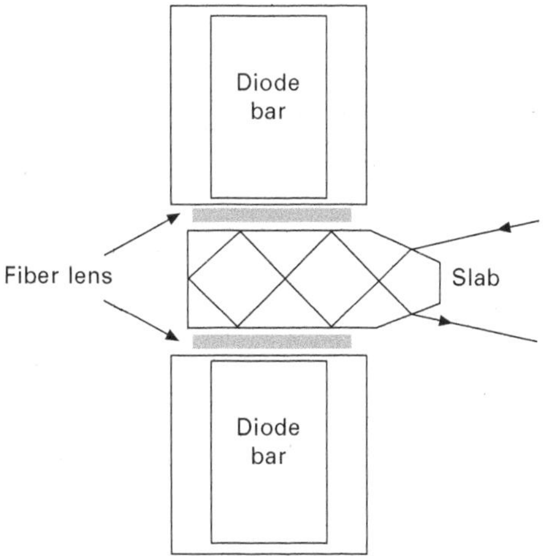

10.2 The coplanar pumped folded slab (CPFS) amplifier design.

© Woodhead Publishing Limited, 2013

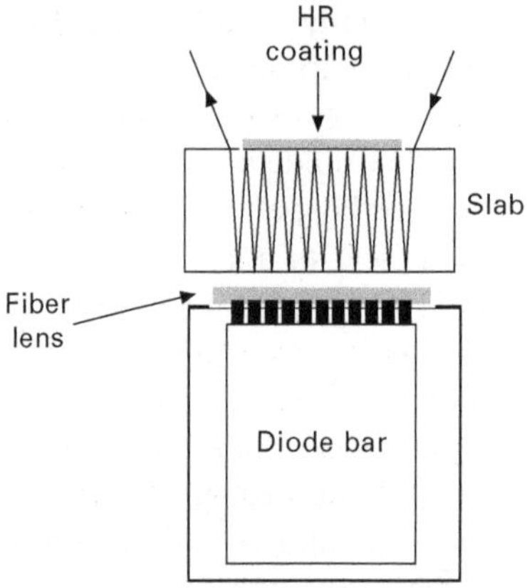

10.3 Amplifier module of the tightly folded resonator (TFR).

All these cavity designs employ resonators as long as several tens of centimeters and are intended for obtaining average output powers in the range of several watts. They turn out not to be suitable when shorter Q-switched pulses are needed, approaching (and even breaking) the nanosecond barrier. To this purpose, a short cavity length is crucial in order to reduce both the round-trip time and the number of round trips needed to build up the pulse. Unfortunately, short resonators capable of producing high brightness emission pose severe design challenges due to a smaller fundamental cavity mode inside the active medium, a significant drawback when trying to up-scale the power without compromising beam quality. Only in the early 2000s did commercial products based on Nd:YVO_4 active medium, fast electro-optical Q-switch devices and very compact resonators become available on the market. With these systems, using either the single oscillator configuration (http://www.brightsolutions.it, ~500 ps 'Wedge' model), or a master oscillator power amplifier (MOPA) (Pearce and Ireland 2003), multi-kHz, ~1 ns-long pulses with average output power of several watts have been demonstrated.

In the last decade significant research efforts have been devoted to scale the average output power of these actively Q-switched laser systems up to several tens of watts. Average power scaling always involves larger active volumes, necessary to distribute the thermal load and avoid thermal stress fracture of crystals. In order to ease the removal of pump-generated heat, the typical active medium shape is a slab with a large aspect ratio and wide top and bottom surfaces conductively cooled with water or Peltier thermo-electric cooler modules. Pump power is provided by either multiple diode bars or stacked diode arrays. One issue concerns how to efficiently couple the pump radiation in a properly shaped active medium. Moreover, the cavity should be properly designed in order either to sustain a large cavity fundamental

© Woodhead Publishing Limited, 2013

mode or to arrange a multi-pass path of a small cavity fundamental mode inside a large active volume.

The only way to produce a large cavity mode in short resonators is to employ unstable rather than stable resonators. One interesting solution, exploited in earlier CO_2 lasers and originally proposed by Du *et al.* (1998, 2003) for DPSSLs, relies on a resonator design in which the cavity is stable in the *y–z* (vertical) plane, and unstable in the *x–z* (horizontal) plane, as shown in Fig. 10.4. In this laser arrangement, the fast-axis emission of the stacked diode array with relatively high beam quality was reshaped and homogenized in a planar waveguide and then focused in the active medium. The slow-axis emission was imaged in the crystal, and the final result was a rather uniform pump sheet with cross-section ~0.5×12 mm^2. The heat deposited in the Nd:YVO_4 slab was removed through the wide surfaces parallel to the *x–z* plane. About 100 W was obtained in cw operation under approximately 250 W pumping with about 55% slope efficiency. The laser was then actively Q-switched with a Pockels cell. At 10 kHz repetition rate, 6.5 ns long pulses with 5.6 mJ pulse energy were obtained. Increasing the frequency up to 50 kHz, 80 W average power with ≈11 ns long pulses were demonstrated with good beam quality ($M^2 \leq 1.5$ after cutting sidelobes with a spatial filter, with power loss ≈ 10%). Employing a similar cavity design with two Nd:YAG slabs each pumped by six stacked diode arrays, Zhu *et al.* (2008) further increased the average output power from the stable/unstable resonator setup. About 190 W in the cw regime and about 170 W in Q-switching operation at 10 kHz repetition rate with similar pulse duration (≈11 ns) and beam quality were obtained. A further proof of the effectiveness of the proposed solution

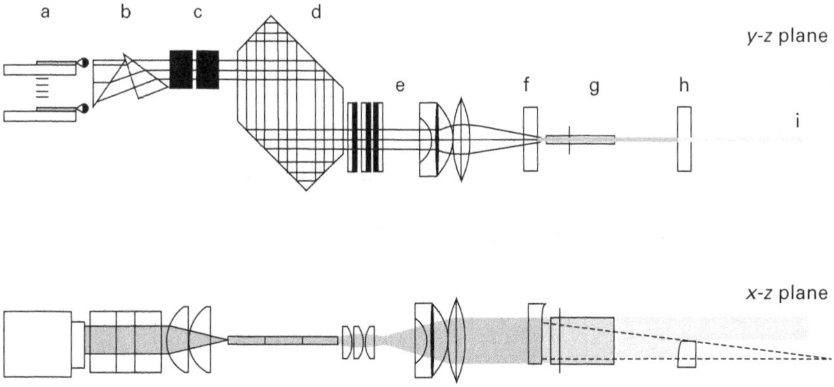

10.4 Pump and cavity setup of the stable/unstable partially end-pumped resonator: a, diode stack; b, anamorphic prisms pair; c, cylindrical lenses; d, planar waveguide; e, imaging group; f, rear mirror M1; g, Nd:YVO_4 slab; h, output coupler M2; i, output laser beam (from Du *et al.*, 2003, with permission).

© Woodhead Publishing Limited, 2013

is the high second-harmonic generation efficiency obtained, slightly lower than 60%, that allowed production of up to 93 W green light at 10 kHz in a Type I critically phase matched, 15 mm long LBO crystal.

The multipass slab (MPS) concept was efficiently exploited, instead, by Dergachev *et al.* (2007). The general design of such a configuration is depicted in Fig. 10.5. The emission of the diode bars is fast-axis collimated and directly coupled in the slab in order to realize a thin gain sheet in the active medium. A dielectric high-reflectivity coating at the pump wavelength back-reflects the unabsorbed pump, maximizing absorption. Efficient energy extraction is provided by the multiple passes of the laser mode through the active volume. To improve extraction efficiency, the overlap of the laser mode and the gain volume can be further optimized by an elliptical shaping of the laser mode inside the active medium. This design proved to be very effective, employing as active media both fluoride materials, such as Nd:YLF, and oxide materials such as Nd:YVO_4. The superior thermo-optical properties of the fluoride matrix ease TEM_{00} operation even at high pump power levels. Up to 35 W with 52% slope efficiency were reported in cw operation of a Nd:YLF slab pumped with 80 W. Nevertheless also efficient Nd:YVO_4 based oscillators were demonstrated with this design. About 14 W average power at 100 kHz repetition rate was obtained in an acousto-optical Q-switched Nd:YVO_4 MPS laser with 16 ns long pulses at 1064 nm. Also efficient operation al 1342 nm was reported, with up to 5 W average power at 50 kHz and 50 ns long pulses.

Bernard and Alcock (1993) proposed another interesting setup employing

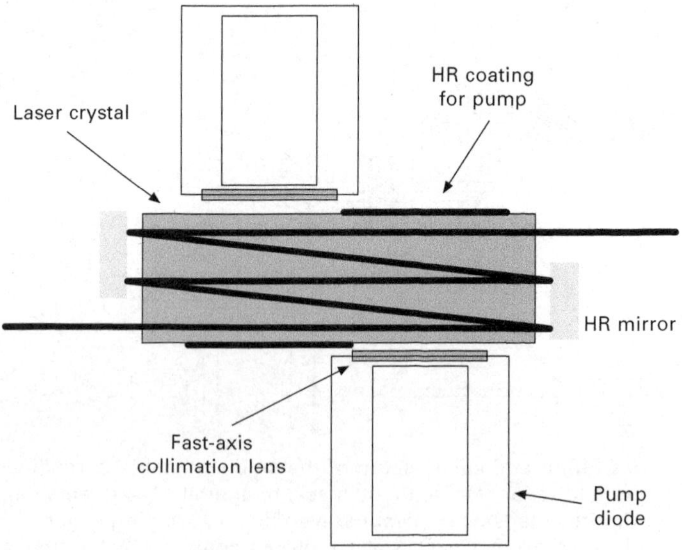

10.5 Multipass slab amplifier conceptual scheme.

a side-pumped slab, which was more deeply exploited a decade later, taking advantage of high power cw diode bars that became available. Indeed, this bounce amplifier or grazing incidence amplifier allowed exceptionally high small signal gain (up to ~30–40 dB) even in cw pumping, owing to the short absorption length of Nd:YVO$_4$ (Fig. 10.6). Oscillators with 23.1 W output power (cw), optical-to-optical efficiency of 58% (TEM$_{00}$) and 68% slope efficiency (multimode) have been demonstrated (Minassian *et al.* 2003). The most limiting aspect of this architecture is perhaps related to the strongly asymmetric (also power-dependent) thermal lensing requiring a relatively complicated resonator design including cylindrical lenses.

Some applications benefit from high-repetition rate systems with constant pulse duration. This kind of performance cannot be realized when cavity losses are modulated by an acousto-optical or electro-optical Q-switch. Indeed, pulse duration is dependent on the initial gain before Q-switching, and such gain modulation decreases with the repetition frequencies. Such a performance can, instead, be realized relying on cavity-dumping Q-switch. In this case, in fact, as far as the electro-optical switch is fast enough, the pulse duration is basically set by the cavity round-trip time. A remarkable example of such a design was reported in McDonagh *et al.* (2006). The resonator was Z-folded, as schematically shown in Fig. 10.7. The 0.5%-doped, 30 mm long Nd:YVO$_4$ crystal was pumped at 888 nm in order to both reduce thermal load and take advantage of the low and isotropic vanadate absorption at this wavelength. The unabsorbed pump power was transmitted through an M$_2$ mirror and re-imaged in the active medium with a lens and a totally reflective mirror at 888 nm (not shown in Fig. 10.7). The polarizing mirror, quarter wave-plate and Pockels cell acted together as a variable output coupler (either almost totally reflective or totally transmitting) for cavity dumping. A maximum average power of 47 W was reached at 50 kHz in Q-switched cavity-dumped operation and an approximately constant pulse duration of 6 ns was measured in the range 30–100 kHz. This constant, short pulse duration increases significantly the peak power at high repetition rates with

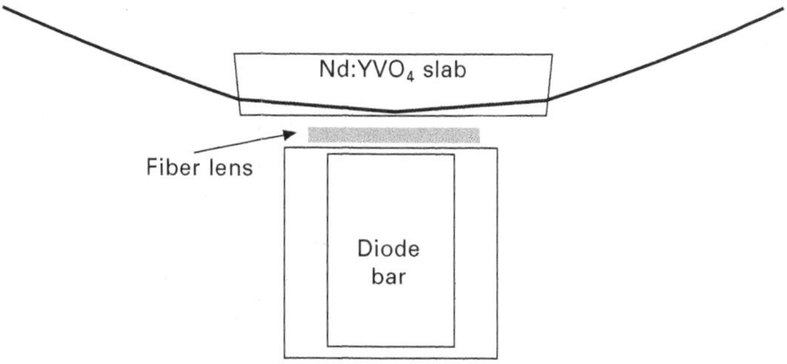

10.6 Grazing incidence amplifier module.

© Woodhead Publishing Limited, 2013

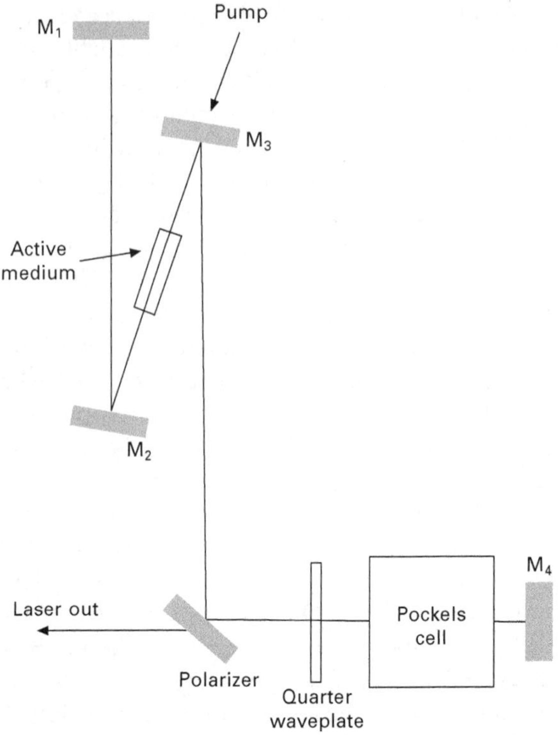

10.7 Schematic depiction of the cavity-dumped Q-switched laser setup.

respect to traditional actively Q-switched lasers, allowing for straightforward efficient harmonic generation. More than 30 W average power at 40 kHz at 532 nm were obtained in a non-critically phase-matched LBO crystal, with a conversion efficiency of 70%.

In many applications it is particularly important to be able to precisely trigger the Q-switched laser pulse emission, hence actively Q-switched laser sources are the only possible choice. But, whenever precise pulse triggering is not of concern, passively Q-switched (PQS) diode-pumped solid-state lasers become a very attractive alternative, offering the advantage of an inherent simplicity, cost-effectiveness and reliability. By choosing the suitable combination of active medium and saturable absorber characteristics, PQS systems can cover a wide range of pulse energies, pulse durations and repetition frequencies. Taking advantage of the relatively high saturation fluence and the excellent thermo-mechanical properties of yttrium aluminum garnet, Cr^{4+}:YAG solid-state saturable absorbers settled as state-of-the-art solutions for PQS lasers operating at 1 micron and delivering high-energy pulses (hundreds of microjoules), up to tens of kilohertz repetition rate. A

© Woodhead Publishing Limited, 2013

representative example of this class of DPSSL is reported in Agnesi *et al.* (1997) where high energy (100–200 μJ), tens of nanoseconds long (15–30 ns), diffraction-limited beam quality ($M^2 \sim 1.1$) pulses, with repetition rates of 6–20 kHz have been demonstrated. In this kind of system, Nd:YAG is the natural candidate as active medium owing to its superior thermo-optical properties, high energy storage and natural predisposition to be passively Q-switched, thanks to its high saturation fluence. This allows for significant cavity design simplification, since no additional intracavity optics are needed to properly focus the cavity mode in the Cr:YAG saturable absorber in order to satisfy the PQS threshold condition (Chen and Tsai 2001),

$$\frac{\ln(1/T_0^2)}{\ln(1/T_0^2) + \ln(1/R) + L} \frac{\sigma_s}{\sigma_L} \frac{A_L}{A_s} >> \frac{\gamma}{1-\beta} \qquad [10.1]$$

as in the case of *a*-cut, Nd:YVO_4. In Eq. 10.1, T_0 is the saturable's small signal transmission, R the output coupler reflectivity, σ_L and A_L the emission cross-section and mode area for the laser medium, σ_S and A_S the absorption cross-section and mode area for the absorber, γ the inversion reduction factor (1 for four-level lasers, 2 for three-level lasers) and β the ratio between excited-state absorption cross-section and σ_S.

However, Nd:YVO_4 offers some intrinsic advantages with respect to Nd:YAG, such as much higher gain, a naturally polarized emission that allows thermally induced birefringence to be overcome, and a broader and stronger absorption line peaked at 808 nm. The consequently shorter absorption length results in a better mode matching with the highly divergent pump beam produced by multi-mode fiber coupled laser diodes and allows for higher optical-to-optical efficiencies. All these features have been exploited in a particular design concept (Agnesi and Dell'Acqua 2003). The resonator layout is depicted in Fig. 10.8. Exploiting the high natural birefringence of yttrium vanadate, even for wedge angles of a few degrees, output polarization can be switched by simply horizontally tilting the output coupler. Indeed,

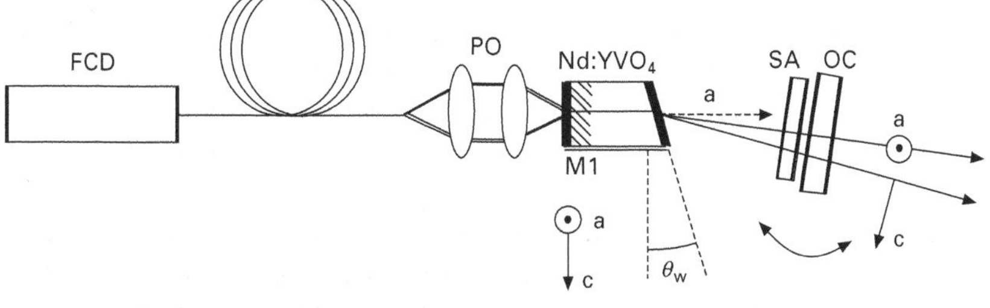

10.8 FCD, fiber-coupled diode; PO, pump optics; M1, pump mirror; SA, Cr:YAG saturable absorber; OC, output coupler; θ_w: wedge cut angle (from Agnesi and Dell'Acqua, 2003, with permission).

© Woodhead Publishing Limited, 2013

since the intracavity beam is normal to the high-reflectivity left mirror, the ordinary and extraordinary waves will experience different refraction and emerge from the crystal at different angles toward the output coupler. Hence the laser can be readily operated in the ordinary polarization exploiting the lower emission cross-section for easy PQS operation in a plane–plane compact resonator. Pulses as short as 6 ns with energies exceeding 150 μJ, corresponding to peak powers higher than 20 kW with good beam quality ($M^2 < 1.4$), were obtained.

The cavity shortening allowed by PQS can be pushed to the limit by designing monolithic or quasi-monolithic resonators, in which the cavity is completely filled by the active medium and the saturable absorber, resulting in intrinsically single longitudinal mode, maintenance-free lasers. One possible solution is to bond a highly doped, thin Nd:YAG crystal with a Cr:YAG saturable absorber, with the minimum length of both crystals basically set by the necessary absorption at pump and laser wavelength, respectively. Very short pulses (~200 ps), with a multi-kilohertz repetition rate and energies of a few tens of microjoules, were first demonstrated by Zayhowski and Dill (1994) employing a microchip Nd/Cr:YAG laser. Microchip laser technology is illustrated in greater detail in Chapter 14. Exploiting the concept of fast spatial mode filtering in unstable resonators, Agnesi *et al.* (2006) demonstrated a diffraction-limited, single-longitudinal-mode (SLM) 0.5 mJ 420 ps Nd:$GdVO_4$ PQS laser with Cr:YAG. The 9 mm long Fabry–Perot resonator (with the active medium) was pumped by a 140 W quasi-cw diode bar up to 200 Hz repetition rate. In order not to destroy optical elements during Q-switching, a large pump area of ~1.4×0.8 mm^2 in the *a*-cut 1%-doped Nd:$GdVO_4$ crystal was chosen. The use of relatively high-loss Cr:YAG (21% transmission) and high output coupling (20% reflectivity) required high peak gain. In turn, this dominated thermal lensing and shaped the mode profile as in classic unstable resonators. Indeed, owing to the slow build-up time of PQS lasers, the seed photons had time to propagate into the far field, thus accomplishing full build-up of the fundamental spatial mode as well, as clearly shown by numerical simulations.

The SLM selection in PQS occurs under some conditions which are often fulfilled even in resonators few centimeters long, owing to the slow build-up that tends to favor the longitudinal mode nearest to the gain peak (Isyanova and Welford 1999). Unidirectional monolithic non-planar ring PQS oscillators were also proved to be excellent performers in this respect (Freitag *et al.* 1997).

In order to further downscale pulse duration, even shorter cavities than those allowed by Nd/Cr:YAG microchip lasers are needed. To this purpose, a key role is played by semiconductor saturable absorber mirrors (SESAMs). These devices offer some advantages with respect to Cr:YAG in designing a microchip PQS laser, in particular:

© Woodhead Publishing Limited, 2013

- They work in reflection and not in transmission mode and hence do not contribute to the resonator optical length, allowing for a significant reduction in cavity round-trip time.
- Their saturation fluence can be adjusted in order to optimize the performance of the PQS microchip laser. Saturation fluence can be made sufficiently low to employ even *a*-cut Nd:YVO$_4$ as active medium. This is particularly important, since the higher intrinsic absorption of vanadate crystals with respect to Nd:YAG allows for a shortening of the active medium to ~100 μm and consequently of the cavity length.
- The modulation depth of the SESAM can be adjusted independently of other saturable absorber parameters. This allows the design of microchip lasers with very different pulse repetition rates and pulse durations.

All these considerations and some practical useful guidelines for the design of microchip lasers based on SESAMs are given in Spühler *et al.* (1999). The shortest pulses ever reported for a Q-switched DPSSL were only 37 ps long, with 8.5 mW output power at 160 kHz repetition frequency (1.4 kW peak power) for a Nd:YVO$_4$/SESAM microchip laser. Further optimization of the optical bonding of SESAM with the active medium by the use of spin-on-glass glue to avoid parasitic etalon effects due to residual air gaps between resonator components, allowed significant enhancement of the stability and average power of the Nd:YVO$_4$/SESAM microchip laser. Nodop *et al.* (2007) demonstrated up to 100 mW average power at 166 kHz repetition rate with 100 ps long pulses and an impressive energy of 1 μJ in 50 ps long pulses (20 kW peak power) at 40 kHz repetition rate.

10.3 Power/energy limitations and oscillator scaling concepts

The most natural way to optimize both operation efficiency and output beam brightness is with end-pumping. However, this is basically limited to low to medium power oscillators, up to a few tens of watts. Indeed, when designing a high-brightness TEM$_{00}$ laser source one must trade off thermal lens effects and practical design constraints, such as the resonator size. First, the thermal fracture limit F (W/cm) yields the pump absorption coefficient α_P (at wavelength λ_P) that a given laser material can withstand at a certain pump level:

$$F = \left.\frac{dP_{abs}}{dz}\right|_{max} = \alpha_P P_P(0) \qquad [10.2]$$

Then the pump and laser mode sizes W_P should match reasonably well within the absorption length of the laser material (with refractive index n) which must be comparable to the pump Rayleigh range z_{RP}:

© Woodhead Publishing Limited, 2013

$$z_{RP} \approx \frac{1}{\alpha_P} \tag{10.3}$$

$$\frac{nW_P^2}{\lambda_P M_P^2} \approx \frac{1}{\alpha_P} \tag{10.4}$$

where M_P^2 is the beam quality of the pump.

$$W_P \approx \sqrt{\frac{\lambda_P M_P^2}{n\alpha_P}} \tag{10.5}$$

The thermal focal length (Innocenzi *et al.* 1990) depends on the pump area and absorbed pump power. In particular, we see that

$$f_{th} \propto \frac{W_P^2}{P_{abs}} \propto \frac{M_P^2}{\alpha_P P_{abs}} \propto \frac{M_P^2}{F} \tag{10.6}$$

Therefore, the focal length is ultimately settled by the thermal fracture limit, and the mode size in the laser material scales as $\sqrt{P_{abs} M_P^2 / F}$. Since in a stable resonator the TEM_{00} mode size basically depends on the square root of the cavity length (Koechner 2006), it is clear that upscaling to high powers soon becomes challenging. Advancements on high-brightness, high-power laser diodes for pumping with lower M_P^2 parameters (Clarkson and Hanna 1996) have somehow alleviated the problem but the practical limit has only been pushed a little forward.

Besides, there are other important issues related to thermal aberrations, nicely reviewed by Clarkson (2001). Indeed, it is well known that the thermal lens is not ideally 'spherical', but rather induces severe non-spherical aberrations and sometimes even astigmatism, owing to the anisotropy of laser materials or to non-uniform thermal contacts between crystal faces and the cooling medium. Flat-top pump beam profiles such as those generally available with laser diodes coupled to multimode fibers allow the best performance in terms of minimization of thermal aberrations, since the lens is quadratic inside the pump cross-section and non-sphericity occurs outside. Thus, a trade-off between operation efficiency and beam quality is often considered: the ratio between the fundamental laser mode radius w_L (measured at $1/e^2$ of peak intensity level) and the radius of the pump beam w_P is generally chosen to be ≈0.8–0.9. Furthermore, the preferred stable resonator design allows the condition $dw_L/df_{th} < 0$ to be met: this means that, since the focal length of the aberrated thermal lens increases radially, the corresponding fundamental eigenmodes have smaller radius, as well as the next few higher-order modes which would start to oscillate in these exterior annular regions of the pumped cross-section. Therefore, such higher-order modes are eventually suppressed by the competition with the TEM_{00} on-axis mode,

© Woodhead Publishing Limited, 2013

even without additional intracavity apertures. However, this allows only a moderate power upscaling, until the onset of spatial modes of even higher order, that need to be suppressed by hard apertures.

In addition to these issues, energy transfer upconversion (ETU) has been demonstrated to increase thermal lensing and the associated aberrations. ETU originates from photon absorption in highly populated upper states, leading to the increase of thermal lensing owing to subsequent multi-phonon decay following the upconversion. This not only happens in Q-switching operation, but even merely when comparing thermal lenses in lasing and non-lasing conditions, since cw oscillation requires that the population inversion be clamped to the threshold level. A quantitative investigation of the effect in lasers of practical importance was carried out by Hardman *et al.* (1999). An immediate consequence is that an accurate characterization of thermal lensing required to start or refine a specific laser design must be done with lasers operating in free-running mode, not merely with the pumped rod without the cavity. The thermal lens can be determined by measuring the output beam size and the M^2 in both transverse directions x–y, for several levels of pump power. Applying a best-fit approach to the resonator model taking into account the dependence of the thermal lens on the absorbed pump power, $f_{th} = C_{x,y}/P_{abs}$, allows one to determine the coefficient C for each coordinate x and y.

To further complicate the matter, induced birefringence must also be considered when pumping at high power levels isotropic materials such as Nd:YAG (Koechner 2006, Clarkson 2001). However, when linearly polarized lasers are required, most often intrinsically birefringent materials are chosen (such as Nd:YVO$_4$ or Nd:YLF) thereby easily offsetting any thermally induced perturbation with their intrinsically high natural birefringence.

It is also worth pointing out some general guidelines in choosing the most appropriate laser material for a given application. Starting from the mode-matching and scaling considerations, Eqs 10.2–10.6, one can summarize the relevant laser parameters as in Table 10.1.

The single-pass small-signal gain g_0 is given by

$$g_0 = \int \sigma_L n(z) \mathrm{d}z = \frac{\lambda_P}{\lambda_L} \frac{P_{abs}}{P_{SL}} \qquad [10.7]$$

Table 10.1 Relevant parameters of the most used Nd and Yb laser materials

	τ_f (μs)	$\sigma_L \times 10^{19}$ (cm^2)	F_{SL} (J/cm^2)	F (W/cm)	g_0 max (a.u.)
Nd:YVO$_4$	100	15	0.125	250	1
Nd:YAG	230	2.8	0.670	550	0.94
Nd:YLF	500	1.8	1.0	100	0.24
Yb:YAG	950	0.20	9.4	550	0.28

© Woodhead Publishing Limited, 2013

where $n(z)$ is the population inversion along the beam propagation direction z, and P_{abs} and P_{SL} are the absorbed pump power and the saturation power, respectively.

From Eq. 10.6 we find that the gain g_0 scales as

$$g_0 \propto \frac{\sigma_L \tau_f P_{abs}}{A} \propto \sigma_L \tau_f \frac{F}{M_P^2} \qquad [10.8]$$

In Table 10.1 we show the gain normalized to the maximum value among those of different crystals, most popular in solid-state lasers. Yb:YAG and Nd:YAG are the most natural candidates for high-power applications, owing to the high fracture limit F. With their long lifetime, Nd:YLF and Yb:YAG yield the highest pulse energy in Q-switching at low repetition frequency. Instead, Nd:YVO_4 is preferred for high-repetition rate Q-switching applications due to the shortest fluorescence time, and with Nd:YAG allows the generation of the shortest nanosecond pulses owing to the highest gain achievable.

Several strategies have been devised to increase the output power of end-pumped lasers without compromising their beam quality. One solution is using ~880 nm laser diodes for pumping directly into the $^4F_{3/2}$ upper laser level thereby reducing significantly the quantum defect. This yields higher intrinsic efficiency and comparable beam quality at higher output power, especially at 1064 nm but also at the quasi-three-level transitions ~915/946 nm (Pavel *et al.* 2008) and reduces the thermal effects in the active medium. A Nd:YVO_4 laser end-pumped at either 808 nm or 880 nm has been shown to improve the slope efficiency dP_{out}/dP_{abs} from 49% to 53% (Pavel *et al.* 2007), with significant reduction of the thermal load, especially useful in high-power lasers. Furthermore, thermal diffractive losses due to coupling of energy into rapidly diverging higher-order modes can be reduced by use of diffusion-bonded undoped end-caps (Chang *et al.* 2008), which cancel the effect of the bulging of the heated surface of the doped crystal, therefore mitigating the resulting aberration on the beam wavefront. In addition, the bonded element provides an improved thermal contact with the heated crystal, reducing the temperature gradient (Fig. 10.9).

While end-pumping is the preferred method for achieving efficient TEM_{00} oscillation, significant power scaling was accomplished traditionally by recurring to either slab geometry (Koechner 2006) or side-pumping, following design rules established for the former class of lamp-pumped lasers (Magni 1986), which obviously still apply.

Owing to the one-dimensional temperature gradient, the slab geometry has been shown to outperform rods when the aspect ratio (width/thickness) is greater than 2 (Koechner 2006), therefore providing a means for power scaling maintaining the absorbed power per unit length safely below the fracture limit. This concept was only partially exploited by Shine *et al.* (1995), who demonstrated a 40 W TEM_{00} Nd:YAG zig-zag slab laser, side-

© Woodhead Publishing Limited, 2013

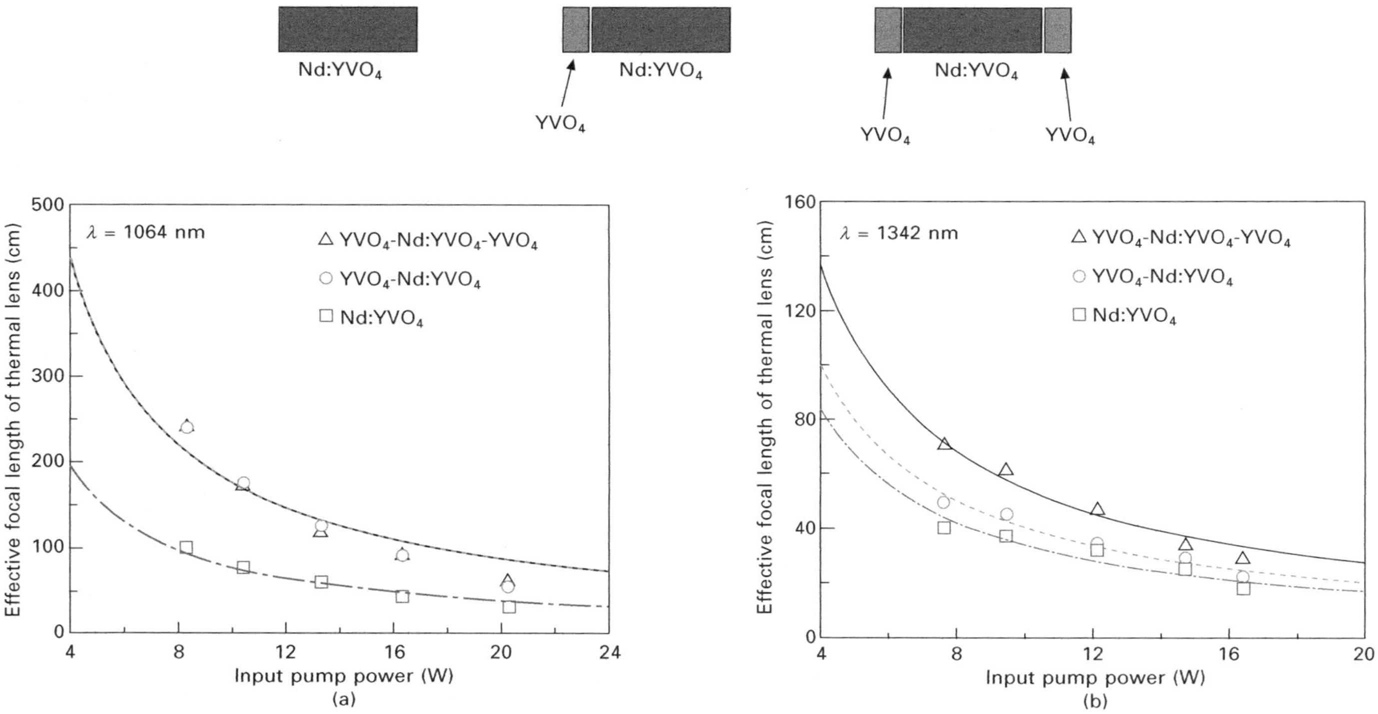

10.9 Comparison of thermal lensing at 1064 and 1342 nm with conventional, single- and double-bonded Nd:YVO_4 crystals (from Chang *et al.*, 2008, with permission).

© Woodhead Publishing Limited, 2013

pumped by fiber-coupled laser diodes delivering 235 W. Up to 72 W were achieved in multimode operation with 36% slope efficiency, a remarkable result for side-pumping. In this case the slab geometry concept was not fully exploited for further upscaling, since the aspect ratio was ≈1; nevertheless this result gives an idea of the potential. The main drawback of the zig-zag design is the imperfect overlap between gain distribution and the resonant beam path, and the necessity of very high quality surface polishing because wavefront aberrations add up on several bounces, affecting the output beam quality more heavily than with the end-pumped rod setup. The asymmetric thermal lens arising from such geometry was effectively compensated by the use of an off-normal-incidence intracavity spherical mirror, introducing the necessary astigmatism.

Notwithstanding the clear difficulty of selectively coupling pump energy into the fundamental mode, direct side-pumping of Nd:YAG rods with laser arrays symmetrically arranged in several units around the rod (Fig. 10.10) and creating a rather uniform internal illumination (and gain distribution) resulted in remarkable achievements in terms of both output power and overall system efficiency: Furuta *et al.* (2005) demonstrated an oscillator delivering 540 W average power in quasi-cw operation and 440 W in Q-switching (burst mode) at 60–100 kHz when pumped by 1180 W average power. The optical-to-optical efficiency was greater than 37% and electrical-to-optical efficiency was above 18%. The beam quality was $M^2 = 9$, typical for this laser architecture. Such performance has been further improved in the latest side-pumped high-power Nd:YAG industrial lasers, where up to 10 kW average power (cw) is now available (Akiyama *et al.* 2002).

An innovative approach for high-power, high-brightness lasers was established in the 1990s, owing to the introduction of thin-disk technology (Giesen and Speisen 2007). The technology makes use of a very thin, highly

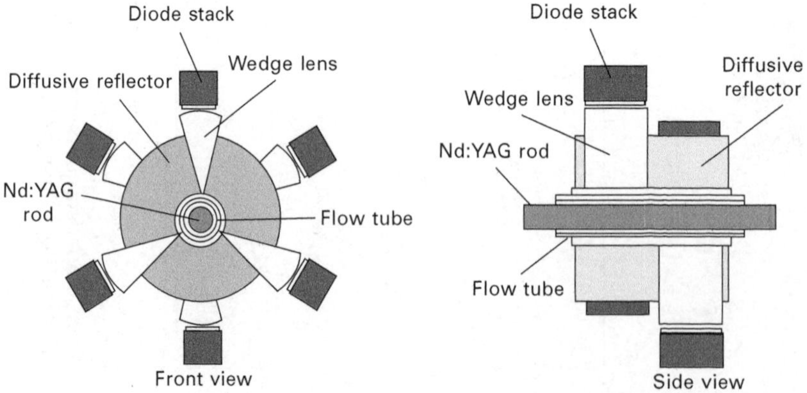

10.10 Example of side-pumped laser head (from Furuta *et al.*, 2005, with permission).

© Woodhead Publishing Limited, 2013

doped crystal (most often Yb:YAG) with a few hundred microns thickness end-pumped with multipass geometry (up to few tens of passes), and takes advantage of longitudinal cooling geometry which virtually eliminates the thermal lens (Fig. 10.11). Therefore, large mode area resonators with no significant thermal lensing were demonstrated, paving the way to outstanding performances in terms of laser efficiency, brightness, and output power up to several kilowatts. Although quasi-three-level materials with very small quantum defects such as Yb^{3+}:host take full advantage of the thin-disk concept and indeed have been exploited to record power levels, Nd:YVO_4 and Nd:YAG were also investigated as quasi-three-level systems around 900 nm, which is of great interest for blue light generation (Pavel *et al.* 2008).

10.4 Power scaling with master oscillator/power amplifier (MOPA) architectures

Notwithstanding the recent development of very high power oscillators, as reviewed in the previous section, a very effective concept for power scaling is the master oscillator/power amplifier (MOPA). This is especially useful when relatively short pulses, high average power and high beam quality

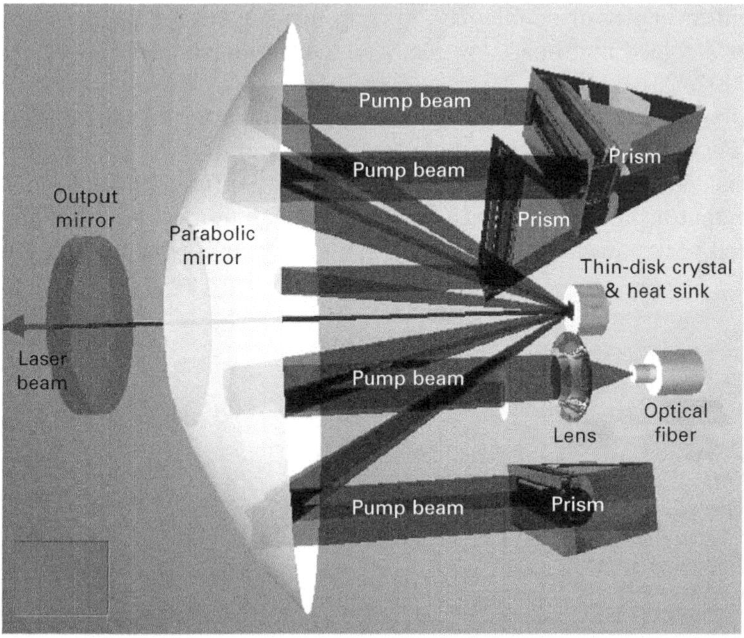

10.11 Representative layout of a thin-disk laser, with multi-pass pump unit and a simple, linear cavity (from Pavel *et al.*, 2007, with permission).

© Woodhead Publishing Limited, 2013

($M^2 < 1.5$) are required. Indeed, short pulses ranging from ~100 ps to a few nanoseconds generally require short (or very short, for microchip oscillators!) resonators which necessarily support small-size TEM_{00} and are therefore not suitable to withstand high power with diffraction-limited output.

In particular, for micro-machining applications and high-resolution laser ranging, MOPA based on short-pulse microchip seeders are of great interest. Passively Q-switched microchip lasers including a Cr:YAG or SESAMs deliver pulses in the range from ~100 ps to 1 ns at repetition rates from a few hundred hertz to hundreds of kilohertz. Therefore, MOPA systems are very attractive alternatives to much more complex mode-locked lasers with cavity-dumping or regenerative amplifiers, as long as the target pulse width is longer than ~100 ps.

A first demonstration of picosecond microchip MOPAs was done by Forget *et al.* (2002), who employed a Nd:YAG/Cr:YAG microchip delivering 800 ps pulses at 28 kHz, with 110 mW average power. A multi-pass amplifier based on an end-pumped 10 mm 0.3% doped Nd:YVO_4 crystal, very effective owing to its exceptionally high gain cross-section, allowed four-pass amplification up to 4 W with nearly TEM_{00} output beam and efficiency ≈27% (the pump power was 15 W). A substantially more complicated six-pass setup allowed up to 5.7 W output power with proportionally higher efficiency.

Simpler MOPA setups have been employed with side-pumped technology, either single- or double-pass grazing incidence slabs (Agnesi *et al.* 2010) or Z-folded multi-pass symmetrically side-pumped slabs (Dergachev *et al.* 2007, 2010). Starting from a 50 μJ, 500 ps Nd:YAG oscillator passively Q-switched with Cr:YAG up to 10 kHz, a single Nd:YVO_4 grazing amplifier (Fig. 10.12) side-pumped by a 40 W diode array collimated by a microlens yielded 5.5 W and 580 ps pulses in a nearly diffraction-limited ($M^2 \approx 1.1$) output using a double pass, with 13% extraction efficiency. Pulse energy and peak power are the highest reported to date for such MOPA solid-state

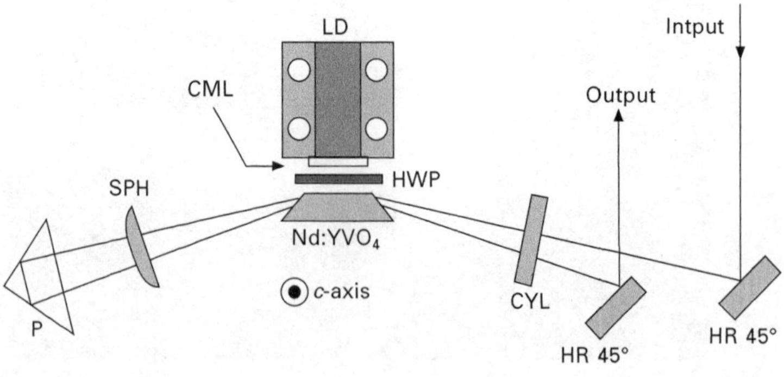

10.12 Grazing incidence double-pass amplifier. LD: laser diode; CML: collimation micro-lens; HWP: half wave-plate; SPH: spherical lens; CYL: cylindrical lens; P: right angle prism.

© Woodhead Publishing Limited, 2013

lasers. These limit basically the extraction efficiency, since the beam could not be further focused inside the amplifier to improve saturation without incurring optical damage (AR coatings are usually safe at intensity up to 500 MW/cm^2 at 10 ns, which scales as the square root of pulse duration, resulting in ≈2 GW/cm^2 at 580 ps, which is comparable to the maximum peak laser intensity).

For example, another MOPA experiment with more ordinary ~ 20–30 ns pulses from an actively Q-switched oscillator delivering 43 W at 0.75–1.7 MHz showed comparatively higher extraction efficiency, up to 45% (Chard and Damzen 2009). With 155 W pump power and 34 W injection, the amplifier yielded 104 W total output power with $M^2 < 1.8$. In this particular realization, the bounce amplifier used a double side-pumped design with a single crystal slab.

Another excellent demonstration of amplification of short pulses with MOPA employed a Z-folded multipass setup based on a Nd:YVO$_4$ amplifier module side-pumped by two laser diode bars yielding up to 60 W absorber pump power, that was used to efficiently boost a 600 mW seeder (Dergachev *et al.*, 2010). The seeder was a microlaser with air gaps between its components, with a SESAM/active medium (Nd:YVO$_4$)/OC combination designed for generation of pulses in the range 0.8–5 ns at a repetition rate of 250–2000 kHz. Therefore, the designer here deliberately chose to trade off pulse width with higher seeding output power, to allow easier saturation of the amplifier and efficient extraction. The MOPA output reached 12 W in the range 20–200 kHz with high beam quality and 20% optical-to-optical efficiency. Also in such a case the seed beam and the gain sheet could be made tighter than in Agnesi *et al.* (2010), allowing efficiency to be optimized since the peak pulse power was much lower (due to the much higher repetition rate).

However, end-pumping geometry remains the preferred choice when the seeder is sufficiently powerful to ensure efficient energy extraction. Smaller gain per module is generally available with single-pass longitudinally pumped DPSS amplifiers. Given the pump rate producing some gain per unit length γ_0, the amplified power P per unit length increases according to

$$\frac{\mathrm{d}P}{\mathrm{d}z} = \frac{\gamma_0}{1 + P/P_{\mathrm{SL}}} P \qquad [10.9]$$

where P_{SL} is the saturation power. Clearly, the higher the launched seed power, the sooner saturation occurs for which efficient extraction is possible:

$$\frac{\mathrm{d}P}{\mathrm{d}z} \approx \gamma_0 P_{\mathrm{SL}} \qquad [10.10]$$

Therefore, most efficient end-pumped MOPAs generally employ high-power oscillators.

For example, Kim *et al.* (2006) amplified a 3.8 W single-frequency cw

© Woodhead Publishing Limited, 2013

oscillator at 1064 nm with a three-stage amplifier. Each stage was made by two 10 mm, 0.1% doped Nd:YVO_4 rods, with relay imaging between them in order to absorb most of the pump power left after each rod into the next rod. The two vanadate crystals were rotated by 90° to cancel thermal lens astigmatism (a half-wave plate was inserted between the crystals to amplify always the right seed polarization in each crystal). The first stage was pumped by two 30 W fiber-coupled laser diodes, whereas each of the next stages was pumped by two 50 W fiber-coupled laser diodes at 808 nm. Optimum design of the amplifier required seeding with a slightly smaller beam diameter than the pump beams, in order to minimize thermal aberrations. Up to 79 W cw with $M^2 < 1.5$ were achieved for a total of 265 W pump power, with a 29% extraction efficiency.

McDonagh *et al.* (2007) demonstrated a 111 W, 33 ps MOPA at 110 MHz based on a 56 W oscillator (cw mode-locked). The amplifier here was a 30 mm, 0.5% doped Nd:YVO_4 rod pumped by a 108 W fiber-coupled diode at 888 nm for lower thermal release, with double-pass pumping (the residual pump after the first pass was back-reflected and re-imaged into the crystal). Therefore, the amplifier was designed for efficient extraction in a single pass (the pump spot diameter was 1.35 mm) delivering 55 W from the 108 W absorbed, maintaining an excellent beam quality parameter M^2 below 1.1.

In yet another example, Yan *et al.* (2008) amplified a Q-switched oscillator delivering 35 W at 500 kHz (pump power was 84 W). Two amplifier stages were chosen, each employing a 16 mm long (end capped), 0.5% doped Nd:YVO_4 crystal end-pumped by two 42 W fiber-coupled diodes at 808 nm, focused to a spot diameter of 0.8 mm. The power at the MOPA output was 108 W, corresponding to an extraction efficiency of 43% (52% for the last stage). The beam quality was better than two times diffraction limited. This further design example confirms the excellent performance in terms of efficiency and beam quality with a really simple setup starting from a relatively high-power oscillator.

A particularly interesting idea for a MOPA architecture was proposed and investigated in Ma *et al.* (2007), where the beam of the pump diode array was partially sampled (≈12%) and used to pump the oscillator, while the remaining part pumped the multipass amplifier (Fig. 10.13). The oscillator and the amplifier share the same $12 \times 1 \times 10$ mm^3, 0.4% doped Nd:YVO_4 slab. Clearly, the design of a relatively low-power Q-switched oscillator is much easier and can be readily optimized to meet the target specifications, especially in terms of pulse width and beam quality. The six-pass amplifier takes advantage of progressive expansion of the beam width at each pass, ensuring safe operation while the peak power increases and allowing good overlapping with the gain volume. The oscillator delivers a maximum 6 W in Q-switching at 100 kHz (15 ns pulse width), whereas the total output power of the MOPA was 28.8 W. The extraction efficiency (total pump to

© Woodhead Publishing Limited, 2013

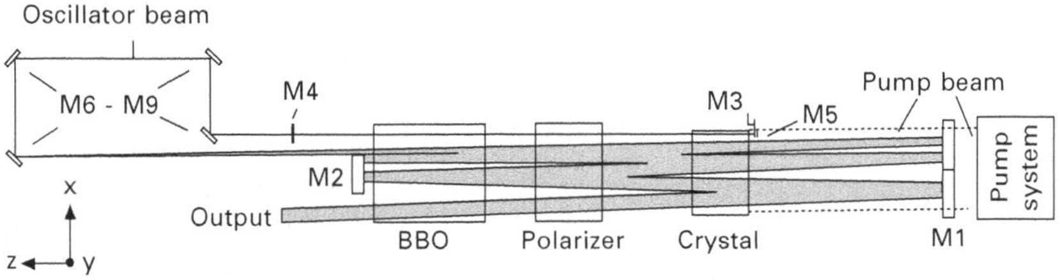

10.13 Schematic of the experimental setup proposed by Ma *et al.* (2007) (from Ma *et al.*, 2007, with permission).

output) in this case was 21%, therefore one could expect further optimization margins.

Among the latest trends in short pulse amplification we would like to mention the use of relatively long Nd:YAG crystal fibers of a few centimeters, allowing better pump absorption and homogenization through guiding; furthermore, it is worth mentioning the spreading use of Yb^{3+}-doped photonics crystal rods. The outlying idea shared by both approaches is to make use of high pump power by distributing it along the fiber in order to avoid any deleterious thermal effect, while the seed beam cross-section stays more or less unchanged along the propagation. Indeed, though one should design the amplifier such that the Rayleigh range matches the absorption length, here the pump beam can be more tightly focused but, being confined through total internal reflections, it will maintain good overlap with the seed transverse mode along the whole crystal length. Interesting and promising results are being achieved: for example, Martial *et al.* (2011) used a 50 mm long, 1 mm diameter, 0.2% doped Nd:YAG fiber rod pumped by a 60 W fiber-coupled laser diode at 808 nm to amplify compact microlasers. With a 1 ns pulse, 42 kHz, 5 W seeder they achieved 20.5 W in single-pass with 26% extraction efficiency and $M^2 < 1.4$. As an energy amplifier, they used instead a 450 ps seeder delivering 80 mW at 1 kHz: the output energy after double-pass amplification was 2.7 mJ with $M^2 < 1.2$, with 34% extraction efficiency. However, an even more sophisticated technology inspired by the same underlying ideas is most likely represented by Yb^{3+}-doped silica photonic rod fibers with large mode area (LMA). In a quite impressive demonstration (Laurila *et al.* 2011), oscillator single-mode operation at 110 W with 210 W pump power at 976 nm was achieved with a 75 cm long photonic rod fiber having 85 μm core diameter. Q-switching provided as much as 104 W at 100 kHz, with 21 ns pulses.

10.5 Future trends

As already mentioned in this review, amplification of short pulses from microchip lasers is becoming a hot topic since there are many attractive applications for such laser systems. New microchip designs and fabrication

© Woodhead Publishing Limited, 2013

techniques have recently allowed practical 100 ps, or even 50 ps, seeders at hundreds of kilohertz to become available as off-the-shelf commercial devices.

Amplification in traditional, either bulk end- or side-pumped amplifiers, as outlined in the previous section, as well as in the newest crystal/fiber rods or even in readily available LMA Yb fibers, provided a convincing demonstration (Steinmetz *et al.* 2011) of the feasibility of high-power picosecond lasers at ~100 kHz to ~1 MHz. This suggests that traditional but more complicated and expensive amplification techniques of DPSS mode-locked lasers such as regenerative amplification would be obsolete. Furthermore, taking advantage of nonlinear optical effects such as self-phase-modulation in the same optical fiber amplifiers, the pulses have been shortened down to few picoseconds.

Many of the future advancements in this field will be likely related to new material developments, of both fibers and crystal rods. For example, it has been shown that crystal rod fibers are very promising for power and energy scaling: further refinements of the micropulling-down technique (Didierjean *et al.* 2006, Veronesi *et al.* 2012) employed to fabricate such fibers will allow researchers to match the laser performance of the most powerful lasers based on traditional crystals and even outperform them when further power scaling to hundreds of watts will be considered, in virtue of the superior thermal handling capability allowed by the more favorable surface/volume ratio of long fibers.

Other particular Nd crystals and geometries are being currently investigated as future candidates for high-power lasers, such as waveguide lasers fabricated with epitaxy, ion-implanting or femtosecond pulses (Lee *et al.* 2002, Siebenmorgen *et al.* 2010) and ceramic lasers (Tokurakawa *et al.* 2012).

Especially attractive also is the option of starting from Nd oscillators, allowing relatively short pulses and high repetition frequency in Q-switching operation, using Yb amplifiers to boost the average power or the pulse energy. Indeed, Yb amplifiers have several advantages:

1. They require pumping at 940–980 nm, thus relying on smaller quantum defect and the inherent thermal load, a big plus when very high power is the target.
2. The laser diode technology at such pump wavelength is inherently more reliable than at 808 nm.
3. Given the broad amplification bandwidth of Yb^{3+} in silica or, more suitable for non-fiber laser architectures, some new crystalline media such as YCOB (Heckl *et al.* 2010), CALGO (Ricaud *et al.* 2011) and sesquioxides (Saraceno *et al.* 2011), the exact wavelength of the seeder is no longer a critical design aspect, as for the thermal shift of the fluorescence peak in Nd:YAG or Nd:YVO_4 amplifiers.

In conclusion, we see a bright future for high-power industrial DPSSLs,

© Woodhead Publishing Limited, 2013

which for several reasons will not soon be replaced entirely by fiber lasers, notwithstanding the aggressive competition. In particular, until very large core fibers demonstrate guiding of clean TEM_{00} modes with very high peak power pulses, bulk amplifiers will likely dominate the scene for a long time.

10.6 Sources of further information and advice

As with all review papers, this one, summarizing the state-of-the-art of neodymium-doped DPSSLs, is necessarily biased by the authors' direct experience and does not presume to cover all the possible aspects in full depth. Whereas most of the latest exciting developments are outlined by the scientific specialist literature (of which we have sampled some of the most representative examples, included in the Reference section), classic advanced textbooks such as Koechner's are also regularly updated to include the latest advancements.

Web-based resources such as Paschotta's *Encyclopedia on Laser Physics and Technology* are also valuable to quickly find references and start a deeper search on a specific subject. Also instructive for an update of the applications and latest market-driven laser developments is the regular reading of commercial journals such as *Laser Focus World* and *Photonics Spectra*, which also provide an extensive referencing to many industrial players, from whose websites one can learn directly of new product details.

Finally, much can be learned even by reading patents, and indeed some top-notch industrial laser research most easily originates from patents rather than scientific papers, so this is definitely worth considering.

We include here a list of useful web links for all these resources:

http://www.osa.org
http://ieeexplore.ieee.org
http://www.springer.com/physics/journal/340
http://www.rp-photonics.com/encyclopedia.html
http://www.laserfocusworld.com/
http://photonics.com/

10.7 References

Agnesi, A. and Dell'Acqua, S. (2003) 'High peak-power diode-pumped passively Q-switched Nd:YVO_4 laser', *Appl. Phys. B* 76, 351–354

Agnesi, A., Dell'Acqua, S., Morello, C., Piccinno, G., Reali, G. C. and Sun, Z. (1997) 'Diode-pumped neodymium lasers repetitively Q-switched by Cr^{4+}:YAG solid-state saturable absorbers', *IEEE J. Selected Topics in Quantum Electron.* 3, 45–52

Agnesi, A., Pirzio, F., Reali, G. and Piccinno, G. (2006) 'Sub-nanosecond diode-pumped passively Q-switched Nd:$GdVO_4$ laser with peak power >1 MW', *Appl. Phys. Lett.* 89, 101120

© Woodhead Publishing Limited, 2013

Agnesi, A., Dallocchio, P., Pirzio, F. and Reali, G. (2010) 'Sub-nanosecond single-frequency 10-kHz diode-pumped MOPA laser', *Appl. Phys. B* 98, 737–741

Akiyama, Y., Takada, H., Yuasa, H. and Nishida, N. (2002) 'Efficient 10 kW diode-pumped Nd:YAG rod laser', *Advanced Solid-State Lasers*, 68, paper WE4

Baer, T. M., Head, D. F., Gooding, P., Kintz, G. J. and Hutchison, S. (1992) 'Performance of diode-pumped Nd:YAG and Nd:YLF lasers in a tightly folded resonator configuration', *IEEE J. Quantum Electron.* 28, 1131–1138

Bernard, J. E. and Alcock, A. J. (1993) 'High-efficiency diode-pumped Nd:YVO_4 slab laser', *Opt. Lett.* 18, 968–970

Chang, Y. T., Huang, Y. P., Su, K. W. and Chen, Y. F. (2008) 'Comparison of thermal lensing effects between single-end and double-end diffusion-bonded Nd:YVO_4 crystals for $^4F_{3/2}\rightarrow^4I_{11/2}$ and $^4F_{3/2}\rightarrow^4I_{13/2}$ transitions', *Opt. Express* 16, 21155–21160

Chard, S. P. and Damzen, M. J. (2009) 'Compact architecture for power scaling bounce geometry lasers', *Opt. Express* 17, 2218–2223

Chen, Y. F. and Tsai, S. W. (2001) 'Simultaneous Q-switching and mode-locking in a diode-pumped Nd:YVO_4–Cr:YAG laser', *IEEE J. Quantum Electron.* 37, 580–586

Clarkson, W. A. (2001) 'Thermal effects and their mitigation in end-pumped solid-state lasers', *J. Phys. D: Appl. Phys.* 34, 2381–2395

Clarkson, W. A. and Hanna, D. C. (1996) 'Two-mirror beam-shaping technique for high-power diode bars', *Opt. Lett.* 21, 375–377

Dergachev, A., Flint, J. H., Isyanova, Y., Pati, B., Slobodtchikov, E. V., Wall, K. F. and Moulton, P. F. (2007) 'Review of multipass slab laser systems', *IEEE J. Selected Topics in Quantum Electron.* 13, 647–660

Dergachev, A., Moulton, P. F., Petrich, G. S., Kolodziejski, L. A. and Kärtner, F. X. (2010) 'Semiconductor Q-switched, short-pulse, high-power, MHz-rate laser', *Advanced Solid-State Photonics*, paper AMC5

Didierjean, J., Castaing, M., Balembois, F., Georges, P., Perrodin, D., Fourmigué, J. M., Lebbou, K., Brenier, A. and Tillement, O. (2006) 'High-power laser with Nd:YAG single-crystal fiber grown by the micro-pulling-down technique', *Opt. Lett.* 31, 3468–3470

Du, K., Wu, N., Xu, J., Giesekus, J., Loosen, P. and Poprawe, R. (1998) 'Partially end-pumped Nd:YAG slab laser with a hybrid resonator', *Opt. Lett.* 23, 370–372

Du, K., Li, D., Zhang, H., Shi, P., Wei, X. and Diart, R. (2003) 'Electro-optically Q-switched NdYVO_4 slab laser with a high repetition rate and a short pulse width', *Opt. Lett.* 28, 87–89

Forget, S., Balembois, F., Georges, P. and Devilder, P.-J. (2002) 'A new 3D multipass amplifier based on Nd:YAG or Nd:YVO_4 crystals', *Appl. Phys. B* 75, 481–485

Freitag, I., Tünnermann, A. and Welling, H. (1997) 'Passively Q-switched Nd:YAG ring lasers with high average output power in single-frequency operation', *Opt. Lett.* 22, 706–708

Furuta, K., Kojima, T., Fujikawa, S. and Nishimae, J. (2005) 'Diode-pumped 1 kW Q-switched Nd:YAG rod laser with high peak power and high beam quality', *Appl. Opt.* 44, 4119–4122

Giesen, A. and Speisen, J. (2007) 'Fifteen years of work on thin-disk lasers: results and scaling laws', *IEEE J. Quantum Electron.* 13, 598–609

Hardman, P. J., Clarkson, W. A., Friel, G. J., Pollnau, M. and Hanna, D. C. (1999) 'Energy-transfer upconversion and thermal lensing in high-power end-pumped Nd:YLF laser crystals', *IEEE J. Quantum Electron.* 35, 647–655

Heckl, O. H., Kränkel, C., Baer, C. R. E., Saraceno, C. J., Südmeyer, T., Petermann, K.,

© Woodhead Publishing Limited, 2013

Huber, G. and Keller, U. (2010) 'Continuous-wave and modelocked Yb:YCOB thin disk laser: first demonstration and future prospects', *Opt. Express* 18, 19201–19208

Innocenzi, M. E., Yura, H. T., Fincher, C. L. and Fields, R. A. (1990) 'Thermal modeling of continuous-wave end-pumped solid-state lasers', *Appl. Phys. Lett.* 56, 1831–1833

Isyanova, Y. and Welford, D. (1999) 'Temporal criterion for single-frequency operation of passively Q-switched lasers', *Opt. Lett.* 24, 1035–1037

Kim, J. W., Yarrow, M. J. and Clarkson, W. A. (2006) 'High power single-frequency continuous-wave Nd:YVO_4 master-oscillator power amplifier', *Appl. Phys. B* 85, 539–543

Koechner, W. (2006) *Solid-State Laser Engineering*, Berlin, Springer

Laurila, M., Saby, J., Alkeskjold, T. T., Scolari, L., Cocquelin, B., Salin, F., Broeng, J. and Lægsgaard, J. (2011) 'Q-switching and efficient harmonic generation from a single-mode LMA photonic bandgap rod fiber laser', *Opt. Express* 19, 10824–10833

Lee, J. R., Baker, H. J., Friel, G. J., Hilton, G. J. and Hall, D. R. (2002) 'High-average-power Nd:YAG planar waveguide laser that is face pumped by 10 laser diode bars', *Opt. Lett.* 27, 524–526

Ma, Z., Li, D., Hu, P., Shell, A., Shi, P., Haas, C. R., Wu, N. and Du, K. (2007) 'Monolithic Nd:YVO_4 slab oscillator–amplifier', *Opt. Lett.* 32, 1262–1264

Magni, V. (1986) 'Resonators for solid-state lasers with large-volume fundamental mode and high alignment stability', *Appl. Opt.* 25, 107–117

Martial, I., Balembois, F., Didierjean, J. and Georges, P. (2011) 'Nd:YAG single-crystal fiber as high peak power amplifier of pulses below one nanosecond', *Opt. Express* 19, 11667–11679

McDonagh, L., Wallenstein, R. and Knappe, R. (2006) '47 W, 6 ns constant pulse duration, high-repetition-rate cavity-dumped Q-switched TEM_{00} Nd:YVO_4 oscillator', *Opt. Lett.* 31, 3303–3305

McDonagh, L., Wallenstein, R. and Nebel, A. (2007) '111 W, 110 MHz repetition-rate, passively mode-locked TEM_{00} Nd:YVO_4 master oscillator power amplifier pumped at 888 nm', *Opt. Lett.* 32, 1259–1261

Minassian, A., Thompson, B. and Damzen, M. J. (2003) 'Ultrahigh-efficiency TEM_{00} diode-side-pumped Nd:YVO_4 laser', *Appl. Phys. B* 76, 341–343

Nighan, W. L., Dudley, D. K., Keirstead, M. S. and Petersen, A. B. (1995) 'Highly efficient, diode-bar-pumped Nd:YVO_4 laser with >13 W TEM_{00} output', *Advanced Solid State Lasers*, 24, paper LA10

Nodop, D., Limpert, J., Hohmuth, R., Richter, W., Guina, M., and Tünnermann, A. (2007) 'High-pulse-energy passively Q-switched quasi-monolithic microchip lasers operating in the sub-100-ps pulse regime', *Opt. Lett.* 32, 2115–2117

Pavel, N., Lünstedt, K., Petermann, K. and Huber, G. (2007) 'Multipass pumped Nd-based thin-disk lasers: continuous-wave laser operation at 1.06 and 0.9 μm with intracavity frequency doubling', *Appl. Opt.* 46, 8256–8263

Pavel, N., Kränkel, C., Peters, R., Petermann, K. and Huber, G. (2008) 'In-band pumping of Nd-vanadate thin-disk lasers', *Appl. Phys. B* 91, 415–419

Pearce, S. and Ireland, C. L. M. (2003) 'Performance of a CW pumped Nd:YVO_4 amplifier with kHz pulses', *Opt. Laser Technol.* 35, 375–379

Ricaud, S., Jaffres, A., Loiseau, P., Viana, B., Weichelt, B., Abdou-Ahmed, M., Voss, A., Graf, T., Rytz, D., Delaigue, M., Mottay, E., Georges, P. and Druon, F. (2011) 'Yb:$CaGdAlO_4$ thin-disk laser', *Opt. Lett.* 36, 4134–4136

Richards, J. and McInnes, A. (1997) 'High brightness 10 kHz diode pumped Nd:YAG laser', *Advanced Solid State Lasers*, 10, paper PS1

© Woodhead Publishing Limited, 2013

Saraceno, C. J., Heckl, O. H., Baer, C. R. E., Golling, M., Südmeyer, T., Beil, K., Kränkel, C., Petermann, K., Huber, G. and Keller, U. (2011) 'SESAMs for high-power femtosecond modelocking: power scaling of an Yb:$LuScO_3$ thin disk laser to 23 W and 235 fs', *Opt. Express* 19, 20288–20300

Shine, R. J., Alfrey, A. J. and Byer, R. L. (1995) '40-W cw, TEM_{00}-mode, diode-laser-pumped, Nd:YAG miniature-slab laser', *Opt. Lett.* 20, 459–461

Siebenmorgen, J., Calmano, T., Petermann, K. and Huber, G. (2010) 'Highly efficient Yb:YAG channel waveguide laser written with a femtosecond-laser', *Opt. Express* 18, 16035–16041

Spühler, G. J., Paschotta, R., Fluck, R., Braun, B., Moser, M., Zhang, G., Gini, E. and Keller, U. (1999) 'Experimentally confirmed design guidelines for passively Q-switched microchip lasers using semiconductor saturable absorbers', *J. Opt. Soc. Am. B* 16, 376–388

Steinmetz, A., Eidam, T., Nodop, D., Limpert, J. and Tünnermann, A. (2011) 'Nonlinear compression of Q-switched laser pulses to the realm of ultrashort durations', *Opt. Express* 19, 3758–3764

Tokurakawa, M., Shirakawa, A., Ueda, K., Yagi, H., Yanagitani, T., Kaminskii, A. A., Beil, K., Kränkel, C. and Huber, G. (2012) 'Continuous wave and mode-locked Yb^{3+}:Y_2O_3 ceramic thin disk laser', *Opt. Express* 20, 10847–10853

Veronesi, S., Zhang, Y., Tonelli, M., Agnesi, A., Greborio, A., Pirzio, F. and Reali, G. (2012) 'Spectroscopy and efficient laser emission of Yb^{3+}:LuAG single crystal grown by micro-PD', *Opt. Commun.* 285, 315–321

Yan, X., Liu, Q., Fu, X., Wang, Y., Huang, L., Wang, D. and Gong, M. (2008) 'A 108 W, 500 kHz Q-switching Nd:YVO_4 laser with the MOPA configuration', *Opt. Express* 16, 3356–3361

Zayhowski, J. J. and Dill, C. III (1994) 'Diode-pumped passively Q-switched picosecond microchip lasers', *Opt. Lett.* 19, 1427–1429

Zhu, P., Li, D., Qi, B., Schell, A., Shi, P., Haas, C., Fu, S., Wu, N. and Du, K. (2008) 'Diode end-pumped high-power Q-switched double Nd:YAG slab laser and its efficient near-field second-harmonic generation', *Opt. Lett.* 33, 2248–2250

© Woodhead Publishing Limited, 2013

11
System sizing issues with diode-pumped quasi-three-level materials

A. JOLLY, Commissariat à l'Energie Atomique, Centre d'Etudes Scientifiques et Techniques d'Aquitaine, France

DOI: 10.1533/9780857097507.2.283

Abstract: This chapter aims to provide an overview of ytterbium-based bulky laser systems in the fields of nanosecond and short pulse generation. After some basics of interest in the understanding of the specific design issues of quasi-three-level materials, and of their operating conditions in a given laser architecture, the generic architectures of sources and amplifiers are described in some detail, referring to simplified sizing equations. Then we review the state-of-the-art of high-performance YAG-based and tungstate-based systems, of evident commercial interest in numerous applications, as an illustration of the former principles previously described.

Key words: ytterbium, quasi-three-level, diode-pumping, nanosecond pulses, ultrashort pulses, Q-switching, regenerative amplification, mode-locking.

11.1 Introduction

Rapid progress has taken place over the past 10 years in the field of diode-pumped solid-state lasers, and more specifically in the emerging area of ytterbium-doped (Yb-doped) laser systems. The arrival at maturity of the technology of InGaAs/GaAs laser diode (LD) bars and stacks has led to the availability of commercial high-power pumps in the range of wavelengths 940–980 nm. This opens the route to the development of highly energetic short-pulse sources with a high pulse repetition frequency (PRF), which implies an increasing number of new applications in industry [1, 2], medicine and science [3, 4]. This chapter starts with some highlights on the fundamentals of three-level materials and their operating conditions. After recalling some basics for the optimisation of the pump and of energy extraction, dedicated to the design of nanosecond and sub-picosecond YAG and tungstate-based sources of interest in this chapter, we focus on critical thermal issues. The second step consists of a review of the generic techniques of widespread use with Yb-based systems for nanosecond and sub-picosecond pulse generation, i.e. Q-switching, mode-locking and regenerative amplification. A few analytical expressions are included for sizing purposes, and to help identify the relevant parameters. In a third step, we propose an overview of the state-of-the-art of high-energy and high-power sources in the field of

© Woodhead Publishing Limited, 2013

YAG and KGW–KYW-based systems, and we show how to discuss orders of magnitude in the related optical performance.

11.2 Ytterbium-doped materials and bulk operating conditions

This first section aims to evidence the operating conditions of Yb-doped materials and, despite the counterpart of hard-to-please pump conditions, their interest for bulky systems dedicated to the production of highly energetic and short pulses. This ensures an appropriate choice of the materials' geometry versus the pump conditions, to reach the required small-signal net gain for efficient energy extraction. In addition, we plan to show why the search for optimised optical performance also requires tight thermal control.

11.2.1 Technological options in the 1 μm range

Ytterbium versus neodymium

The only two realistic options for the realisation of efficient diode-pumped sources in the spectral range of 1 μm are those of neodymium or ytterbium-based configurations [5], according to the following considerations:

- Neodymium-doped materials benefit from a pure four-level scheme [6], with no transparency threshold and quite high-gain capabilities due to elevated emission cross-sections. Saturation intensities in the range 0.1–10 J/cm^2 then ensure good energy extraction capabilities. They remain fairly consistent with the optical damage limitations in the usual materials and coatings. But these materials exhibit narrow amplification bandwidths, thus leading to significant performance limitations in short-pulse amplification, and they suffer from limited energy storage capabilities.
- On the other hand, Yb-doped materials usually benefit from very broad bandwidths. In addition to obvious interests for tunability and short-pulse generation [3, 7], this results in simpler selection criteria for the diode pump, in terms of required spectral width and thermal control. The longer fluorescence lifetimes and larger saturation intensities enable good energy storage capabilities. Furthermore, the low quantum defect and the lack of any possible excited-state absorption, due to the very simple electronic structure of the trivalent ion Yb^{3+} [6], enable superior performance in the presence of a high PRF. Just for a basic reminder, given pump and laser wavelengths λ_P and λ_L, the quantum defect is determined by $\eta_Q = 1 - (\lambda_P/\lambda_L)$. The specific issue of the operating room temperature and, at the same time, of the possibility of taking advantage of low temperature properties may be discussed in terms of benefits and

© Woodhead Publishing Limited, 2013

drawbacks. The more significant drawbacks nevertheless consist of the quasi-three-level energetic scheme and of low emission cross-sections. This applies to the elevated pump intensities (I_p) which are required to overcome the transparency threshold, and to the need to afford rather low small-signal gain coefficients in relationship with plus or minus significant reabsorption effects at λ_L.

11.2.2 Fundamentals for the implementation of bulky ytterbium-doped materials

Quasi-three-level basics

The electronic structure of ytterbium, the last and heaviest rare-earth in the classification of Mendeleev, comprises 67 electrons. Its incomplete outer 4f layer is short of one electron. This completely governs the overall optical properties of the doped material [8, 9], which are determined by the electric-field distribution around and by the Stark effect. Strong splitting effects then occur in the two manifolds of interest in the lasing transitions, namely $^2F_{5/2}$ and $^2F_{7/2}$. The processes of stimulated absorption and emission apply to similar concepts, given the wavelength (λ) considered in the spectral range 900–1100 nm. They are determined by their spectral cross-sections, denoted $\sigma_a(\lambda)$ and $\sigma_e(\lambda)$ respectively. Both spectra exhibit an intense line near 970–985 nm, the so-called 'zero-line', which determines the boundary limit between the processes of emission and absorption. All Yb-doped materials are described by means of quasi-three-level energetic schemes. In the theoretical worst-case of a pure three-level scheme, which implies no Stark sub-level splitting but a binary distribution of the ionic population in the two manifolds, population inversion only starts when at least one half of this ionic population has been excited. On the otherhand, a pure four-level scheme involves lasing transitions towards an unpopulated lower level. This is the most efficient configuration. Yb-doped materials rather correspond to intermediate situations between these two extrema. Depending on the actual thermal distribution of the ionic population in the lower multiplet, and more specifically on that of its higher Stark sub-levels, the related model can either match a nearly four-level scheme or a nearly three-level scheme. Competing Yb-doped materials differ by the internal field's strength around the active ion and the splitting ratios of their Stark sub-levels (Fig. 11.1). This also explains why the operating temperature may lead to severe performance limitations.

The energy storage efficiency is governed by the fluorescence lifetime (τ_F). There exists a relationship with the emission cross-section $\sigma_e(\lambda)$ of the material:

© Woodhead Publishing Limited, 2013

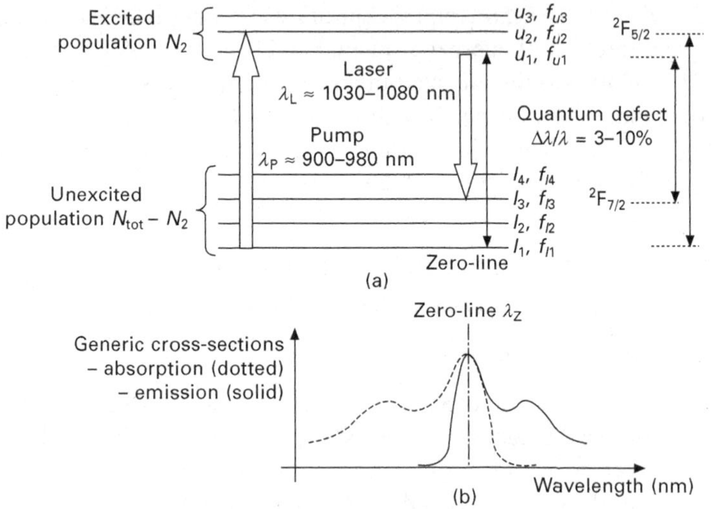

11.1 (a) Electronic transition model of the Yb^{3+} ion, and (b) related spectral distributions of generic absorption and emission cross-sections. The upper and lower multiplets $^2F_{5/2}$ and $^2F_{7/2}$ are split, respectively, in three and four Stark sub-levels.

$$\tau_F = \frac{1}{8\pi n^2 c \int_{\Delta\lambda} \frac{\sigma_e(\lambda)\,d\lambda}{\lambda^4}} \qquad [11.1]$$

where c, n and $\Delta\lambda$ denote the speed of light, the material's refraction index and the gain-bandwidth respectively. Equation 11.1 makes evidence of a quasi constant product $\Delta\lambda \times \tau_F$. The broader is the bandwidth, the smaller the fluorescence lifetime. In most Yb-doped materials, the peak values of $\sigma_e(\lambda)$ and τ_F lie in the ranges of 10^{-20} cm² and of 1 ms, respectively, while $\Delta\lambda$ varies from some tens of nanometres to about 100 nm. These orders of magnitude provide useful numbers regarding generic spectroscopic data, whatever the material under consideration.

Saturated pump absorption

Prior to further considerations about amplification, the suitable conditions for efficient pump absorption need to be discussed with respect to the pump intensity (I_p). Let us then consider a slab of total absorbing surface Σ and of thickness l (Fig. 11.2(a)), pumping at λ_p from the left side, and N_{tot} the doping level (cm^{-3}). Any absorbed pump photon is supposed to intercept the projection of the absorption cross-section (σ_a) in the plane of Σ.

Assuming unsaturated absorption, the absorbed fraction of pump power (ΔI_p) from $z = 0$ to $z = l$ is equal to the total fraction of A occupied by the

© Woodhead Publishing Limited, 2013

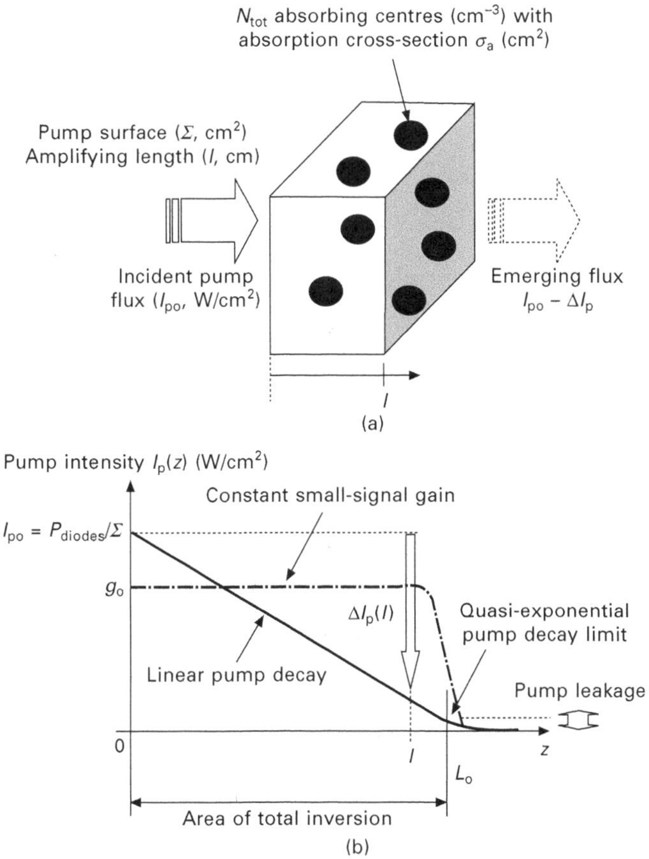

11.2 High-intensity pump absorption across a given length of heavily doped slab, leading to saturated absorption: (a) parameters involved and (b) pump deposition profile.

ionic population. This ensures $\Delta I_p/I_p = -\Delta\Sigma/\Sigma$ for $\Delta\Sigma = N_{tot}\sigma_a Al \cdot N_{tot}$ being low enough to avoid any spatial overlap and significant ionic interaction, we get $\Delta I_p/I_p = -\sigma_a N_{tot} l$. This leads to a differential equation in $I_p(z)$, to be integrated from $z = 0$ to l. The solution takes the well-known form of Biot and Savart's law, given the incident pump-intensity I_{po}:

$$I_p(l) = I_{po}\exp(-\sigma_a N_{tot} l) \qquad [11.2]$$

The related exponential decay implies a low $I_p(0)$ which does not account for saturated pump absorption. Since this is usually required in Yb-doped materials, Eq. 11.2 must be completed by more representative – but complicated – calculations. Lacking the saturated pump absorption, the actual amount of absorbed pump should be over estimated by a large amount. A fairly simple analytical approximation [10] may be used in the case of strongly saturated pump absorption, to get the right orders of magnitude by means of a pure linear decrease. Under the restriction of a collimated pump, $I_p(z)$ can be fitted along the pump axis z (Fig. 11.2(b)) according to:

$$I_p(z) \approx I_{po}\left(1 - \frac{z}{L_o}\right) \quad [11.3]$$

The decreasing slope of $I_p(z)$ determines the actual absorption length, L_o, as a function of two coefficients, $\beta_a(\lambda_p) = \dfrac{\sigma_a(\lambda_p)N_{tot}}{1 + \dfrac{\sigma_e(\lambda_p)}{\sigma_a(\lambda_p)}}$ in units of cm^{-1}, and

dimensionless $\eta_a(\lambda_p) = \dfrac{1}{1 + \dfrac{\sigma_e(\lambda_p)}{\sigma_a(\lambda_p)}}$:

$$L_o(\lambda_p) = \frac{\eta_a(\lambda_p) + \dfrac{I_p}{I_{sat\_abs}}}{\beta_a(\lambda_p)} \quad [11.4]$$

$I_p(z)$ must be considered as a fraction of the saturated pump intensity I_{sat_abs}:

$$I_{sat\_abs}(\lambda_p) = \frac{hc}{\lambda_p \sigma_a(\lambda_p)\tau_F} \quad [11.5]$$

Proper sizing according to Eqs 11.3 and 11.4 prevents any significant reabsorption at λ_L in the situation of single-pass absorption, leading to $l \approx L_o$. With a longer l, the only remaining area still experiencing an exponential decay is located towards the opposite side to the pump.

Small-signal spectral gain and available tuning range versus reabsorption limitations

The characteristic ratio $\sigma_e(\lambda_L)/\sigma_a(\lambda_L)$ provides a useful parameter to discuss the processes of bleaching and of gain saturation in quasi-three-level schemes. In a pure three-level scheme near the zero-line, it is unity. In a pure 'four-level' scheme, it is zero. Quasi-three-level schemes imply values in the range of a few percent to about 10%. Referring to the presence of a transparency threshold in the material and to the population inversion (N_2), the minimum value of the actual ionic fraction $\beta = N_2/N_{tot}$ to be excited at λ_L is given by [3, 8]:

$$\beta_{min} = \frac{N_{2min}}{N_{tot}} = \frac{1}{1 + \dfrac{\sigma_e(\lambda_L)}{\sigma_a(\lambda_L)}} \quad [11.6]$$

© Woodhead Publishing Limited, 2013

β_{min} may vary in quite large proportions versus λ_L. Due to decreasing reabsorption it decreases for large values of λ_L. Given $\lambda_P \ll \lambda_L$, far from zero-line pumping [8], the related pump intensity to achieve the transparency threshold at λ_L is written as:

$$I_{min}(\lambda_P, \lambda_L) \approx \beta_{min}(\lambda_L) I_{sat\_abs}(\lambda_P) \qquad [11.7]$$

The comparison of the usual values of I_{min} with those of the commercial diode pumps evidences the need for strong focusing. For example, in the favourable case of Yb^{3+}:YAG, $\beta_{min} \approx 5.5\%$ and $I_{min} \approx 1.7$ kW/cm^2. The relationship between the excited ionic population N_2 in the upper manifold and the small-signal gain coefficient (g_o) is determined by the conservation of ytterbium concentration, i.e. $g_o(\lambda_L) = \sigma_e(\lambda_L) N_2 - \sigma_a(\lambda_L)(N_{tot} - N_2)$. Given β and β_{min} versus λ_L, then referring to $g_o = \sigma_{gain} N_{tot}$ by means of the so-called effective-gain cross-section $\sigma_{gain}(\lambda_L)$, a usual mode of representation to evaluate the attainable amplification bandwidth makes use of:

$$\sigma_{gain}(\lambda_L) = \beta\sigma_e(\lambda_L) = [1 - \beta]\sigma_a(\lambda_L) = \frac{\beta - \beta_{min}}{1 - \beta_{min}}\sigma_e(\lambda_L) \qquad [11.8]$$

Peak output energy extraction and energy storage

The saturation intensity (I_{sat}) of amplification in the material is among the most important data of interest. It is expressed as:

$$I_{sat} = \frac{hc}{\lambda_L \tau_F} \frac{1}{\sigma_e(\lambda_L) + \sigma_{abs}(\lambda_L)} \qquad [11.9]$$

I_{sat} ranges from about 10 kW/cm^2 for relatively high-gain materials like Yb^{3+}:YAG and tungstates, up to nearly 100 kW/cm^2 for broader-bandwidth materials. In the first-order, at elevated pumping rates, the amount of stored energy at λ_L, E_{sto}, in the pumped volume is related to the small-signal gain coefficient (g_o) by the following approximation:

$$E_{sto} = N_2 \frac{hc}{\lambda_L} = g_o I_{sat} \tau_F \qquad [11.10]$$

Provided the value of $\sigma_e(\lambda_L)$ is in the material under consideration, Eqs 11.8–11.10 settle the relevant criteria to reach a reasonable value of g_o. For bulky materials, 'reasonable' means that g_o essentially needs to compensate for the external optical losses in the lasing configuration involved, thus simply leading to a slightly positive power balance in the amplifying path. I_{sat} provides the ultimate limit for energy extraction. The optimisation of the overall laser efficiency then requires, at the same time, efficient energy storage and the highest possible ratio I_p/I_{min}. This implies strong pump focusing, within the limitations of the efficiency of the *LD* pump and of the optical damage hazards in the material itself or in boundary coatings.

© Woodhead Publishing Limited, 2013

For quasi-continuous-wave (Q-CW) pumping, of interest with PRFs below 1 kHz to benefit from the highest possible pump power, the optimisation of the laser efficiency also involves the use of the optimal pump duration (T_p). Apart from any ASE consideration [3], the energy-storage efficiency is governed by the ratio T_p/τ_F:

$$\eta_{sto} = \frac{E_{sto}(T_p)}{E_{pump}} = \frac{1 - \exp(-T_p/\tau_F)}{T_p/\tau_F} \qquad [11.11]$$

This justifies the usual pump duration, i.e. $T_p = \tau_F$ to 1.5 ξ τ_F.

11.2.3 Temperature effects and related issues

As specified hereafter, the use of bulky Yb-doped materials always implies great care regarding the thermal conditions. Despite a low quantum defect, some part of the absorbed pump power remains transferred to the host lattice in the form of heat [11–13] by phonon processes. Furthermore, the fundamental-level populating rate and the linewidths of the various Stark sub-levels in the two manifolds are determined by the average material temperature, which strongly influences the laser performance. Besides the usual room-temperature effects, the implementation of low-temperature designs using cryogenic means may also evidence some interest in the search for outstanding optical efficiencies and ultimate PRFs.

Material thermal properties

The first connected item of data is the thermal conductivity k, in units of W m^{-1} K^{-1}. The room temperature value of K may vary somewhat with the doping level [14], and it usually experiences a strong increase towards low temperatures. The other involved data are the coefficient of thermal expansion α, in units of K^{-1}, and, referring to Hooke's law [15] dedicated to elasticity which determines the material's internal stresses (σ, Pa), Poisson's ratio ν and Young's modulus E. Elevated thermomechanical stresses and constraints [15, 16] may be experienced at high pump intensity, together with surface or bulk-fracture hazards. These parameters then determine the so-called thermal-shock parameter R_T, a statistical quantity of prime importance in units of W m^{-1} to account for the tolerable surface tensile stress σ_T:

$$R_T = \frac{1 - \nu}{\alpha E} k \sigma_T \qquad [11.12]$$

Internal stresses also generate combined optical effects and geometrical modifications in the material's shape and dimensions. A non-uniform stress distribution obviously results in a non-uniform, and usually anisotropic, spatial distribution of the refraction index. This leads to combined depolarisation

© Woodhead Publishing Limited, 2013

and polarisation–rotation effects, depending on the pump distribution, the nature of the material and the boundary cooling conditions. Depending on the photoelastic tensor [15], the orientation of the optical indicatrix, which characterises the distribution of the linear components of the refraction index, then undergoes significant plus or minus deformations. According to the material's class of symmetry, up to 20 or more than 36 independent coefficients may be needed to identify the actual ellipsoid index at a given location inside. Even assuming comprehensive knowledge of each of these coefficients, appropriate modelling would imply efficient spatial sampling, together with the use of numerical tools using finite elements. Comprehensive calculations usually remain very difficult, or even impossible. More efficient approaches then consist in the coupling of experimental data with approximated analytical formulations.

Generic modelling of the optical effects due to temperature gradients in a cylindrical rod

This configuration takes advantage of analytical expressions [15], to be used in the basic understanding of thermo-optical issues. Let us start from the heat equation, in a suitable form to account for a scalar or vectorial description of the temperature (T), i.e. the material's isotropy or anisotropy, and geometrical symmetries or asymmetries. Let us also consider a rod of length and radius l, r_o with isotropic side-cooling, r being the radial coordinate inside and P_{th} the equilibrium thermal power dissipated inside the rod. With a top-hat pump beam along the pump axis z and an arbitrary axial pump-absorption profile $I_p(z)$, but unsaturated pump, the solution to this heat equation provides the following three-dimensional temperature gradient:

$$T(r, z) - T(r_o, z) = \frac{P_{th}}{4\pi k}\frac{\alpha_{NS}\exp(-\alpha_{NS}z)}{1-\exp(-\alpha_{NS}l)}\begin{cases}\ln\left[\frac{r_o^2}{\omega_p(z)}\right]+1-\frac{r^2}{\omega_p^2(z)}, r \le \omega_p(z)\\ \ln\left(\frac{r_o^2}{r^2}\right), r \ge \omega_p(z)\end{cases} \quad [11.13]$$

where $\omega_p(z)$ figures the circular pump radius along z and α_{NS} the unsaturated pump-absorption coefficient. Equation 11.13 does not account for any heat flux along z, and does not apply to thin disks. But it evidences the sizing criteria of rod-like geometries, to minimise $\Delta T(r, z) = T(r, z) - T(r_{o,z})$. This implies a minimum diameter, the first option within technological limitations due to polishing issues, or the enlargement of the focusing spot, at the expense of laser efficiency. The central gradient, which is governed by the

© Woodhead Publishing Limited, 2013

unsaturated absorption length $1/\alpha_{NS}$, undergoes only slight variations with l and remains proportional to P_{th}/k:

$$\Delta T(r, z) = \frac{P_{th}}{4\pi k} \frac{\alpha_{NS}\exp(-\alpha_{NS}z)}{1 - \exp(-\alpha_{NS}l)} \frac{r^2}{\omega_p^2} \quad [11.14]$$

Anyway, due to the counter-balancing of pump absorption by stimulated amplification, this simplified model cannot accurately fit the actual thermal load in the rod when lasing. Forgetting that the actually absorbed pump power is lower under non-lasing than under lasing, the thermal gradient may be overestimated by a significant amount. In real life, more precise numerical methods are required to include the combined effects of total or partial absorption–saturation, with combined fluorescence and lasing. Easy-to-use analytical calculations such as Eqs 11.13–11.14 only remain valid for the analysis of some orders of magnitude and worst-case limits.

The second modelling step involves the coupling of thermo-optic and thermo-elastic phenomena, using either numerical or experimental means, or using hybrid processes. Up to the end of this section, we just attempt to provide the reader with a basic approach of the way to operate. The simpler calculations take into account thermally induced variations in the optical index dn/dT [6, 15], in the form of a spherical thermal–optical lens of which the focal length is written as a function of the quantum (η_Q) and absorption (η_a) efficiencies:

$$f_{tho} = \frac{2k}{\eta_Q \eta_{abs} I_p \mathrm{d}n/\mathrm{d}T} \quad [11.15]$$

Boundary effects due to parasitic curvature phenomena on the laser faces are not considered in Eq. 11.15, which will nevertheless evidence a true interest in the discussion of YAG and tungstate-based designs to be presented in the last section of this chapter. More complete models take into account the anisotropy of the material itself, and/or asymmetrical temperature gradients [16, 17] in combination with thermal expansion effects, stress and strain fields. These complex phenomena usually involve significant depolarisation losses. A rough estimation, though more precise than that made in Eq. 11.15, applies to the combination of thermo-optic and thermo-elastic effects. Considering cylindrical coordinates and a quadratic approximation of $n(r)$ in the form $n(r) = n_o\left[1 - \frac{Q}{2k}\left(\frac{1}{2n_o}\frac{\mathrm{d}n}{\mathrm{d}T} + \alpha n_o^2 C_{r,\phi}\right)r^2\right]$, with n_o, Q and $C_{r,\phi}$ denoting respectively the room-temperature refraction index of reference, the thermal load in the material and dimensionless functions of the photo-elastic coefficients [17], yields:

$$f_{th} = \frac{k}{\eta_{abs}\eta_Q I_p}\left(\frac{1}{2}\frac{\mathrm{d}n}{\mathrm{d}T} + \alpha\left(C_{r,\phi}n_o^3 + \frac{r_o(n_o - 1)}{l}\right)\right)^{-1} \quad [11.16]$$

© Woodhead Publishing Limited, 2013

The above functions $C_{r,\phi}$ depend on the r- and ϕ-polarised photo-elastic variations of the refraction index, $\Delta n_{r,\phi}$, according to $C_{r,\phi} = -\frac{n_0^3}{2\Delta n_{r,\phi}}\frac{\alpha Q}{k}r^2$. The global thermo-optic coefficient is the sum of three contributions, i.e. dn/dT, the thermal expansion and the photo-elastic terms. The use of Eqs 11.13–11.16 for sizing purposes only makes sense if all the contributing coefficients are identified. Referring to YAG, KGW and KYW matrices, Table 11.1 summarises typical numbers from the literature.

11.3 Overview of Yb-based systems pump architectures and modes of operation

This section provides an overview of generic sources of nanosecond and sub-picosecond pulses in the field. It aims to focus on the basics of Q-switching and mode-locking, and on the process of regenerative amplification dedicated to the production of high-energy pulses using Yb-doped materials. The related concepts and characteristic numbers must be considered as complementary data to Chapter 9, as applied to the discussion of the optical performance of YAG and tungstate-based systems.

11.3.1 Pump architectures

High-power laser diodes and pump geometry

Diode pumps are key components in any Yb-based laser system [18–20]. They make use of high-power LD bars or stacks in the spectral range of 900–980 nm. High-brightness single LD stripes and bars may provide pump intensities (I_p) in excess of 10^5 W/cm^2, within the power limitation (P_p) of a few to some tens of watts, by means of imaging techniques. With the large angular divergence of individual LD emitters being typically 35–40° and 5–10° along the fast and slow axes respectively, this mainly involves efficient fast-axis collimation and beam symmetrisation. Efficient collimating techniques, up to five or ten times the diffraction limit, are based on either upstream close-coupling of an array of micro-optical cylindrical lenses or of a fibre-bundle in front of the emissive surface, or on the use of coupled off-axis stepped mirrors. To get access to higher pump powers, large water-cooled LD stacks can be used. Commercial stacks, which make use of CW LD-bar assemblies up to 50 W and more per bar, or of Q-CW assemblies up to 100 W and more per bar, are available in the range of P_p = 1 to 5–10 kW pump powers.

Given the selected diode pump, the pump architecture characterises the definition of the coupling geometry with the Yb-based piece of material. Generic options refer to end-pumping [6, 13, 21], when the pump and beam

© Woodhead Publishing Limited, 2013

Table 11.1 Room-temperature material data: (*) rough estimation by combining published data, and (**) reliable data not found at the time of publication, even though not so different from *KGW* numbers

	Thermal conductivity	Thermal expansion coefficient	Specific heat	Young's modulus	Poisson's ratio	Knoop hardness	Fracture limit	Tensile strength	Thermal shock parameter	Global thermo-optic coefficient	Refractive index variation
	k ($W.m^{-1}.K^{-1}$)	α ($10^{-6}.K^{-1}$)	C_p ($J.g^{-1}.K^{-1}$)	E (GPa)	ν	(mohs)	K_c (MPa)	(GPa)	R_T (W/m)	χ (10^{-6} K^{-1})	dn/dT ($10^{-6}.K^{-1}$)
YAG	9–11	6.7–8.2	0.59	280	~0.25	8.5	1.4	0.13–0.28	600–790	10	+7.3 to +8.9
KGW	2.5–3.8	1.6–8.5	0.26	92–152	~0.25	4–5	0.3	* 20–180	210–280	7.5	~0.4
KYW	3–4	2–8.5	**	**	~0.25	4–5	**	**	**	**	~0.4

© Woodhead Publishing Limited, 2013

axes are the same, to side-pumping [6, 22] when they are orthogonal, or to mixed designs in the situation of off-axis or resonant configurations. The selected option must account, at the same time, for the optimisation of the pump and laser mode-volume overlap, and for the efficiency and the spatial uniformity in the absorption of the pump.

The end-pump configuration (Fig. 11.3) of a fibre-coupled LD bar or LD stack with a cylindrical rod is one of the most widely used low-power designs [13]. The deposition of suitable coatings on the two sides of the rod, which can be conductively or radially cooled by means of a water jacket, is an option to integrate a fairly compact multimode cavity. Then the rear coating must ensure a good pump transmission at λ_P, but a high reflectance at λ_L, in the whole spectral range of interest. Given the doping level, the optimum length of the rod depends on the pump intensity. It needs to match the pump absorption length, either single-pass or double-pass. Due to the low quantum defect, and depending on the selected options, coatings with sharp spectral transitions may be required. The output coating may be replaced by an external mirror, to built a free-space cavity and enable some control of the polarisation extinction ratio (PER), or Q-switching by means of a waveplate and a Pockels cell. The most critical sizing and pump issues consist in the management of the beam quality in the presence of elevated pump intensities, versus the radial thermal gradients and the related thermo-optical effects (Eqs 11.13–11.16). This configuration takes advantage of some compactness and robustness, with a relative insensitivity to misalignment effects in the presence of environmental perturbations.

The so-called 'thin-disk' geometry [23–25] is another end-pump configuration of reference, which has taken a major place in the field of ultra-high power systems. This configuration enables a strong reduction of the transverse thermal gradients, and of associated thermal-lensing effects. It makes use of a very thin disk of Yb-doped material (Fig. 11.4), to be pumped

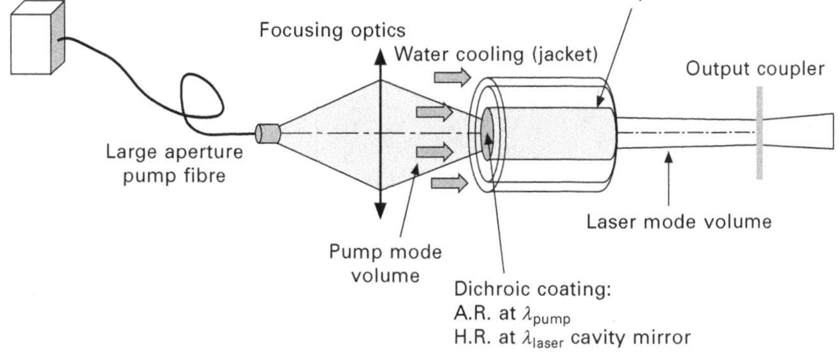

11.3 End-pump basic geometry.

© Woodhead Publishing Limited, 2013

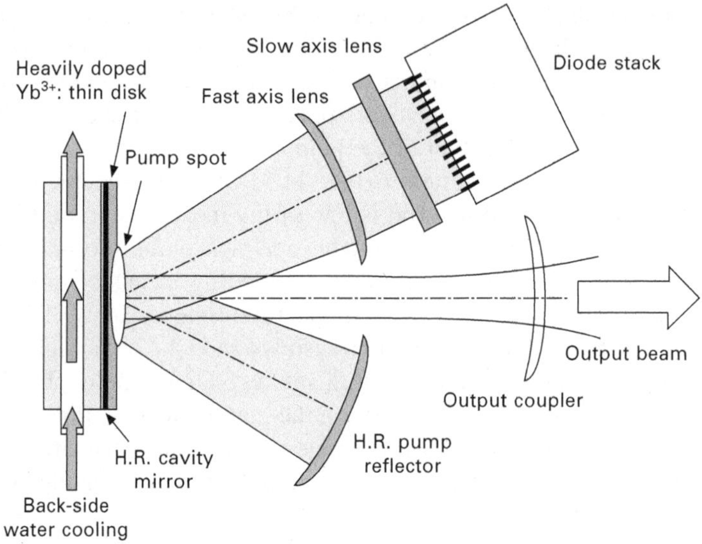

11.4 Thin-disk four-pass resonant pump geometry.

using double-pass or multipass absorption [15, 26, 27]. The material's very low thickness, typically 100 to 300 μm, makes possible efficient and uniform back-side cooling throughout the whole cross-section of the laser beam. The only significant issues are those of packaging, which must be consistent with fragile materials to ensure a good thermal contact on the whole rear surface without any deleterious stress. Sizing criteria [25] mainly involve the needs of efficient pump absorption, relative to the search for a minimum small-signal gain. This explains why a number of highly efficient designs have been demonstrated with YAG and KGW–KYW-based materials, taking advantage of rather large cross-sections and of high doping capabilities. Free-space pumping schemes compete with fibre pumps, which benefit from more spatial uniformity. Apart from the need for rather complex optical designs in the situation of multipass pump absorption, the main limitation to be kept in mind is that of the energy storage and related gain depletion due to transverse ASE [6, 26]. This will be discussed later.

Side-pumping is another efficient option, to get more flexibility in the management of power and gain-scaling compromises while minimising the limitations due to nonlinear effects. Fully adjustable and possibly larger pump surfaces enable the optimisation of the small-signal gain versus the system's requirements and expected output energy, with less ASE. Thanks to the linear geometry of LD bars and stacks, this geometry also benefits from a good consistency with straightforward free-space pumping. It takes advantage of large-sized low-cost optics and simple optical schemes.

Figure 11.5 depicts a relatively simple side-pumped design [22]. By means

© Woodhead Publishing Limited, 2013

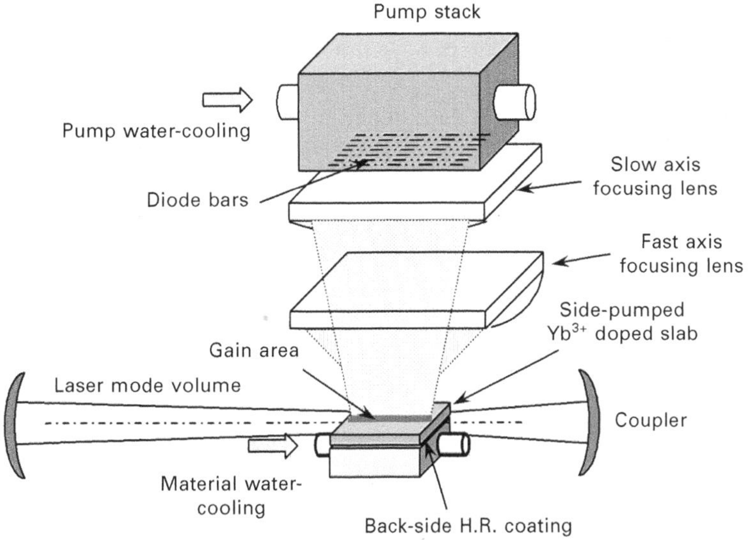

11.5 Two-pass side-pump geometry of a long slab.

of only two plano-cylindrical macro-lenses, it makes use of a unique large-sized, fast-axis collimated LD stack. Providing efficient double-pass pump absorption, this geometry enables the generation of a long and nearly uniform gain stripe in a thin plate of Yb^{3+}: YAG. The large bottom lens, which is aligned along the LD bar's axis, helps to image the emitting surface into the central area of the plate. The vertical position of the orthogonal upper lens determines the suitable optical magnification for relevant length matching. The laser performance may be optimised with respect to the symmetry of the beam, while operating the material with large and fully adjustable small-signal net gains (G_o). To give numbers at P_{pump} = 1.5–2 kW, values of G_o in excess of 2 within an active section of 2–3 mm^2 are easily obtained using a 1 cm long plate. The main counterparts include a poorer spatial overlap between the pump and the laser beams, and stronger astigmatism.

ASE limitations

As already underlined, the topic of ASE limitations must be set within the discussion of pump and geometrical sizing issues. The aim is the maintaining of efficient energy storage. To give orders of magnitude with the generic pump geometry of the slab in Fig. 11.2, given a side aperture denoted by $D = \sqrt{\Sigma}$, let us consider the average density of absorbed pump energy (ρ, J/cm^3). Within the elementary slice $dV = D.l.dy$ of thickness dy somewhere in the slab along one of the transverse axes (y), the spontaneous energy (E_{SP}) produced is determined by $dN_2/dt \approx N_2/\tau_F$, i.e. $dE_{SP}/dt \approx (\rho Dl/\tau_F)\,dy$. Integrating

© Woodhead Publishing Limited, 2013

along the whole pump duration T_p we get, $E_{SP}(y) = (T_p\rho Dl/\tau_F)\,dy$ of which the only fraction to be amplified is $\frac{\Omega}{4\pi} = \frac{1}{2}\left(1 - \cos\left(\frac{\pi}{2} - \theta_{limit}\right)\right) = \frac{n-1}{2n}$. The limiting solid-angle Ω is determined by the half-angle of which the summit value is $\pi/2 - \theta_{limit}$, if $\theta_{limit} = \arcsin(1/n)$ figures as the total-reflexion limit in the presence of the material index n at λ_L. The value of the total ASE energy in the slab is obtained thanks to a second integration step along the y axis, considering the propagation path up to the opposite side of the slab and no lateral reflexion from the large faces. This leads to the following transcendental equation in D:

$$E_{ASE} = \int_0^D E_{SP}(y)\frac{\Omega}{4\pi}\exp(g_o(D-y))\,dy = \frac{T_P}{\tau_F}\frac{\Omega}{4\pi}\rho V\frac{\exp(g_o D) - 1}{g_o D} \quad [11.17]$$

Numerical or graphical solutions of Eq. 11.17 can be obtained as a function of the acceptable ratio $E_{ASE}/\rho V$, to estimate the acceptable geometrical limitations versus g_o whatever the situation of a long gain stripe or a thin disk. Given $E_{ASE}/E_{sto} < 10\%$ in the case of a thin slab, for example, we find $g_o \approx 7$.

11.3.2 Generic nanosecond and sub-picosecond modes of operation

Nanosecond pulse generation and amplification

This section applies to both basic Q-switching and regenerative amplification [6, 27], as involved in most current Yb-based systems dedicated to the production of energetic nanosecond pulses. Recalling Eqs 11.1, 11.11 and 11.17, Q-switching [28] takes advantage of the efficiency of energy storage thanks to a large τ_F, a benefit of ytterbium. This mode of operation involves a lossy cavity (Fig. 11.6(a)) and rapid switching of the Q-factor inside, at the end of the so-called pump cycle which determines the time interval (T_p) of reference. A giant pulse is generated downstream of the output coupler, following T_p. Yb-based bulky systems usually make use of active triggering, thanks to polarisation–rotation using a Pockels cell, a polariser and a quarter-wave plate.

Most often, the exact Q-switched pulse shape is determined with the well-known formalism of rate equations [6, 27], using numerical calculations like Runge–Kutta algorithms. But more straightforward analytical expressions provide a simpler approach to evidence the characteristics of the pulse, of interest for fully parametric sizing purposes. We denote by l the material's amplification length and by V the active volume. The pump ratio over the threshold (PRT) consists of the basic sizing parameter to quantify the optical performance [22]. I_{pTh} denotes the actual threshold pump intensity, and its

© Woodhead Publishing Limited, 2013

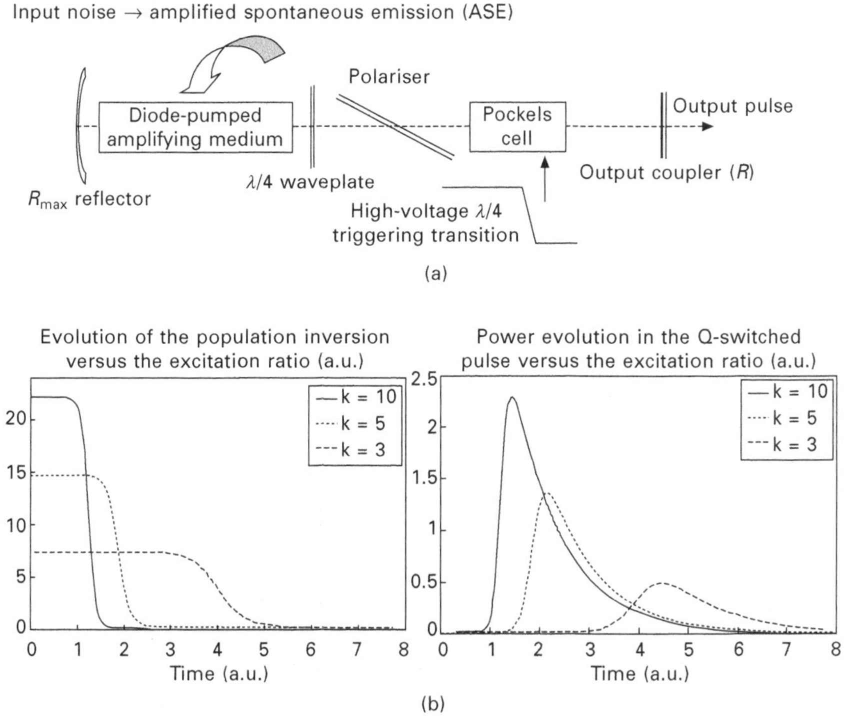

11.6 (a) Actively Q-switched cavity and (b) related features, given the photon lifetime τ_C: temporal decrease of the population inversion and output pulse shape versus *k*, the ratio of pump power over the threshold at the time of triggering.

value may be written as $k_{TH} = I_p/I_{pTh}$ in terms of measurable values, or as $k_{Th} = N_2/N_{2Th}$ for modelling topics. In this situation, k_{Th} refers to the threshold value of the population inversion, of which the generic expression in an Yb-based lossy cavity is:

$$N_{2Th} = \frac{\gamma_{tot} + 2\sigma_a(\lambda_L)N_{tot}l}{2[\sigma_e(\lambda_L) + \sigma_a(\lambda_L)]l} \qquad [11.18]$$

where γ_{tot} figures the optical losses, except those due to the three-level configuration. I_{pTh} then needs to account for the transparency intensity I_{min}, according to Eqs 11.6–11.9, and to include the reabsorption effects at λ_L. Its value may be referred to the well-known expression for the lasing threshold of four-level configurations, $I_{pTh} = [\gamma_{tot} - \ln(R)]hc/2\lambda_L\eta_Q\sigma_e\tau_F$, as a function of the reflectivity (R) of the output coupler. The values of k_{Th} and of the population inversion at the time of triggering completely determine the energy contained in a Q-switched pulse, according to [6]:

© Woodhead Publishing Limited, 2013

$$E_{out} \approx N_{2Th} \frac{V \frac{hc}{\lambda_L}}{1 - \frac{\ln(R)}{\gamma_o}} \frac{k_{Th} + \text{LambertW}\{-k_{Th}\exp(-k_{Th})\}}{\psi} \qquad [11.19]$$

where the material constant ψ varies from 1 up to nearly 2 for 'three-level' conditions, and LambertW expresses the mathematical Lambert's W-function so that LambertW(u).exp[LambertW(u)] = u. The larger is the PRT, the higher the output energy and the shorter the pulse width. The corresponding temporal evolutions of the output pulse shape and of the population inversion are depicted in Fig. 11.6(b), for k_{TH} = 3, 5 and 10.

The technique of regenerative amplification [3, 6, 27, 29], which has been proved of particular interest for high-energy Yb-based systems, takes advantage of the capability of huge amplification net gains by recirculation along a low net-gain path. As distinguished from Q-switching, the amplification process is no longer initiated from the ASE noise, but from the pulse to be amplified, at its time of arrival. This requires relevant synchronisation, using a dual-electrode Pockels cell (Fig. 11.7(a)). As for Q-switching above, the operating conditions are based on the progressive transfer of the complete amount of stored energy (Eq. 11.10) in the active medium towards the circulating pulse. The cavity length is determined by the spatial extension of the pulse inside, to prevent any overlap, and by the transition times of the Pockels cell. Because two independent input and output paths are required to trap and to extract the pulse, the involved optical scheme is more complex. Thanks to rapid $\pi/2$ polarisation switching, the linearly P-polarised input pulse enters the cavity at the location of the input/output polariser. Progressive amplification then operates round-trip per round-trip, up to the saturation of the net round-trip gain. With low-energy input pulses and energy extraction up to the saturation of the net gain, the overall energy gain may reach impressive values, in the range of 10^3 to 10^9.

Figure 11.7(b) describes the evolution of the laser power in the cavity during the whole regenerative sequence, thanks to photodiode monitoring. The envelope of the successive pulses, which are spaced by the round-trip time, obviously resembles the leading edge of a Q-switched pulse. The abrupt rear-side evidences the extraction of the circulating pulse.

An Yb-based regenerative amplifier may also be calculated in a fairly simple way, by considering the ratio of the reabsorption at λ_L to optical losses, $\xi(\lambda_L) = 2\sigma_a(\lambda_L)N_{tot}l/\gamma_o$, and amplification up to net-gain saturation to optimise the overall energy efficiency. This implies that the condition $N_2 = N_{2Th}$ has been reached at the time of pulse extraction. The peak output energy of the pulse to be amplified [22] is proportional to I_{sat}, as defined above in Eq. 11.9:

© Woodhead Publishing Limited, 2013

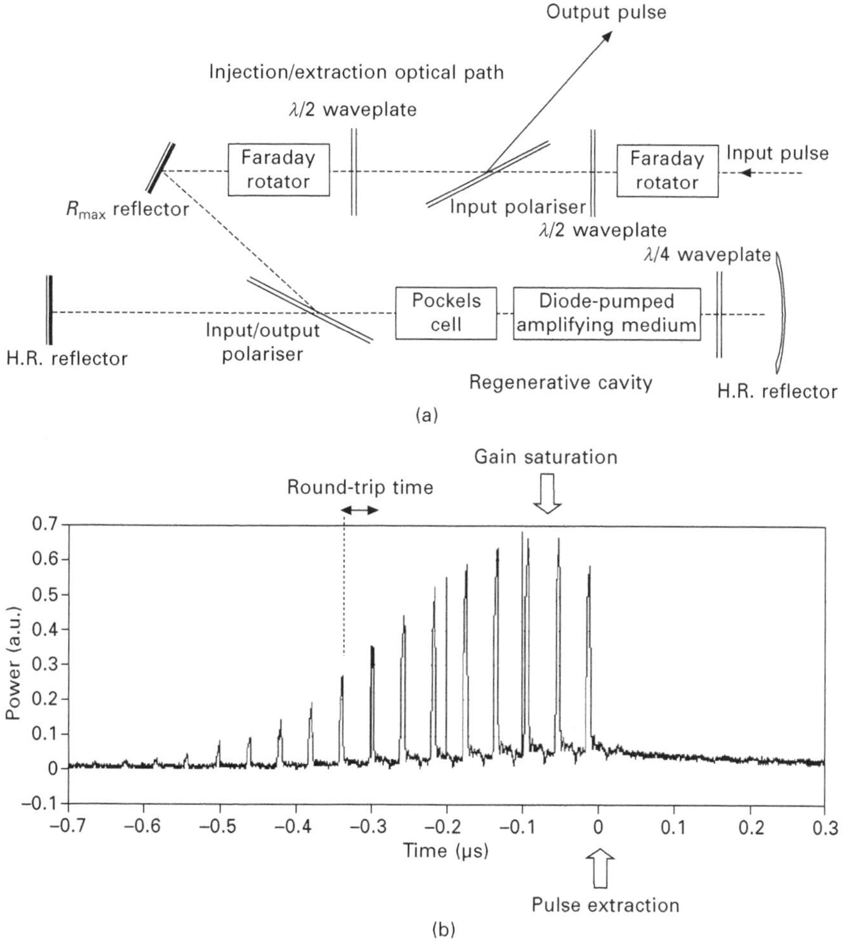

11.7 (a) Regenerative amplifier and (b) power monitoring thanks to back-reflector leakage.

$$E_{out}(\lambda_L) = \gamma_o S_{beam} I_{sat} \tau_F \times (1-\beta_{min})(1+\xi)\left\{k_{Th} - 1 - \frac{1}{1+\xi}\ln[(1+\xi)\cdot k_{Th} - \xi]\right\} \quad [11.20]$$

and the build-up time of the process is:

$$T_c = \frac{2L_{cavity}}{c(\gamma_o + 2\sigma_a N_{tot} l)(k_{Th} - 1)}$$

© Woodhead Publishing Limited, 2013

$$\times \ln\left\{\frac{G}{\gamma_o(1-\beta_{min})(1+\xi)\left\{k_{Th}-1-\dfrac{\ln[(1+\xi)k_{Th}-\xi]}{1+\xi}\right\}}\right\} \qquad [11.21]$$

with G the overall net gain. Towards the pure four-level limit it can be verified that Eq. 11.20 simplifies to $E_{out} = \gamma_o \cdot S_{beam} I_{sat} \tau_F \cdot [k_{Th} - 1 - \ln(k_{Th})]$, as established in [27].

Expressions 11.18–11.21 provide the most important data for basic understanding of Q-switching and regenerative amplification. This completes simple models for the analysis of the optical performance in any Yb-based cavity.

Despite some operating complexity and critical synchronisation issues, Yb-based regenerative amplifiers consist of an efficient tool to produce energetic, either nanosecond or sub-picosecond pulses. Considering pulse trains at a reduced PRF or of a limited duration, the attainable output energy per pulse is governed by the stored energy and the optical losses of the cavity, within optical-damage limitations regarding the peak power inside. These limitations can be partly overcome by the addition of chirp pulse amplification (CPA), which involve upstream stretching and downstream recompression in the presence of a large spectral bandwidth. This implies the preservation of ultimate broad-bandwidth capabilities by means of additional spectral equalisation techniques. The cavity must be operated with a tight compensation for the spectral non-uniformity of the small-signal gain and for the effects of internal group delay dispersion (GDD), near overall zero-dispersion. The issue of GDD will be discussed hereafter in more detail, regarding short-pulse generation. A number of representative works in the field of regenerative amplification will be selected in the last section dedicated to YAG and tungstate-based systems, referring to the equations above in terms of optical performance and of pumping configurations.

Short pulse generation by mode-locking

Mode-locking [30, 31] consists of the unique process to produce periodic trains of ultra-short pulses, of duration ranging from some tens of femtoseconds to several picoseconds. Starting from a large number (N) of frequency modes in a long-length (L) cavity, this process essentially involves phase or intensity locking. Thanks to the stabilisation of internal interferences in the presence of mode competition, a unique set of standing waves is allowed to oscillate in the CW regime. A single transverse-mode cavity is needed, with homogeneous spectral broadening. Mode-locking involves either active or passive means.

© Woodhead Publishing Limited, 2013

Active mode-locking implies loss modulation by electro-optical means, in exact synchronism with the cavity frequency (F_m). In the presence of pure homogeneous broadening across the whole gain bandwidth B_{FWHM} and of a Lorentzian gain profile, which apply to the behaviour of Yb-based materials, and without any GDD or self phase modulation (SPM) effect, the theory of Kuizenga and Siegman settles the attainable minimum pulse-width:

$$\delta t_{FWHM} \approx 0.45\left(\frac{G}{M}\right)^{0.25}(F_m B_{FWHM})^{-0.5} \qquad [11.22]$$

where G and M denote the peak gain and the modulation index, respectively. The value of I_{pump} must be adjusted in such a way that the total cavity losses γ_o are exactly balanced by the pulse-train power at the equilibrium, which leads to [3]:

$$G = \gamma_o + \frac{M}{4}(2\pi F_m \delta t_{FWHM})^2 \qquad [11.23]$$

Given L = 20 cm and a gain bandwidth $B_{FWHM} \approx 20$ nm, for example, to match generic orders of magnitude with Yb-based cavities of widespread use, (23) leads to $\delta t \approx 100 - 150$ fs.

Passive mode-locking refers to self-modulation effects [31–33], due to the nonlinear properties of materials in the presence of a high laser intensity. The usual configurations in the field involve either semiconductor saturable absorber mirrors (SESAM), Kerr-lens mode-locking (KLM) or soliton generation. KLM-based cavities require the addition of a starter inside, for the initiation of the process by mechanical or electro-mechanical means. This is not the case with SESAMs and solitons. KLM essentially implies good adjustment of beam-focusing somewhere inside, at the location of a nonlinear component of which the optical index may be modulated with a high efficiency. The solitonic regime applies to more complicated interactions between the GDD and Kerr effects. This technique takes advantage of very stable pulse trains and of a good resistance to unwanted variations in the PRT. SESAM-based mode-locking makes use of intensity-dependent ultra-fast saturation and desaturation phenomena in the electronic population of semiconductor quantum-well structures. This consists of the most widely used option, which combines self-starting capabilities and fairly stable operation. Denoting by ΔR the modulation index at full saturation of the involved SESAM [3], one obtains:

$$\delta t_{FWHM} \approx \frac{0.9}{B_{FWHM}}\sqrt{\frac{G}{\Delta R}} \qquad [11.24]$$

where G is the peak gain, as above, and $B_{FWHM} = c\Delta\lambda_L/\lambda_L^2$ is the gain bandwidth. If $\Delta\lambda_L = 20$ nm around the peak λ_L, this gives typical values of δt_{FWHM} ~150–250 fs.

© Woodhead Publishing Limited, 2013

In similar terms as those of broadband regenerative amplification above, the search for the attainable minimal δt_{FWHM} according to Eqs 11.22–11.24 requires an exact compensation for the GDD. The GDD characterises the temporal enlargement of a broad-bandwidth pulse after its propagation along a dispersive path. The related wavelength-dependent phase shift $\varphi(\lambda)$ is governed by the Sellmeier coefficients of the average refraction index $n(\lambda)$ in this path. The generic expression to be used in the determination of the λ-dependent time delay is $\delta\varphi(\lambda) = \frac{\lambda_L^3 l}{2\pi c^2}\left[\frac{{}^2 n(\lambda)}{\lambda^2}\Big|_{\lambda=\lambda_L}\right]$ Considering an initially unchirped Gaussian pulse of duration δt_{FWHMo}, the subsequent temporal enlargement becomes noticeable when GDD > δt^2_{FWHMo}. It is written as [3]:

$$\Delta t_{FWHM} = \delta t_{FWHMo}\left[1 + \left(4\ln\left(\frac{2GDD}{\delta t_{FWHMo}}\right)\right)^2\right]^{0.5} \qquad [11.25]$$

Expression 11.25 helps in determining the conditions for the compensation of GDD effects within the entire gain bandwidth of interest, usually to reach nearly-zero global dispersion, or sometimes slightly negative dispersion in the case of solitonic regimes in the anomalous-dispersion domain. Widely used compensation techniques involve Gires–Tournois interferometer mirrors (GTI), which exhibit negative dispersion, in combination with prism pairs. GTIs are fairly compact optical, asymmetrical components comprising a front-side Fabry–Pérot structure and a highly reflective back-side mirror.

Figure 11.8 describes a generic passively mode-locked cavity using SESAMs, with GDD compensation by means of a prism pair and GTIs. It comprises a long single-mode resonator, which encloses all the suitable optical components for fine control of the lasing conditions. This implies separate means for spectral tuning within B_{FWHM} near the peak λ_L, for the selection of polarisation, for the GDD compensation and beam cleaning, for the adjustment of the beam focusing in the SESAM, and for proper long-term stabilisation of the cavity length. This helps to avoid any hazard due to environmental thermo-mechanical or electrical perturbations. An optional starter has been included, if necessary for some operating regimes. A number of variations around this design are compared in the next section.

11.4 YAG–KGW–KYW-based laser systems for nanosecond and sub-picosecond pulse generation

Nowadays high-energy nanosecond sources have initiated a large variety of applications in numerous fields of industry, medicine, physics and R&D.

© Woodhead Publishing Limited, 2013

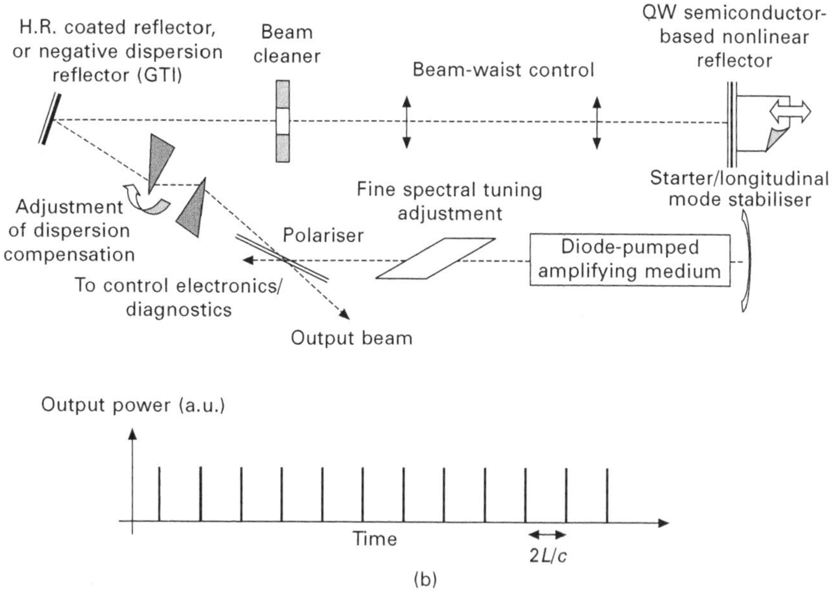

11.8 Passive mode-locking: (a) generic cavity of optical length *L* and (b) output pulse-train (b).

Representative industrial applications are the surface treatment of materials using thermal processes, marking and soldering, cutting and selective ablation. Others in medicine are retina operations, cauterisation of blood vessels, and dental treatment. The search for inertial confinement fusion or the detection of pollutants by LIDARs also takes place among the huge variety of fields of interest in physics and environmental surveillance, as well as the synthesis of ultra-high-temperature materials in chemistry. Most often, basic needs consist of the combination of high-energy and high-PRF, with a good beam quality. This last section then aims to provide an overview of laser systems using Yb-doped YAG and tungstate-based crystals, which are found among well-established commercial solutions. A special focus is proposed on some recent advances with YAG at cryogenic temperatures, for the search of unusual optical-to-optical efficiencies in excess of 50%.

11.4.1 YAG-based sources for nanosecond pulse generation

The isotropic YAG ($Y_3Al_5O_{12}$) composition, in its crystalline form or as ceramics, benefits from an excellent compromise between suitable spectroscopic, thermal and mechanical material properties. The growth of relatively large crystal boules of good optical quality is quite easy. Yb^{3+}:YAG

© Woodhead Publishing Limited, 2013

also takes advantage of large absorption and emission cross-sections (Fig. 11.11(a) below) and of good thermal conductivity. Apart from the sharp zero-line at 980 nm, which does not usually match the best pumping conditions, Yb^{3+}:YAG benefits from a broad absorption bandwidth around 940 nm. The most efficient output wavelength is $\lambda_L = 1030$ nm. These considerations may be evidenced in Fig. 11.9, which shows the room-temperature emission and absorption cross-sections, together with the related spectral gain as defined in Eq. 11.8. The shape of $\sigma_{gain}(\lambda)$ is shown in Fig. 11.11(b) below, for the sake of comparison with Fig. 11.9 for YAG, to evidence the broader bandwidth capabilities of the crystal and explain its longer than usual λ_L, which may

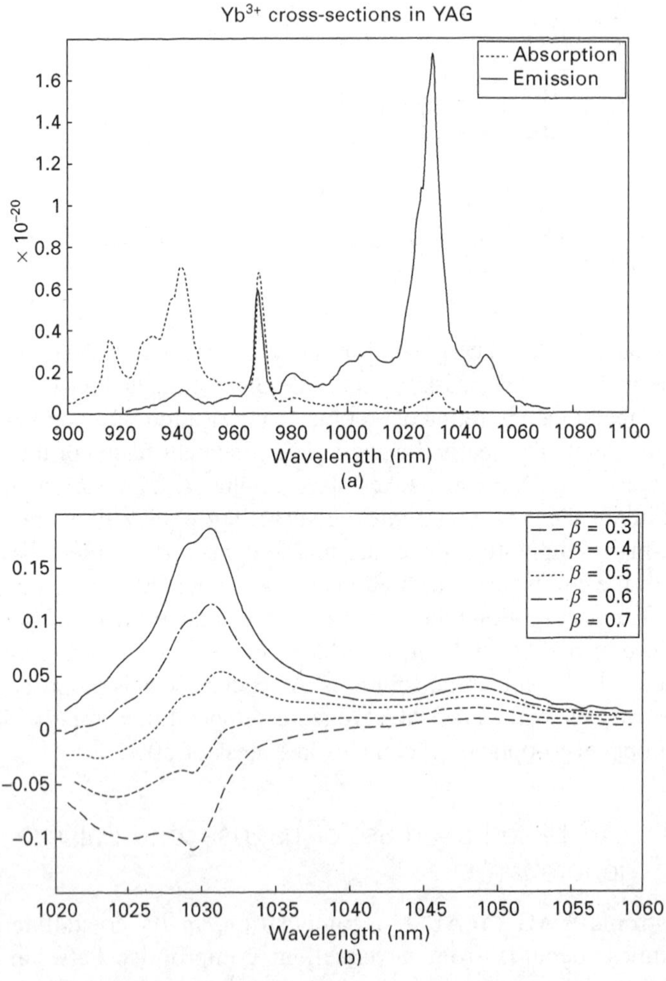

11.9 (a) Unpolarised absorption and emission room-temperature cross-sections in Yb^{3+}:YAG and (b) spectral gain with 2% doping.

© Woodhead Publishing Limited, 2013

be typically shifted by 10 to 30 nm. This provides basic data of interest to size the usual YAG-based systems in relationship to Eqs 11.2–11.11 and 11.17–11.21.

However, when the material's temperature is lowered, these spectra have been shown to undergo strong modifications. Spectral broadening effects then tend to be reduced by a large amount [34, 35]. Recent works in the range of cryogenic temperatures (*T*) have established the following:

- A significant increase of σ_e at λ_L = 1030 nm, by a factor of about 7 when *T* decreases from 300 to 10 K.
- A small increase of σ_a, by a factor of 1.5 to less than 2, depending on the value of λ_P, together with a nearly constant absorption bandwidth around 940 nm.
- The absence of any reabsorption effects, due to the depopulation of the upper Stark levels in the lower manifold. This needs to be referred to the fact that, at 300 K, the reabsorption at λ_L essentially occurs between the lower Stark levels of the upper manifold and the upper Stark levels of the lower manifold. Below *T* = 100 K, the crystal then behaves like a pure 'four-level' material.
- An increase of the thermal conductivity more than tenfold compared to its room-temperature value, i.e. from 10 W m^{-1} K^{-1} to approximately 500 W m^{-1} K^{-1} at the temperature of liquid nitrogen. YAG then looks like a metal. The peak thermal conductivity is experienced near *T* = 30–60 K.

The corresponding data, as presented in Fig. 11.10 from room temperature down to cryogenic temperatures, indicate that the material tends to behave like a highly conductive 'four-level' metal. Despite some added complexity and more critical operating conditions, these tremendous variations in the thermal and laser properties necessarily imply new factors at work. The energy storage and extraction efficiency data involved in Eqs 11.10, 11.11, 11.19 and 11.20 may be optimised for quite specific conditions. Thermal limitations and focal length effects, as described in Eqs 11.13–11.16, can be decreased by a large amount. Representative works in the field will be discussed hereafter.

On the basis of combined high-energy and high-efficiency selection criteria, Table 11.2 provides an overview of the attainable laser performance with some representative YAG-based designs from the open literature, considering either room-temperature [36–38] or low-temperature [39–41] conditions. These works reflect the progress of this technology since earlier research in the 1990s, and help to illustrate the state-of-the-art:

- Side-versus end-pumping issues always need to be discussed, in any case. This assumes a number of compromises. More or less complex optical designs have been demonstrated by means of slabs and rods, or

© Woodhead Publishing Limited, 2013

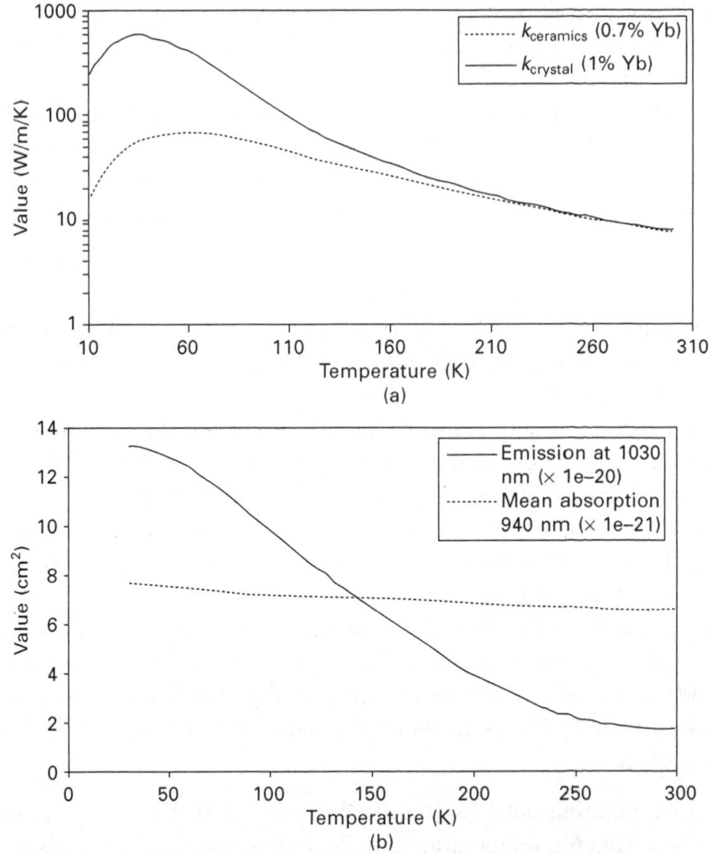

11.10 Cryogenic operations with Yb^{3+}:YAG: (a) variation of the thermal conductivity in doped crystal and ceramics, and (b) variation of the cross-sections of interest for diodepumping and amplification (peak emission at 1030 nm and averaged absorption over 10 nm).

thin disks and plates, from single-pass up to 4-pass, 6-pass and 32-pass pump absorption.

- Nowadays, room-temperature systems have been shown to have optical efficiencies up to about 30%, with nanosecond-pulse energies from some millijoules to nearly 3 J, at 1 Hz to 1 kHz PRF. λ_L = 1030 nm remains the unique wavelength of interest for efficient Q-switching.
- Cryogenic designs have been proved to be the most efficient, enabling pulse energies as high as 20–40 mJ, with up to 5 kHz PRFs. Despite their poor wall-plug efficiency and obvious integration issues, the pump-to-laser efficiency may exceed 60%.

A number of sizing data of particular interest in the use of the basic equations from Sections 11.2 and 11.3 have been collected in Table 11.2,

© Woodhead Publishing Limited, 2013

for the discussion of the optical performance of YAG-based systems. The first step in the analysis makes use of Eqs 11.3–11.8, regarding absorption and gain data, and of Eqs 11.9–11.11 and 11.13–11.16 regarding the energy transfer and thermal issues. The second step, to look at the consistency of the results with the expected output performance, applies to Eqs 11.19–11.21 in terms of pulse data.

As an example, let us refer to the low-temperature operation of a Q-switched YAG disk [40]. Applying Eqs 11.3 and 11.4 to the data provided, we calculate L_o = 1.1 mm. This is consistent with nearly total double-pass pump absorption in the stipulated thickness of 1 mm. The data depicted in Fig. 11.10 are taken into account by multiplying the usual value of $\sigma_e(\lambda_L)$ at 1030 nm in Yb^{3+}:YAG by a factor of 4 and by updating the pump reamplification ratio $\mu = \sigma_e(\lambda_P)/\sigma_a(\lambda_P)$. Rather than the usual value of about 20% for λ_P = 940 nm at T = 300 K, we consider μ = 5%. Starting from β_{min} = 2% in Eq. 11.8 and β = 8.5%, with g_o = 5.5, the related value E_{out} = 42 mJ in the chapter is verified using Eq. 11.19 with R = 72%, under the realistic assumption γ_o = 4%. The interested reader will also verify that these calculations apply with a net gain G_o = 0.07, at k_{Th} = 7. The thermal focal lengths in relationship with Eqs 11.15 and 11.16 are 8–9 m and 1.8 m. Similar verifications can be made the same way referring to [41], in the conditions of a high-PRF system, of which the optical efficiency reaches nearly 50%.

11.4.2 Tungstate-based sources for ultra-short pulse generation

Due to its moderate gain bandwidth, Yb-doped YAG suffers from some limitations in the field of ultra-short pulse generation. Since compact and robust sources producing 100–500 fs pulses at a high PRF consist of a well-established need for more and more industrial and scientific applications, other material options must be considered. This results from the basic difference between the involved femto-picosecond and nanosecond laser-matter interaction processes, i.e. the need to deal with dielectric breakdown effects or thermal deposition. A fairly good option to overcome the spectral limitations with YAG may be the selection of a tungstate-based design, using Yb^{3+}:KGd $[WO_4]_2$ (KGW) and/or KY$[WO_4]_2$ (KYW), which exhibit well-established broad-bandwidth capabilities. These highly anisotropic crystals [42–44] belong to the family of the monoclinic space group. Despite a lower τ_F, from 350 to 600 μs, they exhibit a set of quite attractive properties, both optical and thermal [45–48]. The large bandwidth broadening comes from a large splitting ratio of the sub-Stark levels in the energy diagram, and the anisotropy results from a highly asymmetric electric field in the closed environment of the dopant ion. KGW and KYW benefit from elevated emission cross-sections, in the same range as for Yb^{3+}: YAG for lasing wavelengths of interest. In addition, they can be easily doped up to a very high ratio.

© Woodhead Publishing Limited, 2013

Table 11.2 Nanosecond-pulse generation: comparison of some representative YAG-based designs with specified operating conditions (*), and referring to basic extrapolation by means of the sizing equations (**)

Selected design		Thin-disk, room-temperature and high PRF, Q-switching versus cavity-dumping and regenerative amplification	Long slab, room-temperature and low PRF, high-energy multipass MOPA system	Short slab, low-temperature and low PRF, Q-switched cavity	Long slab, low temperature and high PRF, high-efficiency, Q-switched cavity
System design	YAG material/doping	Crystal	Crystal/1.4%	Ceramics/9.8%	Crystal/1%
	Lasing conditions	Compared modes of operation, varying the pump spot in a thin disk	Slab-based amplifier with four-pass angular multiplexing, 6.4 ns–200 mJ input from a Q-switched MOPA	Back-cooled, end-pumped thick plate in a V-shaped cavity	Bulk-cooled, high-gain dichroic pump design through undoped endcap
Diode-pumped laser head	Pump source	60–190 W	Free-space, two Q-CW water-cooled, fast-axis collimated modules of 16 LD bar stacks, 2 × 13 kW at 940 nm	Fibred, single CW LD bar, 150 W at 938 nm	Fibred, single CW LD bar, 250 W at 940 nm
	Pump absorption		Single-pass, symmetric edge-pump, polarisation coupling and imaging optics, T_p = 0.8–1.5 ms	Double-pass, tilted pump, T_p = 1–4 ms	Single-pass, axial pump
	Material geometry	Disk	16 × 16 × 3.3 mm^3 slab	4 × 4 × 1 mm^3 brick	5 × 5 × 23 mm^3 brick
	Pump spot diameter or size (mm)	1.2–3.1 *	3.3 × 16 *	7 *	1.2–1.5 *
	I_p (kW/cm^2)		7.5 **	0.4 **	**
	g_o (cm^{-1}) (cf. Eq. 11.9)	Up to 0.73 *	**	5.5 **	**

© Woodhead Publishing Limited, 2013

Thermal data	$F_{thermal\ gradient}$ (cm)			800–900 **	**
	$F_{thermal\ lens}$ (cm)			180 **	**
Q-switching conditions	Beam diameter in the material (mm)	1.2 **		5–7 **	1.2 *
	$R_{coupler}$ (%)	90 *		72 *	10 *
	γ_o (%)	<2% **		**	**
	Length (cm)	85 *		175 *	43 *
	E_{out} (mJ)	Up to 18 *		Up to 42 *	Up to 23 **
	FWHM (ns)	250–550 *		Max. 200 *	16 *
Regenerative/multipass amplification	Number of round-trips/passes	40 to 1000 round trips, four passes per round trip *	Four passes *		
	γ_o (%)	10 *			
	G_o	Up to 0.18 per pass *	Up to 160 *		
	k_{Th}	<1.2 **	5 *	7.5 **	6.25 **
	E_{out} (mJ)	1 * (unsaturated G)	2900 *		
	FWHM (ns)	8 *	6.4 *		
Pulse output data	PRF (kHz)	1–13 *	0.01 *	0–0.35 *	5 *
	Wavelength (nm)	1025–1030 *	1030 *	1030 *	1030 *
	P_{out} (W)	12–64 *	10 *	17 **	100–114 *
	Optical efficiency (%)	10–34 *	10–15 *	11 **	46 **
Reference		[34]	[36]	[40]	[41]

© Woodhead Publishing Limited, 2013

KGW benefits from rather good thermal properties thanks to existing athermal crystalline orientations [9, 49]. The selection of one of these suitable orientations, which correspond to the nearly-exact compensation of the thermo-optic and thermal-expansion effects, then helps to reduce the thermal-length effects. Figure 11.11 depicts the characteristic spectral data to be kept in mind against the polarisation of interest, together with so-called 'athermal' directions. The largest emission and absorption cross-sections correspond to the so-called 'm-cut' orientation, for beam propagation along the crystallographic axis *m* (N_m). Referring to Eqs 11.22 and 11.24, the gain cross-section of KGW is consistent with the production of pulses of duration as short as 100 fs, with a large efficiency in the range λ_L = 1030–1080 nm. The anisotropy also concerns the thermal conductivity, with respective values of k = 2.8–2.5 and 3 W/m/K, along the crystalline orientations <100>< 010> and <001>. For the sake of a compromise between optical and thermal requirements, the usual propagation directions are those of the dotted axes in the '*m–g*' plane or in the '*g–p*' plane, which correspond to respective polarisation states parallel to N_p or to N_m. Despite average k of about one-third that of YAG, the use of athermal directions enables highly efficient KGW laser designs at elevated output power. A number of thermal-lensing measurement results have been published with KGW in a number of pumping and lasing configurations. They indicate rather large discrepancies and a strong dependence on the experimental conditions, from nearly similar values for focal-length as in YAG [15] down to about one-third of those values [42].

One can use Eq. 11.16 to relate the experimental data [15] with the expected orders of magnitude in the situation of CW end-pumping. This concerns the focal length of the induced thermal lens at I_p = 1 to 2 kW/cm^2, considering a pump spot diameter of ~200 μm, which decreases down to about 17 cm and 7 cm for 5–6 W of absorbed pump power, respectively, in 2 mm-thick and 3 mm-thick lasing pieces of $Yb^{3+}{}_{8\%}$:YAG and $Yb^{3+}{}_{5\%}$:KGW. The former 'lasing' condition is important, since the absorption of the same power at the same value of I_p in non-lasing pieces usually involves shorter focal-lengths.

KYW is closed to KGW in terms of crystalline structure and opto-mechanical properties, but its τ_F of about 600 μs enables more efficient energy storage. As shown in Fig. 11.12, its spectral data are not so different. Among the two selected polarisation directions, which are the most efficient in KYW, the axis *a* evidences the largest gain in the range 1000–1100 nm. Considering the crystalline orientations, a number of modes of representation are depicted in the literature [42–44]. In some of them, the crystalline orientation of KYW may be referenced to main two axes of interest, denoted '*a*' and '*b*' in the triplet of orthogonal vectors of reference '*a–b–c*'. The interested reader will find in references [9], [11] and [16] the exact values of the tilts between the

© Woodhead Publishing Limited, 2013

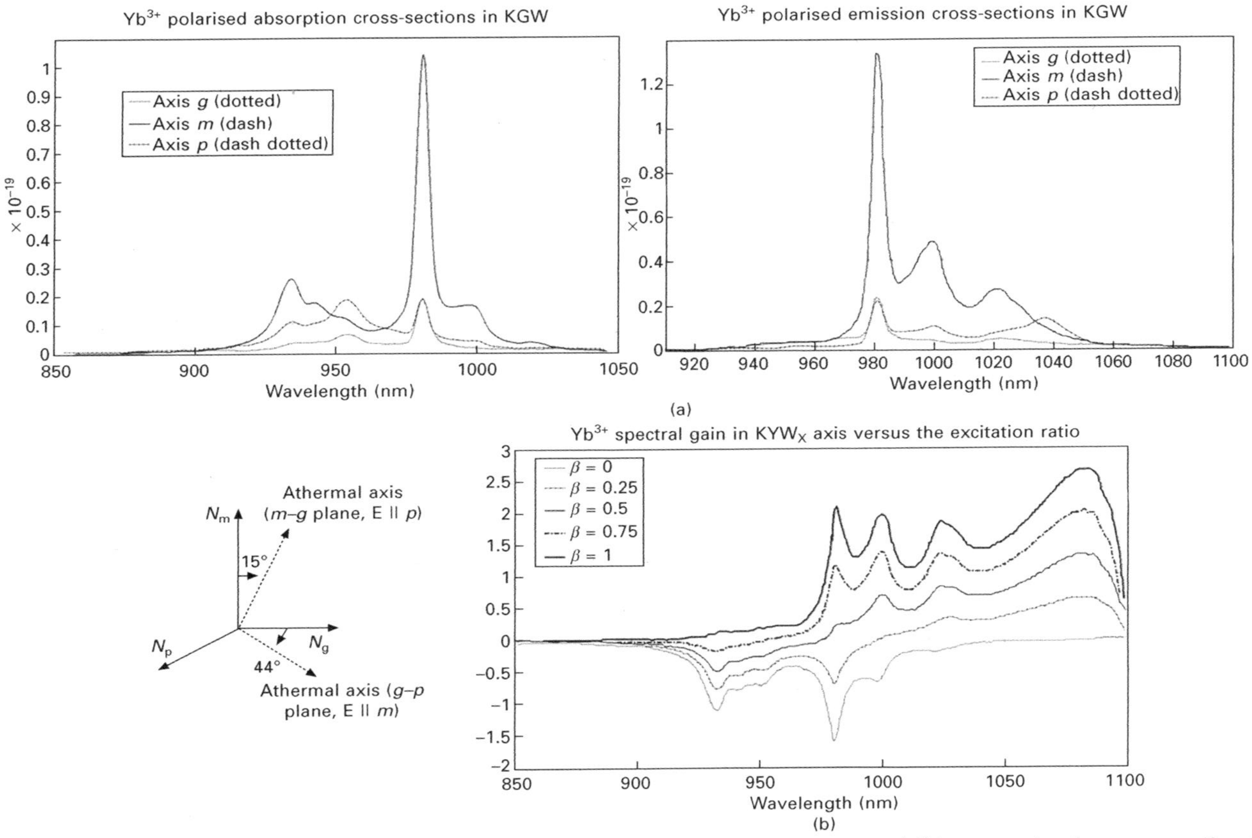

11.11 (a) Polarised room-temperature absorption and emission cross-sections and (b) spectral gain cross-section in Yb^{3+}: KGW.

© Woodhead Publishing Limited, 2013

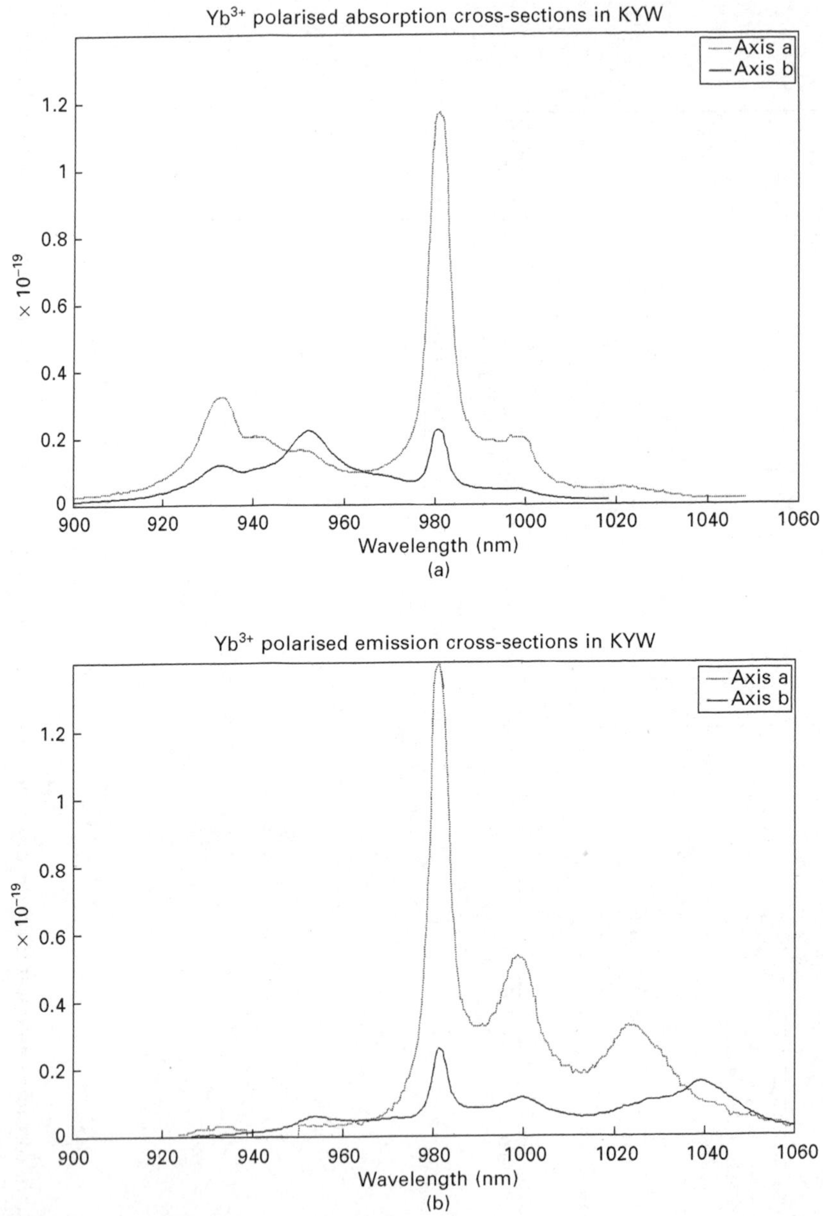

11.12 Polarised room-temperature (a) absorption and (b) emission cross-sections in Yb^{3+}: KYW.

different axes of the involved conventional triplets '*a*–*b*–*c*' and 'N_p–N_g–N_m', or the so-called *X*–*Y*–*Z* axes, with much detail concerning the spectroscopy and crystallography of the two materials.

© Woodhead Publishing Limited, 2013

Table 11.3 summarises the specifications and the optical performance of a number of representative Yb^{3+}:KGW and KYW-based systems, which may be considered as the state-of-the-art. As was done in Table 11.2 for nanosecond systems, the data in the table also help to make the relationship with the architectures of the generic systems from Figs 11.8 and 11.10 and with the related sizing equations. To give some numbers in relation to representative examples, cavity designs enabling output powers in the range of P_{out} = 7–10 W [50] have been demonstrated at δt_{FWHM} = 290 fs, for λ_L = 1039 nm and PRF = 45 MHz, or [53] at δt_{FWHM} = 416 fs for λ_L = 1048 nm and PRF = 1 MHz. KYW-based regenerative amplifiers [52] have been proven to give access to P_{out} = 10–50 W and $E_{out} \approx$ 10 μJ per pulse for λ_L = 1028 nm and PRF = 25 MHz, or [51] E_{out} = 500 μJ for λ_L = 1034 nm and PRF = 20 kHz, up to E_{out} = 5.5 mJ [47] at λ_L = 1024 nm, and even E_{out} = 27 mJ [54] at λ_L = 1030 nm and PRF = 100 Hz. The first two configurations enable quite short δt_{FWHM} = 200–240 fs, while the latter imply longer δt_{FWHM} = 340–670 fs.

As was done above for YAG systems, we can illustrate the interest of the basic equations from Sections 11.2 and 11.3 to discuss KYW-based regenerative amplification [51]. Referring to Eq. 11.4, we get $L_o \sim$ 2.3 mm. This value is consistent with the concerned 24-pass pump configuration, for nearly total absorption throughout the 100 μm-thick disk, with 5% doping. Then we assume nearly total population inversion and consider the properties of Yb^{3+}:KYW, i.e. I_{sat_abs} (980 nm) ≈ 1.4 kW/cm^2 and I_{sat} (1034 nm) ≈ 18 kW/cm^2 for the polarisation of interest. In addition, due to the lack of any specification, we make use of a major sizing assumption that γ_o = 10%. Looking at the involved optical configuration and combination of broadband reflectors, this should provide a realistic estimation. Provided β_{min} = 12%, following Eq. 11.6, $I_{min} \sim$ 170 W/cm^2, following Eq. 11.7, and reabsorption losses are in the range of 2%, total inversion yields k_{TH} = 1.23–1.27 and g_o ~ 6–7. With P_p ranging from 35 to 75 W, as specified, it can be shown that the net-gain coefficient G_o varies from about ~0.2 to 0.1. These low values evidence the need for a very large number of amplification passes, to ensure efficient energy extraction from the thin disk. Referring to Eq. 11.21, up to 300–500 passes are required. These conditions enable an overall net gain G up to 5.5 × 10^4 and E_{out} = up to 500 μJ. The interested reader will scan the parameters of interest in Eqs 11.20 and 11.21 to prove how critical the values of γ_o may be in this kind of design. He will verify that γ_o > 20% should correspond to the actual order of magnitude of the ultimate limit for possible lasing.

From a more generic viewpoint, in any situation, compromises must be made to optimise given data at the expense of the others, regarding the overall optical performance and system design. This may concern the energy or peak power per pulse, the average power and PRF, the beam quality, or

© Woodhead Publishing Limited, 2013

Table 11.3 Short-pulse sources: comparison of some representative tungstate-based designs with specified operating conditions (*), and referring to basic extrapolation by means of the sizing equations (**)

Selected design		High PRF, ultra-short pulse generation (first demonstration)	Ultra-low PRF, CPA-based, low-power, short-pulse amplification	High PRF, high-power, ultra-short pulse generation	Medium PRF, CPA-based, ultra-high-power, short-pulse amplification	Medium PRF, high-power, high net-gain by cascading, short-pulse generation
System design	Material/doping	KGW/5%	KYW/10%	KGW/1.5%	KYW/10%	KYW/5%
	Mode-locking	SPM/GDD CW cavity + SESAM	200 fs input pulses at 1053 nm	KLM CW cavity + SESAM	200 fs–9 nJ–50 MHz input pulses at 1034 nm	Cavity dumping + Herriot cell + SESAM
	Regenerative amplification		800 ps upstream stretching, downstream compression		31 ps upstream stretching, 12-pass/ round trip, 53 round trips max., downstream compression	
	GDD compensation	Pair prism		GTI		GTI
Laser head	Pump source	Free-space, two LD stripes, 2 × 3 W at 980 nm	Free-space, coupled and single LD stripes, 2 × 1.6 W at 940 nm, 1 × 20 W at 980 nm	Free-space, two LD bars, 2 × 23 W at 981 nm, symmetric	Fibred, single LD bar, 78 W at 981 nm	Fibred, two LD bars, 30 W at 980 nm (core 200 μm)
	Pump absorption	Axial, single-pass, bidirectional, 10°C thermal regulation, 3 mm absorption depth	Axial, single-pass and side, double-pass, 18°C thermal regulation, 4 mm absorption depth	Axial, single-pass, bidirectional, 1.4 mm absorption depth	24-pass, 24 × 100 μm absorption depth	Axial, single-pass, undirectional, 2 mm absorption depth
	Material geometry	Single, 3 mm-thick plate	Single, 4 mm-thick, long thin slab	Single, 1.4 × 10 × 10 mm^2, thick plate	Single, 7 mm-diameter 100 μm thin disk	Two cascaded, 4 × 1.5 × 2 mm^3 slabs
	Pump spot diameter (μm)		115 × 30 *	100–200 *	750 *	160 *

© Woodhead Publishing Limited, 2013

	I_p (kW/m^2)	**	**	**	8–19 *	**
	I_{sat_abs} (W/cm^2)				1400 **	
	g_o (cm^{-1})	**	**	**	6–7 **	**
Cavity	$R_{coupler}$ (%)	96–97 *	98 *	78–84 *		
	γ_o (%)	**	**	**	10 **	**
	k_{Th}	**	**	**	1.2–1.3 max. **	**
Output data	PRF (MHz)	86 *	0.001 *	45 *	0.02 *	1 *
	Wavelength (nm)	1032–1054 *	1040–1045 *	1039 *	1028–1034 *	1048 *
	Pulse width (fs)	112–224 *	400–460 *	134–433 *	185 *	416 *
	E_{out} per pulse (nJ)	7–14 **	40,000–65,000 *	200 *	500,000 *	7000 *
	P_{peak_pulse} (MW)	0.064 *	90–160 **	0.5–1.5 **	2700 **	12 *
	Average P_{output} (W)	0.18–0.82 *	0.04–0.065 **	10 *	10 *	7 *
Reference		[45]	[46]	[47]	[51]	[53]

© Woodhead Publishing Limited, 2013

even the robustness and compactness of the system. Table 11.3 also aims to identify these compromises and helps to point out the current trends in the field:

- Mode-locking by means of SESAMs or their derivatives is a technique in widespread use, in combination with dispersive prism pairs and GTIs to ensure proper GDD compensation.
- Tungstate-based amplifying laser heads generally involve axial pumping by means of single or multiple LD stripes, either free-space or fibred. The usual net gain being less than 2, rather low-loss cavities are required. Downstream regenerative amplification in combination with CPA-like techniques can provide nearly millijoule pulse energies.
- Referring to the usual data for basic cavities with a moderate pulse energy, i.e. some tens to some hundreds of nanojoules with a PRF in the range of 50–100 MHz, the decrease of this PRF below 1 Hz or less provides a good option to increase the pulse energy while taking advantage of a fairly simple and robust architecture.
- The attainable minimum pulse widths lie in the range of 100–200 fs, thanks to attainable $B_{FWHM} \approx$ 10–20 nm. This enables peak powers in excess of about 1 MW, from a basic CW cavity, or up to 10 MW by combining cavity-dumping and mode-locking, and more than 30 MW by implementing low-PRF input pulse clipping prior to regenerative amplification. A peak of more than 2 GW may be reached by adding CPA-like techniques.

Beyond the results presented here in the range λ_L = 1030–1054 nm, it is worth underlining that tuning capabilities from 1025 to 1060 nm have also been demonstrated with good optical performance.

11.5 Conclusion and future trends

This chapter was dedicated to the generic issues of interest in the understanding of ytterbium-based laser systems using YAG and KGW–KYW materials, which are widely used in the field of high-energy and high-power sources. The interest of basic equations for sizing purposes is underlined by means of a few representative examples from the literature, evidencing the critical contribution of the pumping-head design to the overall optical performance.

The pumping architecture may involve different options to be discussed against the output requirements and the system's operating conditions. These options include rods, slabs and thin disks, using either end- or side-pumping. Slab-based systems have been demonstrated up to pulse energies in the range of 0.5 to nearly 3 J in the nanosecond domain. Rods and slabs take advantage of quite flexible optical schemes and robust architectures. But nowadays, they need to compete with thin disks, which have been proven to be consistent

© Woodhead Publishing Limited, 2013

with higher average powers, at least up to energies below ~300 mJ. A number of demonstrations are available up to about 6 kW, together with 4 to 32 multiple-pass pumping schemes in order to overcome small-signal gain limitations and take advantage of reduced thermal-lensing effects.

Regenerative amplification is the most efficient process to get access to high-energy pulses, starting from a low-energy input. This mode of operation helps one to afford a low small-signal gain, as usually experienced in bulky Yb-based pumping-heads, while still enabling a very high net gain to be reached. A number of thin-disk, YAG or KYW-based amplifiers have been demonstrated in combination with upstream mode-locking, for the generation of short pulses at 100–800 fs at more than 20 μJ of energy, with 80–100 W average power and 1–20 kHz PRFs.

Future trends in the field of nanosecond sources imply a number of challenging routes. To combine high energy and high output power, the first aim refers to the competition between room-temperature multipass operation and cryogenic options. This aim consists of the optimisation of energy extraction, either affording a low small-signal gain at the expense of a rather complex optical design, or attempting to enhance it by means of a higher emission cross-section when lowering the temperature of the Yb-based material. The related issues, which are not the same as for short-pulse sources, essentially involve system considerations like robustness and compactness. A second direction is that of new Yb-doped crystals and ceramics. Ceramics take advantage of improved thermo-mechanical properties [55, 56] and of possible material combinations for the realisation of composite micro-structures [57]. This evidences some interest in improved pump confinement and the implementation of integrated optical functions. Representative examples may be the integration of thermally conductive bonding layers and scalable periodic optical structures. A third direction refers to OPA pump sub-systems, as parts of complex laser systems. This implies more specific needs, such as those of a good beam quality, of single-mode operation and temporal shaping.

In the field of short and ultra-short pulse generation, future trends also suggest a number of competing routes. Yb-based systems will be able to produce multi-millijoule sub-picosecond pulses in the near future, then opening the route to the development of fairly compact and efficient sources. The development of ultra-broadband new materials, either crystalline or ceramics, is the first direction. The search for highly disordered crystals or ceramics still involves active R&D, as it has done since the 2000s. Ceramics may offer the opportunity for previously inconceivable chemical compositions, thanks to solid-state sintering temperatures prior to phase transitions in the materials. A second direction concerns rod-like, large mode-area fibres [58]. They provide intermediate solutions with respect to bulky systems, of interest for overcoming their thermal limitations due to optical-guidance capabilities

© Woodhead Publishing Limited, 2013

and large thermal-exchange surfaces. When stacked using beam-combining techniques, such as Talbot designs, they may lead to new architectures of interest in the search for intermediate energies at a high PRF. As a third direction, it is worth underlining the combination of ultra-high PRFs, in the GHz domain, and of high-power passive mode-locking techniques for industrial needs. This will benefit, for example, applications which involve scanning needs. Other directions are towards spectral-shaping capabilities in preamplifiers dedicated to cascaded laser architectures, for the search for very high pulse energies and ultra-high powers. This exhibits some interest, for example, in coupling applications to titanium–sapphire systems or to optical-parametric-CPA (OPCPA) techniques. Likewise, these complex configurations not only require properly shaped ultra-broadband sources, but also make use of high-energy nanosecond pumps, mainly YAG-based nowadays, of the same kind as those previously evoked.

11.6 References

[1] J. Lopez *et al.*, 'Ultrafast laser with high energy and high average power for industrial micromachining: comparison ps–fs', LMF Session 4, *Proc. ICALEO* (2011)

[2] W. Horn, 'High power diode lasers for industrial applications', Paper 502, *Proc. ICALEO* (2007)

[3] M.E. Fremann *et al.* (ed), *Ultrafast Lasers, Technology and Applications*, Marcel Dekker, New York (2001)

[4] W. Krupke, 'DPSSLs: Status and prospects for materials processing', preprint UCRL-JC-135913, *3rd Symposium on Advanced Photon Processing and Measurement Technologies*, Tokyo, Japan, 7–8 November 1999

[5] T.Y. Fan, 'Diode-pumped solid state lasers', *Lincoln Laboratory Journal*, 3(3), 413–425 (1990)

[6] W. Koechner, 'Solid state laser engineering', in *Optical Sciences*, 5th edn, Springer (1999)

[7] B. Faircloth, 'High-brightness high-power fiber coupled diode laser system for material processing and laser pumping', invited paper, in *High-Power Diode Laser Technology and Applications*, M. S. Zediker (ed.), *Proc. SPIE*, Vol. 4973 (2003)

[8] A. Courjaud, 'Sources laser femtosecondes pompées par diode basées sur l'ion ytterbium', thesis THE 036064 01, Université Bordeaux 1 (2001)

[9] P. Klopp, 'New Yb^{3+}-doped laser materials and their application in continuous-wave and mode-locked lasers', Dissertation-thesis, Humboldt-Universität, Berlin (2006)

[10] H. Coïc, 'Analytic modelling of high-gain ytterbium-doped fibre amplifiers', *J. Pure Appl. Opt.*, 4, 120–129 (2002)

[11] J.F. Nye, *Physical Properties of Crystals*, Clarendon Press, Oxford (1985)

[12] T.Y. Fan, 'Heat generation in Nd:YAG and Yb:YAG', *IEEE J. Quantum Electron.*, 29(6), June 1993

[13] Y.F. Chen *et al.*, 'Optimization in scaling fiber-coupled laser-diode end-pumped

© Woodhead Publishing Limited, 2013

lasers to higher power: Influence of thermal effect', *IEEE J. Quantum Electron.*, 33(8), August 1997

[14] R. Gaume *et al.*, 'A simple model for the prediction of thermal conductivity in pure and doped insulating crystals', *Appl. Phys. Lett.*, 83(7), August 2003

[15] S. Chesnay, 'Nouveaux matériaux laser dopés à l'ytterbium: performances en pompage par diode et étude des effets thermiques', thesis, Université Paris Sud (2002)

[16] S. Biswal *et al.*, 'Thermo-optical parameters measured in ytterbium-doped potassium gadolinium tungstate', *Appl. Opt.*, 44, 3093–3097 (2005)

[17] D. Stučinskas, 'Thermal lens diagnostics and mitigation in diode end pumped lasers', Doctoral dissertation, Vilnius University (2010)

[18] M. Behringer, High-Power Diode Laser Technology and Characteristics, in *Springer Series in Optical Sciences*, 128, 5–74 (2007)

[19] S. Heinemann and L. Leininger, 'Fiber coupled diode lasers and beam-shaped high-power stacks', invited paper, *Proc. SPIE*, 3267 • 0277-786X/98

[20] H.G. Treusch *et al.*, 'Fiber-coupling technique for high-power diode laser arrays', *Proc. SPIE*, 3267 • 0277-786X/98

[21] A. Giesen *et al.*, 'Scalable concept for diode-pumped high-power solid-state lasers', *Appl. Phys. B*, 58 (1994)

[22] A. Jolly and E. Artigaut, 'Theorical design for the optimization of a material's geometry in diode-pumped high-energy Yb^{3+}:YAG lasers and its experimental validation at 0.5–1 J', *Appl. Opt*, 43(32), 6016–6022 (2004)

[23] E. Innerhofer *et al.*, '60-W average power in 810-fs pulses from a thin disk Yb:YAG laser', *Opt. Lett.*, 28(5), 367–369 (2003)

[24] J. Neuhaus *et al.*, 'Subpicosecond thin disk laser oscillator with pulse energies of up to 25.9 microjoules by use of an active multipass geometry', *Opt. Express*, 16, 20530–20539 (2008)

[25] D. Kouznetsov *et al.*, 'Scaling laws of disk lasers', *Opt. Mater.*, 31, 754–759 (2009)

[26] D. Albach *et al.*, 'Influence of ASE on the gain distribution in large size, high gain Yb^{3+}:YAG slabs', *Opt. Express*, 17(5), 3792 (2009)

[27] L. Tarassov, 'Physique des processus dans les générateurs de rayonnement optique cohérent', in *Russian to French Translation from the Original Book by V. Kolimeev*, MIR, Moscow (1981)

[28] A. Hofer *et al.*, 'Fully analytical simulation of Q-switched lasers', *Laser Phys. Lett.*, 1(6), 282–284 (2004)

[29] H. Liu *et al.*, 'Directly diode-pumped millijoule subpicosecond Yb:glass regenerative amplifier', *Opt. Lett.*, 24(13), 917–919 (1999)

[30] C. Rullière (ed), *Femtosecond Laser Pulses – Principles and Experiments*, 2nd edn, Springer (2004)

[31] R. Paschotta and U. Keller, 'Passively mode-locked solid-state lasers', in *Solid-State Lasers and Applications, Optical Science and Engineering*, A. Sennaroglu (ed.), CRC Press, Taylor & Francis (2007)

[32] C. Honninger *et al.*, 'Diode-pumped thin disk Yb:YAG regenerative amplifier', *Appl. Phys. B*, 65, 423 (1997)

[33] T. Südmeyer *et al.*, 'High-power ultrafast thin disk laser oscillators and their potential for sub-100-femtosecond pulse generation', *Appl. Phys. B*, 97, 281–295 (2009)

[34] J. Dong and P. Deng, 'Temperature dependent emission cross-section and fluorescence lifetime of Cr, Yb: YAG crystals', *J. Phys. Chem. Sol.*, 64, 1163–1171 (2003)

© Woodhead Publishing Limited, 2013

[35] J. Dong and M. Bass, 'Dependence of the Yb^{3+} emission cross section and lifetime on temperature and concentration in yttrium aluminum garnet', *J. Opt. Soc. Am. B*, 20(9), September 2003

[36] F. Butze *et al.*, 'Nanosecond-pulsed thin disk Yb:YAG lasers', *Proc. ASSPP Int. Conf. OSA* (2004)

[37] A. Jolly *et al.*, 'Generation of variable width pulses from an Yb^{3+}:YAG Integrated Dumper – Regenerative Amplifier', *Opt. Express*, 15(2), 466 (2007)

[38] M. Siebold *et al.*, 'High-energy, diode-pumped, nanosecond Yb:YAG MOPA system', *Opt. Express*, 16(6), 3674 (2008)

[39] S. Tokita *et al.*, 'Efficient high-average-power operation of Q-switched cryogenic Yb:YAG laser oscillator', *Jap. J. Appl. Phys.*, 44(50), 1529–1531 (2005)

[40] J. Kawanaka *et al.*, '42-mJ Q-switched active-mirror laser oscillator with a cryogenic Yb:YAG ceramics', paper, *ASSP 2007*

[41] J.G. Manni *et al.*, '100-W Q-switched cryogenically cooled Yb:YAG laser', *IEEE J. Quantum Electron.*, 46(1) (2010)

[42] O. Musset and J. P. Boquillon, 'Flashlamp-pumped Nd:KGW laser at repetition rates up to 50 Hz', *Appl. Phys. B*, 65, 13–18 (1997)

[43] M.C. Pujol *et al.*, 'Crystalline structure and optical spectroscopy of Er^{3+}-doped $KGd(WO_4)_2$ single crystals', *Appl. Phys. B*, 68, 187–197 (1999)

[44] L. Tang *et al.*, 'Phase diagram, growth and spectral characteristic of Yb^{3+}:$KY(WO_4)_2$ crystal', *J. Crystal Growth*, 282, 376–382 (2005)

[45] F. Brunner *et al.*, 'Diode-pumped femtosecond Yb:$KGd(WO_4)_2$ laser with 1.1-W average power', *Opt. Lett.*, 25(15) (2000)

[46] H. Liu *et al.*, 'Directly diode-pumped Yb:$KY(WO_4)_2$ regenerative amplifiers', *Opt. Lett.* 27(9) May 2002

[47] F. Brunner *et al.*, '240-fs pulses with 22-W average power from a mode-locked thin-disk Yb:$KY(WO_4)_2$ laser', *Opt. Lett.*, 27(13), 1162–1164 (2002)

[48] D. Nickel *et al.*, 'Ultrafast thin-disk Yb:$KY(WO_4)_2$ regenerative amplifier with a 200-kHz repetition rate', *Opt. Lett.*, 29(23), December 2004

[49] J.H. Hellstrom *et al.*, 'Laser performance and thermal lensing in high-power diode-pumped Yb:KGW with athermal orientation', *Appl. Phys. B*, 83, 55–59 (2006)

[50] G.R. Holtom, 'Mode-locked Yb:KGW laser longitudinally pumped by polarization-coupled diode bars', *Opt. Lett.*, 31(18), 2719–2721 (2006)

[51] K. Ogawa *et al.*, 'Multi-millijoule, diode-pumped, chirped-pulse Yb:$KY(WO_4)_2$ regenerative amplifier', *Opt. Express*, 15(14), 8598 (2007)

[52] U. Buenting *et al.*, 'Regenerative thin disk amplifier with combined gain spectra producing 500 μJ sub 200 fs pulses', *Opt. Express*, 17, 8046–8050 (2009)

[53] G. Palmer *et al.*, '12 MW peak power from a two-crystal Yb:KYW chirped-pulse oscillator with cavity-dumping', *Opt. Express*, 18(18) 19095 (2010)

[54] D.N. Papadopoulos *et al.*, 'Broadband high-energy diode-pumped Yb:KYW multipass amplifier', *Opt. Lett.*, 36(19), 3816–3818 (2011)

[55] T. Taira, 'Recent advances in crystal optics/Avancées récentes en optique cristalline – Ceramic YAG lasers', *C. R. Physique*, 8, 138–152 (2007)

[56] V. Cardinali, 'Matériaux lasers dopés à l'ion ytterbium: Performances lasers en pompage par diodes lasers et étude des propriétés thermo-optiques à des températures cryogeniques', thesis, Ecole Polytechnique-Paristech (2011)

[57] K. Ueda *et al.*, 'Scalable ceramic lasers', *Laser Physics*, 15(7), 927–938 (2005)

[58] L. Lago *et al.*, 'High-energy temporally shaped nanosecond-pulse master-oscillator power amplifier based on ytterbium-doped single-mode microstructured flexible fiber', *Opt. Lett.*, 36(5), 734–736 (2011)

© Woodhead Publishing Limited, 2013

12
Neodymium doped lithium yttrium fluoride ($Nd:YLiF_4$) lasers

N. U. WETTER, Centro de Lasers e Aplicações – IPEN/SP-CNEN, Brazil

DOI: 10.1533/9780857097507.2.323

Abstract: As a laser material, the neodymium doped lithium yttrium fluoride laser shows some very important and favorable characteristics when it comes to laser beam quality, efficient high-energy pulsed operation and parametric processes, amongst other applications. In this chapter we will discuss several approaches that have resulted in efficient, high-quality laser beams usually through decreasing the influence of the structural drawbacks of the YLF host whilst taking advantage of its favorable optical properties.

Key words: laser material, solid-state lasers, rare earth lasers, neodymium lasers, diode-pumped lasers, laser resonators.

12.1 Introduction

Since the first construction of a neodymium doped lithium yttrium fluoride laser ($Nd:YLiF_4$ or more briefly Nd:YLF) in 1981, its main advantages with respect to the already well-known and widespread oxide laser Nd:YAG (yttrium aluminum garnet, $Y_3Al_5O_{12}$) have been recognized (Pollak *et al.*, 1982). These are large upper state lifetime, weak thermal lensing and natural birefringence (Vanherzeele, 1988). Thermal conductivity, absorption and emission cross-section are roughly half those of the YAG host at similar doping level (Ryan and Beach, 1992; Pfistner *et al.*, 1994). The birefringence causes two emission lines, one at 1047 nm and another at 1053 nm that matches the wavelength transition of Nd^{3+} glass amplifiers commonly used in high pulse energy systems and laser fusion facilities. The 1047 nm emission is obtained for light polarized parallel to the crystal *c*-axis (π-polarization) and the 1053 nm emission for light polarized perpendicular to the *c*-axis (σ-polarization). Both transitions originate from the $^4F_{3/2}$ upper laser level and terminate at the second Stark splitting of the $^4I_{11/2}$ lower laser level (see Fig. 12.1). Nd:YLF may be efficiently diode pumped around 800 nm presenting two main absorptions at 792 nm and 797 nm and a roughly 50% smaller absorption at 806 nm. The 797 nm peak absorption coefficient is higher for π-polarization than for σ-polarization and of the order of 5–10 cm^{-1} for 1 mol% of neodymium doping.

The large upper state lifetime of Nd:YLF measures more than half a

© Woodhead Publishing Limited, 2013

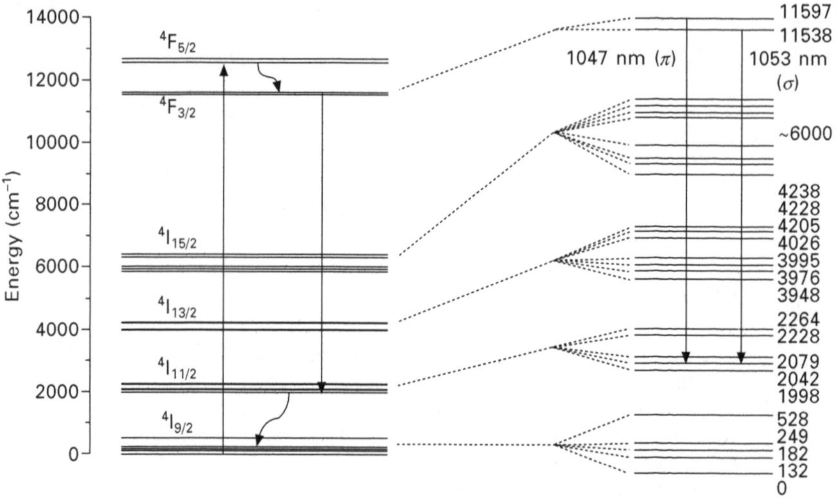

12.1 Energy level diagram of Nd:YLF (Kaminskii, 1996).

millisecond, showing potential for very large energy storage which is good for diode pumping, high peak powers and large pulse energies. However, it has been observed that under Q-switched operation, especially at low repetition rates, the system performance deteriorates (Beach *et al.*, 1993). This behavior has been attributed to quenching of the upper laser level lifetime. The strong yellow fluorescence caused by upconversion under non-lasing operation is easily observed by the naked eye and a measure for this quenching process indicating a bigger heat load, which is followed by a series of generally unwanted consequences (Fan *et al.*, 1986). Heat generation during the non-lasing transitions in Nd:YLF lasers pumped by powerful high-brightness diodes is approximately twice as high as during lasing and 30% higher than in Nd:YAG under similar pumping conditions. This is the downside of a long upper laser level lifetime in Nd:YLF: non-linear processes originating from the upper laser level become stronger (Pollnau *et al.*, 1998).

The increased heat load inside the crystal is accompanied by an adverse decrease in thermal conductivity that causes an altogether much higher temperature at the center of the pump face, thus strongly increasing stress and strain in this area. If the stress is above the sample's fracture limit, damage to the crystal might occur during instances of non-lasing operation. When compared to YAG, YLF has an approximately five times lower fracture limit. In fact it is so small that in many cases fracture in Nd:YLF has been observed simultaneously with the occurrence of thermal lensing, which makes it difficult to experimentally foresee this catastrophic damage (Bollig *et al.*, 2010).

The weak lensing in YLF can be observed under lasing conditions when

© Woodhead Publishing Limited, 2013

upconversion loss is minimized and is a consequence of two simultaneously occurring effects: the negative refractive index change with increasing temperature and the positive lens created by end-face bulging of the laser crystal. Both contributions tend to cancel each other but the former contribution is stronger in the π-polarization, causing an altogether negative thermal lens, whereas for the σ-polarization the contribution of the negative index change is less, causing a small positive lens (Cerullo *et al.*, 1992). It is this characteristic of YLF that has brought a renewed interest to this laser material, because with today's high brightness diodes the focal length of the thermal lens generated inside the crystal rapidly becomes less than the cavity length and may turn the resonator unstable. Especially in longitudinally pumped laser designs, the onset of resonator instability puts a limit to the maximum achievable output power. Therefore, thermal lensing normally sets the upper limit in terms of absorbed pump power for high-power, diode-pumped solid-state laser systems.

The low fracture limit and thermal lensing of Nd:YLF call for efficient cooling and other measures to decrease heat load and resultant strain and stress in the host material. This can be done by using a low cooling temperature to increase thermal conductivity and also crystal geometries with large cooling surface to sample volume ratios, such as long and thin rods or thin slabs. Additionally, the thermal gradient inside the rod and the resultant stress and thermal lens can be decreased by keeping the dopant concentration low to decrease upconversion and by detuning of the pump wavelength from the absorption peak, or by using larger pump spot sizes to dilute the absorption over a larger volume. However, these measures are only effective whenever high pump intensity is not an issue, such as in Q-switched laser operation.

12.2 Pumping methods of Nd:YLF lasers

12.2.1 Lamp-pumped Nd:YLF lasers

If the pump light is deposited over a relatively large volume, such as in lamp-pumped lasers, the heat gradient is automatically smaller and relatively high pulse energies and continuous (cw) powers can be obtained but efficiency and spatial beam quality are limited. A series of cavity designs have been employed that partly overcome these efficiency and spatial beam quality drawbacks of lamp-pumped lasers (Hanna *et al.*, 1981). A relatively simple method to achieve fundamental mode operation in lamp-pumped laser rods is the mode-filling technique (Magni, 1986). By choosing appropriate resonator mirrors and distances between mirrors and rod, the cross-section of the fundamental mode inside the rod can be increased to a size where diffraction losses at the borders of the rod become too large for the next higher transverse mode to oscillate. Nd:YLF is especially suited for this method

© Woodhead Publishing Limited, 2013

because of its natural birefringence and permits fundamental mode operation in standard, one-quarter-inch diameter rods; meanwhile, thermally induced birefringence limits the use of this technique in YAG rods to diameters of a couple of millimeters (Cerullo *et al.*, 1993; Wetter *et al.*, 1993). Forty watts of continuous fundamental mode output power at 1047 nm and 35 W for 1053 nm have been obtained for 7 kW of lamp pump power using this technique together with a 6.35 millimeter diameter Nd:YLF rod (Cerullo *et al.*, 1992). Eventually, for very large fundamental-mode diameters, the resonator becomes unstable as a function of TEM_{00} diameter fluctuations, a problem that can be mitigated by joining the stability zones using, for example, resonator arms of the same length, as shown in Fig. 12.2 (Cerullo *et al.*, 1993; Wetter *et al.*, 2008a).

As pointed out in the introduction, lamp-pumped Nd:YLF lasers are used for laser applications that need high power, high energy and good beam quality. One famous example is the OMEGA facility at the University of Rochester's Laboratory for Laser Energetics (LLE). One of the many amplifier stages used to achieve the petawatt powers and kilojoule pulse energies necessary to drive the laser fusion and high-energy-density physics experiments is an optical parametric chirped pulse amplifier (OPCPA) whose final amplification stage consists of a four-pass ring Nd:YLF laser using two flashlamp-pumped, one-inch-diameter rods. This front end produces 2.4 ns pulses of 2 J at 5 Hz (Kelly *et al.*, 2006) before the pulses get further amplified by a chain of Nd:glass amplifiers.

12.2.2 End-pumped Nd:YLF lasers

When efficiency and spatial beam quality are an issue, as in most of today's laser designs, then high pump intensity is necessary and the range of possible cavity setups is limited. High pump intensity can be achieved under diode pumping using high brightness diodes (Wetter, 2001). For example, a 50 W diode bar coupled to a 100 μm fiber of 0.22 NA that is focused into a 300 μm focal spot size has a confocal length of four millimeters and a pump intensity of 70 kW/cm^2. The small dimensions of the round pump-spot size inside the host can then be matched to the fundamental mode inside the resonator, allowing for high beam quality. This end-pumping scheme is by far the most used cavity setup for efficient and high beam-quality lasers.

Pump saturation is normally not a problem at these high pump intensities, because as soon as the lasing threshold is overcome, laser operation clamps the upper laser level population so that upon increasing the pump level the effective upper laser level lifetime decreases and the active ions can be cycled ever more quickly, increasing thereby the lasers' output power. This, in principle, permits one to decrease pump and laser mode size within the active media, permitting lower threshold and better slope efficiency as

© Woodhead Publishing Limited, 2013

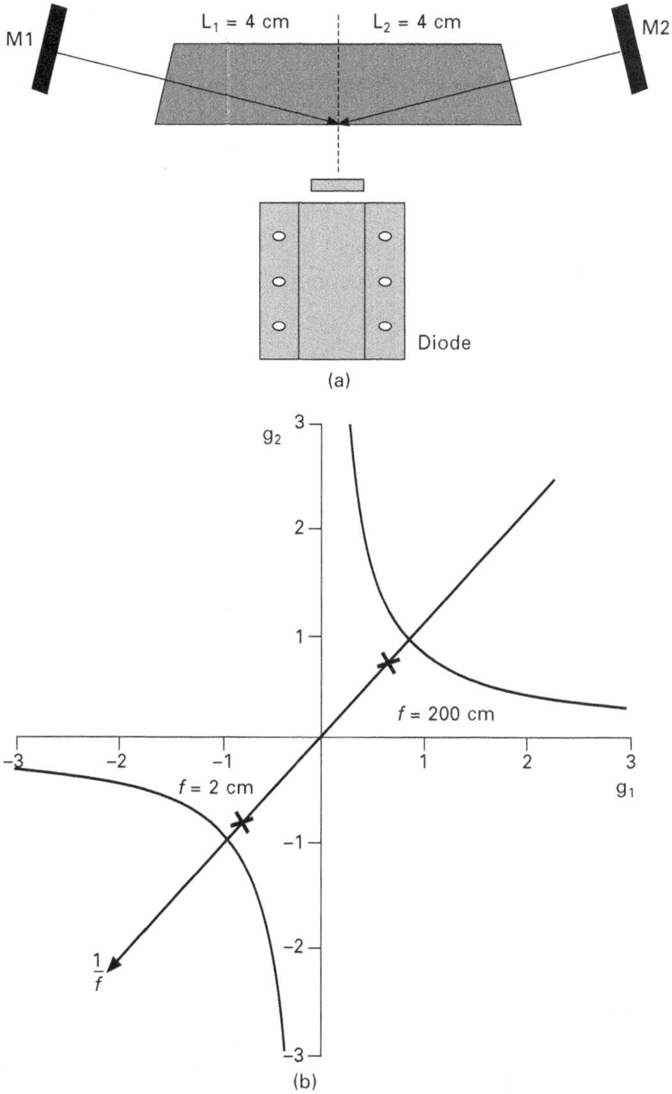

12.2 Example of joint stability zones in a side-pumped laser. (a) Simple high-efficiency resonator with equal distance between crystal and mirror M1 and M2, respectively. (b) Joined stability zones of this resonator, permitting stable operation with a thermal lens ranging from 200 cm to 2 cm as a function of pump power (Wetter *et al.*, 2008a).

long as good overlap between both beams is still guaranteed. Although the thermally induced gradient does increase with smaller pump beam waists, thermal lensing is not necessarily increased because the smaller laser beam waists are less affected by this gradient. A turning point is achieved when

© Woodhead Publishing Limited, 2013

the beam quality of the pump beam starts to decrease the overlap efficiency between pump and laser inside the absorption region. This problem can be circumvented in part if the laser is operated much above threshold, because in such a case one may depart from the optimum relation between laser mode to pump mode of 1 to 1 and choose a smaller pump mode such that the ratio becomes 2:1 at the center of the gain media, in which case good efficiency is still guaranteed for most practical pump setups. For even higher ratios, thermally induced diffraction loss becomes a problem because the thermal gradient departs from a parabolic shape across the laser beam's cross-section. It should be noted, however, that for high-gain lasers, which are operated several times above threshold, the benefit of smaller pump waists results generally only in a minor gain in efficiency even if pump overlap can be maintained. Additionally, the higher thermal gradient increases the probability of thermally induced catastrophic crystal fracture, which is particularly low in Nd:YLF (fracture limit of 40 MPa). In fact, the common 'roll-over' effect in the input–output power curve, associated with a thermal lens that is strong enough to turn the laser cavity unstable, is hardly visible with Nd:YLF crystals and most often its onset occurs practically together with the crystal's fracture. As a result, most diode end-pumped neodymium doped lasers oscillating at the fundamental wavelength operate generally only a few times above threshold, using beam waists between 200 μm and 400 μm.

The first demonstration of an end-pumped Nd:YLF laser in 1986 used a 30 mW GaAlAs single stripe diode and achieved already 38% slope efficiency at 1047 nm during cw operation (Fan *et al.*, 1986). For higher pump powers the width of the diode stripe has to be increased and the pump beam gets increasingly elliptical. It is still possible to achieve very high efficiency if these elliptical beams are correctly matched to the cavity mode. Using cylindrical cavity mirrors, an ellipticity factor of 12 has been achieved in a Brewster cut Nd:YLF crystal, matching the cavity mode to the pump beam of the 500 μm wide diode emitter (Zehetner, 1995). The authors achieved in excess of 1 W of fundamental mode cw output power at 1047 nm with 71% slope efficiency and 41.8% optical-to-optical efficiency (see Fig. 12.3).

For even higher output power, it is no longer possible to increase the width of the stripe for cooling reasons, and several stripes, generally between 10 and 60, are grown in parallel on a 1 cm wide diode array, also called a diode bar. The ellipticity of this emitter is 1:10,000 and only rather complex devices can transform this pump beam into a beam with approximately equal dimensions and quality factors in the horizontal and vertical directions (Wetter, 2001). For this reason it is generally preferred to use fiber-coupled diode bars. At one watt of output power in fundamental mode, an optical-to-optical efficiency of 50% with respect to absorbed pump power has been achieved for the 1047 nm emission (Lue *et al.*, 2010b) when pumped at 806 nm.

© Woodhead Publishing Limited, 2013

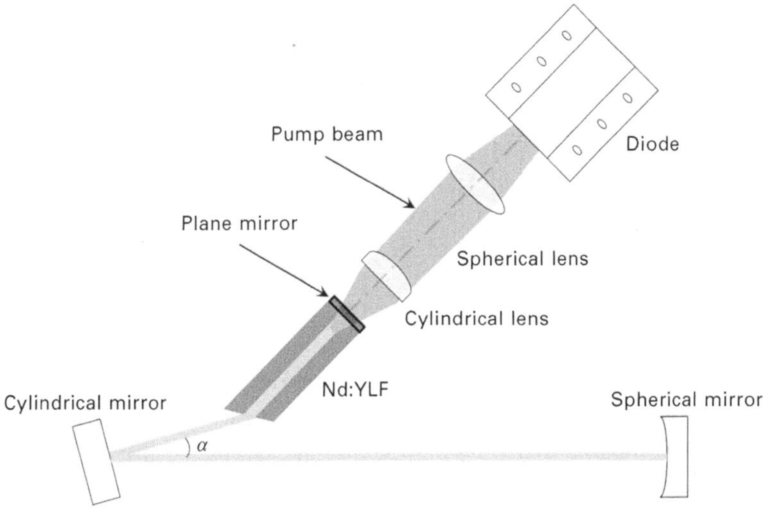

12.3 End-pumped laser cavity scheme. This folded cavity uses a cylindrical mirror in order to match laser and pump beam inside the Nd:YLF crystal (Zehetner, 1995).

An interesting scheme for high-power diode pumping is the doubly end-pumped tightly folded cavity. In this setup the gain media are pumped from both ends, preferably with fiber-coupled diodes for symmetrical beam quality. In order to place the pump-focusing optics close to the gain crystal, the cavity is folded by two mirrors that are placed close to the end-faces. Using two 100 W diodes, coupled into 600 μm diameter fibers, an optical-to-optical efficiency of 40% has been obtained for quasi-cw emission at 1053 nm (Babushkin and Seka, 1998). Given the pump beam quality factor which generally is several tens of times worse than the fundamental mode, the overlap that can be achieved in the axial direction is usually limited to no more than 1 cm. This confocal length, which equals twice the Rayleigh length, has to be enough to absorb a considerable fraction of the pump power and achieve good absorption efficiency. Therefore, the doping concentration should be enough so that the confocal range equals about three absorption lengths, which results in 95% of absorbed pump power. As a consequence, there is a large pump-induced heat load that has to be removed. By far the most common way to cool end-pumped lasers is by cooling the side faces or the barrel surface of the crystal using a rectangular or cylindrical geometry, respectively. This edge-cooling method is responsible for a radial temperature gradient that in turn causes a thermal lens which in most cases can be approximated by a parabolic profile at the center of the pump spot (Clarkson, 2001). The fundamental mode size should be less than or at most equal to the pump spot size, as outside the pump spot the lens profile departs

© Woodhead Publishing Limited, 2013

from the parabolic shape and therefore causes aberration losses (Pfistner *et al.*, 1994).

For higher output powers, the fundamental mode size inside the crystal should be increased, eventually accompanied by lower doping concentration and detuning of the pump wavelength. A careful balance of these measures resulted in 60 W cw output power at 1053 nm, 44% optical- to-optical efficiency with respect to absorbed pump power (38% overall optical efficiency) and good beam quality without observing lifetime quenching during Q-switched operation (Bollig *et al.*, 2010). The authors used a doping level of 0.5% of Nd^{3+} and the pump wavelength and spot size were 806 nm and 1 mm, respectively. Using an acousto-optic modulator, Q-switched operation was achieved with repetition rates ranging from 5 kHz (pulse energy of 10.4 mJ, duration of 76 ns, 52 W of average output power) to 30 kHz.

The relatively small thermal conductivity and the small tensile strength of Nd:YLF limit the maximum output power that can be achieved with end-pumped rod designs because increase in pump power absorption means stronger thermal gradients. Increasing further the pump spot size permits higher output powers but at the expense of beam quality and laser efficiency (Beach *et al.*, 1993), given the relatively low pump absorption and small stimulated emission cross-section. In order to increase pump power absorption density, a geometry has to be employed that increases the pump volume but at the same time keeps the distance from the pump spot to the cooled surface small. This can be achieved with thin, end-pumped slabs. Using a diode stack that was focused into a 0.4 mm × 12 mm pump spot size inside two 1 mm thin slabs, 127 W of cw multimode output power at 1047 nm was achieved inside a stable cavity (Li *et al.*, 2007). Using an unstable cavity, shown in Fig. 12.4, the same authors achieved almost diffraction-limited cw output of 74 W with 37% optical-to-optical efficiency (Li *et al.*, 2008).

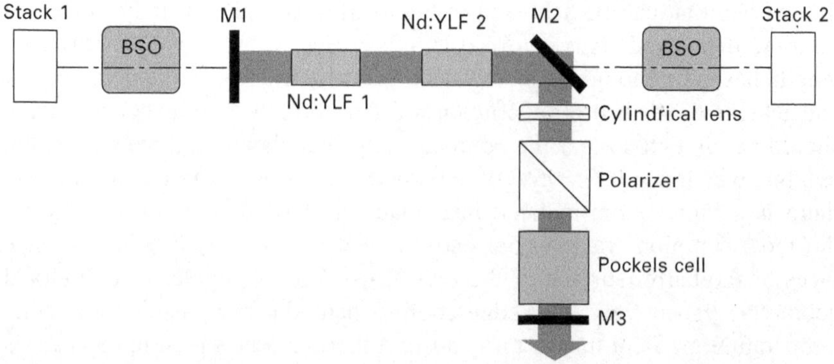

12.4 Unstable folded resonator design with diffraction-limited output employing two Nd:YLF crystals pumped by two diode stacks and beam-shaping optics (BSO) (Li *et al.*, 2008).

© Woodhead Publishing Limited, 2013

The same authors tested electro-optic (EO) Q-switched operation and achieved 39% optical-to-optical efficiency at 10 kHz repetition rate for the fundamental wavelength of 1047 nm. The highest multimode pulse energy and duration were 25.4 mJ and 5.9 ns at 1 kHz repetition rate, whilst diffraction-limited pulse energy and duration were only slightly less, 24.2 mJ and 7 ns, respectively. Optical-to-optical efficiency at 5 kHz and 1 kHz were approximately 29% and 12% respectively.

End-pumping is also traditionally used for ultra-short pulse generation with mode-locked Nd:YLF lasers that tends to give slightly shorter pulse duration because of its broader linewidth of 1.45 nm when compared to Nd:YAG. Using an antiresonant semiconductor Fabry–Pérot saturable absorber (SESAM), 3.3 ps duration pulses with 700 mW of average output power for 2 W of pump power at 220 MHz repetition rate have been achieved (Keller *et al.*, 1992). Even broader linewidths of up to 1.9 nm are possible with mixed host crystals, as for example Nd:LuYLF that contains a mixture of 50% lutetium and 50% yttrium, generating shorter pulses than with pure Nd:YLF (Maldonado *et al.*, 2001).

12.2.3 Direct pumping of Nd:YLF lasers

The energy difference from the $^4F_{5/2}$ upper pump level in Nd:YLF, populated when using standard 805 nm diodes, to the upper laser level is 995 cm^{-1} and the energy difference from the lower $^4I_{11/2}$ laser level to the ground level is 1997 cm^{-1} (Pollnau *et al.*, 1998). Directly pumping into the $^4F_{3/2}$ emitting level therefore causes a 33% reduction in heat load caused by pump absorption. Three possible pump wavelengths are adequate for this purpose, 863 nm, 872 nm and 880 nm, corresponding to the first, second and third Stark splitting of the ground level, respectively (Lue *et al.*, 2010b). Sometimes this is also referred to as 'thermally boosted pumping' when the pump absorption transition originates from the second or third Stark level (see Fig. 12.1). However, pump absorption at these wavelengths is small, of the order of 1–3 cm^{-1} and therefore laser-to-pump beam overlap efficiency is not optimal and a large amount of pump power usually passes through the crystal without being absorbed. The latter effect can be mitigated by using clever arrangements that reimage the pump beam into the crystal, thereby achieving a second pass of the pump beam through the crystal. A higher doping level beyond 1 mol% is not an option because it causes increasing scattering losses and decrease of upper level lifetime. A slope efficiency of 76% (optical-to-optical efficiency of approximately 66%) at 1047 nm and 0.8 watt of cw output power was obtained in a small linear cavity containing a 8 mm long Nd:YLF crystal pumped at 880 nm (Lue *et al.*, 2010b). 9.5 W of cw output power in fundamental mode were obtained at 1053 nm with 71% slope efficiency and 63% of optical efficiency with

© Woodhead Publishing Limited, 2013

respect to absorbed pump power using a 10 mm long crystal pumped also at 880 nm in a V-shaped cavity using a polarization preserving pump setup (Schulz and Kracht, 2009). In both cases the output power was limited by the available pump power and not by onset of thermal effects. As a last resource, one can combine direct upper laser level pumping with a pump source of excellent beam quality such as a Ti:sapphire laser. This permits the use of long crystals that well absorb the pump power with good pump-to-laser beam overlap. This scheme has been employed in a simple linear cavity by Zhang *et al.* (2011). They achieved a slope efficiency of almost 64%, limited mainly by the short Nd:YLF crystal of only 8 mm. So far the maximum cw output power of commercial Ti:sapphire lasers is about 10 W, posing a limit to the maximum output power that can be achieved using this technique. Additionally, at this output power level, diode pumping is economically much more viable.

12.2.4 Side-pumping of Nd:YLF lasers

High-power diode lasers generally come in the form of a linear array of emitters and are diffraction limited perpendicular to the array and highly multimode in the direction of the array. This makes it very simple to pump lasers laterally with diode bars by placing them along the intracavity beam path inside the gain medium. In this configuration the diode bar generates a thin gain sheet inside the YLF crystal that is passed once or several times by the intracavity fundamental mode in the perpendicular direction. One of the major concerns with side pumping is the pump-to-laser beam overlap, given the relatively low pump absorption of Nd:YLF. It is therefore of interest to have large intracavity fundamental modes. Alternatively, one may reflect the pump beam several times back to the location of the laser mode as in cylindrical pump cavities. A large amount of work has been done on multipass-slab resonators that use a combination of these methods (Dergachev *et al.*, 2007). In this setup, a gain crystal of length 28 mm and cross-section 6 mm × 2 mm ($W \times H$) is pumped by two laterally offset fast-axis collimated diode bars. The pump radiation is reflected once from the facet opposite to the pump face and the fundamental beam undergoes five passes through the gain sheet. At up to 64 W of cw pump power the authors obtained 20 W and 30 W of output power at 1053 nm and 1047 nm, corresponding to an optical efficiency of 31% and 43%, respectively (Snell *et al.*, 2000). Using the same technology, further improvement was achieved by direct pumping into the upper laser level using a wavelength of 863 nm, resulting in cw multimode operation at 47% optical efficiency and TEM_{00} operation with 37% efficiency at 1047 nm (Pati and Rines, 2009).

The same group investigated pulsed operation with Nd:YLF in an oscillator–amplifier system that consisted of a single-frequency master

© Woodhead Publishing Limited, 2013

oscillator delivering 1 mJ and 10 ns pulses that were subsequently amplified by three pre-amplifiers (using the same design outlined above) and one main amplifier to generate 55 mJ at 1047 nm with 1 kHz repetition rate (Isyanova and Moulton, 2007). Their main amplifier used a diode stack with a total of 285 W of pump power.

An optical-to optical efficiency of 47% in fundamental mode with 16.7 W of the peak output power during quasi-cw operation has been achieved at the wavelength of 1053 nm by a pumping method that can be envisioned as a mix of side- and longitudinal pumping (Wetter *et al.*, 2009; Deana and Wetter, 2012). The design makes use of the high inversion density created by π-polarized absorption and located in a shallow region near the pumped surface as with end-pumping and takes advantage of the 1053 nm transition to benefit from the weaker thermal lens. The fundamental mode enters the quadratically shaped slab of 13 mm side-length and 3 mm height at the Brewster angle and undergoes total internal reflection at 56° at the pump facet (see Fig. 12.5). The gain sheet is then double-passed by the fundamental beam in a controlled manner to ensure high beam quality (Wetter *et al.*, 2008b).

Good efficiency in Q-switched operation has been obtained in a novel setup that used a 3 mm diameter rod cut in half and polished along the optical axis (Pati *et al.*, 2008) with a cross-section that is a half-circle. The 792 nm pump wavelength is incident upon the AR-coated cylindrical surface and reflected back by the HR-coated large plane surface which also serves for conduction cooling. Using a Cr:YAG saturable absorber, the authors achieved close to diffraction-limited 7 mJ pulses at 1053 nm with 17% of optical efficiency and less than 10 nanosecond pulses.

A good example of power scalability has been presented by Hirano *et al.* (2000). A conduction-cooled, 78 mm long Nd:YLF rod was embedded in a MgF_2 confinement cavity of approximately triangular cross-section and pumped by 18 five-bar cw stacks with 168 W of average power. The large lateral surfaces of this high thermal conductivity pump chamber were HR coated and served as heat sink whilst the flattened ridges were AR coated for the pump radiation. A total output power of 72 W with an M^2 value of 8 and 43% optical efficiency were measured at 1053 nm.

Lamp-pumped Nd:YLF amplifiers used in laser fusion and high-energy-density experiments have been substituted in many facilities by diode-pumped amplifiers because of the stringent requirements in terms of energy and intensity stability that cannot be met by flashlamps. One such example is the photo-injector of the linear collider nuclear test facility at the European organization for nuclear research (CERN) (Petrarca *et al.*, 2011). The photo-injector produces first 8 ps pulses of 6.6 nJ at a repetition rate of 1.5 GHz with 10 W of average power using a commercial Nd:YLF MOPA system. A 400 μs long burst of pulses then gets amplified in two Nd:YLF diode-

© Woodhead Publishing Limited, 2013

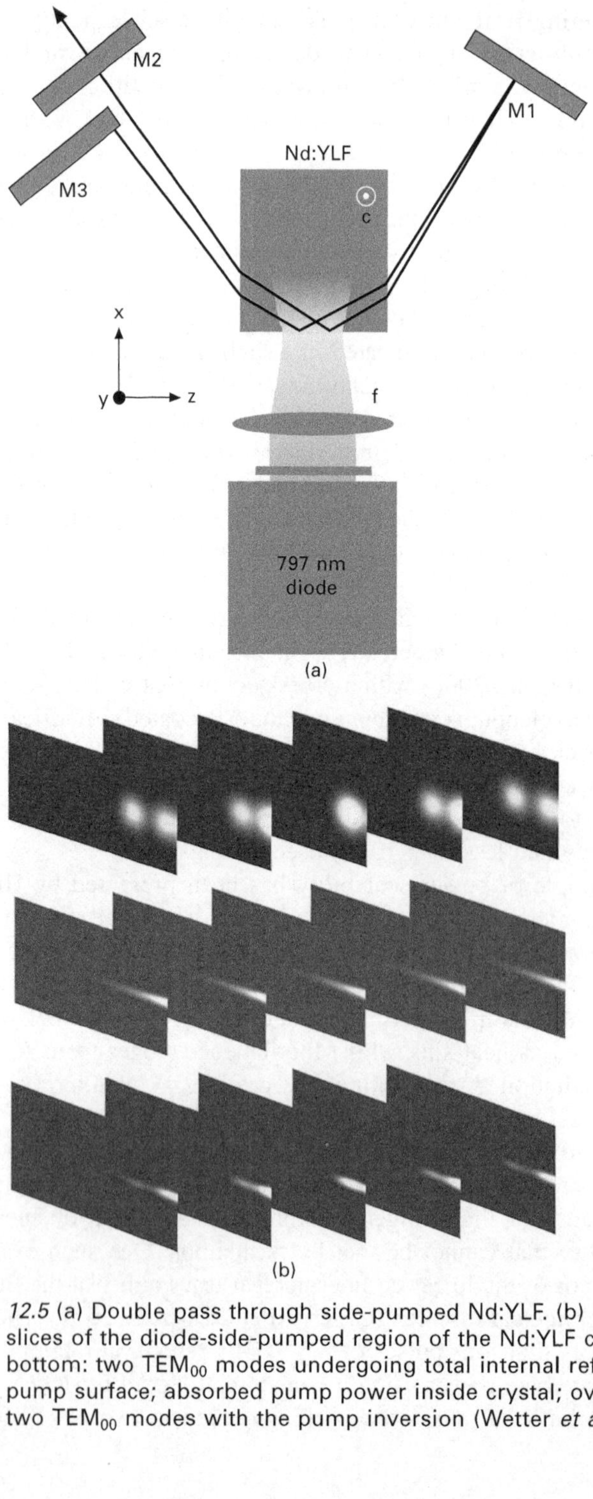

12.5 (a) Double pass through side-pumped Nd:YLF. (b) Five *xy*-slices of the diode-side-pumped region of the Nd:YLF crystal. Top to bottom: two TEM_{00} modes undergoing total internal reflection at the pump surface; absorbed pump power inside crystal; overlap of the two TEM_{00} modes with the pump inversion (Wetter *et al.*, 2009).

© Woodhead Publishing Limited, 2013

side-pumped amplifier stages with 5 Hz repetition rate (Ross *et al.*, 2003). At pump powers of 15 kW and 17.8 kW in the first and second amplifier stages, the pulses get amplified to 5.5 μJ. Another such system that employs a Nd:YLF amplifier in its front end is the HALNA (High Average power Laser for Nuclear fusion Application) system at Osaka University in Japan (Yasuhara *et al.*, 2008). The diode-pumped Nd:YLF regenerative ring double amplifier system increases the output energy from 13 nJ to 380 mJ in eight round-trips (Sekine *et al.*, 2007).

12.3 Alternative laser transitions

12.3.1 Laser transitions to the ground level

Amongst the most commonly used neodymium doped laser hosts such as YAG, YLF, YVO_4 and $GdVO_4$, the energy of the $^4F_{3/2}$ level is highest for Nd:YLF. Consequently, transitions to the uppermost Stark level of the $^4I_{9/2}$ ground state have the shortest wavelength in Nd:YLF, specifically 908 nm and 903 nm for the σ- and π-transition, respectively. This is of interest for second harmonic generation (SHG) to the deep blue region (Jonas Jakutis *et al.*, 2010). Using direct pumping into the emitting level and a special mirror coating with high transmission at 880 nm and high reflection at 908 nm, 4.7 W of cw output power at 11.8 W of absorbed pump power were achieved (Liang *et al.*, 2011). Given the lower emission cross-section at 903 nm, this emission is much harder to obtain (Spiekermann and Laurell, 2000).

The natural birefringence and the small thermal lens of Nd:YLF are of paramount importance whenever non-linear parametric processes are required, such as SHG or SFG (sum-frequency generation), and have resulted in a strong revival of Nd:YLF as a laser material. By reimaging the 35 W fiber-coupled 880 nm pump beam into a 15 mm long Nd:YLF crystal inside a Z-cavity, 4.3 W of cw blue output power at 454 nm have been obtained using SHG in a 10 mm long LBO crystal (Lue *et al.*, 2010a).

12.3.2 Laser transitions to the $^4I_{13/2}$ energy level

Upon emission into the $^4I_{13/2}$ energy level, Nd:YLF generates two strong lines at 1313 nm (σ-polarization) and 1321 nm (π-polarization). Both lines have been efficiently brought to lasing action in fundamental mode by using direct end-pumping into the emitting level and a crystal whose laser facet opposite to the pump facet was HR coated for the pump wavelength to achieve better pump absorption. An optical efficiency of 49% and 3.6 W of cw output power has been achieved at 1321 nm (Lue *et al.*, 2010c). Using a c-cut 16 mm long crystal, an optical efficiency of 30% with an output power of 3.1 W has been obtained (Li *et al.*, 2011). Both wavelengths can

© Woodhead Publishing Limited, 2013

be efficiently frequency doubled into the red spectral region as has been demonstrated using tunable single-frequency ring cavities. More than one watt of single-frequency output power in a diffraction limited mode has been obtained at 657 nm and 660 nm using a periodically poled $KTiOPO_4$ crystal for SHG inside a ring cavity (Zondy *et al.*, 2010; Sarrouf *et al.*, 2007). Pump wavelength was off-center at 806 nm and the doping level of the Nd:YLF crystal was only 0.7 at% to dilute the absorption and keep upconversion low. Unidirectional operation was achieved using a Brewster-cut TGG rod and a zero-order half-wave plate (see Fig. 12.6). Fine tuning of the wavelength was obtained with a thin fused intracavity etalon.

These ring-cavities permit tunable, mode-hop free operation up to the maximum output power because of their absence of spatial hole-burning and are therefore ideally suited for cooling transitions of atomic species in clocks such as silver, calcium and lithium atoms (Camargo *et al.*, 2010).

12.4 Future trends

As has been pointed out in this chapter, almost all major drawbacks of Nd:YLF with respect to other neodymium doped hosts can be overcome if the heat is properly removed. Use of long and thin Nd:YLF rods with low doping levels is a solution but efficient pumping is still limited by incomplete pump overlap or expensive pump sources with low output powers such as Ti:sapphire. Recently a very high pump power of the order of 12 W has been achieved with tapered diodes in a MOPA (Master Oscillator Power Amplifier) configuration at almost diffraction-limited beam quality (Sumpf *et al.*, 2009). If these devices become commercial then this should prove a route to efficient Nd:YLF lasers with output powers achieved so far only

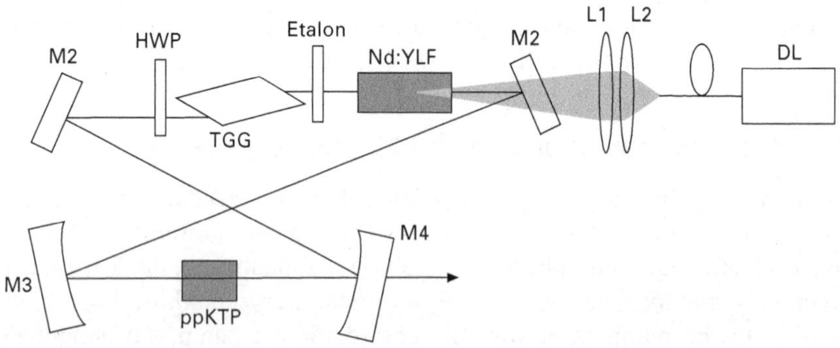

12.6 Single-frequency, tunable ring-cavity for efficient SHG conversion to the red (Camargo *et al.*, 2009). M1–M4 are the ring-cavity mirrors, HWP is the half-wave-plate that together with the TGG makes the optical diode, L1 and L2 are the two doublet lenses that focus the fiber-coupled diode laser (DL) into the Nd:YLF crystal.

© Woodhead Publishing Limited, 2013

by oxide hosts whilst maintaining the excellent beam quality inherent to this material.

12.5 References

Babushkin, K. & Seka, W. 1998. Efficient 1053-nm Nd:YLF laser end pumped by a 100-W quasi-cw diode array. *Lasers and Electro-Optics, 1998, CLEO 98*. Technical Digest. Summaries of papers presented at the Conference on 3–8 May 1998, pp. 180–181.

Beach, R., Reichert, P., Benett, W., Freitas, B., Mitchell, S., Velsko, A., Davin, J. & Solarz, R. 1993. Scalable diode-end-pumping technology applied to a 100-MJ Q-switched Nd^{3+}-YLF laser-oscillator. *Optics Letters*, 18, 1326–1328.

Bollig, C., Jacobs, C., Esser, M. J. D., Bernhardi, E. H. & Von Bergmann, H. M. 2010. Power and energy scaling of a diode-end-pumped Nd:YLF laser through gain optimization. *Optics Express*, 18, 13993–14003.

Camargo, F., Zanon-willette, T., Sarrouf, R., Badr, T., Wetter, N. U. & Zondy, J. J. 2009. 1.3 Watt single-frequency Nd:YLF/ppKTP red laser. *2009 Conference on Lasers and Electro-Optics and Quantum Electronics and Laser Science Conference (Cleo/Qels 2009)*, Vols 1–5, 1241–1242.

Camargo, F. A., Zanon-willette, T., Badr, T., Wetter, N. U. & Zondy, J. J. 2010. Tunable single-frequency Nd:YVO(4)BiB/(3)O(6) ring laser at 671 nm. *IEEE Journal of Quantum Electronics*, 46, 804–809.

Cerullo, G., Desilvestri, S. & Magni, V. 1992. High-efficiency, 40-W cw NdYLF laser with large TEM_{00} mode. *Optics Communications*, 93, 77–81.

Cerullo, G., Desilvestri, S., Magni, V. & Svelto, O. 1993. Output power limitations in cw single transverse-mode Nd:YAG lasers with a rod of large cross-section. *Optical and Quantum Electronics*, 25, 489–500.

Clarkson, W. A. 2001. Thermal effects and their mitigation in end-pumped solid-state lasers. *Journal of Physics D – Applied Physics*, 34, 2381–2395.

Deana, A. M. & Wetter, N. U. 2012. High-efficiency, Q-switched and diffraction-limited Nd:YLF side-pumped laser. *In*: Graf T. *et al.* (ed.), *Laser Sources and Application Proc SPIE*, Vol. 8433, 84330B.

Dergachev, A., Flint, J. H., Isyanova, Y., Pati, B., Slobodtchikov, E. V., Wall, K. F. & Moulton, P. F. 2007. Review of multipass slab laser systems. *IEEE Journal of Selected Topics in Quantum Electronics*, 13, 647–660.

Fan, T. Y., Dixon, G. J. & Byer, R. L. 1986. Efficient GaAlAs diode-laser-pumped operation of Nd:YLF at 1.047-μm with intracavity doubling to 523.6 nm. *Optics Letters*, 11, 204–206.

Hanna, D. C., Sawyers, C. G. & Yuratich, M. A. 1981. Telescopic resonators for large-volume TEM_{00} mode operation. *Optical and Quantum Electronics*, 13, 493–507.

Hirano, Y., Yanagisawa, T., Ueno, S., Tajime, T., Uchino, O., Nagai, T. & Nagasawa, C. 2000. All-solid-state high-power conduction-cooled Nd:YLF rod laser. *Optics Letters*, 25, 1168–1170.

Isyanova, Y. & Moulton, P. F. 2007. Single-frequency, 55 W average power, 1-kHz pulse rate Nd:YLF MOPA system. *CLEO2007, Conference on Lasers and Electro-Optics/Quantum Electronics and Laser Science Conference and Photonic Applications Systems Technologies*. Optical Society of America, CTuD7.

Jonas Jakutis, N., Fabíola, A. C. & Niklaus Ursus, W. 2010. Deep blue Nd:$LiYF_4$ laser in quasi-continuous and continous operation. Optical Society of America, ATuA12.

© Woodhead Publishing Limited, 2013

Kaminskii, A. A. 1996. *Crystalline Lasers: Physical Processes and Operating Schemes*. Boca Raton, FL, CRC Press.

Keller, U., Miller, D. A. B., Boyd, G. D., Chiu, T. H., Ferguson, J. F. & Asom, M. T. 1992. Solid-state low-loss intracavity saturable absorber for Nd:YLF lasers – an antiresonant semiconductor Fabry-Pérot saturable absorber. *Optics Letters*, 17, 505–507.

Kelly, J. H., Waxer, L. J., Bagnoud, V., Begishev, I. A., Bromage, J., Kruschwitz, B. E., Kessler, T. J., Loucks, S. J., Maywar, D. N., McCrory, R. L., Meyerhofer, D. D., Morse, S. F. B., Oliver, J. B., Rigatti, A. L., Schmid, A. W., Stoeckl, C., Dalton, S., Folnsbee, L., Guardalben, M. J., Jungquist, R., Puth, J., Shoup, M. J., Weiner, D. & Zuegel, J. D. 2006. OMEGA EP: High-energy petawatt capability for the OMEGA laser facility. *Journal de Physique IV*, 133, 75–80.

Li, C. L., Zhang, X. H., Liang, W. & Zhao, Z. M. 2011. Diode-pumped continuous-wave Nd:YLF laser at 1313 nm. *Laser Physics*, 21, 340–342.

Li, D., Ma, Z., Haas, R., Schell, A., Simon, J., Diart, R., Shi, P., Hu, P. X., Loosen, P. & Du, K. M. 2007. Diode-pumped efficient slab laser with two Nd:YLF crystals and second-harmonic generation by slab LBO. *Optics Letters*, 32, 1272–1274.

Li, D., Ma, Z., Haas, R., Schell, A., Zhu, P., Shi, P. & Du, K. 2008. Diode-end-pumped double Nd:YLF slab laser with high energy, short pulse width, and diffraction-limited quality. *Optics Letters*, 33, 1708–1710.

Liang, W., Zhang, X. H., Liang, Z. L., Liu, Y. Q. & Liang, Z. 2011. Efficient continuous-wave 908 nm Nd:YLF laser emission under direct 880 nm pumping. *Laser Physics*, 21, 320–322.

Lue, Y., Zhang, X., Cheng, W. & Xia, J. 2010a. All-solid-state cw frequency-doubling Nd:$YLiF_4$/LBO blue laser with 4.33 W output power at 454 nm under in-band diode pumping at 880 nm. *Applied Optics*, 49, 4096–4099.

Lue, Y.-F., Zhang, X.-H., Zhang, A.-F., Yin, X.-D. & Xia, J. 2010b. Efficient 1047 nm CW laser emission of Nd:YLF under direct pumping into the emitting level. *Optics Communications*, 283, 1877–1879.

Lue, Y. F., Xia, J., Zhang, X. H., Zhang, A. F., Wang, J. G., Bao, L. & Yin, X. D. 2010c. High-efficiency direct-pumped Nd:YLF laser operating at 1321 nm. *Applied Physics B – Lasers and Optics*, 98, 305–309.

Magni, V. 1986. Resonators for solid-state lasers with large-volume fundamental mode and high alignment stability. *Applied Optics*, 25, 107–117.

Maldonado, E. P., Barbosa, E. A., Wetter, N. U., Courrol, L. C., Ranieri, I. M., Morato, S. P. & Vieria, N. D. 2001. Mode-locking operation of Nd:LuYLF. *Optical Engineering*, 40, 1573–1578.

Pati, B. & Rines, G. A. 2009. Direct-pumped Nd:YLF laser. *Advanced Solid-State Photonics*, Optical Society of America.

Pati, B., Wall, K. F., Isyanova, Y. & Moulton, P. F. 2008. Passively Q-switched Nd:YLF laser in a D-rod configuration. *2008 Conference on Lasers and Electro-Optics and Quantum Electronics and Laser Science Conference*, Vols 1–9, 120–121.

Petrarca, M., Martyanov, M., Divall, M. C. & Luchinin, G. 2011. Study of the powerful Nd:YLF laser amplifiers for the CTF3 photoinjectors. *IEEE Journal of Quantum Electronics*, 47, 306–313.

Pfistner, C., Weber, R., Weber, H. P., Merazzi, S. & Gruber, R. 1994. Thermal beam distortions in end-pumped Nd:YAG, Nd:GSGG, and Nd:YLF rods. *IEEE Journal of Quantum Electronics*, 30, 1605–1615.

Pollak, T. M., Wing, W. F., Grasso, R. J., Chicklis, E. P. & Jenssen, H. P. 1982. Cw laser operation of Nd:YLF. *IEEE Journal of Quantum Electronics*, 18, 159–163.

© Woodhead Publishing Limited, 2013

Pollnau, M., Hardman, P. J., Kern, M. A., Clarkson, W. A. & Hanna, D. C. 1998. Upconversion-induced heat generation and thermal lensing in Nd:YLF and Nd:YAG. *Physical Review B*, 58, 16076–16092.

Ross, I. N., Csatari, M. & Hutchins, S. 2003. High-performance diode-pumped Nd:YLF amplifier. *Applied Optics*, 42, 1040–1047.

Ryan, J. R. & Beach, R. 1992. Optical-absorption and stimulated-emission of neodymium in yttrium lithium-fluoride. *Journal of the Optical Society of America B – Optical Physics*, 9, 1883–1887.

Sarrouf, R., Sousa, V., Badr, T., Xu, G. & Zondy, J.-J. 2007. Watt-level single-frequency tunable Nd:YLF/periodically poled KTiOPO$_4$ red laser. *Optics Letters*, 32, 2732–2734.

Schulz, B. & Kracht, D. 2009. Nd:YLF laser pumped at 880 nm. *Advanced Solid-State Photonics*, Optical Society of America.

Sekine, T., Matsuoka, S., Kawashima, T., Kan, H., Kawanaka, J., Tsubakimoto, K., Nakatsuka, M. & Izawa, Y. 2007. High order wavefront correction for high-energy Nd:YLF rod amplifier by phase conjugate plate. *CLEO2007, Conference on Lasers and Electro-Optics/Quantum Electronics and Laser Science Conference and Photonic Applications Systems Technologies*. Optical Society of America, CFA3.

Snell, K. J., Lee, D., Wall, K. F. & Moulton, P. F. 2000. Diode-pumped, high-power CW and modelocked Nd:YLF lasers. *Advanced Solid State Lasers (ASSL)*, Davos, Switzerland: Optical Society of America.

Spiekermann, S. & Laurell, F. 2000. Continuous wave and Q-switched operation of diode pumped quasi-three level Nd:YLF lasers. *Advanced Solid State Lasers, Proceedings*, 34, 60–62.

Sumpf, B., Hasler, K.-H., Adamiec, P., Bugge, F., Dittmar, F., Fricke, J., Wenzel, H., Zorn, M., Erbert, G. & Traenkle, G. 2009. High-brightness quantum well tapered lasers. *IEEE Journal of Selected Topics in Quantum Electronics*, 15, 1009–1020.

Vanherzeele, H. 1988. Thermal lensing measurement and compensation in a continuous-wave mode-locked Nd:YLF laser. *Optics Letters*, 13, 369–371.

Wetter, N. U. 2001. Three-fold effective brightness increase of laser diode bar emission by assessment and correction of diode array curvature. *Optics and Laser Technology*, 33, 181–187.

Wetter, N. U., Maldonado, E. P. & Vieira, N. D. 1993. Enhanced efficiency of a continuous-wave mode-locked Nd:YAG laser by compensation of the thermally-induced, polarization-dependent bifocal lens. *Applied Optics*, 32, 5280–5284.

Wetter, N. U., Camargo, F. A. & Sousa, E. C. 2008a. Mode-controlling in a 7.5 cm long, transversally pumped, high power Nd:YVO$_4$ laser. *Journal of Optics A – Pure and Applied Optics*, 10.

Wetter, N. U., Sousa, E. C., Camargo, F. D., Ranieri, I. M. & Baldochi, S. L. 2008b. Efficient and compact diode-side-pumped Nd:YLF laser operating at 1053 nm with high beam quality. *Journal of Optics A – Pure and Applied Optics*, 10.

Wetter, N. U., Sousa, E. C., Ranieri, I. M. & Baldochi, S. L. 2009. Compact, diode-side-pumped Nd^{3+}:YLiF$_4$ laser at 1053 nm with 45% efficiency and diffraction-limited quality by mode controlling. *Optics Letters*, 34, 292–294.

Yasuhara, R., Kawashima, T., Sekine, T., Kurita, T., Ikegawa, T., Matsumoto, O., Miyamoto, M., Kan, H., Yoshida, H., Kawanaka, J., Nakatsuka, M., Miyanaga, N., Izawa, Y. & Kanabe, T. 2008. 213 W average power of 2.4 GW pulsed thermally controlled Nd:glass zigzag slab laser with a stimulated Brillouin scattering mirror. *Optics Letters*, 33, 1711–1713.

© Woodhead Publishing Limited, 2013

Zehetner, J. 1995. Highly efficient diode-pumped elliptic mode Nd:YLF laser. *Optics Communications*, 117, 273–276.

Zhang, F. D., Zhang, X. H., Liang, W. & Li, C. L. 2011. Thermally-boosted pumping of Nd:LiYF(4) using Ti:sapphire laser. *Laser Physics*, 21, 639–642.

Zondy, J. J., Camargo, F. A., Zanon, T., Petrov, V. & Wetter, N. U. 2010. Observation of strong cascaded Kerr-lens dynamics in an optimally-coupled cw intracavity frequency-doubled Nd:YLF ring laser. *Optics Express*, 18, 4796–4815.

© Woodhead Publishing Limited, 2013

13
Erbium (Er) glass lasers

B. I. DENKER, B. I. GALAGAN and S. E. SVERCHKOV, A. M. Prokhorov General Physics Institute, Russian Academy of Sciences, Russia

DOI: 10.1533/9780857097507.2.341

Abstract: The interest in Yb-sensitized Er-activated glass lasers is mostly associated with the eye-safety of their wavelength (λ ~1.53–1.56 µm) corresponding to $^4I_{13/2} - {}^4I_{15/2}$ Er^{3+} ion transition. The chapter first discusses the design of flashlamp pumped Er:glass lasers. Then the typical configurations of diode pumped Er lasers are considered. Means of Q-switching for Er glass lasers and applications of Er glass lasers are reviewed. In conclusion erbium glass lasers are compared with crystalline ones utilizing the same Er^{3+} transition and with Raman lasers.

Key words: eyesafe lasers, Co:spinel, phosphate laser glass, Yb:Er crystal.

13.1 Introduction

The attractive properties of ~1.54 micron Er:glass laser radiation are excellent transparency of the earth's atmosphere and most common optical materials in combination with the presence of sensitive room-temperature Ge and InGaAs photodetectors. But its most noticeable feature is its relative eye-safety. The maximal permissible exposure of pulsed 1.54-micron laser radiation hitting the naked human eye is five orders of magnitude higher than that for ~1.06-micron neodymium lasers. According to ANSI Z136-1-2007, single-pulse Maximum Permissible Exposure (MPE_{SP}) for nanosecond pulses at 1.5–1.8 µm is 1 J/cm^2 vs. 5×10^{-6} J/cm^2 at 1 µm.

Outstanding eye-safety is associated with the appropriately high (tens of cm^{-1}) extinction of liquid water at ~1.5 microns. Laser radiation of this wavelength is totally absorbed by the transparent eye tissues and cannot reach the retina. In contrast to high absorption in liquid water, there is practically no absorption in water vapor (and moist air). It is also well known that the Er:glass laser wavelength coincides well with the spectral minimum of losses in silica waveguides, though this chapter is devoted to non-fiber configurations of Er:glass lasers. Fiber erbium lasers are discussed in Chapter 15. In non-fiber configurations weak and narrow Er^{3+} ion absorption bands in combination with their three-level lasing scheme in glasses inevitably require sensitization by Yb^{3+} ions (see the energy level scheme in Fig. 13.1). Yb^{3+} sensitizing ions absorb the near infrared (900–1000 nm) pump radiation

© Woodhead Publishing Limited, 2013

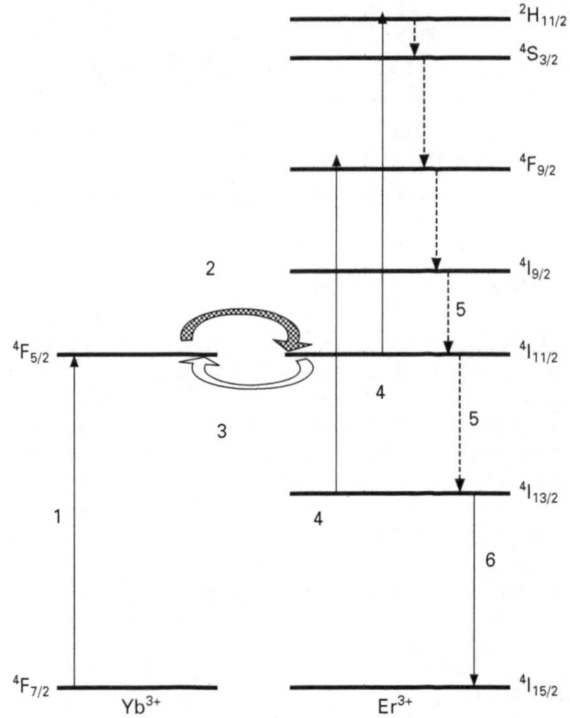

13.1 The principal energy level scheme and energy transformation processes in Yb, Er doped media: 1 optical pumping into Yb absorption band; 2, 3 direct and back Yb-Er energy transfer; 4 up-conversion losses; 5 (→) multiphonon relaxation; 6 lasing.

provided by flashlamps or laser diodes (LDs) and transfer the energy to the lasing Er^{3+} ions. So the main subject of this chapter is bulk Yb–Er glass lasers. Some information about these glasses can be found in Chapter 5 of this book. In the present chapter we just give some minimal information for the proper choice of glass for specific types of lasers.

The 1.54-micron lasing effect in Yb and Er co-doped glass was first observed by E. Snitzer and R. Woodcock in 1965.[1] Nevertheless for a long time the efficiency of Yb–Er glass lasers was below any practical requirements. The fundamental studies reported by Gapontsev *et al.*[2] and Alekseev *et al.*[3] showed the unique properties of phosphate glass as a host for efficient Yb–Er lasers. As already mentioned in Chapter 5, phosphate glass up till now still remains the only host that combines long lifetime (~7 ms) of the $^4I_{13/2}$ Er^{3+} upper laser level with the low (2–3 μs) lifetime of the $^4I_{11/2}$ Er^{3+} level that is in resonance with the Yb^{3+} $^2F_{5/2}$ excited state. Rapid $^4I_{11/2} \rightarrow {}^4I_{13/2}$ multiphonon relaxation is required to prevent reverse energy transfer to Yb ions. The peculiarities of the Yb^{3+} and Er^{3+} laser ion system led to the unusual situation that until now there have been no commercially available crystalline Yb–Er laser materials. Many years of continuous searching for such a host has brought only limited success. Some crystalline Yb–Er, resonantly pumped Er lasers and Raman lasers will also be briefly mentioned in this chapter as alternative sources of 1.5-micron laser radiation. Considerable progress in flashlamp pumped Er glass lasers in the mid 1980s was associated with the suggestion of Cr^{3+} ions as a second sensitizer[4] and the development of the first commercially available Cr–Yb–Er laser glasses. Cr^{3+} (and sometimes small amounts of Nd^{3+}) sensitizing ions utilize the visible-range flashlamp light and transfer the energy to Yb^{3+} ions. It should be mentioned that the presence of chromium and neodymium in Yb–Er laser glasses can cause a number of parasitic effects and for this reason their presence in all diode pumped lasers (and in certain cases in flashlamp pumped lasers also) should be avoided.

The obvious drawback of glass in comparison to crystalline laser hosts is its low thermal conductivity, limiting the laser average power. Despite this drawback Yb–Er and Cr–Yb–Er glass lasers remain the most simple and widespread source of coherent radiation in the 'eyesafe' spectral window near 1.5 μm. Their advantages are as follows:

- The possibility of operating in quite different regimes ranging from continuous wave (CW) to high-energy pulses with duration ranging from milliseconds to femtoseconds
- The possibility of being efficiently pumped either by xenon flash lamps or by InGaAs laser diodes (LDs).

In this chapter we shall focus our attention on the energy possibilities and operation modes of Yb–Er glass lasers of different configurations. We shall start with the design of flashlamp pumped lasers. As for diode pumped lasers, their configurations can be roughly divided into two quite different groups: transverse and longitudinal diode pumped lasers. Transversely pumped lasers are typically powered by pulsed laser diode arrays and stacks. Longitudinally pumped lasers are typically tiny microchips powered by a single or a few CW diodes. After that we will review Q-switching means for Er glass lasers. The next part of this chapter will be devoted to applications of Er glass lasers. In conclusion we shall show the advantages and the drawbacks of erbium glass lasers in comparison to Yb–Er doped crystalline lasers, resonantly pumped Er doped crystal lasers and Raman lasers.

13.2 Flashlamp pumped erbium (Er) glass lasers

Flashlamp pumped operation may seem an anachronism now, but in the case of Er:glass lasers it remains the simplest and the cheapest way to obtain high energy (tens of joules in the free-run regime and joules in the Q-switch

© Woodhead Publishing Limited, 2013

regime[5,6]) 1.54 μm pulses. The free-run efficiencies of flashlamp pumped Cr–Yb–Er glass lasers can reach several percent[4,5,7,8] while a threshold value can be rather low. For example, a 5 J threshold laser with 1.8% slope efficiency was demonstrated.[9] This laser had a 2.5 × 35 mm rod of LGS glass from the Institute of Radio Engineering and Electronics, Moscow, Russia. As was already mentioned, the average power of an Er glass laser is limited by its low thermal conductivity, causing easy thermal fracture of the glass. For many years, laser glass developers and manufacturers tried to enhance this value. The record high average output power (up to 20 W from a ∅5 × 152 mm laser rod[10]) was obtained using a commercially available QX-Er glass from Kigre, Inc. (USA) with surface strengthening by ion exchange. Nevertheless the typical average power of free-running Er glass lasers lies within the range of 0.3–0.5 W per 1 cm of active rod length. For this reason reaching the threshold of CW lasing in arc lamp pumped Er glass lasers is very difficult. We know only one successful example[11] where CW laser action was obtained using a Xe-filled arc lamp and a thin (2.7 mm) Yb–Er glass rod with minimized (10^{19} cm^{-3}) Er content.

The design of Er glass flashlamp lasers has some peculiarities. Ignoring the following few simple recommendations may result in dramatic efficiency losses:

- Er glass is a pure three-level laser medium. For this reason practically all the glass active element should be pumped. The non-pumped zones are a source of laser light absorption losses. In the case of the traditional rod configuration the non-pumped rod ends should be as short as possible.
- The optical excitation of the Yb–Er glasses (especially chromium-free ones) takes place mostly via Yb ions. The only Yb absorption band lies in the near infrared spectral range (900–1000 nm). Water extinction in this spectral range has the order of 1 cm^{-1}. For this reason the water-cooled laser efficiency can be drastically decreased due to pump light absorption in the water layer surrounding the flashlamp and the laser rod. The possible problem solutions are (1) a pump cavity design with narrow (~1 mm) gaps for water flow around the laser rod and the flashlamp, or (2) using a coolant with low extinction in the near IR. There are some suitable special organic liquids that, however, may suffer from low thermal stability or low photostability. An absolutely stable coolant with excellent transparency in the Yb absorption band is deuterized water (D_2O).
- Yb concentration in all Yb–Er laser glasses for flashlamp pumping is well optimized and lies within a narrow range of (1.5–2.5) × 10^{21} ions per cm^3. Fixed Yb content determines a rather narrow reasonable range for the flashlamp pumped laser rod thickness: 2–10 mm.
- Efficient Xe lamp operation in the infrared (in the Yb absorption band)

© Woodhead Publishing Limited, 2013

requires low plasma temperature and thus low current density. For this reason the optimal pump pulse duration in Yb–Er glass lasers is several times longer than that usual for Nd lasers. Typical values vary from ~1 ms to 5 ms. Millisecond-long pump pulse durations are also well matched with the long (~7 ms) upper laser level lifetime and the relatively slow (~1 ms) Yb→Er energy transfer rate.

- The efficiency of flashlamp pumped Cr–Yb–Er glasses is typically 2–4 times higher than that of Cr-free (Yb–Er and Nd–Yb–Er) glasses. But the room-temperature Cr–Yb energy transfer quantum yield does not exceed 60%. Higher heat dissipation in Cr-containing glass lasers results in their easier thermal fracture and thus ~1.5 times lower attainable average power.
- The efficiency of Cr-free (Yb–Er and Nd–Yb–Er) glasses can be considered to be independent of temperature in a wide temperature range from ~ –40°C or lower up to the glass softening temperature (~400°C). In contrast to this, the efficiency of chromium sensitization rapidly decreases with temperature.[12] Cr–Yb–Er glasses stop lasing typically at 100–150°C. On the other hand, low temperatures should increase the Cr–Yb–Er glass laser efficiency.

13.3 Laser diode (LD) pumped erbium (Er) glass lasers

All the possible (non-waveguide) configurations of diode pumped Yb–Er glass lasers can be divided into two quite different main groups: transversely and longitudinally pumped lasers.

The transversely pumped lasers are to some extent an updated analog of flashlamp pumped ones. All of them utilize the pulsed (pulse duration in the millisecond range) operating regime of the pumping laser diodes. High pump power (100 W at least) requires using not single diodes but single or multiple laser arrays or array stacks. The active element of a cylindrical (or sometimes rectangular) rod shape has a thickness measured in millimeters and a length measured in centimeters. In contrast, the longitudinally pumped lasers are much more compact. Typically they are excited by a single (or a few) laser diodes and the pumping area diameter is measured in tens (or a few hundreds) of microns. The small diameter of the pumping area makes it possible to lower the lasing threshold and to obtain a continuous wave (CW) regime easily. The length of the gain medium is determined by the pump light penetration depth, which is dependent on Yb concentration and on the exact pump wavelength. Typically it lies within the 0.5–5 mm range.

In principle, any Er glass initially designed for flashlamp excitation (preferably with no Cr or Nd additives) and with proper Er content is suitable for LD pumping. Nevertheless some glasses have also been designed especially

© Woodhead Publishing Limited, 2013

for LD pumped lasers. For example, an Yb–Er glass[13] with enhanced Yb content (4×10^{21} cm^{-3}) was developed at the General Physics Institute (GPI), Moscow, Russia. High Yb concentration has made it possible to show remarkable parameters in miniature transversely pumped lasers (see next section). Also a special Yb–Er laser glass with lowered thermal expansion and enhanced mechanical strength for high-power longitudinally pumped microchip lasers was developed at the same Institute.[14]

13.3.1 Transversely pumped lasers

The pioneering work on transversely pumped Er glass lasers was made by T. Allik and J. A. Hutchinson[15] in 1992. They pumped a cylindrical glass rod (∅2 × 10 mm) from two opposite sides with a pair of ~100 W laser diode arrays emitting at ~970 nm, i.e. close to the peak of Yb ion absorption in the glass. The pump pulse duration was about 1 ms. The free-running output energy reached 20 mJ at the slope efficiency of 14%. The laser glass was QE-7 from Kigre, Inc. doped with 0.13 wt% Er_2O_3 and 15 wt% Yb_2O_3. The glass was designed for flashlamp pumping and was co-doped with 0.26 wt% Nd_2O_3 which was useless for diode pumping. After this work it became clear that diode array pumping of Yb–Er glass would become a new technology competing with flashlamp operation. The following several examples demonstrate the potential of LD pumped lasers.

An extensive approach to laser design was demonstrated in the paper of Wu *et al.*[16] The ∅3 × 35 mm Yb–Er glass rod was surrounded by 48 (!) 1.2 cm long laser diode arrays providing total pump energy of 17.5 joules at 929 nm with a 4 nm full-width-at-half-maximum (FWHM) line width in millisecond pulses. The slope efficiency of the lasers was measured at about 9.5%. The maximum output energy reached 1.2 joules. Gagarskii *et al.*[17] demonstrated a much simpler design. The laser had a ∅2 × 10 mm rod that was placed near the 100 W array and parallel to it at a distance of about 100 μm. Part of the cylindrical surface of the active element, not directly illuminated by the diodes, was covered with a silver foil, ensuring a back-reflection into the element of the pump radiation transmitted through it. High pumping efficiency was obtained due to high Yb concentration in the laser glass (4×10^{21} cm^{-3}). The absorption coefficient in the glass reached 10 cm^{-1} at the pump wavelength of ~940 nm. Full power pumping with 4.8 ms long pulses produced output energy of 23 mJ, i.e. about the same as in Allik and Hutchinson[15] but using only a single 100 W LD array.

Georgiou *et al.*[18,19] also described a well-optimized laser based on the same glass with high Yb concentration. The laser rod (∅3 × 12 mm) was mounted on a heat sink – a hollow cylindrical gilded brass holder – and pumped by three multi-bar laser diode arrays, symmetrically arranged at 120° angles around the rod (Fig. 13.2). Three slotted side openings on the

© Woodhead Publishing Limited, 2013

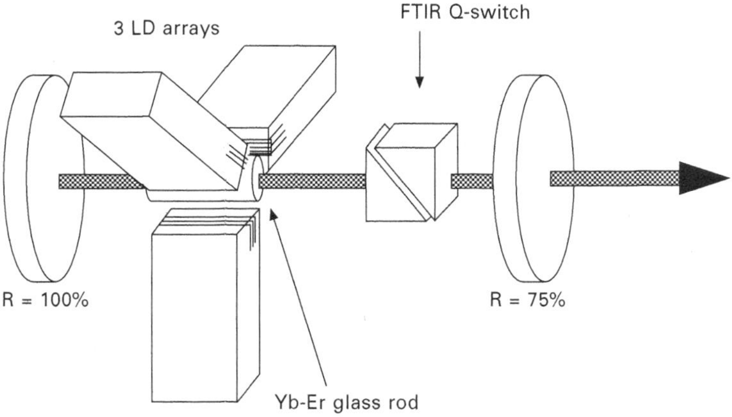

13.2 Optical scheme of the LD array pumped Yb–Er glass laser Q-switched by a frustrated total internal reflection (FTIR) shutter.

rod holder allowed efficient direct optical coupling of the pump radiation to the laser-active material. For the same purpose, the LD emitting facets were positioned at <1 mm distance from the laser rod, while the divergent, scattered or unabsorbed pump radiation was reflectively recycled thanks to the rod holder gold plating. This design effectively improved the optical coupling efficiency and homogenized pump distribution over the rod cross-section. Three 959 nm diode arrays were able to deliver up to 2.65 J in 5.5 ms long pulses. Such a design[18,19] provided up to 360 mJ and up to 1.2 J when the pump pulses were extended to ~20 ms. The slope efficiency reached a record high value of 20.5%. The removal of the waste thermal energy from the LD arrays and from the rod holder was effected by their mounting on thermoelectrically cooled heat sinks via thin indium foil. The Q-switched operation of the same laser will be considered below.

13.3.2 Longitudinally pumped microlasers

Longitudinal pumping of thin (in the order of a few millimeters) Er–Yb glass plates is the widespread configuration of laser diode pumped erbium glass lasers. Laser resonator mirrors can be deposited directly on the glass plate or can be separated a few millimeters away from it. This configuration can utilize either single or a few laser diodes or laser diode arrays (with the help of some optical radiation concentrators). In such a configuration CW and high pulse repetition rate regimes are easily obtained due to the low volume of the pumped active media and thus low thresholds. For a comprehensive description of the physics and technology of microchip and miniature lasers, see Chapter 14 of this book. Some remarks have to be made about the choice of the active and sensitizing ion concentration in the glass for longitudinally

© Woodhead Publishing Limited, 2013

pumped microchip lasers. In principle, Er concentration in Yb–Er glasses can be varied in the rather wide range of about (1–15) × 10^{19} cm^{-3}. At the lower limit the efficiency of Yb→Er energy transfer is decreased. The upper end is determined by the undesirable interaction of excited Er ions reducing the energy storage time.[20] In small-sized pumped chips the reasonable Er content lies within approximately (0.5–1.5) × 10^{20} cm^{-3} in order to provide sufficient gain over the distance of about 1 mm. (Er content in flashlamp pumped lasers is usually close to the minimum of ~(1.5–2) × 10^{19} cm^{-3} in order to reduce the lasing threshold.) As for Yb concentration, the glass for longitudinally LD pumped microlasers should contain as much Yb ions as technologically possible for the given glass composition. The first longitudinally InGaAs LD pumped bulk Er–Yb glass laser[21] was reported in 1990. The authors used pulsed pumping of Er–Yb glass plate into the peak of the Yb absorption spectrum. The slope efficiency was 7.1%, and the threshold was as low as 0.9 mJ. In the first demonstration[22] of a CW diode pumped Er glass microlaser its power did not exceed 2 mW. High output parameters were demonstrated[23] in a 0.3 × 1 × 1 mm^3 Er–Yb glass microchip with both sides directly coated to make the resonator mirrors. The threshold was as low as 9 mW of absorbed power, the CW output reached 220 mW, and the slope efficiency reached 38% with regard to incident pump power.

Operation of CW microchip lasers is strongly affected by the thermal properties of the glass. Enhancing the microlaser power requires using the special strong Yb–Er glasses. For example, it was shown[24] that the output power of a chip made of QE-7 Yb–Er glass (with customized Er content) from Kigre, Inc. was limited to less than 50 mW due to glass fracture. The usage of a QX–Er glass[25] with a higher thermal loading limit allows the output power to be increased to 300 mW. Experiments with the GPI Yb–Er concentrated glass[26] have allowed a CW output power of 350 mW to be reached.

13.4 Means of Q-switching for erbium (Er) glass lasers

13.4.1 Active Q-switching devices

In principle, the active Q-switching devices for Er glass lasers are based on the same principles as those for other lasers. Of course, the material of the Q-switcher should be transparent at the lasing wavelength. For this reason hydrogen (and deuterium)-containing crystals for Pockels cells (like KH_2PO_4–KDP) are excluded. The suitable electrooptical crystals that can be used in Er glass lasers are lithium niobate and β-barium borate (BBO). These devices suffer from a number of drawbacks. Their driving voltage is high (several kilovolts) and roughly proportional to the laser wavelength.

© Woodhead Publishing Limited, 2013

Their performance decreases in high average power lasers due to additional losses caused by laser light depolarization in the thermally stressed laser glass. In addition, the lithium niobate optical damage threshold is usually lower than that of other laser optical components.

It is important to remember that a low gain in Er–Yb laser glasses makes the laser output very sensitive to all kinds of optical losses in a cavity. At the same time a low gain means a relatively long giant pulse buildup time and thus much softer demands for the Q-switching rate. That is why relatively slow optomechanical Q-switching devices are of interest for Er glass lasers. The spinning 90° roof prism is the oldest device of this sort which should not be neglected. At optimized spinning rate it introduces low losses and enables Q-switch pulses to form. Other advantages are polarization independence and a high optical damage threshold. The temperature insensitivity of this device is an important peculiarity for field applications. An inconvenience factor is that the pump pulse should be matched to the prism spinning phase and not vice versa.

Frustrated total internal reflection (FTIR) shutters may seem a convenient Q-switching means for Er glass lasers due to low losses, a high enough switching rate, a high optical damage threshold and the possibility of switching the light of any polarization. Such a shutter (schematically shown in Fig. 13.2) consists of a pair of fused silica prisms rigidly connected with each other. The ~0.5–1 μm gap between the prisms makes the system weakly transparent for laser light. This gap can be rapidly collapsed (in hundreds of nanoseconds) and the shutter opened by applying the properly shaped voltage pulse at piezoelectric cells mounted on the prisms. Usually the driving voltage does not exceed ~300–400 V, which is much less than the voltage required for electrooptical cells.

Comparison of FTIR shutters with BBO Pockels cells in LD array pumped Er glass lasers was made by Wu *et al.*[27] These tests showed the evident advantage of FTIR shutters. FTIR performance was investigated in a large variety of erbium glass lasers including flashlamp pumped,[6] single diode CW pumped,[17] and pulsed diode array pumped ones.[17,18] In the last of these three papers, the FTIR Q-switched a diode array pumped laser based on a ∅3 × 12 mm rod of the glass which emitted up to 50 mJ Q-switched pulses with an efficiency of 5.4%. To our knowledge this output energy is the highest ever value obtained for a diode pumped Q-switched Er glass laser. A serious restriction for field applications of FTIR shutters is their limited temperature range of operation. At low temperatures the shutter prisms tend to 'stick' and the shutter remains open. At elevated temperatures the gap between the prisms tends to increase together with the required driving voltage. Another limitation originates from their limited lifespan of ~10^6 pulses. For most purposes this is quite enough, but in the case of high pulse repetition rate lasers with CW pumping their resource can be exhausted in minutes.[17]

© Woodhead Publishing Limited, 2013

13.4.2 Passive Q-switchers for Er glass lasers

Mass applications require cheap, compact, simple and efficient Q-switching devices for 1.54 μm lasers. The bleaching filters for solid-state lasers with shorter wavelengths were found long ago, but the search for efficient passive Q-switching materials for Er glass lasers brought practical results only about a decade ago. Uranium-doped CaF_2 and similar crystals[28] were the first bleaching materials for erbium glass lasers, though their Q-switching efficiency was moderate. A number of Co^{2+} doped crystalline hosts have been proposed for Q-switching Er glass lasers. Co^{2+} ions in tetrahedral coordination have an intensive absorption band at ~1.5 μm. Co-doped La–Mg–aluminate – $LaMgAl_{11}O_{19}$ (Co^{2+}:LMA) crystal[29] – is a birefrigent and dichroic crystal having a high enough absorption cross-section for e-polarization only. Thus an additional polarizing element in the laser cavity is required. Cobalt-doped ZnSe crystal[30] has no such problem due to its cubic structure. It should be noted that the bleached state lifetime in this material reaches 400 μs. Thus the Q-switched pulse repetition rate (in high-power CW pumped lasers) cannot be much higher than that value. The most widespread cobalt-doped material for passive Q-switching of erbium glass lasers is magnesium–aluminum spinel, $MgAl_2O_4$, also referred to as MALO.[14,31–35] This is a cubic crystal with excellent mechanical properties and high optical damage threshold. A high enough absorption cross-section of cobalt ions at the laser wavelength in spinel (~3×10^{-19} cm^2, see Fig. 13.3) is combined with sufficient bleached state lifetime (~350 ns). Laser characterization of this material showed that

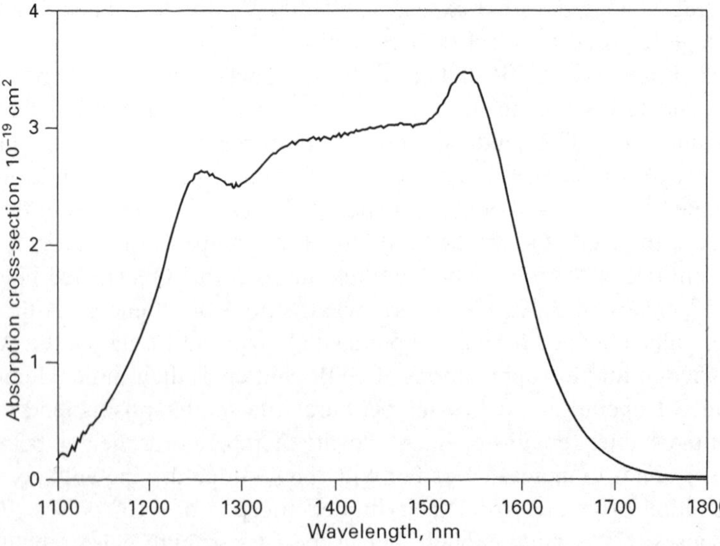

13.3 Absorption cross-section of Co^{2+} ions in $MgAl_2O_4$ crystal.

© Woodhead Publishing Limited, 2013

Co:MALO Q-switched laser efficiency surely overwhelms the efficiency of uranium-doped CaF_2 and is about 80% in comparison with the FTIR Q-switched laser.[32,33]

The cubic structure of the $Co:MgAl_2O_4$ crystal and its favorable spectroscopic parameters have made it a very convenient material for passive Q-switching of microchip lasers. For example, TEM_{00} mode pulses as short as 2.3 ns, peak power up to 2 kW and optical-to-optical efficiency up to 18% were obtained[35] using Co:MALO. High mechanical strength and high thermal conductivity of the crystal together with its thermal expansion value close to that of the laser glass has made it possible to use Co-doped spinel plate simultaneously as a Q-switcher and an efficient heat sink for a microchip laser.[14] The laser average output power reached 180 mW in a stable train of 5 ns pulses in this paper.

Passive Q-switching of low-gain Er glass lasers has some peculiarities. Single transverse mode operation is usually quite stable so that in the case of uniform pump distribution over the laser aperture no special mode selection is required. But, in contrast to higher gain Nd:glass it is quite tricky to obtain stable multiple transverse mode single pulses from a passively Q-switched Er glass laser. Typically the multimode output includes two or three different pulses with separation ranging from tens of nanoseconds to hundreds of microseconds, each pulse having its own mode structure.[36] Usually the fundamental mode emits first, and the following pulses have another mode structure. There was only one article[31] about stable multimode output of a passively Q-switched laser emitting 250 mJ ~100 ns pulses. This became possible due to a relatively high gain in glass, high outcoupling ($R = 40\%$) and as short a resonator as possible.

13.5 Applications of erbium (Er) glass lasers

Erbium glass lasers find applications in such fields as the following:

- Eye-safe rangefinders and LIDARs. In addition to eye-safety, the advantages of ~1.5-micron radiation for this purpose are high air transparency and the presence of sensitive room-temperature detectors. A typical Er:glass rangefinder emits Q-switched 3–10 mJ pulses and is capable of measuring distances up to ~10 km. For flashlamp pumped devices without forced cooling, the time interval between the pulses is a few seconds. Diode array pumped rangefinder emitters can have a repetition rate up to 10–20 Hz.
- Surgery. Millisecond-long pulsed or CW fiber-delivered 1.5-micron radiation is a delicate tool for delicate tissue coagulation and bloodless cutting. The typical output energy can vary in a wide range from ~0.1 to ~10 J. It should be noted that in this field of application erbium fiber lasers compete with pulsed free-run Yb–Er bulk glass lasers.

© Woodhead Publishing Limited, 2013

- Laser Induced Breakdown Spectroscopy (LIBS). A Q-switched laser pulse with energy ~10 mJ is focused down to target, creating a dielectric breakdown or 'plasma spark'. The excited atoms decay, resulting in a narrow 'fingerprint' elemental emission line spectra. The main advantage of Er glass lasers for this purpose is the insensitivity of the silicon charge-coupled device (CCD) arrays of the spectrum analyzers.[37]

In conclusion we would like to mention the advantages and the drawbacks of erbium glass lasers in comparison to other sources of ~1.5 μm eyesafe laser radiation: Yb–Er doped crystalline lasers, resonantly pumped Er doped crystal lasers, optical parametric oscillator (OPO) and Raman lasers.

13.6 Crystal lasers emitting at about 1.5 microns: advantages and drawbacks

As was mentioned in the introduction, Yb–Er glass lasers are the most efficient and simple laser sources of ~1.5-micron eyesafe radiation. Nevertheless there are other types of solid-state lasers emitting in approximately the same spectral range.

13.6.1 Yb–Er crystal lasers

Attempts to develop a crystalline Yb–Er laser medium are rather numerous. The high thermal conductivity of practically any crystal in comparison to glass gives hope to enhance the average power of the laser. In particular, laser action under LD or Ti:sapphire pumping has been obtained at different wavelengths in the 1.53–1.64 μm spectral range in such crystals as YAG,[38,39] Y_2SiO_5,[40] $Ca_2Al_2SiO_7$,[41] and YVO_4.[42,43] Nevertheless, the highest slope efficiencies of the crystalline lasers mentioned did not exceed ~5–6%, i.e. several times less than that of LD pumped glass lasers. As a result, the CW or average output power of all of them did not exceed the values attainable for glass lasers. The reason is low efficiency of the upper laser level population due to reverse energy transfer from the long-living $^4I_{11/2}$ Er^{3+} level back to Yb ions.

Laser action has also been obtained in a number of Yb–Er doped borate crystals: in calcium yttrium oxoborate $Ca_4YO(BO_3)_3$,[44, 45] calcium gadolinium oxoborate $Ca_4GdO(BO_3)_3$,[46] $LaSc_3(BO_3)_4$ (LSB),[47] and $YAl_3(BO_3)_4$.[48] Short (<1 μs) $^4I_{11/2}$ Er^{3+} level lifetime in borate crystals excludes the reverse energy transfer to Yb ions and the slope efficiencies of the lasers mentioned reached almost 30%, i.e. they were about the same as for typical glass lasers. But the lowered (~1.2 ms or less in comparison to ~7 ms in phosphate glass) upper laser level lifetime in all these crystals resulted in lasing thresholds almost an order of magnitude higher than in glass lasers. Another big disadvantage

© Woodhead Publishing Limited, 2013

of calcium lanthanide oxoborate crystals is their easy cleavage. For these reasons the average power of these lasers was also no higher than that of glass lasers. The only crystalline Yb–Er laser material that can demonstrate at present an average output exceeding the power of glass lasers of similar geometry is yttrium aluminum borate (YAB), $YAl_3(BO_3)_4$. High thermal conductivity and the good mechanical properties of the crystal have made it possible to obtain 1 W of continuous output from a YAB chip at the wavelength of 1.555 microns. This value is at least three times the power of a glass laser that can be expected in the same configuration. The slope efficiency was also very high (up to 35%). Unfortunately the very low (0.325 ms) upper laser level lifetime in this material is the cause of the high lasing threshold (0.5–1 W in the case of the pump beam waist ~100 μm). And the main drawback of this material is its high cost in comparison to glass: YAB crystals can be grown by the flux method only and the crystal size does not exceed a few millimeters.

13.6.2 Resonantly pumped Er:YAG lasers

There is one more way to develop high power ~1.5 μm crystal lasers – direct optical pumping of Er ions in non-sensitized crystals. It may make it possible to enhance the average power by more than an order of magnitude, although this method meets some difficulties connected with the low extinction factor of such crystals in comparison to Yb-sensitized materials.

First, the lower level of the considered $^4I_{13/2} \rightarrow {}^4I_{15/2}$ Er laser transition is either a thermally populated Stark component of the $^4I_{15/2}$ manifold located just a few hundred cm^{-1} above the ground level, or the ground state, which means lasing through a three-level lasing scheme. For this reason the $^4I_{13/2}$ level population required to reach inversion at room temperature is ~15–50%. Second, the concentration of Er^{3+} ions in the host should be limited to about 10^{20} $ions/cm^3$ in order to avoid an intensive up-conversion process. The Er concentration limitation means that the transverse pumping of a thin rod-shaped Er doped laser crystal (in contrast to Yb–Er glass) is not realistic, due to low extinction, and the only possibility is the longitudinal pumping of a crystal.

$^4I_{15/2} \rightarrow {}^4I_{13/2}$ transition is the strongest among Er absorption lines. The reasonable extinction length of a crystal with ~10^{20} cm^{-3} Er^{3+} can be as low as ~20–30 mm. Resonant pumping means very low quantum defect between the pumping and the lasing wavelengths and thus very low heat dissipation in the laser crystal. But in this case an inversion at the transitions terminating at the ground level is impossible. Lasing can be achieved only at the transitions terminating at excited Stark components of the $^4I_{15/2}$ manifold. Er:YAG crystal combines excellent thermomechanical properties and large ground-state splitting of the $^4I_{15/2}$ Er level. For this reason almost all of the

© Woodhead Publishing Limited, 2013

papers concerning resonantly pumped erbium lasers are devoted to the YAG host having emission peaks at ~1.61 and ~1.64 μm. There are two kinds of powerful pump sources for this material: InGaAsP/InP LDs (λ ~ 1.4–1.5 μm) and Er:glass fiber lasers (λ ~ 1.53 μm). Both of them were successfully utilized for Er:YAG resonant pumping.

Garbusov *et al.* published the pioneering paper[49] concerning direct InGaAsP/InP LD pumping of Er:YAG, demonstrated a free-running laser action. The output power in these experiments was low, but the photon-to-absorbed-photon efficiency reached already a noticeable value of 26%. The next paper from the same authors[50] is an impressive demonstration of power scaling. They used two-dimensional stacks of either 1470 nm or 1530 nm InGaAsP/InP LD with power reaching 200 W and 300 W respectively. The optical-to-optical slope efficiencies in the free-run regime reached 45%. The free-running energy per pulse reached an impressive value of 1 J.

An alternative approach in the design of resonantly pumped Er:YAG lasers is to use a hybrid laser configuration in which an Er fiber source pumps 'in-band' the Er bulk crystalline laser. This scheme combines the advantages of efficient CW high-power generation of cladding pumped fiber lasers with the energy storage and high pulse energy capabilities of bulk solid-state laser crystals. In contrast to high power LD arrays, high-power fiber laser beam divergence can be more or less diffraction limited and its lasing spectrum can be made narrow and can be tuned to provide exact matching with the strongest Er:YAG absorption peak at 1532 nm. Output power of 26 W was obtained[51] at 75 W pump using a 0.25 at% 58 mm long Er:YAG rod.

Thus, the average output power of resonantly pumped Er:YAG lasers can be much higher than that available from Yb,Er:glass lasers but their design is noticeably more complicated.

13.6.3 Raman lasers

In these lasers ~1.0 or ~1.3 μm high peak power Nd laser pulses are transformed into the ~1.5 μm eye-safe spectral region.[52–56] Their main advantage over Er:glass lasers is a much higher repetition rate of neodymium crystal lasers and the obvious drawback is the complexity of their overall optical scheme.

The systems discussed actually represent a family of solid-state bulk lasers operating in the ~1.5 μm region.

13.7 References

1 Snitzer E, Woodcock R (1965), 'Yb^{3+}–Er^{3+} glass laser', *Applied Physics Letters*, 6, 45–46.

2 Gapontsev V P, Matitsin S, Izyneev A, Kravchenko V (1982), 'Erbium glass lasers and their applications', *Optics and Laser Technology*, 14, 189–196.

© Woodhead Publishing Limited, 2013

3 Alekseev N E, Gapontsev V P, Zhabotinskii M E, Kravchenko V B, Rudnitskii Yu P (1983), *Laser Phosphate Glasses*, Moscow, Nauka Publishing House.
4 Lunter S G, Murzin A G, Tolstoi M N, Fedorov Yu K, Fromzel V A (1984), 'Energy parameters of lasers utilizing erbium glasses sensitized with ytterbium and chromium', *Quantum Electronics*, 14, 66–70.
5 Denker B I, Maksimova G V, Osiko V V, Sverchkov S E, Sverchkov Yu E (1991), 'Investigation of the lasing capabilities of new erbium glasses', *Quantum Electronics*, 21, 964–966.
6 Denker B I, Osiko V V, Sverchkov S E, Sverchkov Yu E, Fefelov A P, Khomenko S I (1992), 'Highly efficient erbium glass lasers with Q-switching based on frustrated total internal reflection', *Quantum Electronics*, 22, 500–503.
7 Izyneev A A, Sadovskii P I (1997), 'New highly efficient LGS-KhM erbium-doped glass for uncooled miniature lasers with a high pulse repetition rate', *Quantum Electronics*, 27, 771–775.
8 Byshevskaya-Konopko L O, Vorob'ev I L, Izyneev A A, Sadovskii P I, Sergeev S N (2001), 'Optimisation of the pumping parameters of a repetitively pulsed erbium laser', *Quantum Electronics*, 31, 861–863.
9 Gapontsev V, Gromov A, Izyneev A, Sadovskii P, Stavrov A, Tipenko Yu, Schkadarevich A (1989), 'A low threshold erbium glass mini-laser', *Quantum Electronics*, 19, 447–449.
10 Jiang S, Hamlin S, Myers J, Rhonehouse D, Myers M (1996), 'High average power 1.54 μm Er–Yb doped phosphate glass laser', *OSA Technical Digest Series, Optical Society of America*, 9, 380–381.
11 Alekseev N E, Byshevskaya-Konopko L O, Vorob'ev I L, Izyneev A A, Sadovskii P I (2003), 'Continuous wave lasing at 1.54 μm in a flashlamp pumped ytterbium–erbium-doped glass', *Quantum Electronics*, 33, 1062–1064.
12 Galagan B I, Danileiko Yu K, Denker B I, Osiko V V, Sverchkov S E (1998), 'Nature of the temperature dependence of the lasing efficiency of erbium laser glasses and the mechanism of the influence of sensitizers on this efficiency', *Quantum Electronics*, 28, 313–315.
13 Maksimova G V, Sverchkov S E, Sverchkov Yu E (1991), 'Lasing tests on new ytterbium–erbium laser glass pumped by neodymium laser', *Quantum Electronics*, 21, 1324–1325.
14 Karlsson G, Laurell F, Tellefsen J, Denker B, Galagan B, Osiko V, Sverchkov S (2002), 'Development and characterization of Yb–Er laser glass for high average power laser diode pumping', *Applied Physics B*, 75, 1–6.
15 Allik T, Hutchinson J A (1992), 'Diode array pumped Er,Yb:phosphate glass laser', *Applied Physics Letters*, 60, 1424–1426.
16 Wu R, Hamlin S, Myers M, Myers J, Hutchinson J, Marshall L (1996), '1.2 J high energy diode pumped 1535 nm Er^{3+}, Yb^{3+}:glass laser', *Conference on Lasers and Electro-Optics Europe '96 (CLEO/Europe)*, http://www.kigre.com/files/er10.pdf.
17 Gagarskii S V, Galagan B I, Denker B I, Korchagin A A, Osiko V V, Prikhod'ko K V, Sverchkov S E (2000), 'Diode-pumped ytterbium–erbium glass microlasers with optical Q-switching based on frustrated total internal reflection', *Quantum Electronics*, 30, 10–12.
18 Georgiou E, Musset O, Boquillon J-P, Denker B, Sverchkov S (2001), '50 mJ/30 ns FTIR Q-switched diode pumped Er:Yb:glass 1.54 μm laser', *Optics Communications*, 198, 147–153.
19 Georgiou E, Musset O, Boquillon J-P (2002), 'Diode pumped bulk Yb:Er:glass 1.54

μm pulsed laser with 1.2 J and 25% optical efficiency', *OSA TOPS*, 68, *Advanced Solid State Lasers*, 226.

20 Denker B, Galagan B, Osiko V, Sverchkov S (2001), 'Peculiarities of energy storage and relaxation in Yb–Er glasses with enhanced Er content', *Advanced Solid State Lasers, Technical Digest*, 389–394.

21 Hutchinson J A, Caffey D P, Scaaus C F (1990), 'Diode pumped eyesafe erbium glass laser', *CLEO'90 Technical Digest, Anaheim*, paper CPDP1.

22 Laporta P, De Silvestri S, Magni V, Svelto O (1991), 'Diode-pumped cw bulk Er:Yb:glass laser', *Optics Letters*, 16, 1952–1954.

23 Thony P, Molva E (1996), '1.55 μm wavelength CW microchip lasers', *OSA TOPS*, 1, *Advanced Solid State Lasers*, 296–300.

24 Laporta P, Taccheo S, Longhi S, Svelto O, Svelto C (1999), 'Erbium–yitterbium microlasers: Optical properties and lasing charicteristics', *Optical Materials*, 11, 269–288.

25 Taccheo S, Sorbello G, Laporta P, Karlsson G, Laurell F (2001), '230-mW diode-pumped single-frequency Er:Yb laser at 1.5 μm', *IEEE Photonics Technology Letters*, 13, 19–21.

26 Huber G, Danger T, Denker B I, Galagan B I, Sverchkov S E (1998), 'Diode-pumped cw laser around 1.54 μm using Yb,Er-doped silico-boro-phosphate glass', *OSA Technical Digest Series, Optical Society of America*, 6, 181.

27 Wu R, Myers J, Hamlin S (1998), 'Comparative results of LD pumped Er glass lasers Q-switched with BBO Pockels cell and FTIR methods', *OSA TOPS*, 19, *Advanced Solid State Lasers*, 159–161.

28 Stultz R, Camargo M, Montgomery S, Birnbaum M, Spariosu K (1994), 'U^{4+}:SrF_2 efficient saturable absorber Q-switch for the 1.54 μm erbium:glass laser', *Applied Physics Letters*, 64, 948.

29 Thony P, Ferrand B, Molva E (1998), '1.55 μm passive Q-switched microchip laser', *OSA TOPS*, 19, *Advanced Solid State Lasers*, 150.

30 Birnbaum M, Camargo M, Lee S, Unlu F, Stultz R (1997), 'Co^{2+}:ZnSe saturable absorber Q-switch for 1.54 μm Er:Yb:glass laser', *OSA TOPS*, 10, *Advanced Solid State Lasers*, 148–151.

31 Denker B, Galagan B, Godovikova E, Meilman M, Osiko V, Sverchkov S, Kertesz I (1999), 'The efficient saturable absorber for 1.54 μm Er glass lasers', *OSA TOPS*, 26, *Advanced Solid State Lasers*, 618–620.

32 Galagan B I, Godovikova E A, Denker B I, Meil'man M L, Osiko V V, Sverchkov S E (1999), 'Efficient bleachable filter based on Co^{2+}:$MgAl_2O_4$ crystals for Q-switching of λ = 1.54 μm erbium glass lasers', *Quantum Electronics*, 29, 189–190.

33 Wu R, Myers J D, Myers M J, Denker B I, Galagan B I, Sverchkov S E, Hutchinson J A, Trussel W (2000), 'Co:$MgAl_2O_4$ crystal passive Q-switch performance at 1.34, 1.44 and 1.54 microns', *OSA TOPS*, 34, *Advanced Solid State Lasers*, 254–256.

34 Mikhailov V, Yumashev K, Denisov I, Prokoshin P, Posnov N, Moncorgé R, Vivien D, Ferand E, Guyot Y (1999), 'Passive Q-switch performance at 1.3 μm (1.5 μm) and nonlinear spectroscopy of Co^{2+}:$MgAl_2O_4$ and Co^{2+}: $LaMgAl_{11}O_{19}$ crystals', *OSA TOPS*, 26, *Advanced Solid State Lasers*, 317–324.

35 Karlsson G, Pasiskevichius V, Laurell F, Tellefsen J A, Denker B, Galagan B, Osiko V, Sverchkov S (2000), 'Diode-pumped Er–Yb:glass laser passively Q-switched by use of Co^{2+}:$MgAl_2O_4$ as a saturable absorber', *Applied Optics*, 39, 6188–6192.

36 Wu R, Chen T, Myers J, Myers M, Hardy C (2003), 'Multi-pulses behavior in a erbium glass laser Q-switched by cobalt spinel', *AeroSense 2003, SPIE Vol. 5086, Laser Radar Technology and Applications VIII*, Orlando, FL, 21–25 April 2003.

© Woodhead Publishing Limited, 2013

37 Myers M J, Myers J D, Sarracino J T, Hardy C R, Guo B, Christian S M, Myers J A, Roth F, Myers A (2010), 'LIBS system with compact fiber spectrometer, head mounted spectra display and hand held eye-safe erbium glass laser gun', *SPIE Photonics West 2010, Solid State Lasers XIX: Technology and Devices, Conference La101, #7578-87,* 26 January 2010.

38 Schweizer T, Jensen T, Heumann E, Huber G (1995), 'Spectroscopic properties and diode pumped 1.6 μm laser performance in Yb-codoped Er:$Y_2Al_5O_{12}$', *Optics Communications*, 118, 557.

39 Georgiou E, Kiriakidi F, Musset O, Boquillon J-P (2005), '1.65-μm Er:Yb:YAG diode-pumped laser delivering 80-mJ pulse energy', *Optical Engineering*, 44, 64202.

40 Li C, Moncorgé R, Souriau J, Borel C, Wyon C (1994), 'Room-temperature CW laser action of Y_2SiO_5:Yb,Er at 1.57 μm', *Optics Communications*, 107, 61–64.

41 Simondi-Teisseire B, Viana B, Lejus A M, Benitez J-M, Vivien D, Borel C, Templier R, Wyon C (1996), 'Room temperature CW laser operation at 1.55 μm (eye-safe range) of Yb:Er:Ce:$Ca_2Al_2SiO_7$ crystals', *IEEE Journal of Quantum Electronics*, 32, 2004–2009.

42 Sokolska I, Heumann E, Kuck S, Lukasiewicz T (2000), 'Laser oscillations of Er:YVO_4 and Er,Yb, YVO_4 crystals in the spectral range around 1.6 μm', *Applied Physics B*, 71, 893–896.

43 Tolstik N A, Troshin A E, Kurilchik S V, Kisel V E, Kuleshov N V, Matrosov V N, Matrosova T A, Kupchenko M I (2007), 'Spectroscopy, continuous-wave and Q-switched diode-pumped laser operation of Er^{3+},Yb^{3+}:YVO_4 crystal', *Applied Physics B*, 86, 275–278

44 Burns P, Dawes J, Dekker P, Piper J, Jiang H, Wang J (2002), 'CW diode-pumped microlaser operation at 1.5–1.6 μm in Er,Yb:YCOB', *IEEE Photonics Technology Letters*, 14, 1677–1679.

45 Burns P, Dawes J, Dekker P, Piper J, Jiang H, Wang J (2003), '250 mW continuous-wave output from Er,Yb:YCOB laser at 1.5 μm', *Advanced Solid State Photonics 2003, Technical Digest*, 8–12.

46 Denker B, Galagan B, Ivleva L, Osiko V, Sverchkov S, Voronina I, Hellström J, Karlsson G, Laurell F (2004), 'Luminescent and laser properties of Yb,Er-activated $GdCa_4O(BO_3)_3$ – a new crystal for eyesafe 1.5 micrometer lasers', *Applied Physics B*, 79, 577–581.

47 Lebedev V A, Pisarenko V F, Selina N V, Perfilin A A, Brik M G (2000), 'Spectroscopic and luminescent properties of Yb,Er:$LaSc_3(BO_3)_4$ crystals', *Optical Materials*, 14, 121–126.

48 Tolstik N A, Kurilchik S V, Kisel V E, Kuleshov N V, Maltsev V V, Pilipenko O V, Koporulina E V, Leonyuk N I (2007), 'Efficient 1 W continuous-wave diode-pumped Er,Yb:$YAl_3(BO_3)_4$ laser', *Optics Letters*, 32, 3233–3235.

49 Garbusov D, Kudryashov I, Dubinskii M (2005), 'Resonantly diode laser pumped 1.6 μm erbium doped yttrium aluminum garnet solid state laser', *Applied Physics Letters*, 86, 131115.

50 Garbusov D, Kudryashov I, Dubinskii M (2005), '110 W (0.9 J) pulsed power from resonantly diode-laser-pumped 1.6 μm Er:YAG laser', *Applied Physics Letters*, 87, 121101–121103.

51 Shen D Y, Sahu J K, Clarkson W A (2006), 'Electrooptically Q-switched Er:YAG laser in-band pumped by an Er–Yb fiber laser', *Advanced Solid State Photonics 2006, Technical Digest*, paper WD4.

© Woodhead Publishing Limited, 2013

52 Bruns D G, Bruesselbach H W, Stowall H D, Rockwell D A (1982), 'Scalable visible Nd:YAG pumped Raman laser source', *IEEE Journal of Quantum Electronics*, QE-18, 246–1252.

53 Stultz R, Nieuwsma D, Gregor E (1991), 'Eyesafe high pulse rate laser progress at Hughes', *Proc. SPIE Int. Soc. Opt. Eng.*, 1419, paper 08.

54 Murray J T, Powell R C, Peyghambarian N, Smith D, Austin W, Stolzenberger R A (1995), 'Generation of 1.5 μm radiation through intracavity solid-state Raman shifting in $Ba(NO_3)_2$ nonlinear crystals', *Optics Letters*, 20, 1017.

55 Murray J T, Powell R C, Peyghambarian N, Smith D, Austin W (1995), 'Eyesafe solid-state intracavity Raman laser', *OSA Proc. ASSL'95*, 24, 267–269.

56 Zverev P, Basiev T, Prokhorov A (1999), 'Stimulated Raman scattering in Raman crystals', *Optical Materials*, 11, 335–352.

© Woodhead Publishing Limited, 2013

14
Microchip lasers

J. J. ZAYHOWSKI, Massachusetts Institute of Technology, USA

DOI: 10.1533/9780857097507.2.359

Abstract: Microchip lasers are a rich family of solid-state lasers defined by their compact size, robust integration, reliability, and potential for low-cost mass production. CW microchip lasers cover a wide range of wavelengths, often operate single frequency in a near-ideal mode, and can provide a modest amount of tunability. *Q*-switched microchip lasers provide the shortest output pulses of any *Q*-switched solid-state laser, with peak powers up to several hundred kilowatts.

Key words: solid-state laser, diode-pumped laser, microchip laser, *Q*-switched laser, passively *Q*-switched microchip laser.

14.1 Introduction

Most solid-state lasers are built from discrete optical components that must be carefully assembled and critically aligned. Laser assembly is typically performed by trained technicians, and is time consuming and therefore expensive. As a result, the cost of most solid-state lasers makes them unattractive for a wide range of applications for which they would otherwise be well suited. Additional characteristics that have historically impeded the widespread use of solid-state lasers include their size and reputation for being fragile and unreliable. This was even truer in the early 1980s, the infancy of microchip laser development, than it is today. Microchip lasers were developed to overcome these limitations – cost, size, robustness, and reliability – and thereby become viable components for a variety of large-volume applications. The term 'microchip laser' was coined at MIT Lincoln Laboratory in the early 1980s to draw an analogy between this new class of device and semiconductor electronic microchips with their inherent small size, reliability, and low-cost mass production.

Consider the simplest of all possible microchip lasers, a small piece of solid-state gain medium polished flat and parallel on two opposing sides, with dielectric cavity mirrors deposited directly onto the polished faces, as

This work was sponsored by the Department of the Air Force under Air Force Contract #F19628-95C-0002. Opinions, interpretations, conclusions and recommendations are those of the author and are not necessarily endorsed by the United States Government.

© Woodhead Publishing Limited, 2013

shown in Fig. 14.1 (Zayhowski and Mooradian, 1989a; this device will be discussed in detail in Section 14.2). Fabrication of the laser starts with a large boule of gain material, such as Nd:YAG, which may cost several thousand dollars. The boule is sliced into wafers about 0.5 mm thick. The wafers are then polished and dielectrically coated before they are diced into 1-mm-square pieces, with each piece being a complete laser. One 250-mm-long boule can produce up to 250 wafers, and one 125-mm-diameter wafer can produce up to 6000 lasers, making the cost per laser very small. And, throughout the fabrication process, the lasers never need to be handled independently.

To complete a microchip laser system, the laser must be coupled to a pump source, to energize the gain medium. Microchip lasers are pumped with semiconductor diode lasers. The 1980s were a time of rapid development in diode lasers. The amount of power that was available from commercial diode lasers was rapidly increasing and the cost per watt of output power was quickly decreasing, with projections of extremely inexpensive, high-power diode lasers in the near future. As a result, diode lasers fit nicely into the picture of low-cost microchip laser systems. To keep the cost of the system low, it is important that the coupling of the diode to the microchip laser be performed inexpensively. The use of a flat–flat laser cavity eliminates any critical alignment between the diode and the laser and makes the assembly of the system quick and simple, with the potential for inexpensive automation.

The use of simple, small, monolithic, mass-produced lasers noncritically coupled to low-cost semiconductor diode lasers gives microchip laser systems their defining characteristics – small size, robust integration of components, reliability, and the potential for low-cost mass production – and differentiates them from other miniature laser systems that are designed and constructed using more conventional techniques.

Since their early development, microchip lasers have evolved into a rich family of devices with capabilities that often exceed those of conventional lasers. The first microchip lasers developed operated continuous wave (cw),

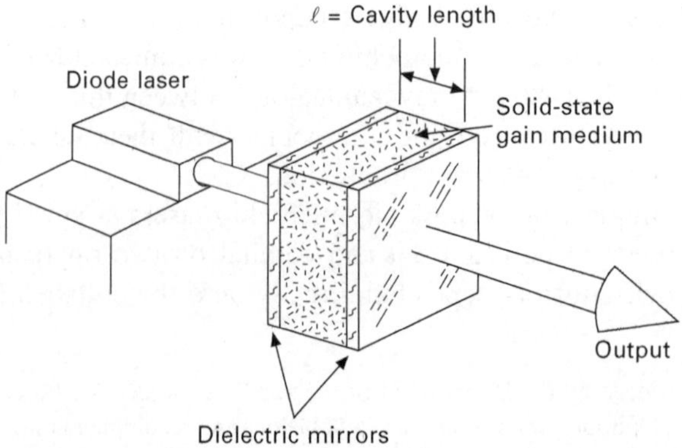

14.1 Monolithic microchip laser.

© Woodhead Publishing Limited, 2013

and a variety of cw microchip lasers were quickly demonstrated, covering a wide range of wavelengths. Some of the early applications required a modest amount of tunability, and several tuning mechanisms were incorporated into the laser structure. It was not long before researchers realized that the short cavity lengths inherent in microchip lasers gave them tremendous potential as pulsed devices. This led to the development of actively Q-switched microchip lasers, quickly followed by the most successful variation of microchip laser, the passively Q-switched microchip laser, which has demonstrated the shortest output pulse of any Q-switched solid-state laser.

14.1.1 Chapter scope and organization

Section 14.2 gives an overview of the types of microchip lasers that have been demonstrated, including monolithic and composite-cavity devices, and briefly discusses the different pump configurations that are employed. Section 14.3 discusses several mechanisms that define the transverse modes of microchip lasers. Section 14.4 talks about their spectral (longitudinal-mode) characteristics, including their single-frequency performance, linewidth, frequency stability, and tuning. The polarization of microchip lasers is briefly discussed in Section 14.5. One of the more interesting aspects of microchip lasers is their ability to produce short, high-peak-power pulses. Actively and passively Q-switched microchip lasers are discussed in Section 14.6, along with a brief overview of gain-switched and mode-locked devices. The ability to extend the wavelength coverage of microchip laser systems using nonlinear optical techniques has proven to be very important for many applications and is briefly covered in Section 14.7. Section 14.8 looks quickly at amplified systems. Finally, Section 14.9 takes a look at where this technology may be headed.

The goal of this chapter is to give the reader a broad overview of the issues associated with microchip lasers, the physics that define their characteristics, and their capabilities. It is assumed that the reader has some familiarity with lasers and there is no attempt to make the discussion completely self-contained. There are no theoretical derivations, and equations are kept to a minimum. To complement this, there is extensive referencing to allow the interested reader to quickly find more detailed and formal material. Alternatively, a single-source, comprehensive, formal treatment of the material is given by Zayhowski *et al.* (2007).

It is worth noting that the literature contains several reports of 'microchip lasers' that do not satisfy the defining characteristics listed above – small size, robust integration of components, reliability, and the potential for low-cost mass production. In many cases, these lasers employ a cavity fabricated from discrete optical components with a 'microchip' gain element. They will not be discussed further in this chapter.

© Woodhead Publishing Limited, 2013

14.2 Microchip lasers: a broadly applicable concept

14.2.1 Monolithic microchip lasers

In its simplest embodiment a microchip laser consists of a small piece of solid-state gain medium polished flat and parallel on two sides. Cavity mirrors are dielectrically deposited directly onto the polished surfaces and the laser is longitudinally pumped with a semiconductor diode laser. Microchip lasers that consist of a single dielectrically coated material are referred to as monolithic.

The earliest microchip lasers were monolithic devices based on optical transitions in the Nd^{3+} ion near 1.06 or 1.32 μm, using a variety of gain media including Nd:YAG (Zayhowski and Mooradian, 1989a), NPP (Zayhowski and Mooradian, 1989a), LNP (Zayhowski and Mooradian, 1989c; Dixon *et al.*, 1989), Nd:GSGG (Zayhowski and Mooradian, 1989a), Nd:YVO_4 (Taira *et al.*, 1991), Nd:$LaMgAl_{11}O_{19}$ (Mermilliod *et al.*, 1991), Nd:YCeAG (Gavrilovic *et al.*, 1992), Nd:YLF (Leilabady *et al.*, 1992), $Nd_xY_{1-x}Al_3(BO_3)_4$ (Amano, 1992), Nd:La_2O_2S (Zarrabi *et al.*, 1993), Nd:MgO:$LiNbO_3$ (MacKinnon *et al.*, 1994), and Cr,Nd:YAG (Yao *et al.*, 1994). It was not long, however, before other active ions were investigated and monolithic microchip lasers were constructed at a variety of wavelengths. Cr:LiSAF microchip lasers were demonstrated in the 0.8- to 1.0-μm spectral region (Sutherland *et al.*, 1995); Yb:YAG was used in microchip lasers operating at 1.05 μm (Fan, 1994); Yb,Er:glass devices were demonstrated at 1.5 μm for applications in fiber-optic communications (Laporta *et al.*, 1993); Tm (Zayhowski *et al.*, 1995a) and Ho (Storm and Rohrbach, 1989; Harrison and Martinsen, 1994a) microchip lasers operating near 2 μm were built for remote sensing; Cr^{2+}-doped chalcogenide devices were demonstrated near 2.3 and 2.5 μm (Mirov *et al.*, 2002); and 3-μm microchip lasers were investigated for medical applications (Harrison and Martinsen, 1994b).

Of the gain media demonstrated in microchip lasers, Nd:YAG and Nd:YVO_4 are the most developed and the most commonly used, but Yb:YAG is becoming more popular, particularly in passively *Q*-switched devices because of its longer upper-state lifetime. All of the lasers listed above operated in a near-ideal TEM_{00} mode, and several exhibited single-frequency (single-longitudinal-mode) performance with a narrow linewidth. The $Nd_xY_{1-x}Al_3(BO_3)_4$ microchip laser reported by Amano (1992) had another very interesting characteristic. In addition to being a good gain medium, $Nd_xY_{1-x}Al_3(BO_3)_4$ has a second-order nonlinearity that allows it to perform harmonic generation, and the microchip laser was 'self-frequency-doubled' to produce green output at 531 nm. The Nd:MgO:$LiNbO_3$ microchip laser reported by MacKinnon *et al.* (1994) was electro-optically tunable. The doubly doped Cr,Nd:YAG laser reported by Yao *et al.* (1994) is the only one of the lasers listed above that was not cw – it was self-passively *Q*-switched.

© Woodhead Publishing Limited, 2013

14.2.2 Composite-cavity microchip lasers

The search for multifunctional gain media – gain media that simultaneously act as harmonic converters, electro-optical material, or passive *Q*-switches – has led to several interesting monolithic devices. However, the multifunctional media are often less than ideal in one or both of their functions, and/or are difficult or expensive to grow. Higher-performance devices can be obtained by combining two or more specialized materials within the same composite cavity. In composite-cavity microchip lasers the constituent materials are bonded together to form a quasi-monolithic device, with dielectric mirrors deposited (or bonded) on the outer surfaces.

Several issues must be considered before committing to a composite-cavity microchip laser design. Different materials have different refractive indices. Does the interface between the two materials need to be antireflecting or is some level of reflectivity acceptable, or even desirable? Should the interface be tilted with respect to the optic axis of the cavity to avoid coupled-cavity effects, and if so does that have an undesirable impact on the potential to mass-produce the lasers? Different materials have different thermal expansion coefficients. How will the materials be bonded together? If a bonding agent is used it must have acceptable optical, mechanical, and thermal properties. If contact or diffusion bonding is used, will it be robust enough to hold up over the range of operating and storage conditions the device will be exposed to? If the elements are simply placed in close proximity to each other and held in place by an external fixture, will that be robust and reliable enough for the intended application, and does it allow for cost-effective production? If the answers to all of these questions are not satisfactory, the less-than-ideal performance of a monolithic device, if it can be constructed, may be the best solution.

Composite cavities have been successfully employed in harmonically converted green (MacKinnon and Sinclair, 1994) and blue (Matthews *et al.*, 1996) cw microchip lasers, coupled-cavity single-frequency microchip lasers using gain media with a broad gain profile (Sutherland *et al.*, 1995), electro-optically tunable microchip lasers (Zayhowski *et al.*, 1993), actively *Q*-switched microchip lasers (Zayhowski and Dill, 1992, 1995), passively *Q*-switched microchip lasers (Zayhowski and Dill, 1994; Zayhowski, 2000), and Raman-shifted passively *Q*-switched microchip lasers (Grabtchikov *et al.*, 2002; Demidovich *et al.*, 2003).

14.2.3 Pumping microchip lasers

Microchip lasers are pumped by semiconductor diode lasers. The simplest configuration places the microchip laser in close proximity to the output facet of the diode with no intervening optics (Zayhowski and Mooradian, 1989a).

© Woodhead Publishing Limited, 2013

As the amount of pump power increases, the diameter of the oscillating mode in a microchip laser usually decreases (see Section 14.3) and, for this proximity-coupled configuration, the typically large divergence of the diode output overfills the oscillating mode volume, resulting in inefficient operation or multi-transverse-mode oscillation. The situation can be improved by putting a lens between the diode and the microchip cavity, which is common practice in moderate- or high-power microchip laser systems (output powers in excess of several tens of milliwatts).

An alternative configuration for pumping microchip lasers uses a fiber-coupled diode laser. This configuration decouples the diode-laser system from the microchip cavity and offers several practical advantages. It facilitates an extremely compact laser head which can fit into small places, coupled to the rest of the system by a single, flexible optical fiber. This was an essential feature when an ultraviolet microchip laser was used as part of a cone penetrometer system to probe subsurface soil contamination (Zayhowski and Johnson, 1996; Bloch *et al.*, 1998). High-power microchip lasers are often pumped by beam-combined diode-laser arrays. In this case, fiber coupling allows a convenient separation of the engineering of the pump subsystem from the microchip cavity, and the fiber serves the secondary function of shaping the combined output of the diode-laser array. Fiber coupling also makes it easy to independently control the temperature of the diodes and the microchip cavity, simplifies handling of the diode by the person assembling the system, and makes it easy to change diodes in case of failure. The costs of fiber coupling generally include some loss of optical power at the input to the fiber, an increase in the pump étendue, and an increased complexity of the pump source.

14.3 Transverse mode definition

The devices discussed in Section 14.2 all use a flat–flat cavity design. The eigenmodes of a flat–flat cavity are plane waves, yet all of the lasers operated in a near-ideal, TEM_{00} mode with a well-defined mode radius. What is it that determines the transverse mode dimensions of a flat–flat microchip laser? The mode-defining mechanisms vary depending on the gain medium, other media within a composite-cavity device, and pump power.

14.3.1 Thermal guiding

Thermal waveguiding/lensing

For many microchip lasers the transverse mode is determined primarily by thermal effects (Zayhowski, 1991a). When a microchip laser is longitudinally pumped, the pump beam deposits heat as it pumps the gain medium. For microchip lasers that are short relative to the absorption length of the pump

© Woodhead Publishing Limited, 2013

light, the heat is deposited nearly uniformly along the cavity length and generally results in a radially symmetric temperature distribution. In materials with a positive change in refractive index with temperature, such as Nd:YAG, this creates a thermal waveguide. Weak waveguides support a single mode, the HE_{11} mode (which is closely matched to the TEM_{00} free-space mode), with a relatively large mode diameter. As the gain medium is pumped harder waveguiding becomes stronger, the diameter of the HE_{11} mode decreases, and additional transverse modes are supported. However, as long as most of the pump power is deposited within the HE_{11} mode volume, the laser will continue to oscillate in the fundamental mode. At high-enough pump powers, the diameter of the fundamental mode becomes significantly smaller than the pump diameter and multi-transverse-mode operation sets in.

When the absorption length for the pump light is short compared to the cavity length, it is more natural to think of the thermally induced index change as creating a lens at the pump end of the cavity. For cavities with a large Fresnel number, as is the case with microchip lasers, the result is the same; a system with a lens can be modeled as a waveguide and vice versa.

Thermal strain

The heat deposited by the pump also induces curvature of the cavity mirrors when they are deposited directly on the gain medium (or other absorbing media in a composite-cavity device). For materials with a positive thermal expansion coefficient this contributes to the stabilization of the transverse mode (MacKinnon and Sinclair, 1992). In some materials, such as Nd:YLF, this term dominates and can lead to stable-transverse-mode operation in an otherwise flat–flat cavity despite a negative change in index. Nonuniform thermal expansion of the gain medium also results in strain-induced index variations. This effect tends to be less important than others in determining the cavity's transverse mode characteristics and can usually be ignored.

14.3.2 Aperture guiding

In three-level or quasi-three-level lasers there may be significant absorption of the oscillating radiation in unpumped regions of the gain medium. For longitudinally pumped devices, this creates a radially dependent loss (aperture) that can restrict the transverse dimensions of the lasing mode, resulting in a smaller mode radius than predicted by thermal effects alone. Aperture guiding can be important in Yb:YAG microchip lasers (Fan, 1994). A saturable loss at the oscillating wavelength has a similar effect. In passively *Q*-switched devices it results in a dynamic aperture that opens as the *Q*-switched pulse forms and is an important mode-defining mechanism (Yao *et al.*, 1994; Arvidsson, 2001).

© Woodhead Publishing Limited, 2013

14.3.3 Gain guiding

The absence of gain in the unpumped regions of the gain medium can be sufficient to define a stable transverse mode (Kogelnik, 1965). This effect is similar to aperture guiding. It is usually insignificant compared to thermal guiding, but can be important in low-duty-cycle lasers where thermal effects are minimized. It can also be important in lasers operating near threshold (Kemp *et al.*, 1999).

14.3.4 Gain-related index guiding

Optical gain provides dispersion. Laser modes that are spectrally detuned from the center of the gain profile will see a refractive index profile that is a function of their detuning. As a result of gain-related index guiding, each of the longitudinal modes of a microchip laser has a slightly different spatial profile, a different amount of overlap with the pump, and a different far-field divergence. Gain-related index guiding has been shown to play an important role in $Nd:YVO_4$ microchip lasers (Kemp *et al.*, 1999), and can lead to interesting effects such as self *Q*-switching (Conroy *et al.*, 1998).

14.3.5 Fabrication tolerances

For a flat–flat laser cavity to create a circularly symmetric fundamental transverse mode, parallelism between the cavity mirrors is critical. How parallel the mirrors must be is directly related to the strength of the guiding mechanisms discussed above. When thermal mechanisms dominate, as is usually the case for cw devices, the strength of the guiding increases with pump power (Zayhowski, 1991a). Low-power flat–flat microchip lasers must have extremely parallel faces on the cavity; the parallelism requirement is greatly relaxed for higher-power devices.

14.3.6 Curved mirrors

Methods consistent with low-cost mass fabrication have been developed to put curved mirrors on microchip lasers (Zayhowski, 1990b; Rabarot *et al.*, 1997). Curved mirrors can stabilize the transverse mode of the laser when the mechanisms discussed above are not strong enough to do so. In practice, they can reduce the threshold of cw microchip lasers and make low-power operation more consistent from device to device, at the expense of increased fabrication cost and tighter tolerances on pump alignment. For a large variety of gain media they are not needed, especially when the laser is pumped with medium- or high-power diodes (typically more than several tens of milliwatts).

© Woodhead Publishing Limited, 2013

14.3.7 Pump considerations

To ensure oscillation in the fundamental transverse mode, it is necessary for the fundamental mode to use most of the gain available to the laser. If the radius of the fundamental mode is much smaller than the radius of the pumped region of the gain medium, higher-order transverse modes will oscillate.

14.4 Spectral properties

14.4.1 Single-frequency operation

For most solid-state lasers the free spectral range of the laser cavity is much less than the gain bandwidth of the active medium. For example, the free spectral range of a 10-cm-long empty cavity is 1.5 GHz; the gain bandwidth of commonly used solid-state laser media is greater than 100 GHz. As a result, solid-state lasers tend to oscillate at several frequencies simultaneously. Part of the original microchip laser concept involved making the laser cavity sufficiently short that the free spectral range of the cavity was comparable to the gain bandwidth and single-frequency cw operation could be obtained.

Spatial hole burning and energy diffusion

Lasers will typically oscillate in a single longitudinal mode just above threshold. In a laser with a homogeneously broadened gain medium and uniform intracavity optical field, the first mode to oscillate will deplete the optical gain and clamp the inversion at its threshold value, such that no other mode should reach threshold. Multimode oscillation occurs when the optical field in the cavity is not uniform. Microchip lasers typically use a standing-wave cavity. In a standing-wave cavity light traveling toward and away from the output coupler coherently superimpose to form a sinusoidal intensity distribution. At the maxima of the intensity distribution there is strong gain saturation and the population inversion is depleted. However, at nulls in the optical field the oscillating mode is unable to deplete the inversion. As a result, the inversion density is not uniform, but has 'holes' at the positions corresponding to the peaks in the optical intensity and large buildups at its nulls. This phenomenon is known as spatial hole burning. The gain at the nulls in the optical field will continue to increase as the gain medium is pumped harder. Because other cavity modes have a different longitudinal profile than the first mode to oscillate and can use the population inversion at these positions, this can lead to multimode operation.

Spatial hole burning can be mitigated by energy diffusion. Energy diffusion moves some of the inverted population away from the nulls in the optical field toward the maxima, where it can be effectively depleted. How far above threshold a standing-wave laser will operate in a single longitudinal mode

© Woodhead Publishing Limited, 2013

is determined by the optical length of the cavity, the effects of spatial hole burning and energy diffusion, and the spectral characteristics of the gain medium (Zayhowski, 1990a, 1990c).

Methods of obtaining single-frequency operation

Detailed analysis shows that how far above threshold a cw laser will oscillate in a single longitudinal mode is a strong function of cavity length (Zayhowski, 1990a, 1990c) and that robust single-mode operation can be achieved well above threshold even when the free spectral range of the cavity is considerably less than the gain bandwidth if one of the cavity modes is near the center of the gain peak and only a single gain transition needs to be considered.

Most gain media have several gain lines. If the laser is intended to operate on the strongest gain line, it may be possible to ignore the others. If not, it is often possible to discriminate between gain lines with properly designed dielectric cavity mirrors. For closely spaced gain transitions of comparable strength, the length of the cavity can be chosen such that if a cavity mode falls at the peak of the desired transition, other cavity modes straddle the undesired transitions, reducing the gain cross-section of those transitions for those modes. Finally, if there are no better options, a coupled-cavity composite microchip laser can provide the needed frequency discrimination (Sutherland *et al.*, 1995); or a 'twisted-mode' laser, which uses a quarterwave plate on either side of the gain medium in a composite two-mirror cavity, can be used to eliminate spatial hole burning (Wallmeroth, 1990).

Robust single-frequency cw operation still requires that one of the cavity modes falls near the peak of the desired gain transition. This necessitates precise, subwavelength control of the cavity's optical length, and is most often accomplished by thermal control of the cavity with the appropriate feedback.

Single-frequency operation of monolithic microchip lasers often requires a compromise of device efficiency. In many gain media, including Nd:YAG, the cavity length required to obtain robust single-frequency operation is comparable to or shorter than the absorption length of the gain medium at the pump wavelength. Materials with very short pump absorption lengths have a twofold advantage in obtaining efficient single-frequency operation, since for a given length of gain media they absorb more pump power, and a short absorption length mitigates the effects of spatial hole burning such that a longer cavity can be used for the same level of discrimination against multimode operation (Zayhowski, 1990a, 1990c).

Single-frequency operation of Q-switched microchip lasers

When *Q*-switched lasers operate in more than one longitudinal mode the output pulse is intensity modulated as a result of mode beating; single-

© Woodhead Publishing Limited, 2013

longitudinal-mode lasers produce pulses with a smoothly varying temporal profile. Smooth temporal profiles are desirable for many applications in high-resolution ranging (Abshire *et al.*, 2000; Afzal *et al.*, 1997; Degnan, 1993, 1999), altimetry (Degnan *et al.*, 2001), and light detection and ranging (LIDAR). They also minimize the potential for field-induced optical damage in the laser.

It is easier to achieve single-longitudinal-mode operation for *Q*-switched microchip lasers than for cw devices. This is because the mode with the largest gain will build up more quickly than competing modes, and once the pulse forms it will drive the optical inversion well below its cw threshold. In cw operation, the oscillating modes leave the inversion at its cw threshold. Often, each pulse of a *Q*-switched microchip laser is single frequency, but at high repetition rates (when the interpulse timing is short compared to the relaxation time of the gain medium) spatial hole burning creates a situation where consecutive pulses oscillate in different longitudinal cavity modes. For the same reason, the afterpulse (see Section 14.6.1) associated with a given *Q*-switched pulse is often in a different longitudinal mode.

In an actively *Q*-switched laser all potential lasing modes start to build up when the cavity *Q* is switched. All modes start from spontaneous emission and the amount of gain seen by the mode determines how quickly it builds up. In a passively *Q*-switched laser, the cavity *Q* is not switched until after a mode starts to build up; it is the light in the first mode that causes *Q*-switching to occur. As a result, when the cavity *Q* switches, that initial mode already contains a significant photon population and has a head start in the race to form a pulse. Since the initial mode has the highest net gain, the number of photons in the first mode at any time during the pulse buildup will exceed the number in any other mode by a factor greater than the number of photons in the first mode at the time a second mode reaches threshold. As a result, single-frequency pulses are easily achieved in passively *Q*-switched systems.

14.4.2 Fundamental linewidth

Spontaneous emission

One contribution to the spectral width of all lasers is the coupling of spontaneous emission to the oscillating mode (Schawlow and Townes, 1958; Lax, 1966). This contribution has a Lorentzian power spectrum with a width that scales inversely with the output power of the laser, and spectral broadening occurs on a time scale corresponding to the relaxation frequency of the laser. For many ultrastable lasers this contribution alone determines the fundamental linewidth, and it is common practice to obtain the linewidth of such lasers by fitting the tails of the measured power spectrum to a Lorentzian curve.

© Woodhead Publishing Limited, 2013

Fundamental thermal fluctuations

In very small lasers, such as microchip lasers, there is a second important contribution to the fundamental linewidth: thermal fluctuations of the cavity length at a constant temperature (Jaseja *et al.*, 1963). These fluctuations result in a Gaussian power spectrum that, for monolithic devices, scales as the inverse of the oscillating mode volume (Zayhowski, 1990b). This type of spectral broadening occurs on a time scale consistent with the acoustic resonances of the laser (or its components). In lasers with a small mode volume the contribution due to thermal fluctuations can be much larger than that due to spontaneous emission. However, because a Gaussian curve decays more quickly than a Lorentzian curve, the tails of the power spectrum still correspond to the Lorentzian contribution. Nd:YAG microchip lasers with a cavity length of ~1 mm typically have a Gaussian spectral profile with a linewidth between 5 and 7 kHz, with spectral tails corresponding to a Lorentzian contribution of only a few hertz (Zayhowski, 1990b).

Technical noise

Other factors that contribute to the linewidth of microchip lasers include fluctuations of the pump power, optical feedback to the laser cavity, mechanical vibrations, and temperature variations. These contributions are less fundamental, however, and in principle can be controlled. In addition, they tend to occur on a longer time scale than spontaneous emission and fundamental thermal fluctuations. Although they are often not important factors in attempts to measure the fundamental linewidth of a laser, they are important contributions to frequency fluctuations of microchip lasers on time scales of practical interest.

Transform-limited pulses

Single-longitudinal-mode *Q*-switched microchip lasers typically have a Fourier-transform-limited optical spectrum that is much broader than the fundamental linewidth of cw devices.

14.4.3 Frequency tuning

Frequency tuning of a laser can occur in one of two ways. If the longitudinal mode spacing of the laser cavity is less than the gain bandwidth, the cavity is capable of supporting several modes at different frequencies. A single frequency can be selected through the insertion of one or more frequency-selective elements. Typically, a small repositioning of the frequency-selective element results in the selection of a different longitudinal mode at a new

© Woodhead Publishing Limited, 2013

operating frequency. The frequency-selective element is used to select one of several cavity modes and discrete tuning is obtained. Rapid tuning can be obtained through the use of electro-optic components. This method of frequency tuning is common in large cavities composed of discrete components, but much less common in microchip lasers.

The other way a laser can be tuned is to change the frequency of a given cavity mode by changing the cavity's optical length (length times refractive index). Because the optical length can be changed continuously, this leads to continuous tuning. This type of tuning is limited by the cavity's free spectral range, except under transient conditions. Once the change in the cavity's optical length shifts the modes by a full free spectral range an adjacent cavity mode is positioned at the frequency where the initial mode started. For the same reasons that the initial mode was originally favored, the adjacent mode is now favored and the laser will have a tendency to mode hop (the longitudinal mode number will change by one), returning to the original frequency.

The optical length of the elements in a laser cavity can be changed using a variety of techniques including thermal tuning, stress tuning, and electro-optic tuning. Pump-power modulation represents a special case of thermal tuning. Each of these techniques allows continuous frequency modulation of a single longitudinal cavity mode and is discussed below.

Thermal tuning

Changing the temperature of an element in a laser cavity, or the entire cavity, is often the simplest way to tune a laser. A change in temperature usually results in a change to both the physical length of the component and its refractive index, both of which affect the optical length. In multi-element cavities the total change in the cavity's optical length is the sum of the changes in all intracavity elements. The response time of the cavity is limited by a thermal diffusion time, which is typically several milliseconds or more. Thermal tuning, or control, is often necessary to maintain single-longitudinal-mode operation of cw microchip lasers, but is often too slow when applications require the dynamic control of the laser's oscillating frequency.

Stress tuning

The cavity modes of a resonator will also tune as elements within the cavity are squeezed. For squeezing transverse to the resonator's optical axis, the main effect is usually an elongation of the material along the axis (Owyoung and Esherick, 1987). Superimposed on this is the stress-optic effect. In crystals with cubic symmetry, the stress-optic effect can split the frequency degeneracy of orthogonally polarized optical modes because the stress-optic

© Woodhead Publishing Limited, 2013

coefficients are different for light polarized parallel to and perpendicular to the applied stress. For squeezing along the optical axis of the cavity there is a compression of the squeezed elements and the frequency degeneracy of orthogonally polarized modes remains unchanged.

By using a piezoelectric transducer to squeeze a monolithic Nd:YAG microchip laser, tuning was obtained at modulation frequencies up to 20 MHz, although nonresonant response was limited to ~80 kHz (Zayhowski and Mooradian, 1989b). The nonresonant tuning response was 300 kHz V^{-1}.

Electro-optic tuning

For applications including frequency-modulated optical communications and chirped coherent laser radar, high rates of tuning are required. These rates can only be achieved electro-optically. The frequency response of a laser whose optical length is varied is well understood. When a linear voltage ramp is applied to an intracavity electro-optic crystal, the frequency of the laser undergoes a series of steps whose spacing in time is the cavity round-trip time (Genack and Brewer, 1978). When the rise time of the voltage is long compared to the cavity round-trip time, if we assume that the change in the optical length of the cavity tracks the applied voltage, the frequency has an approximately linear chirp. Because two steps are required to define a modulation frequency (one step up and one step down), the maximum response frequency of the cavity is the inverse of twice the round-trip time.

The sensitivity of electro-optic voltage-to-frequency conversion increases linearly with the fraction of the cavity length occupied by the electro-optic element. For high-sensitivity tuning, it is desirable to fill the cavity with as large a fraction of electro-optic material as possible. However, it is often still important to keep the total cavity length as small as possible, for the reasons discussed above. In addition, when the electric field is applied transverse to the cavity's optical axis, as the length of the electro-optic crystal increases, the capacitance of the electrodes on the crystal increases, resulting in higher energy requirements and slower electrical response.

Most electro-optic crystals are piezoelectrically active. As a result, when a voltage is applied to the crystal, stress and refractive indices are modulated. Although the effects of stress may normally be small compared to the index modulation, a free-standing crystal can act as a high-*Q* acoustic cavity. At the acoustic resonant frequencies the piezoelectric effect can cause a greatly enhanced frequency-modulation response of the laser. Depending on the dimensions of the crystal, acoustic resonances can fall between a few kilohertz and several megahertz. To eliminate these resonances, the electro-optic material can be bonded to materials with a similar acoustic impedance

© Woodhead Publishing Limited, 2013

to transmit the electrically excited acoustic waves out of the crystal (Schulz and Henion, 1991).

For the reasons discussed above, the high-speed tuning of a laser is most linear when the shortest possible laser cavity is used. Short cavity length offers an advantage in tuning range as well, because the amount of frequency tuning that can be obtained without mode hopping is limited by the cavity's free spectral range. As a result, the short length of electro-optically tuned microchip lasers makes them attractive for applications that require high sensitivity, a large continuous tuning range, and high tuning rates. Composite-cavity electro-optically tuned microchip Nd:YAG/$LiNbO_3$ lasers have been continuously tuned over a 30-GHz range with a tuning sensitivity of ~14 MHz V^{-1} (Zayhowski *et al.*, 1993). The tuning response was relatively flat for tuning rates from dc to 1.3 GHz. Monolithic Nd:MgO:$LiNbO_3$ (MacKinnon *et al.*, 1994) and Nd:$LiNbO_3$ (Vieira *et al.*, 1997, 2001) electro-optically tuned microchip lasers have also been demonstrated.

Pump-power modulation

Changes in pump power induce frequency changes in the output of solid-state lasers. As the pump power increases more heat is deposited in the gain medium, causing the temperature to rise and changing both the refractive index and length. Because frequency tuning via pump-power modulation relies on thermal effects, it is often thought to be too slow for many applications. In addition, modulating the pump power has the undesirable effect of changing the amplitude of the laser output. However, for small lasers significant frequency modulation can be obtained at relatively high modulation rates with little associated amplitude modulation (Zayhowski and Keszenheimer, 1992). For example, pump-power modulation of a 1.32-μm microchip laser has been used to obtain 10-MHz frequency modulation at a 1-kHz rate and 1-MHz frequency modulation at a 10-kHz rate, with an associated amplitude modulation of less than 5%. This technique has been employed to phase-lock two lasers and introduced less than 0.1% amplitude modulation on the slave laser (Keszenheimer *et al.*, 1992). When it can be used, pump-power modulation has advantages over other frequency modulation techniques since it requires very little power, no high-voltage electronics, no special mechanical fixturing, and no additional intracavity elements.

Whether or not frequency modulation is desired, it is important to understand the frequency response of a microchip laser to changes in pump power, because pump-power fluctuations are often responsible for a large portion of the frequency fluctuations in the laser output (Zayhowski and Keszenheimer, 1992).

© Woodhead Publishing Limited, 2013

14.5 Polarization control

It is often desirable for a laser to oscillate in a single linear polarization. This is easily achieved if the gain medium in the laser has a preferred polarization, but can be problematic for monolithic lasers with isotropic gain media. A common approach in larger cavities is to include a polarizing element, such as a Brewster plate, within the cavity. In single-frequency devices, the presence of any tilted optical element may be sufficient to polarize a laser since the reflection coefficient of surfaces is often different for s- and p-polarized waves. In monolithic microchip lasers with an isotropic gain medium it may be possible to induce polarized operation by creating anisotropic structure on one of the cavity mirrors, but a successful example of this is not readily available in the literature.

For microchip lasers with isotropic gain media, the polarization degeneracy of the gain medium can often be removed by applying uniaxial transverse stress. If there is little polarization selectivity within a laser cavity, feedback from external surfaces may determine the polarization of the oscillating mode. Although this effect is usually undesirable, it has been used to controllably switch the polarization of microchip lasers (Zayhowski, 1991b). Finally, in the absence of any strong polarizing mechanism the polarization of the pump light may determine the polarization of the laser output.

14.6 Pulsed operation

Pulsed output has been obtained from microchip lasers using a variety of techniques, including active *Q*-switching, passive *Q*-switching, gain switching, and mode locking.

14.6.1 Actively *Q*-switched microchip lasers

The shortest output pulse that can be obtained from a *Q*-switched laser has a full width at half maximum of 8.1 times the round-trip time of light in the laser cavity divided by the natural logarithm of the round-trip gain (Zayhowski and Kelley, 1991). Since microchip lasers are physically very short, they have very short cavity lifetimes. This leads to the possibility of producing very short *Q*-switched output pulses.

Optimization of a *Q*-switched microchip laser for minimum pulse width, maximum pulse energy, or maximum peak power is similar to optimization of any other laser for the same parameters. It typically corresponds to an output coupling in the high-*Q* state that results in the laser being between 3 and 3.6 times its cw threshold inversion density when the pulse begins to form, depending on the characteristic to be optimized and cavity losses (Degnan, 1989; Zayhowski and Kelley, 1991).

© Woodhead Publishing Limited, 2013

Coupled-cavity Q-switched microchip lasers

The trick with a microchip laser is to get the *Q*-switching mechanism into the cavity without sacrificing the short cavity length or compromising other desirable properties associated with a microchip laser. Coupled-cavity electro-optically *Q*-switched microchip lasers (illustrated in Fig. 14.2) are one solution to this problem. They have, to date, demonstrated the shortest *Q*-switched pulses obtained from an actively *Q*-switched solid-state laser. From a Nd:YAG device, pulse widths of 270 ps have been obtained and repetition rates up to 500 kHz were demonstrated (Zayhowski and Dill, 1992). From a Nd:YVO_4 microchip laser, pulse widths of 115 ps and repetition rates of 2.25 MHz were obtained, with peak output powers as high as 90 kW when pumped with a 1-W laser diode (Zayhowski and Dill, 1995).

The principle behind the operation of a coupled-cavity *Q*-switched laser is that an etalon containing an electro-optic element serves as a variable-reflectivity output coupler for a gain cavity defined by two reflective surfaces adjacent to the gain medium. The reflectivity of the etalon for the potential lasing frequencies of the device is controlled by a voltage applied to a pair of electrodes in contact with the electro-optic material. The potential lasing modes are determined primarily by the gain cavity. In the low-*Q* state, the etalon must have high transmission for all potential lasing modes so that none can reach threshold. In the high-*Q* state, the reflectivity of the etalon is high for the desired mode and a *Q*-switched output pulse develops. To ensure that all potential modes of the gain cavity can be simultaneously suppressed, the optical length of the etalon must be nearly an integral multiple of the optical length of the gain cavity. The higher the *Q* of the etalon, the tighter the tolerance on its length. For an isotropic gain medium such as Nd:YAG,

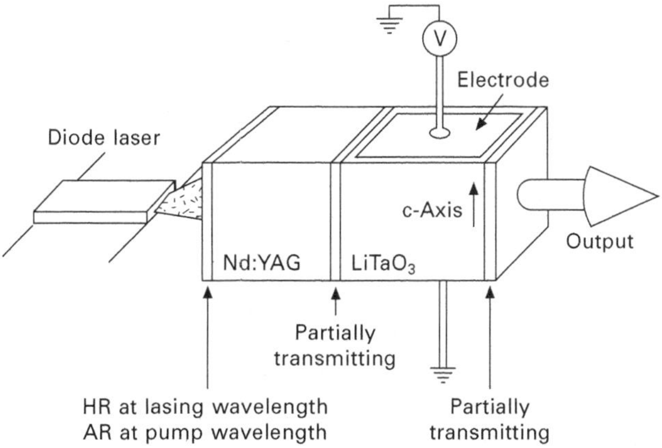

14.2 Coupled-cavity electro-optically *Q*-switched microchip laser; HR = highly reflective, AR = antireflective.

the length tolerance imposes a restriction on the birefringence of the electro-optic material, because in the low-Q state oscillation must be suppressed for modes of both polarizations.

Afterpulsing

The first output pulse that forms after a laser cavity is Q-switched will not extract all of the energy stored in the gain medium. This is particularly true for single-frequency standing-wave cavities because of spatial hole burning. The residual gain left by the first output pulse can contribute to the continued development of a pulse in a competing longitudinal mode, resulting in a second pulse that is delayed in time with respect to the first pulse and typically much smaller in magnitude. This type of afterpulse is minimized when there is a large difference in the gain of the primary and competing modes, which is most easily accomplished in cavities with a large mode spacing (short cavity length) and when the primary mode is positioned at the peak of the gain profile. In microchip lasers, the latter can often be accomplished by controlling either the pump power to the laser or its temperature, or by applying a controlled amount of stress to one of the intracavity elements (see Section 14.4.3). As the primary pulse gets closer to the peak of the gain profile the time delay between the primary pulse and the afterpulse increases, and the magnitude of the afterpulse decreases.

There is an additional mechanism for afterpulsing when the thermalization times of the laser manifolds are long compared to the Q-switched output pulse. Thermalization of the manifolds can rapidly pump the laser transition and lead to gain-switched pulses following the primary Q-switched pulse. These gain-switched pulses can be in the same longitudinal mode as the primary pulse and can be very stable in amplitude and time.

In actively Q-switched lasers the afterpulse can be prevented by rapidly decreasing the cavity Q after the first output pulse exits the system.

Pulse bifurcation

In Q-switched lasers operating at high pulse repetition rates, when the pulse-to-pulse spacing is comparable to or shorter than the upper-state lifetime of the gain medium, some of the energy not extracted by one pulse (during one Q-switching cycle) will still be present in the laser cavity when the next pulse forms (during the next cycle). As a result of spatial hole burning the residual gain may be greater at positions where it favors the development of a pulse in a different longitudinal mode than the initial pulse. This effect will be strongest in high-Q standing-wave cavities and when the difference between the thresholds of the two modes is small. The second pulse will then preferentially deplete the gain most available to it, creating a situation

© Woodhead Publishing Limited, 2013

where the first mode may again be favored in the subsequent pulse. As a result, it is not uncommon to see pulse bifurcation in high-repetition-rate *Q*-switched lasers, with pulses alternating between two longitudinal modes. At still higher pulse repetition rates more modes can come into play, creating trains of three or more modes occurring in a regular sequence or, depending on the stability of the system, chaotically.

When the gain differential between modes is extremely small, even small amounts of spatial hole burning can cause pulse bifurcation. The effect can be significant even when the pulse-to-pulse spacing is several times the spontaneous lifetime of the gain medium. In systems with small mode-gain differentials operating at high repetition rates, time-averaged spectra will show the presence of multiple oscillating modes – this should not be interpreted as proof of multimode pulses; each individual pulse may still be single frequency with a smooth temporal profile, free of mode beating. Because of gain-related index-guiding effects (see Section 14.3.4), pulses in different modes will have slightly different energies and transverse profiles.

14.6.2 Passively *Q*-switched microchip lasers

Passive *Q*-switching provides an attractive alternative to active *Q*-switching for many applications. A passively *Q*-switched laser contains a gain medium and a saturable absorber. As the gain medium is pumped it accumulates energy and emits photons. Over many round trips in the resonator the photon flux sees gain, fixed loss, and saturable loss. If the gain medium saturates before the saturable absorber, the laser will tend to oscillate cw. On the other hand, if the photon flux builds up to a level that saturates, or bleaches, the saturable absorber first, the resonator will see a dramatic reduction in intracavity loss and the laser will *Q*-switch, generating a short, intense pulse of light, without the need for any high-voltage or high-speed switching electronics.

To manage thermal effects, high-power passively *Q*-switched microchip lasers are often pulse pumped. A common implementation uses an external clock signal to turn the pump diodes on and a signal generated by the *Q*-switched output pulse to turn them off. By operating high-power pump diodes at a low duty cycle, larger amounts of energy can be stored in the gain medium of the laser with reduced thermal loading. This results in a larger oscillating mode diameter (see Section 14.3) and more energetic pulses.

In addition to simplicity of implementation, the advantages of a passively *Q*-switched laser include the generation of pulses with a well-defined energy and duration that are insensitive to pumping conditions. In contrast to actively *Q*-switched lasers, the population inversion at the time the cavity *Q*-switches is fixed by material parameters and the design of the laser. In the absence of pulse bifurcation it is identical from pulse to pulse. This can result in extremely stable pulse energies and pulse widths, with measured stabilities

© Woodhead Publishing Limited, 2013

of better than 1 part in 10^4. The stability of the pulse energy and pulse width is achieved at the expense of pulse timing stability. Pulse formation does not occur until the population inversion reaches the proper value. Fluctuations in the pump source are the primary cause of pulse-to-pulse timing jitter in many passively *Q*-switched lasers.

Optimization of passively *Q*-switched microchip lasers for minimum pulse width, maximum pulse energy, or maximum peak power is similar to optimization of actively *Q*-switched devices when the gain cross-section of the gain medium is much smaller than the absorption cross-section of the saturable absorber. As the gain cross-section of the gain medium approaches the absorption cross-section of the saturable absorber, the minimum possible pulse width gets larger, both the maximum pulse energy and the maximum peak power get smaller, and the optimal output coupling decreases (Degnan, 1995). When the gain cross-section of the gain medium is larger than the absorption cross-section of the saturable absorber the laser will tend to oscillate cw, unless the diameter of the oscillating mode is smaller in the saturable absorber, which is typically not the case in a microchip laser.

Bulk saturable absorbers

The most commonly used saturable absorber for passive *Q*-switching of microchip lasers is Cr^{4+}:YAG. It has been used to *Q*-switch Nd:YAG microchip lasers operating at 1.064 μm (Zayhowski and Dill, 1994), 946 nm (Zayhowski *et al.*, 1996), and 1.074 μm (Zayhowski *et al.*, 2000); Nd:YVO_4 microchip lasers operating at 1.064 μm (Jaspan *et al.*, 2000); Nd:$GdVO_4$ microchip lasers operating at 1.062 μm (Liu *et al.*, 2003a); and Yb:YAG microchip lasers operating at 1.03 μm (Zhou *et al.*, 2003; Dong *et al.*, 2006).

The combination of Cr^{4+}:YAG and a doped-YAG gain medium, such as Nd:YAG, is particularly attractive from the point of view of an extremely robust device. Since both materials use the same host crystal, YAG, they can be diffusion-bonded to each other in a way that blurs the distinction between a monolithic and a composite-cavity device. Both materials have the same thermal and mechanical properties and the same refractive index, and the bond between them can be sufficiently strong that the composite device acts in all ways as if it were a single crystal. An early, low-power, diffusion-bonded Nd:YAG/Cr^{4+}:YAG microchip laser is shown in Fig. 14.3. As an alternative to diffusion bonding, Nd:YAG can be epitaxially grown on Cr^{4+}:YAG and vice versa, using nonequilibrium growth techniques that can produce Nd or Cr^{4+} concentrations that could not otherwise be achieved. Finally, Cr^{4+},Nd:YAG can be grown as a single crystal, although this approach precludes some of the device optimization that can be achieved when the two materials are physically distinct (Zayhowski and Wilson, 2003a).

Typically Nd:YAG/Cr^{4+}:YAG passively *Q*-switched microchip lasers are

© Woodhead Publishing Limited, 2013

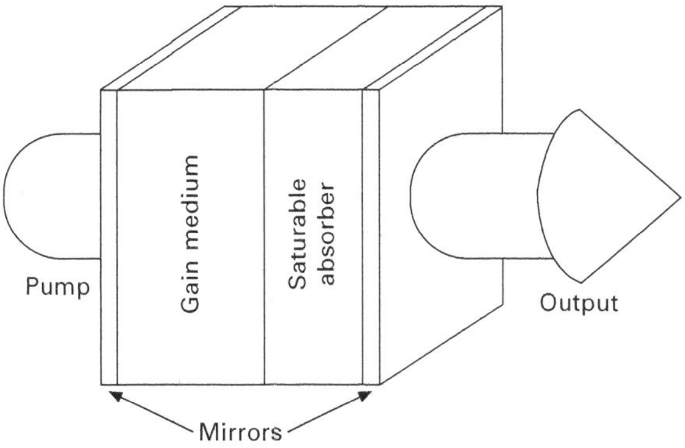

(a)

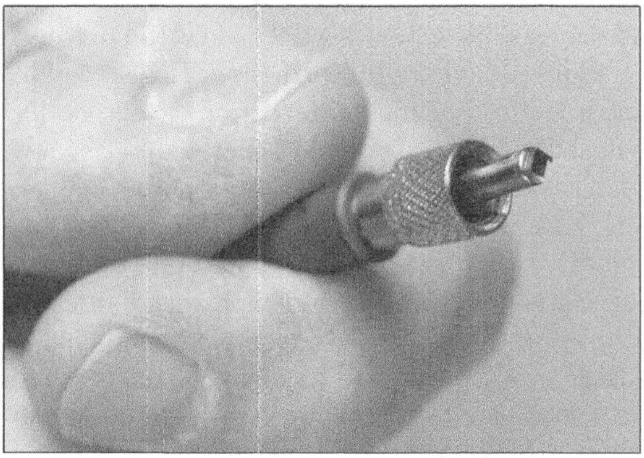

(b)

14.3 Composite-cavity Nd:YAG/Cr^{4+}:YAG passively *Q*-switched microchip laser: (a) schematic and (b) photograph of laser bonded to ferrule of pump fiber.

operated with an output coupling equal to the single-pass loss of the saturable absorber. The minimum pulse width that can be obtained is limited by the average inversion density that can be achieved within the oscillating mode volume, pump-induced bleaching of the saturable absorber, and a reduction in the gain cross-section caused by heating of the gain medium as the laser is pumped harder (Zayhowski and Wilson, 2003a; Jaspan *et al.*, 2004). Lasers using Cr^{4+},Nd:YAG as both the gain medium and the saturable absorber have demonstrated 1.064-μm output pulses with durations as short as 290 ps (Wang *et al.*, 1995). Pump-induced bleaching of the saturable absorber is reduced in composite-cavity Nd:YAG/Cr^{4+}:YAG microchip lasers. Devices containing a Nd:YAG gain medium diffusion bonded to a Cr^{4+}:YAG saturable absorber have generated pulses as short as 200 ps when pumped with a cw 1-W diode (Zayhowski and Dill, 1994; Zayhowski *et al.*, 1995b). Pulse-pumping with

© Woodhead Publishing Limited, 2013

high-power diode-laser arrays allows a further reduction in pulse width, and has resulted in pulses as short as 148 ps (Zayhowski and Wilson, 2003a).

The maximum pulse energy and peak output power that can be obtained from a passively *Q*-switched Nd:YAG/Cr^{4+}:YAG microchip laser are dependent on pump power, pulse repetition rate, and cavity length. Increasing the pump power in cw-pumped devices results in thermal effects that lead to a smaller oscillating-mode diameter, limiting the amount of stored energy the fundamental mode can access and increasing the fluence of the oscillating mode for a given pulse energy, which ultimately leads to damage. If the diameter of the fundamental oscillating mode becomes much smaller than that of the pumped volume, multi-transverse-mode operation occurs. The superposition of the electric fields from the multiple modes causes spatial hotspots in the laser beam and effectively reduces the damage threshold of the device. The decrease in mode diameter with pump powers can be mitigated by using a longer cavity (at the expense of pulse width and peak power), but at some point the benefits of a microchip laser are lost.

Passively *Q*-switched Nd:YAG/Cr^{4+}:YAG microchip lasers cw pumped with a 1-W diode laser typically generate 1.064-μm pulses with energies up to ~15 μJ, with peak powers up to ~30 kW, at pulse repetition rates of ~10 kHz (Zayhowski and Dill, 1994; Zayhowski *et al.*, 1995b). Less energetic pulses, with lower peak powers, have been demonstrated at repetition rates up to 110 kHz (Gong *et al.*, 2006). By pulse pumping microchip lasers with high-power diode-laser arrays, pulse energies up to ~30 μJ, with peak powers up to ~100 kW, are achieved at repetition rates of ~10 kHz (Zayhowski and Wilson, 2003a); pulse energies of ~250 μJ, with peak powers up to ~500 kW, are obtained at repetition rates of ~1 kHz (Zayhowski, 1998); and pulse energies of several millijoules, with peak powers of several megawatts, are achieved at repetition rates up to ~100 Hz (Liu *et al.*, 1998; Tsunekane *et al.*, 2010; Pavel *et al.*, 2011).

The combination of Yb:YAG, with a gain transition at 1.03 μm, and Cr^{4+}:YAG has the same potential as Nd:YAG/Cr^{4+}:YAG for monolithic and quasimonolithic integration (Zhou *et al.*, 2003; Dong *et al.*, 2006). The longer upper-state lifetime of Yb:YAG (compared to Nd:YAG) is attractive for low-repetition-rate systems since it allows the gain medium to accumulate energy for a longer time, making it possible to use lower-power, less expensive diode pumps. A disadvantage of Yb:YAG is its much lower gain cross-section, which dramatically increases the amount of stored energy required to obtain a short pulse width and increases the saturation fluence. Although this increases the pulse energy and peak power for a given pulse width, it can lead to optical damage. Nonetheless, impressive short-pulse, high-energy performance has been reported; pulses as short as 237 ps have been generated with peak powers up to 720 kW at 3.5-kHz repetition rate (Dong *et al.*, 2007a).

© Woodhead Publishing Limited, 2013

To date, the best results in the eye-safe spectral region were obtained from passively *Q*-switched microchip lasers that use Yb,Er:glass as the gain medium and Co^{2+}:$MgAl_2O_4$ (Co^{2+}:MALO) as the saturable absorber. When pumped with a 1-W 975-nm diode laser, these devices have produced output pulses of ~5 ns duration at repetition rates up to 25 kHz, with peak powers as high as 1.6 kW and average output powers up to 150 mW (Denker *et al.*, 2002; Karlsson *et al.*, 2002). At low duty cycles, pulses as short as 880 ps, pulse energies as high as 110 μJ, and peak powers in excess of 35 kW have been demonstrated (Chinn and King, 2006).

Other combinations of gain medium and bulk solid-state saturable absorber have been used in *Q*-switched microchip lasers operating at a variety of wavelengths. Table 14.1 lists several gain-medium/bulk-saturable-absorber combinations that have been demonstrated in miniature or microchip lasers. Cr^{4+}:YAG has been used with numerous gain media at wavelengths between 0.91 and 1.08 μm. The Cr^{5+}-vanadates are a relatively new family of saturable absorbers that are used in the same spectral region. Although lasers based on them have not yet demonstrated the short pulse widths or high peak powers that are obtained with Cr^{4+}:YAG, they offer the potential for quasimonolithic and monolithic (Yu *et al.*, 2008) integration with vanadate gain media. V^{3+}:YAG has been used over the spectral range from 0.93 to 1.44 μm. It is most useful, however, in the long-wavelength portion of this range, where Cr^{4+}:YAG is not an option. Co^{2+}:$LaMgAl_{11}O_{19}$ (Co^{2+}:LMA) and Co^{2+}:MALO extend the coverage of solid-state saturable absorbers into the eye-safe spectral region.

Bulk semiconductor saturable absorbers have been used to *Q*-switch miniature solid-state lasers at wavelengths from 1 to 2 μm (Tsou *et al.*, 1993; Kajave and Gaeta, 1996; Gu *et al.*, 2000; Lan *et al.*, 2010; Yao *et al.*, 2010), but have a much lower damage threshold than the solid-state saturable absorbers described above and are therefore rarely used. They also have very different thermal and mechanical properties than solid-state gain media, making robust integration of miniature devices challenging.

Semiconductor saturable-absorber mirrors

Semiconductor saturable-absorber mirrors (SESAMs) (Kärtner *et al.*, 1995; Keller *et al.*, 1996; Spühler *et al.*, 1999) consist of an antiresonant semiconductor Fabry–Pérot etalon formed by a semiconductor layer grown on top of a highly reflecting semiconductor Bragg mirror and covered by a dielectric reflector. The semiconductor layer typically consists of absorptive quantum-well layers in an otherwise transparent medium. The bandgap of the quantum wells can be engineered to provide saturable absorption at a wide variety of wavelengths. In a cavity containing a SESAM, the SESAM is used as one of the cavity mirrors, with the oscillating light incident on the

© Woodhead Publishing Limited, 2013

Table 14.1 Passively *Q*-switched laser gain-medium/bulk-saturable-absorber combinations

Wavelength (μm)	Gain medium	Saturable absorber	References
0.916	Nd: $LuVO_4$	Cr^{4+}:YAG	He *et al.*, 2009
0.933, 0.936	Nd:GGG	V^{3+}:YAG	Huang *et al.*, 2010
0.935	Nd:CNGG	Cr^{4+}:YAG	Li *et al.*, 2009
0.946	Nd:YAG	Cr^{4+}:YAG	Zayhowski *et al.*, 1996
1.017	Yb:YVO_4	Cr^{5+}:YVO_4	Zolotovskaya *et al.*, 2007
1.02	Yb:$NaGd(WO_4)_2$	Cr^{4+}:YAG	Liu *et al.*, 2007b
1.03	Yb:YAG	Cr^{4+}:YAG	Zhou *et al.*, 2003; Dong *et al.*, 2006
1.03	Yb:LuAG	Cr^{4+}:YAG	Dong *et al.*, 2007b
1.03	Yb:$Y_3Sc_2Al_3O_{12}$	Cr^{4+}:YAG	Dong *et al.*, 2008
1.04	Yb:KGW	V^{3+}:YAG	Lagatsky *et al.*, 2000
1.04	Yb:YGAB	Cr^{4+}:YAG	Brenier *et al.*, 2007
1.059, 1.062	Nd:CNGG	Cr^{4+}:YAG	Yu *et al.*, 2009
1.06	Nd:$Gd_xY_{1-x}VO_4$	Cr^{4+}:YAG	Liu *et al.*, 2003b
1.06	Nd:$LuVO_4$	Cr^{4+}:YAG	Liu *et al.*, 2007a
1.06	Nd:$Lu_xGd_{1-x}VO_4$	Cr^{4+}:YAG	Yu *et al.*, 2007a
1.06	Nd:$Lu_xGd_{1-x}VO_4$	Cr^{5+}:$GdVO_4$	Yu *et al.*, 2007b
1.061	Nd:$BaGd_2(MoO_4)_4$	Cr^{4+}:YAG	Zhu *et al.*, 2008
1.062	Nd:$GdVO_4$	Cr^{4+}:YAG	Liu *et al.*, 2003a; Forget *et al.*, 2005
1.062	Nd:GAGG	Cr^{4+}:YAG	Zhi *et al.*, 2010
1.064	Nd:YAG	Cr^{4+}:YAG	Zayhowski and Dill, 1994
1.064	Nd:YVO_4	Cr^{4+}:YAG	Bai *et al.*, 1997; Jaspan *et al.*, 2000
1.064	Nd:$Lu_xY_{1-x}AG$	Cr^{4+}:YAG	Xu *et al.*, 2010
1.064	Nd:$Gd_xY_{1-x}VO_4$	Cr^{4+}:YAG	Zhuang *et al.*, 2011
1.074	Nd:YAG	Cr^{4+}:YAG	Zayhowski *et al.*, 2000
1.080, 1.081	Nd:$CaYAlO_4$	Cr^{4+}:YAG	Li *et al.*, 2010a
1.33	Nd:GAGG	V^{3+}:YAG	Zhang *et al.*, 2010
1.34	Nd:$YAlO_3$	Co^{2+}:LMA	Yumashev *et al.*, 1999
1.34	Nd:YAG	V^{3+}:YAG	Jabczynski *et al.*, 2001; Sulc *et al.*, 2005
1.34	Nd:YAP	V^{3+}:YAG	Jabczynski *et al.*, 2001
1.34	Nd:YVO_4	V^{3+}:YAG	Jabczynski *et al.*, 2001
1.34	Nd:YVO_4	Co^{2+}:LMA	Huang *et al.*, 2007
1.34	Nd:$LuVO_4$	V^{3+}:YAG	Liu *et al.*, 2008b
1.34	Nd:$Gd_xY_{1-x}VO_4$	V^{3+}:YAG	Huang *et al.*, 2009; Li *et al.*, 2010b
1.34	Nd:$Lu_xY_{1-x}VO_4$	V^{3+}:YAG	Zhang *et al.*, 2011
1.44	Nd:YAG	V^{3+},YAG	Kuleshov *et al.*, 2000
1.5	Yb,Tm:YLF	Co^{2+}:LMA	Braud *et al.*, 2000
1.54	Er:glass	Co^{2+}:LMA	Thony *et al.*, 1999; Yumashev *et al.*, 1999
1.54	Er:glass	Co^{2+}:MALO	Wu *et al.*, 2000; Denker *et al.*, 2002

© Woodhead Publishing Limited, 2013

dielectric reflector. SESAMs have been used to generate *Q*-switched pulses with durations less than 100 ps, and have been used with a variety of gain media operating over a wide range of wavelengths. However, the onset of optical damage typically limits the pulse energies to the nanojoule regime (Braun *et al.*, 1996, 1997).

SESAMs offer both advantages and limitations when compared to bulk saturable absorbers. The advantages include the fact that they can be engineered to operate at a wide variety of wavelengths, making SESAMs particularly interesting where good combinations of gain medium and bulk saturable absorber have not been identified. In addition, with SESAMs the physical length of the saturable-absorber region is small and its contribution to the round-trip time of light in the laser cavity is negligible, resulting in the shortest possible *Q*-switched pulses. One of the main limitations of SESAMs is their relatively low damage threshold. Typical values for the damage fluence of the semiconductor materials are ~10 mJ cm^{-2}. The damage fluence of the laser can be increased beyond this value by using highly reflective dielectric layers between the gain medium and the SESAM. This, however, effectively reduces the saturable loss, reduces the pulse energy, and increases the pulse duration. Typical values for the saturable loss of SESAMs are in the range of 10% or less. The largest pulse energy reported for a passively *Q*-switched microchip laser using a SESAM is 4 μJ, with fractions of a microjoule being more typical.

A practical limitation of SESAMs is that they cannot be used as input or output couplers. Also, their thermal expansion coefficients are not matched to the gain media, precluding the possibility of bonding the gain medium to the saturable absorber in a robust, quasi-monolithic fashion, although recent progress has been made using spin-on-glass glues for lasers of modest power in a laboratory environment (Nodop *et al.*, 2007). Finally, the carrier recombination time in semiconductors (the upper-state lifetime of the SESAM) can be quite short, ranging from ~10 ps to ~1 ns, depending on the material and the growth conditions. When the upper-state lifetime is comparable to the output pulse width, it can introduce a significant parasitic loss. On the other hand, it effectively closes the *Q* switch and can prevent the formation of afterpulses. It also allows SESAMs to be operated at extremely high pulse repetition rates.

As a result of their advantages and limitation, SESAMs are most attractive in applications requiring short, low-energy pulses, and where requirements on the system's robustness can be relaxed. In this regime, they can be operated at very high repetition rates, can produce extremely short pulses, and can be engineered to work with gain media at many different wavelengths.

At wavelengths near 1.06 μm, pulses as short as 180 ps have been obtained by *Q*-switching $Nd:LaSc_3(BO_3)_4$ microchip lasers with SESAMs (Braun *et al.*, 1996); pulses as short as 37 ps were obtained in a $Nd:YVO_4$

© Woodhead Publishing Limited, 2013

laser (Spühler *et al.*, 1999). Although these devices operate at lower pulse energies and peak powers than Cr^{4+}:YAG-based devices, they have been pulsed at repetition rates up to 7 MHz. SESAMs have also been used to *Q*-switch microchip lasers operating near 1.03 μm (Spühler *et al.*, 2001), 1.34 μm (Keller *et al.*, 1996; Fluck *et al.*, 1997), and 1.5 μm (Fluck *et al.*, 1998). Guidelines specific to the design of microchip lasers passively *Q*-switched by SESAMs are available in the literature (Spühler *et al.*, 1999).

Afterpulsing

The same mechanisms that cause afterpulsing and pulse bifurcation in actively *Q*-switched lasers (Section 14.6.1) cause afterpulsing and pulse bifurcation in passively *Q*-switched devices as well. In passively *Q*-switched lasers that use bulk saturable absorbers, the recovery time of the saturable absorber is typically too slow to prevent the formation of an afterpulse. In this case, afterpulsing that results from competing longitudinal modes is minimized by positioning the saturable absorber near the center of the microchip cavity, at the possible expense of increased pump-induced bleaching of the saturable absorber (Zayhowski and Wilson, 2003a). Because of the standing-wave nature of the intracavity optical intensity distribution, there is spatial hole burning in the saturable absorber and the first mode to lase does not completely bleach the saturable absorber for the second mode. The differential loss for adjacent modes is greatest when the saturable absorber is located in the center of the cavity.

For applications where afterpulsing cannot be tolerated and the afterpulse is in a different longitudinal mode than the primary pulse, it may be possible to eliminate the afterpulse with spectral filtering (an appropriately tuned etalon) outside the laser cavity, if the laser is bifurcation free. In uncontrolled environments it may be necessary for the filter to actively track the laser, or vice versa.

Pulse bifurcation and pulse-to-pulse amplitude stability

The amplitude and pulse width of the output from a passively *Q*-switched laser are determined by material characteristics and cavity design. In the absence of pulse bifurcation (see Section 14.6.1) they can be extremely stable. Passively *Q*-switched microchip lasers operating on a single longitudinal mode have demonstrated pulse-to-pulse amplitude stabilities better than 1 part in 10^4 at pulse repetition rates of 7 kHz.

In high-repetition-rate passively *Q*-switched lasers pulse bifurcation usually results in alternating strong and weak pulses. The timing interval between the pulses typically varies in accordance with the amplitudes; the period preceding a weak pulse is shorter than the period preceding a strong pulse.

© Woodhead Publishing Limited, 2013

At higher pulse repetition rates, as the pulse train subdivides into pulses of more longitudinal modes, there will be a greater variation in pulse amplitudes and pulse-to-pulse timing.

For some applications, pulse-to-pulse amplitude stability is critical and pulse bifurcation cannot be tolerated. To maximize the pulse repetition rate at which bifurcation-free operation is achieved, the gain differential between pulses must be maximized. Techniques that reduce afterpulsing also encourage bifurcation-free operation. Thus, observations of the afterpulse can be useful in tuning a passively *Q*-switched laser for maximum pulse-to-pulse stability.

Pulse-to-pulse timing stability

The pulse-to-pulse timing stability of passively *Q*-switched lasers is generally worse than that of actively *Q*-switched lasers. In bifurcation-free systems it is often limited by noise in the pump source and is typically ~1% of the pump duration. The timing stability is improved in pulse-pumped systems where the diode is operated at as high a peak power as possible for just long enough to produce an output pulse. This approach has the added benefit of minimizing thermal effects in the passively *Q*-switched laser, since it minimizes efficiency losses due to spontaneous decay of the gain medium, leading to additional benefits in performance.

In a system that uses a pulsed pump source, the pulse-to-pulse timing jitter of the passively *Q*-switched laser gets smaller as the pump period decreases. The minimum possible pump duration is limited by the power (or cost) of the available pump diodes. One approach to minimize this limitation is to cw pump the laser just below threshold and then quickly raise the pump power to its maximum value just before an output pulse is desired. This minimizes timing jitter at the expense of overall system efficiency.

Other techniques used to reduce the pulse timing jitter include hybrid active/passive loss modulation (Arvidsson *et al.*, 1998) and external optical synchronization (Dascalu *et al.*, 1996). These approaches have resulted in timing jitter as low as 20 ps (Hansson and Arvidsson, 2000; Steinmetz *et al.*, 2010), at the expense of added system complexity and some sacrifice in the amplitude and pulse-width stability of a true passively *Q*-switched system.

14.6.3 Gain-switched microchip lasers

Gain-switched lasers are an attractive alternative to *Q*-switched devices when the gain medium has a short upper-state lifetime, giving it a limited capacity to accumulate the energy from low-power cw or quasi-cw pump sources; or when energetic, pulsed pump sources are readily available. They

© Woodhead Publishing Limited, 2013

may also be an attractive option for passive systems when good saturable absorbers have not been identified for the operating wavelength or gain medium employed.

Diode-pumped, 1.3-μm Nd:YAG microchip lasers, pumped with a peak power of 1 W, have been gain-switched to produce a 100-kHz train of pulses with durations of 170 ns and peak powers of 1.8 W (Zayhowski, 1990b). In this case, the microchip lasers were resonantly pumped at their relaxation frequency to minimize both the pulse width and the fluctuations in pulse amplitude.

A more interesting variation of a gain-switched laser system uses a passively *Q*-switched microchip laser as the pump. This approach is particularly attractive when the absorption of the gain medium is well matched to the fundamental or one of the harmonics of Nd:YAG or Yb:YAG. It can result in efficient, compact, and robust sources of short-pulsed laser radiation at diverse wavelengths. Optimal performance is obtained when the inversion density in the gain-switched laser cavity is about three times above threshold at the time the pulse forms. If the goal is to have a single output pulse extract the maximum amount of energy from each pump pulse, the duration of the pump pulse should typically be less than 3–10 times the output pulse width. Longer pump pulses can lead to multiple output pulses.

Miniature Cr^{4+}:YAG gain-switched lasers, pumped by the fundamental output of a passively *Q*-switched Nd:YAG microchip laser, have produced 4-ns output pulses at a wavelength of 1.44 μm, with a pulse energy of 2 μJ (Zayhowski *et al.*, 2001). Using a frequency-doubled passively *Q*-switched microchip laser as the pump source, gain-switched Ti:Al_2O_3 lasers have been demonstrated with output pulses as short as 350 ps at wavelengths near 800 nm, with tuning over 65 nm, linewidths as narrow as 0.05 nm, and pulse energies up to 20 μJ (Zayhowski *et al.*, 2001; Zayhowski and Wilson, 2002). Miniature gain-switched Ce^{3+}:$LiCaAlF_6$ lasers, pumped by the fourth harmonic of passively *Q*-switched Nd:YAG and Nd:YVO_4 microchip lasers, have demonstrated 500-ps ultraviolet pulses tunable over the range from 283 to 314 nm, with pulse energies of several microjoules (Spence *et al.*, 2006; Liu *et al.*, 2008a). These lasers were in turn used to pump miniature gain-switched Ce^{3+}:$LiLuF_4$ lasers that produced 500-ps ultraviolet pulses tunable from 306 to 308 nm, with pulse energies around 1 μJ (Liu *et al.*, 2008a).

14.6.4 Mode-locked microchip lasers

Although many microchip lasers are designed to operate in a single longitudinal mode, they need not be. One compelling reason to build a multimode microchip laser would be for mode-locked operation. In this case, multimode operation is essential to the performance of the device and the optical length of the cavity determines the pulse repetition rate. Optical-

© Woodhead Publishing Limited, 2013

domain generation of high-quality millimeter-wave signals for fiber radio has led to the development of monolithic electro-optically mode-locked 1.085-μm Nd:$LiNbO_3$ microchip lasers with pulse widths as short as 18.6 ps and repetition rates up to 20 GHz (3.48-mm cavity length) (Vieira *et al.*, 1997, 2001). Here, the Nd:$LiNbO_3$ served as both the gain medium and the electro-optic material, and the repetition rate of the laser is the carrier frequency of the radio-frequency signal.

14.7 Nonlinear frequency conversion

Nonlinear frequency generation is an important adjunct to laser technology, since it accesses frequency domains that are important for a variety of applications but are not covered by good laser gain media.

14.7.1 Nonlinear frequency conversion with cw microchip lasers

Intracavity second-harmonic generation has been used in both monolithic and composite-cavity microchip lasers to convert the fundamental infrared wavelength to the green (Amano, 1992; MacKinnon and Sinclair, 1994) and blue (Matthews *et al.*, 1996) regions of the spectrum. However, since frequency doubling introduces nonlinear loss into the laser cavity, care must be taken to obtain stable output. Instabilities can result from the coupling of the cavity modes through the doubling process and result in chaotic amplitude fluctuations. This is commonly referred to as the green problem (Baer, 1986). The easiest solution, if stable output is required, is to ensure that only one cavity mode will oscillate, even in the presence of efficient frequency conversion (see Section 14.4.1).

14.7.2 Nonlinear frequency conversion with *Q*-switched microchip lasers

The high peak powers obtained from *Q*-switched microchip lasers enable a variety of miniature nonlinear optical devices. Harmonic generation, frequency mixing, parametric conversion, and stimulated Raman scattering have been used with passively *Q*-switched microchip lasers for frequency conversion to wavelengths covering the entire spectrum from 213 nm to 4.3 μm in extremely compact optical systems.

Harmonic conversion

Because of its high peak intensity, the output of even the low-average-power Nd:YAG/Cr^{4+}:YAG passively *Q*-switched microchip lasers can be efficiently

© Woodhead Publishing Limited, 2013

frequency doubled by simply placing a 5-mm-long piece of $KTiOPO_4$ (KTP) near the output facet of the laser, with no intervening optics. With 1-W-pumped devices, doubling efficiencies as high as 70% have been demonstrated (Zayhowski and Dill, 1994), although more typical numbers are between 45% and 60%. Since the resulting green output has a high peak intensity and good mode quality, it is possible to frequency-convert it into the ultraviolet by placing an appropriate nonlinear material, such as β-BaB_2O_4 (BBO) adjacent to the output facet of the KTP, as shown in Fig. 14.4. With this approach, a 1-W-pumped 1.064-μm Nd:YAG/Cr^{4+}:YAG passively *Q*-switched microchip laser has been frequency-converted to produce up to 7 μJ of green, 1.5 μJ of third-harmonic, 1.5 μJ of fourth-harmonic, and 50 nJ of 213-nm (fifth-harmonic) light at a typical pulse repetition rate of 10 kHz (Zayhowski *et al.*, 1995b; Zayhowski, 1996a, 1996b, 1997c). The wavelength diversity offered by harmonic conversion opens up numerous applications. To address those applications, microchip laser systems are commercially available at 532, 355, and 266 nm. Higher-power pulse-pumped passively *Q*-switched Nd:YAG/Cr^{4+}:YAG microchip lasers have been harmonically converted to produce higher-power green and ultraviolet output (Zayhowski, 1998; Zayhowski *et al.*, 1999), including more than 1 μJ of 215-nm light at the fifth harmonic of the 1.074-μm transition in Nd:YAG (Zayhowski *et al.*, 2000).

An interesting variation of the harmonically converted passively *Q*-switched microchip laser uses a self-frequency-doubling material, Nd:$GdCa_4O(BO_3)_3$,

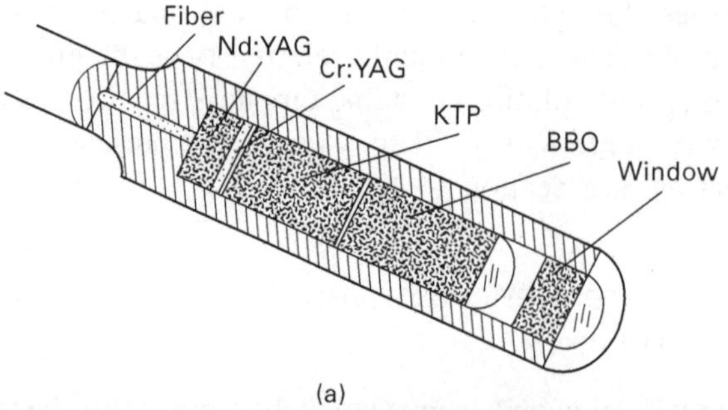

(a)

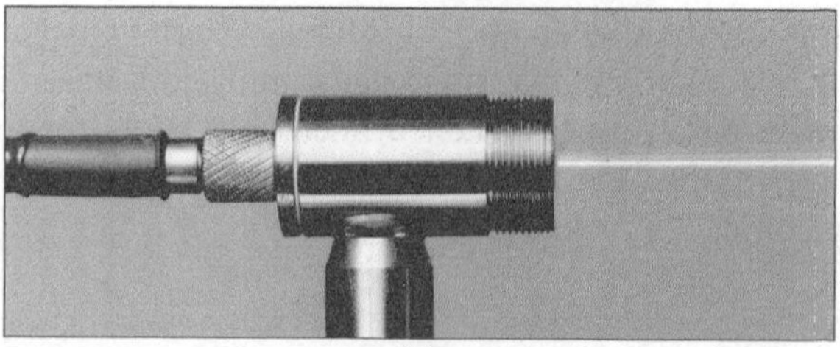

(b)

14.4 Fiber-coupled frequency quadrupled passively *Q*-switched microchip laser: (a) schematic and (b) photograph.

© Woodhead Publishing Limited, 2013

as the gain medium and Cr^{4+}:YAG as the saturable absorber (Zhang *et al.*, 2001). Systems using a single material as the gain medium, saturable absorber, and nonlinear crystal have been proposed but not yet reduced to practice.

Optical parametric conversion

Passively *Q*-switched microchip lasers can also be used to pump optical parametric amplifiers (OPAs), oscillators (OPOs), and generators (OPGs). The unfocused 1.064-μm output of a passively *Q*-switched microchip laser pumped with a 10-W diode-laser array has been used to efficiently drive several periodically poled lithium niobate (PPLN) OPGs, covering the spectral range from 1.4 to 4.3 μm (Zayhowski, 1997a; Zayhowski and Wilson, 2000). A photograph of an OPG system is shown in Fig. 14.5. A similar range of wavelengths has been accessed with microchip-laser-pumped OPOs (Capmany *et al.*, 2001).

The wavelength coverage of microchip-laser-based systems has been pushed further into the infrared by using the output of a microchip-laser-pumped PPLN OPA to pump a zinc germanium phosphate (ZGP) OPA (Zayhowski, unpublished). This system was demonstrated at signal and idler wavelengths of 3.4 and 8.1 μm, respectively, and more than 250 nJ per pulse was obtained at 8.1 μm. The same approach could be used to access wavelengths through the mid-infrared to beyond 10 μm.

KTP OPOs have been pumped with the second harmonic of a passively *Q*-switched microchip laser, and operated at signal and idler wavelengths between 700 and 2000 nm (Zayhowski, 1997b).

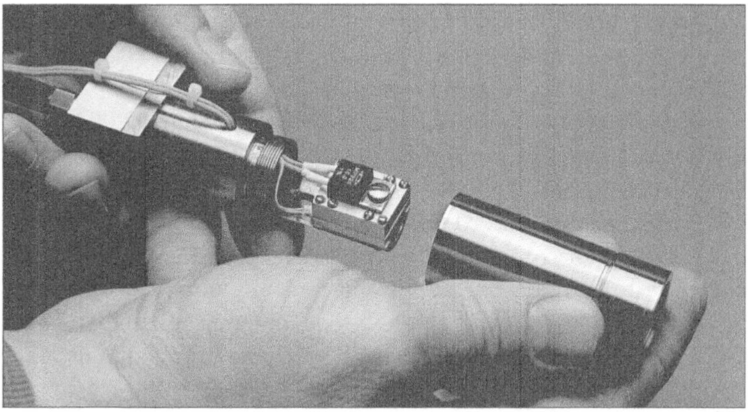

14.5 Optical parametric generator pumped by a passively *Q*-switched microchip laser. The microchip laser used to pump the OPG is contained in the first, covered section of the housing. An oven containing the OPG is in the second section, and is shown uncovered in the photo.

© Woodhead Publishing Limited, 2013

Raman conversion

Intracavity Raman frequency conversion has been demonstrated in composite-cavity Nd:YAG/Cr^{4+}:YAG and Nd:LSB/Cr^{4+}:YAG passively *Q*-switched microchip lasers using $Ba(NO_3)_2$ as the Raman medium (Demidovich *et al.*, 2003). Up to 1.2 μJ of energy was obtained at the Stokes wavelength of 1.196 μm, corresponding to 8% conversion efficiency. The pulse duration, at 1.196 μm, was as short as 118 ps.

Supercontinuum generation

Raman combined with other nonlinear optical effects can be used to create extremely broadband, spatially coherent optical continua in microchip-laser-pumped optical fibers (Agrawal, 1995; Zayhowski, 1997c, 1999, 2000; Johnson *et al.*, 1999). Properly engineered microstructured fibers can increase the efficiency of the processes involved (Ranka *et al.*, 2000; Provino *et al.*, 2001; Wadsworth *et al.*, 2004; Champert *et al.*, 2004) and result in high-power white-light systems (Champert *et al.*, 2002).

14.8 Microchip amplifiers

Microchip lasers can produce desirable waveforms and beam profiles, but often at the expense of output power or, in the case of pulsed systems, pulse energy. For example, passively *Q*-switched microchip lasers excel in generating subnanosecond, single-frequency pulses, but the small volume of the gain medium limits the stored-energy capacity.

As the required output power of the laser system increases, the need for amplification becomes apparent, and a wide variety of amplifiers have been used to increase the output power of microchip lasers. These include amplifiers based on solid-state gain media (Druon *et al.*, 1999; Isyanova *et al.*, 2001; Zayhowski and Wilson, 2002, 2003b; Manni, 2005) and on large-mode-area fibers (Di Teodoro *et al.*, 2002; Di Teodoro and Brooks, 2005a, 2005b; Brooks and Di Teodoro, 2006; Schrader *et al.*, 2006; Kir'yanov *et al.*, 2009; Zhao *et al.*, 2011). In most of these oscillator/amplifier systems the design of the oscillator is independent of the amplifier, which is later added on to increase the power and typically dominates the size, weight, power consumption, and complexity of the laser system. The exception to this is the energy-scavenging amplifier (Zayhowski and Wilson, 2004).

14.8.1 Energy-scavenging amplifiers

Many microchip lasers are inefficient, in part due to inefficient absorption of the pump light owing to the short length of the gain medium. In some

© Woodhead Publishing Limited, 2013

systems the unabsorbed pump light can be collected (scavenged) and used to pump an in-line amplifier, as shown in Fig. 14.6, resulting in increased system power and efficiency, with minimal added cost, size, or complexity. For example, the efficiency of short-pulse, passively Q-switched Nd:YAG/Cr^{4+}YAG microchip laser systems has been more than tripled with energy-scavenging Nd:YVO_4 amplifiers (Zayhowski and Wilson, 2004). The use of energy-scavenging amplifiers, where applicable, offers an additional degree of freedom to laser designers, since it can free them of efficiency concerns

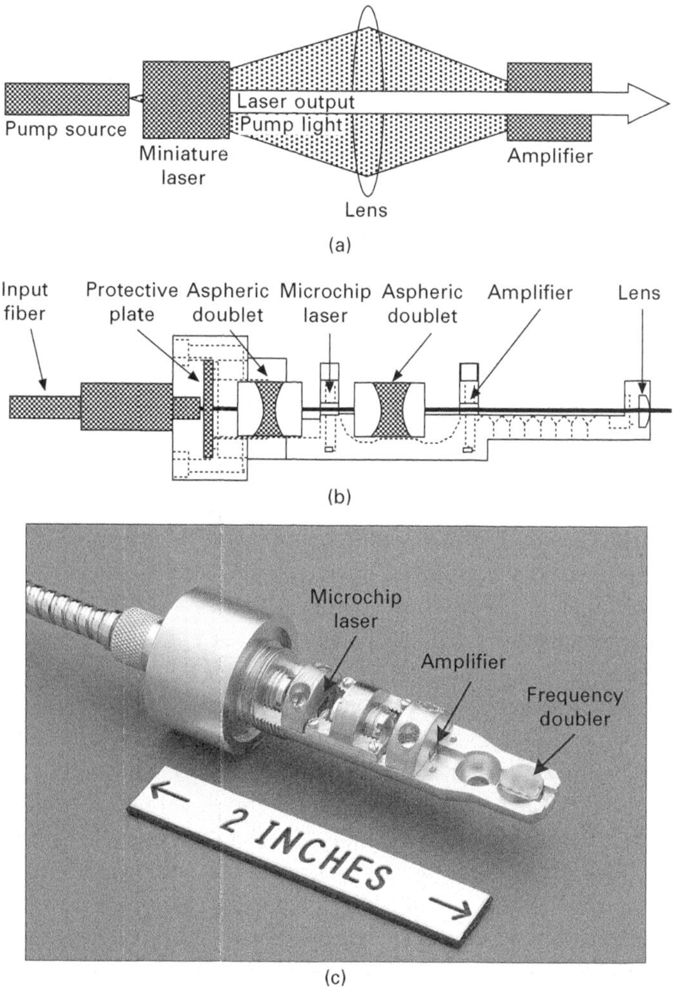

14.6 Energy-scavenging amplifier: (a) illustration of concept, (b) schematic of a current embodiment, and (c) photograph of frequency-doubled version.

© Woodhead Publishing Limited, 2013

when designing the oscillator, knowing that the system efficiency and power can be recovered at little expense with the amplifier.

14.9 Future trends

Microchip lasers are well understood, and have been thoroughly explored theoretically and experimentally. Within their range of capabilities they are the simplest, most compact, and most robust implementation of solid-state laser. CW microchip lasers face strong competition from diode lasers and fiber lasers. Diode lasers are more efficient, smaller, and simpler, and are available at a greater variety of wavelengths. Fiber lasers also tend to be more efficient and can produce higher output powers in a fundamental transverse mode. To compete against either of these technologies, cw microchip lasers need to find a niche application that exploits their unique spectral properties. Nonetheless, within the solid-state laser development community the cw microchip geometry seems to have established itself as a testing ground for newly developed gain media.

On the other hand, *Q*-switched microchip lasers provide capabilities that cannot be matched by semiconductor devices or fiber lasers. Semiconductor diode lasers have a very limited capacity to store energy, and their facets are damaged at very modest optical intensities. Fiber lasers have much longer cavities, which prevent them from producing short *Q*-switched pulses and make them susceptible to the onset of undesirable nonlinear effects at peak powers well below those demonstrated with *Q*-switched microchip lasers. Fiber amplifiers can be used to increase the peak power from pulsed semiconductor diodes up to the level of *Q*-switched microchip lasers (Cheng *et al.*, 2005; Galvanauskas *et al.*, 2007), but this requires several stages of amplification and results in more complicated devices. Additionally, the output of high-peak-power fiber amplifiers typically has a broader bandwidth and interpulse amplified spontaneous emission (ASE) that may be detrimental for some applications.

As a result of their small size, robust construction, reliability, and relatively low cost, coupled with their ability to produce energetic, diffraction-limited, Fourier-transform-limited, subnanosecond pulses, passively *Q*-switched microchip lasers have been embraced for applications in high-resolution time-of-flight three-dimensional imaging. Because their high peak output intensity allows for the efficient generation of ultraviolet light in very compact and reliable formats, they have also been well accepted in the field of ultraviolet fluorescence spectroscopy, and are an integral part of numerous fielded spectroscopic instruments. In addition, passively *Q*-switched microchip lasers have started to make inroads in applications involving laser scribing and marking, laser breakdown spectroscopy, and most recently laser ignition. The continued development of the technology will be driven by

© Woodhead Publishing Limited, 2013

applications. The full potential of passively *Q*-switched microchip lasers as low-cost mass-produced components will not be realized until they are accepted into large-volume markets. Once that happens, these devices should be attractive for a wide variety of additional applications, where the current price of high-performance lasers is an overwhelming obstacle to their use.

14.10 Sources of further information and advice

A single-source, comprehensive, formal treatment of miniature solid-state lasers, including microchip lasers, with adequate background for those readers new to the area can be found in the following reference: Zayhowski J J, Welford D, and Harrison J (2007), 'Miniature solid-state lasers', in Gupta M C and Ballato J, *The Handbook of Photonics,* 2nd edn, Boca Raton, FL, CRC Press, Ch. 10.

Readers whose primary interest is in passively *Q*-switched microchip lasers can find a similar treatment, restricted to passively *Q*-switched microchip lasers but supplemented with a discussion of early applications of the devices, in: Zayhowski J J (2007), 'Passively *Q*-switched microchip lasers', in Sennaroglu A, *Solid-State Lasers and Applications,* Boca Raton, FL, CRC Press, Ch. 1.

14.11 References

Abshire J, Ketchum E, Afzal R, Millar P, and Sun X (2000), 'The geoscience laser altimeter system (GLAS) for the ICEsat mission', *Conf. Lasers Electro-Optics, Tech. Dig.*, 602–603.

Afzal R S, Yu A W, Zayhowski J J, and Fan T Y (1997), 'Single-mode, high-peak-power passively *Q*-switched diode-pumped Nd:YAG laser', *Opt. Lett.*, 22, 1314–1316.

Agrawal G P (1995), *Nonlinear Fiber Optics*, 2nd edn, San Diego, CA, Academic Press, Ch. 8.

Amano S (1992), 'Microchip NYAB green laser', *Rev. Laser Eng.*, 20, 723–727 (in Japanese).

Arvidsson M (2001), 'Far-field timing effects with passively *Q*-switched lasers', *Opt. Lett.*, 26, 196–198.

Arvidsson M, Hansson B, Holmgren M, and Lindstrom C (1998), 'A combined actively and passively Q-switched microchip laser', *SPIE*, 3265, 106–113.

Baer T (1986), 'Large-amplitude fluctuations due to longitudinal mode coupling in diode-pumped intracavity-doubled Nd:YAG lasers', *J. Opt. Soc. Am. B*, 3, 1175–1180.

Bai Y, Wu N, Zhang J, Li J, Li S, Xu J, and Deng P (1997), 'Passively *Q*-switched Nd:YVO_4 laser with a Cr^{4+}:YAG crystal saturable absorber', *Appl. Opt.*, 36, 2468–2472.

Bloch J, Johnson B, Newbury N, Germaine J, Hemond H, and Sinfield J (1998), 'Field test of a novel microlaser-based probe for in situ fluorescence sensing of soil contamination', *Appl. Spectrosc.*, 52, 1299–1304.

Braud A, Girard S, Doualan J L, and Moncorgé R (2000), 'Wavelength tunability and passive *Q*-switching of a (Yb,Tm):YLF laser operating around 1.5 μm', *Conf. Lasers Electro-Optics, Tech. Dig.*, 463–464.

© Woodhead Publishing Limited, 2013

Braun B, Kärtner F X, Keller U, Meyn J-P, and Huber G (1996), 'Passively Q-switched 180-ps Nd:$LaSc_3(BO_3)_4$ microchip laser', *Opt. Lett.*, 24, 405–407.

Braun B, Kärtner F X, Zhang G, Moser M, and Keller U (1997), '56-ps passively *Q*-switched diode-pumped microchip laser', *Opt. Lett.*, 22, 381–393.

Brenier A, Tu C, Zhu Z, and Li J (2007), 'Diode pumped passive *Q* switching of Yb^{3+}-doped $GdAl_3(BO_3)_4$ nonlinear laser crystal', *Appl. Phys. Lett.*, 90, 71103–71105.

Brooks C D, and Di Teodoro F (2006), 'Multimegawatt peak-power, single-transverse-mode operation of a 100 μm core diameter, Yb-doped rodlike photonic crystal fiber amplifier', *Appl. Phys. Lett.*, 89, 111119-1–111119-3.

Capmany J, Bermudez V, Callejo D, and Dieguez E (2001), 'Microchip OPOs operate in the infrared', *Laser Focus World*, June, 143–148.

Champert P A, Popov S V, and Taylor J R (2002), 'Generation of multiwatt, broadband continua in holey fibers', *Opt. Lett.*, 27, 122–124.

Champert P A, Couderc V, Leproux P, Février S, Tombelaine V, Labonté L, Roy P, and Froehly C (2004), 'White-light supercontinuum generation in normally dispersive optical fiber using original multi-wavelength pumping system', *Opt. Express*, 12, 4366–4371.

Cheng M Y, Chang Y C, Galvanauskas A, Mamidipudi P, Changkakoti R, and Gatchell P (2005), 'High-energy and high-peak-power nanosecond pulse generation with beam quality control in 200-μm core highly multimode Yb-doped fiber amplifiers', *Opt. Lett.*, 30, 358–360.

Chinn S R, and King V (2006), 'Subnanosecond (Er,Yb) glass *Q*-switched microlasers: 3-D transient modeling and experiments', *IEEE J. Quantum Electron.*, 42, 1128–1136.

Conroy R S, Lake T, Friel G J, Kemp A J, and Sinclair B D (1998), 'Self-*Q*-switched Nd:YVO_4 microchip lasers', *Opt. Lett.*, 23, 457–459.

Dascalu T, Pavel N, Lupei V, Philipps G, Beck T, and Weber H (1996), 'Investigation of a passive *Q*-switched, externally controlled, quasicontinuous or continuous pumped Nd:YAG laser', *Opt. Eng.*, 35, 1247–1251.

Degnan J J (1989), 'Theory of the optimally coupled Q-switched laser', *IEEE J. Quantum Electron.*, 25, 214–220.

Degnan J J (1993), 'Millimeter accuracy satellite laser ranging: a review', *Contributions of Space Geodesy to Geodynamics: Technology, AGU Geodynamics Series*, 25, 133–162.

Degnan J J (1995), 'Optimization of passively *Q*-switched lasers', *IEEE J. Quantum Electron.*, 31, 1890–1901.

Degnan J J (1999), 'Engineering progress on the fully automated photon-counting SLR2000 satellite laser ranging station', *SPIE*, 3865, 76–82.

Degnan J J, McGarry J, Zagwodzki T, Dabney P, Geiger J, Chabot R, Steggerda C, *et al.* (2001), 'Design and performance of an airborne multikilohertz photon-counting microlaser altimeter', *Int. Arch. Photogrammetry Remote Sensing*, XXXIV-3/W4, 9–16.

Demidovich A A, Apanasevich P A, Batay L E, Grabtchikov A S, Kuzmin A N, Lisinetskii V A, Orlovich V A, *et al.* (2003), 'Sub-nanosecond microchip laser with intracavity Raman conversion', *Appl. Phys. B*, 76, 509–514.

Denker B I, Galagan B I, Osiko V V, and Sverchkov S E (2002), 'Materials and components for miniature diode-pumped 1.5 μm erbium glass lasers', *Laser Phys.*, 12, 697–701.

Di Teodoro F, and Brooks C D (2005a), '1.1 MW peak-power, 7 W average-power, high-spectral-brightness, diffraction-limited pulses from a photonic crystal fiber amplifier', *Opt. Lett.*, 30, 2694–2696.

© Woodhead Publishing Limited, 2013

Di Teodoro F, and Brooks C D (2005b), 'Multistage Yb-doped fiber amplifier generating megawatt peak-power, subnanosecond pulses', *Opt. Lett.*, 30, 3299–3301.

Di Teodoro F, Koplow J P, Moore S W, and Kliner D A V (2002), 'Diffraction-limited, 300-kW peak-power pulses from a coiled multimode fiber amplifier', *Opt. Lett.*, 27, 518–520.

Dixon G J, Lingvay L S, and Jarman R H (1989), 'Properties of close-coupled, monolithic lithium neodymium tetraphosphate laser', *SPIE*, 1104, 107–112.

Dong J, Shirakawa A, and Ueda K (2006), 'Sub-nanosecond passively Q-switched Yb:YAG/Cr^{4+}:YAG sandwiched microchip laser', *Appl. Phys. B*, 85, 513–518.

Dong J, Ueda K, Shirakawa A, Yagi H, Yanagitani T, and Kaminskii A A (2007a), 'Composite Yb:YAG/Cr^{4+}:YAG ceramics picosecond microchip lasers', *Opt. Express*, 15, 14516–14523.

Dong J, Ueda K, and Kaminskii A A (2007b), 'Efficient passively *Q*-switched Yb:LuAG microchip laser', *Opt. Lett.*, 32, 3266–3268.

Dong J, Ueda K, and Kaminskii A A (2008), 'Continuous-wave and Q-switched microchip laser performance of Yb:$Y_3Sc_2Al_3O_{12}$ crystals', *Opt. Express*, 16, 5241–5251.

Druon F, Balembois F, Georges P, and Brun A (1999), 'High-repetition-rate 300-ps pulsed ultraviolet source with a passively *Q*-switched microchip laser and a multipass amplifier', *Opt. Lett.*, 24, 499–501.

Fan T Y (1994), 'Aperture guiding in quasi-three-level lasers', *Opt. Lett.*, 19, 554–556.

Fluck R, Braun B, Gini E, Melchoir H, and Keller U (1997), 'Passively *Q*-switched 1.34-μm Nd:YVO_4 microchip laser with semiconductor saturable-absorber mirrors', *Opt. Lett.*, 22, 991–993.

Fluck R, Häring R, Paschotta R, Gini E, Melchior H, and Keller U (1998), 'Eyesafe pulsed microchip laser using semiconductor saturable absorber mirrors', *Appl. Phys. Lett.*, 72, 3273–3275.

Forget S, Druon F, Balembois F, Georges P, Landru N, Fève J-P, Lin J, and Weng Z (2005), 'Passively Q-switched diode-pumped Cr^{4+}:YAG/Nd^{3+}:$GdVO_4$ monolithic microchip laser', *Opt. Commun.*, 259, 816–819.

Galvanauskas A, Cheng M Y, Hou K C, and Liao K H (2007), 'High peak power pulse amplification in large-core Yb-doped fiber amplifiers', *IEEE J. Sel. Topics Quantum Electron.*, 13, 559–566.

Gavrilovic P, O'Neill M S, Meehan K, Zarrabi J H, and Singh S (1992), 'Temperature-tunable, single frequency microcavity lasers fabricated from flux-grown YCeAG:Nd', *Appl. Phys. Lett.*, 60, 1652–1654.

Genack A Z, and Brewer R G (1978), 'Optical coherent transitions by laser frequency switching', *Phys. Rev. A*, 17, 1463–1473.

Gong M, Wang Y, Wang D, and Liao Y (2006), 'Stable 100 kHz operation of passively Q-switched microchip laser', *Electron. Lett.*, 42, 760–762.

Grabtchikov A S, Kuzmin A N, Lisinetskii V A, Orlovich V A, Demidovich A A, Danailov M B, Eichler H J, *et al.* (2002), 'Laser operation and Raman self-frequency conversion in Yb:KYW microchip laser', *Appl. Phys. B*, 75, 795–797.

Gu J, Tam S C, Lam Y L, Chen Y, Kam C H, Tan W, Xie W J, Zhao G, and Yang H (2000), 'Novel use of GaAs as a passive Q-switch as well as an output coupler for diode-pumped infrared solid state lasers', *SPIE*, 3929, 222–235.

Hansson B, and Arvidsson M (2000), 'Q-switched microchip laser with 65ps timing jitter', *Electron. Lett.*, 36, 1123–1124.

Harrison J, and Martinsen R J (1994a), 'Thermal modeling for mode-size estimation in

© Woodhead Publishing Limited, 2013

microlasers with application to linear arrays in Nd:YAG and Tm,Ho:YLF', *IEEE J. Quantum Electron.*, 30, 2628–2633.

Harrison J, and Martinsen R J (1994b), 'Operation of linear microlaser arrays near 1 μm, 2 μm and 3 μm', in Fan T Y, and Chai B H T, *OSA Proceedings on Advanced Solid-State Lasers* 20, Washington, DC, OSA, 272–275.

He K, Gao C, Wei Z, Li D, Zhang Z, Zhang H, and Wang J (2009), 'Diode-pumped passively Q-switched Nd:LuVO4 laser at 916 nm', *Opt. Commun.*, 282, 2413–2416.

Huang H T, He J L, Zuo C H, Zhang H J, Wang J Y, Liu Y, and Wang H T (2007), 'Co^{2+}:LMA crystal as saturable absorber for a diode-pumped passively Q-switched Nd:YVO_4 laser at 1342 nm', *Appl. Phys. B*, 89, 319–321.

Huang H T, Zhang B T, He J L, Yang J F, Xu J L, Yang X Q, Zuo C H, and Zhao S (2009), 'Diode-pumped passively Q-switched Nd:$Gd_{0.5}Y_{0.5}VO_4$ laser at 1.34 μm with V^{3+}:YAG as the saturable absorber', *Opt. Express*, 17, 6946–6951.

Huang H T, He J L, Zhang B T, Yang J F, Xu J L, Zuo C H, and Tao X T (2010), 'V^{3+}:YAG as the saturable absorber for a diode-pumped quasi-three-level dual-wavelength Nd:GGG laser', *Opt. Express*, 18, 3352–3357.

Isyanova Y, Manni J G, Welford D, Jaspers M, and Russell J A (2001), 'High-power, passively Q-switched microlaser – power amplifier system', in Marshal C, *OSA TOPS* 50, *Advanced Solid-State Lasers*, Washington, DC, OSA, 186–190.

Jabczynski J K, Kopczynski K, Mierczyk Z, Agnesi A, Guandalini A, and Reali G C (2001), 'Application of V^{3+}:YAG crystals for *Q*-switching and mode-locking of 1.3-μm diode-pumped neodymium lasers', *Opt. Eng.*, 40, 2802–2812.

Jaseja T S, Javan A, and Townes C H (1963), 'Frequency stability of He–Ne masers and measurements of length', *Phys. Rev. Lett.*, 10, 165–167.

Jaspan M A, Welford D, Xiao G, and Bass M (2000), 'Atypical behavior of Cr:YAG passively *Q*-switched Nd:YVO_4 microlasers at high-pumping rates', *Conf. Lasers Electro-Optics, Tech. Dig.*, 454.

Jaspan M, Welford D, and Russell J A (2004), 'Passively *Q*-switched microlaser performance in the presence of pump-induced bleaching of the saturable absorber', *Appl. Opt.*, 43, 2555–2560.

Johnson B, Joseph R, Nischan M, Newbury A, Kerekes J P, Barclay H T, Willard B, and Zayhowski J J (1999), 'A compact, active hyperspectral imaging system for the detection of concealed targets', *SPIE*, 3710, 144–153.

Kajave T T, and Gaeta A L (1996), 'Q switching of a diode-pumped Nd:YAG laser with GaAs', *Opt. Lett.*, 21, 1244–1246.

Karlsson G, Laurell F, Tellefsen T, Denker B, Galagan B, Osiko V, and Sverchkov S (2002), 'Development and characterization of Yb-Er laser glass for high average power laser diode pumping', *Appl. Phys. B*, 75, 41–46.

Kärtner F X, Brovelli L R, Kopf D, Kamp M, Calasso I, and Keller U (1995), 'Control of solid state laser dynamics by semiconductor devices', *Opt. Eng.*, 34, 2024–2026.

Keller U, Weingarten K J, Kärtner F X, Kopf D, Braun B, Jung I D, Fluck R, Hönninger C, Matuschek N, and Aus der Au J (1996), 'Semiconductor saturable absorber mirrors (SESAMs) for femtosecond to nanosecond pulse generation in solid-state lasers', *IEEE J. Sel. Topics Quantum Electron.*, 2, 435–453.

Kemp A J, Conroy R S, Friel G J, and Sinclair B D (1999), 'Guiding effects in Nd:YVO_4 microchip lasers operating well above threshold', *IEEE J. Quantum Electron.*, 35, 675–681.

Keszenheimer J A, Balboni E J, and Zayhowski J J (1992), 'Phase locking of 1.32-μm microchip lasers through the use of pump-diode modulation', *Opt. Lett.*, 17, 649–651.

© Woodhead Publishing Limited, 2013

Kir'yanov A V, Klimentov S M, and Mel'nikov I V (2009), 'Specialty Yb fiber amplifier for microchip Nd laser: Towards ~1-mJ/1-ns output at kHz-range repetition rate', *Opt. Commun.*, 282, 4759–4764.

Kogelnik H (1965), 'On the propagation of Gaussian beams of light through lens-like media including those with a loss or gain variation', *Appl. Opt.*, 4, 1562–1569.

Kuleshov N V, Podlipensky A V, Yumashev K V, Kretschmann H M, and Huber G (2000), 'V:YAG saturable absorber as a *Q*-switch for diode-pumped Nd:YAG-lasers at 1.44 μm and 1.34 μm', *Conf. Lasers Electro-Optics, Tech. Dig.*, 228.

Lagatsky A A, Abdolvand A, and Kuleshov N V (2000), 'Passive Q switching and self-frequency Raman conversion in a diode-pumped Yb:KGd$(WO_4)_2$ laser', *Opt. Lett.*, 25, 616–618.

Lan R, Pan L, Utkin I, Ren Q, Zhang H, Wang Z, and Fedosejevs R (2010), 'Passively Q-switched Yb^{3+}:NaY$(WO_4)_2$ laser with GaAs saturable absorber', *Opt. Express*, 18, 4000–4005.

Laporta P, Taccheo S, Longhi S, Svelto O, and Sacchi G (1993), 'Diode-pumped microchip Er-Yb laser', *Opt. Lett.*, 18, 1232–1234.

Lax M (1966), 'Quantum noise V: Phase noise in a homogeneously broadened maser', in Kelley P K, Lax B, and Tannenwald P E, *Physics of Quantum Electronics*, New York, McGraw-Hill, 735–747.

Leilabady P A, Anthon D W, and Gullicksen P O (1992), 'Single-frequency Nd:YLF cube lasers pumped by laser diode arrays', *Conf. Lasers Electro-Optics, Tech. Dig.*, 54.

Li Q, Feng B, Zhang D, Zhang Z, Zhang H, and Wang J (2009), '*Q*-switched 935 nm Nd:CNGG laser', *Appl. Opt.*, 48, 1898–1903.

Li D, Xu X, Meng J, Zhou D, Xia C, Wu F, and Xu J (2010a), 'Diode-pumped continuous wave and *Q*-switched operation of Nd:CaYAlO_4 crystal', *Opt. Express*, 18, 18649–18654.

Li X, Li G, Zhao S, Xu C, Li Y, Liu H, and Zhang H (2010b), 'Enhancement of passively Q-switched performance at 1.34 μm with a class of Nd:$Gd_xY_{1-x}VO_4$ crystals', *Opt. Express*, 18, 21552–21556.

Liu H, Zhou S H, and Chen Y C (1998), 'High-power monolithic unstable-resonator solid-state laser', *Opt. Lett.*, 23, 451–453.

Liu J, Ozygus B, Yang S, Erhard J, Seelig U, Ding A, Weber H, *et al.* (2003a), 'Efficient passive *Q*-switching operation of a diode-pumped Nd:$GdVO_4$ laser with a Cr^{4+}:YAG saturable absorber', *J. Opt. Soc. Am. B*, 20, 652–661.

Liu J, Wang Z, Meng X, Shao Z, Ozygus B, Ding A, and Weber H (2003b), 'Improvement of passive *Q*-switching performance reached with a new Nd-doped mixed vanadate crystal Nd:$Gd_{0.64}Y_{0.36}VO_4$', *Opt. Lett.* 28, 2330–2332.

Liu F Q, Xia H R, Pan S D, Gao W L, Ran D G, Sun S Q, Ling Z C, *et al.* (2007a), 'Passively Q-switched Nd:$LuVO_4$ laser using Cr^{4+}:YAG as saturable absorber', *Opt. Laser Technol.*, 39, 1449–1453.

Liu J, Petrov V, Zhang H, Wang J, and Jiang M (2007b), 'Efficient passively *Q*-switched laser operation of Yb in the disordered NaGd$(WO_4)_2$ crystal host', *Opt. Lett.*, 32, 1728–1730.

Liu H, Spence D J, Coutts D W, Sato H, and Fukuda T (2008a), 'Broadly tunable ultraviolet miniature cerium-doped LiLuF lasers', *Opt. Express*, 16, 2226–2231.

Liu F, He J, Zhang B, Xu J, Dong X, Yang K, Xia H, and Zhang H (2008b), 'Diode-pumped passively Q-switched Nd:$LuVO_4$ laser at 1.34 μm with a V^{3+}:YAG saturable absorber', *Opt. Express*, 16, 11759–11763.

MacKinnon N, and Sinclair B D (1992), 'Pump power induced cavity stability in

© Woodhead Publishing Limited, 2013

lithium neodymium tetraphosphate (LNP) microchip lasers', *Opt. Commun.*, 94, 281–288.

MacKinnon N, and Sinclair B D (1994), 'A laser diode array pumped, Nd:YVO_4/KTP, composite material microchip laser', *Opt. Commun.*, 3–4, 183–187.

MacKinnon N, Norrie C J, and Sinclair B D (1994), 'Laser-diode-pumped, electro-optically tunable Nd:MgO:$LiNbO_3$ microchip laser', *J. Opt. Soc. Am. B*, 11, 519–522.

Manni J G (2005), 'Amplification of microchip oscillator emission using a diode-pumped wedged-slab amplifier', *Opt. Commun.*, 252, 117–126.

Matthews D G, Conroy R S, Sinclair B D, and MacKinnon N (1996), 'Blue microchip laser fabricated from Nd:YAG and $KNbO_3$', *Opt. Lett.*, 21, 198–200.

Mermilliod N, François B, and Wyon C (1991), '$LaMgAl_{11}O_{19}$:Nd microchip laser', *Appl. Phys. Lett.*, 59, 3519–3520.

Mirov S B, Fedorov V V, Graham K, Moskalev I S, Badikov V V, and Panyutin V (2002), 'Erbium fiber laser-pumped continuous-wave microchip Cr^{2+}:ZnS and Cr^{2+}:ZnSe lasers', *Opt. Lett.*, 27, 909–911.

Nodop D, Limpert J, Hohmuth R, Richter W, Guina M, and Tünnermann A (2007), 'High-pulse-energy passively *Q*-switched quasi-monolithic microchip lasers operating in the sub-100-ps pulse regime', *Opt. Lett.*, 32, 2115–2117.

Owyoung A, and Esherick P (1987), 'Stress-induced tuning of a diode-laser-excited monolithic Nd:YAG laser', *Opt. Lett.*, 12, 999–1001.

Pavel N, Tsunekane P, and Taira T (2011), 'Composite, all-ceramics, high-peak power Nd:YAG/Cr^{4+}:YAG monolithic micro-laser with multiple-beam output for engine ignition', *Opt. Express*, 19, 9378–9384.

Provino L, Dudley J M, Maillotte H, Grossard N, Windeler R S, and Eggleton B J (2001), 'Compact broadband continuum source based on microchip laser pumped microstructured fibre', *Electron. Lett.*, 37, 558–560.

Rabarot M, Fulbert L, Molva E, Thony P, Marty V, and Dastouet S (1997), 'Fibre coupling of microchip lasers with silica microlenses', *Pure Appl. Opt.*, 6, 699–705.

Ranka J K, Windeler R S, and Stentz A J (2000), 'Visible continuum generation in air–silica microstructure optical fibers with anomalous dispersion at 800 nm', *Opt. Lett.*, 25, 25–27.

Schawlow A L, and Townes C H (1958), 'Infrared and optical masers', *Phys. Rev.*, 12, 1940–1949.

Schrader P E, Farrow R L, Kliner D A V, Fève J P, and Landru N (2006), 'High-power fiber amplifier with widely tunable repetition rate, fixed pulse duration, and multiple output wavelengths', *Opt. Express*, 14, 11528–11538.

Schulz P A, and Henion S R (1991), 'Frequency-modulated Nd:YAG laser', *Opt. Lett.*, 16, 578–580.

Spence D J, Liu H, and Coutts D W (2006), 'Low-threshold miniature Ce:LiCAF lasers', *Opt. Commun.*, 262, 238–240.

Spühler G J, Paschotta R, Fluck R, Braun B, Moser M, Zhang G, Gini E, and Keller U (1999), 'Experimentally confirmed design guidelines for passively *Q*-switched microchip lasers using semiconductor saturable absorbers', *J. Opt. Soc. Am. B*, 16, 376–388.

Spühler G J, Paschotta R, Kullberg M P, Graf M, Moser M, Mix E, Huber G, *et al.* (2001), 'A passively Q-switched Yb:YAG microchip laser', *Appl. Phys. B*, 72, 285–287.

Steinmetz A, Nodop D, Martin A, Limpert J, and Tünnermann A (2010), 'Reduction of timing jitter in passively *Q*-switched microchip lasers using self-injection seeding', *Opt. Lett.*, 35, 2885–2887.

Storm M E, and Rohrbach W W (1989), 'Single-longitudinal-mode lasing of Ho:Tm:YAG at 2.091 μm', *Appl. Opt.*, 28, 4965–4967.

© Woodhead Publishing Limited, 2013

Sulc J, Jelinkova H, Nejezchleb K, and Skoda V (2005), 'Nd:YAG/V:YAG microchip laser operating at 1338 nm', *Laser Phys. Lett.*, 2, 519–524.

Sutherland J M, Ruan S, Mellish R, French P M W, and Taylor J R (1995), 'Diode-pumped, single-frequency, Cr:LiSAF coupled-cavity microchip laser', *Opt. Commun.*, 113, 458–462.

Taira T, Mukai A, Nozawa Y, and Kobayashi T (1991), 'Single-mode oscillation of laser-diode-pumped Nd:YVO_4 microchip lasers', *Opt. Lett.*, 16, 1955–1957.

Thony P, Fulbert L, Besesty P, and Ferrand B (1999), 'Laser radar using a 1.55-μm passively Q-switched microchip laser', *SPIE*, 3707, 616–623.

Tsou Y, Garmire E, Chen W, Birnbaum M, and Asthana R (1993), 'Passive *Q* switching of Nd:YAG lasers by use of bulk semiconductors', *Opt. Lett.*, 18, 1514–1516.

Tsunekane M, Inohara T, Ando A, Kido N, Kanehara K, and Taira T (2010), 'High peak power, passively *Q*-switched microlaser for ignition of engines', *IEEE J. Quantum Electron.*, 46, 277–284.

Vieira A J C, Herczfeld P R, and Contarino V M (1997), '20 GHz mode-locked Nd:$LiNbO_3$ microchip laser', *Conf. Lasers Electro-Optics, Tech. Dig.*, 141–142.

Vieira A J C, Herczfeld P R, Rosen A, Ermold M, Funk E E, Jemison W D, and Williams K J (2001), 'A mode-locked microchip laser optical transmitter for fiber radio', *IEEE Trans. Microwave Theory Tech.*, 49, 1882–1887.

Wadsworth W J, Joly N, Knight J C, Birks T A, Biancalana F, and Russell P St J (2004), 'Supercontinuum and four-wave mixing with Q-switched pulses in endlessly single-mode photonic crystal fibers', *Opt. Express*, 12, 299–309.

Wallmeroth K (1990), 'Monolithic integrated Nd:YAG laser', *Opt. Lett.*, 15, 903–905.

Wang P, Zhou S-H, Lee K K, and Chen Y C (1995), 'Picosecond laser pulse generation in a monolithic self-*Q*-switched solid-state laser', *Opt. Commun.*, 114, 439–441.

Wu R, Myers J D, Myers M J, Denker B I, Galagan B I, Sverchkov S E, Hutchinson J A, and Trussel W (2000), 'Co^{2+}:$MgAl_2O_4$ crystal passive Q-switch performance at 1.34, 1.44, and 1.54 micron', *SPIE*, 3929, 42–45.

Xu X, Cheng S, Meng J, Li D, Zhou D, Zheng L, Xu J, *et al.* (2010), 'Spectral characterization and laser performance of a mixed crystal Nd:$(Lu_xY_{1-x})_3Al_5O_{12}$', *Opt. Express*, 18, 21370–21375.

Yao G, Lee K K, Chen Y C, and Zhou S (1994), 'Characteristics of transverse mode of diode-pumped self-Q-switched microchip laser', in Fan T Y, and Chai B H T, *OSA Proc. Advanced Solid-State Lasers* 20, Washington, DC, OSA, 28–31.

Yao B, Tian Y, Li G, and Wang Y (2010), 'InGaAs/GaAs saturable absorber for diode-pumped passively Q-switched dual-wavelength Tm:YAP lasers', *Opt. Express*, 18, 13574–13579.

Yu H, Zhang H, Wang Z, Wang J, Yu Y, Shao Z, and Jiang M (2007a), 'Enhancement of passive Q-switching performance with mixed Nd:$Lu_xGd_{1-x}VO_4$ laser crystals', *Opt. Lett.*, 32, 2152–2154.

Yu H, Zhang H, Wang Z, Wang J, Yu Y, Gao W, Tao X, *et al.* (2007b), 'Cr^{5+}:$GdVO_4$ as a saturable absorber for a diode-pumped Nd:$Lu_{0.5}Gd_{0.5}VO_4$ laser', *Opt. Express*, 15, 11679–11684.

Yu H, Zhang H, Wang Z, Wang J, Yu Y, Gao W, Tao X, and Jiang M (2008), 'Growth and passively self-Q-switched laser output of new Nd^{3+},Cr^{5+}:$GdVO_4$ crystal', *Opt. Express*, 16, 3320–3325.

Yu H, Zhang H, Wang Z, Wang J, Yu Y, Shi Z, Zhang X, and Jiang M (2009), 'High-power dual-wavelength laser with disordered Nd:CNGG crystals', *Opt. Lett.*, 34, 151–153.

© Woodhead Publishing Limited, 2013

Yumashev K V, Denisov I A, Posnov N N, Mikhailov V P, Moncorgé R, Vivien D, Ferrand B, and Guyot Y (1999), 'Nonlinear spectroscopy and passive *Q*-switching operation of a Co^{2+}:$LaMgAl_{11}O_{19}$ crystal', *J. Opt. Soc. Am. B*, 16, 2189–2194.

Zarrabi J H, Gavrilovic P, Williams J E, O'Neill M S, and Singh S (1993), 'Single-frequency, diode-pumped, neodymium-doped lanthanum oxysulfide microchip laser', *Conf. Lasers Electro-Optics, Tech. Dig.*, 588–589.

Zayhowski J J (1990a), 'Limits imposed by spatial hole burning on the single-mode operation of standing-wave laser cavities', *Opt. Lett.*, 15, 431–433.

Zayhowski J J (1990b), 'Microchip lasers', *Lincoln Lab. J.*, 3, 427–446.

Zayhowski J J (1990c), 'The effects of spatial hole burning and energy diffusion on the single-mode operation of standing-wave lasers', *IEEE J. Quantum Electron.*, 26, 2052–2057.

Zayhowski J J (1991a), 'Thermal guiding in microchip lasers', in Jenssen H P, and Dubé G, *OSA Proc. Advanced Solid-State Lasers* 6, Washington, DC, OSA, 9–13.

Zayhowski J J (1991b), 'Polarization-switchable microchip lasers', *Appl. Phys. Lett.*, 58, 2746–2748.

Zayhowski J J (1996a), 'Microchip lasers create light in small places', *Laser Focus World*, April, 73–78.

Zayhowski J J (1996b), 'Ultraviolet generation with passively *Q*-switched microchip lasers', *Opt. Lett.*, 21, 588–590 ('Ultraviolet generation with passively *Q*-switched microchip lasers: errata', *Opt. Lett.*, 21, 1618).

Zayhowski J J (1997a), 'Periodically poled lithium niobate optical parametric amplifiers pumped by high-power passively *Q*-switched microchip lasers', *Opt. Lett.*, 22, 169–171.

Zayhowski J J (1997b), 'Microchip optical parametric oscillators', *IEEE Photon. Technol. Lett.*, 9, 925–927.

Zayhowski J J (1997c), 'Covering the spectrum with passively *Q*-switched picosecond microchip laser systems', *Conf. Lasers Electro-Optics, Tech. Dig.*, 11, 463–464.

Zayhowski J J (1998), 'Passively *Q*-switched microchip lasers and applications', *Rev. Laser Eng.*, 26, 841–846.

Zayhowski J J (1999), 'Passively *Q*-switched microchip lasers find real-world application', *Laser Focus World*, August, 129–136.

Zayhowski J J (2000), 'Passively *Q*-switched Nd:YAG microchip lasers and applications', *J. Alloys Compounds*, 303–304, 393–400.

Zayhowski J J, and Dill III C (1992), 'Diode-pumped microchip lasers electro-optically *Q* switched at high pulse repetition rates', *Opt. Lett.*, 17, 1201–1203.

Zayhowski J J, and Dill III C (1994), 'Diode-pumped passively *Q*-switched picosecond microchip lasers', *Opt. Lett.*, 19, 1427–1429.

Zayhowski J J, and Dill III C (1995), 'Coupled-cavity electro-optically *Q*-switched Nd:YVO_4 microchip lasers', *Opt. Lett.*, 20, 716–718.

Zayhowski J J, and Johnson B (1996), 'Passively *Q*-switched microchip lasers for environmental monitoring', in *Laser Applications to Chemical, Biological and Environmental Analysis, Tech. Dig.* 3, Washington, DC, OSA, 37–39.

Zayhowski J J, and Kelley P L (1991), 'Optimization of *Q*-switched lasers', *IEEE J. Quantum Electron.*, 27, 2220–2225 ('Corrections to optimization of *Q*-switched lasers', *IEEE J. Quantum Electron.*, 29, 1239).

Zayhowski J J, and Keszenheimer J A (1992), 'Frequency tuning of microchip lasers using pump-power modulation', *IEEE J. Quantum Electron.*, 28, 1118–1122.

Zayhowski J J, and Mooradian A (1989a), 'Single-frequency microchip Nd:YAG lasers', *Opt. Lett.*, 14, 24–26.

© Woodhead Publishing Limited, 2013

Zayhowski J J, and Mooradian A (1989b), 'Frequency-modulated Nd:YAG microchip lasers', *Opt. Lett.*, 14, 618–620.

Zayhowski J J, and Mooradian A (1989c), 'Microchip lasers', in Shand M L, and Jenssen H P, *OSA Proceedings on Advanced Solid State Lasers* 5, Washington, DC, OSA, 288–294.

Zayhowski J J, and Wilson A L (2000), 'Miniature sources of subnanosecond 1.4–4.0-μm pulses with high peak power', in Keller U, Injeyan H, and Marshall C, *OSA TOPS* 34, *Advanced Solid-State Lasers*, Washington, DC, OSA, 308–311.

Zayhowski J J, and Wilson Jr A L (2002), 'Miniature, pulsed Ti:sapphire laser system', *IEEE J. Quantum Electron.*, 38, 1449–1454.

Zayhowski J J, and Wilson Jr A L (2003a), 'Pump-induced bleaching of the saturable absorber in short-pulse Nd:YAG/Cr^{4+}:YAG passively *Q*-switched microchip lasers', *IEEE J. Quantum Electron.*, 39, 1588–1593.

Zayhowski J J, and Wilson Jr A L (2003b), 'Miniature, high-power 355-nm laser system', in Zayhowski J J, *OSA TOPS* 83, *Advanced Solid-State Photonics*, Washington, DC, OSA, 357–362.

Zayhowski J J, and Wilson Jr A L (2004), 'Energy-scavenging amplifiers for miniature solid-state lasers', *Opt. Lett.*, 29, 1218–1220.

Zayhowski J J, Schulz P A, Dill III C, and Henion S R (1993), 'Diode-pumped composite-cavity electrooptically tuned microchip laser', *IEEE Photon. Technol. Lett.*, 5, 1153–1155.

Zayhowski J J, Harrison J, Dill III C, and Ochoa J (1995a), 'Tm:YVO_4 microchip laser', *Appl. Opt.*, 34, 435–437.

Zayhowski J J, Ochoa J, and Dill III C (1995b), 'UV generation with passively *Q*-switched picosecond microchip lasers', *Conf. Lasers Electro-Optics, Tech. Dig.*, 15, 139.

Zayhowski J J, Fan T Y, Cook C, and Daneu J L (1996), '946-nm passively *Q*-switched microlasers', *M.I.T. Lincoln Lab., Solid State Res., Quarterly Tech. Rep.* 1996:3, ESC-TR-96-096, 5–6.

Zayhowski J J, Dill III C, Cook C, and Daneu J L (1999), 'Mid- and high-power passively *Q*-switched microchip lasers', in Fejer M M, Injeyan H, and Keller U, *OSA TOPS* XXVI, *Advanced Solid-State Lasers*, Washington, DC, OSA, 178–186.

Zayhowski J J, Cook C C, Wormhoudt J, and Shorter J H (2000), 'Passively *Q*-switched 214.8-nm Nd:YAG/Cr^{4+}:YAG microchip-laser system for the detection of NO', in Keller U, Injeyan H, and Marshall C, *OSA TOPS* 34, *Advanced Solid-State Lasers*, Washington, DC, OSA, 409–412.

Zayhowski J J, Buchter S C, and Wilson A L (2001), 'Miniature gain-switched lasers', in Marshall C, *OSA TOPS* 50, *Advanced Solid-State Lasers*, Washington, DC, OSA, 462–469.

Zayhowski J J, Welford D, and Harrison J (2007), 'Miniature solid-state lasers', in Gupta M C, and Ballato J, *The Handbook of Photonics,* 2nd edn, Boca Raton, FL, CRC Press, Ch. 10.

Zhang X, Zhao S, Wang Q, Zhang S, Sun L, Liu X, Zhang S, and Chen H (2001), 'Passively *Q*-switched self-frequency-doubled Nd^{3+}:$GdCa_4O(BO_3)_3$ laser', *J. Opt. Soc. Am. B*, 18, 770–779.

Zhang B, Yang J, He J, Huang H, Liu S, Xu J, Liu F, *et al.* (2010), 'Diode-end-pumped passively Q-switched 1.33 μm Nd:$Gd_3Al_xGa_{5-x}O_{12}$ laser with V^{3+}:YAG saturable absorber', *Opt. Express*, 18, 12052–12058.

Zhang S, Huang H, Xu L, Wang M, Chen F, Xu J, He J, and Zhao B (2011), 'Continuous wave and passively Q-switched Nd:$Lu_xY_{1-x}VO_4$ laser at 1.34 μm with V^{3+}:YAG as the saturable absorber', *Opt. Express*, 19, 1830–1835.

© Woodhead Publishing Limited, 2013

Zhao J, Yan P, Shu J, and Ruan S (2011), '1 MW peak power, sub-nanosecond master oscillator fiber power amplifier', *2011 Symposium on Photonics and Optoelectronics (SOPO),* Piscataway, NJ, 1–3.

Zhi Y, Dong C, Zhang J, Jia Z, Zhang B, Zhang Y, Wang S, *et al.* (2010), 'Continuous-wave and passively *Q*-switched laser performance of LD-end-pumped 1062 nm Nd:GAGG laser', *Opt. Express*, 18, 7584–7589.

Zhou Y, Thai Q, Chen Y C, and Zhou S (2003), 'Monolithic Q-switched Cr,Yb:YAG laser', *Opt. Commun.*, 219, 365–367.

Zhu H, Chen Y, Lin Y, Gong X, Luo Z, and Huang Y (2008), 'Efficient quasi-continuous-wave and passively *Q*-switched laser operation of a $Nd^{3+}:BaGd_2(MoO_4)_4$ cleavage plate', *Appl. Opt.*, 47, 531–535.

Zhuang S, Yu H, Wang Z, Zhang H, Wang J, Guo L, Chen L, *et al.* (2011), 'Passively Q-switched $Nd:Gd_{0.63}Y_{0.37}VO_4/Cr^{4+}:YAG$ microchip laser', *J. Crystal Growth*, 318, 691–694.

Zolotovskaya S A, Yumashev K V, Kuleshov N V, Matrosov V N, Matrosova T A, and Zupchenko M I (2007), 'Absorption saturation properties and laser Q-switch performance of Cr^{5+}-doped YVO_4 crystal', *Appl. Phys. B*, 86, 667–671.

© Woodhead Publishing Limited, 2013

15
Fiber lasers

B. SAMSON, Nufern, USA and L. DONG, Clemson University, USA

DOI: 10.1533/9780857097507.2.403

Abstract: Fiber lasers are arguably the most disruptive laser technology to emerge in the last decade, spanning kilowatt-class CW lasers, pulsed lasers from nanoseconds down to ultrafast, and wavelengths from 1 μm to 2 μm. This chapter covers the basic laser physics of these systems and the waveguide designs to optimize device performance along with the spectroscopy and nonlinear limitations of fibers. Many of the key experimental results of the last decade are detailed, across all key fiber laser operating regimes, and spanning commercially relevant fiber laser systems through to the latest cutting-edge research using novel fiber designs.

Key words: fiber laser, large mode area fiber, kW laser, Yb-doped fiber, single-mode laser.

15.1 Introduction and history

The period of time from 2000 till 2010 saw an unprecedented advance in fiber laser technology. As shown in Fig. 15.1, the CW power from a single Yb-doped fiber laser operating around 1 μm and delivering nearly single-mode beam quality increased from ~100 W [1] to 10 kW [2]. This rapid increase in output power was enabled by several factors but it is important to note that many of these high-power results were limited only by the available pump power rather than by any engineering or physical limit of the fiber itself. To this extent, progress on scaling output power from a single-mode Yb-doped fiber was closely tied to the available laser diode power and specifically to the progress on suitable high brightness diodes that could be efficiently coupled into a single Yb-doped fiber.

The first demonstration of a fiber laser dates as far back as 1964 by Keoster and Snitzer [3]. However, the key enabling advances for high-power fiber lasers were pioneered in the late 1980s at Polaroid Corporation, when a rare-earth-doped double-clad fiber, the key enabling technology for high-power cladding-pumped fiber lasers, was first described. This concept, that a pump waveguide surrounding a doped core of the fiber is formed by a second cladding region, allows the use of high power, low brightness laser diode pumps, and has been proven to be a major enabling factor for power scaling of fiber laser technology. It was subsequently demonstrated in a cladding-

© Woodhead Publishing Limited, 2013

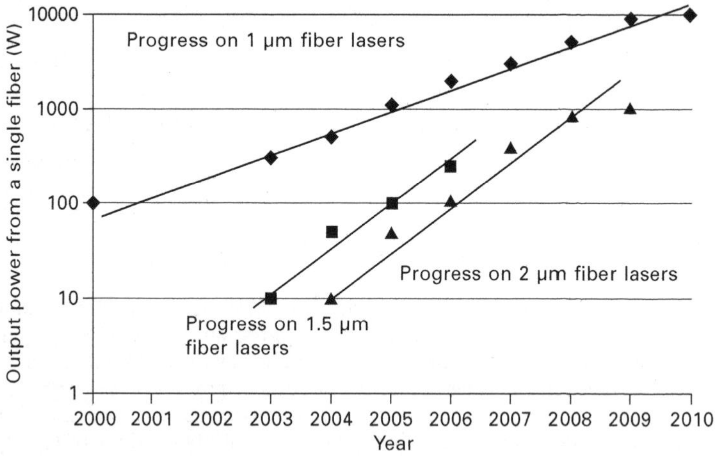

15.1 Progress on power scaling of near-single-mode fiber lasers at wavelengths 1 μm, 1.5 μm and 2 μm.

pumped fiber laser in 1989 [4] by the same group. This cladding-pumped geometry reduces the pump absorption by a factor approximately determined by the ratio of the core to clad areas, but allows for a significantly larger waveguide for coupling pump light. This large multimode pump waveguide is a much better match to the beam from high power multimode laser diodes. The ability to form the second pump waveguide by using low refractive index fluoro-acrylate polymer coatings proved to be another valuable innovation. It enabled a high power diode to efficiently couple to the fiber, since the typical numerical apertures (NAs) formed in these polymer-coated fibers exceed 0.46NA, far higher than the typical NA of a multimode laser diode deliver fiber, which is in the range 0.15–0.2NA.

In some of these initial experiments in Fig. 15.1, the adoption of very large cladding diameters was common (up to 800 μm in some cases), allowing a greater amount of diode power to be coupled into the cladding. However, it is fair to say that the fiber laser industry has settled on a more modest 400 μm clad diameter as the standard over the last five years. This 400 μm clad diameter offers a good compromise between large diameter to couple pump light and the practical handling and coiling of the fiber, which is more difficult in larger fibers, together with additional considerations of the optimum pump absorption and fiber length. In the early part of the last decade, several theoretical studies on limits of fiber lasers further enhanced the realization that high power CW lasers are very practical in rare-earth-doped silica fibers [5, 6]. The dramatic improvements in diode laser technology during this same time played an equally important part in the scaling of output powers of fiber lasers, first through the improved diode efficiency, enabling higher power and lower thermal load, and later through improved

© Woodhead Publishing Limited, 2013

facet damage thresholds enabling longer lifetime of high power diodes. Other improvements, including higher powers from high-brightness diode bars and stacks [7] and wavelength combining as means to improve overall pump brightness, have enabled kilowatts of pump power to be coupled into the cladding of the target fiber. These diode improvements were critical to the power scaling of fiber lasers as shown in Fig. 15.1.

With the adoption of large clad diameter fibers, it also became important to develop fibers with larger core diameters. This requirement was two-fold: firstly to scale up the mode field of the signal but also to reduce fiber length through an increase in the cladding absorption for the pump light. The ability of the fiber to maintain single-mode beam quality when the core diameter is increased to the 20–30 μm diameter range gave rise to the nomenclature of large mode area (LMA) fibers, implying a lower NA [8] and hence fewer core modes, as described in the next section. Through a variety of techniques researchers have shown it is possible to operate these few-moded fibers with good output beam quality over the lengths required to make practical laser devices [9, 10]. With the development of larger core diameter fibers, the energy storage and pulsed performance of fiber lasers are also increased, making pulsed fiber devices, operating anywhere from nanoseconds down to femtoseconds, highly attractive. The ease of scaling average power in the fiber geometry has made high average power pulsed fiber systems relatively cost-effective when compared with the DPSSL counterparts. The commercial success of pulse fiber lasers in the 1 mJ regime delivering 10–50 W average power (nominally 10 kW peak power and 100 ns pulse duration) is significant. They arguably have become the dominant laser for industrial marking and engraving applications.

As the scaling of pulse energy progressed, it became clear that limitation to the LMA technology still existed, with difficulty scaling core size beyond 30 μm diameter without sacrificing beam quality. In the last five years or so, a great deal of fiber research has concentrated on more sophisticated fiber designs for scaling the mode field, inducing innovations such as photonic crystal fibers (PCF), chirally coupled core (CCC) fibers, and leakage channel fibers (LCF) amongst others (for a recent review of these fibers see reference 11). The underlying theme of much of this fiber research involves scaling core diameter into the 50–100 μm regime in Yb-doped fibers for high peak power pulsed fiber lasers. In addition to the scaling of fiber lasers at 1 μm using Yb-doped fibers, progress has been made more recently on scaling at wavelengths around 1.5 μm and 2 μm using erbium and thulium doped silica fibers respectively. Here the greatest advances have been in the efficiency of the fibers. Power scaling of fiber lasers at these 'eyesafer' operating wavelengths has now reached the kilowatt level, as shown in Fig. 15.1.

As the technology of fiber lasers has advanced in the last 10 years, so have the commercial applications of the technology. Annual sales of fiber

© Woodhead Publishing Limited, 2013

lasers exceeded \$500m, dominated by industry leader IPG Photonics. The commercial impact of fiber lasers now spans the complete range of industrial products from kilowatt-class CW lasers for cutting and welding, through nanosecond pulsed fiber lasers for marking and engraving, to ultrafast fiber lasers for super-continuum generation and materials processing. The overall market share for fiber lasers as a total of all industrial lasers is now approaching 10%, and growing at a CAGR of 20%, taking market share from CO_2 and DPSSLs in almost all market segments [12].

15.2 Principle of fiber lasers

15.2.1 Overview

One primary benefit of fiber lasers over other solid state and gas lasers is superior mode quality and robustness against thermal effects. The optical waveguide can be made to support only a single Gaussian-like optical mode. Single-mode optical fibers are widely used in telecommunications. This high level of mode control is hard to achieve through any other means and is the primary reason why fiber lasers were first studied in the 1960s [3, 13]. The second key benefit of fiber lasers is the confinement of low brightness pump lights in a double-clad design. The double-clad design, illustrated in Fig. 15.2, confines pump light in a larger multimode optical waveguide around a smaller rare-earth-doped central core. This enables the use of high power, lower brightness diodes as pump and, as a consequence, creates great potential to scale output power. The confinement of pump light enables unmatched high efficiency, as pump light is forced repeatedly to interact with the active core along the length of the fiber. A fiber laser can be seen

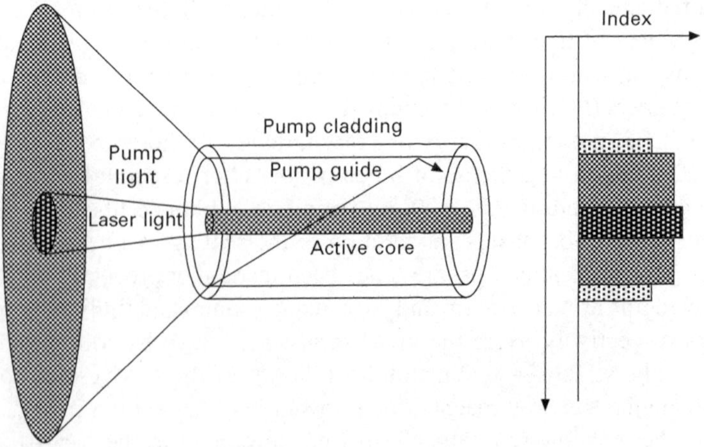

15.2 Illustration of a double-clad optical fiber.

© Woodhead Publishing Limited, 2013

as a very efficient brightness converter. The lower overlap between the pump light and active medium in the core does lead to lower pump absorption. This is, however, well compensated by the possibility of using a long length of optical fiber. The third benefit of fiber lasers is the superior thermal load handling due to the long and slender geometry which allows efficient heat transfer from the active core radially to the fiber surface just a fraction of a millimeter away. An additional benefit of fiber lasers is robustness and low maintenance as a consequence of the flexible optical fiber construction. This is very important for the use of fiber lasers in manufacturing environments where they can be operated by unskilled personnel with minimum downtime and maintenance.

15.2.2 Basics of optical fibers

A conventional optical fiber is illustrated in Fig. 15.3. It consists of a circular core of radius ρ and refractive index n_{co}, surrounded by a cladding of refractive index n_{cl}. It is typically coated by a high index polymer to protect the glass surface and to strip off any lights in the cladding. For single-mode optical fibers, the core refractive index n_{co} is only slightly higher than the cladding refractive index n_{cl}. A useful parameter is numerical aperture (NA), which is closely related to the acceptance angle θ_a of the incident cone illustrated in Fig. 15.2.

$$\mathrm{NA} = \sqrt{n_{co}^2 - n_{cl}^2} = n_0 \sin(\theta_a)$$

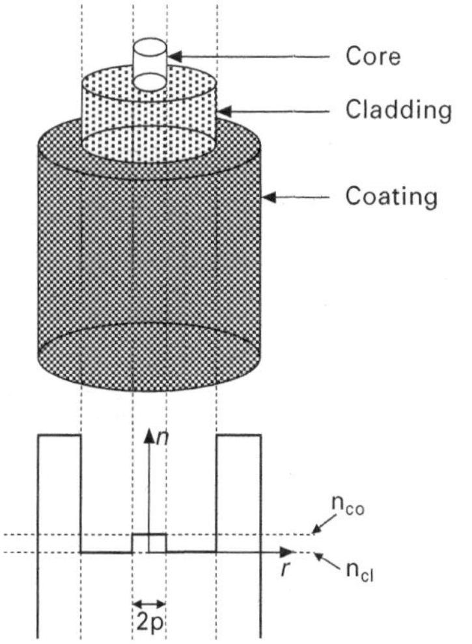

15.3 Illustration and parameter definitions for a conventional optical fiber.

© Woodhead Publishing Limited, 2013

where n_0 is the refractive index of the medium outside the fiber and is usually 1 for air. Another very useful fiber parameter is normalized frequency, defined at vacuum wavelength λ as

$$V = \frac{2\pi}{\lambda} \rho \mathrm{NA}$$

The normalized frequency V takes into account both the dimensional and refractive index scalability of an optical waveguide and describes the modal properties of any optical fiber at the defined V value. A guided mode has a constant transverse field distribution over the length of a fiber with a well-defined propagation constant,

$$\beta = \frac{2\pi}{\lambda} n_{\mathrm{eff}}$$

where n_{eff} is the effective index of the guided mode and falls between core and cladding index n_{co} and n_{cl}. It describes the guiding strength of a mode. In general, a mode is strongly guided when its effective index n_{eff} is well above n_{cl} and is cut off when $n_{\mathrm{eff}} = n_{\mathrm{cl}}$. In weakly guided optical fiber, modes are well represented by the approximate linearly polarized modes $\mathrm{LP}_{\mathrm{lm}}$, where l and m are azimuthal and radial mode numbers respectively. Modes in step-index fibers are given in Fig. 15.4 for $V < 10$. The normalized propagation constant is defined as

$$b = \frac{n_{\mathrm{eff}}^2 - n_{\mathrm{cl}}^2}{n_{\mathrm{co}}^2 - n_{\mathrm{cl}}^2}$$

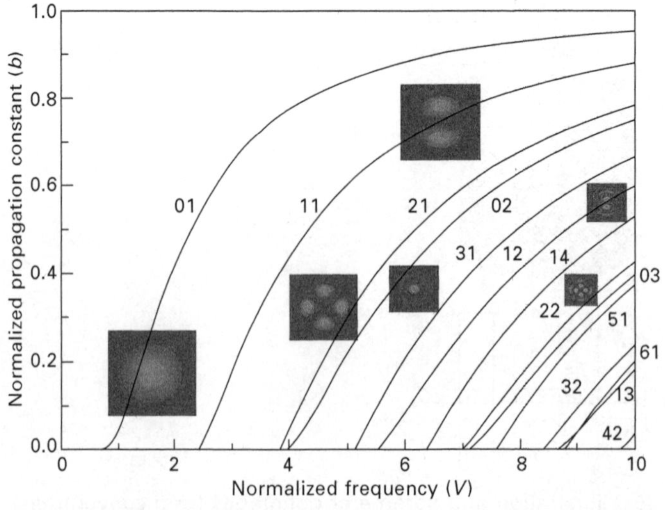

15.4 Normalized propagation constants for a step-index fiber.

with b varying from 0 to 1, corresponding to $n_{\text{eff}} = n_{\text{cl}}$ and n_{co} respectively.

It can be seen in Fig. 15.4 that effective mode indexes generally increase with V, indicating stronger mode guidance at larger V values. Single-mode operation exists only when $V < 2.405$ for step-index fibers, where the second mode (LP_{11}) cuts off. Since the V value is determined by the product of core radius ρ and NA, many combinations will provide the same effective mode index. The V value also increases towards shorter wavelengths due to the $1/\lambda$ dependence. The total number of guided modes can be estimated by

$$N = \frac{V^2}{2}$$

Dispersion of an optical fiber is the sum of material dispersion D_{m} and waveguide dispersion D_{w}, $D = D_{\text{m}} + D_{\text{w}}$. Material dispersion of silica is shown in Fig. 15.5(a) with a zero-dispersion wavelength (ZDW) at ~1.3 µm. Waveguide dispersion can be written as

$$D_{\text{w}} = -\frac{n_{\text{co}} - n_{\text{cl}}}{\lambda c} V \frac{\text{d}^2(Vb)}{\text{d}V^2} = \frac{n_{\text{co}} - n_{\text{cl}}}{\lambda c} D(V)$$

where c is the speed of light in vacuum and $D(V)$ is a dimensionless function dependent only on V and shown in Fig. 15.5(b), which can be viewed as the normalized waveguide dispersion. Waveguide dispersion is proportional to the index difference between core and cladding and inversely proportional to wavelength. It is also worth noting that waveguide dispersion is entirely normal in the single-mode regime and becomes anomalous in the multimode regime only when $V > 2.68$ for step-index fiber. It is common to expand the propagation constant $\beta(\omega)$ at a reference frequency ω_0:

$$\beta(\omega) = \beta_0 + \beta_1(\omega - \omega_0) + \frac{\beta_2}{2}(\omega - \omega_0)^2 + \frac{\beta_3}{6}(\omega - \omega_0)^3 + \ldots$$

where β_1 is group delay and β_2 describes first-order dispersion.

$$\beta_1 = \left(\frac{\text{d}\beta}{\text{d}\omega}\right)_{\omega=\omega_0} = \frac{1}{c}\left(n + \omega\frac{\text{d}n}{\text{d}\omega}\right) = \tau_{\text{g}}$$

$$\beta_2 = \left(\frac{\text{d}\beta_1}{\text{d}\omega}\right)_{\omega=\omega_0} = \frac{1}{c}\left(2\frac{\text{d}n}{\text{d}\omega} + \omega\frac{\text{d}^2 n}{\text{d}\omega^2}\right)$$

D and β_2 are related through

$$D = \frac{\text{d}\tau_{\text{g}}}{\text{d}\lambda} = \frac{\text{d}\beta_1}{\text{d}\lambda} = -\frac{2\pi c}{\lambda^2}\beta_2$$

The optical pulse broadens during propagation in dispersive optical fibers

© Woodhead Publishing Limited, 2013

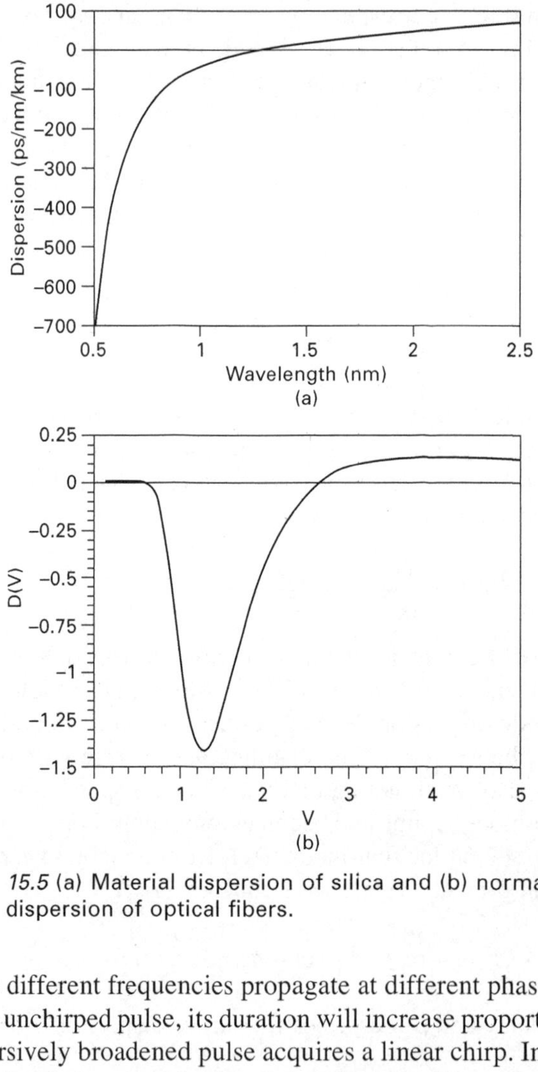

15.5 (a) Material dispersion of silica and (b) normalized waveguide dispersion of optical fibers.

because different frequencies propagate at different phase velocities. For an initially unchirped pulse, its duration will increase proportional to dispersion. A dispersively broadened pulse acquires a linear chirp. In optical fibers with normal dispersion, $D < 0$, $\beta_2 > 0$, low frequency travels to the leading edge of the pulse and high frequency to the trailing edge of the pulse, leading to a positive chirp. The situation is reversed in optical fibers with anomalous dispersion, $D > 0$, $\beta_2 < 0$.

15.2.3 Rare-earth-doped silica glass

Techniques for rare-earth doping in silica fibers were developed during the 1980s. There are two key techniques. The first is the solution-doping method [14]. Silica soot is deposited first using any one of the established vapor phase deposition techniques. This is typically done by reacting a vapor such

© Woodhead Publishing Limited, 2013

as $SiCl_4$ with oxygen at a temperature below the sintering temperature. A rare-earth compound, typically a chloride, is dissolved in a solution. The porous silica soot is then immersed in the solution. Rare-earth ions are attached to the porous soot. The solution is eventually drained and the soot dried before sintering. The extent of rare-earth incorporation can be controlled by solution concentrations, soaking time and soot porosity. The key benefit of this technique is ease of implementation, which largely explains its wide use. Soluble compounds of all relevant ions are readily available.

The second technique for rare-earth incorporation is to introduce the rare earth via the vapor phase [15, 16]. Organometallic compounds are typically used for their relatively high vapor pressures. The key benefit of this technique is its compatibility with the vapor phase techniques already used in fiber preform fabrication, which typically offer good controllability, less impurity, and possibly good doping uniformity. The key drawback is complexity in its implementation. This is a consequence of the low vapor pressure of the precursors. It typically requires the use of elevated temperatures with heated delivery lines.

Silica, with its rigid closed structure, is not known for good rare-earth solubility. Other glasses such as phosphate have more open structures and are much better at incorporation of rare-earth ions. Clustering and phase separation can happen at high doping levels, which can lead to ion-ion interactions and often a degradation of performance. This is not typically a problem for erbium-doped fiber amplifiers for telecommunications, where it is mitigated by using low doping levels and, consequently, long amplifier lengths of a few tens of meters. For high power fiber lasers, a much shorter length is often desired to minimize nonlinear effects which typically limit power scaling. Furthermore, the double-clad design reduces the overlap of the pump light and the rare-earth ions in the single-mode core by typically well over two orders of magnitude, requiring much higher doping levels. Typically modifiers such as aluminum are co-doped to allow much higher rare-earth doping levels by creating more open sites. Phosphorus can be co-doped to achieve similar effects.

An example of ion-ion interaction is cooperative up-conversion of two excited ions to a single excited ion at a higher energy level, where the higher energy level has an energy which is approximately the sum of the two lower energy levels. This typically leads to energy loss when the higher level is de-excited. This de-excitation can also lead to more energetic photons, which can cause photo-darkening by the creation of color centers in the fiber. A useful ion-ion interaction is the case of the Tm^{3+} system, where the higher level $^3H_4$ is excited first, which when de-excited, leading to two excited Tm^{3+} ions at the lasing levels $^3H_6$ in the so-called two-for-one process (see Fig. 15.6).

Many rare-earth ions have been investigated in fiber amplifiers and lasers.

© Woodhead Publishing Limited, 2013

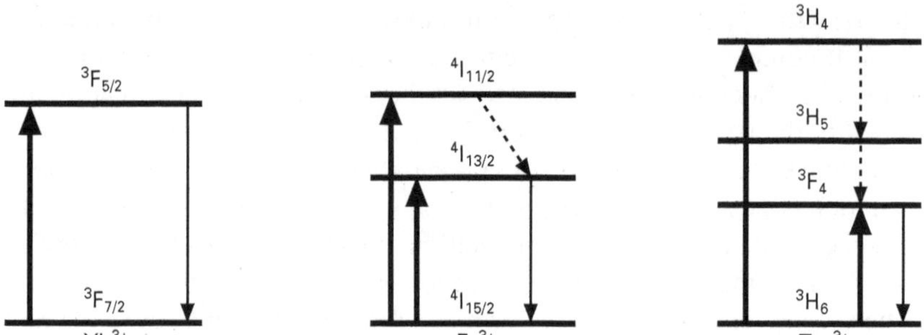

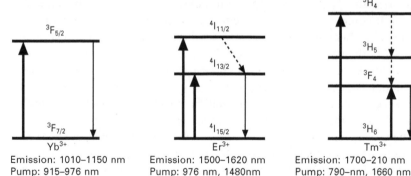

15.6 Illustration of relevant energy levels of Yb^{3+}, Er^{3+} and Tm^{3+} commonly used in fiber lasers.

Yb^{3+} and Tm^{3+} doped silica fibers are the only systems which have operated at CW power levels over 1 kW in fiber laser or amplifier configurations. Fiber lasers of ~400 W have been demonstrated in an Er^{3+} system (see Fig. 15.1). One major constraint for power scaling is the availability of high power pump diodes. Currently, they are mostly at ~976 nm, ~915 nm and ~800 nm, most suitable for pumping Yb^{3+} and Er^{3+} systems. One earlier driver for high power diode developments was the need for Er^{3+} pumps for fiber amplifiers in telecommunications. The relevant energy levels of Yb^{3+}, Tm^{3+} and Er^{3+} are illustrated in Fig. 15.6. Yb^{3+} ions have a very simple system of two energy levels consisting of three sub-levels and four sub-levels in the upper and lower branches respectively, as illustrated in Fig. 15.7(a) [17, 18]. Despite little interest early on, the Yb^{3+} system soon caught on once it was realized that it offers broad gain bandwidths and can lase as a four-level system at a very low threshold at the long-wavelength end of the gain spectrum (see Fig. 15.7). The simple two-level system excludes many ion-and-ion interactions existing in other rare-earth ions, enabling ytterbium doping levels of well over an order of magnitude higher than those of erbium. This allows shorter fiber lengths, a highly desirable feature for short pulse fiber lasers requiring generation of high peak powers. The low quantum defect was also very advantageous for power scaling. These combined benefits soon established the Yb^{3+} system as the dominant force for high power fiber lasers. With a tandem pumping scheme using ytterbium fiber lasers as pump at 1017 nm to further lower quantum defect, IPG Photonics demonstrated record 10 kW single-mode fiber lasers [19].

The Tm^{3+} system offers very broad gain at ~2 μm (see Fig. 15.8). This wavelength falls in one of the transmission windows of the atmosphere. It is blocked by the cornea, which prevents damage to the retina, making it safer for the eyes. The main drive is for defense applications in recent years. The availability of high power pumps at ~790 nm and the low quantum defect

© Woodhead Publishing Limited, 2013

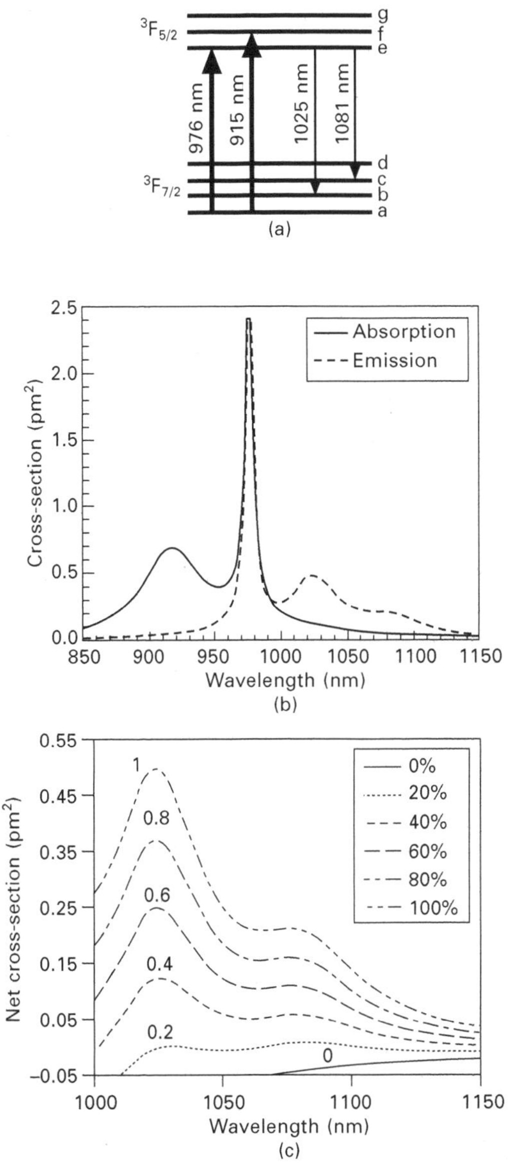

15.7 (a) Yb^{3+} energy level diagram, (b) absorption and emission cross-sections of Yb^{3+} ions in aluminosilicate glass and (c) net cross-section of Yb^{3+} at various inversions.

in the two-for-one process [20] are keys to the demonstration of 1 kW Tm^{3+} fiber lasers (see a later section of this chapter for more details). The laser wavelength at ~2 μm also makes it an ideal pump source for the generation of mid-IR wavelengths which have a wide range of spectroscopic and sensing

© Woodhead Publishing Limited, 2013

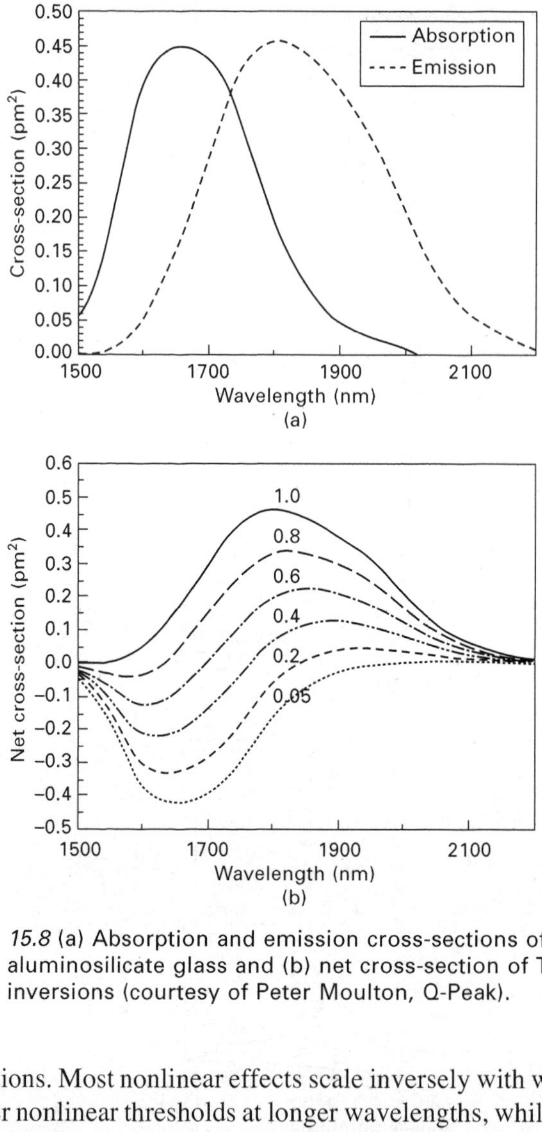

15.8 (a) Absorption and emission cross-sections of Tm^{3+} ions in aluminosilicate glass and (b) net cross-section of Tm^{3+} at various inversions (courtesy of Peter Moulton, Q-Peak).

applications. Most nonlinear effects scale inversely with wavelength, leading to higher nonlinear thresholds at longer wavelengths, while waveguide scales linearly with wavelength, leading to lower optical intensity due to the large effective mode area. These lead to the potential for much higher optical powers from Tm^{3+} fiber lasers.

Er^{3+}-doped fiber lasers were among the first fiber lasers to be intensively studied (see Fig. 15.9). This was largely due to the wide availability of Er^{3+}-doped fibers, the necessary pumps, isolators and other components as a result of the development of Er^{3+} fiber amplifiers. The much lower doping levels and lack of pumps for in-band pumping led to slow progress in power levels. High power Er^{3+} fiber lasers have been demonstrated exclusively by

© Woodhead Publishing Limited, 2013

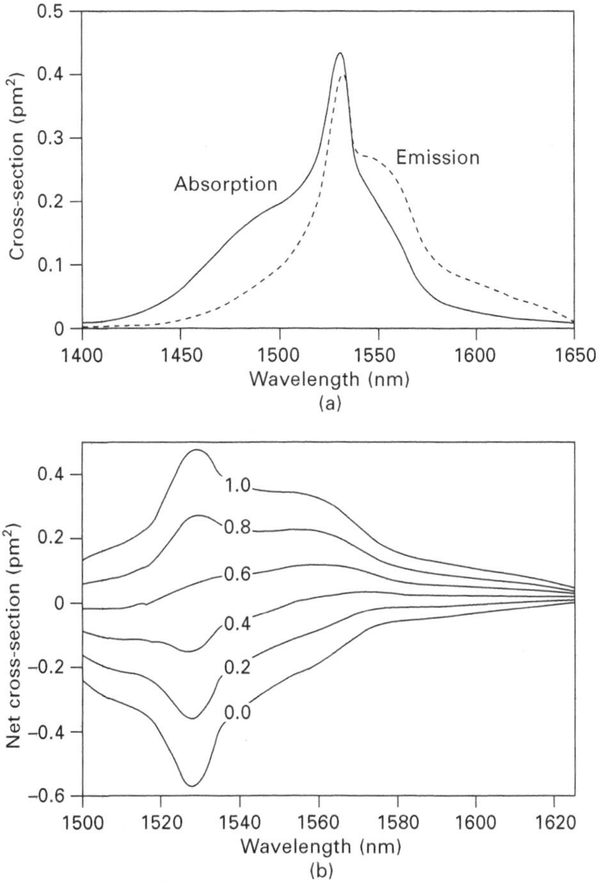

15.9 (a) Absorption and emission cross-sections of Er^{3+} ions in aluminosilicate glass and (b) net cross-section of Er^{3+} at various inversions (courtesy of Martin Fermann, IMRA).

using Yb^{3+} co-doping to enhance pump absorption at ~976 nm. An excited Yb^{3+} ion at the $^3F_{5/2}$ level can lead to the excitation of an Er^{3+} ion at the $^4I_{11/2}$ level through an energy transfer process involving phonons. This effectively makes shorter double-clad fibers possible for the Er^{3+} system. The large quantum defect of this process largely explains the lower demonstrated powers compared to Yb^{3+} and Tm^{3+} doped fiber lasers.

15.2.4 Nonlinear effects

High optical intensity in a small core and long interaction length in the order of a few meters to a few tens of meters in fiber lasers can lead to severe limitations due to nonlinear effects. Typically nonlinear processes relevant to

© Woodhead Publishing Limited, 2013

fiber lasers are stimulated Brillouin scattering, stimulated Raman scattering and self-phase modulations.

In stimulated Brillouin scattering (SBS), an intense beam can interact with an acoustic wave in the medium, leading to a weak reflected beam. The reflected beam, generally referred to as a Stokes wave, and the original beam can then interfere to amplify the acoustic wave through electrostriction. This process can start from thermal noise and is self-stimulating due to the positive feedback. SBS can lead to a significant fraction of the original beam being reflected at high power levels. Energy and momentum conservation require the frequencies ν_a, ν_p and ν_s and the wave vectors k_a, k_p and k_s, where a, p and s stand for acoustic, pump and Stokes waves respectively, to satisfy

$$\nu_a = \nu_p - \nu_s$$

$$k_a = k_p - k_s$$

The acoustic frequency involved in SBS is ~16 GHz at an optical wavelength of ~1 μm in silica fibers, with a spectral bandwidth of about a few tens of megahertz. The acoustic spectral bandwidth is referred to as the SBS gain bandwidth. Any back-propagating optical wave with a frequency separation from the Stokes wave less than the SBS gain bandwidth will experience the SBS gain. The SBS gain bandwidth is determined by the acoustic damping of the silica glass. The SBS threshold in an optical fiber can be determined for an optical wave with a spectral width significantly less than the SBS spectral width by

$$P_{SBS} = 21 \frac{A_{eff}}{g_B L_{eff}}$$

where g_B is the peak Brillouin gain (= 3–5 × 10^{-11} m/W in a silica fiber, mostly wavelength independent); A_{eff} is the effective mode area; and L_{eff} is the effective nonlinear length. The effective mode area A_{eff} for an optical mode with spatial electric distribution of $E(x, y)$ is given by

$$A_{eff} = \frac{\left(\iint_{A\infty} |E(x, y)|^2 \, dxdy\right)^2}{\iint_{A\infty} |E(x, y)|^4 \, dxdy}$$

where the integration is across the entire cross-sectional area. The effective length for an amplifier with output power P_{out}, length L and power distribution $P(z)$ is given by

$$L_{eff} = \int_0^L \frac{P(z)}{P_{out}} dz$$

SBS typically has the lowest threshold among all nonlinear effects for an optical wave with narrow spectral width and can also be a major limitation

© Woodhead Publishing Limited, 2013

in optical fiber telecommunication systems. Since it is a highly coherent process, SBS can be significantly suppressed by spectral broadening to beyond the SBS gain bandwidth. In telecommunications, the optical carrier wave is typically modulated to effectively suppress SBS. For the same reason, SBS is usually not the major nonlinear limit for pulse systems with sub-nanosecond durations. In recent years, the drive to significantly scale power in single-frequency fiber lasers for directed energy weapons in defense has led to strong interest in SBS suppression techniques. One technique that has attracted much interest in recent years is to design a transverse acoustic velocity profile in optical fibers. Since transverse acoustic power distribution is determined by acoustic modes, a transverse acoustic velocity profile can be engineered to minimize overlap of acoustic and optical powers in the fiber [21, 22]. It has been pointed out recently that significant SBS suppression by this method is limited due to the difficulty of engineering both guided and leaky acoustic modes in an optical fiber [23]. By bandwidth broadening to over 10 GHz, over 1 kW of power has been demonstrated recently for single-mode fiber lasers.

Stimulated Raman scattering (SRS) is a light scattering process due to the excitation of optical phonons via the direct interaction of an electric field and matter in an optical fiber. The scattered Stokes wave is in the forward direction. The optical phonons involved have a much higher frequency than the acoustic phonons involved in the SBS process. Energy conservation requires the scattered Stokes wave to have a lower frequency. Since momentum conservation is automatically satisfied to a large extent, the SRS gain bandwidth is significantly larger than that of SBS and is over 40 THz with a peak at ~13 THz in silica fibers. The SRS threshold is determined by

$$P_{SRS} = 16 \frac{A_{eff}}{g_R L_{eff}}$$

The peak Raman gain coefficient $g_R = \sim 1 \times 10^{-13}$ m/W at a pump wavelength of 1 μm in silica fibers. For other pump wavelengths, g_R scales inversely with pump wavelength.

Self-phase modulation (SPM) originates from the time-dependent nonlinear phase when an optical pulse propagates in an optical fiber. The time-dependent phase leads to frequency generation at the leading and trailing edges of an optical pulse. Continuous red-shift at the leading edge and blue shift at the trailing edge occur, resulting in a positive chirp (see Fig. 15.10). The chirp is linear near the center of a Gaussian pulse. SPM is more pronounced in ultrafast pulses with durations less than a few tens of picoseconds. The nonlinear phase ϕ_{NL} induced by optical power P is given by

$$\phi_{NL} = \gamma P L_{eff}$$

© Woodhead Publishing Limited, 2013

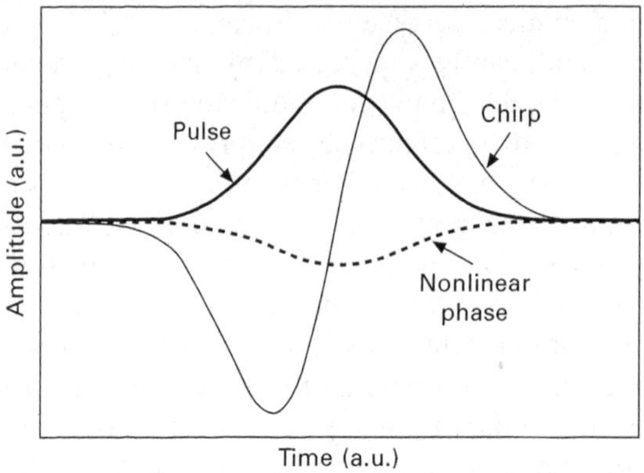

15.10 SPM induced chirp in an optical pulse.

where the nonlinear coefficient γ at optical wavelength λ is given by

$$\gamma = \frac{2\pi n_2}{\lambda_{A_{\text{eff}}}}$$

n_2 = ~2.6×10^{-20} m^2/W is the Kerr nonlinear coefficient for silica fibers. The peak nonlinear phase is commonly referred to as the B-integral, which is an important parameter to consider when designing an ultrafast laser. The B-integral is typically kept below π to minimize pulse distortion by SPM in a laser. A useful parameter is the dispersion length z_{nl}:

$$z_{\text{nl}} = \frac{1}{\gamma_{\text{P}}}$$

We can then write the nonlinear phase as

$$\phi_{\text{NL}} = \frac{L_{\text{eff}}}{z_{\text{nl}}}$$

With the right pulse shape, the positive chirp arising from SPM can be compensated by the negative chirp as a result of propagation in an optical fiber with anomalous dispersion. This can be precisely achieved when a pulse has a $\text{sech}^2(t/\tau)$ intensity profile without chirp. These pulses are referred to as optical solitons, which can propagate with pulse shape unchanged in optical fibers with anomalous dispersion and SPM. The FWHM pulse width is given by $\Delta\tau = 1.763\tau$. τ is also referred to as the soliton pulse width. Soliton requires the precise balance of SPM and dispersion and its power is defined by

$$P_{\text{s}} = \frac{|\beta_2|}{\gamma\tau^2} = \frac{3.11|\beta_2|}{\gamma\Delta\tau^2}$$

© Woodhead Publishing Limited, 2013

with pulse energy

$$E_s = \frac{3.53|\beta_2|}{\gamma\Delta\tau}$$

Similarly to bulk medium, Kerr nonlinearity can lead to self-focusing in the optical fibers. Self-focusing can collapse optical beam, leading to catastrophic breakdown when the nonlinear self-focusing effect overcomes diffraction. Since the nonlinear coefficient and diffraction scale similarly to waveguide dimension, resulting in a critical power, not intensity, for self-focusing,

$$P_{SF} = \frac{1.86}{k_0 n_0 n_2}$$

where $k_0 = 2\pi/\lambda$ is the wave vector and n_0 is the effective index of the optical mode. The critical power for nonlinear self-focusing is estimated to be 4–6 MW in silica optical fibers.

15.2.5 Thermal limits

An analytic solution can be found for heat diffusion in a cylindrical optical fiber. Assuming heat is generated only in the active core in the center, the solution indicates a quadratic temperature distribution in the core and a logarithmic distribution in the cladding. In typical silica fibers, it takes only a few tens of microseconds for thermal equilibrium to be established. The surface temperature rise is dominated by heat transfer [24]:

$$\Delta T \approx \frac{Q_0 \rho^2}{2Rh}$$

where Q_0 is the heat density generated in the core, ρ is the core radius, and R is the fiber outer diameter. The heat transfer coefficient h increases with ΔT. The temperature rise ΔT is plotted in Fig. 15.11 for convective cooling for fibers with outer diameter $2R = 250$ μm at various heat loads W_h (W/m) deposited in the core (h is estimated using $h = 60.64 + 15.5\Delta T^{0.25}$ W/m²/K [24]).

15.2.6 Optical damage

For a perfect surface, the optical damage threshold at the surface is close to that of the bulk medium in theory. In practice, the damage threshold at the surface can be much lower than that of the bulk. Recently, it has been shown that the optical damage threshold at ~1064 nm is close to 480 GW/cm² in bulk silica due to electron avalanche and independent of pulse width for

© Woodhead Publishing Limited, 2013

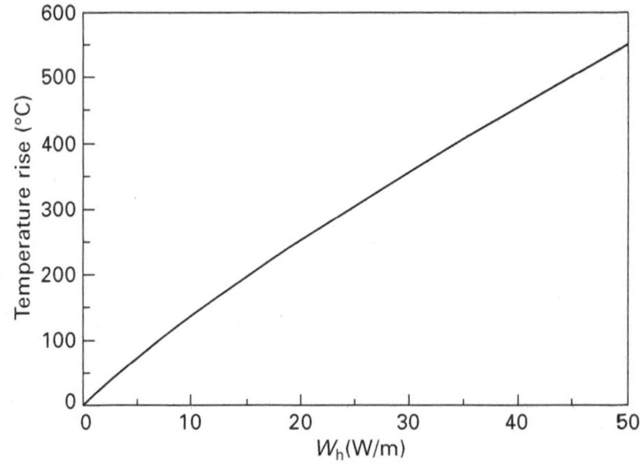

15.11 Temperature rise in 250 μm silica fiber using only convective heat removal.

pulse durations above 100 ps [25]. For shorter pulses, the avalanche evolves more slowly than the pulse envelope, leading to a slightly higher threshold [25]. In practice, surface damage has a much lower damage threshold and can be avoided in fiber systems by splicing as many fibers and components as possible and avoiding the use of free-space components. At the output from the fiber laser where the intensity is usually highest, it is common to use glass endcaps which allow the beam to expand to a safe intensity before reaching the glass/air interface.

15.3 High power continuous wave (CW) fiber lasers

The first published near-single-mode, 1 kW CW fiber laser [26] used free-space pumping of an Yb-doped fiber laser from both ends of the laser cavity with high brightness 972 nm diode stacks as shown in Fig. 15.12. The large clad diameter of the fiber (600 μm) used in this experiment was essential to allow coupling of high power diode stacks at 1 kW into each end of the Yb-doped fiber. In order to maintain a reasonable fiber length the core diameter was also rather large at 43 μm, resulting in a few-moded output and measured beam quality M^2~3.4. The resulting slope efficiency of 80% and CW output power of 1.01 kW, limited only by the available pump power, showed the potential for the technology to compete with existing CO_2 and diode-pumped solid state lasers.

However, since those early results, it has become clear that the use of free-space optics in the fiber laser cavity is not ideal, requiring precision alignment of the external mirrors to the core of the fiber during the operational lifetime of the laser. The high intensity of the intra-cavity power that is

© Woodhead Publishing Limited, 2013

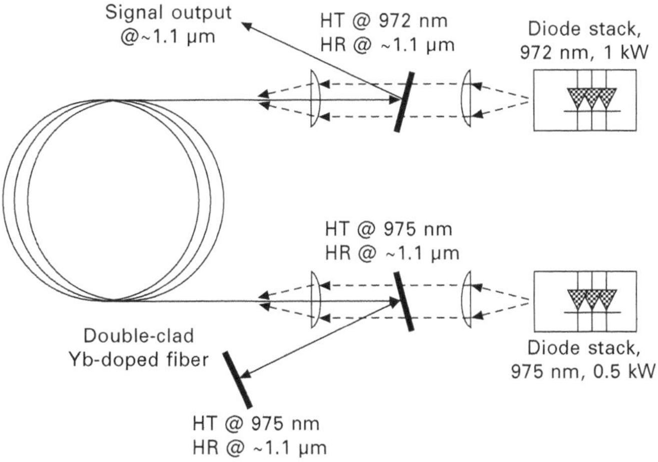

15.12 Experiment demonstrating the world's first near single mode 1 kW CW fiber laser (courtesy of ORC, Southampton University).

focused into a small core can lead to surface damage and reliability issues as well as additional intra-cavity losses.

A possible solution to this problem is the adoption of monolithic fiber laser devices, where the free-space mirrors are replaced with a pair of fiber Bragg gratings (FBGs) acting as the high reflector and output coupler of the laser cavity. The challenge is to operate these monolithic cavities at the kilowatt CW power level in LMA fibers such as 20/400 (20 μm diameter core and 400 μm diameter inner cladding) without premature failures at the intra-cavity splice points and the gratings. Examples of CW FBG cavities operating at power levels >1 kW (Fig. 15.13) and delivering single-mode beam quality [27] are in the literature, as are examples of monolithic multimode fiber laser cavities. Once optimized, these monolithic cavities do not need realignment during the life of the device, removing any cleaning and degradation of the fiber surface from the maintenance schedule of the laser. However, the packaging and optimization of the splices and FBGs themselves are non-trivial at these power levels as are the matching of the various fibers, since the composition of the Yb-doped fiber and various passive and photosensitive fibers needs to be carefully matched across the entire fiber laser chain [28].

Options for replacing the free-space diode stacks in Fig. 15.12 as the pump source of the kilowatt fiber laser started to emerge after 2007. Specifically, commercially available fiber coupled bars started to increase in brightness and become available with 200 μm and 400 μm delivery fiber, rather than the larger diameter fibers of the early diode bars. In addition to this, improvements in diode material efficiency allowed for higher output power per bar and more available power from 200 μm and 400 μm delivery fibers. One of the first

© Woodhead Publishing Limited, 2013

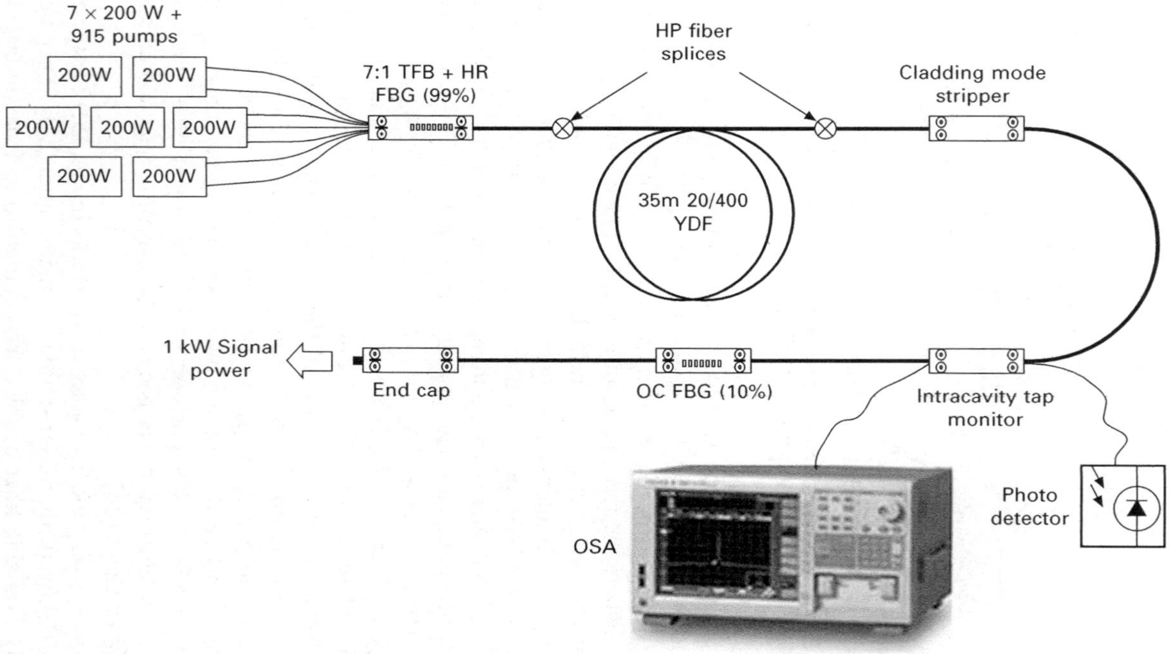

15.13 Schematic of a monolithic fiber laser cavity operating at >1 kW CW power using fiber Bragg gratings acting as the laser cavity mirrors (courtesy of ITF Labs).

© Woodhead Publishing Limited, 2013

demonstrations of a monolithic master oscillator power amplifier (MOPA) configuration at the kilowatt power level was demonstrated in 2007 [29] and used a co-pumped MOPA scheme shown in Fig. 15.14, by combining 6 × 200 W pump modules at 976 nm through in a (6 + 1) to 1 combiner to deliver ~1.2 kW of 976 nm pump into a 400 μm diameter double-clad fiber (0.46NA). This monolithic pumping scheme is ideal for MOPA configurations where a signal port is required for coupling into the power amplifier stage. One advantage of the MOPA scheme is the lack of any FBGs in the high power section of the fiber device.

Some of the options for pump combining using industry-standard diode delivery fibers (105, 200 and 400 μm) into standard double-clad fibers (125 μm, 250 μm and 400 μm with 0.46NA) are summarized in Table 15.1 [30]. These correspond to the most common inner clad diameters for the commercially available Yb-doped double-clad fibers [31]. The current state of the art for (6 + 1) to 1 couplers into 400 μm DC fiber is now around 2 kW. A typical high power coupler for these multi-kilowatt power levels is shown in Fig. 15.15 and is often water-cooled at these power levels. Insertion

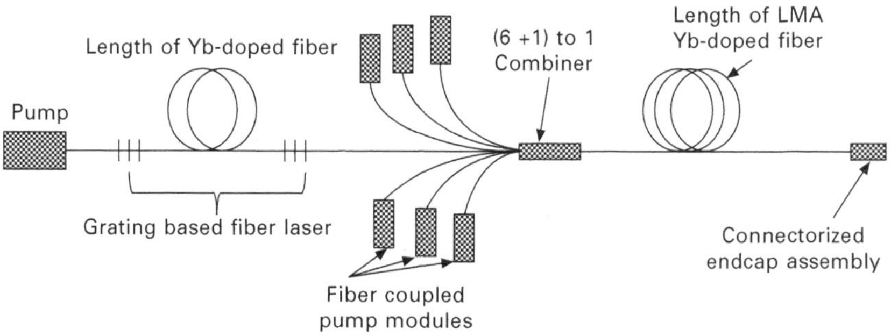

15.14 Monolithic all-fiber MOPAs are an alternative to high power laser cavities for CW power scaling, eliminating the need for high power Bragg gratings and offering flexibility of seed source.

Table 15.1 Examples of high-power coupler designs indicating the maximum number of pump fibers, assuming fully filled NA for these fibers and the three industry standard output double-clad fiber diameters

Input fibers/output fiber	125 μm DCF, NA = 0.46	250 μm DCF, NA = 0.46	400 μm DCF, NA = 0.46
105/125 μm, NA 0.15	7 × 1	19 × 1	61 × 1
105/125 μm, NA 0.22	4 × 1	7 × 1	37 × 1
200/220 μm, NA 0.22	1 × 1	4 × 1	7 × 1
400/440 μm, NA 0.22	N/A	1 × 1	3 × 1

Source: Gonthier *et al.*, 2005 [30].

© Woodhead Publishing Limited, 2013

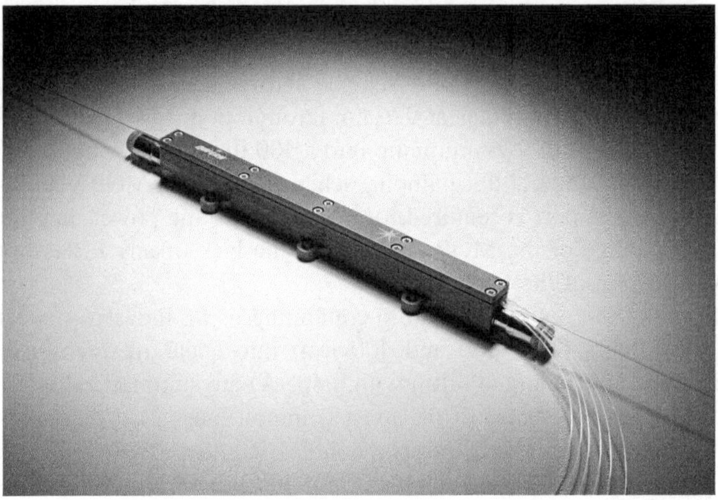

15.15 High power (6 + 1) to 1 coupler technology is now suitable for operating at 2 kW CW with very low insertion losses, suitable for co-pumped MOPAs (courtesy of Nufern).

losses for couplers operating at these power levels must be in the range 2–4% maximum for a co-pumped coupler and lower for a counter-pumped device because of the higher signal power level and therefore additional heating. A typical pump diode for splicing to these couplers is shown in Fig. 15.16, and is based on fiber coupled bars with an SMA connector [32]. The advantage of fiber connectorized pump diodes is the field replacement, which can be important in industrial applications where diode replacement over the lifetime of the laser has to be considered. Typical pump diodes delivering 200 W and, more recently, 350 W are commercially available coupled into 200 μm diameter fiber (0.22NA). Additional features may be readily combined within the pump module, including protective filters against fiber laser back-reflections at 1064 nm (which may damage the diode) and wavelength locking of the diode to narrow the output linewidth, making the pump diode better matched to the narrow 976 nm absorption line shape of the Yb-doped fiber.

The alternative to using fiber-coupled diode bars as the pump is based on multiple high-power single-emitter diodes as the building block. State-of-the-art single emitters operate at 10–12 W and may be used individually as the pump source or by combining multiple diodes into a single delivery fiber. These multi-emitter modules deliver between 60 W and 100 W and are coupled into 105 μm diameter fiber (0.15NA), making this an ideal high brightness pump source for fiber lasers [33–35]. These multi-emitter pump modules are commercially available as a reliable high-power source at a range of wavelengths between 800 and 976 nm. Furthermore, this

© Woodhead Publishing Limited, 2013

15.16 One attractive pump option for high power CW fibers are fiber coupled bars, which now deliver several hundred watts power coupled into 200 mm 0.22NA fiber (photo courtesy of DILAS).

pump module can be combined in a 19:1 combiner (Table 15.1) to allow a monolithic pump source exceeding 2 kW of power into 400 μm diameter double-clad fiber, enabling a monolithic fiber laser easily exceeding 1 kW output power, as shown in Fig. 15.17 [33]. The advantages of the single emitter versus the diode bar pump can depend on the application, since the cost, lifetime/reliability and cooling/drive current requirements can all be different. Obviously the splicing and handling of 19 pump delivery fibers into the combiner requires some manpower that can affect the overall cost, since the option of a connectorized delivery fiber is not common for the single-emitter based pump modules, requiring direct splicing to the pump combiner.

One of the major advantages of fiber lasers is the potential for single-mode beam quality at high power levels. However, the ability of Yb-doped LMA fibers to operate at the multiple-kilowatt power level with good beam quality has become a challenge, as physical effects at high output powers affect the modal properties of the laser/amplifier [36, 37]. An example of this is shown in Fig. 15.18 [38], where the measured beam quality for two fiber amplifiers is shown at varying power levels. Mode instability associated with the mode coupling phenomena appears at a certain threshold power level that is currently thought to have its roots in the thermal gradient across the core of the fiber. The power level for the onset of this effect appears to be

© Woodhead Publishing Limited, 2013

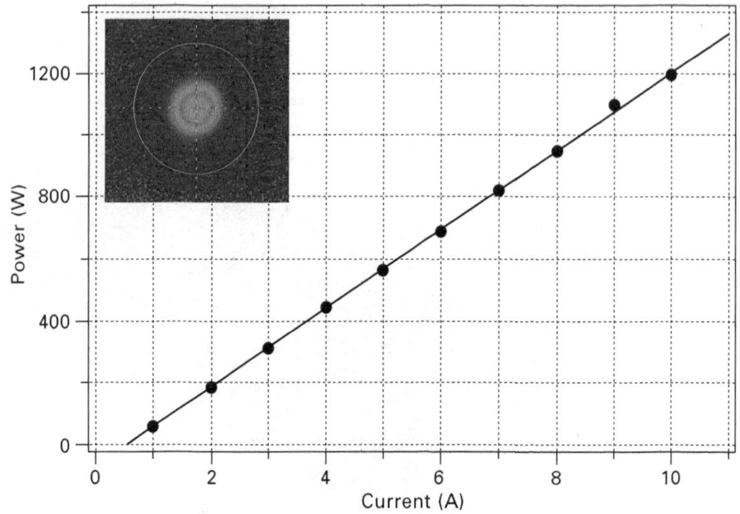

15.17 Fiber laser output power showing near-single-mode beam quality (M^2 = 1.3) at 1.2 kW, based on 18 single emitter based pump diode modules combined in a 19 × 1 coupler, each module producing ~120 W pump power (data courtesy of JDSU).

strongly fiber and device (inversion) dependent, with some of the highest-power cladding-pumped MOPA results showing 2 kW output power in LMA fibers without the modal effects shown in Figs 15.19 and 15.20 [28]. To date the full physics of this phenomenon of mode instability is unclear and remains a topic for further research.

In some applications the beam quality does not have to be single mode, and commercial components that combine the single-mode output from multiple fiber lasers into a multi-kilowatt multimode beam are available, and are based on fused silica combiner technology [39]. These multimode fiber lasers in the 2–6 kW regime are ideal for cutting and welding as a reliable replacement for DDPSL and CO_2 lasers, and are commercially available from many suppliers, often as part of a complete system with beam switch and various delivery fiber options.

Other limitations to the available output power from single-mode fiber lasers are the intrinsic nonlinear thresholds for the optical fiber amplifier in combination with the delivery fiber. The threshold for SBS in particular is a major problem for high power single-frequency fiber amplifiers, occurring around 50–100 W in practical LMA fiber amplifier systems using state-of-the-art Yb-doped fibers and typical amplifier pump schemes [40]. Although schemes to suppress the SBS gain have been theoretically investigated and some have been experimentally validated, no 1 kW single-frequency fiber amplifier has been demonstrated to date with single-mode beam quality. Broadening of the linewidth can raise the SBS threshold into the kilowatt regime and a

© Woodhead Publishing Limited, 2013

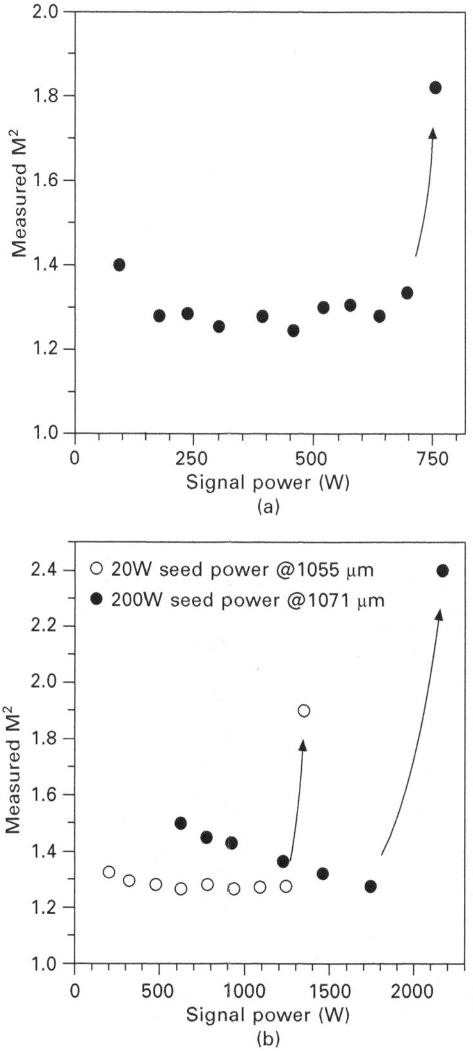

15.18 Examples of mode instabilities measured in two different high-power fibers and identified as a possible limiting factor in scaling high power fibers: (a) a 33 mm core PCF fiber and (b) a 30 mm core step index fiber (data courtesy of Friedrich Schiller University, Jena, Germany).

monolithic fiber amplifier delivering 1.5 kW with ~10 GHz linewidth has been demonstrated [41] using standard LMA fibers and components. Such high-power, narrow-linewidth amplifiers are useful building blocks for beam combining (coherent or spectral) into the tens of kilowatts power level with excellent efficiency and beam quality [42–44]. Military applications for these

© Woodhead Publishing Limited, 2013

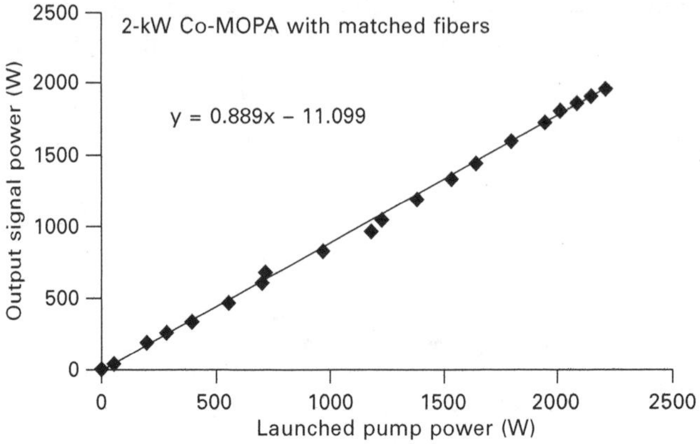

15.19 2 kW output power from a co-pumped MOPA system using a series of matched fibers that enable single mode operation at high power levels (data courtesy of Nufern).

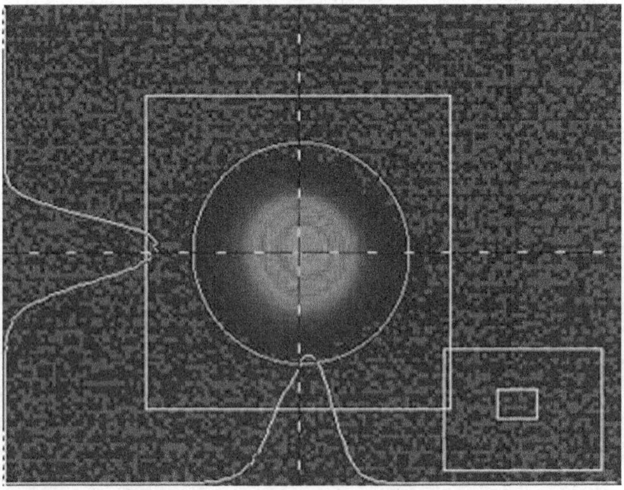

15.20 Measured beam quality from 20/400 MOPA at 2 kW without observing mode instabilities (courtesy of Nufern).

power levels are being investigated and consist of directed energy weapons in shipboard defense and missile defense, amongst other applications [45]. The attraction of fiber based devices in this application would be the high E-O efficiency, robust all-fiber platform, compact/lightweight packaging, excellent beam quality, and flexible design including remote amplifier head which separates the heat-generating elements such as diodes and power supplies from the fiber amplifier, as shown in Fig. 15.21.

Although many of these specific amplifier and laser designs are non-polarization maintaining, kilowatt-level PM systems have been demonstrated

© Woodhead Publishing Limited, 2013

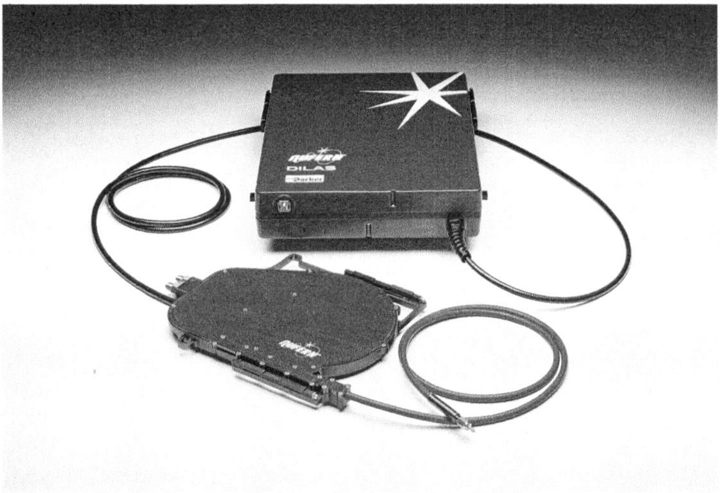

15.21 Photograph of 1.5 kW monolithic amplifier, based on a remote fiber amplifier head for overcoming nonlinear limitation associated with SBS in narrow linewidth systems (photo courtesy of Nufern).

[46]. Many of the same building blocks including combiners and Yb-doped LMA fibers are made with PANDA-style PM fibers, typically producing 13–17 dB polarization extinction ratio (PER) depending on the detail. In many cases, the use of a non-PM fiber amplifier with external polarization control works well, but the behavior of PM fibers in an amplifier can have different SBS thresholds and mode-instability thresholds compared with non-PM fiber amplifiers at high power levels [47].

Fundamental limitations to power scaling of CW fiber lasers have been summarized in several papers, including the work of the Lawrence Livermore National Laboratory (LLNL) [48], which investigated the nonlinear limits along with the thermal limits and self-focusing limits of various cladding-pumped fiber designs. The highest-power single-mode CW fiber device demonstrated is the 10 kW system from IPG in 2010 [19]. The novel method of pumping this final amplifier at 1018 nm, using fiber laser pumps, significantly reduces the thermal load on the fiber. The high-brightness nature of the pump source can reduce the inner clad size for the pump waveguide. This, in turn, can reduce the fiber length needed, despite the low absorption cross-section for Yb-doped silica at 1018 nm. The schematic and experiment results for a 10 kW single-mode fiber laser are shown in Fig. 15.22.

15.4 Pulsed fiber lasers

Some of the early work investigating LMA fibers focused on the advantages in the pulsed regime, making use of the energy storage of the larger core

© Woodhead Publishing Limited, 2013

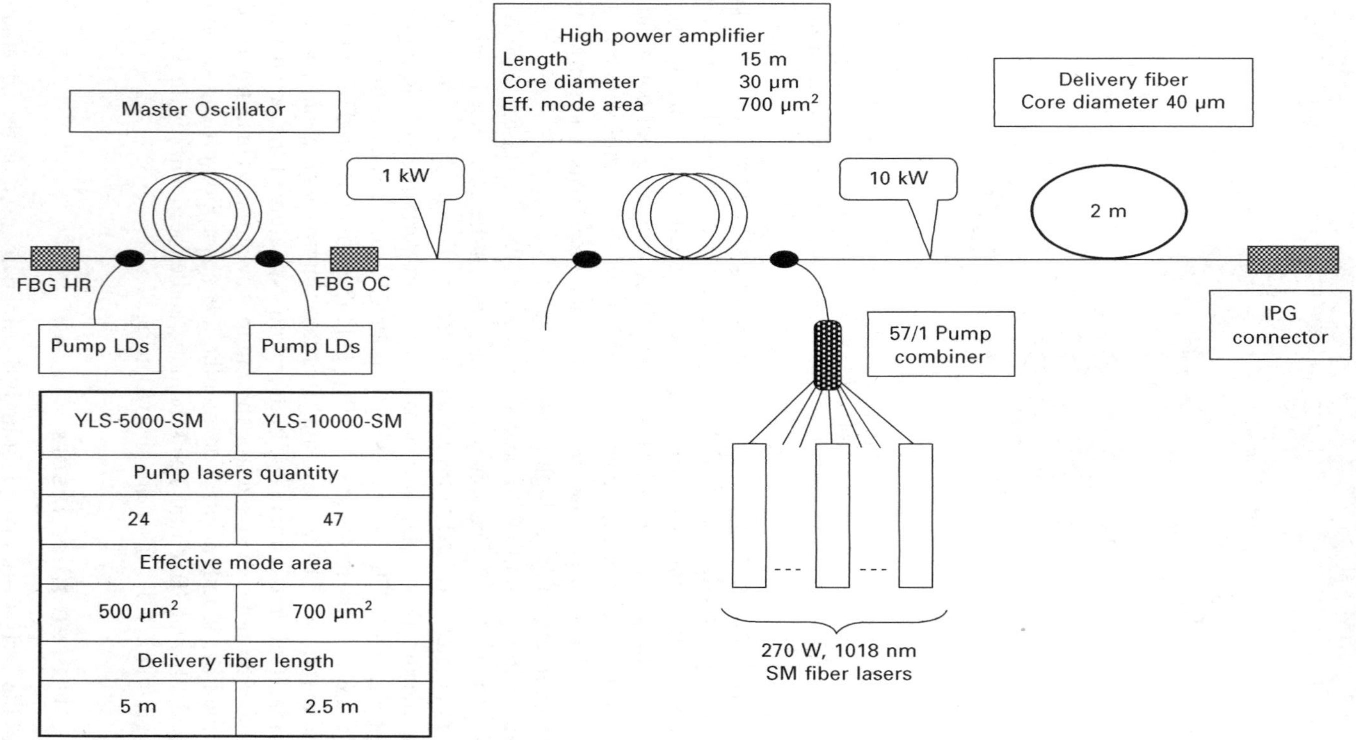

YLS-5000-SM	YLS-10000-SM
Pump lasers quantity	
24	47
Effective mode area	
500 μm^2	700 μm^2
Delivery fiber length	
5 m	2.5 m

15.22 Schematic of the 10 kW CW fiber laser first demonstrated in 2009 (courtesy of IPG Photonics).

© Woodhead Publishing Limited, 2013

combined with the increased nonlinear threshold from the increased mode field and shorter fiber length [49, 50]. In addition to highlighting improvements in terms of fiber design, the early work highlighted the ease of producing multistage MOPA systems in fiber form including managing inter-stage ASE filtering to allow higher overall gain [51] and increase pulse energy. Theoretical papers investigating the limits to scaling pulse energy of fiber lasers dated from the same period and included fiber designs to increase pulse energy at the extreme wavelengths of the Yb-gain spectrum [52].

A resurgence of the topic started in 2002 when the group at the Naval Research Lab (NRL) [53] amplified a microchip laser using a single-stage LMA fiber amplifier. In addition to producing record peak powers (300 kW) from a 20 μm core diameter (0.1NA) Yb-doped LMA fiber, the paper highlighted the excellent beam quality from the LMA fiber amplifier, attributed to coiling-induced filtering of the higher order modes in a multimode fiber amplifier, as shown in Fig. 15.23 and 15.24 [9].

Variations on this LMA based MOPA design have been investigated by numerous groups since then. In particular alternative seed sources other than microchip lasers, which are limited in terms of repetition rates, include direct Q-switched fiber lasers which can also be made from all polarization-

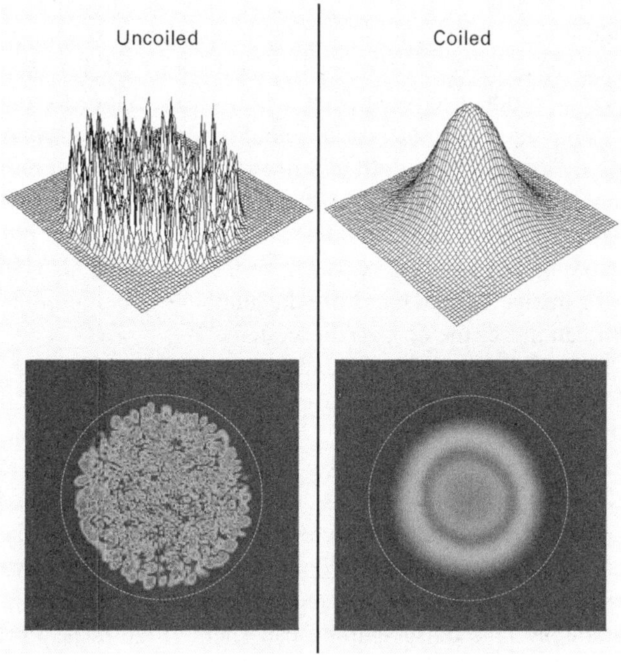

15.23 Examples of the beam quality from multimode fiber before and after coiling the fiber to suppress higher order modes (photo courtesy of Dahv Kliner).

© Woodhead Publishing Limited, 2013

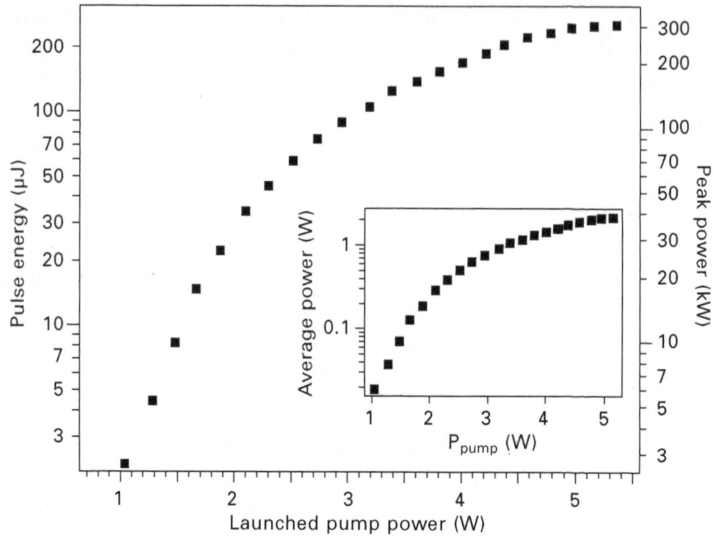

15.24 Pulse energy scaling in LMA fiber amplifier using a coiled fiber amplifier (Courtesy Dahv Kliner).

maintaining components and fibers [54], directly modulated laser diodes and CW diode lasers combined with external modulation schemes [55–59]. There are advantages and disadvantages to the various MOPA schemes and many commercial products covering a wide range of designs have also been launched by companies over the last decade. Some schemes may be suitable for producing the narrow linewidth and linear polarized output required for frequency doubling, whilst others offer maximum flexibility and pulse control, a useful feature for optimizing the pulse shape before the final high-power fiber amplifier stage. The output power from the seed laser also determines the amount of gain needed and therefore the number of fiber amplifier stages, which in turn can affect the cost of the system.

The use of a modulated laser diode as a seed source is particularly attractive, offering a highly flexible laser system that can operate from CW down to sub-nanosecond durations [57]. By pre-shaping the pulse shape from the seed source (Fig. 15.25), it is possible to compensate for saturation effects on the pulse shape in the subsequent fiber amplifier stages [58, 59]. The use of an external modulator in combination with the seed laser diode source, adds a further level of pulse control that is very attractive in fiber systems, allowing shaping of the pulse by the diode drive current and the external modulator. This pulse shaping can enable high peak powers from the final amplifier stage and offers a level of pulse flexibility in the final application that is unmatched by Q-switched lasers. Additionally, these pulse-tailored fiber lasers are of interest in materials processing applications

© Woodhead Publishing Limited, 2013

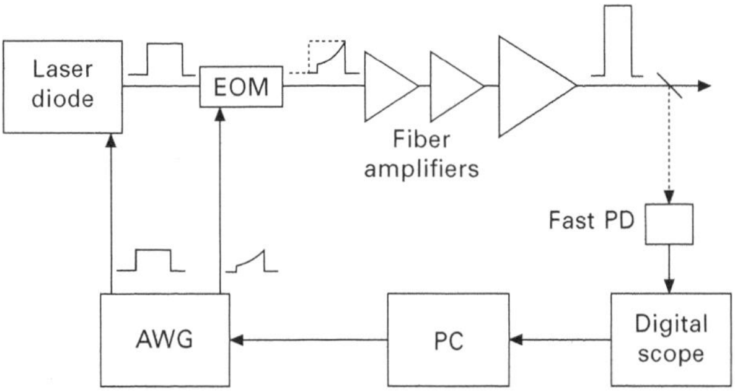

15.25 Schematic of the adaptive pulse shaping technique used to compensate for gain saturation in the fiber amplifier stage (courtesy of ORC, Southampton University).

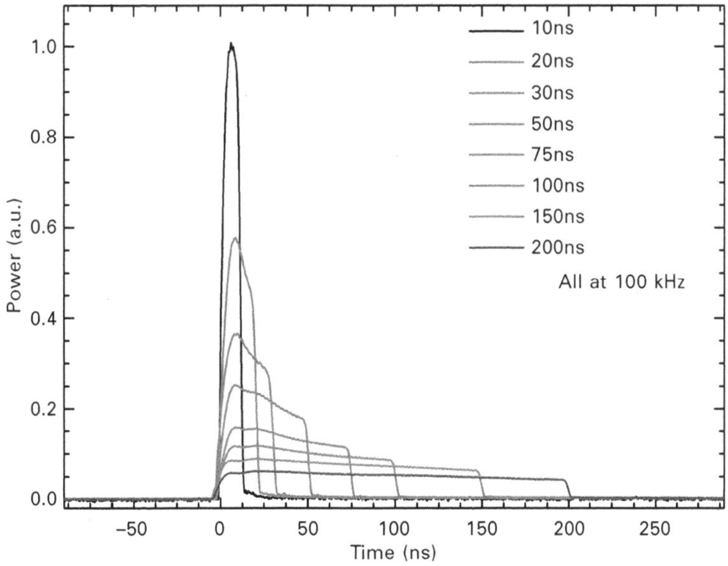

15.26 Examples of controlled pulse profiles from a pulsed fiber MOPA system. All pulses are of 100 mJ energy with varying pulse duration from the same pulsed laser source (courtesy of Multiwave).

where the material response can be affected by the pulse shape. Examples of commercially available systems are shown in Figs 15.26 and 15.27, [60], where pulse-shape tailoring is shown to offer a range of marking effects in stainless steel.

Pulsed fiber laser systems offer advantages compared with diode-pumped solid-state lasers in flexibility of repetition rates and pulse duration. Importantly, this is combined with excellent beam quality at high average power levels in the range of 50 W. However, compared with solid-state lasers, most of the

© Woodhead Publishing Limited, 2013

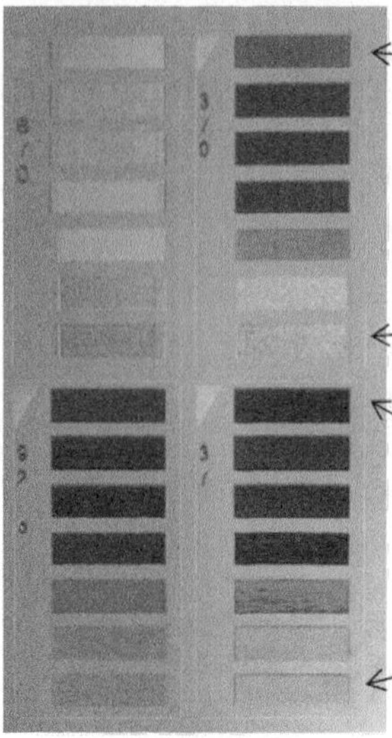

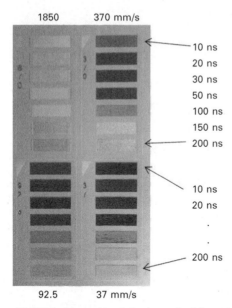

15.27 Marking of stainless steel with various combinations of pulse duration (10–200 ns) and four different scan speeds (1850–37 mm/s) all with a fixed average power of 13 W from a fiber laser with tailored pulses shown in Fig. 15.26 (courtesy of Multiwave).

commercial fiber systems are limited to a modest 1 mJ pulse energy under these pulsed conditions. However, many of these attributes are ideal for marking and engraving systems, making fiber lasers a highly attractive and cost-effective laser source in materials processing worldwide. The monolithic all-fiber nature of fiber lasers makes for reliable and stable fiber laser sources which are strong alternatives to DPSSL based systems. In addition, the flexibility of a fiber delivery system can open up system designs that would be difficult to achieve in a cost-effective manner otherwise. Typical systems operate with 2–5 m of delivery fiber.

The limiting factor in scaling pulse energy in the final amplifier stage depends on the pulse duration/peak power and the required beam quality. The benefit of using LMA fibers is the ability to deliver single-mode beam quality, but is limited to fibers with core size typically less than 30 μm in diameter. Fibers with larger mode field and core diameter can amplify to greater pulse energy and peak power, but only if the application can utilize multimode output and deteriorated beam quality. In many applications this is not acceptable. Research into new fiber technology to scale the core size and mode field of Yb-doped fibers is ongoing [61–64] and a detailed discussion of this topic is covered in a later section of this chapter. The adoption of photonic crystal fibers in particular made a large impact in scaling pulse energy in fiber lasers. To date, the highest reported Q-switched fiber laser

© Woodhead Publishing Limited, 2013

is shown in Fig. 15.28 and corresponds to 26 mJ pulse energy (<60 ns pulse duration) with near-single-mode output from a custom PCF fiber based on a three-stage MOPA system with final amplifier stage using a PCF with ~90 μm MFD [65].

Frequency doubling of the output of pulsed fiber systems is readily done using external nonlinear crystals, providing the linewidth of the fiber laser and polarization are optimized. Control of the polarization is achieved by using PM fiber and components throughout the system. The linewidth from these pulsed MOPA systems is somewhat dependent on the pulse duration, but can be easily scaled to the tens of kilowatt peak power with linewidths suitable for LBO and other standard nonlinear crystals. An example is shown in Figs 15.29 and 15.30, where a high repetition rate fiber laser system (10 MHz, 5 ns) producing 110 W of IR with 20 pm linewidth is converted to 60 W of green with 54% efficiency [66]. External modulation rather than direct modulation of the diode is an often-used technique for maintaining the linewidth of the seed source, which can broaden in pulsed diodes.

Fifth-harmonic generation [67] for gas sensing and generation of extreme UV wavelengths down to 13.5 nm [68] based on simple Yb-doped fiber MOPA schemes has been demonstrated, enabled by the megawatt peak powers produced at 1064 nm from very large mode area fiber systems. Flexible pulse formats are also a useful tool for pulsed LIDAR systems, where the advantages of fiber lasers are also the ability to operate at eye-safer wavelengths around 1.5 μm as well as to provide compact, lightweight systems with excellent beam quality [69]. These seeded MOPA systems have also been demonstrated at 2 μm wavelengths based on Tm-doped fibers. Pulsed sources at these

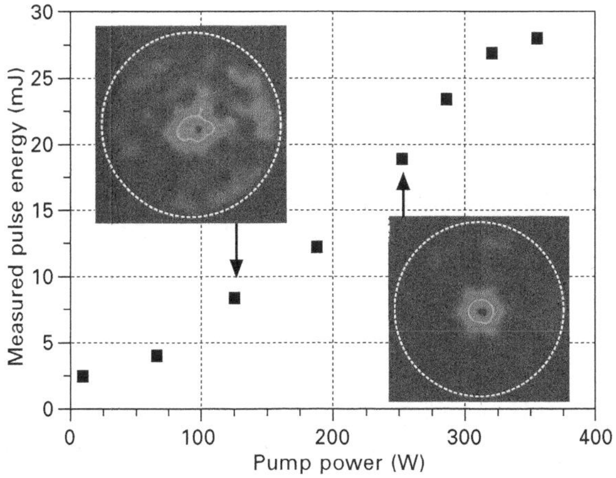

15.28 Highest Q-switched pulse energy reported to date from a near-single-mode fiber amplifier is 26 mJ, with sub-60 ns pulse duration using a custom PCF fiber design (data courtesy of Friedrich Schiller University, Jena, Germany).

© Woodhead Publishing Limited, 2013

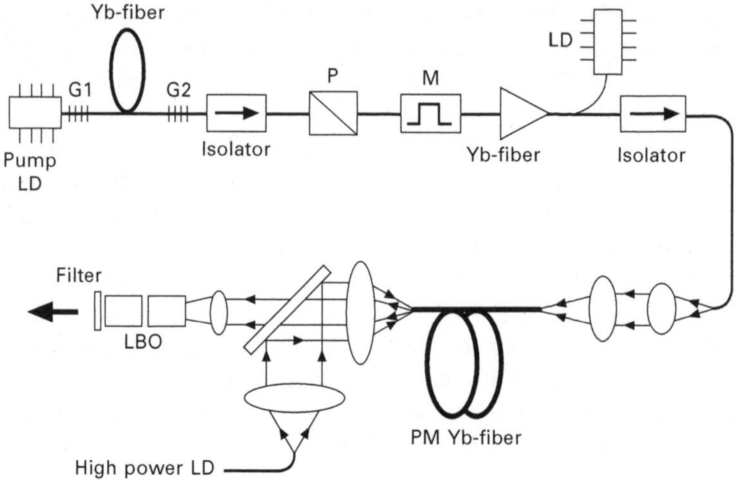

15.29 Experiment to generate high repetition rate, high average power picosecond pulses for frequency conversion to green (courtesy of Aculight).

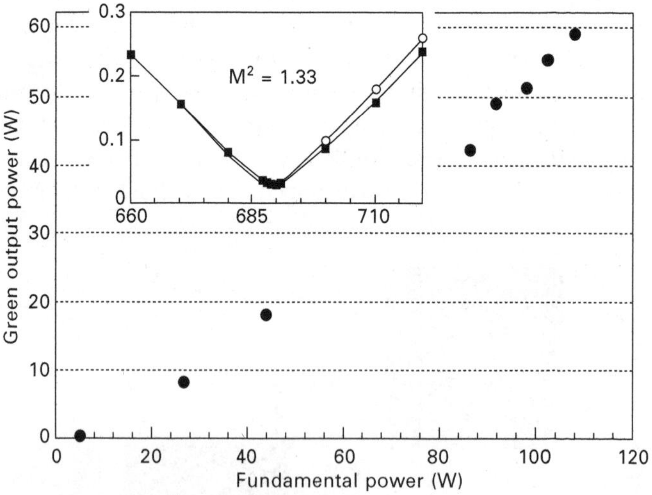

15.30 Results of the experiment shown in Fig. 15.29 to generate 60 W of green (courtesy of Aculight).

wavelengths are a useful source for generating mid-IR wavelengths based on nonlinear conversion with commercially available materials. Supercontinuum generation based on nonlinear conversion of high repetition rate, picosecond fiber lasers has also emerged in the last five years as an excellent broadband source for visible and near-UV wavelengths. Applications for such sources include biomedical imaging. Mid-IR supercontinuum generation using soft

© Woodhead Publishing Limited, 2013

glass fibers and micro-structured fibers [70, 71] has also been demonstrated and commercialization of fiber based supercontinuum sources has been highly successful in the marketplace with multiple suppliers selling systems.

Although the majority of this chapter is focused on silica glass fibers and the lasers produced by these fibers, non-silica fibers are playing an increasingly large role in niche laser markets where silica does not perform well, mostly at wavelengths outside the transmission band of silica (>2.1 μm) and where it is easy to justify the additional cost and complexity of using non-silica glass. In addition, the rather poor solubility of rare-earth ions in silica glass compared with other glasses has led to commercialization of phosphate-based glass fiber lasers, for DBR and DFB devices which require high doping levels and short cavity lengths [72]. The extension of non-silica glass fibers to micro-structured designs is a particularly novel research topic currently, enabled by a much wider range of material (optical and thermal) properties compared with silica glass [73].

15.5 Ultrafast fiber lasers

15.5.1 Introduction

Due to strong inhomogeneous broadening effects associated with the random nature of glass, large gain bandwidth can be obtained in rare-earth-doped optical fibers. Gain bandwidths of around 50 nm, 45 nm and 200 nm are possible in ytterbium, erbium and thulium fiber lasers respectively, leading to possible pulse widths down to 30 fs. Mode-locked fiber lasers generating pulses from a few tens of femtoseconds to tens of picoseconds with repetition rates from megahertz to gigahertz have been demonstrated. Initial studies in the 1990s were driven by applications in telecommunications, two-photon microscopy, terahertz generation and LIDAR. In the last decade, the focus has shifted to materials processing including high precision micro-matching in manufacturing of semiconductor devices, medical devices and engine parts, largely due to the increased pulse energies from fiber chirped pulse amplification (FCPA) systems. Mode-locked fiber lasers have also been used in laser-Assisted in situ Keratomileusis (LASIK) eye surgery procedures in recent years. Recently precision frequency combs using fiber mode-locked lasers have attracted significant interest for a wide range of applications from optical clocks to spectroscopy.

Compared to the performance of solid-state counterparts, mode-locked fiber lasers are significantly limited by SPM due to the tightly confined optical modes. Recently, with advances in optical fibers with large effective mode areas, pulse energies up to ~1 mJ have been achieved in FCPA systems with significant stretching ratios. In additional to the robustness provided by flexible optical fibers, one significant benefit of mode-locked fiber lasers over their

© Woodhead Publishing Limited, 2013

solid-state counterparts is the potential to scale average powers. Average powers in the hundreds of watts have been demonstrated recently. This is very important for applications in manufacturing where high throughputs are always very much desired.

15.5.2 Actively mode-locked lasers

Active mode-locking produces well-synchronized optical pulses which can be used in high-speed optical communications. Both amplitude modulation (AM) and frequency modulation (FM) mode-locking at repetition rate of tens of gigahertz can be achieved using integrated $LiNbO_3$ modulators. Pulse durations in actively mode-locked fiber lasers are, however, limited to >1 ps.

Typical modulators operate at a period of the cavity round-trip time. The mechanism can be understood in both the time and the frequency domain. In the time domain, modulation forces pulses to pass the modulator at its high transmission cycle. This effectively produces optical pulses equally spaced in time. In the frequency domain, modulation produces new frequency components on both sides of the original optical frequency at spacings which equal integer multiples of the modulation frequency. When the modulation frequency is chosen to be the cavity mode spacing, this effectively leads to coupling between adjacent cavity modes and locks them in phase. In FM, the optical phase is modulated instead of amplitude. Pulses from actively mode-locked lasers are Gaussian, $A_0\exp(-t^2/2\tau^2)$, with their pulse duration given by the Kuizenga–Siegman formula [73]:

$$\tau^4 = \frac{2g}{M\Omega_{\mathrm{m}}^2\,\Omega_{\mathrm{g}}^2}$$

where g is peak gain, M is modulation depth, Ω_{m} is modulation frequency, and Ω_{g} is gain bandwidth. Pulse duration in an actively mode-locked laser is fundamentally limited by modulation frequency. Assuming FWHM of spectral bandwidth $\Delta\nu$ and pulse width $\Delta\tau$, the time–bandwidth product is $\Delta\nu\Delta\tau = 0.63$ for bandwidth-limited Gaussian pulses.

15.5.3 Passively mode-locked lasers

Passively mode-locked fiber lasers produce the shortest pulses [74]. A record of 28 fs was demonstrated from an ytterbium oscillator recently [75]. A saturable absorber is typically used to encourage the formation of optical pulses. An optical pulse can often start from noise. An oscillating pulse gets narrower for every round trip through the saturable absorber. Assuming the temporal response of the saturable absorber is fast enough, the time gate

© Woodhead Publishing Limited, 2013

as a result of the saturable absorber action gets narrower with steeper sides when the pulse narrows. Eventually, the pulse width is limited by the gain bandwidth available. The temporal slope of modulation is much steeper with saturable absorbers in passive locking than with those in active mode-locking, which is essentially limited by the modulator frequency, resulting in much shorter pulses.

Passively mode-locked fiber lasers typically operate in an anomalous regime with a $\text{sech}^2(t/\tau)$ intensity profile, also referred to as a soliton laser. The hyperbolic secant has an exponential tail. The system behaves linearly at the pulse tails due to the low optical intensity, and the governing second-order differential equation requires an exponential solution in this linear regime.

Early mode-locked fiber lasers used a nonlinear amplifying loop mirror as the saturable absorber. An example is shown on the right-hand side of Fig. 15.31 in a figure-8 laser, named after its configuration [76]. The nonlinear loop mirror consists of a 3 dB coupler and a loop of fiber with an erbium-doped fiber amplifier (EDFA) at one end. The counter-clockwise propagating optical power is amplified at the beginning of the loop and accumulates more nonlinear phase than clockwise-propagating optical power. A polarization controller is required to ensure polarization is aligned for the two optical waves at the coupler. At low powers the nonlinear loop mirror is reflecting, and at high power the phase difference between counter-clockwise and clockwise waves can approach π due to the nonlinear phase, resulting in light coming out of the second port in transmission. A figure-8 laser can be constructed using the nonlinear mirror as shown in Fig. 15.31. An isolator is required to ensure unidirectional propagation. An output coupler is used for coupling out the output power. The polarization controller can also be used to adjust the powers of both counter-clockwise and clockwise waves in the two polarization states. This leads to the possibility of achieving a

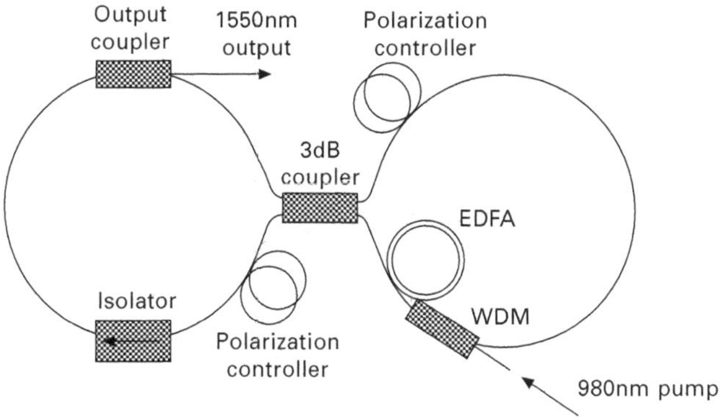

15.31 A typical figure-8 fiber laser.

© Woodhead Publishing Limited, 2013

phase difference of π even at lower powers. In this case, light is reflected only at high powers. One disadvantage of a nonlinear amplifying loop mirror is its polarization sensitivity. Saturable absorbers based on nonlinear polarization rotation are often used in later implementations of mode-locked fiber lasers.

An example of a saturable absorber based on nonlinear elliptical polarization rotation is shown in Fig. 15.32. Considering both linear and nonlinear phase-including SPM and cross-phase modulation terms, the phases for the x and y polarizations with intensity of I_x and I_y after a propagation of L in distance are

$$\phi_x = [\beta_x + \gamma I_x + \frac{2}{3}\gamma I_y]$$

$$\phi_y = [\beta_y + \gamma I_y + \frac{2}{3}\gamma I_x]$$

This will cause a rotation of the elliptic polarization, leading to intensity-dependent transmission after the polarizer. A saturable absorber based on nonlinear elliptic polarization rotation can also be implemented in reflection using a mirror. Nonlinear saturable absorbers based on the Kerr effect have a very fast response. The drawback is complexity in their implementations.

The semiconductor saturable absorber mirror (SESAM) is a compact integrated device which typically incorporates a semiconductor Bragg mirror and a single quantum well absorber layer based on interband transition. Recovery time is a lot slower than for Kerr nonlinearity, typically in the few to hundreds of picoseconds. As a result of its simplicity and ease of use, the SESAM is widely used in fiber mode-locked lasers and especially in commercial products. Carbon nanotubes and graphene are also used in fiber mode-locked lasers as saturable absorbers in recent years. They have the benefits of a faster response time than SESAM and can be directly deposited on the fiber ends. Issues related to reliability and power handling are still to be fully understood.

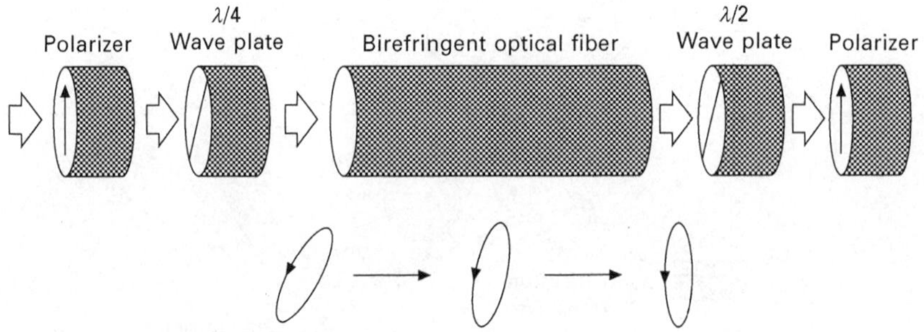

15.32 An example of a saturable absorber based on nonlinear elliptic polarization rotation.

© Woodhead Publishing Limited, 2013

Early mode-locked fibers were mostly implemented using erbium-doped optical fibers. In addition to the wide availability of integrated components developed for the telecommunications markets, desired optical fibers with anomalous dispersion are also readily available for erbium-doped lasers operating around 1550 nm. This is because material dispersion of silica is anomalous in this wavelength regime (Fig. 15.5). As interest grew in ytterbium fiber lasers, considerable interest developed in ytterbium mode-locked fiber lasers. Material dispersion is normal at ytterbium wavelengths of ~1 μm. Bearing in mind that waveguide dispersion is also normal in the single-mode regime (see Fig. 15.5), it is not impossible to obtain single-mode fibers with anomalous dispersion for the ytterbium wavelength. Early ytterbium mode-locked lasers used intra-cavity bulk grating pairs to achieve overall anomalous dispersion in the cavity. Later on, chirped fiber Bragg gratings were also used for more integrated and compact implementations. An example of this is illustrated in Fig. 15.33, with pulse energies up to 10 nJ and pulse widths down to 100 fs [77].

Soliton fiber lasers can typically tolerate more SPM due to the balance of SPM by dispersion. One issue is the formation of Kelly side bands in the pulse spectrum. When soliton is periodically perturbed due to the loss and amplification in a mode-locked fiber, energy is shed into a broad continuum. This process is enhanced when the continuum and soliton is phase-matched from pulse to pulse, leading to draining of energy from the soliton and appearance of sharp narrow spectral peaks in the pulse spectrum.

The master mode-locking equation can be solved to obtain various pulse parameters for mode-locked lasers to understand various regimes of operation. This was done by Haus [78]. The solution is a chirped pulse with a field $a(t)$ given by

$$a(t) = A_0 sech^{1+i\beta}\left(\frac{t}{\tau}\right)$$

The normalized pulse width τ/τ_0 and chirp β versus normalized round-trip dispersion D_n are given in Fig. 15.34 for various normalized nonlinearity values δ_n:

$$D_n = \frac{\beta_2 L}{2}\Omega_g^2 \qquad \delta_n = \frac{\gamma\tau_0 W}{6}\Omega_g^2$$

where Ω_g is the gain bandwidth, W is the pulse energy, and τ_0 is the pulse width for $D_n = 0$ and $\delta_n = 0$. It can be seen from Fig. 15.34(a) that the minimum pulse width is achieved at zero dispersion, $D_n = 0$, for a system without nonlinearity, $\delta_n = 0$. For nonlinear systems with $\delta_n > 0$, the minimum pulse width is achieved at a small amount of anomalous dispersion $D_n < 0$. This is due to the balance of SPM with anomalous dispersion in the cavity. Pulse width increases significantly in the normal dispersion regime

© Woodhead Publishing Limited, 2013

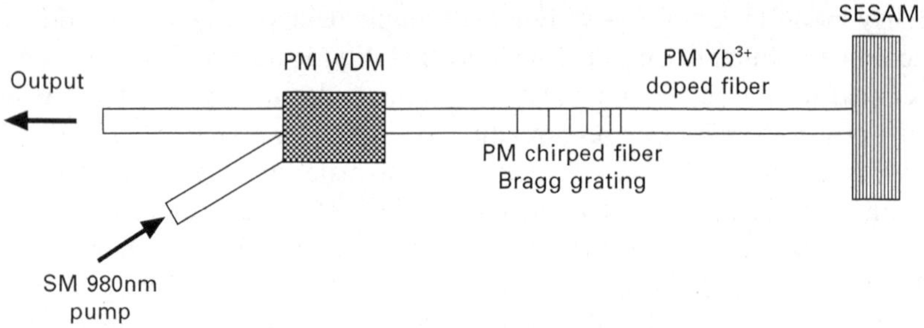

15.33 Set-up of a robust and compact passively mode-locked ytterbium fiber laser.

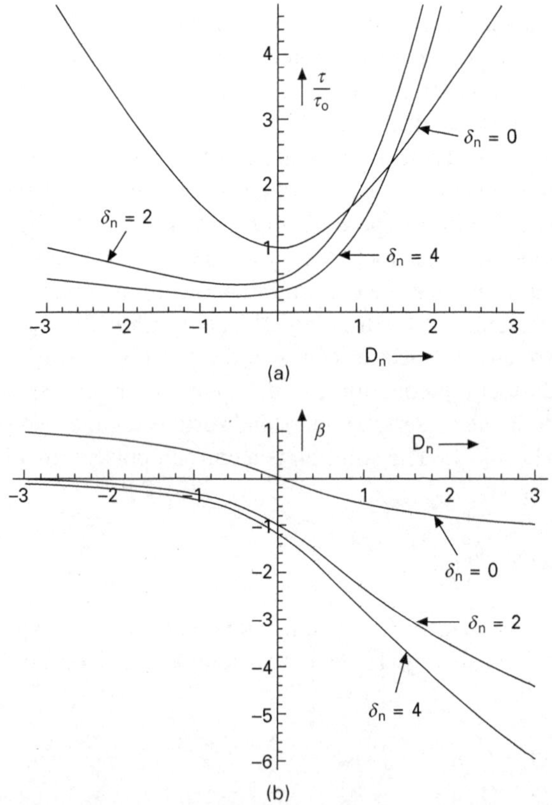

15.34 Normalized (a) pulse duration and (b) chirp from mode-locked lasers (from Haus [78]).

accompanied by a large chirp. Recently mode-locked fiber lasers in the large normal dispersion regime have been demonstrated, giving a broad chirped pulse. This was largely predicted by Haus [78]. For small saturable absorption action and weak filtering in the anomalous regime, a soliton-like solution

© Woodhead Publishing Limited, 2013

can be found from the master mode-locking equation, which is chirp-free and has a hyperbolic secant profile. The pulse is formed through the balance of SPM and dispersion. It can be viewed as a soliton weakly perturbed by amplification, saturable absorbing action and filtering.

Stretched-pulse fiber lasers can scale pulse energy by 100-fold beyond soliton fiber lasers [79]. An example is shown in Fig. 15.35. The dispersion is almost balanced by the selection of two fibers with anomalous and normal dispersion respectively. Nonlinear elliptical polarization rotation is used as saturable absorber in this case. The pulse circulating in the ring is stretched and compressed by a factor of 20 in one round trip. This leads to significant reduction of nonlinearity. The pulse is Gaussian. Since this is no longer a soliton laser, Kelly sides are absent in this case.

15.5.4 Pulse energy and power scaling of ultrafast fiber lasers

Further pulse energy scaling requires a fiber chirped pulse amplification (FCPA) system. A schematic of a FCPA system is illustrated in Fig. 15.36. Sometimes, bulk grating stretchers are used instead of fiber stretchers. A stretching ratio up to three orders of magnitude is sometimes used to significantly lower the pulse peak powers. Typically, bulk gratings are used for the pulse compression to remove the chirp. The reduction in the nonlinear effect by the much lower optical intensity in the optical amplifier leads to the possibility of much higher peak powers after compression. Pulse energies up to 1 mJ and average powers in the hundreds of watts have been demonstrated using large core photonic crystal fibers.

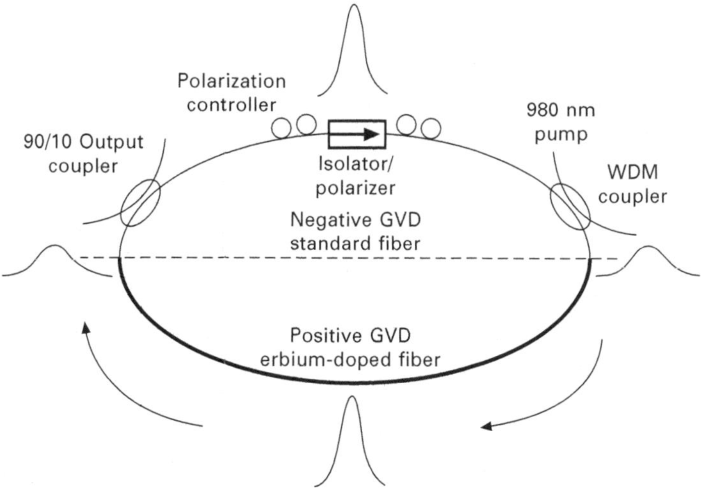

15.35 A setup for the all-fiber stretched-pulse laser.

© Woodhead Publishing Limited, 2013

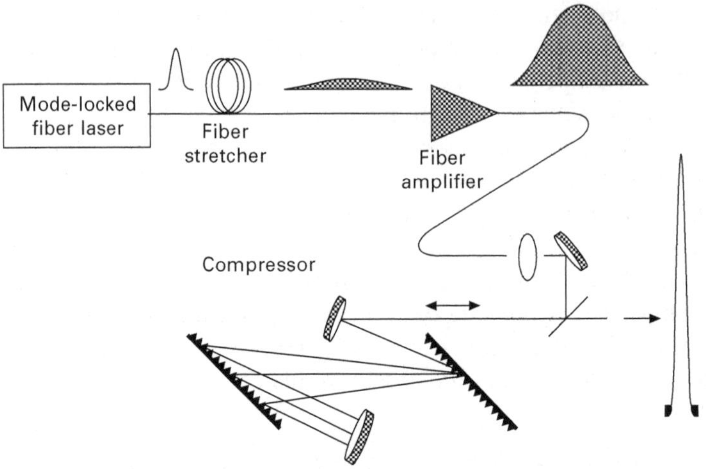

15.36 Schematic of a fiber chirped pulse amplification system.

15.6 Continuous wave (CW) and pulsed fiber lasers at alternative wavelengths

Power scaling of fiber lasers at wavelengths around 1.5 μm has been achieved in the Er:Yb co-doped fiber system described in the earlier section. The advantage of this system is the suitability for cladding pumping, where the Yb absorption is high enough to allow fiber lengths in the 3–10 m range using traditional fiber designs. Efficient energy transfer between Yb ions and Er ions in the glass is a well-understood process from telecommunication fiber amplifiers, where it is widely used to make CATV amplifiers at the 1-2 W power level. With optimization of the composition and waveguide, typical slope efficiencies for these fibers are 35–40% and the dramatic consequence of high pump rates on a non-optimized fiber is shown in Fig. 15.37, where the slope efficiency is seen to roll over at high power levels, due to parasitic 1 μm ASE [80]. The stringent requirements on the glass host make the design of LMA fibers very challenging for this system; in particular the low NA requirements are not compatible with the phosphosilicate fiber compositions required for efficient Er:Yb co-coped fibers [81]. The adoption of a low index pedestal region around the core has been a successful technique to lower the effective NA of the core. Both LMA and PM-LMA Er:Yb doped fibers have been shown to operate at >100 W without ASE and enable high peak powers, although not at the levels of the Yb-doped systems. They can be fabricated in PM designs with suitable birefringence for fiber lasers and amplifiers [82]. The likelihood of these Er:Yb fibers ever operating at the kilowatt power level seems remote, given the current state of the technology.

© Woodhead Publishing Limited, 2013

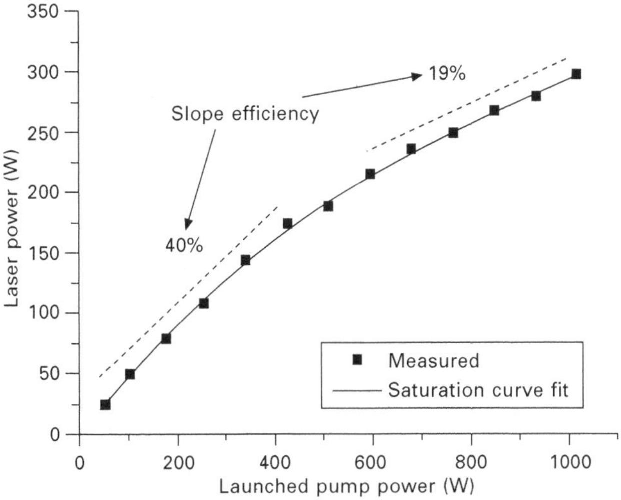

15.37 Power-scaling fiber lasers at 1.5 μm using Er:Yb co-doped fibers has been hampered by the parasitic 1 μm ASE generated from the Yb ions that limits the available power (data courtesy of ORC, Southampton University).

Because of the relatively low efficiency of the Er:Yb system and the resulting high thermal load on the fiber, little progress has been made on power scaling fibers at 1.5 μm beyond the few hundred watts level. Resonant pumping of erbium-doped fibers at wavelengths between 1480 and 1530 nm is a potentially interesting method [83] for power scaling fiber lasers, by lowering the quantum defect and thermal load on the doped fiber. However, the lack of suitable high power, high brightness pump diodes has limited the experimental success of this scheme. To date the highest power and slope efficiency demonstrated are around 100 W and 80% respectively for a 1530 nm cladding pumped erbium fiber device [84]. Some of this most recent power scaling work is utilizing high power Er:Yb doped fiber lasers as the pump source, so the overall optical-to-optical efficiency is low and the overall cost of the pump diodes too high for most commercial applications. Furthermore, progress on developing high power, high brightness diodes at 1530 nm for direct cladding pumping erbium fibers is still hampered by the low E-O efficiency of the diode material.

In contrast, progress on power scaling Tm-doped fibers at 2 μm has progressed rapidly over the last five years, with the demonstrated single-mode output power from fiber increasing from 10 W CW output power in 2002 [85] to the 1 kW level in 2010 [86, 87], as shown in Figs 15.38 and 15.39. The major factor in this power scaling work was the optimization of the Tm-doped glass composition to increase the efficiency of 790 nm cladding pumped fibers through the cross-relaxation of Tm ions (so-called

© Woodhead Publishing Limited, 2013

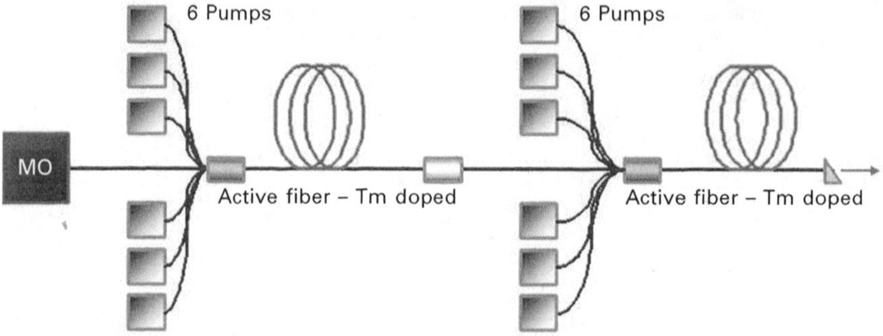

15.38 Schematic of the two-stage co-pumped amplifier used to demonstrate the first 1 kW CW fiber laser at 2 μm, based on 790 nm pumping of Tm-doped LMA fibers (courtesy of Q-peak).

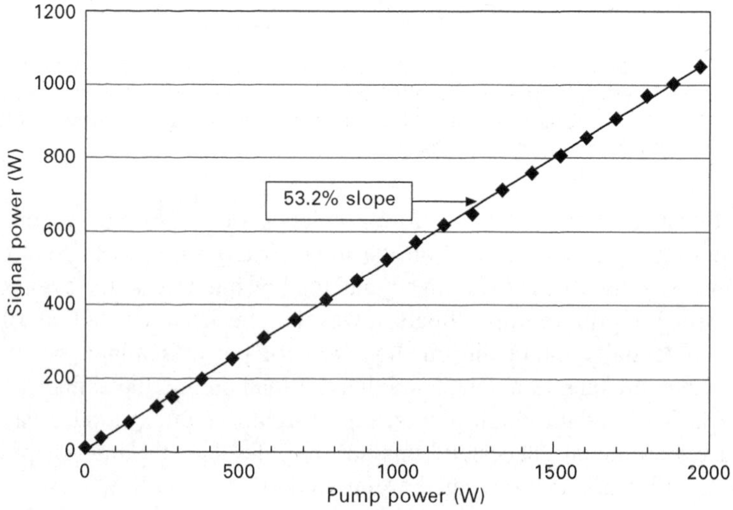

15.39 1 kW 2 μm fiber MOPA system operating at >50% slope efficiency and pumped at 793 nm (data courtesy of Q-peak).

2:1 cross-relaxation), described in the earlier section. Optimization of the composition, requiring relatively high Tm concentrations, has shown that this cross-relaxation process can increase the slope efficiency from ~35% for 790 nm pumping to 65% and that this efficiency is maintained at high power levels, as seen in Fig. 15.39. However, the availability of high power diode pumps at 790 nm is also a major factor in the rapid development of these high power 2 μm fiber lasers.

Another major feature of Tm-doped fibers is the extended tuning wavelength for this system, operating from <1.9 to over 2.1 μm, with optimization of the fiber length (Fig. 15.40) [88]. This wavelength region spans important

© Woodhead Publishing Limited, 2013

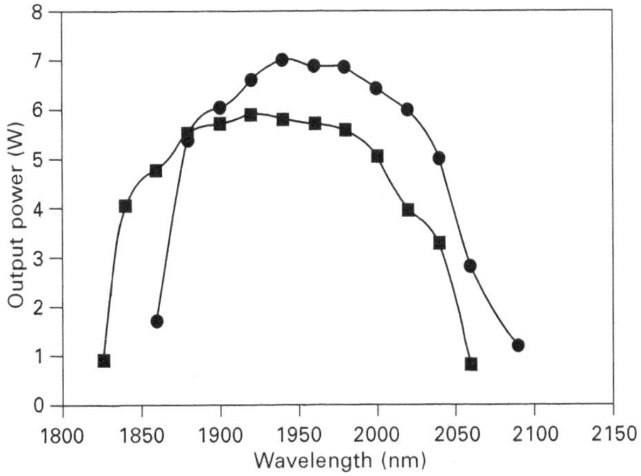

15.40 Tuning curve of double-clad Tm fiber lasers can extend from 1850 to 2100 nm and depends on the fiber length used (data courtesy of ORC, Southampton University).

atmospheric transmission windows as well as gas sensing wavelengths and is an excellent match to the water absorption peak in the human body at 1940 nm, which has enabled a growing medical laser market for Tm-doped fiber lasers operating between 10 and 100 W [89].

Similar to the Er:Yb system, the optimization of the core glass composition leads to a relatively high NA waveguide, making LMA fiber designs challenging. The adoption of pedestal fibers as a method to lower the core effective NA has been highly successful and is another key enabling factor in the demonstrated performance of these fibers over the last five years. Finally, the high efficiency of the Tm-doped fibers combined with the widespread availability of low-cost, high-power diode bars at 790 nm makes the overall cost of developing a commercial high power 2 μm fiber laser realistic and likely to expand in the next 5–10 years.

Pulsed 2 μm systems have been demonstrated based on the Q-switched and seeded MOPA design, similar to the 1 μm system, with the exception that the seed diode technology is not as mature at the longer operating wavelengths. The generation of high pulse energies at wavelengths around 2 μm is attractive for mid-IR conversion, and multistage Tm-doped fiber amplifiers have generated high pulse energies combined with high average powers, as shown in Figs 15.41 and 15.42, based on an amplified gain-switched fiber laser as the signal source [90, 91]. Frequency doubling of pulsed and CW 2 μm sources to 950 nm has been demonstrated [92], targeting wavelengths around 450 nm for sensing and underwater applications. Material processing of plastics is also emerging [93], using Tm-doped fiber lasers as the source which also has the additional benefit of an 'eyesafer' operating wavelength.

© Woodhead Publishing Limited, 2013

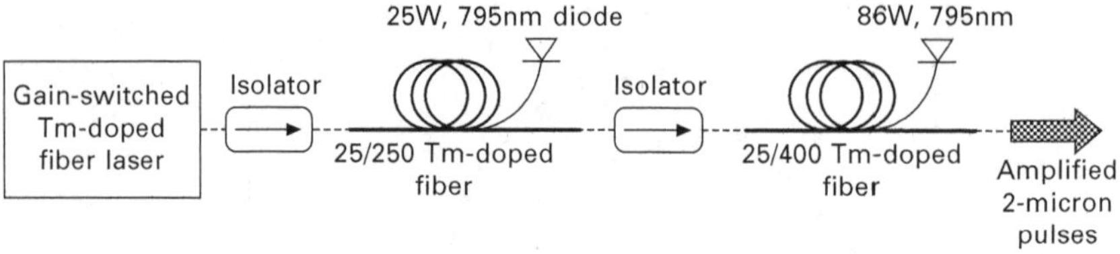

15.41 Pulsed MOPA system operating at 2 μm and based on Tm-doped fibers (courtesy of BAE Systems).

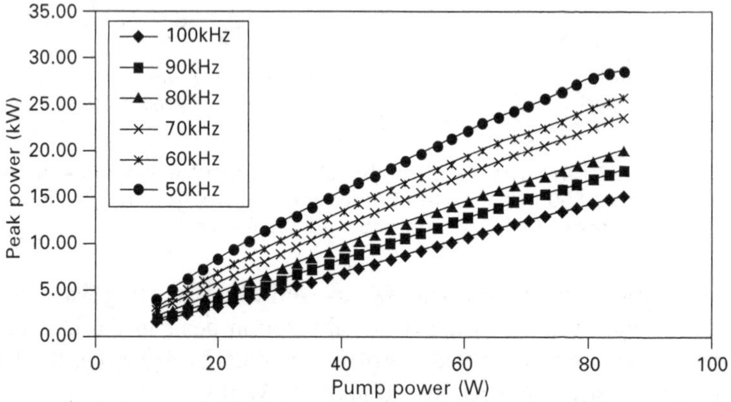

15.42 Results of the experiment shown in Fig. 15.14 with 13 ns pulse duration (courtesy of BAE Systems).

Further power scaling of CW cladding pumped Tm-doped fibers using 793 nm pumps, much beyond the 1 kW level, seems unlikely due to the thermal load on the fiber, and much of the recent research has focused on the resonant pumping of Ho-doped fibers to reduce the thermal load. This system has a theoretical slope efficiency around 90% and has demonstrated efficiencies exceeding 70% in non-silica fibers [94]. Despite the very new nature of the research on high power holmium fiber lasers, progress has already demonstrated ~140 W CW lasing at 2.13 μm using Tm-silica fiber lasers (~1950 nm) as the pump source for Ho-doped fibers [95], as shown in Fig. 15.43. A scheme for power scaling of Ho-fiber MOPA is shown in Fig. 15.44. Another advantage of the Ho-fiber is the longer operating wavelength of > 2.1 μm which is interesting for access to some of the important atmospheric transmission windows and for possible gas sensing.

15.7 Emerging fiber technologies for fiber lasers

Overcoming nonlinear limits is key to further power scaling of fiber lasers. The most effective method is to scale the effective mode area of optical fibers

© Woodhead Publishing Limited, 2013

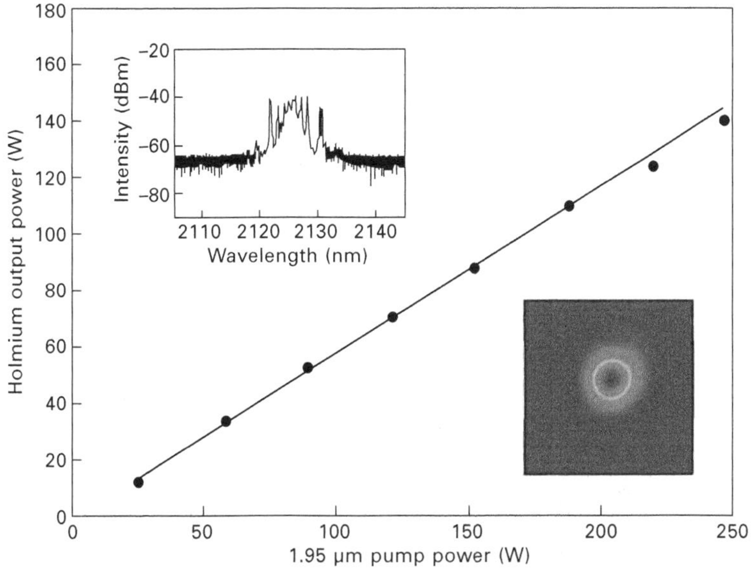

15.43 Resonant pumping of Ho-doped silica fibers at 1.95 μm using a Tm-fiber laser has been demonstrated to operate at ~150 W CW power at a wavelength 2.13 μm (data courtesy of DSTO Australia).

while maintaining robust single-mode operation. In conventional fibers, more modes are guided when the core diameter is increased. Various techniques have been studied to overcome this limit. They typically fall into three categories: (1) introduction of differential mode losses to suppress higher-order mode propagation, (2) operating within or near the single-mode regime by lowering the NA of the optical fiber, and (3) use of higher-order modes.

Conventional large mode area (LMA) fibers fall into the first category. Coiling is used to introduce differential mode losses to suppress high-order modes [96] along with careful selective launch of the fundamental mode [97]. The fundamental mode is better guided than higher-order modes and, consequently, most robust against coiling. As core size increases, more modes are guided in the core with much reduced modal spacing in the effective refractive index. This makes it harder to suppress higher-order modes without significantly affecting fundamental modes. As modes are better guided in larger cores, tighter coils are necessary, which can cause mode coupling among modes. This eventually limits scalability of LMA to ~30 μm core diameter. A typical PM-LMA fiber is shown in Fig. 15.45. A non-circular pump guide is typically used to minimize the propagation of skew rays which do not intersect the active core and, therefore, to increase pump absorption, and polarization-maintaining stress elements are positioned either side of the core to induce birefringence across the core.

© Woodhead Publishing Limited, 2013

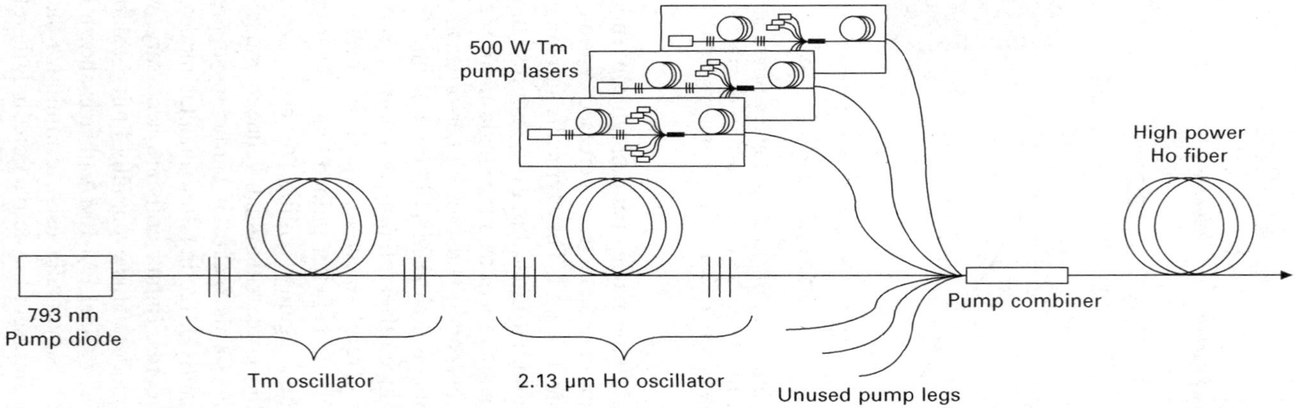

15.44 Schematic of system for power-scaling a resonantly pumped Ho-doped MOPA system.

© Woodhead Publishing Limited, 2013

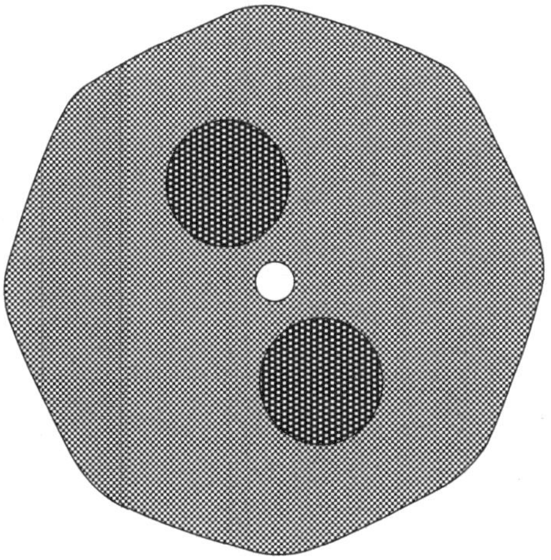

15.45 Illustration of a PM-LMA fiber.

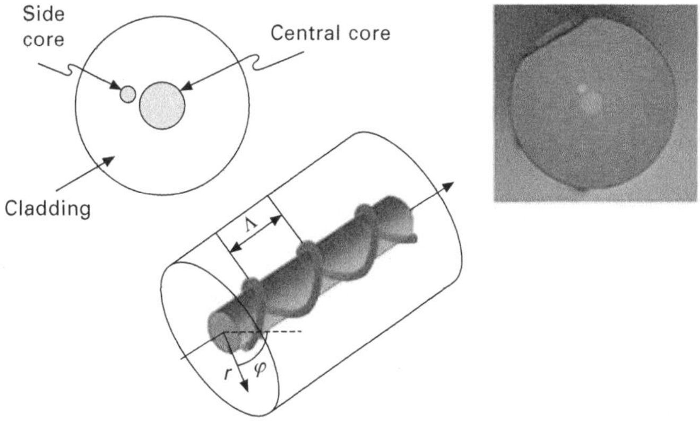

15.46 Illustration of a chirally coupled core fiber.

Designs based on resonant coupling have also been studied [98]. Typically, an additional waveguide is created adjacent to the active core. The modes in the side core are designed to resonantly couple with higher-order modes in the active core. A notable example is chirally coupled core (CCC) fiber (see Fig. 15.46) [99], where a single or multiple side cores are placed very close to an active central core. The preform is spun during fiber drawing, leading to the side cores being wrapped around the central active core in a helical fashion. The second-order mode in the active core can be made to be in resonance with the fundamental mode in the side core and, consequently, has higher losses due to the high bending loss of the side cores in the tight helix. The resonance can be tuned to the laser wavelength by adjusting the pitch of the helix. Recently, a core diameter of ~50 μm has been demonstrated

© Woodhead Publishing Limited, 2013

[100]. Strong resonant coupling needs both phase-matching and good spatial overlap of the modes involved. When the core diameter increases, all modes are more confined to the core center, leading to less overlap with the side cores. This eventually can limit the scalability of all approaches based on resonance coupling. The resonance typically has a narrow spectral width. This can also constrain the usefulness of this approach, where the wavelength of the resonance can be affected by coiling or temperature changes in fiber lasers.

Differential mode losses can also be obtained in leakage channel fibers (LCF) where an open cladding replaces the closed core–cladding boundary in a conventional fiber (see Fig. 15.47) [101–104]. The open cladding structures in combination with the fact that the fundamental mode has a larger spatial size lead to the possibility of designs with negligible fundamental mode loss and significant higher-order mode losses. The effect is not resonance-based, leading to weak wavelength dependence. Higher-order modes extend into the open cladding of leakage channel fibers even in very large cores, leading to significantly improved core diameter scalability. Single-mode operation with core diameters of ~180 μm has been demonstrated.

Recently, it has been shown that all-solid photonic bandgap fibers can provide very high differential mode losses even at very large cores [105–107]. All-solid photonic bandgap fibers guide lights by anti-resonance of the lattice of high index nodes in the cladding, i.e. photonic bandgaps (see Fig. 15.48), not the total internal reflections in conventional optical fibers. Nominally, no modes guide in the defect core in the center. The guidance effect in photonic bandgap fibers has strong mode dependence. It can be designed to selectively guide just a single mode and, therefore, provides significant higher-order mode suppression. Core diameter of ~50 μm has been demonstrated [108].

Since *V* value scales with the product of core diameter and NA, it is easy

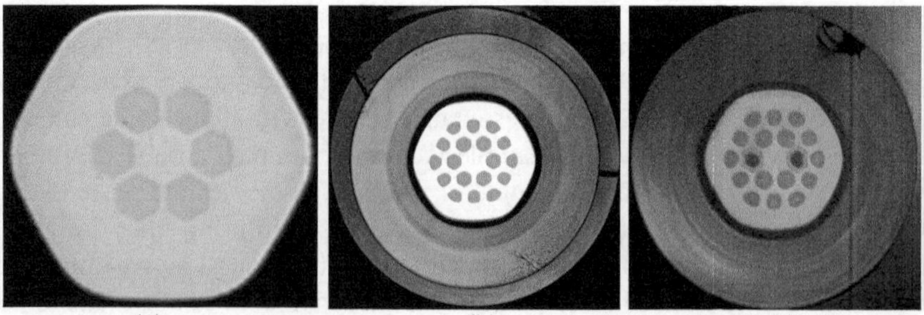

(a) (b) (c)

15.47 Illustration of various types of all-glass leakage channel fibers (a) with low index polymer as pump cladding, (b) with fluorine-doped silica as pump cladding, (c) with stress elements for maintaining polarization.

© Woodhead Publishing Limited, 2013

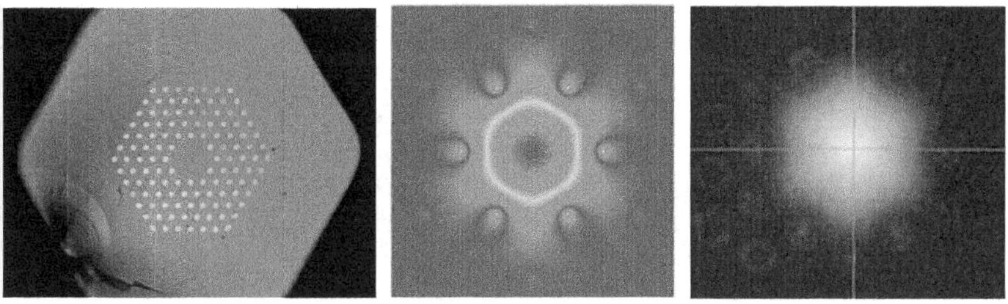

(a) (b) (c)

15.48 (a) A fabricated all-solid photonic bandgap fiber with core diameter of 50 μm, (b) simulated fundamental mode and (c) measured fundamental mode.

to see that a smaller NA can be used to compensate the increase of core diameter while maintaining single-mode operation. A number of approaches fall into this second category. One fundamental issue with this approach is that the optical mode becomes increasingly sensitive to bends and other perturbations when NA becomes small. In theory, the NA of a conventional fiber can be lowered to allow very large cores. In practice, it is very hard to fabricate fibers with NA <0.06 due to index controllability of the fabrication process.

Photonic crystal fibers (PCF) fall in this second category (see Fig. 15.49) [109–112]. The cladding of a PCF can be viewed as a composite material consisting of glass and air. The index of the composite can be accurately controlled by adjusting the dimensions of the air holes in the cladding. Coilable PCFs can be made with core diameter below ~40 μm. PCFs with over 40 μm core diameter have to be used in short straight sections. Core diameter up to ~100 μm has been fabricated, although with much less higher-order mode suppression. PCFs typically have a ring of air holes for pump cladding. High pump NA exceeding 0.6 is quite common. This is very useful for pump coupling. The presence of air holes also makes splicing very difficult, as air holes can collapse unlike conventional solid optical fibers.

It should also be noted that the effective mode area typically decreases with a reduction of coil diameter (see Fig. 15.50) [113]. This is due to the fact that a coiling can be seen as equivalent to the introduction of a linear refractive index across the fiber. This makes the part of the core towards the center of the coil have an effectively higher index, which becomes the effective waveguide supporting a smaller mode. This effect leads to mode compression in coiled fiber.

The third category of approaches is to use a higher-order mode, typically one of the LP_{0n} modes instead of the fundamental mode [114, 115]. It is argued that perturbations in fibers have anti-symmetry as they typically arise from bending effects. This prevents them from coupling to symmetrical modes.

© Woodhead Publishing Limited, 2013

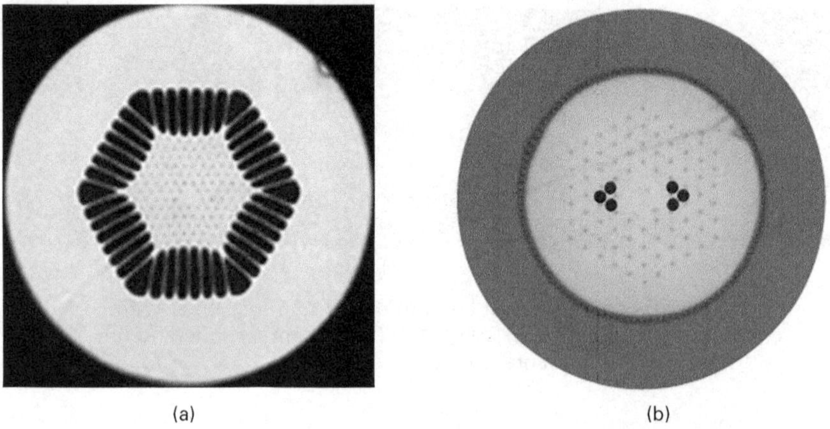

15.49 Illustration of (a) regular and (b) polarization-maintaining photonic crystal fibers.

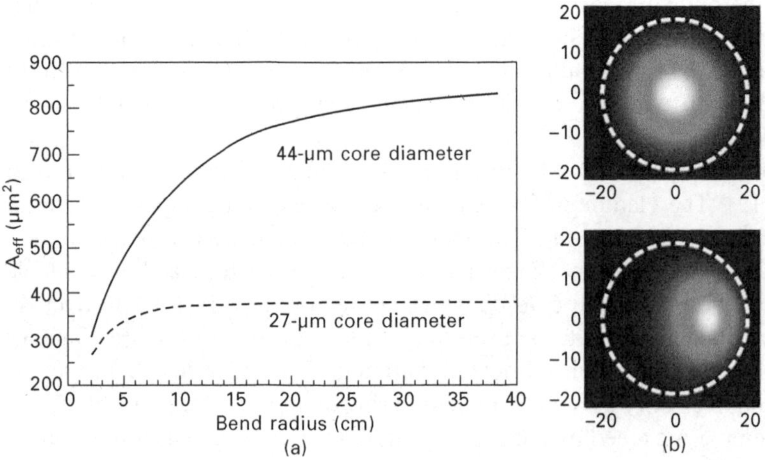

15.50 Mode area compression in coiled fibers (courtesy of John Fini, OFS).

The nearest modes with anti-symmetry are further away for a higher-order LP_{0n} mode than the fundamental mode, leading to more robust propagation of such a higher-order mode. The higher-order mode has a very large effective mode area and does not suffer much from mode area compression when coiled. Mode converters will have to be used at the input and output ends to convert lights from the fundamental mode to the higher-order mode and then back again after amplification. One major issue is that the fiber supports many modes, leading to significantly higher amplified spontaneous emission (ASE), which limits maximum gain. Recently it has been shown

© Woodhead Publishing Limited, 2013

that this can be overcome by using a pump in the same mode to suppress ASE in other modes [116]. A further issue is that not all perturbations are strictly anti-symmetrical in practice, leading to some coupling to other nearby symmetrical modes.

Another technique to further improve mode quality is to design a rare-earth doping profile to overlap better with the desired mode and less well with undesired modes [117]. This can be used in any of the approaches discussed so far, and has been used widely in practice.

15.8 Conclusion and future trends

Despite the rapid pace of progress in fiber laser technology in the past decade or so, we expect continued innovations in the field for the next decade, with some of the new research topics covered in this chapter possibly emerging from the laboratory and making it into commercial products. In particular, novel fiber designs for scaling the mode field and managing nonlinear limitation continues to be a robust topic for research with numerous innovations in the field of micro-structured fiber still to be uncovered. The impact on fiber laser device performance will be mostly in the pulsed fiber laser domain, where the current device performance lags significantly behind that of more traditional solid-state laser technologies. Improvement in longer wavelength fiber lasers will continue and commercial success of the new wavelengths such as 2 μm will start to emerge as the technology becomes cheaper to deploy and the applications become more visible. There are also opportunities for non-silica fibers, such as fluorides and chalcogenides, to excel given the limited transmission properties of silica glass in the mid-IR. New commercial opportunities will emerge across the full range of markets, spanning industrial, medical, military and sensing applications and driving further innovation through increased R&D budgets.

15.9 References

1. Dominic, V., MacCormack, S., Waarts, R., Sanders, S., Bickness, S., Dohle, R., Wolak, E., Yeh, E. and Zucker, E., '110W Fiber laser', OSA, Conference on Lasers and Electro-Optics (CLEO), 1999.
2. Fomin, V., Abramov, M., Ferin, A., Abramov, A., Mochalov, D., Platonov, N., and Gapontsev, V., '10 kW single-mode fiber laser', presented at 5th International Symposium on High-Power Fiber Lasers and Their Applications, St Petersburg, 28 June–1 July, 2010.
3. Keoster, C. J. and Snitzer, E., 'Amplification in a fiber laser', Appl. Opt., 13: 1182–1186, 1964.
4. Po, H., Snitzer, E., Tumminelli, R., Zenteno, L., Hakimi, F., Cho, N. M., and Haw, T., 'Double-clad high brightness Nd fiber laser pumped by GaAlAs phased array', Optical Fiber Communication Conference (Optical Society of America, Washington, DC, 1989), pp. PD7.

© Woodhead Publishing Limited, 2013

5. Brown, D. C., and Hoffman, H. J., 'Thermal, stress, and thermo-optic effects in high average power double-clad silica fiber lasers', IEEE J. Quant. Electron., 37, 2, 2001.
6. Galvanauskas, A., 'High power fiber lasers', Optics and Photonics News: 42–47, July 2004.
7. Wolf, P., Köhler, B., Rotter, K., Hertsch, K., Kissel, H., and Biesenbach, J., 'High-power, high-brightness and low-weight fiber coupled diode laser device', SPIE Photonics West 2011, Paper 7918–23.
8. Taverner, D., Richardson, D. J., Dong, L., Caplen, J. E., Williams, K., and Penty, K. V., '158-μJ pulses from a single-transverse-mode, large-mode-area erbium-doped fiber amplifier', Opt. Lett., 22: 378–380, 1997.
9. Koplow, J., Kliner, D., and Goldberg, L., 'Single-mode operation of a coiled multimode fiber amplifier', Opt. Lett., 25: 442–444, 2000.
10. Fermann, M., 'Single-mode excitation of multimode fibers with ultrashort pulses', Opt. Lett., 23: 52–54, 1998.
11. Ahmed, M. A. and Voss, A., 'Optical fibers for high power single-mode beam delivery', Optik and Photonik, 2: 38–43, May 2012.
12. See, for example, Laser Focus World, vol. 48, January 2012.
13. Snitzer, E., 'Proposed fiber cavities for optical masers', J. Appl. Phys., 23: 36–39, 1961.
14. Townsend, J. E., Poole, S. B., and Payne, D. N., 'Solution doping technique for fabrication of rare earth doped optical fibers', Electron. Lett., 23: 329–331, 1987.
15. Poole, S. B., Payne, D. N., and Fermann, M. E., 'Fabrication of low loss optical fibers containing rare earth ions', Electron. Lett., 21: 737–738, 1985.
16. Sire, J. E., Dubois, J. T., Eisentraut, K. J., and Sievers, R. E., 'Volatile lanthanide chelates: II Vapor pressures, heats of vaporization, and heats of sublimation', J. Am. Chem. Soc., 91: 3476–3481, 1969.
17. Pask, H. M., Carman, R. J., Hanna, D. C., Tropper, A. C., Mackechnie, C. J., Baber, P. R., and Dawes, J. M., 'Ytterbium-doped silica fiber lasers: versatile sources for the 1–1.2 μm region', IEEE J. Sel. Top. Quantum Electron., 1: 2–13, 1995.
18. Paschotta, R., Nilsson, J., Tropper, A. C., and Hanna, D. C., 'Ytterbium-doped fiber amplifiers', IEEE J. Quantum Electron., 33: 1049–1056, 1997.
19. Shiner, B., 'Recent progress in high power fiber lasers', presented at ALAW 2009.
20. Jackson, S. D. and Mossman, S., 'Efficiency dependence on the Tm^{3+} and A^{l+} concentrations for Tm^{3+}-doped silica double clad fiber lasers', Appl. Opt., 42: 2702–2707, 2003.
21. Kobyakov, A., Kumar, S., Chowdhury, D. Q., Ruffin, A. B., Sauer, M., and Bickham, S. R., 'Design concept for optical fibers with enhanced threshold', Opt. Express, 13: 5338–5346, 2005.
22. Gray, S., Liu, A., Walton, D. T., Wang, J., Li, M. J., Chen, X., Ruffin, A. B., DeMeritt, J. A., and Zenteno, L. A., '502 Watt, single transverse mode, narrow linewidth, bidirectionally pumped Yb-doped fiber amplifier', Opt Express, 15: 17044–17050, 2007.
23. Dong, L., 'Limits of stimulated Brillouin scattering suppression in optical fibers with transverse acoustic waveguide designs', IEEE. J. Lightwave Technol., 28: 3156–3161, 2010.
24. Davis, M. K., Digonnet, M. J. F., and Pantell, R. H., 'Thermal effects in doped fibers', IEEE J. Lightwave Technol., 16: 1013–1023, 1998.

© Woodhead Publishing Limited, 2013

25. Smith, A. V., Do, B. T., and Söderlund, M. J., 'Optical damage limits to pulse energy from fibers', IEEE J. Sel. Top. Quantum Electron., 15: 153–158, 2009.
26. Jeong, Y., Sahu, J. K., Payne, D. N., and Nilsson, J., 'Ytterbium-doped large-core fiber laser with 1 kW continuous-wave output power', OSA *ASSP 2004*, New Mexico, 1–4 February 2004, PD1 (Postdeadline).
27. Xiao, Y., Brunet, F., Kanskar, M., Faucher, M., Wetter, A., and Holehouse, N., '1-Kilowatt CW all-fiber laser oscillator pumped with wavelength-beam-combined diode stacks', Opt. Express, 20: 3296–3301, 2012.
28. Oulundsen, G., Farley, K., Abramczyk, J., and Wei, K., 'Fiber for fiber lasers: Matching active and passive fibers improves fiber laser performance', Laser Focus World, Vol. 48, January 2012.
29. Edgecumbe, J., *et al.*, 'Kilowatt level, monolithic fiber amplifiers for beam combining applications at 1 micron', Proc. DEPS SSDLTR, 2007.
30. Gonthier, F., 'All-fiber pump coupling techniques for double-clad fiber amplifiers and lasers', TFII1-3 CLEO Europe, 2005.
31. www.nufern.com
32. www.DILAS.com
33. Yu, H. *et al.*, '1.2 kW single-mode fiber laser based on 100-W high brightness pump diodes', Proc SPIE, 8237: 82370G1–6, January 2012.
34. Karlsen, S., Price, R. K., Reynolds, M., Brown, A., Mehl, R., Patterson, S., and Martinsen, R. J., '100-W, 105-μm, 0.15NA fiber coupled laser diode module', SPIE Photonics West 2009, Vol. 7198-29, 2009.
35. Gapontsev, V., Moshegov, N., Trubenko, P., Komissarov, A., Berishev, I., Raisky, O., Strougov, N., Chuyanov, V., Maksimov, O., and Ovtchinnikov, A., 'High-brightness 9XXnm pumps with wavelength stabilization', Proc. SPIE, 7583: 75830A–75830A-9, 2010.
36. Jauregui, C., Eidam, T., Limpert, J., and Tünnermann, A., 'Impact of modal interference on the beam quality of high-power fiber amplifiers', Opt. Express, 19: 3258–3271, 2011.
37. Smith, A. V., and Smith, S. J., 'Mode instability in high power fiber amplifiers', Opt. Express, 19: 10180–10192, 2011.
38. Eidam, T., Wirth, C., Jauregui, C., Stutzki, F., Jansen, F., Otto, H., Schmidt, O., Schreiber, T., Limpert, J., and Tünnermann, A., 'Experimental observations of the threshold-like onset of mode instabilities in high power fiber amplifiers', Opt. Express, 19: 13218–13224, 2011.
39. Wettera, A., Fauchera, M., Lovelady, M., and Séguina, F., 'Tapered fused-bundle splitter capable of 1 kW CW operation', Proc. SPIE, 6453, 64530I, 2007.
40. Chen, X., Li, M., Wang, J., Walton, D. T., and Zenteno, L. A., 'Comprehensive modeling of single frequency fiber amplifiers for mitigating stimulated Brillouin scattering', IEEE J. Lightwave Technol., 27: 2189–2198, 2009.
41. Edgecumbe, J., *et al.*, 'kW-Class, narrow-linewidth counter pumped fiber amplifiers', Proc. SSDLTR, 2010.
42. Goodno, G. D., McNaught, S. J., Rothenberg, J. E., McComb, T. S., Thielen, P. A., Wickham, M. J., and Weber, M. E., 'Active phase and polarization locking of a 1.4 kW fiber amplifier', Opt. Lett., 35: 1452–1544, 2010.
43. Wirth, C., Schmidt, O., Tsybin, I., Schreiber, T., Eberhardt, R., Limpert, J., Tünnermann, A., Ludewigt, K., Gowin, M., Have, E., and Jung, M., 'High average power spectral beam combining of four fiber amplifiers to 8.2 kW', Opt. Lett., 36: 3118–3120, 2011.

© Woodhead Publishing Limited, 2013

44. Rothenberg, J., 'Limits of power scaling by fiber laser amplifier combination', Optical Fiber Communications (OFC) 2009 High Power Fiber Laser Workshop, 23 March 2009.
45. See, for example, Hecht, J., Laser Focus World Podcast, October 2011, 'High-power solid-state lasers for military applications'.
46. Machewirth, D. P., Wang, Q., Samson, B., Tankala, K., O'Connor, M., and Alam, M., 'Current developments in high-power monolithic polarization maintaining fiber amplifiers for coherent beam combining applications', SPIE Proc., Fiber Lasers IV: Technology, Systems, and Applications: Proceedings, Vol. 6453, February 2007.
47. Smith, A.V., and Smith, J. J., 'Mode competition in high power fiber amplifiers', Opt. Express, 19: 11318–11329, 2011.
48. Dawson, J. W., Messerly, M. J., Beach, R. J., Shverdin, M. Y., Stappaerts, E. A., Sridharan, A. K., Pax, P. H., Heebner, J. E., Siders, C.W., and Barty, C. J. P., 'Analysis of the scalability of diffraction-limited fiber lasers and amplifiers to high average power', Opt. Express, 16: 13240–13260, 2008.
49. Alvarez-Chavez, J. A., Offerhaus, H. L., Nilsson, J., Turner, P. W., Clarkson, W. A., and Richardson, D. J., 'High-energy, high-power ytterbium-doped Q-switched fiber laser', Opt. Lett., 25: 37–39, 2000.
50. Offerhaus, H. L., Alvarez-Chavez, J. A., Nilsson, J., Turner, P. W., Clarkson, W. A., and Richardson, D. J., 'Multi-mJ, multi-watt Q-switched fiber laser', CLEO 1999, Baltimore, MD, paper CPD10.
51. Desthieux, B., Laming, R. I., and Payne, D. N., '110 kW (0.5 mJ) pulse amplification at 1.5 mm using a gated cascade of three erbium-doped fiber amplifiers', Appl. Phys. Lett., 63: 586–588, 1993.
52. Nilsson, J., Minelly, J. D., Paschotta, R., Tropper, A. C., and Hanna, D. C., 'Ring-doped cladding-pumped single-mode three-level fiber laser', Opt. Lett., 23: 355–357, 1998.
53. Teodoro, F., Koplow, J. P., Moore, S. W., and Kliner, D. A. V., 'Diffraction-limited, 300-kW peak-power pulses from a coiled multimode fiber amplifier', Opt. Lett., 27: 518–520, 2002.
54. Khitrov, V., Samson, B., Machewirth, D., Yan, D., Tankala, K., and Held, A., 'High-peak power-pulsed single-mode linearly-polarized LMA fiber amplifier and Q-switch laser', Proc. SPIE, Fiber Lasers IV: Technology, Systems, and Applications, Vol. 6453, 2007.
55. Dupriez, P., Piper, A., Malinowski, A., Sahu, J. K., Ibsen, M., Thomsen, B. C., Jeong, Y., Hickey, L. M. B., Zervas, M. N., Nilsson, J., and Richardson, D. J., 'High average power, high repetition rate, picosecond pulsed fiber master oscillator power amplifier source seeded by a gain-switched laser diode at 1060 nm', IEEE Photon. Technol. Lett., 18: 1013–1015, 2006.
56. Ryser, M., Neff, M., Pilz, S., Burn, A., and Romano, V., 'Gain-switched laser diode seeded Yb-doped fiber amplifier delivering 11-ps pulses at repetition rates up to 40-MHz', Proc. SPIE, Fiber Lasers IX: 8237, paper 82373I, 2012.
57. Lauterborn, T., Heinemann, S., and Galvanauskas, A., 'System integration aspects of pulsed fiber lasers in MOPA configuration', Presented at DEPS SSDLTR 2006, Paper Fiber 1-1.
58. Malinowski, A., Vu, K. T., Chen, K. K., Nilsson, J., Jeong, Y., Alam, S., Lin, D., and Richardson, D. J., 'High power pulsed fiber MOPA system incorporating electro-optic modulator based adaptive pulse shaping', Opt. Express, 17: 20927–20937, 2009.

© Woodhead Publishing Limited, 2013

59. Schimpf, D. N., Ruchert, C., Nodop, C., Limpert, J., Tünnermann, A., and Salin, F., 'Compensation of pulse-distortion in saturated laser amplifiers', Opt. Express, 16: 17637–17646, 2008.
60. Hendow, S. T., and Shakir, S. A., 'Structuring materials with nanosecond laser pulses', Opt. Express, 18: 10188–10199, 2010.
61. Liu, C.-H., Chang, G., Litchinitser, N., Guertin, D., Jacobson, N., Tankala, K., and Galvanauskas, A., 'Chirally coupled core fibers at 1550-nm and 1064-nm for effectively single-mode core size scaling', CLEO, Paper CTuBB3, 2007.
62. Schrader, P. E., Fève, J. P., Farrow, R. L., Kliner, D. A. V., Schmitt, R. L., and Do, B. D., 'Power scaling of fiber-based amplifiers seeded with microchip lasers', Proc. SPIE, 6871: 68710T-68710T-11, 2008.
63. Limpert, J., Deguil-Robin, R., Manek-Hönninger, I., Salin, F., Röser, F., Liem, A., Schreiber, T., Nolte, S., Zellmer, H., Tünnermann, A., Broeng, J., Petersson, A., and Jakobsen, C., 'High-power rod-type photonic crystal fiber laser', Opt. Express, 13: 1055–1058, 2005.
64. Dong, I., McKay, H. A., Fu, L., Ohta, M., Marcinkevicius, A., Suzuki, S., and Fermann, M. E., 'Ytterbium-doped all glass leakage channel fibers with highly fluorine-doped silica pump cladding', Opt. Express, 17: 8962–8969, 2009.
65. Stutzki, F., Jansen, F., Liem, A., Jauregui, C., Limpert, J., and Tünnermann, A., '26 mJ, 130 W *Q*-switched fiber-laser system with near-diffraction-limited beam quality', Opt. Lett., 37: 1073–1075, 2012.
66. Liu, A., Norsen, M. A., and Mead, R. A., '60-W green output by frequency doubling of a polarized Yb-doped fiber laser', Opt. Lett., 30: 67–69, 2005.
67. Schrader, P. E., Farrow, R. L., Kliner, D. A. V., Fève, J., and Landru, N., 'High-power fiber amplifier with widely tunable repetition rate, fixed pulse duration, and multiple output wavelengths', Opt. Express, 14: 11528–11538, 2006
68. Mordovanakis, A. G., Hou, K.-C., Chang, Y.-C., Cheng, M.-Y., Nees, J., Hou, B., Maksimchuk, A., Mourou, G., and Galvanauskas, A., 'Demonstration of fiber-laser-produced plasma source and application to efficient extreme UV light generation', Opt. Lett., 31: 2517–2519, 2006.
69. Torruellas, W., 'High peak power fiber lasers for Lidar/Ladar applications', IEEE-LEOS, Baltimore, MD, 2007.
70. Dudley, J. M., and Taylor, J. R., *Supercontinuum Generation in Optical Fibers*, Cambridge University Press, 2010.
71. Chen, K. K., Alam, S., Price, J. H. V., Hayes, J. R., Lin, D., Malinowski, A., Codemard, C., Ghosh, D., Pal, M., Bhadra, S. K., and Richardson, D. J., 'Picosecond fiber MOPA pumped supercontinuum source with 39 W output power', Opt. Express, 18: 5426–5432, 2010.
72. Leigh, M., Shi, W., Zong, J., Yao, Z., Jiang, S., and Peyghambarian, N., 'High peak power single frequency pulses using a short polarization maintaining phosphate glass fiber with a large core', Appl. Phys. Lett., 92: 181108-1, 2008.
73. Schulzgen, A., Li, L., Temyanko, V. L., Suzuki, S., Moloney, J. V., and Peyghambarian, N., 'Single-frequency fiber oscillator with watt-level output power using photonic crystal phosphate glass fiber', Opt. Express, 14: 7087–7092, 2006.
74. Kuizenga, D. I., and Siegman, A. E., 'Modulator frequency detuning of effects in the FM mode-locked laser', IEEE J. Quantum Electron., QE-6: 803–808, 1970.
75. Fermann, M. E., 'Ultrashort pulse sources based on single-mode rare-earth-doped fibers', J. Appl. Phys. B, B58: 197, 1994.

© Woodhead Publishing Limited, 2013

76. Zhou, X., Yoshitomi, D., Kobayashi, Y., and Torizuka, K., 'Generation of 28-fs pulses from a mode-locked ytterbium fiber oscillator', Opt. Express, 16: 7055, 2008.
77. Duling, I. N., *Compact Sources of Ultra-short Pulses*, Cambridge Studies in Modren Optics, Cambridge University Press, 1995.
78. Haus, H. A., 'Mode-locking of lasers', IEEE J. Sel. Top. Quantum Electron., 6: 1173–1185, 2000.
79. Hartl, I., Imeshev, G., Dong, L., Cho, G., and Fermann, M. E., 'Ultra-compact dispersion compensated femtosecond fiber oscillator and amplifiers', Conf. on Lasers and Electro-optics, paper CThG1, 2005.
80. Jeong, Y., Yoo, S., Codemard, C. A., Nilsson, J., Sahu, J. K., and Payne, D. N., 'Erbium:ytterbium co-doped large-core fiber laser with 297 W continuous-wave output power', IEEE J. Sel. Top. Quantum Electron., 13: 573–579, 2007.
81. Carter, A., *et al.*, 'Robustly single-mode polarization maintaining Er/Yb co-doped LMA fiber for high power applications', Proc. OSA CLEO 2007, paper CTuS6, 2007.
82. Khitrov, V., Ding, J., Samson, B., Machewirth, D., and Tankala, K., '1.0 μm-ASE-free, high power, large-mode-area erbium and ytterbium co-doped amplifier', SPIE Fiber Lasers VI: Technology, Systems, and Applications, Vol. 7195, 2009.
83. Zhang, J., Fromzel, V., and Dubinskii, M., 'Resonantly cladding-pumped Yb-free Er-doped LMA fiber laser with record high power and efficiency', Opt. Express, 19: 5574–5578, 2011.
84. Lim, E., Alam, S., and Richardson, D. J., 'The impact of pair-induced quenching on the power scaling of in-band pumped erbium doped fiber amplifiers', Opt. Express, 20: 13886–13895, 2012.
85. Frith, G., Lancaster, D. G., and Jackson, S. D., '85W Tm^{3+}-doped silica fibre laser', Electron. Lett., 41: 1207–1208, 2005.
86. Ehrenreich, T., Leveille, R., Majid, I, Tankala, K., Rines, G., and Moulton, P., '1-kW, all-glass Tm:fiber laser', SPIE Photonics West 2010: LASE Fiber Lasers VII: Technology, Systems, and Applications, Conference 7580, Session 16: Late-Breaking News.
87. Moulton, P. F., Rines, G. A., Slobodtchikov, E. V., Wall, K. F., Frith, G., Samson, B., and Carter, A. L. G., 'Tm-doped fiber lasers: Fundamentals and power scaling', IEEE J. Sel. Top. Quantum Electron., 15: 85–92, 2009.
88. Clarkson, W. A., Barnes, N. P., Turner, P. W., Nilsson, J., and Hanna, D. C., 'High-power cladding pumped Tm-doped silica fiber laser with wavelength tuning from 1860 to 2090 nm', Opt. Lett., 27: 1989–1991, 2002.
89. Fried, N. M., 'Thulium fiber laser lithotripsy: An in vitro analysis of stone fragmentation using a modulated 110-watt thulium fiber laser at 1.94 mm', Lasers in Surgery and Medicine, 37: 53–58, 2005.
90. Creeden, D., Budni, P. A., and Ketteridge, P. A., 'Pulsed Tm-doped fiber lasers for mid-IR frequency conversion', SPIE Fiber Lasers VI: Technology, Systems, and Applications, Vol. 7195, 2009.
91. Jiang, M., and Tayabati, P., 'Stable 10 ns, kW peak-power pulse generation from gain-switched thulium-doped fiber laser', Opt. Lett., 32: 1797–1799, 2007.
92. Frith, G., McComb, T., Samson, B., Torruellas, W., Dennis, M., Carter, A., Khitrov, V., and Tankala, K., 'Frequency doubling of Tm-doped fiber lasers for efficient 950 nm generation', Proc. OSA ASSP 2009, Denver, CO, Fiber Lasers and Bulk Solid-State Lasers: Paper WB.

© Woodhead Publishing Limited, 2013

93. Smock, D., 'Fiber laser process targets medical plastics welding', Plastics Today, May 2012.
94. Guhur, A., and Jackson, S. D., 'Efficient holmium-doped fluoride fiber laser emitting 2.1 μm and blue upconversion fluorescence upon excitation at 2 μm', Opt. Express, 18: 20164–20169, 2010.
95. Hemming, A., Bennetts, S., Simakov, J., and Carter, A., 'A 140 W large mode area double clad holmium fibre laser', SPIE Fiber Lasers IX: Technology, Systems, and Applications, Vol. 8237, 2012, post-deadline paper, and Proc. SPIE, Vol. 8237, 82371J-9.
96. Koplow, J. P., Kliner, D. A. V., and Goldberg, L., 'Single-mode operation of a coiled multimode fiber amplifier', Opt. Lett., 25: 442–444, 2000.
97. Fermann, M. E., 'Single-mode excitation of multimode fibers with ultra-short pulses', Opt. Lett., 23: 52–54, 1998.
98. Fini, J. M., 'Design of solid and microstructure fibers for suppression of higher-order modes', Opt. Express, 13: 3477–3490, 2005.
99. Liu, C. H., Chang, G., Litchinitser, N., Galvanauskas, A., Guertin, D., Jacobson, N., and Tankala, K., 'Effectively single-mode chirally-coupled core fiber', Advanced Solid-State Photonics, paper ME2, OSA Technical Digest Series (CD), Optical Society of America, 2007.
100. Hu, I. N., Ma, X., Zhu, C., Liu, C. H., Sosnowski, T., and Galvanauskas, A., 'Experimental demonstration of SRS suppression in chirally coupled core fibers', Advanced Solid-State Photonics, paper AT1A.3, 2012.
101. Dong, L., McKay, H. A., and Fu, L., 'All-glass endless single-mode photonic crystal fibers', Opt. Lett., 33: 2440–2442, 2008.
102. Dong, L., Peng, X., and Li, J., 'Leakage channel optical fibers with large effective area', J. Opt. Soc. Am. B, 24: 1689–1697, 2007.
103. Dong, L., Wu, T. W., McKay, H. A., Fu, L., Li, J., and Winful, H. G., 'All-glass large-core leakage channel fibers', IEEE J. Sel. Top. Quantum Electron., 15: 47–53, 2009.
104. Fu, L., McKay, H. A., Suzuki, S., Ohta, M., and Dong, L., 'All-glass pm leakage channel fibers with up to 80 μm core diameters for high gain and high peak power fiber amplifiers', post-deadline paper MF3, Advanced Solid State Photonics, Denver, CO, February 2009.
105. Egorova, O. N., Semjonov, S. L., Kosolapov, A. F., Denisov, A. N., Pryamikov, A. D., Gaponov, D. A., Biriukov, A. S., Dianov, E. M., Salganskii, M. Y., Khopin, V. F., Yashkov, M. V., Gurianov, A. N., and Kuksenkov, D. V., 'Single-mode all-silica photonic bandgap fiber with 20 μm mode field diameter', Opt. Express, 16: 11735–11740, 2008.
106. Kashiwagi, M., Saitoh, K., Takenaga, K., Tanigawa, S., Matsuo, S., and Fujimaki, M., 'Practically deployable and effectively single-mode all-solid photonic bandgap fiber with large effective area', CLEO, paper CThM3, 2011.
107. Saitoh, K., Murao, T., Rosa, L., and Koshiba, M., 'Effective area limit of large-mode-area solid-core photonic bandgap fibers for fiber laser applications', Opt. Fiber Technol., 16: 409–418, 2010.
108. Dong, L., Saitoh, K., Kong, F., Foy, P., Hawkins, T., and McClane D., 'Mode Area Scaling for high power fiber lasers with all-solid photonic bandgap fibers', SPIE Defense Security and Sensing, paper 8381-5, 2012.
109. Limpert, J., Liem, A., Reich, M., Schreiber, T., Nolte, S., Zellmer, H., Tünnermann, A., *et al.*, 'Low-nonlinearity single-transverse-mode ytterbium-doped photonic crystal fiber amplifier', Opt. Express, 12: 1313–1319, 2004.

© Woodhead Publishing Limited, 2013

110. Limpert, J., Deguil-Robin, N., Manek-Hönninger, I., Salin, F., Röser, F., Liem, A., Schreiber, T., *et al.*, 'High-power rod-type photonic crystal fiber laser', Opt. Express, 13: 1055–1058, 2005.
111. Limpert, J., Schmidt, O., Rothhardt, J., Röser, F., Schreiber, T., and Tünnermann, A., 'Extended single-mode photonic crystal fiber', Opt. Express, 14: 2715–2719, 2006.
112. Brooks, C. D., and Di Teodoro, F., 'Multi-megawatt peak power, single-transverse-mode operation of a 100 μm core diameter, Yb-doped rod-like photonic crystal fiber amplifier', Appl. Phys. Lett., 89: 111119–111121, 2006.
113. Fini, J. M., 'Bend-resistant design of conventional and microstructure fibers with very large mode area', Opt. Express, 14: 69–81, 2006.
114. Ramachandran, S., Nicholson, J. W., Ghalmi, S., Yan, M. F., Wisk, P., Monberg, E., and Dimarcello, F. V., 'Light propagation with ultra large modal areas in optical fibers', Opt. Lett., 31: 1797–1799, 2006.
115. Ramachandran, S., 'Spatially structured light in optical fibers, applications to high power lasers', paper MD1, Advanced Solid State Photonics, Denver, CO, February 2009.
116. Headley, C., Phillips, J., Fini, J., Gonzales, E., Ghalmi, S., Yan, M., Nicholson, J., Wisk, P., Fleming, J., Monberg, E., Dimarcello, F., Windeler, R. S., and Fishteyn, M., 'Amplification of a large-mode area single higher order mode in a fiber amplifier', SPIE Photonics West, paper 8237–59, 2012.
117. Marciante, J. R., 'Gain filtering for single-spatial-mode operation of large-mode-area fiber amplifiers', IEEE J. Sel. Top. Quantum Electron., 15: 30–36, 2009.

© Woodhead Publishing Limited, 2013

16
Mid-infrared optical parametric oscillators

M. HENRIKSSON, Swedish Defence Research Agency, Sweden

DOI: 10.1533/9780857097507.2.463

Abstract: This chapter gives an overview of recent advances in the field of generation of mid-infrared (2–15 μm) coherent radiation using optical parametric oscillators (OPO). The content includes time scales from continuous wave over nanosecond to ultrafast pulses and both single frequency devices for spectroscopy and high power devices for defence and medical applications. The chapter discusses advances in second order nonlinear optical crystals for the mid-infrared region, novel cavity configurations and more general improvements allowed by the development of laser sources for OPO pumping.

Key words: coherent mid-infrared sources, optical parametric oscillators, second-order optical nonlinearity, nonlinear optical crystals.

16.1 Introduction

This chapter will try to give an overview of mid-infrared (mid-IR) coherent sources based on optical parametric oscillators (OPO). The mid-IR is here defined as approximately the 2–15 μm wavelength range, with the main focus on the 4–9 μm range where much of the interesting development is happening. When writing a text like this, one is always limited by time and space and the content will by necessity not be comprehensive in its coverage of the subject. I will here try to give an overview of the field of optical parametric oscillators, and especially focus on the, in my view, most interesting developments during the last decade.

The importance of the mid-IR wavelength range stems from the many interesting applications including spectroscopy, laser surgery and military applications such as infrared countermeasures. At the same time there has historically been a lack of practically useful laser sources in many parts of this spectral region. Even though the first OPO was demonstrated in 1965 (Giordmaine and Miller, 1965), they were long of limited usefulness. This is no longer the case, mainly because of new nonlinear crystal materials and improved pump lasers. At the same time other mid-IR laser sources such as quantum cascade lasers (QCL) and transition metal lasers (e.g. Cr:ZnSe, Fe:ZnSe) are being developed. This means that the mid-IR spectrum that has

© Woodhead Publishing Limited, 2013

historically only been reached through some gas lasers is getting better and better coverage. Direct laser sources and OPOs have different characteristics in several ways and thus which kind of source is the best will vary from application to application.

The absorption of molecules in the mid-IR is important in several different ways. The first is that the major components of the atmosphere (oxygen, nitrogen, water and carbon dioxide) all have low absorption in large parts of this spectral range. This means that there are several spectral windows with high transmission, as seen in Fig. 16.1, allowing long range optical applications. The second interesting feature of molecule absorption is that most molecules have very distinct absorption lines in the mid-IR, providing a fingerprint that makes it possible to identify them using optical spectroscopy. The varying absorption spectra of different molecules also allow selective excitation and heating. This has great potential for minimizing collateral damage during laser surgery.

This chapter will start with a short introduction to the physics of nonlinear optics and OPOs and then review some of the more important developments in mid-IR OPOs that have appeared during the last decade. The OPOs are divided in three groups depending on the characteristics of the produced radiation, tuneable single longitudinal mode OPOs for spectroscopy, high power and high energy nanosecond pulse OPOs and synchronously pumped

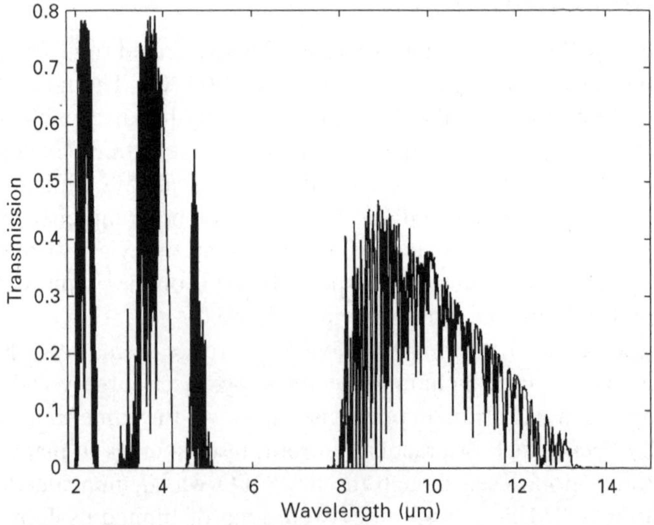

16.1 Atmospheric transmission in the mid-IR over 10 km in sub-arctic summer conditions as generated by the MODTRAN5 model (www.modtran.org). The transmission in some wavelength regions varies with atmospheric conditions, especially the aerosol distribution, but the transmission windows remain fundamentally the same.

© Woodhead Publishing Limited, 2013

ultrashort pulse OPOs. This division is suitable as there are important distinctions between these three types of OPOs. At the end of the chapter there is a short discussion on other types of laser sources in the mid-IR and finally an outlook on where the field might be heading in the future.

16.2 Nonlinear optics and optical parametric devices

As an introduction to the review of important progress in the field of OPOs we will here recapitulate some of the main points about OPO theory and discuss some of the more recent additions to the knowledge of the physics of OPOs. There is still a lack of a book that describes all the theory behind OPOs. There are, however, some book chapters and review papers that contain much useful information on the physics and technology of OPOs. Some examples are Harris (1969), Sutherland (2003), Debuisschert *et al.* (2000) and Svelto *et al.* (2007).

When the electronic response of a material is not linear to the driving electromagnetic field there will be frequency components in the material polarization that are not present in the incident field. For sufficiently high field strengths all materials will behave nonlinearly. Here we are mainly interested in the second order nonlinearity that describes interactions between three electromagnetic waves. The second order nonlinearity, and all higher even orders of nonlinearity, is zero in symmetric materials, and thus only present in some crystal materials. The nonlinear coefficient $\chi^{(2)}$ is a third order tensor relating the three polarization components of the material polarization to the polarization components of the two incident electromagnetic waves. After the propagation direction and the wavelengths and polarization planes of all three interacting waves have been fixed, an effective nonlinearity d_{eff} for that specific interaction can be calculated and used as a scalar number.

A parametric interaction is a way to use nonlinear optics to divide the energy of a photon between two new photons of longer wavelength. Convention says that the incoming beam with the highest photon energy is called the pump. The pump beam, through a second order nonlinear optical difference frequency generation (DFG) process, will generate two new beams, the signal and the idler, as illustrated in Fig. 16.2. The shorter wavelength beam of these two is usually called the signal and the longer wavelength beam the idler. In a parametric interaction the state of the nonlinear material where the conversion is performed is always the same before and after the process. This means that the photon energy is conserved and

$$\omega_{\text{p}} = \omega_{\text{s}} + \omega_{\text{i}} \qquad [16.1]$$

where ω_{p}, ω_{s} and ω_{i} are the angular frequencies of the pump, the signal and the idler, respectively.

© Woodhead Publishing Limited, 2013

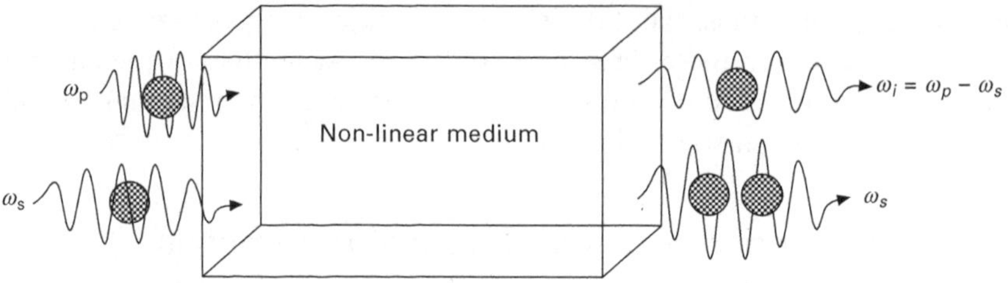

16.2 In a parametric amplification process, difference frequency generation between pump ω_p and signal ω_s will remove one pump photon and add another signal photon while generating an idler photon at the difference frequency ω_i.

The difference frequency process demands that there is always some initial electromagnetic field at either the signal or the idler frequency. If no seed beam is provided this electromagnetic field will be provided by the vacuum fluctuations in a process called optical parametric generation (OPG) or parametric superfluorescence (PSF). The case where there is a seed beam is called DFG or optical parametric amplification (OPA). When there is some energy at the signal or idler frequency, either supplied as a seed beam or from amplification of the vacuum fluctuations, DFG of the pump and the signal will amplify the field at the idler frequency while at the same time DFG of the pump and the idler will amplify the signal. In this way more and more energy is transferred from the pump to the signal and the idler as the beams propagate through the nonlinear crystal. There is no absolute distinction between devices that are said to operate through DFG and OPA, but in general it is called DFG when the nonlinear contribution to the material polarization, and thus the gain, at the seed wavelength through mixing of the pump and the difference frequency is small. In an OPA the photon numbers of the signal and idler are of similar magnitude after passage through the crystal and they are mutually amplifying each other.

The OPO puts the nonlinear crystal inside a cavity to provide feedback of the generated radiation, making the parametric process seed itself at a later time. If the gain in a single passage through the nonlinear crystal is larger than the losses in the feedback, the field oscillating in the cavity will grow and, if the pump is applied during enough time, reach macroscopic intensities even for moderate pump intensities. The parametric process will then deplete the pump intensity, limiting further growth of the peak intensity. To be useful the OPO cavity losses should include outcoupling of a beam for use in the intended application.

The wavelength pair that experience gain in a parametric process is determined by phase matching. The mixing of pump and signal beams at one position will generate an idler beam with a certain phase compared to the two generating fields. When the beams propagate through the nonlinear

© Woodhead Publishing Limited, 2013

crystal they will travel at slightly different speeds because of the chromatic dispersion in the material. This means that when the pump and signal fields are mixed at a later time in a new position of the crystal the phase of the material polarization generated at the idler frequency will not fit the phase of the idler beam that was generated before and has travelled through the crystal. Adding two fields at the same frequency that are more than $\pi/2$ out of phase with each other will cause them to cancel out and the energy is transferred back to the pump by sum frequency generation (SFG) of the signal and the idler. There are in principle two ways to get around the phase mismatch caused by the chromatic dispersion in bulk devices. The first way is to use the fact that the refractive index and thus the phase velocity depend on the polarization direction in birefringent materials. This is called birefringent phase-matching (BPM). The other way is to periodically change the phase of the generated beam by inverting the crystal structure and thus the sign of D_{eff}. This is called quasi phase-matching (QPM).

The phase matching condition can be described by the equation

$$\frac{n_{\text{p}}}{\lambda_{\text{p}}} - \frac{n_{\text{s}}}{\lambda_{\text{s}}} - \frac{n_{\text{i}}}{\lambda_{\text{i}}} \pm \frac{1}{\Lambda_{\text{g}}} = 0 \qquad [16.2]$$

where λ means wavelength and n refractive index with the indices p, s and i signifying pump, signal and idler. The inversion period of the nonlinear coefficient d_{eff} is given as Λ_g. For BPM this period is infinite and the contribution to the phase matching relation zero. A more thorough discussion on phase matching and the ways of reaching it is provided in Chapter 6 of this book.

So far we have assumed discrete wavelengths for the three beams. However, they have certain spectral widths, where the phase matching relation is perfectly fulfilled for the centre wavelengths, and the phase mismatch grows as the wavelengths change. In fixed wavelength nonlinear interactions, such as second harmonic generation (SHG), acceptance intervals are given in temperature or angle for the nonlinear crystal. This can be relevant in an OPO in the case where the signal or idler is fixed by seeding or by a wavelength-selective cavity. In the general case the signal and idler wavelengths will, however, adapt to the optimum phase matching. It is therefore more relevant to discuss the spectral bandwidths that fulfil phase matching.

For fixed pump wavelength and crystal parameters the phase matching bandwidth of the signal and idler can be shown to be

$$\Delta\omega_{\text{s,i}} = \frac{2C}{|t_{\text{w,si}}|} \qquad [16.3]$$

where C is a constant that is often chosen as 0.886, and the temporal walk-off between signal and idler is

© Woodhead Publishing Limited, 2013

$$t_{w,si} = L \frac{n_{g,s} - n_{g,i}}{c} \qquad [16.4]$$

This expression is strictly only valid for moderate pump intensities and without depletion of the pump. There are several factors that may influence the bandwidth, especially for higher pump intensities. The generated bandwidth will, for example, in the high depletion regime also be affected by the temporal walk-off between the pump and the generated beams, as for low walk-off a certain time slice of the signal will always propagate together with the same time slice of the pump and the generated intensity is thus limited by the pump intensity. For larger temporal walk-off an initial spike in the signal and idler will have access to more energy from the pump and the exponential growth is not limited as quickly. Larger temporal walk-off between pump and signal will thus lead to larger temporal modulation of the signal and hence wider bandwidth (Arisholm *et al.*, 2001). An important factor in this is that the bandwidth will also influence the back-conversion, allowing operation higher above threshold with wide bandwidths.

There is a principal difference between the behaviour of type I, where signal and idler have the same polarization, and type II, where signal and idler have different polarizations, phase matching close to degeneracy that has implications for mid-IR OPOs. For type I phase matching the refractive indices of the signal and idler are equal at degeneracy. This will cause a rapid change of the wavelength for a very small change of the refractive index and a wide gain bandwidth as illustrated in Fig 16.3. In a type II OPO there is a significant difference in refractive index of the two generated waves even when their wavelengths are degenerate because of the birefringence of the material (a few examples of isotropic nonlinear materials exist). Because of this the tuning curves cross at degeneracy, the tuning speed is no different than at other wavelengths, as seen in Fig. 16.4, and the gain bandwidth stays narrow. This means that wavelength stability close to degeneracy is easier to achieve with type II interaction, while type I interaction can allow very wide tuning using a wavelength-selective cavity.

A similar relation to eq. 16.3 can be introduced for the maximum acceptable pump bandwidth if the signal or idler frequency is fixed. When both the signal and idler are free to adapt, the situation is more complicated and not suitable for analytical evaluation. The spectral intensity of the pump will, however, influence the conversion efficiency as the parametric gain will be distributed over a larger part of the spectrum and hence be weaker at a certain wavelength. Perrett *et al.* (2004) have shown experimentally how the slope efficiency of a ZGP OPO is reduced as the pump bandwidth increases. The author has investigated the same thing numerically, showing that there is no pump acceptance bandwidth below which the efficiency is unaffected by further decrease in pump bandwidth. Instead the OPO efficiency is decreasing

© Woodhead Publishing Limited, 2013

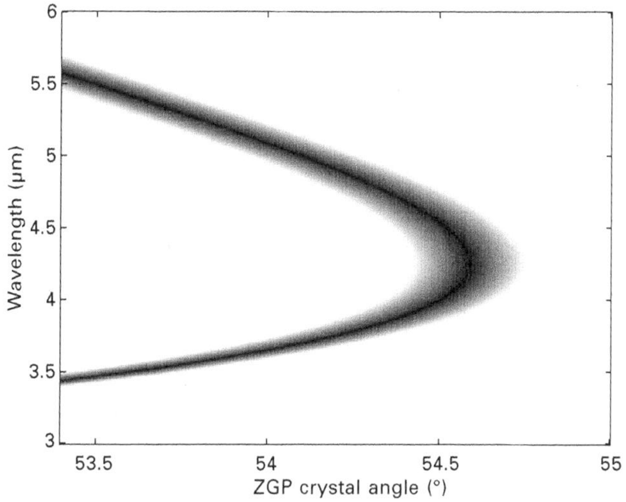

16.3 Phase mismatch for type I phase matching in ZGP with 2128 nm pump wavelength. The colour scale is from zero phase mismatch for black colour to more than π phase mismatch over 10 mm propagation for white colour. Close to degeneracy a large range of wavelengths have low phase mismatch for the same crystal angle and the generated bandwidth will be large in an unconstrained OPO.

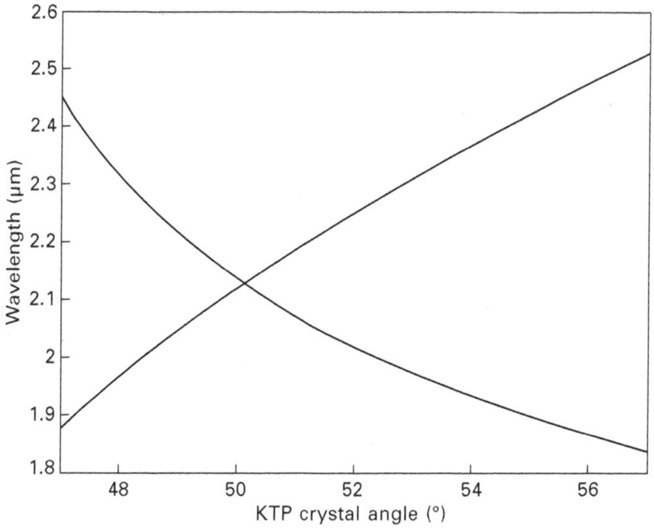

16.4 Phase mismatch for type II phase matching in KTP (*xz*-plane) with 1064 nm pump wavelength. The colour scale is from zero phase mismatch for black colour to more than π phase mismatch over 10 mm propagation for white colour. The two branches have different polarization, and this together with the birefringence of the material causes the spectral bandwidth to be narrow and almost constant during tuning.

© Woodhead Publishing Limited, 2013

almost linearly with increasing pump bandwidth until the OPO no longer reaches the oscillation threshold (Henriksson *et al.*, 2009).

The OPO consists of three important components, the pump, the nonlinear crystal and the feedback cavity. An optical cavity will only allow those wavelengths to oscillate where the roundtrip optical length of the cavity corresponds to an integer number of wavelengths. This causes the existence of longitudinal modes, separated in frequency by the inverse of the roundtrip time.

An important distinction between different cavity types is whether it is a singly or doubly resonant OPO (SRO or DRO), that is, whether only one or both of the signal and idler is resonant in the cavity. The difference in feedback may cause the pump intensity oscillation threshold for the DRO to be several orders of magnitude lower than for the SRO. The double resonance, however, also means that both the signal and idler spectra contain longitudinal modes. Energy conservation demands that the frequencies of the longitudinal modes of the signal and idler add up exactly to the longitudinal modes in the pump spectrum. The chromatic dispersion of different elements in the OPO will cause a small difference in the mode separations of the signal and idler. Because of this only certain mode pairs will add up to energy conservation. At the same time thermal and mechanical fluctuations cause the longitudinal modes to move in frequency. The result in a narrow gain bandwidth OPO is that the output jumps around between different modes and large intensity fluctuations as the phase mismatch of the mode pairs that fulfil energy conservation varies. For larger gain bandwidths there will always be some modes, and usually several, that fulfil energy conservation with low phase mismatch. In this case the spectrum will be very unstable and consist of multiple peaks, but the output energy may be relatively stable. To generate a stable frequency from a DRO the cavity has to be actively stabilized to compensate for unavoidable fluctuations in the roundtrip time of the cavity.

In the SRO the spectrum of the non-resonant wave is free to adapt to fluctuations of the frequencies of the pump and cavity modes, allowing stable oscillation. The price is, however, a much higher threshold than in the DRO. There have also been triple resonant and pump enhanced cavities where the pump is also resonant to further reduce the needed pump intensity. With improving pump lasers these are now of limited interest because of very high demands on stabilization of the cavity to obtain stable oscillation.

In addition to the cavity mirrors the cavity roundtrip length also has a large influence on an OPO. An increased roundtrip time will decrease the number of passes through the nonlinear crystal per time unit, reducing the effective gain. The longer roundtrip time will in this way increase the build-up time for the pulse, thus reducing the conversion efficiency and increasing the threshold of a pulsed OPO as shown by Brosnan and Byer (1979). For this

© Woodhead Publishing Limited, 2013

reason it is important to keep the cavity short in pulsed OPOs. Another cavity length effect is that a short cavity will have a higher Fresnel number

$$N_F = \frac{D^2}{L\lambda} \quad [16.5]$$

where D is the beam diameter, L the cavity length and λ the wavelength. A high Fresnel number reduces the spatial filtering and may allow higher order spatial modes to oscillate, causing bad beam quality. A third effect is that matching of the roundtrip time has an effect for nanosecond pulses when using multi-longitudinal mode pump sources just as it has for pico- and femtosecond pulses as seen in Fig. 16.5. The Reduced threshold of DROs when the lengths were matched was discussed already in 1966 (Harris, 1966). An increase in output energy and spectral clustering from a DRO appears around every length where the ratios of pump and OPO roundtrip time are fractions of small integers (Arisholm *et al.*, 2000). The effect of increased output energy and lowered threshold for roundtrip time matched cavities is present also in a narrowband SRO (Henriksson *et al.*, 2010). The cause

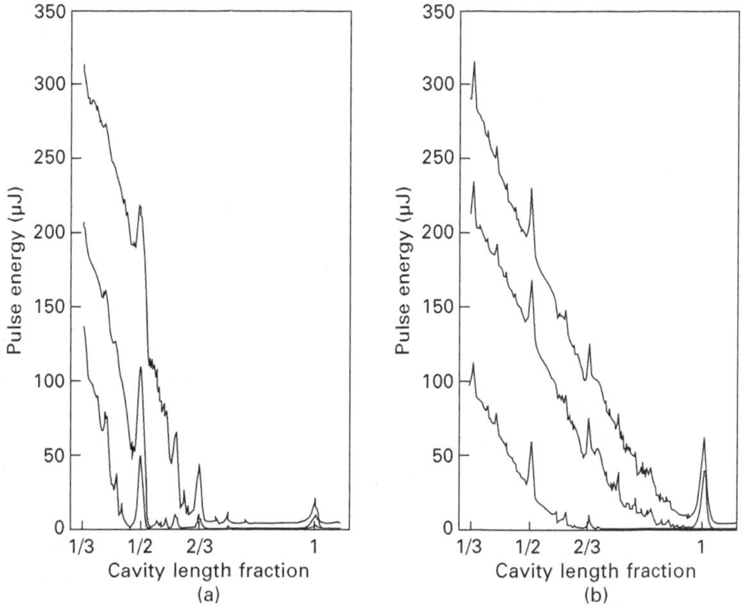

16.5 The output power from an OPO pumped by a multi-longitudinal mode laser is significantly increased when the ratio of the roundtrip times for the pump laser and OPO cavities forms a fraction of low integers. There are qualitative differences between (a) singly resonant OPOs and (b) doubly resonant OPOs, where the enhancement decreases further above threshold in an SRO but stays the same in a DRO. Adapted from Henriksson *et al.* (2010).

© Woodhead Publishing Limited, 2013

of these effects is that the electric field of a multi-longitudinal mode pump laser is periodic with the roundtrip time even if phases for the modes of the pump are random but constant.

There are many aspects of the OPO that cannot be described by analytical formulas, which makes numerical modelling tools interesting. Especially noteworthy is Arlee Smith's freely available SNLO (http://www.as-photonics.com/SNLO) that includes both simple calculations of phase matching and numerical modelling of OPO performance in many regimes.

16.3 Nonlinear optical materials for the infrared region

The nonlinear optical crystal is the most important component of the OPO. Selection of a suitable material depends on several parameters. The first is that the material must have second order nonlinearity. This is only possible in non-centrosymmetric materials, removing almost all pure elements, all non-structured materials such as gases, liquids and glasses and also many crystals. The material should also preferably have a high nonlinear figure of merit, defined as

$$\eta_{\mathrm{NL}} = \frac{d_{\mathrm{eff}}^2}{n^2} \qquad [16.6]$$

This figure of merit contains all material parameters in the conversion efficiency of a non-depleted pump calculation. Another demand of the crystal is that it has to be possible to phase-match the interaction, either by the crystal being birefringent or by fabrication of periodic structures that allow quasi phase-matching. Second order non-linear optical materials and phase-matching technology are discussed in more detail in Chapter 6. To be useful for an OPO the crystal material must also be transparent for all involved waves. Low loss, from bulk absorption, bulk scattering and surface reflection, is needed to keep down the threshold of the OPO. If the device is to operate at high average power, transparency is not enough, but the crystal has to have low absorption to avoid thermal problems. Low thermal conductivity of the material will worsen the problem of absorbed power. Too high absorption may also lead to crystal damage. Crystal damage, and especially surface damage, is a limiting factor in many nonlinear devices, making damage threshold one of the selection criteria for nonlinear crystals. Another issue is crystal availability. Some promising materials have never been grown as large low-loss monocrystals, and can thus not be used for devices.

The most common nonlinear materials for the visible and near-infrared are oxide-based crystals. Some of them are also useful in the mid-infrared. The most common ones are lithium niobate (LN), lithium tantalate (LT), potassium titanyl phosphate (KTP) and its isomorphs. The transmission curves

© Woodhead Publishing Limited, 2013

of some crystals are shown in Fig. 16.6. The longest wavelength transmission is available with LN, but even here high average power operation is limited to approximately 4 μm.

These materials may be used in bulk form with BPM. The nonlinearity is, however, much higher with all fields parallel in the crystal and thus QPM is very attractive. Periodic poling (PP), where the ferroelectric polarization is reversed by application of high voltage in periodic domains, allows this possibility, providing the necessary inversion of the effective nonlinearity. This was first demonstrated for bulk materials in PPLN (Webjorn *et al.*, 1994). Periodically poled materials are now the dominating nonlinear medium for low pulse energy OPOs, and basically the only one for CW and ultrashort pulse operation.

Some non-oxide nonlinear crystal materials are transparent at longer wavelengths, especially ternary and quaternary chalcopyrite semiconductors. Petrov (2012) reviews all non-oxide nonlinear crystals. Traditionally the best material for generation of wavelengths longer than 4.5 μm with 1 μm pumping has been $AgGaS_2$ (AGS), but this material suffers from a low damage threshold and bad thermal properties. Recently there has been a surge in material development producing some promising crystals where OPOs have been realized, for example $CdSiP_2$ (Petrov *et al.*, 2009) and $LiInSe_2$ (Zondy *et al.*, 2005), but their practical usefulness is not yet clear.

If the demand for pumping at 1 μm is relaxed and longer wavelength pump lasers are allowed, ZGP becomes a possibility. ZGP needs a pump wavelength at longer than 2 μm because of residual absorption, but has otherwise

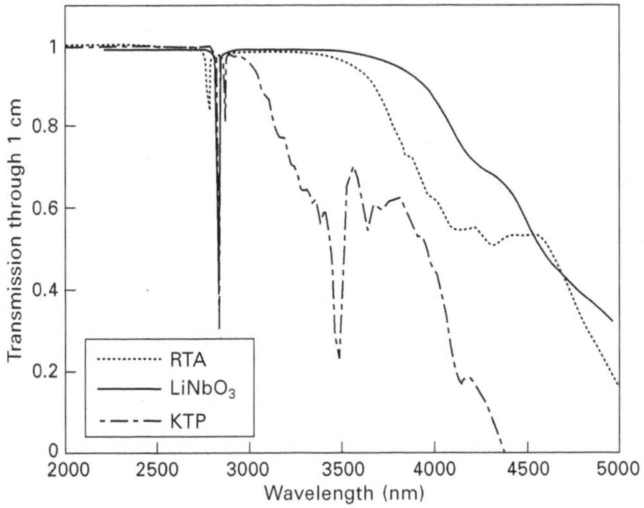

16.6 Transmission curves for $KTiOPO_4$ (KTP), $RbTiOAsO_4$ (RTA) and $LiNbO_3$.

© Woodhead Publishing Limited, 2013

attractive performance and has been the dominating crystal for longer than 4 μm OPOs during the last decade. The first use of ZGP for nonlinear optics was published in 1971 (Boyd *et al.*, 1971), but many problems remained to be solved. Further development went on in the Soviet Union during the 1980s, and was restarted again in the USA in the 1990s. Absorption, crystal size and damage threshold are all still improving (Zawilski *et al.*, 2008).

Several other nonlinear crystals of this group have been produced over the years. Most of them have remained curiosities with only some experimental use. A few, such as AGSe, CdSe and GaSe, see some use, primarily for generation of infrared wavelengths beyond 10 μm through DFG.

Binary isotropic semiconductor materials can have high nonlinearities, good thermal and mechanical properties and transparency reaching far into the mid-IR. The problem was for a long time that they are not birefringent and cannot be periodically poled by application of high voltages. There was for that reason no effective phase-matching method available. Efforts were made to produce QPM materials by stacking and bonding of thin plates with alternating crystal directions, but residual interface losses and expensive production methods limited their use. The development of orientation patterned growth (OP) of GaAs, where the domains are grown with alternating crystal orientation using epitaxial techniques (Eyres *et al.*, 2001), has changed this, even though crystals are not yet commercially available. Further development of the growth process now allows at least 1.48 mm thick QPM structures (Gonzalez *et al.*, 2011), and thicker samples are expected in the future. GaAs needs a pump wavelength longer than 1.8 μm because of two-photon absorption, but has shown good performance with 2 μm pumping. For the future, larger bandgap materials such as GaP may allow possibilities for pumping with Nd-lasers while still maintaining transmission over a large part of the mid-IR (Schunemann and Setzler, 2011).

16.4 Tuneable single frequency optical parametric oscillators (OPOs) for spectroscopy

Many spectroscopic techniques for gas sensing demand tuneable laser sources with narrow spectral bandwidth, preferably only a single longitudinal cavity mode. In the mid-IR one of the main ways to generate this radiation has been and is with OPOs.

For continuous wave (CW) operation the single longitudinal mode pump sources have historically been very limited in output power. Hence, the OPOs have been doubly resonant or have used pump enhanced cavities to reach threshold. These OPOs demanded stabilized pump lasers and strict control of cavity lengths and thermal effects to reach stable operation while still producing low output powers, and their usefulness was thus limited (see, e.g., Smith, 1973).

© Woodhead Publishing Limited, 2013

Although the first CW SRO was demonstrated using bulk non-critically phase-matched KTP (Yang *et al.*, 1993), it was when PPLN started to be used (Bosenberg *et al.*, 1996) that they started to be practically useable. The availability of long high nonlinearity PPLN crystals together with improved pump lasers has revolutionized the tuneable single frequency OPOs, from overconstrained DROs with multiparameter tuning (Eckardt *et al.*, 1991) to relatively easily controllable SROs. An important factor behind the success using PPLN is also that QPM allows walk-off free operation at any wavelength, something that is necessary with the tight focusing and long interaction length that is needed for low intensity CW pump lasers.

Another big development during recent years has been the availability of higher power CW tuneable single frequency pump lasers in the near-IR, mainly DFB-diode laser seeded fibre amplifiers and fibre lasers with Bragg grating feedback but also power increases in the Ti:sapphire technology and semiconductor lasers. This has allowed CW single frequency SROs with mode hop free tuning of the idler within the parametric gain bandwidth by pump tuning (Klein *et al.*, 2000; Lindsay *et al.*, 2005). Single resonant operation reduces the instability compared to DROs, but passive mode selection with etalons (Henderson and Stafford, 2006), surface gratings in Littrow configuration (Vainio *et al.*, 2009) or volume Bragg gratings (VBG) (Vainio *et al.*, 2010) are still often used to avoid the OPO operation switching between different longitudinal cavity modes on a longer time scale. The surface grating is very interesting close to degeneracy where the parametric gain curve is flat, as it allows wide mode-hop tuning with mode selectivity over the whole gain region of the parametric process. The Bragg grating is very limited in tuning, but has the advantage that it can handle higher powers and allows the OPO to be operated higher above threshold without multimode operation (Vainio *et al.*, 2011). Another way to allow stable operation at higher powers is to reduce the thermal loading of the nonlinear crystal. This can be done by introducing some signal output coupling in the cavity. A MgO:PPLN OPO with signal output coupling generated 7.7 W near 3 μm and at the same time 9.8 W of signal (Chaitanya Kumar *et al.*, 2011).

The main cause of the mode hopping in SROs is thermal fluctuations in the nonlinear crystal, and stabilization of the crystal temperature to around 10 mK has been shown to allow stable operation with pump tuning over 140 GHz without any intracavity frequency selective elements (Vainio *et al.*, 2008). Discontinuous tuning over longer wavelengths is still possible by selection of grating period and crystal temperature.

By stabilizing the signal wavelength to an external wavelength reference by controlling the OPO cavity length, it was possible to extend the mode-hop free idler tuning range of a pump tuned SRO to 500 GHz at 3300 nm (Andrieux *et al.*, 2011). This wavelength was chosen because it provides an

© Woodhead Publishing Limited, 2013

unusually large flat gain bandwidth when pumping with a 1066 nm external cavity diode laser amplified in an Yb-fibre amplifier, so the same wide tuning cannot be reproduced at all wavelengths.

So far CW OPOs are limited to near infrared pump wavelengths and PPLN crystals. Wavelength tuning up to 4.8 μm has been reported (van Herpen *et al.*, 2003), but the idler power is reduced by idler absorption above 4.2 μm. For longer wavelengths the absorption in PPLN will raise the oscillation threshold too much for CW operation. Longer wavelengths have primarily been produced with DFG in non-oxide materials, but this produces only very limited output power.

One improvement in the field of DROs in recent years has been the entangled cavity OPO (ECOPO) where the cavity lengths for signal and idler are controlled separately (Drag *et al.*, 2002; Berrou *et al.*, 2010). With continuous monitoring of the wavelength and computer control of the cavity length, long term stability and over 100 GHz continuous tuning has been achieved. Further development of the design has produced a configuration dubbed the NesCOPO using only three reflective surfaces instead of five and producing a pulsed OPO with 2.5 μJ threshold and spectroscopic beam quality (Hardy *et al.*, 2011).

For high pulse energy spectroscopic applications the traditional way is to use a seeded OPO, although an OPA amplifying the output from a low energy OPO is also a possibility. An ECOPO was amplified in a MOPA setup to generate up to 11 mJ in spectroscopic quality pulses near 2 μm (Raybaut *et al.*, 2009). A seeded high energy OPO uses a larger beam diameter. In this way a seeded OPO is not limited to non-critical phase-matching and can use BPM.

High spectral quality OPOs at longer wavelengths are not often reported, but are possible in the pulsed regime. A ZGP OPO pumped by a Q-switched Ho-laser and seeded by the idler from a tuneable PPLN OPO allowed narrow linewidth oscillation in the 4 to 10 μm wavelength range (Lee *et al.*, 2006).

16.5 High power and high energy nanosecond pulselength systems

16.5.1 1 μm pumped single OPO

Generation of nanosecond pulse length high repetition rate radiation at 2 to 4 μm wavelength using OPOs with various nonlinear crystals is an established technology. A standard two-mirror linear cavity OPO with a PPLN, PPKTP, bulk KTP or bulk KTA nonlinear crystal pumped by a Nd:YAG or $Nd:YVO_4$ laser will in general provide high conversion efficiency and good beam quality in this regime. As an example of the power levels that it is possible

© Woodhead Publishing Limited, 2013

to reach, a MgO:PPLN OPO can generate 16.7 W at 3.84 μm and at the same time 46 W at 1.47 μm (Peng *et al.*, 2009).

When tuning a PPLN OPO to idler wavelengths around 4 μm there is significant absorption. In addition to the increased threshold and reduced efficiency, in high average power setups this will also cause heating of the nonlinear crystal. The heating will change the phase-matching conditions and reduce efficiency. This can partially be compensated by using an oven with a longitudinal temperature gradient (Mason and Wood, 2004). In this way the idler output power can be increased by more than 30% (Godard *et al.*, 2010), as shown in Fig. 16.7.

To reach longer wavelengths than are possible with oxide crystals it is necessary to move to new materials. Promising results around 6 μm have been reached with CSP in the last few years (Petrov *et al.*, 2009). This is the first chalcogenide material where the thermal properties allow a 1 kHz repetition rate (Petrov *et al.*, 2010), all earlier experiments having been at a low repetition rate. The high nonlinear figure of merit of CSP also makes it possible to use very short crystals. In this way sub-nanosecond pulses have been produced in an OPO for the first time (Petrov *et al.*, 2010). Temperature tuning between 6.1 and 6.5 μm has been shown in a non-critical phase-matching cut. Tuning to even longer wavelengths has been shown using a $LiInSe_2$ OPO where idler wavelengths from 4.7 to 8.7 μm have been demonstrated

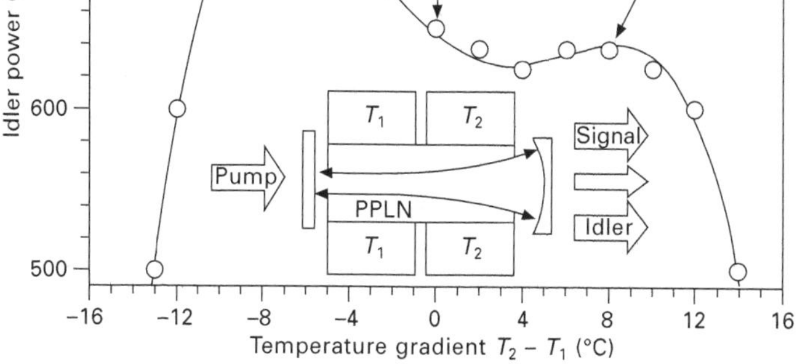

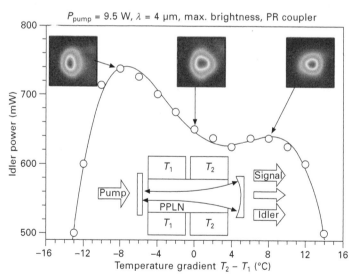

16.7 A two-zone oven with a temperature gradient can enhance the output in a PPLN OPO with large idler absorption. Reprinted from Godard A, Raybaut M, Schmid T, Lefebvre M, Michel A-M and Péalat M (2010), 'Management of thermal effects in high-repetition-rate pulsed optical parametric oscillators', Opt Lett, 35, 3667–3669.

© Woodhead Publishing Limited, 2013

at low average power (Marchev *et al.*, 2009). The record tuning from a 1 μm pumped OPO is, however, still the 3.9 to 11.3 μm achieved in $AgGaS_2$ (Vodopyanov *et al.*, 1999). Both CSP and LISe have better thermo-mechanical properties than AGS, making higher average powers possible and potentially making them more practically useful than AGS.

16.5.2 1 μm pumped tandem OPO

As direct conversion from 1.06 μm to wavelengths longer than 4 μm at high average powers is a problem with available nonlinear crystals, other solutions have been sought. There are several nonlinear materials that allow conversion starting at wavelengths around 2 μm, primarily ZGP and OP-GaAs. One way of generating the 2 μm pump source is to start with a Nd-laser at 1.06 μm and use an OPO to generate the intermediate wavelength. The setup where one OPO is used to pump a second OPO is here called a tandem OPO.

In addition to pulse energy, important properties of the pump of the second OPO, and hence demands on the output from the first OPO, are high beam quality and narrow spectral bandwidth. A problem is that the OPO that converts from 1.06 μm to close to 2 μm is working close to spectral degeneracy. The spectral selectivity of coated mirrors is not high enough to achieve an SRO cavity this close to degeneracy, causing power instability because of the overconstrained DRO system. In addition for a type I OPO where the signal and idler have the same polarization the gain bandwidth of the nonlinear interaction is very wide. In a free-running mirror OPO this will cause the bandwidth to be too wide for efficient pumping of the second OPO. Type II interaction avoids this problem because the refractive indices are different for the signal and idler, causing the phase matching error to grow more rapidly with changing wavelength. Type II interactions can, however, not access as high nonlinear coefficients as type I interaction in quasi phase-matched materials, reducing the efficiency of the devices, especially in low pulse energy applications. An additional advantage is that since the signal and idler from the first OPO have the same polarization they can both be used to pump the second OPO stage, in principle doubling the available pump power.

Tandem OPOs at kilohertz repetition rates using bulk type II KTP OPOs for the first step and a ZGP OPO to convert to the 3.5 to 5 μm wavelength range have been available for some time (Phua *et al.*, 1998; Cheung *et al.*, 1999). With careful optimization, conversion efficiency of 14% from 1 μm to the mid-IR has been reached (Arisholm *et al.*, 2002). The use of PPLN or PPKTP crystals in the first step can reduce the threshold because of the higher nonlinearity, but the type I operation demands some form of bandwidth-reducing cavity in the first OPO. Experiments have shown an increase in the ZGP OPO slope efficiency from 9 to 33% when the pump

© Woodhead Publishing Limited, 2013

bandwidth was reduced from 8 to 2.3 nm (Perrett *et al.*, 2004). The two cases were both PPLN OPOs using an L-cavity with a surface relief grating to force narrowband operation around degeneracy. The problem with this type of cavity is that by necessity it becomes long and in addition includes significant losses. A partial solution to this problem is the use of a MOPA structure to amplify the high beam quality, narrow bandwidth radiation from an OPO using a surface grating or an etalon (Saikawa *et al.*, 2006). The disadvantage with a MOPA system is the increased complexity of the system and the higher number of optical components needed.

An alternative method of bandwidth narrowing is to use a volume Bragg grating (VBG) as cavity mirror in the OPO. The VBG is a bulk component with a periodically varying refractive index, causing a narrow bandwidth reflection and thereby limiting the spectrum of the generated radiation as shown in Fig. 16.8. Using a VBG allows a short cavity length and introduces very low extra cavity losses. In this way a reasonable efficiency PPKTP OPO with a spectral output close to degeneracy so that both signal and idler can be used to pump a ZGP OPO at the same time is possible (Henriksson *et al.*, 2007). A tandem OPO setup has generated up to 3.2 W in the 3.5 to 5 μm wavelength range from 26 W of 1 μm pump power at 20 kHz PRF (Henriksson *et al.*, 2008). Using a large aperture PPLN crystal, bandwidth narrowing with a VBG has been proven to work well at higher pulse energy, generating up to 61 mJ pulse energy with 7.6 MW peak power in a 1.4 nm region around degeneracy at 2128 nm (Saikawa *et al.*, 2007). The short cavity with large beams of this setup, however, gave a relatively bad beam quality with $M^2 \approx 22$, which still is an improvement by the spatial filtering of the VBG from the $M^2 \approx 27$ measured with an ordinary mirror.

The first OP-GaAs OPO was realized in 2004 using a tandem OPO setup where a PPLN OPO with an intracavity etalon produced a tuneable pump wavelength in the 1.8 to 2 μm range. This was used to pump an OP-GaAs OPO that produced tuneable mid-IR radiation from 2.28 to 9.14 μm. The threshold was 16 μJ and the OPO generated up to 3 μJ idler energy at 7.9 μm in an $M^2 \approx 2$ beam (Vodopyanov *et al.*, 2004).

16.5.3 2 μm laser pumped OPOs

Direct laser pumping could provide a simpler and more efficient solution for a 2 μm pumped OPO than a tandem OPO setup. Lasers doped with thulium or holmium emit wavelengths around 2 μm. Tm-lasers are, however, not suitable for Q-switched operation, while holmium lacks absorption lines in regions where pump diodes are available. The early solution was co-doped crystal where energy was absorbed by the Tm ions and then transferred to Ho ions that lased. This, however, suffered from large thermal problems.

The revolution in 2 μm lasers for OPO pumping was started by the

© Woodhead Publishing Limited, 2013

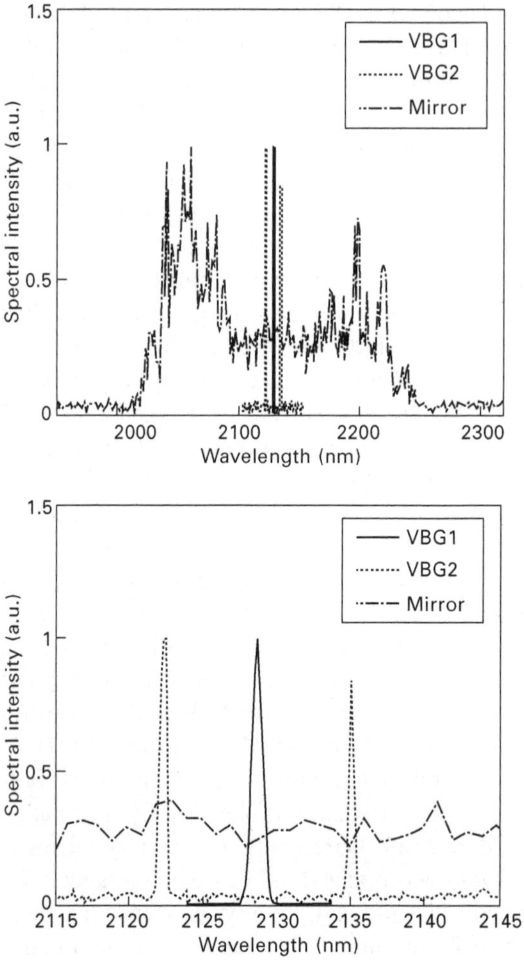

16.8 The use of a VBG as a wavelength-selective cavity mirror will drastically reduce the spectral bandwidth of the output from a degenerate PPKTP OPO. Two spectra using different VBGs with resonance wavelengths almost exactly at degeneracy and slightly off degeneracy are shown together with a reference spectrum collected using a standard outcoupling mirror. Adapted from Henriksson *et al.* (2007).

introduction of in-band pumping of holmium lasers by separate thulium lasers. This has been done both intracavity by Tm:YAG (Bollig *et al.*, 1998) and Tm:YLF lasers (Schellhorn *et al.*, 2003) and extracavity (Budni *et al.*, 2000). A further improvement was the introduction of Tm-fibre lasers for pumping the holmium lasers (Lippert *et al.*, 2003, 2006). The fibre laser geometry handles the generated heat in the Tm laser better and allows problem-free operation. This allowed 5 W of ZGP OPO output from a 15 W CW Tm-fibre

© Woodhead Publishing Limited, 2013

laser with an intermediate 9.8 W Ho:YAG laser Q-switched at 20 kHz with $M^2 \approx 1.8$ (Lippert *et al.*, 2006).

Further power scaling has been done using a three-mirror ring oscillator ZGP OPO. The ring oscillator removes the risk for back-reflection into the pump laser and thus allows operation without an optical isolator. Faraday rotators able to handle high powers at 2 μm are not available and thus isolators are a big problem. The ZGP OPO delivered up to 22 W output power in the 3 to 5 μm region using 37.7 W pump power from the Tm-fibre laser pumped Ho:YAG laser (Lippert *et al.*, 2010) seen in Fig. 16.9. A Ho-laser pumped ZGP OPO produced 0.95 W near 8 μm and was tuneable up to 9.8 μm (Lippert *et al.*, 2008).

Performance reached with OP-GaAs is very similar to what can be reached with ZGP in the same setup (Kieleck *et al.*, 2009). The limited aperture of OP-GaAs, however, limits the power from OP-GaAs OPOs. The highest average power published so far is 4.9 W with a 50 kHz Q-switched Ho:YAG laser as pump source (Kieleck *et al.*, 2010). At 20 kHz PRF the same setup produced 147 μJ pulse energy in the mid-IR. A beam quality of $M^2 \approx 1.5$ was measured slightly below maximum output power.

A ZGP OPO directly pumped by a gain-switched Tm-fibre laser with a large mode area fibre amplifier has reached up to 2 W average power in the 3.5–5 μm region (Creeden *et al.*, 2008a, 2008b). The conversion efficiency was limited by the relatively low intensity of the pump, which forced tight

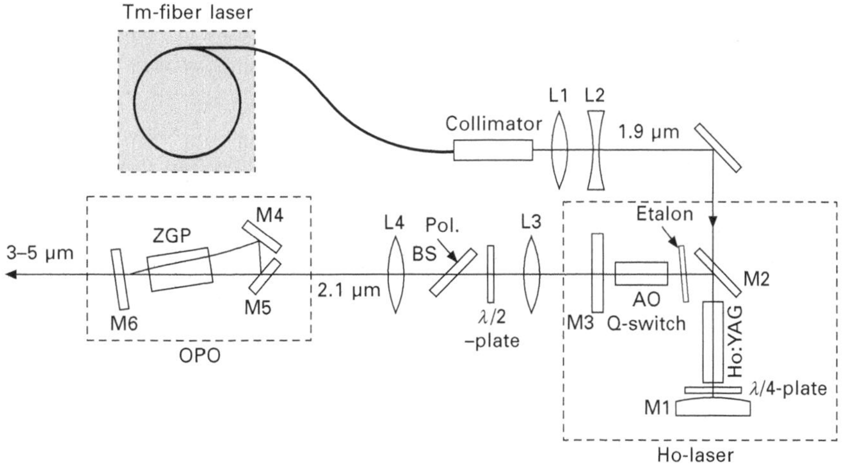

16.9 Experimental setup for a high power mid-infrared source with a V-shaped three-mirror ring ZGP OPO pumped by a Tm-fiber-laser-pumped Ho:YAG laser. Reprinted from Lippert E, Fonnum H, Arisholm G and Stenersen K (2010), 'A 22-watt mid-infrared optical parametric oscillator with V-shaped 3-mirror ring resonator', Opt Express, 18, 26475–26483.

© Woodhead Publishing Limited, 2013

focusing with resulting thermal problems and walk-off effects. Pumping an OP-GaAs OPO with the output from a Tm:Ho:silica fibre laser, an average power of 2.2 W at 40 kHz repetition rate was reached (Kieleck *et al.*, 2010). High average power from fibre-source pumped OP-GaAs OPOs is also claimed without details in Schunemann and Setzler (2011). With further development of Tm and Ho-doped pulsed fibre lasers and amplifiers, which is a research topic at the moment (Eichhorn, 2010), it is probable that this type of setup may reach much higher average powers in the near future.

16.5.4 High energy pulses

To scale OPOs to high pump energy the diameter of the beams needs to be increased to avoid damage to the nonlinear crystals. At the same time an increase in the cavity roundtrip time of a nanosecond OPO will reduce the conversion efficiency and thus the cavity needs to be kept short. A short cavity with wide beams will have a high Fresnel number and thus produce bad beam quality. A number of methods have been used to produce high energy pulses with good beam quality. High energy mid-IR sources at longer wavelengths are often two-stage setups starting at 1 μm, and thus one of the major problems is to generate the 2 μm pump that is used in the second stage, often a ZGP OPO.

Wavelengths up to 3.5 μm can be generated by Nd-laser pumped OPOs in bulk KTA. As an example an intracavity pumped bulk KTA OPO produced 31 mJ at 3.5 μm at 10 Hz repetition frequency (Zhong *et al.*, 2010). The pulse length of 5.5 ns means a peak power higher than 5 MW. The use of PPLN is limited by the available apertures and the damage threshold. Using elliptical beams the energy from the thin but wide crystals can be scaled at the cost of reduced beam quality (Missey *et al.*, 2000). There also exist some hero samples of PPLN up to 5 mm thickness that have allowed up to 110 mJ output near 2.1 μm (Saikawa *et al.*, 2007). The high Fresnel number resonator led to a bad beam quality with $M^2 \approx 27$. The KTP family of ferroelectric materials also allow poled crystals several millimetres thick, but the wavelength does not reach further than for the bulk crystals.

Arisholm *et al.* (2004) used a master oscillator power amplifier (MOPA) setup where a low energy KTP OPO beam was expanded and amplified in a KTP OPA to produce high energy pulses near 2 μm. Using the low energy idler from a KTA OPO as a seed beam at 3.7 μm or 2.8 μm in a ZGP OPA pumped by the high energy beam made further conversion to the mid-IR possible. In this way they were able to produce high energy pulses with good beam quality at 4.7 or 8.8 μm wavelength (Haakestad *et al.*, 2008). The output energy was 8 mJ at 8 μm and 33 mJ for the combined beams at 3.7 and 4.7 μm. The beam quality was measured to be $M^2 \approx 2$–4 in both wavelength regions. Dividing the OPA in several stages and removing either

© Woodhead Publishing Limited, 2013

the signal or the idler between the crystals can reduce back-conversion and further improve the beam quality (Arisholm *et al.*, 2004). The MOPA configuration can be used also to achieve narrow linewidth operation by using spectral narrowing in the master OPO (Saikawa *et al.*, 2006).

The walk-off between signal and idler in a type II critically phase-matched OPO will introduce a spatial filtering of the beams that improves the beam quality in the walk-off plane. Walk-off is generally seen as a negative effect as it reduces the conversion efficiency and makes the beam asymmetric. With the large beam diameters necessary in high energy OPOs, significant walk-off can be tolerated with insignificant loss in conversion efficiency. The RISTRA (rotated image singly-resonant twisted rectangle) OPO is a ring cavity that is folded in three dimensions (Fig. 16.10) and uses the spatial walk-off to improve the beam quality (Smith and Armstrong, 2002). Ho:YLF pumped operation of a ZGP RISTRA has generated up to 10 mJ at 3.4 μm with $M^2 < 1.8$ (Dergachev *et al.*, 2007). This shows that the RISTRA works also for type I interaction. An alternative way to get spatial filtering in two dimensions from spatial walk-off is to use two different crystal materials with the walk-off in different planes, e.g. KTA and BBO, in the same linear cavity to spatially filter the beam in both directions (Farsund *et al.*, 2010). This has been shown to generate good beam quality with $M^2 < 2$ from a high Fresnel number cavity in the near-IR with negligible loss of conversion efficiency.

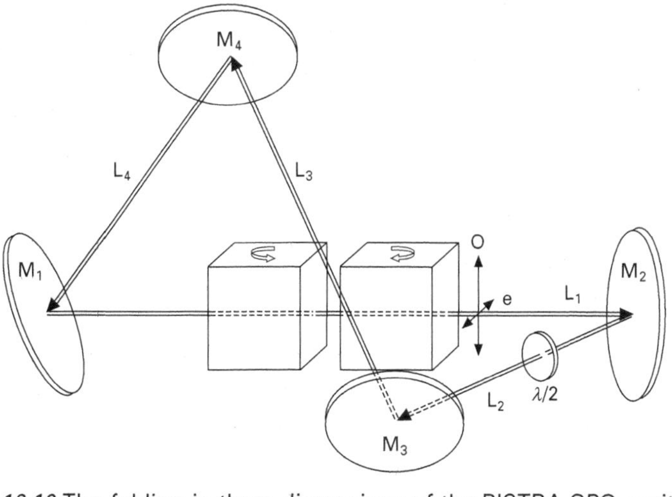

16.10 The folding in three dimensions of the RISTRA OPO cavity will produce an image rotation that together with the walk-off in the birefringent crystal spatially filters the beam and improves the beam quality. Reprinted from Smith A V and Armstrong D J (2002), 'Nanosecond optical parametric oscillator with 90° image rotation: design and performance', J Opt Soc Am B, 19, 1801–1814.

© Woodhead Publishing Limited, 2013

16.6 Ultrashort pulse systems

OPOs pumped by ultrashort pulse lasers have more in common with CW OPOs than they do with nanosecond pulse OPOs. As the femto- or picosecond pump pulse is much shorter than any possible cavity roundtrip, synchronous pumping where the roundtrip time is matched to the pump laser pulse repetition frequency is necessary. This means that for each pass through the nonlinear crystal the IR pulse will propagate together with a new pump pulse. The high repetition rate limits the intensity, and accurate control of the mode size is necessary, just as in the CW OPO.

PPLN and PPLT OPOs at wavelengths up to 4 μm and pulse lengths from a few hundred femtoseconds to a few picoseconds pumped by various mode-locked lasers, e.g. Ti:sapphire, Nd-lasers, Yb-fibre lasers and Er-fibre lasers, are today well known. The use of CW mode-locked Yb-fibre lasers as a pump source (O'Connor *et al.*, 2002) was a great advance in the practical usability of the sources compared to earlier systems mainly based on Nd:YAG lasers. The maximum mid-IR power reported was 7.8 W idler power at 3.57 μm generated using a Yb:YAG thin-disk laser to pump a PPLT OPO with fibre feedback (Südmeyer *et al.*, 2004). Using a more conventional PPLN OPO cavity and an 81.1 MHz PRF, 20.8 ps pulse length Yb-fibre laser pump 4.6 W average power was generated at 3.33 μm (Kokabee *et al.*, 2010). For longer wavelengths the crystal absorption raises the threshold and reduces the efficiency. The high pump intensity with ultrashort pulses, however, makes the single pass gain much higher than in the CW case. For an ultrashort pulse the energy will generally more than double and may increase up to several hundred times in a single pass through the parametric amplification stage, while the gain for a CW parametric amplifier is only a few per cent even if many crystals are used. In this way much higher idler losses can be tolerated with ultrashort pulses and tuning of a PPLN OPO out to 7.3 μm has been shown at low average power (Watson *et al.*, 2002). At this long wavelength the absorption is high and all the idler output is generated in a thin slice by the output facet of the crystal. By optimizing the crystal length and mode size to maximize the ratio between nonlinear gain and linear loss, up to 1.1 W has been reached at 4.5 μm (Ruebel *et al.*, 2011).

In precision metrology applications it is not enough to have a train of short pulses with high power, but the frequency comb of longitudinal modes must also be stabilized in phase. A PPLN OPO synchronously pumped by an amplified Yb-fibre laser has provided up to 1.5 W idler power and idler tunability from 2.8 to 4.8 μm. Electronic locking of the frequency comb makes this a suitable source for spectroscopy (Adler *et al.*, 2009). A degenerate OPO is an overconstrained system that when the cavity length is stabilized to provide oscillation will be phase locked to the pump laser. In this way a frequency comb with a carrier envelope offset (CEO) which follows that of

© Woodhead Publishing Limited, 2013

the Er-fibre pump laser has been generated with a 20 dB bandwidth from 2500 to 3800 nm (Leindecker *et al.*, 2011). Changing to OP-GaAs as the nonlinear medium and a Cr:ZnSe pump laser, this technology has been moved to a centre wavelength of 4.9 μm (Vodopyanov *et al.*, 2011).

Using pulsed or chopped mode-locked lasers to produce microsecond macropulses comprising pulse trains of picosecond laser pulses, the average powers have been reduced enough to operate 1 μm synchronously pumped OPOs in AGS and CSP to generate longer idler wavelengths, where OPO operation is otherwise inhibited by thermal damage to the nonlinear crystals. The CSP OPO already shows better performance than the AGS OPOs and, with improved crystal growth which eliminates non-intrinsic absorption near the band-gap, the 15 mW average power near 6.4 μm can be scaled to higher powers (Peremans *et al.*, 2009). Because of the high intensity and thus high gain with ultrashort pulses, OPG is also an alternative, and 154 mW average power has been generated at 6.2 μm using a CSP crystal without any cavity feedback (Chalus *et al.*, 2010).

Using a tandem OPO setup where a picosecond mode-locked Nd:YLF laser pumps a PPLN OPO that in turn pumps a CdSe OPO, a tuning range from 9.1 to 9.7 μm was reached. To ensure efficient operation and allow pump tuning of the CdSe OPO, a surface relief grating in a Littrow configuration was used in the PPLN cavity to select a part of the gain spectrum (Watson *et al.*, 2003). Using a similar setup to pump an idler resonant ZGP OPO, up to 800 mW was generated with signal tunability from 3.8 to 4.5 μm (Dherbecourt *et al.*, 2010). Longer-wavelength ultrashort pulses are otherwise normally generated by DFG, which can be reasonably efficient because of the high peak intensities that are available with ultrashort pulses.

Further OPA and optical parametric chirped pulse amplification (OPCPA) stages to create high energy ultrashort pulses are possible in the mid-IR, but are outside the scope of this chapter.

16.7 Sources of further information and advice

There are several only slightly overlapping communities of scientists who are interested in mid-IR laser sources. The different interests include, for example, defence, atmospheric spectroscopy and laser surgery. There are also several barely related technologies for generating laser radiation in the wavelength range of interest, including QCL, direct solid-state and gas lasers in addition to the nonlinear optical down-conversion discussed here.

An effort to bring all of these groups together is the Mid-Infrared Coherent Sources (MICS) conference series. The conferences so far have been in Cargèse, Corsica, France in 1998, St Petersburg, Russia in 2001, Barcelona, Spain in 2005 and Trouville, France in 2009, and a continuation of the

© Woodhead Publishing Limited, 2013

series is planned. In addition to the conference proceedings, the community got together and produced two review books after the conferences in 2001 (Sorokina and Vodopyanov, 2003) and 2005 (Ebrahim-Zadeh and Sorokina, 2008) that give a good overview of the area.

Other recent reviews of mid-IR sources include one focused on CW sources (Ebrahim-Zadeh, 2010) and one on ultrafast OPOs (Reid *et al.*, 2011). Petrov (2012) discusses the use of non-oxide materials for mid-IR nonlinear optics and their use in parametric devices for all temporal ranges. Godard (2007) gives a broader overview of all types of laser sources available in the mid-IR.

Transition metal lasers, particularly Cr:ZnSe in the 2 to 3 μm range and Fe:ZnSe around 4 μm are one alternative to OPOs (Mirov *et al.*, 2010). Another important possibility is the semiconductor quantum cascade lasers (QCL) that have narrowband tuneability from each device and are starting to reach higher powers in an increasing part of the mid-IR spectrum (Capasso, 2010; Rao and Karpf, 2011). When looking for a laser source in mid-IR one should also not forget gas lasers which, even if they have their problems, have been workhorses in many applications for a long time, the CO_2 laser at 10.6 and 9.6 μm being the best known and most used.

16.8 Future trends

The future can only be speculated about, but there are some ongoing trends that are bound to continue. The first is that the pump sources will improve further. The main performance driver at the moment is the large mode area fibre amplifier that allows amplification of CW lasers to higher power levels without detrimental nonlinear effects. There is also ongoing work on pulsed fibre amplifiers that might make fibre lasers the dominating pump source for high average power OPOs in the future. The main use of fibre sources for direct OPO pumping has so far been Yb-fibre sources near 1 μm. In the future Tm and Ho doping for 2 μm pump sources may become important for generation of wavelengths beyond 4 μm.

Another pump laser development that might be of future interest is that Ho lasers directly pumped by 1.9 μm diode lasers are starting to appear (Scholle and Fuhrberg, 2008; Newburgh *et al.*, 2011). This can provide more compact and higher wall plug efficiency Ho lasers than the current practice of pumping with Tm-fibre lasers. The potential for this will rise as 1.9 μm diode stacks become mature.

During recent years there has been progress in nonlinear crystals in mainly two areas, orientation patterned semiconductors, such as OP-GaAs, and new crystal materials that can be pumped at 1 μm and generate idler wavelengths beyond 5 μm, e.g. CSP and LISe. Further refining of the production technology that makes these two types of crystals commercially

© Woodhead Publishing Limited, 2013

available can have a large impact on which OPO architectures are possible for longer wavelengths.

One interesting possibility for the future is tuneable single longitudinal mode CW OP-GaAs OPOs pumped by Tm-fibre lasers. This setup has the possibility of becoming for longer wavelengths what the Yb-fibre pumped PPLN OPO is for wavelengths in the 2 to 4 μm range. This opens great possibilities for spectroscopy at longer wavelengths.

There is a drive towards larger aperture QPM materials for high energy applications. This is the case both for periodically poled materials, such as PPLN, PPSLT and PPRTA, and for OP-GaAs. This should be most interesting for OPA stages as the beam quality suffers too much in non-critically phase-matched high-Fresnel-number oscillators.

In summary there is a good possibility that mid-IR sources based on OPOs can become more compact by the use of fibre-based pump sources. The shorter wavelength range up to 4 μm will also benefit from larger aperture QPM materials and new cavity designs to produce high energy pulses with good beam quality.

By the improvement of emerging nonlinear optical materials the OPO performance available today up to 4 μm can be extended to longer wavelengths with 1 μm pump sources. There is a parallel development of better 2 μm pump sources for pumping the currently available long wavelength nonlinear crystals and it will be a competition on which technique shows the best progress.

16.9 References

Adler F, Cossel K C, Thorpe M J, Hartl I, Fermann M E and Ye J (2009), 'Phase-stabilized, 1.5 W frequency comb at 2.8–4.8 μm', Opt Lett, 34, 1330–1332.

Andrieux E, Zanon T, Cadoret M, Rihan A and Zondy J-J (2011), '500 GHz mode-hop-free idler tuning range with a frequency-stabilized singly resonant optical parametric oscillator', Opt Lett, 36, 1212–1214.

Arisholm G, Lippert E, Rustad G and Stenersen K (2000), 'Effect of resonator length on a doubly resonant optical parametric oscillator pumped by a multilongitudinal-mode beam', Opt Lett, 25, 1654–1656.

Arisholm G, Rustad G and Stenersen K (2001), 'Importance of pump-beam group velocity for backconversion in optical parametric oscillators', J Opt Soc Am, 18, 1882–1890.

Arisholm G, Lippert E, Rustad G and Stenersen K (2002), 'Efficient conversion from 1 to 2 μm by a KTP-based ring optical parametric oscillator', Opt Lett, 27, 1336–1338.

Arisholm G, Nordseth Ö and Rustad G (2004), 'Optical parametric master oscillator and power amplifier for efficient conversion of high-energy pulses with high beam quality', Opt Express, 12, 4189–4197.

Berrou A, Raybaut M, Godard A and Lefebvre M (2010), 'High-resolution photoacoustic and direct absorption spectroscopy of main greenhouse gases by use of a pulsed entangled cavity doubly resonant OPO', Appl Phys B, 98, 217–230.

© Woodhead Publishing Limited, 2013

Bollig C, Hayward R A, Clarkson W A and Hanna D C (1998), '2-W Ho:YAG laser intracavity pumped by a diode-pumped Tm:YAG laser', Opt Lett, 23, 1757–1759.

Bosenberg W R, Drobshoff A, Alexander J I, Myers L E and Byer R L (1996), 'Continuous-wave singly resonant optical parametric oscillator based on periodically poled $LiNbO_3$', Opt Lett, 21, 713–715.

Boyd G D, Buehler E and Storz F G (1971), 'Linear and nonlinear optical properties of $ZnGeP_2$ and CdSe', Appl Phys Lett, 18, 301–304.

Brosnan S J and Byer R L (1979), 'Optical parametric oscillator threshold and linewidth studies', IEEE J Quantum Electron, QE-15, 415–431.

Budni P A, Lemons M L, Mosto J R and Chicklis E P (2000), 'High-power high-brightness diode-pumped 1.9-μm thulium and resonantly pumped 2.1-μm holmium lasers', IEEE J Sel Top Quantum Electron, 6, 629–635.

Capasso F (2010), 'High-performance midinfrared quantum cascade lasers', Opt Eng, 49, 111102.

Chaitanya Kumar S, Das R, Samanta G K and Ebrahim-Zadeh M (2011), 'Optimally-output-coupled, 17.5 W, fiber-laser-pumped continuous-wave optical parametric oscillator', Appl Phys B, 102, 31–35.

Chalus O, Schunemann P G, Zawilski K T, Biegert J and Ebrahim-Zadeh M (2010), 'Optical parametric generation in $CdSiP_2$', Opt Lett, 35, 4142–4144.

Cheung E, Palese S, Injeyan H, Hoefer C, Ho J, Hilyard R, Komine H, Berg, J and Bosenberg W (1999) 'High power conversion to mid-IR using KTP and ZGP OPOs', in *Advanced Solid State Lasers*, M. Fejer, H Injeyan and U Keller, eds., Vol. 26 of OSA Trends in Optics and Photonics, Optical Society of America, 1999, paper WC1. http://www.opticsinfobase.org/abstract.cfm?URI=ASSL-1999-WC1.

Creeden D, Ketteridge P A, Budni P A, Setzler S D, Young Y E, McCarthy J C, Zawilski K, Schunemann P G, Pollak T M, Chicklis E P and Jiang M (2008a), 'Mid-infrared $ZnGeP_2$ parametric oscillator directly pumped by a pulsed 2 μm Tm-doped fiber laser', Opt Lett, 33, 315–317.

Creeden D, Ketteridge P A, Budni P A, Zawilski K, Schunemann P G, Pollak T M and Chicklis E P (2008b), 'Multi-watt mid-IR fiber-pumped OPO', in Conference on Lasers and Electro-Optics/Quantum Electronics and Laser Science Conference and Photonic Applications Systems Technologies, Optical Society of America, 2008, CTuII2.

Debuisschert T, Raffy J, Dupont J-M and Pocholle J-P (2000) 'Nanosecond optical parametric oscillators', C R Acad Sci, 1, 561–583.

Dergachev A, Armstrong D, Smith A, Drake T and Dubois M (2007), '3.4-μm ZGP RISTRA nanosecond optical parametric oscillator pumped by a 2.05-μm Ho:YLF MOPA system', Opt Express, 15, 14404–14413.

Dherbecourt J-B, Godard A, Raybaut M, Melkonian J-M and Lefebvre M (2010), 'Picosecond synchronously pumped $ZnGeP_2$ optical parametric oscillator', Opt Lett, 35, 2197–2199.

Drag C, Desormeaux A, Lefebvre M and Rosencher E (2002), 'Entangled-cavity optical parametric oscillator for mid-infrared pulsed single-longitudinal-mode operation', Opt Lett, 27, 1238–1240.

Ebrahim-Zadeh M (2010), 'Continuous-wave optical parametric oscillators', Chapter 17 in *OSA Handbook of Optics, Vol. IV, Optical Properties of Materials, Nonlinear Optics, Quantum Optics*, New York, McGraw-Hill.

Ebrahim-Zadeh M and Sorokina I T (2008), *Mid-infrared Coherent Sources and Applications*, Dordrecht, Springer.

© Woodhead Publishing Limited, 2013

Eckardt R C, Nabors C D, Kozlovsky W J and Byer R L (1991), 'Optical parametric oscillator frequency tuning and control', J Opt Soc Am B, 8, 646–667.

Eichhorn M (2010), 'Pulsed 2 μm fiber lasers for direct and pumping applications in defence and security', Proc SPIE, 7836, 78360B.

Eyres L A, Tourreau P J, Pinguet T J, Ebert C B, Harris J S, Fejer M M, Becouarn L, Gerard B and Lallier E (2001), 'All-epitaxial fabrication of thick, orientation-patterned GaAs films for nonlinear optical frequency conversion', Appl Phys Lett, 79, 904–906.

Farsund Ø, Arisholm G and Rustad G (2010), 'Improved beam quality from a high energy optical parametric oscillator using crystals with orthogonal critical planes', Opt Express, 18, 9229–9235.

Giordmaine J A and Miller R C (1965), 'Tunable coherent parametric oscillation in $LiNbO_3$ at optical frequencies', Phys Rev Lett, 14, 973–976.

Godard A (2007), 'Infrared (2–12 μm) solid-state laser sources: a review'. C R Phys, 8, 1100–1128.

Godard A, Raybaut M, Schmid T, Lefebvre M, Michel A-M and Péalat M (2010), 'Management of thermal effects in high-repetition-rate pulsed optical parametric oscillators', Opt Lett, 35, 3667–3669.

Gonzalez L P, Upchurch D C, Barnes J O, Schunemann P G, Mohnkern L and Guha S (2011), 'Frequency doubling of a CO_2 laser using orientation patterned GaAs', Proc SPIE, 7917, 79171H.

Haakestad M W, Arisholm G, Lippert E, Nicolas S, Rustad G and Stenersen K (2008), 'High-pulse-energy mid-infrared laser source based on optical parametric amplification in $ZnGeP_2$', Opt Express, 16, 14263–14273.

Hardy B, Berrou A, Guilbaud S, Raybaut M, Godard A and Lefebvre M (2011), 'Compact, single-frequency, doubly resonant optical parametric oscillator pumped in an achromatic phase-adapted double-pass geometry', Opt Lett, 36, 678–680.

Harris S E (1966), 'Threshold of multimode parametric oscillators', IEEE J Quantum Electron, 2, 701.

Harris S E (1969), 'Tunable optical parametric oscillators', Proc IEEE, 57, 2096–2113.

Henderson A and Stafford R (2006), 'Low threshold, singly-resonant CW OPO pumped by an all-fiber pump source', Opt Express, 14, 767–772.

Henriksson M, Tiihonen M, Pasiskevicius V and Laurell F (2007), 'Mid-infrared ZGP OPO pumped by near degenerate narrowband type-I PPKTP parametric oscillator', Appl Phys B, 88, 37–41.

Henriksson M, Sjoqvist L, Strömqvist G, Pasiskevicius V and Laurell F (2008), 'Tandem PPKTP and ZGP OPO for mid-infrared generation', Proc SPIE, 7115, 71150O.

Henriksson M, Sjöqvist L, Pasiskevicius V and Laurell F (2009), 'Spectrum of multi-longitudinal mode pumped near-degenerate OPOs with volume Bragg grating output couplers', Opt Express, 17, 17582–17589.

Henriksson M, Sjöqvist L, Pasiskevicius V and Laurell F (2010), 'Cavity length resonances in a nanosecond singly resonant optical parametric oscillator', Opt Express, 18, 10742–10749.

Kieleck C, Eichhorn M, Hirth A, Faye D and Lallier E (2009), 'High-efficiency 20–50 kHz mid-infrared orientation-patterned GaAs optical parametric oscillator pumped by a 2 μm holmium laser', Opt Lett, 34, 262–264.

Kieleck C, Hildenbrand A, Eichhorn M, Faye D, Lallier E, Gérard B and Jackson S D (2010), 'OP-GaAs OPO pumped by 2 μm Q-switched lasers: Tm;Ho:silica fiber laser and Ho:YAG laser', Proc SPIE, 7836, 783607.

© Woodhead Publishing Limited, 2013

Klein M E, Laue C K, Lee D H, Boller K J and Wallenstein R (2000), 'Diode-pumped singly resonant continuous-wave optical parametric oscillator with wide continuous tuning of the near-infrared idler wave', Opt Lett, 25, 490–492.

Kokabee O, Esteban-Martin A and Ebrahim-Zadeh M (2010), 'Efficient, high-power, ytterbium-fiber-laser-pumped picosecond optical parametric oscillator', Opt Lett, 35, 3210–3212.

Lee H R, Yu J, Barnes N P and Bai Y (2006), 'An injection-seeded narrow linewidth singly resonant ZGP OPO', in *Advanced Solid-State Photonics*, Technical Digest, Optical Society of America, 2006, paper MC1.

Leindecker N, Marandi A, Byer R L and Vodopyanov K L (2011), 'Broadband degenerate OPO for mid-infrared frequency comb generation', Opt Express, 19, 6296–6302.

Lindsay I, Adhimoolam B, Groß P, Klein M and Boller K (2005), '110 GHz rapid, continuous tuning from an optical parametric oscillator pumped by a fiber-amplified DBR diode laser', Opt Express, 13, 1234-1239.

Lippert E, Arisholm G, Rustad G and Stenersen K (2003), 'Fiber laser pumped mid-IR source', in *Advanced Solid State Photonics*, J J Zayhowski, ed., Vol. 83 of OSA Trends in Optics and Photonics Series, Optical Society of America, 2003, 292–297.

Lippert E, Nicolas S, Arisholm G, Stenersen K and Rustad G (2006), 'Midinfrared laser source with high power and beam quality', Appl Opt, 45, 3839–3845.

Lippert E, Rustad G, Arisholm G and Stenersen K (2008), 'High power and efficient long wave IR $ZnGeP_2$ parametric oscillator', Opt Express, 16, 13878–13884.

Lippert E, Fonnum H, Arisholm G and Stenersen K (2010), 'A 22-watt mid-infrared optical parametric oscillator with V-shaped 3-mirror ring resonator', Opt Express, 18, 26475–26483.

Marchev G, Tyazhev A, Vedenyapin V, Kolker D, Yelisseyev A, Lobanov S, Isaenko L, Zondy J-J and Petrov V (2009), 'Nd:YAG pumped nanosecond optical parametric oscillator based on $LiInSe_2$ with tunability extending from 4.7 to 8.7 μm', Opt Express, 17, 13441–13446.

Mason P D and Wood N J (2004), 'A high-repetition-rate PPLN mid-infrared optical parametric oscillator source', Proc SPIE, 5620, 308–317.

Mirov S, Fedorov V, Moskalev I, Martyshkin D and Kim C (2010), 'Progress in Cr^{2+} and Fe^{2+} doped mid-IR laser materials', Laser & Photonics Reviews, 4, 21–41.

Missey M, Dominic V, Powers P and Schepler K L (2000), 'Aperture scaling effects with monolithic periodically poled lithium niobate optical parametric oscillators and generators', Opt Lett, 25, 248–250.

Newburgh G A, Word-Daniels A, Michael A, Merkle L D, Ikesue A and Dubinskii M (2011), 'Resonantly diode-pumped Ho^{3+}:Y_2O_3 ceramic 2.1 μm laser', Opt Express, 19, 3604–3611.

O'Connor M V, Watson M A, Shepherd D P, Hanna D C, Price J H V, Malinowski A, Nilsson J, Broderick N G R, Richardson D J and Lefort L (2002), 'Synchronously pumped optical parametric oscillator driven by a femtosecond mode-locked fiber laser', Opt Lett, 27, 1052–1054.

Peng Y, Wang W, Wei X and Li D (2009), 'High-efficiency mid-infrared optical parametric oscillator based on PPMgO:CLN', Opt Lett, 34, 2897–2899.

Peremans A, Lis D, Cecchet F, Schunemann P G, Zawilski K T and Petrov V (2009), 'Noncritical singly resonant synchronously pumped OPO for generation of picosecond pulses in the mid-infrared near 6.4 μm', Opt Lett, 34, 3053–3055.

Perrett B J, Terry J A C, Mason P D and Orchard D A (2004), 'Spectral line narrowing in PPLN OPO devices for 1-μm wavelength doubling', Proc SPIE, 5620, 275–283.

© Woodhead Publishing Limited, 2013

Petrov V (2012), 'Parametric down-conversion devices: The coverage of the mid-infrared spectral range by solid-state laser sources', Opt Mater, 34, 536–554.

Petrov V, Schunemann P G, Zawilski K T and Pollak T M (2009), 'Noncritical singly resonant optical parametric oscillator operation near 6.2 μm based on a $CdSiP_2$ crystal pumped at 1064 nm', Opt Lett, 34, 2399–2401.

Petrov V, Marchev G, Schunemann P G, Tyazhev A, Zawilski K T and Pollak T M (2010), 'Subnanosecond, 1 kHz, temperature-tuned, noncritical mid-infrared optical parametric oscillator based on $CdSiP_2$ crystal pumped at 1064 nm', Opt Lett, 35, 1230–1232.

Phua P B, Lai K S, Wu R F and Chong T C (1998), 'Coupled tandem optical parametric oscillator (OPO): an OPO within an OPO', Opt Lett, 23, 1262–1264.

Rao G N and Karpf A (2011), 'External cavity tunable quantum cascade lasers and their applications to trace gas monitoring', Appl Opt, 50, A100–A115.

Raybaut M, Schmid T, Godard A, Mohamed A K, Lefebvre M, Marnas F, Flamant P, Bohman A, Geiser P and Kaspersen P (2009), 'High-energy single-longitudinal mode nearly diffraction-limited optical parametric source with 3 MHz frequency stability for CO_2 DIAL', Opt Lett, 34, 2069–2071.

Reid D T, Sun J, Lamour T P and Ferreiro T I (2011), 'Advances in ultrafast optical parametric oscillators', Laser Phys Lett, 8, 8–15.

Ruebel F, Anstett G and L'huillier J A (2011), 'Synchronously pumped mid-infrared optical parametric oscillator with an output power exceeding 1 W at 4.5 μm', Appl Phys B, 102, 751–755.

Saikawa J, Fujii M, Ishizuki H and Taira T (2006), '52 mJ narrow-bandwidth degenerated optical parametric system with a large-aperture periodically poled $MgO{:}LiNbO_3$ device', Opt Lett, 31, 3149–3151.

Saikawa J, Fujii M, Ishizuki H and Taira T (2007), 'High-energy, narrow-bandwidth periodically poled Mg-doped $LiNbO_3$ optical parametric oscillator with a volume Bragg grating', Opt Lett, 32, 2996–2998.

Schellhorn M, Hirth A and Kieleck C (2003), 'Ho:YAG laser intracavity pumped by a diode-pumped Tm:YLF laser', Opt Lett, 28, 1933–1935.

Scholle K and Fuhrberg P (2008), 'In-band pumping of high-power Ho:YAG lasers by laser diodes at 1.9 μm', in Conference on Lasers and Electro-Optics/Quantum Electronics and Laser Science Conference and Photonic Applications Systems Technologies, OSA Technical Digest (CD), Optical Society of America, 2008, CTuAA1.

Schunemann P G and Setzler S D (2011), 'Future directions in quasi-phasematched semiconductors for mid-infrared lasers', Proc SPIE, 7917, 79171F.

Smith A V and Armstrong D J (2002), 'Nanosecond optical parametric oscillator with 90° image rotation: design and performance', J Opt Soc Am B, 19, 1801–1814.

Smith R G (1973), 'A study of factors affecting the performance of a continuously pumped doubly resonant optical parametric oscillator', IEEE J Quantum Electron, QE-9, 530–541.

Sorokina I and Vodopyanov K L (2003), *Solid-State Mid-Infrared Laser Sources*, Heidelberg, Springer.

Südmeyer T, Innerhofer E, Brunner F, Paschotta R, Usami T, Ito H, Kurimura S, Kitamura K, Hanna D C and Keller U (2004), 'High-power femtosecond fiber-feedback optical parametric oscillator based on periodically poled stoichiometric $LiTaO_3$', Opt Lett, 29, 1111–1113.

Sutherland R L (2003), *Handbook of Nonlinear Optics*, London, Marcel Dekker.

Svelto O, Longhi S, Della Valle G, Kück S, Huber G, Pollnau M, Hillmer H, Hansmann

© Woodhead Publishing Limited, 2013

S, Engelbrecht R, Brand H, Kaiser J, Peterson A B, Malz R, Steinberg S, Marowsky G, Brinkmann U, Lot D, Borsutzky A, Wächter H, Sigrist M W, Saldin E, Schneidmiller E, Yurkov M, Midorikawa K, Hein J, Sauerbrey R and Helmcke J (2007), 'Lasers and coherent light sources' in F Träger (ed.), *Springer Handbook of Lasers and Optics*, New York, Springer.

Vainio M, Peltola J, Persijn S, Harren F J M and Halonen L (2008), 'Singly resonant cw OPO with simple wavelength tuning', Opt Express, 16, 11141–11146.

Vainio M, Siltanen M, Peltola J and Halonen L (2009), 'Continuous-wave optical parametric oscillator tuned by a diffraction grating', Opt. Express, 17, 7702–7707.

Vainio M, Siltanen M, Hieta T and Halonen L (2010), 'Continuous-wave optical parametric oscillator based on a Bragg grating', Opt Lett, 35, 1527–1529.

Vainio M, Siltanen M, Peltola J and Halonen L (2011), 'Grating-cavity continuous-wave optical parametric oscillators for high-resolution mid-infrared spectroscopy', Appl Opt, 50, A1–A10.

van Herpen M M J W, Bisson S E and Harren F J M (2003), 'Continuous-wave operation of a single-frequency optical parametric oscillator at 4–5 μm based on periodically poled $LiNbO_3$', Opt Lett, 28, 2497–2499.

Vodopyanov K L, Maffetone J P, Zwieback I and Rudermann W (1999), '$AgGaS_2$ optical parametric oscillator continuously tunable from 3.9 to 11.3 μm', Appl Phys Lett, 75, 1204–1206.

Vodopyanov K L, Levi O, Kuo P S, Pinguet T J, Harris J S, Fejer M M, Gerard B, Becouarn L and Lallier E (2004), 'Optical parametric oscillation in quasi-phase-matched GaAs', Opt Lett, 29, 1912–1914.

Vodopyanov K L, Sorokin E, Sorokina I and Schunemann P G (2011), 'Mid-IR frequency comb source spanning 4.4–5.4 μm based on subharmonic GaAs optical parametric oscillator', Opt Lett, 36, 2275–2277.

Watson M A, O'Connor M V, Lloyd P S, Shepherd D P, Hanna D C, Gawith C B E, Ming L, Smith P G R and Balachninaite O (2002), 'Extended operation of synchronously pumped optical parametric oscillators to longer idler wavelengths', Opt Lett, 27, 2106–2108.

Watson M A, O'Connor M V, Shepherd D P and Hanna D C (2003), 'Synchronously pumped CdSe optical parametric oscillator in the 9–10 μm region', Opt Lett, 28, 1957–1959.

Webjorn J, Pruneri V, Russell P S J, Barr J R M and Hanna D C (1994), 'Quasiphase-matched blue light generation in bulk lithium niobate, electrically poled via periodic liquid electrodes', Electron Lett 30, 894–895.

Yang S T, Eckardt R C and Byer R L (1993), 'Continuous-wave singly resonant optical parametric oscillator pumped by a single-frequency resonantly doubled Nd:YAG laser', *Optics Letters*, 18(12), 971–973.

Zawilski K T, Schunemann P G, Setzler S D and Pollak T M (2008), 'Large aperture single crystal $ZnGeP_2$ for high-energy applications', J Crystal Growth, 310, 1891–1896.

Zhong K, Yao J Q, Xu D G, Wang J L, Li J S and Wang P (2010), 'High-pulse-energy high-efficiency mid-infrared generation based on KTA optical parametric oscillator', Appl Phys B, 100, 749–753.

Zondy J-J, Vedenyapin V, Yelisseyev A, Lobanov S, Isaenko L and Petrov V (2005), '$LiInSe_2$ nanosecond optical parametric oscillator', Opt Lett, 30, 2460–2462.

© Woodhead Publishing Limited, 2013

17
Raman lasers

H. M. PASK and J. A. PIPER, Macquarie University, Australia

DOI: 10.1533/9780857097507.2.493

Abstract: Solid-state crystalline Raman lasers continue to evolve as practical and efficient laser sources, efficiently generating output with a diverse range of temporal formats and at wavelengths from the UV to the infrared. In this chapter, we provide the background information required to design efficient crystalline lasers, give an overview of the field, and highlight exciting new developments.

Key words: solid-state laser, Raman laser, stimulated Raman scattering, visible laser, infrared laser, frequency conversion.

17.1 Introduction

Stimulated Raman scattering (SRS) of intense light in crystalline materials has been known since shortly after the invention of the laser, but it is only in the past decade that devices using this phenomenon for practical, efficient generation of infrared, visible and ultraviolet light have emerged in a substantial way.

The potential for efficient wavelength conversion of the 1064 nm fundamental of pulsed (Q-switched) Nd:YAG lasers further out to the infrared using intracavity SRS, and to new regions of the visible using intracavity sum-frequency or second-harmonic generation (SFG/SHG) of the Stokes emission, was first demonstrated in the late 1970s by Ammann (Ammann, 1979; Ammann and Decker, 1977). However, it was not until 15 years later that practical devices using SRS in crystal materials to make efficient, compact pulsed infrared lasers emitting in the eye-safe region (~1.5 μm) were reported (Murray *et al.*, 1995; Zverev *et al.*, 1993), and another five years before efficient all-solid-state pulsed visible sources based on intracavity SHG of Stokes lines generated by SRS in crystals were demonstrated (Pask and Piper, 1999). Subsequently a succession of studies of all-solid-state pulsed Raman lasers has culminated in demonstrated average powers as high as 24.5 W (Feve *et al.*, 2011) and optical conversion efficiencies (diode pump-to-Stokes emission) approaching 24% (Chen *et al.*, 2008), in the near-infrared. Intracavity SHG of Stokes lines in pulsed crystalline Raman lasers has generated visible (yellow) average powers approaching 10 W with diode-yellow optical conversion efficiencies as high as ~18% (Cong *et al.*,

© Woodhead Publishing Limited, 2013

2009, 2010). After 40 years of study of SRS in crystalline Raman materials, the potential for development of new sources appears to have been been realised in the form of efficient *pulsed* Raman lasers.

But the first demonstrations in 2005 by Pask (2005) and Demidovich *et al.* (2005) of efficient, continuous-wave (CW) crystalline intracavity Raman lasers have opened an entirely new opportunity for exploitation of SRS to develop new sources. Thus CW crystalline Raman lasers with multiwatt output powers in the infrared and in the visible (by way of intracavity SFG/SHG) have been demonstrated with optical conversion efficiencies over 20% (Lee *et al.*, 2010c) . The unique cascading nature of SRS has also been exploited to achieve wavelength-selectable output spanning the green-to-orange at multiwatt powers (Lee *et al.*, 2010d). Crystalline Raman laser sources giving multiwatt CW yellow output are now in commercial production for application in ophthalmic surgery; and miniature crystalline Raman lasers generating CW output powers of 100 mW in the visible for diode-pump powers of only 1 W offer promise for widespread use in biomedical diagnostics.

Results of empirical research and associated theoretical modelling undertaken over the past five years have also established fairly clearly the criteria for selection of the best crystalline Raman materials for specific configurations and power regimes, with the standard vanadates ($GdVO_4$, YVO_4) in particular emerging as preferred materials in many situations. The recent availability of synthetic diamond of high quality and appropriate dimensions has added an exciting new material to the list, with highly efficient pulsed and CW diamond Raman lasers operating on Stokes lines in the near-infrared already demonstrated (Feve *et al.*, 2011; Lubeigt *et al.*, 2011; Sabella *et al.*, 2010).

This chapter aims to give an overview of the key principles of operation for crystalline Raman lasers including basic theory of SRS, properties of crystalline Raman materials and introduction to the different Raman laser configurations. We focus on the most recent advances; excellent reviews of SRS in crystalline materials and developments of pulsed-mode crystalline Raman lasers over the first 40 years have been given by Basiev and Powell (2003), Černý *et al.* (2004) and Pask (2003). A more recent review by the current authors (Piper and Pask, 2007) focused on developments of crystalline Raman lasers post-2000, and most recently Pask *et al.* (2008a) have reviewed advances in visible and UV sources based on crystalline Raman lasers. Despite the comparatively short interval since this last review, we will see that developments of crystalline Raman lasers, especially CW visible lasers and diamond-Raman lasers, have already surpassed the expectations we had at that time.

© Woodhead Publishing Limited, 2013

17.2 Raman lasers

17.2.1 Spontaneous Raman effect

The Raman effect involves inelastic scattering of light in an optical medium, simultaneously depositing energy into an excited state of the medium, so that the scattered light has photon energy reduced by the energy deposited in the medium. Accordingly the scattered light has frequency

$$\omega_{S1} = \omega_P - \omega_R$$

where ω_P is the frequency of the incident (pump) light and $\hbar\omega_R$ matches the energy of an electronic, vibrational or rotational energy state of the Raman medium. In the case of crystalline Raman media, $\hbar\omega_R$ matches a vibrational mode of the crystal. Consistent with the above identity, the scattered light has lower frequency and thus longer wavelength than the incident light, and is referred to as the first Stokes emission. In some situations the first Stokes emission may itself undergo spontaneous Raman scattering to generate the second Stokes frequency:

$$\omega_{S2} = \omega_{S1} - \omega_{R.}$$

17.2.2 Stimulated Raman scattering

The phenomenon of SRS comes from the third-order nonlinear polarisability $P_3 = \varepsilon_0\chi_3E^3$ of the optical medium, where ε_0 is the dielectric constant, χ_3 the third-order nonlinear susceptibility, and E the amplitude of the incident optical field. Specifically SRS relates to the imaginary part of χ_3 which is proportional to the square of the normal-mode derivative of the molecular polarisability tensor $\delta\alpha/\delta q$. In accordance with the theory of Boyd (1992) or Penzkofer *et al.* (1979), the integrated Raman scattering cross-section is defined as

$$\frac{\partial\sigma}{\partial\Omega} = \frac{\omega_S^4\mu_S}{c^4\mu_P}\frac{\hbar}{2m\omega_R}\left(\frac{\partial\alpha}{\partial q}\right)^2$$

where μ_s and μ_P are the refractive indices at the Stokes and pump wavelengths, and m is the reduced mass of the molecular oscillator. Γ is the linewidth of the Raman transition, equal to the inverse of the dephasing time T_R for the final state of the transition ($\Gamma = T_R^{-1}$). When T_R is short compared to the duration of the light pulse incident on the Raman medium, the *steady-state* regime applies, and the growth of Stokes intensity $I_S(z)$ along the incident light direction, z, is

$$I_s(z) = I_s(0)\exp(g_R I_P z)$$

where the steady-state Raman gain coefficient (in units of cm/GW) is

© Woodhead Publishing Limited, 2013

$$g_R = \frac{8\pi c^2 N}{\hbar \mu_S^2 \omega_S^3 \Gamma} \left(\frac{\partial \alpha}{\partial \Omega} \right)$$

where N is the number of Raman-active molecules.

Thus the steady-state Raman gain coefficient g_R is higher for shorter Stokes wavelength (larger ω_S), higher Raman scattering cross-section (larger $\delta\alpha/\delta q$) and smaller Raman linewidth (smaller Γ). The Raman scattering cross-section and the polarisation of the Stokes-scattered light depend on the orientation of the crystal with respect to the polarisation of the incident (pump) light. Generation of Stokes light with polarisation orthogonal to that of the pump is effectively precluded by assumptions in the theory, and while this is not strictly true, in practice the strongest Stokes scattering (and therefore gain) is obtained for Stokes polarisation parallel to the pump polarisation.

In the absence of an injected Stokes signal, SRS grows from the spontaneous Stokes noise $I_S(0)$ scattered into solid angle $\Delta\Omega$, given by

$$I_S(0) = \frac{\hbar \omega_S^3 \mu_S^3}{(2\pi)^2 c^2} \Delta\Omega$$

The value of $I_S(0)$ is ~10^{-15}, so that the exponent $g_R I_P z$ governing growth of the SRS field generally needs to have a value of at least 30 for the SRS 'threshold' (usually defined as 1% depletion of the incident light field) to be exceeded. For a crystal with steady-state gain coefficient g_R ~ 10 cm/GW and length 30 mm, the pump intensity I_P necessary to reach the threshold is ~1 GW/cm^2.

The dephasing time T_R for most crystal oscillation modes of interest here is usually ~10 ps, thus the *steady-state* SRS regime applies when the incident light pulse duration is longer than ~1 ns, which is true for most Q-switched solid-state lasers and of course for CW laser sources. For pump light sources with pulse duration τ_P in the picosecond regime, that is where the pulse duration is small compared to the dephasing time T_R, the *transient* Raman regime applies, wherein the Stokes signal grows as

$$I_S(z) = I_S(0) \exp\left(-\frac{\tau_P}{T_R} \right) \exp\left[2\left(\frac{\tau_P g_R I_P z}{T_R} \right)^{1/2} \right]$$

Since g_R depends on Γ^{-1} and $\Gamma^{-1} T_R = 1$, it follows that Stokes growth in the transient regime is independent of Raman linewidth, and the exponent depends on $z^{1/2}$ rather than z as for the steady-state case.

17.3 Solid-state Raman materials

Raman activity has been a key focus of study in analysing the properties of crystalline and glassy optical materials over nearly five decades, since the

© Woodhead Publishing Limited, 2013

first observations of SRS in crystal materials (Eckhardt *et al.*, 1963) only three years after the first demonstration of the ruby laser. Significantly, although diamond was one of the first crystalline Raman materials reported, it is only in the past three years that diamond has emerged as a Raman material with genuine potential for practical application in Raman wavelength conversion devices, owing to the recent availability of optical-quality single-crystal synthetic diamond grown by CVD techniques. In a similar way, the development of solid-state devices using SRS has tracked the availability of high-quality Raman-active materials, with practical devices emerging only over the past 20 years. Foremost in exploration of new Raman-active crystals have been the Basiev and Kaminskii groups from Russia, the former working closely with the Powell group in the US. Summaries of the extensive work undertaken on Raman-active crystals are given in the reviews by Kaminskii (1996) and Basiev and Powell (2003).

Although 100 or more solid-state optical materials have been evaluated for Raman activity, only a few have the potential to be used in practical devices. The primary requirement appears at first to be high Raman gain and a useful Raman Stokes shift, but the ability to make samples of high optical quality, sufficient size (length) and adequate resistance to optical damage is of critical practical importance. SRS results in deposition of heat energy in the Raman medium, that is, Raman heating, thus the thermal properties of Raman crystals are also of critical importance for practical devices. This is especially true in the age of all-solid-state (that is, diode-pumped) lasers, where the optical power densities in laser and Raman crystals can be very high (for example, circulating CW powers of kilowatts in optical mode sizes of 100 microns or less). It follows that high thermal conductivity and low thermo-optic coefficient are also key properties in selection of practical Raman materials.

Table 17.1 gives summary data for a number of solid-state optical materials of practical importance at the present time. As noted above, diamond was one of the first crystalline materials for which SRS was observed. It

Table 17.1 Summary data for selected solid-state Raman materials

Material	Raman shift (cm^{-1})	Raman gain at 1.06 μm (cm/GW)	Thermal conductivity ($W\ m^{-1}\ K^{-1}$)	Thermo-optic coefficient ($10^{-6}\ K^{-1}$)
Diamond	1332	13.5	2000	9.6
$Ba(NO_3)_2$	1047	11	1.17	−20
Fused silica	440	3.2×10^{-3}	1.38	9.2
$KGd(WO_4)_2$	768, 901	4.4, 3.5	2.6, 3.4	−0.8, −5.5
$BaWO_4$	926	8.5	3	–
$GdVO_4$	885	4.5	5	3
KTA	234	2–2.5	1.8	11–16

© Woodhead Publishing Limited, 2013

has a large Raman shift, and the highest steady-state Raman gain of all known crystal materials, resulting from high integrated Raman scattering cross-section and narrow Raman linewidth (~2 cm^{-1}). Equally important, diamond has exceptionally high thermal conductivity, which is a significant advantage in situations where Raman heating is substantial (for example at high average optical powers, or at mid-infrared wavelengths where $\hbar\omega_R$ becomes significant in comparison with $\hbar\omega_P$). However, while synthetic single-crystal diamond of adequate sample size ~10 mm (long) is now available, there remain a number of practical problems with this material: suppliers are few, samples are expensive, the quality is variable, and the material is susceptible to optical damage at the surface, either uncoated or coated.

Early studies of SRS in crystalline barium nitrate $Ba(NO_3)_2$ (Eremenko *et al.*, 1980) demonstrated that this material also has very high steady-state Raman gain, due largely to very narrow Raman linewidth (0.4 cm^{-1}). There have been many reports of the use of $Ba(NO_3)_2$ in pulsed Raman generators and Raman lasers, and indeed in the first-reported CW crystalline Raman laser (Grabtchikov *et al.*, 2004); however, $Ba(NO_3)_2$ is hygroscopic, has a low optical damage threshold (~0.4 GW/cm^2) and has poor thermal conductivity, so suffers serious limitations as a practical material. Likewise lithium iodate, LiO_3, was used to demonstrate the first intracavity crystalline Raman laser based on a flashlamp-pumped Nd^{3+} laser crystal (Ammann and Falk, 1975) but has an even lower optical damage threshold (~0.1 GW/cm^2).

It is something of a paradox that the Raman material that has been most used for practical application is fused silica. This has much lower Raman gain (~10^{-3} cm/GW) than the competing crystalline materials (due to the very broad Raman linewidth of fused silica) but can be produced with extremely high optical quality in very long samples, namely in the form of silica optical fibres. Silica fibre Raman amplifiers are widely used in optical communications; fibre Raman lasers and amplifiers are also capable of producing very high CW powers (tens of watts, see below), wavelength-tunable over a broad range (Chamorovskiy *et al.*, 2011; Agrawal, 1995; Rini *et al.*, 2000).

The emergence of tungstate crystals for SRS applications followed the first reports by Andryunas *et al.* (1985). The 'double' tungstate crystals KGW and KYW have quite high integrated cross-sections (half that of diamond) but comparatively broad Raman linewidths (6–8 cm^{-1}) resulting in moderate Raman gains (4–5 cm/GW). They do, however, have very high optical damage thresholds (>10 GW/cm^2) which means they can tolerate very high pump power densities, and they are available with good optical quality in samples 50 mm long and more (Mochalov, 1997; Wang *et al.*, 2006). KGW/KYW can also be doped comparatively easily and, with Nd^{3+} doping, were the first crystal materials (Andryunas *et al.*, 1985) used to demonstrate

© Woodhead Publishing Limited, 2013

self-Raman action (that is to simultaneously generate the fundamental laser wavelength and the first Stokes Raman wavelength). KGW and KYW also have the useful property of having two distinct Raman shifts with similar steady-state gains for incident polarisation of the incident light aligned along different (mutually perpendicular) crystal axes. Thus KGW has Raman shifts of 768 cm^{-1} and 901 cm^{-1} which can be accessed by rotating the crystal by 90° with respect to the incident light polarisation, allowing different Stokes wavelengths to be accessed for the one crystal. However, KGW and KYW also have the disadvantage of rather poor thermal conductivities, and worse, different values for different crystal axes (Mochalov, 1997). These materials are therefore susceptible to quite strong thermal lensing resulting from Raman heating at high average pump powers, and strongly asymmetric thermal lensing at that, a property that is hard to deal with in simple optical resonators.

Calcium tungstate, $CaWO_4$, had previously been used by Murray *et al.* (1998) in early demonstrations of all-solid-state Raman lasers operating in the eyesafe infrared around 1.5 μm but it is susceptible to optical and mechanical damage. Barium tungstate ($BaWO_4$), lead tungstate ($PbWO_4$) and strontium tungstate ($SrWO_4$) have subsequently all been used effectively for Raman lasers, though $BaWO_4$ is arguably the best of these, with double the steady-state Raman gain of KGW or KYW and a similar optical damage threshold, though with similarly disadvantageous thermal properties (Basiev *et al.*, 2000; Wang *et al.* 2006). Note also that a similar range of crystalline molybdates possess strong Raman gains but relatively poor thermal conductivities.

Although the crystal vanadates $GdVO_4$ and YVO_4 had been widely used as laser materials for some years, it was not until the reports of Kaminskii *et al.* (2001) that their potential as Raman-active materials was realised. Subsequently Nd:$GdVO_4$ and Nd:YVO_4 have been used as self-Raman materials, especially for the CW regime where using a single self-Raman laser crystal has the considerable advantage of reducing surface losses over configurations where separate laser and intracavity Raman crystals are used. The steady-state Raman gains for these vanadate crystals are similar to those of KGW and KYW but their thermal conductivities are superior and their thermo-optic coefficients isotropic. The vanadates are also widely available with high optical quality and at low cost. In order to minimise the effects of waste-pump and Raman heating, and to increase Raman gain, it is common to employ Nd:$GdVO_4$ and Nd:YVO_4 crystals with lower Nd^{3+} doping (0.3 at% vs 1.0 at%) and greater length (10 mm vs 1–3 mm) than is standard for lasers designed to maximise fundamental (1064 nm) and second-harmonic (532 nm) outputs.

The well-known nonlinear (χ_2) crystals KTA and KTP also have quite good Raman gain, so that simultaneous SFG or SHG and SRS may be achieved in the one crystal, so minimising surface losses in intracavity configurations.

© Woodhead Publishing Limited, 2013

The Raman shifts for these are, however, much smaller than for the tungstates and vanadates (Chen, 2005; Liu *et al.*, 2009a).

Finally we note that crystalline silicon has high steady-state Raman (~7.5 cm/GW) and excellent thermal conductivity, offering some potential for use as a practical Raman laser material for the mid-infrared. Indeed silicon was used in the first demonstration of a semiconductor silicon Raman laser operating on the 1686 nm first Stokes of the 1550 nm fundamental gain (Rong *et al.*, 2005, 2006). Despite some excitement following the first announcements of these results, practical devices have yet to emerge.

17.4 Raman generators, amplifiers and lasers

17.4.1 Raman generators and amplifiers

The simplest devices by which SRS is used for wavelength (frequency) conversion are *Raman generators,* involving single-pass conversion of high-peak-power incident light with pulse durations usually in the 10–100 ps (but sometimes also in the nanosecond) ranges, where the Stokes field grows from spontaneous Raman noise, or *Raman amplifiers*, involving the same geometry and pump pulse characteristics but where the Stokes field is grown from an injected optical signal at the Stokes wavelength. As noted above, reaching Raman threshold for crystalline materials typically requires incident pump-light pump intensities of ~1 GW/cm^2, where optical damage to the crystal and the onset of competing nonlinear processes (especially self-focusing) need to be taken into consideration.

Summaries of results for wavelength conversion of high-power picosecond and nanosecond pulses in crystalline Raman generators are given in preceding reviews (Basiev and Powell, 2003; Pask, 2003). Attention is drawn to the more recent comprehensive study reported by Černý *et al.* (2004) involving detailed measurements of Raman thresholds, gains and conversion efficiencies for a range of tungstate crystals with high-power picosecond and nanosecond pulse pumping at 1064, 532 and 355 nm. The highest SRS conversion efficiency yet reported for a Raman generator is a peak-power efficiency of 85% (effectively at the quantum limit) for SRS of 35 ps, 532 nm pump pulses to the first Stokes at 560 nm in a $BaWO_4$ crystal configured for a double pass of the pump light (Černý and Jelínková, 2002).

In the CW regime the most developed form of Raman amplifier is the fibre Raman amplifier, where the Raman medium is a silica glass fibre. Usually pumped in the near-infrared, fibre Raman amplifiers may be kilometres long, offering high gain even though the incident pump intensity is comparatively low (~50 MW/cm^2). High-power, narrowband CW output can be obtained by injection seeding of a Raman fibre amplifier. For example, Taylor *et al.* (2010) report three narrowband injection-seeded silica Raman fibre amplifiers

© Woodhead Publishing Limited, 2013

(30 m length) operating at 1178 nm, each with tens of watts output which, when coherently combined and frequency-doubled in an external cavity, generate 50 W on the sodium D line at 589 nm for Laser Guide Star.

17.4.2 External-resonator Raman lasers

When a Raman amplifier of length l is positioned within a simple optical resonator (two mirrors with reflectivities R_1 and R_2 at the Stokes wavelength), the Raman laser threshold is reached when the round-trip losses at the Stokes wavelength are exceeded by the round-trip Raman gain, that is when

$$R_1 R_2 \exp(2 g_R I_P l) > 1$$

For a fairly standard Raman crystal of length 50 mm and g_R ~5 cm/GW, the incident light intensity I_P required to reach the Raman laser threshold is easily reduced to 10 MW/cm^2 or less for a high-Q cavity. Raman lasers for which the pump light from a separate source is incident on a Raman medium within a (Stokes) resonant cavity, as illustrated in Fig. 17.1(a), are generally called *external-resonator* or *extracavity* Raman lasers.

External-resonator Raman lasers are typically very simple, comprising an input and an output mirror bounding the Raman medium. Typically the input mirror is a dichroic to admit the pump beam and resonate the Stokes orders of interest, while the output mirror typically reflects unabsorbed pump and output couples the Stoke(s) orders of interest. Mirror curvatures and resonator length are selected to match the focal spot of the pumping laser beam, as well as to accommodate any thermal lensing in the Raman crystal, while appropriate choice of mirror coatings can enable selection of a particular Stokes wavelength. A unique aspect of the external resonator is that it can be configured for simultaneous operation at several Stokes wavelengths, the spectral content of the output depending on the resonator losses for each wavelength. Numerical modelling of external-resonator Raman lasers pumped with nanosecond pulses has been carried out by Ding *et al.* (2006).

External-resonator crystalline Raman lasers are quite widely used for an efficient wavelength conversion of Q-switched solid-state lasers with moderate peak power (100 kW to MW) and average powers of 1–10 W. Wavelength conversion from the 1.06 μm output of standard Q-switched Nd^{3+} crystalline solid-state lasers into the eye-safe near-infrared around 1.5 μm is amongst the most common of missions, together with cascaded Stokes shifting in the visible to generate multi-wavelength visible outputs from frequency-doubled (green) Nd^{3+} crystalline solid-state lasers. Sum-frequency or second-harmonic generation of visible Stokes lines by addition of a phase-matched nonlinear crystal (e.g. BBO) internal to the crystalline Raman laser resonator has also been used to generate a broad range of ultraviolet wavelengths (Mildren *et al.*, 2007).

© Woodhead Publishing Limited, 2013

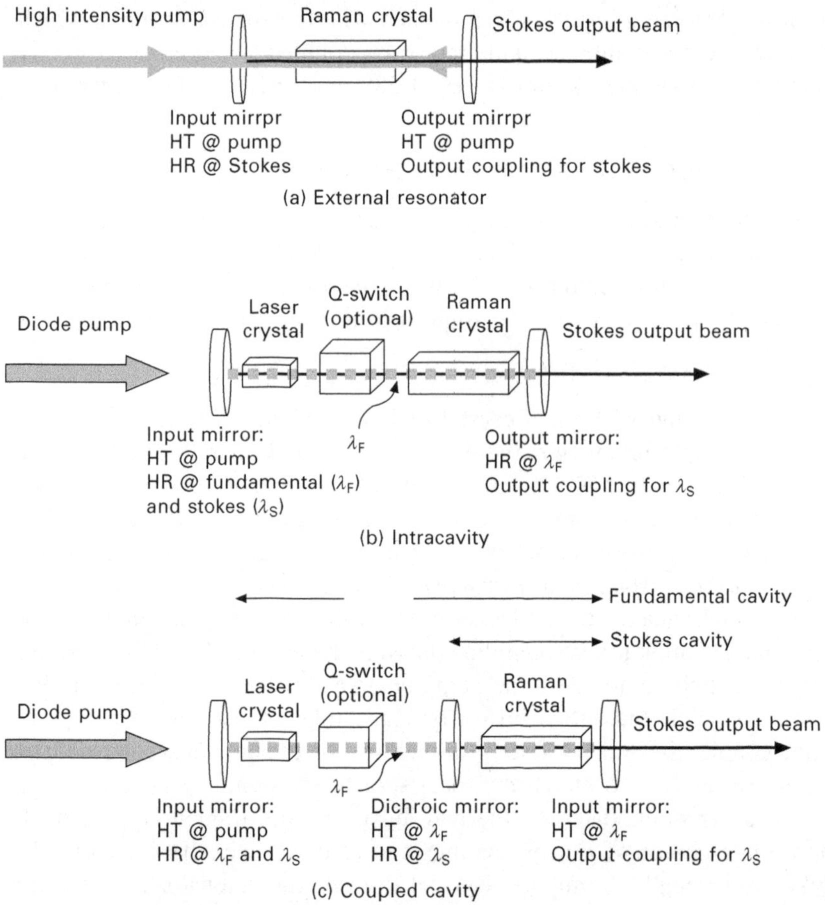

17.1 Design configurations for Raman lasers: (a) external resonator, (b) intracavity and (c) coupled cavity.

A more extensive description and further details of the operating characteristics of external-resonator crystalline Raman lasers are given in Section 17.5 below. Note that fibre Raman lasers typically use external-resonator configurations where the resonator is formed by fibre Bragg gratings bounding either end of a fibre Raman amplifier. The state-of-the-art for fibre Raman lasers is represented by the recent report from Feng *et al.* (2009) of a single-mode, silica glass fibre Raman laser delivering over 150 W CW output at 1120 nm when pumped by a 200 W Yb fibre laser at 1070 nm.

17.4.3 Intracavity Raman lasers

Intracavity Raman lasers include those where the Raman-active crystal is placed within the resonator of a conventional laser, the mirror reflectivities

© Woodhead Publishing Limited, 2013

generally being chosen to give high-Q at the 'fundamental' laser wavelength and appropriate output coupling at the Stokes wavelength (Fig. 17.1(b)). Intracavity configurations encompass also various forms of coupled cavities as in Fig. 17.1(c). Intracavity configurations provide for high circulating intensity at the pump wavelength, and are highly effective in enabling low threshold and high conversion efficiency. They may also be applicable where the pump intensity would otherwise be too low to ensure sufficient Raman gain to reach the Raman laser threshold. In practice intracavity configurations are ideal for diode-pumped crystalline solid-state devices where the fundamental output power (e.g. of a diode-pumped Nd^{3+}:vanadate laser) might only be a few tens or hundreds of milliwatts. Such an arrangement was used for the first all-solid-state crystalline Raman laser (Pask and Piper, 2000) wherein an Nd^{3+} laser crystal pumped by a CW diode formed the fundamental laser medium, the single resonator incorporating both a multi-kilohertz Q-switch and the Raman-active crystal.

The very high circulating powers at the fundamental wavelength generated in intracavity devices using ultra-high-Q resonators also provide the foundation for CW crystalline Raman lasers, first reported in 2005 (Demidovich *et al.*, 2005; Pask, 2005). Even for diode pump powers of only a few watts, circulating powers at the fundamental as high as a kilowatt can be achieved, enabling the CW Raman laser threshold to be reached easily for small mode sizes of ~50 microns diameter in standard Raman crystals. Addition to the resonator of a nonlinear crystal for SFG of the fundamental and first Stokes, or for SHG of the first Stokes, has also resulted in efficient generation of CW output in the visible (lime, yellow) (Lee *et al.*, 2010c; Li *et al.*, 2011a).

The dynamics of intracavity Raman lasers can be very complex, since there are multiple optical processes occurring within the resonator. Strong thermal lensing in the laser crystal (Innocenzi *et al.*, 1990) frequently limits the power and beam quality available from diode-pumped crystalline lasers, including intracavity Raman lasers. In the case of Raman lasers thermal lensing in the Raman crystal due to Raman heating can be an issue too. As we have shown previously (Pask, 2003) the thermal load in the Raman crystal is proportional to the average first Stokes power. Accordingly thermal properties as well as nonlinear properties of Raman crystals must be taken into account when selecting the Raman crystal for an intracavity laser. Many combinations of laser and Raman crystals have been reported, including self-Raman lasers, in which a Raman-active laser crystal performs the dual functions of generating the fundamental and the first Stokes fields. $Nd{:}YVO_4$, $Nd{:}GdVO_4$ and $Nd{:}KGd(WO)_4)_2$ have all been used successfully.

The design of the laser resonator is very important in optimising the performance of the intracavity Raman laser. Once the thermal characteristics of the laser and Raman crystals are known, the resonator layout, cavity length, mirror curvatures and output coupling can be chosen so as to optimise

© Woodhead Publishing Limited, 2013

the optical mode sizes in the crystals. Optimal spot sizes enable efficient extraction of energy from the laser gain crystal and efficient frequency conversion via SRS to occur simultaneously. In a simple intracavity Raman laser the fundamental and Stokes fields oscillate within the same resonator. Coupled-resonator configurations can also be used where the fundamental and Stokes fields exist in separate resonators that overlap in the Raman crystal.

Other design choices include the pump source for the fundamental laser, e.g. diode, arclamp or flashlamp, the pumping geometry, e.g. end-pumped or side-pumped, and whether the laser is pulsed or CW. In the case of pulsed operation, acousto-optic, electro-optic and passive Q-switching have all been used successfully.

Details of design and performance characteristics of intracavity crystalline Raman lasers, Q-switched and CW, are given below in Section 17.5. Extension to the visible by way of intracavity SFG or SHG, in either the Q-switched or the CW mode of operation, is described in Section 17.6.

17.5 Crystalline Raman lasers: performance review

A very diverse range of crystalline Raman lasers have been reported. This section is divided into extracavity and intracavity Raman lasers, with the former including synchronously pumped external-resonator Raman lasers. An overview of the field is given; however, the emphasis of this section is on topical and recent work, especially in the past five years, and with a primary focus on all-solid-state (i.e. diode-pumped) devices.

17.5.1 External-resonator Raman lasers

The potential for efficient frequency conversion using external-resonator crystalline Raman lasers was first established in 1980 by Eremenko *et al.* (1980) wherein the second harmonic of a Nd laser was shifted to the yellow with >25% efficiency using SRS in $Ba(NO_3)_2$. Subsequently He and Chyba (1997) demonstrated optical efficiencies of 40% for wavelength conversion from a 532 nm pump to the first Stokes at 563 nm or the second Stokes at 599 nm, or similar powers at both wavelengths simultaneously, using an external-resonator $Ba(NO_3)_2$ Raman laser. Since then there have been numerous studies involving a variety of different pump lasers and Raman crystals.

Recently Lisinetskii *et al.* (2007b) have used a flashlamp-pumped 300 mJ 1064 nm laser to pump an external-resonator $Ba(NO_3)_2$ Raman laser to demonstrate, with different output couplings, highly efficient wavelength conversion and over 100 mJ of output at each of the first Stokes (1197 nm, 66% efficiency), second Stokes (1369 nm) and third Stokes (1599 nm). An

© Woodhead Publishing Limited, 2013

all-solid-state system with high conversion efficiencies (43%) to the first Stokes (1197 nm) was reported in Pask *et al.* (2003) for pumping with 3 W at 1064 nm from a diode-pumped 1064 nm pulsed laser operating at 5 kHz. For pumping at 532 nm, Mildren *et al.* (2004) reported 45% conversion to the second Stokes at 589 nm using KGW, generating 1.05 W output at 5 kHz. This same system was also configured using a broadband output coupler having 50% transmission to operate simultaneously at five different wavelengths between 578 nm and 621 nm, the unconverted pump light emerging collinearly with the Raman light. The total average output power was 1.68 W at an SRS optical conversion efficiency of 71%.

In 2008, Mildren *et al.* (2008) reported first and second Stokes output from an external cavity CVD-diamond Raman laser pumped by a high-pulse-rate all-solid-state 532 nm laser. Subsequently much higher powers (1.2 W) and efficiencies (63.5%) have been reported (Mildren and Sabella, 2009), clearly demonstrating the potential of diamond to match the performance of other, more conventional Raman media.

While most reported external-resonator Raman lasers have been pumped at 1064 nm or 532 nm, pumping at longer wavelengths has also been demonstrated. The eyesafe region has been targeted using $BaWO_4$ as the Raman-active crystal for wavelength conversion from a 1.3 μm laser pump to the 1.5 μm first Stokes with output pulse energies of 8.5 mJ at 47% conversion efficiency (Zong *et al.*, 2009; Wang *et al.*, 2008). Zong has demonstrated wavelength conversion from a pulsed 4.14 W Nd:YLF pump laser at 1.3 μm to 1.5 μm with 15% efficiency, but also pulse compression from 220 ns to 40 ns. Most recently, Jelínková (2011) reported an external-resonator CVD diamond Raman laser pumped at 1.34 μm and lasing at 1.63 μm. By pumping at 1.9 μm or 1.56 μm, mid-infrared output at 2.31 μm, 2.75 μm and 3.7 μm has been achieved using $BaWO_4$ as the Raman-active crystal (Basiev *et al.*, 2006).

$Ba(NO_3)_2$ has been perhaps the most widely used crystal for high-power pulsed Raman lasers in external resonators, but the relatively poor thermal properties of this material can limit performance (Takei *et al.*, 2002). KGW has also been utilised successfully for high energy and high power pulsed Raman lasers as reported by Basiev *et al.* (2004). Recently, diamond has attracted considerable attention for application to SRS in the near- and mid-infrared. Average powers of 2 W at 1240 nm have been reported for a pulsed external-resonator diamond Raman laser pumped with 3.3 W from a diode-pumped Nd:YAG laser operating at 5 kHz (Sabella *et al.*, 2010). The slope efficiency of 84% was very close to the quantum limit of 86%, and the overall conversion efficiency (1064 nm to 1240 nm) was 61%. In 2011, Feve *et al.* (2011) reported by far the highest average power to date for any crystalline Raman laser, again using diamond. Paying particular attention to thermal management of the diamond crystal, they obtained a maximum

© Woodhead Publishing Limited, 2013

first Stokes (1193 nm) average output power of 24.5 W (from two ends) at a repetition rate of 40 kHz. Output power was limited by optical damage to the AR coatings on the diamond crystal. The pump source was a Q-switched Yb-doped YAG laser, delivering up to 340 W, and the overall conversion efficiency was up to 13%. Notably the Raman laser threshold was very high (~110 W), the laser operating only ~1.5 times above threshold.

While the vast majority of external-resonator Raman lasers have been pulsed lasers, there is one notable exception. This was an external-resonator $Ba(NO_3)_2$ laser pumped by a CW Ar^{3+} laser at 514 nm (Grabtchikov *et al.*, 2004) and the first reported CW crystalline Raman laser. 164 mW at the first Stokes wavelength of 543 nm was generated, from 5 W pump power. More recently, CW operation at the second Stokes was also reported (Grabchikov *et al.*, 2009). Quasi-CW operation of an external-resonator Raman laser using diamond was reported very recently (Kitzler *et al.*, 2011) with output powers of 7.5 W in 6.5 ms pulses at 16% duty cycle.

Most external-resonator Raman lasers utilise simple two mirror cavities, though a variety of alternative architectures have been used. These include unstable resonator designs (Ermolenkov *et al.*, 2005; Karpukhin and Stepanov, 1986) and non-collinear Raman lasers (Mahoney *et al.*, 2006; Mildren, 2011). Adding to the diversity of this group of laser devices is a Raman ring laser operating at 1538 nm (Dashkevich and Orlovich, 2011) and an anti-Stokes Raman laser at 508 nm (Mildren *et al.*, 2009).

There are several applications in biophotonics for which ultrashort pulses at 550–700 nm are required (Girkin and McConnell, 2005; Palero *et al.*, 2005) and which conventional tunable Ti:sapphire lasers cannot access. Raman conversion of picosecond laser sources has been widely explored using Raman generators (Černý *et al.*, 2004), but these generally suffer from high pump threshold, poor output beam quality and difficulties in controlling the spectral content of the output. *Synchronous pumping*, in which the repetition rate of the pump laser is matched to the round-trip time of an external Raman resonator, overcomes these disadvantages, enabling an intense circulating pulse to build up over many pump pulses. Hence synchronous pumping offers the same benefits to picosecond lasers that are available to nanosecond lasers using conventional external-resonator Raman lasers.

In 2009, Granados *et al.* (2009) reported a synchronously pumped KGW Raman laser operating at 559 nm. Pumped with 2 W at 532 nm (pulse duration 10 ps and repetition frequency 80 MHz) from a frequency-doubled CW mode-locked $Nd:YVO_4$, this laser generated a maximum average power of 410 mW with a slope efficiency of 42%, and the overall conversion efficiency was 26%. Pulse compression due to different group velocities of the fundamental and first Stokes emission was investigated, and was a strong function of the cavity length detuning between the fundamental and first Stokes cavities. Maximum first Stokes output power did not correspond exactly to

© Woodhead Publishing Limited, 2013

maximum pulse compression, and for the case where the shortest pulses of duration 3.2 ps were achieved, the maximum output power was 290 mW. The following year Spence *et al.* (2010) reported a diamond Raman laser, synchronously pumped with up to 7.5 W at 532 nm and generating up to 2.2 W output at 573 nm with pulse duration 21 ps. A numerical model was used to explain the dependence of conversion efficiency and pulse duration on cavity length.

Cascaded conversion of a CW mode-locked Raman laser operating at 7.5 W at 532 nm has also been reported in Granados *et al.* (2010), for synchronous pumping of a KGW Raman resonator that comprised two separate cavities for the first and second Stokes wavelengths. 2.4 W at 559 nm and 1.4 W at 589 nm were obtained, with slope efficiencies up to 52% for both Stokes wavelengths. The output pulses were compressed from 28 ps at 532 nm down to 6.5 ps at 559 nm (first Stokes) and to 5.5 ps at 589 nm (second Stokes). The peak power of the generated pulses was almost as high as the pump pulses as a consequence of pulse shortening.

17.5.2 Intracavity Raman lasers

Pulsed intracavity Raman lasers

The majority of pulsed intracavity Raman lasers reported have operated in the near-infrared band between 1.1–1.2 microns. Several researchers have targeted the so-called 'eyesafe' region around 1.5 microns which is important for remote-sensing applications, and two distinct approaches have been taken. The first is by Raman conversion of a 1.3 micron fundamental to the first Stokes at 1.5 microns (Murray *et al.*, 1995; Chen, 2004; Huang *et al.*, 2007). For example in 2004, Chen (2004)reported a diode-pumped Q-switched Nd:YVO_4 self-Raman laser giving 1.2 W output at 1525 nm from a 13.5 W pump. Most recently Huang *et al.* (2007) have reported a very compact, short pulse Nd:KGW self-Raman laser generating 2 ns, 32 mJ pulses at the first Stokes wavelength (1522 nm) at 10 Hz. The second approach for reaching the eyesafe region has been to optimise production of third Stokes output from a 1064 nm fundamental. Such an approach was reported by Dashkevich *et al.* (2009) who from a pump source pulse energy of 6 J achieved 14.7 mJ pulse energies at 1.5 microns. The highest pulse energy reported for a Raman laser operating in the eyesafe band at 1522 nm is 50 mJ reported at 4% efficiency from the fundamental using an intracavity Nd:YAG and KGW combination (Mahoney *et al.*, 2006).

The highest average power yet reported from a pulsed intracavity crystalline Raman laser is 10.5 W at a first Stokes wavelength of 1180 nm (Chen *et al.* 2009). This device was a diode side-pumped AO-Q-switched Nd:YAG/$SrWO_4$ laser and achieved a diode to Stokes efficiency of 6.4%. In Liu

© Woodhead Publishing Limited, 2013

et al. (2009b) 4.55 W at 1091 nm was generated from a side-pumped, AO-Q-switched Nd:YAG laser using KTA as the Raman crystal giving a diode to first Stokes efficiency of 7.5%. Other reports of multi-watt output powers can be found in Pask (2003) and Pask and Piper (2000). The highest optical conversion efficiency reported to date, 23.8%, was achieved for a diode-pumped Q-switched Nd:YAG/$SrWO_4$ laser operating at 1180 nm (Chen *et al.*, 2008). A high efficiency of 17% was also reported by *Chen et al.* (2005), who used a 10 W diode to pump a Nd:YAG/$BaWO_4$ intracavity Q-switched Raman laser, giving 1.6 W at 1181 nm.

Further power scaling of intracavity Raman lasers is certainly possible in the future and will require attention to thermal loading of the Raman and laser crystals using a variety of strategies such as in-band pumping of the fundamental laser at 880 nm (Sato *et al.*, 2003), the use of end-capped crystals (Chen *et al.*, 2011) and careful selection of laser and Raman crystals in respect of thermal and thermo-optic properties.

Pulsed intracavity Raman lasers have also been demonstrated to work efficiently under conditions of modest diode pumping, i.e. up to a few watts. In 1999, Grabtchikov *et al.* (1999) reported the first diode-pumped self-Raman laser, a miniature Nd:KGW laser delivering 4.8 mW at 1181 nm when pumped by a 1 W CW laser diode and passively Q-switched by a Cr^{4+} YAG saturable absorber. Shortly afterwards, Lagatsky *et al.* (2000) reported a miniature Yb:KGW laser that delivered 7 mW with 2% efficiency, and Simons *et al.* (2004) reported a miniature Nd:YVO_4/$Ba(NO_3)_2$ laser that delivered 11 mW with 6.9% efficiency when pumped with a 1.6 W diode.

As mentioned previously, synthetic diamond has emerged as a promising Raman crystal with several attractive features that include a large Raman gain coefficient (13.5 cm/GW), a large Raman shift (1332 cm^{-1}) and very high thermal conductivity (2000 W $m^{-1}K^{-1}$). In 2010, Lubeigt *et al.* (2010) reported pulsed operation of an intracavity Nd:YVO_4 Raman laser in which diamond was the Raman-active crystal. Up to 375 mW average power at 1240 nm was generated for 9.5 W absorbed diode power; birefringence and loss were identified as factors limiting the optical conversion efficiency from diode pump to around 4%. High efficiency (13%) CW Raman laser output has since been achieved (Lubeigt *et al.*, 2011).

The dynamics of pulsed intracavity Raman lasers are complex, and give rise to some interesting and useful features such as pulse shortening and Raman beam cleanup. An overview of these effects can be found in Murray *et al.* (1999) and Pask (2003).

CW intracavity Raman lasers

Even when an intracavity configuration is used to boost the circulating power of the fundamental, reaching the threshold for CW operation is challenging,

© Woodhead Publishing Limited, 2013

usually requiring very small resonator mode diameters and very-high-Q cavities. Meeting these requirements has been greatly helped by two recent technological advances. The first is the availability of high brightness laser diodes at 880 nm, giving small pump modes and allowing in-band pumping of Nd^{3+}, thereby reducing waste-energy heating and thus reducing thermal lensing. The second is the availability of ion-beam sputtered coatings which can provide very high reflectivity coatings at the fundamental and first Stokes wavelengths, and high quality AR-coatings on the laser and Raman crystals. These advances, combined with improved understanding of the underlying physics, have contributed to the consistent improvement in performance of CW intracavity Raman lasers.

Spence *et al.* (2007) reported a numerical model that simulates CW intracavity Raman lasers that operate at the first Stokes wavelength, or at its second harmonic through intracavity doubling. The diode pump power required to reach threshold can be written as (Spence *et al.*, 2007)

$$P_P = \frac{\pi r^2 \lambda_F (T_S + L_S)(T_F + L_F)}{4 g_R l_R \lambda_P}$$

where λ_F is the fundamental wavelength (nm), λ_P is the pump wavelength, g_R and l_R are the Raman gain and crystal length = 3 mm, r is the radius of the cavity mode in the Raman crystal, T_F and T_S are the transmission of the output coupler at the fundamental and Stokes wavelengths, respectively, and L_F and L_S are the losses for the fundamental and Stokes fields, respectively. Clearly it is important to minimise losses at the fundamental and first Stokes wavelengths. In this context the use of self-Raman materials is highly advantageous.

CW intracavity Raman lasers have mainly operated at the first Stokes wavelength in the near-infrared (typically 1.1–1.2 microns); however, there has been one report of a Nd:$GdVO_4$ self-Raman laser operating at the second Stokes wavelength (Lee *et al.*, 2010a) in which 0.95 W output was obtained at 1308 nm. A wide variety of Raman-active crystals have been used in CW intracavity Raman lasers, with Nd:$GdVO_4$, Nd:YVO_4 and Nd:KGW being popular self-Raman crystals, and $SrMoO_4$, $BaWO_4$, $PbWO_4$, $LuWO_4$, $SrWO_4$, KGW and diamond all having been used as Raman crystals in the CW regime.

The highest power reported from a CW intracavity Raman laser is 5.1 W at 1217 nm using diamond as the Raman-active medium (Savitski *et al.*, 2011a). These results were achieved using a diode side-pumped Nd:YLF laser. Much better beam quality was achieved using diamond, compared to KGW, and the brightness of the diamond Raman laser was shown to be 48 times higher than that of the basic pump laser operating with 20% output coupling and 18.4 W output at 1047 nm. This was attributed to Raman beam cleanup and the superior thermal properties of diamond. While the diode to

© Woodhead Publishing Limited, 2013

first Stokes efficiency of this laser was only ~4%, there was no indication that thermal effects were limiting the output, and further power scaling may be possible. CW output powers of 3.36 W at 1180 nm were reported by Fan *et al.* (2009) for a Nd:YVO_4 laser crystal (pumped with an 808 nm diode laser) and intracavity $BaWO_4$ Raman crystal; diode to first Stokes conversion efficiency was 13.2%. Multi-watt powers have also been reported in Dekker *et al.* (2007b), Yu *et al.* (2011) and Savitski *et al.* (2011b). A major factor limiting the performance of CW intracavity Raman lasers has been the strong thermal lensing occurring in the Nd: laser crystal with thermal lenses as strong as 20 mm reported by Dekker *et al.* (2007b). Several strategies for mitigating this strong thermal lensing have been explored, including in-band pumping at ~880 nm (Yu *et al.* 2011; Lee *et al.* 2008), composite (endcapped crystals) (Fan *et al.*, 2010; Omatsu *et al.*, 2009) disk laser geometry (Lubeigt *et al.*, 2011) and the use of Nd:YLF which typically exhibits weak thermal lensing (Bu *et al.*, 2011). Each of these approaches has been shown to be effective in enabling power scaling.

As is the case with pulsed intracavity Raman lasers, CW Raman lasers have been operated efficiently for very modest diode pumping powers. In fact the highest efficiency yet reported for a CW intracavity Raman laser is 13.85% achieved by a Nd:KGW/KGW Raman laser pumped by only 2 W of output power (Lisinetskii *et al.*, 2007a). The use of a single 50 mm long composite crystal contributed to efficient operation by decreasing intracavity losses, especially crystal surface losses.

While the vast majority of CW intracavity Raman lasers have involved Nd^{3+} laser gain media, an important recent development (Parrotta *et al.*, 2011) has been a KGW Raman crystal pumped within the cavity of a CW diode-pumped InGaAs semiconductor disk laser. The Raman threshold was reached for 5.6 W absorbed diode pump power, and output powers up to 0.8 W at 1143 nm were reported with 7.5% diode to first Stokes efficiency. This laser was also tunable between 1133 nm and 1157 nm by tuning the optically pumped semiconductor laser (OPSL).

Much of the research into CW intracavity Raman lasers has focused on exploring new combinations of fundamental laser and Raman crystal materials, improved resonator and thermal design, and power scaling. However, there are now several papers that explore the complexities of these lasers in more detail. Spence *et al.* (2007) have reported a theoretical numerical model for CW intracavity Raman lasers, with and without intracavity frequency doubling. Numerical calculations enable the trade-offs between Raman gain, resonator losses and mode sizes to be investigated, leading to optimised design for maximum laser efficiency and power. Lin *et al.* (2010) have modelled the relaxation oscillations in a CW Raman laser and demonstrated a method for deducing Raman gain and resonator losses. There remains considerable scope for investigating the spectral and temporal characteristics of CW Raman lasers.

© Woodhead Publishing Limited, 2013

17.6 Wavelength-versatile Raman lasers

One of the most exciting directions for development of solid-state Raman lasers has been the development of new laser sources in the yellow and orange spectral regions, and also the UV. Green lasers such as frequency doubled Nd: lasers can be wavelength-shifted to generate yellow-orange output using external-resonator Raman lasers; alternatively the output from near-infrared Raman lasers can be extracavity frequency-doubled. Both of these approaches are fairly simple to implement, especially for pulsed sources, though overall energy conversion efficiency to the visible is typically fairly low (Pask *et al.*, 2008a). Intracavity frequency doubling of the intense Stokes light that is generated in an intracavity or external-resonator Raman laser permits high wavelength conversion efficiencies for both pulsed and CW sources. In this section we describe how intracavity frequency mixing processes such as sum frequency generation (SFG) and second harmonic generation (SHG) can be incorporated into a Raman resonator for efficient wavelength conversion to the visible or UV. The cascaded nature of the SRS process gives rise to a unique mode of operation, which we term 'wavelength-selectable', because the operator can select between various wavelengths simply by configuring the intracavity nonlinear SFG crystal appropriately.

17.6.1 Design principles

Figure 17.2 depicts simple configurations for (a) external resonator and (b) intracavity Raman lasers with intracavity SFG. For the intracavity Raman laser in Fig. 17.2(b), the resonator input and output mirrors are typically coated to admit the pump light from a laser diode (when end-pumping is used), and to resonate the fundamental and first Stokes optical fields. Higher-order Stokes can be resonated if desired; however, obtaining high-reflecting coatings of sufficiently broad bandwidth and high damage thresholds can be difficult. The output coupler typically has high transmission in the visible, and an intracavity dichroic mirror can be use to redirect the back-propagating visible light through the output coupler. Alternatively, a folded resonator configuration can be used to couple the visible output into a single output beam. A Q-switch may also be present, in the case of pulsed lasers. Suitable nonlinear crystals include LBO, BBO, BiBO and KTP. Performing the frequency mixing intracavity leads to high efficiencies because the high intracavity Stokes powers are accessed. Effectively, the frequency mixing acts as a nonlinear output coupling for the Stokes field.

This type of laser involves three optical processes occurring in a single laser resonator, and the key to efficient laser operation is to ensure that the laser parameters are simultaneously optimised for the fundamental laser generation, for conversion to the first Stokes, and for frequency mixing which typically involves SHG of the first Stokes or SFG of the fundamental and

© Woodhead Publishing Limited, 2013

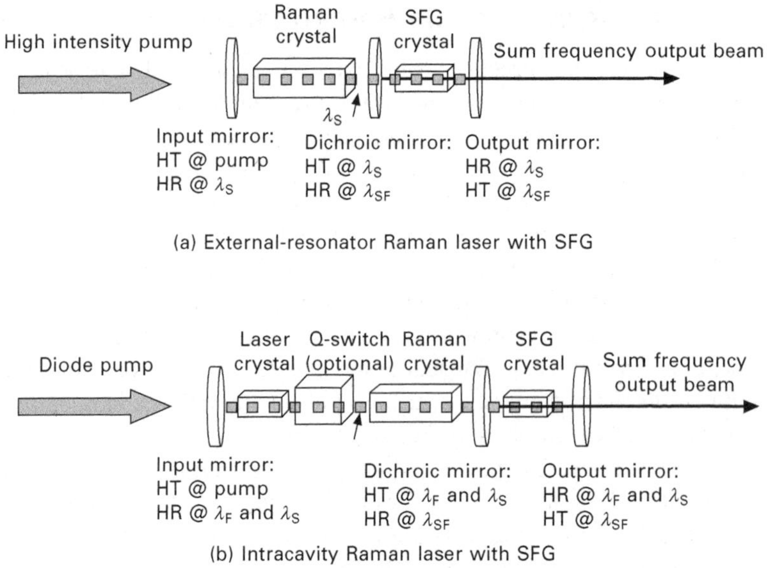

17.2 Simple configurations for (a) external resonator and (b) intracavity Raman lasers incorporating intracavity SFG.

first Stokes. Optimisation is achieved through careful selection of the laser, Raman and doubling crystals, and careful resonator design to optimise the mode sizes in each of these crystals. Importantly, the resonator design must account for the thermal effects which are typically present and frequently strong in the laser and Raman crystals.

The concept for the wavelength-selectable laser is shown in Fig. 17.3. A Raman resonator is employed that is capable of resonating the fundamental, first-order and (if desired) higher-order Stokes wavelengths, together with an intracavity frequency mixing crystal that can be configured to phase-match one of several possible output wavelengths. In principle, if the nonlinear crystal is phase-matched to frequency-double the fundamental at say 1064 nm, the fundamental field will be depleted and no first Stokes will be generated. If the nonlinear crystal is detuned so that there is no SHG of the fundamental, then the first Stokes optical field will build up and the nonlinear crystal can be tuned to allow either SFG of the fundamental and first Stokes, or SHG of the first Stokes. If higher-order Stokes lines are resonated, then the number of lines that can be generated by SFG or SHG is expanded.

17.6.2 Fixed-wavelength Raman lasers incorporating intracavity frequency mixing

A diverse range of pulsed intracavity frequency-doubled Raman lasers operating at yellow-orange wavelengths has been reported: for a thorough

© Woodhead Publishing Limited, 2013

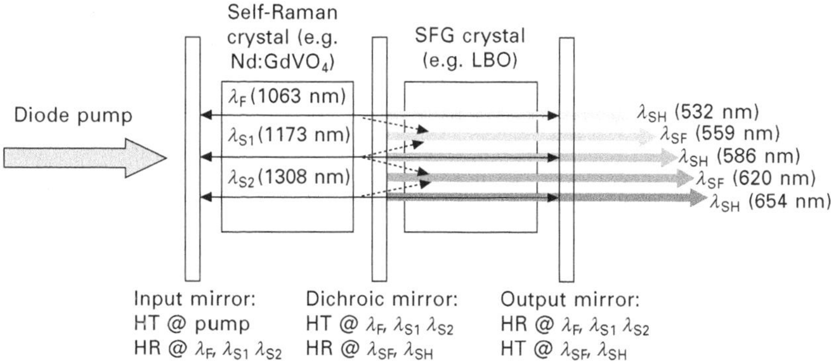

17.3 Concept for a wavelength-selectable Raman laser.

review of developments up to 2008 see Pask *et al.* (2008a). Diode end-pumping and side-pumping (arclamps or diodes) geometries, and many different types of laser, Raman and doubling crystals have been employed to generate many output wavelengths from 579 nm to 599 nm. Output pulse durations can be in the range 3–45 ns, mostly at multi-hertz repetition rates. In terms of output powers and efficiencies, the lasers vary from a low power (5.5 mW) 599 nm laser pumped by a 1.6 W laser diode (Simons *et al.*, 2004) to an 8.3 W, 590 nm laser pumped by 125.8 W diode pump power and Q-switched with 15 kHz pulse repetition frequency (Cong *et al.*, 2010). For the latter, the corresponding optical conversion efficiency from diode laser to yellow laser is 6.57%. Very high efficiency (18%) was reported by the same authors (Cong *et al.*, 2009) for an intracavity frequency-doubled Raman laser constructed using a Nd:YAG ceramic gain medium, a $SrWO_4$ Raman medium, and a KTP frequency-doubling medium. Average power output of 2.93 W was demonstrated at 590 nm for incident pump power of 16.2 W and a pulse repetition frequency of 20 kHz.

While most papers have targeted the second harmonic of the first Stokes wavelength to generate yellow-orange wavelengths, an increasing number of authors have employed sum frequency generation of the fundamental and first Stokes to generate lime wavelengths. Chang *et al.* (2008) have reported an efficient Q-switched Nd:YVO_4 self-Raman laser with intracavity SHG that operated simultaneously at the first Stokes (1176 nm) wavelength with an average power of 0.53 W and 1.67 W at the sum frequency wavelength (1064+1176= 559 nm). Diode pump power was 17.5 W, and pulse repetition rate was 100 kHz.

Also of interest is the use of crystals that have both second- and third-order nonlinearity, and thus can be used for simultaneous SRS and SFG. As well as the $LiIO_3$ crystal used by Ammann (1979), KTP has been used in a Q-switched Nd:YAG laser to generate 1.03 W at 1129 nm and 0.25 W at the sum frequency wavelength of 548 nm, for a diode pump power of 10 W and 10 kHz pulse rate (Chen, 2005).

Realisation of high circulating powers (several hundred watts) in the first demonstrations of CW crystalline intracavity Raman lasers (Demidovich *et al.*, 2005; Pask, 2005) raised the prospect that efficient intracavity SFG and SHG might be achieved in the CW regime. Indeed this was subsequently reported by Dekker *et al.* (2007a) for a Nd:$GdVO_4$ laser with intracavity SRS in KGW and intracavity SHG in LBO. The input and output mirrors had very high reflectivity (>99.99%) at the fundamental and first Stokes wavelengths, but were 95% transmitting for the 588 nm second harmonic of the first Stokes. A maximum CW power of 704 mW at 588 nm was reported for a diode pump power of 13.7 W, above which the output power decreased due to resonator instability. For quasi-CW operation with 50% duty cycle, 1.57 W was achieved for quasi-CW pumping with 20 W diode power, demonstrating that thermal loading was limiting power scaling to higher CW powers. The same 10 mm-long Nd:$GdVO_4$ crystal was also used in a self-Raman configuration to reduce cavity losses and allow a shorter cavity that could accommodate a stronger thermal lens. CW output of 678 mW at 586.5 nm was obtained for a maximum diode pump power of 16.3 W, limited by thermal loading of the Nd:$GdVO_4$.

Since this first demonstration a great deal of effort has been made to achieve efficient multi-watt CW yellow laser output, mostly by reducing resonator losses and managing thermal loading, e.g. by in-band high brightness pumping, and using short resonators. CW Raman lasers with intracavity doubling have been the subject of numerical modelling by Spence *et al.* (2007) where the behaviour of the laser is described as a function of coupling parameters for the Raman and SFG processes. This work is particularly useful as a design tool and for understanding the trade-offs between the many resonator and material parameters.

The highest output power in the yellow reported to date is 4.3 W at 586 nm, with an efficiency of ~17% (Lee *et al.*, 2010d). This was a self-Raman Nd:$GdVO_4$ laser, with frequency doubling in LBO. Multi-watt yellow outputs have also been reported for Nd:$GdVO_4$ lasers incorporating $SrMoO_4$ (Yu *et al.*, 2011) and $BaWO_4$ (Lee *et al.*, 2010b) Raman crystals, and for Nd:$LuVO_4$ in a self-Raman configuration, with LBO for intracavity SHG (Lü *et al.*, 2010a). In all cases, the maximum yellow output power is limited by thermal lensing in the laser crystal. In each of these laser systems a dichroic intracavity mirror is used to the optimise collection of yellow light into a single output beam; this mirror is also effective in preventing the absorption of yellow light in the laser crystal, which otherwise causes a significant contribution to the thermal load.

CW visible output at other wavelengths has also been reported. Lime wavelengths can be obtained by phase-matching the nonlinear crystal for

© Woodhead Publishing Limited, 2013

SFG of the fundamental and Stokes. In fact the highest reported visible output power from a visible Raman laser-based source was 5.3 W at 559 nm (Lee *et al.*, 2010c). This was achieved for a Nd:GdVO$_4$ self-Raman laser incorporating intracavity SFG in LBO; diode-visible conversion efficiency was ~21%. Sum frequency generation at 620 nm from the first and second Stokes lines in a Nd:GdVO$_4$ self-Raman laser has also been reported by Lee *et al.* (2010a) with a diode-visible conversion efficiency of 4.9%.

Efficient generation of visible laser output from CW intracavity Raman lasers with intracavity SFG has been shown to downscale very efficiently to low powers. In particular, the work by Li *et al.* (2011a, b) has used low power diode pumping (3.8 W) to obtain output powers of 320 mW at 588 nm and 660 mW at 559 nm, with corresponding overall efficiencies of 8.4% and 17% respectively. It is noteworthy that these efficiencies are quite comparable to those achieved for pumping with 20 W or more. This miniature visible all-solid-state source featured small pump and optical mode volumes (radii ~80 μm), 1% Nd doping in a short (3 mm long) Nd:YVO$_4$ crystal, a 5 mm long LBO crystal and short cavity length (11 mm), with very low loss coatings on all cavity elements. While the laser is very simple physically, the laser dynamics are complex. Experimental and theoretical studies (Li *et al.*, 2011a, b) show how it is necessary to balance the nonlinear processes of SRS and SFG in order to maximise the efficiency for lime generation. Several other papers (Xia *et al.*, 2011; Lü *et al.*, 2010b; Kananovich *et al.*, 2010) have reported laser operation at 559–561 nm, with typical efficiency in the range 2–13%.

As with conventional green lasers, intracavity frequency-doubled Raman lasers typically exhibit high levels of amplitude noise in the yellow output, analogous to the well-known 'green problem'. Lin *et al.* (2011) have investigated the amplitude noise in green and yellow lasers both experimentally and numerically, using rate equations. In contrast to the green laser, a regime was found at lower powers, where the yellow laser naturally has low amplitude noise. This low-noise 'window' makes it possible to build very simple yellow lasers that do not suffer from amplitude instability.

17.6.3 Wavelength-selectable Raman lasers

Pulsed UV wavelength-selectable laser operation from external-resonator Raman laser

In the UV, pulsed wavelength-selectable laser operation has been reported using an external-resonator Raman laser pumped by frequency-doubled Nd lasers at 532 nm (Mildren *et al.*, 2007). Performance was evaluated using two different pump lasers: one for generating high pulse energies at 10 Hz, and one for generating high average powers at 5 kHz. The Raman resonator

© Woodhead Publishing Limited, 2013

contained a $KGd(WO_4)_2$ crystal and a pair of mirrors that were highly reflective for the first three Stokes wavelengths (559–660 nm). A BBO crystal was used inside the resonator for second harmonic and sum frequency mixing of the three resonating Stokes beams, and by angle-tuning the BBO up to eight different UV lines were obtained between 266 nm and 320 nm. In the high pulse energy regime, seven new frequencies were generated, six of these having pulse energies in the range 88–223 μJ corresponding to conversion efficiencies between 1.1 and 2.4%. In the high average power regime, five new frequencies were generated, four of these with output powers between 11 and 47 mW, corresponding to conversion efficiencies between 0.85% and 2.7%. The UV wavelengths generated sit between the 266 nm and 355 nm harmonics of Nd lasers, where there are few alternative laser sources. Conversion efficiencies higher than 10% appear to be possible in a fully optimised arrangement. Additional UV lines were demonstrated by orienting the KGW crystal so as to access Raman gain for both 768 and 901 cm^{-1}.

Pulsed visible wavelength-selectable laser using intracavity SRS

As early as 1979, Ammann and co-workers (Ammann, 1979) reported wavelength-selectable visible output from an intracavity Raman laser based on simultaneous SRS and SFG/SHG in the one crystal ($LiIO_3$). Over 25 years later a diode-pumped, pulsed, wavelength-selectable intracavity Raman laser was reported (Mildren *et al.*, 2005), where the Nd:YAG fundamental at 1064 nm was Raman shifted to 1159 nm and 1272 nm and intracavity SHG or SFG was used to obtain 1.7 W at 532 nm, 0.95 W at 555 nm or 1.8 W at 579 nm. Temperature tuning across four wavelengths – 532 nm, 555 nm, 579 nm and 606 nm – was also demonstrated with powers between 0.25 W and 1.5 W. Subsequently Pask *et al.* (2008b) reported wavelength-selectable output by angle-tuning BBO to select the output wavelength. BBO is the crystal of choice for angle tuning, as a small rotation of only ~3° enables a green, yellow or red wavelength to be selected. Two sets of output wavelengths could be generated by orienting the KGW crystal so as to access the 768 cm^{-1} or 901 cm^{-1} Raman gain.

CW visible wavelength-selectable laser using an intracavity Raman laser

Wavelength-selectable operation in the visible has been particularly successful, in the regimes of both high power (30 W diode) pumping (Lee *et al.*, 2010d) and low power (3.8 W) pumping (Li *et al.*, 2011b). Physically, these lasers are exceptionally simple, comprising $Nd:GdVO_4$ and LBO together with high-Q resonator end mirrors and (optionally) an intracavity dichroic mirror to reflect visible light back through the output coupler. In both reports, operation was

© Woodhead Publishing Limited, 2013

confined to three-colour operation: green (532 nm), lime (559 nm) and yellow (586 nm); however, in principle the scheme can be extended to include red wavelengths by resonating the second Stokes wavelength.

In Lee *et al.* (2010d), the output power at each wavelength was first maximised individually, by optimising the various resonator parameters (e.g. mirror curvatures, resonator length, etc.), and then wavelength-selectable operation was optimised by choosing the one resonator that gave the best all-round performance at the three wavelengths. The output powers obtained are summarised in Table 17.2, where it is seen that there is a modest trade-off involved in output powers and efficiency for the wavelength-selectable case. Simulations using a rate equation model (Spence *et al.*, 2007) were used to identify the conditions that were optimal for each wavelength.

In Li *et al.* (2011b), wavelength-selectable operation at lime (559 nm) and yellow (588 nm) has been achieved for a miniature self-Raman laser with intracavity SFG/SHG in LBO pumped by only 3.8 W from an 808 nm diode. By changing the LBO temperature in the range 40–100°C, 320 mW could be obtained at 588 nm or 660 mW at 559 nm, with corresponding conversion efficiencies of 8.4% and 17%. Laser output in the visible, as well as the intracavity fundamental and Stokes fields, were monitored for laser resonators with 5 mm LBO, with 10 mm LBO and without LBO, and these data were analysed to show that resonator losses were dominated by bulk crystal losses. Complex laser behaviour arising from the interplay between SRS and SFG was also identified, and is a new consideration for the optimum design of wavelength-selectable CW intracavity Raman lasers.

17.7 Conclusion and future trends

In this chapter we have provided the background information required to design efficient pulsed and CW crystalline Raman lasers, and reviewed the main directions in the development of crystalline Raman lasers. It should be clear that there is a great diversity in the types of Raman laser that have been reported, and that Raman lasers can be tailored to suit the output powers,

Table 17.2 Performance of wavelength-selectable Raman laser reported in Lee *et al.* (2010d). Shown are the maximum output powers that could be achieved at each wavelength, and the wavelength-selectable output powers

Wavelength	LBO temperature	Maximum output power	Maximum diode-visible efficiency	Wavelength-selectable power	Wavelength-selectable efficiency
Green: 532 nm	155°C	6.9 W	24.0%	3.8 W	14.6%
Lime: 559 nm	95°C	5.3 W	20.4%	4.9 W	16.9%
Yellow: 586 nm	45.5°C	4.3 W	17.1%	3.5 W	12.1%

© Woodhead Publishing Limited, 2013

wavelengths and temporal formats required for a wide range of applications. There remains great scope for evaluating new crystalline materials, expanding the wavelength coverage and scaling up the output powers to higher levels, or indeed to scaling down the output power to achieve efficient operation in the hundred-milliwatt range.

Promising new crystalline Raman media continue to appear, and in particular diamond is currently creating great excitement. With its unmatched thermal conductivity, high Raman gain, large frequency shift and broad transparency, diamond is an iconic material. However, there remain several challenges to be overcome in terms of growth, supply, high cost and coating technology, Therefore it is likely to remain a 'high-end' material in the near future, but could play an important role in filling niche opportunities such as Raman conversion to the mid-infrared where the heat loading becomes comparable to the pump photon energies, or scaling to very high power.

As this field matures, and as we consolidate our understanding of the physics that underpins the devices, new capabilities will continue to emerge. There are many exciting new directions that are still in their infancy, for example applying Raman conversion to new laser platforms such as VECSELs and ultrafast lasers. Accessing the blue-green spectral region, achieving low amplitude noise and spectral control of Raman lasers are all areas where there are opportunities to carry out new research and build devices with expanded capabilities.

From a practical point of view, crystalline Raman lasers offer a very simple way of greatly extending the capability of well-established all-solid-state lasers with very little loss of efficiency and at very low cost. Indeed, but for the rather specialised optical coatings needed, many Raman lasers have exactly the same components as 'conventional' solid-state lasers, making them very amenable to commercialisation.

17.8 References

Agrawal, G. P. 1995. *Nonlinear Fiber Optics*. Second edition, Academic Press, New York.

Ammann, E. O. 1979. Simultaneous stimulated Raman scattering and optical frequency mixing in lithium iodate. *Applied Physics Letters*, 34, 838–840.

Ammann, E. O. & Decker, C. D. 1977. 0.9-W Raman oscillator. *Journal of Applied Physics*, 48, 1973–1975.

Ammann, E. O. & Falk, J. 1975. Stimulated Raman scattering at kHz pulse repetition rates. *Applied Physics Letters*, 27, 662–664.

Andryunas, K., Vischchakas, Y., Kabelka, V., Mochalov, I. V., Pavlyuk, A. A., Petrovskii, G. T. & Syrus, V. 1985. SRS-self conversion of Nd laser emission in tungstate crystals^{3+}. *JETP Letters*, 42, 410–412.

Basiev, T. T. & Powell, R. C. 2003. Solid state Raman lasers. In *Handbook of Laser Technology and Applications*, ed. C.E. Webb and J.D.C. Jones, Institute of Physics, 469–497.

© Woodhead Publishing Limited, 2013

Basiev, T. T., Sobol, A. A., Voronko, Y. K. & Zverev, P. G. 2000. Spontaneous Raman spectroscopy of tungstate and molybdate crystals for raman lasers. *Optical Materials*, 15, 205–216.

Basiev, T. T., Danileiko, Y. K., Doroshenko, M. E., Osiko, V. V., Fedin, A. V., Gavrilov, A. V. & Smetanin, S. N. 2004. Powerful $BaWO_4$ Raman laser pumped by self-phase-conjugated Nd:YAG and Nd:GGG Lasers. *Tech. Dig. Adv. Solid State Photon. Conf.*

Basiev, T. T., Basieva, M. N., Doroshenko, M. E., Fedorov, V. V., Osiko, V. V. & Mirov, S. B. 2006. Stimulated Raman scattering in the mid IR range 2.31–2.75–3.7 μm in a $BaWO_4$ crystal under 1.9 And 1.56 μm pumping. *Tech. Dig. Adv. Solid State Photon. Conf.*

Boyd, R. W. 1992. *Nonlinear optics*. First edition, Academic Press, New York.

Bu, Y. K., Tan, C. Q. & Chen, N. 2011. Continuous-wave yellow light source at 579 nm based on intracavity frequency-doubled Nd:YLF/$SrWO_4$/LBO Raman laser. *Laser Physics Letters*, 8, 439–442.

Černý, P. & Jelínková, H. 2002. Near-quantum-limit efficiency of picosecond stimulated raman scattering in $BaWO_4$ crystal. *Optics Letters*, 27, 360–362.

Černý, P., Jelínková, H., Zverev, P. G. & Basiev, T. T. 2004. Solid state lasers with raman frequency conversion. *Progress in Quantum Electronics*, 28, 113–143.

Chamorovskiy, A., Rautiainen, J., Rantamaki, A. & Okhotnikov, O. G. 2011. Raman fiber oscillators and amplifiers pumped by semiconductor disk lasers. *IEEE Journal of Quantum Electronics*, 47, 1201–1207.

Chang, Y. T., Huang, Y. P., Su, K. W. & Chen, Y. F. 2008. Diode-pumped multi-frequency Q-switched laser with intracavity cascade raman emission. *Optics Express*, 16, 8286–8291.

Chen, X., Zhang, X., Wang, Q., Li, P., Li, S., Cong, Z., Jia, G. & Tu, C. 2008. Highly efficient diode-pumped actively Q-switched Nd: YAG-$SrWO_4$ intracavity Raman laser. *Optics Letters*, 33, 705–707.

Chen, X. H., Zhang, X. Y., Wang, Q. P., Li, P., Li, S. T., Cong, Z. H., Liu, Z. J., Fan, S. Z. & Zhang, H. J. 2009. Diode side-pumped actively Q-switched Nd:YAG/$SrWO_4$ Raman laser with high average output power of over 10 W at 1180 nm. *Laser Physics Letters*, 6, 363–366.

Chen, X. H., Zhang, X. Y., Wang, Q. P., Li, P., Liu, Z. J., Cong, Z. H., Li, L. & Zhang, H. J. 2011. Highly efficient double-ended diffusion-bonded Nd:YVO_4 1525-nm eye-safe Raman laser under direct 880-nm pumping. *Applied Physics B: Lasers and Optics*, 1–4.

Chen, Y. F. 2004. Compact efficient self-frequency raman conversion in diode-pumped passively Q-switched Nd:$GdVO_4$ laser. *Applied Physics B: Lasers and Optics*, 78, 685–687.

Chen, Y. F. 2005. Stimulated Raman scattering in a potassium titanyl phosphate crystal: Simultaneous self-sum frequency mixing and self-frequency doubling. *Optics Letters*, 30, 400–402.

Chen, Y. F., Su, K. W., Zhang, H. J., Wang, J. Y. & Jiang, M. H. 2005. Efficient diode-pumped actively Q-switched Nd:YAG/$BaWO_4$ intracavity Raman laser. *Optics Letters*, 30, 3335–3337.

Cong, Z., Zhang, X., Wang, Q., Liu, Z., Li, S., Chen, X., Fan, S., Zhang, H. & Tao, X. 2009. Efficient diode-end-pumped actively Q-switched Nd:YAG/$SrWO_4$/KTP yellow laser. *Optics Letters*, 34, 2610–2612.

Cong, Z., Zhang, X., Wang, Q., Liu, Z., Chen, X., Fan, S., Zhang, H., Tao, X. & Li, S.

© Woodhead Publishing Limited, 2013

2010. Theoretical and experimental study on the Nd:YAG/$BaWO_4$/KTP yellow laser generating 8.3 W output power. *Optics Express*, 18, 12111–12118.

Dashkevich, V. I. & Orlovich, V. A. 2011. Ring solid-state Raman laser at 1538 nm. *Laser Physics Letters*, 8, 661–667.

Dashkevich, V. I., Orlovich, V. A. & Shkadarevich, A. P. 2009. Intracavity Raman laser generating a third Stokes component at 1.5 μm. *Journal of Applied Spectroscopy*, 76, 685–691.

Dekker, P., Pask, H. M. & Piper, J. A. 2007a. All-solid-state 704 mW continuous-wave yellow source based on an intracavity, frequency-doubled crystalline raman laser. *Optics Letters*, 32, 1114–1116.

Dekker, P., Pask, H. M., Spence, D. J. & Piper, J. A. 2007b. Continuous-wave, intracavity doubled, self-Raman laser operation in Nd:$GdVO_4$ at 586.5 nm. *Optics Express*, 15, 7038–7046.

Demidovich, A. A., Grabtchikov, A. S., Lisinetskii, V. A., Burakevich, V. N., Orlovich, V. A. & Kiefer, W. 2005. Continuous-wave raman generation in a diode-pumped Nd^{3+}: $KGd(WO_4)_2$ laser. *Optics Letters*, 30, 1701–1703.

Ding, S., Zhang, X., Wang, Q., Jia, P., Zhang, C. & Liu, B. 2006. Numerical optimization of the extracavity raman laser with barium nitrate crystal. *Optics Communications*, 267, 480–486.

Eckhardt, G., Bortfeld, D. P. & Geller, M. 1963. Stimulated emission of Stokes and anti-Stokes Raman lines from diamond, calcite, and sulfur single crystals. *Applied Physics Letters*, 3, 137–138.

Eremenko, A. S., Karpukhin, S. N. & Stepanov, A. I. 1980. Stimulated Raman scattering of the second harmonic of a neodymium laser in nitrate crystals. *Soviet Journal of Quantum Electronics*, 10, 113–114.

Ermolenkov, V. V., Lisinetskii, V. A., Mishkel, Y. I., Grabtchikov, A. S., Chaikovskii, A. P. & Orlovich, V. A. 2005. A radiation source based on a solid-state Raman laser for diagnosing tropospheric ozone. *Journal of Optical Technology* (a translation of *Opticheskii Zhurnal*), 72, 32–36.

Fan, L., Fan, Y. X., Li, Y. Q., Zhang, H., Wang, Q., Wang, J. & Wang, H. T. 2009. High-efficiency continuous-wave raman conversion with a $BaWO_4$ Raman crystal. *Optics Letters*, 34, 1687–1689.

Fan, L., Fan, Y. X. & Wang, H. T. 2010. A compact efficient continuous-wave self-frequency Raman laser with a composite YVO_4/Nd:YVO_4/YVO_4 crystal. *Applied Physics B: Lasers and Optics*, 101, 493–496.

Feng, Y., Taylor, L. R. & Calia, D. B. 2009. 150 W highly-efficient raman fiber laser. *Optics Express*, 17, 23678–23683.

Feve, J. P. M., Shortoff, K. E., Bohn, M. J. & Brasseur, J. K. 2011. High average power diamond Raman laser. *Optics Express*, 19, 913–922.

Girkin, J. M. & McConnell, G. 2005. Advances in laser sources for confocal and multiphoton microscopy. *Microscopy Research and Technique*, 67, 8–14.

Grabtchikov, A. S., Kuzmin, A. N., Lisinetskii, V. A., Orlovich, V. A., Ryabtsev, G. I. & Demidovich, A. A. 1999. All solid-state diode-pumped raman laser with self-frequency conversion. *Applied Physics Letters*, 75, 3742–3744.

Grabtchikov, A. S., Lisinetskii, V. A., Orlovich, V. A., Schmitt, M., Maksimenka, R. & Kiefer, W. 2004. Multimode pumped continuous-wave solid-state raman laser. *Optics Letters*, 29, 2524–2526.

Grabtchikov, A. S., Lisinetskii, V. A., Orlovich, V. A., Schmitt, M., Schluecker, S., Kuestner, B. & Kiefer, W. 2009. Continuous-wave solid-state two-stokes Raman laser. *IEEE Journal of Quantum Electronics*, 39, 624–626.

© Woodhead Publishing Limited, 2013

Granados, E., Pask, H. M. & Spence, D. J. 2009. Synchronously pumped continuous-wave mode-locked yellow Raman laser at 559 nm. *Optics Express*, 17, 569–574.

Granados, E., Pask, H. M., Esposito, E., McConnell, G. & Spence, D. J. 2010. Multi-wavelength, all-solid-state, continuous wave mode locked picosecond raman laser. *Optics Express*, 18, 5289–5294.

He, C. & Chyba, T. H. 1997. Solid-state barium nitrate Raman laser in the visible region. *Optics Communications*, 135, 273–278.

Huang, J., Lin, J., Su, R., Li, J., Zheng, H., Xu, C., Shi, F., Lin, Z., Zhuang, J., Zeng, W. & Lin, W. 2007. Short pulse eye-safe laser with a stimulated raman scattering self-conversion based on a Nd:KGW crystal. *Optics Letters*, 32, 1096–1098.

Innocenzi, M. E., Yura, H. T., Fincher, C. L. & Fields, R. A. 1990. Thermal modeling of continuous-wave end-pumped solid-state lasers. *Applied Physics Letters*, 56, 1831–1833.

Jelínková, H. 2011. Diamond Raman laser in eye safe region. *Proc. SPIE*, 8306, 83060G.

Kaminskii, A. A. (Ed.) 1996. *Laser Crystals: Their physics and properties*, 4th edn, Springer, New York.

Kaminskii, A. A., Ueda, K. I., Eichler, H. J., Kuwano, Y., Kouta, H., Bagaev, S. N., Chyba, T. H., Barnes, J. C., Gad, G. M. A., Murai, T. & Lu, J. 2001. Tetragonal vanadates YVO_4 and $GdVO_4$ – New efficient $\chi(3)$-materials for Raman lasers. *Optics Communications*, 194, 201–206.

Kananovich, A., Demidovich, A., Danailov, M., Grabtchikov, A. & Orlovich, V. 2010. All-solid-state quasi-CW yellow laser with intracavity self-Raman conversion and sum frequency generation. *Laser Physics Letters*, 7, 573–578.

Karpukhin, S. N. & Stepanov, A. I. 1986. SRS oscillation in the cavity in Ba(nNO), NaNo and CaCO crystals. *Soviet Journal of Quantum Electronics*, 16, 1027–1031.

Kitzler, O., McKay, A. & Mildren, R. P. 2011. CW diamond laser architecture for high power Raman beam conversion. *IQEC/CLEO Pacific Rim*, Sydney.

Lagatsky, A. A., Abdolvand, A. & Kuleshov, N. V. 2000. Passive Q switching and self-frequency Raman conversion in a diode-pumped Yb:$KGd(WO_4)_2$ laser. *Optics Letters*, 25, 616–618.

Lee, A. J., Pask, H. M., Dekker, P. & Piper, J. A. 2008. High efficiency, multi-watt CW yellow emission from an intracavity-doubled self-Raman laser using Nd:$GdVO_4$. *Optics Express*, 16, 21958–21963.

Lee, A. J., Lin, J. & Pask, H. M. 2010a. Near-infrared and orange-red emission from a continuous-wave, second-stokes self-Raman Nd:$GdVO_4$ laser. *Optics Letters*, 35, 3000–3002.

Lee, A. J., Pask, H. M., Piper, J. A., Zhang, H. & Wang, J. 2010b. An intracavity, frequency-doubled $BaWO_4$ Raman laser generating multi-watt continuous-wave, yellow emission. *Optics Express*, 18, 5984–5992.

Lee, A. J., Pask, H. M., Spence, D. J. & Piper, J. A. 2010c. Efficient 5.3 W cw laser at 559 nm by intracavity frequency summation of fundamental and first-Stokes wavelengths in a self-Raman Nd:$GdVO_4$ laser. *Optics Letters*, 35, 682–684.

Lee, A. J., Spence, D. J., Piper, J. A. & Pask, H. M. 2010d. A wavelength-versatile, continuous-wave, self-Raman solid-state laser operating in the visible. *Optics Express*, 18, 20013–20018.

Li, X., Lee, A. J., Pask, H. M., Piper, J. A. & Huo, Y. 2011a. Efficient, miniature, cw yellow source based on an intracavity frequency-doubled Nd:YVO_4 self-Raman laser. *Optics Letters*, 36, 1428–1430.

© Woodhead Publishing Limited, 2013

Li, X., Pask, H. M., Lee, A. J., Huo, Y., Piper, J. A. & Spence, D. J. 2011b. A miniature wavelength-selectable Raman laser: New insights for optimising performance. *Optics Express*, 19, 25623–25631.

Lin, J., Pask, H. M., Lee, A. J. & Spence, D. J. 2010. Study of relaxation oscillations in continuouswave intracavity Raman lasers. *Optics Express*, 18, 11530–11536.

Lin, J., Pask, H. M., Lee, A. J. & Spence, D. J. 2011. Study of amplitude noise in a continuous-wave intracavity frequency-doubled Raman laser. *IEEE Journal of Quantum Electronics*, 47, 314–319.

Lisinetskii, V. A., Grabtchikov, A. S., Demidovich, A. A., Burakevich, V. N., Orlovich, V. A. & Titov, A. N. 2007a. Nd:KGW/KGW crystal: efficient medium for continuous-wave intracavity Raman generation. *Applied Physics B: Lasers and Optics*, 88, 499–501.

Lisinetskii, V. A., Grabtchikov, A. S., Khodasevich, I. A., Eichler, H. J. & Orlovich, V. A. 2007b. Efficient high energy 1st, 2nd or 3rd Stokes Raman generation in IR region. *Optics Communications*, 272, 509–513.

Liu, Z., Wang, Q., Zhang, X., Chang, J., Wang, H., Zhang, S., Fan, S., Sun, W., Jin, G., Tao, X. & Zhang, H. 2009a. A $KTiOAsO_4$ Raman laser. *Applied Physics B: Lasers and Optics*, 94, 585–588.

Liu, Z., Wang, Q., Zhang, X., Zhang, S., Chang, J., Cong, Z., Sun, W., Jin, G., Tao, X. & Sun, Y. 2009b. A diode side-pumped $KTiOAsO_4$ Raman laser. *Optics Express*, 17, 6968–6974.

Lü, Y., Zhang, X., Li, S., Xia, J., Cheng, W. & Xiong, Z. 2010a. All-solid-state cw sodium D2 resonance radiation based on intracavity frequency-doubled self-Raman laser operation in double-end diffusion-bonded Nd^{3+}:$LuVO_4$ crystal. *Optics Letters*, 35, 2964–2966.

Lü, Y. F., Cheng, W. B., Xiong, Z., Lu, J., Xu, L. J., Sun, G. C. & Zhao, Z. M. 2010b. Efficient cw laser at 559 nm by intracavity sum-frequency mixing in a self-Raman Nd:YVO_4 laser under direct 880 nm diode laser pumping. *Laser Physics Letters*, 7, 787–789.

Lubeigt, W., Bonner, G. M., Hastie, J. E., Dawson, M. D., Burns, D. & Kemp, A. J. 2010. An intra-cavity raman laser using synthetic single-crystal diamond. *Optics Express*, 18, 16765–16770.

Lubeigt, W., Savitski, V. G., Bonner, G. M., Geoghegan, S. L., Friel, I., Hastie, J. E., Dawson, M. D., Burns, D. & Kemp, A. J. 2011. 1.6 W continuous-wave Raman laser using low-loss synthetic diamond. *Optics Express*, 19, 6938–6944.

Mahoney, K., Hwang, D., Oien, A. L., Bennett, G. T., Kukla, M., Burgio, K. & Anderson, C. 2006. Compact short pulse eyesafe solid state Raman laser. *Tech. Dig. Adv. Solid State Photon. Conf.*, 29 January 2006.

Mildren, R. P. 2011. Side-pumped crystalline Raman laser. *Optics Letters*, 36, 235–237.

Mildren, R. P. & Sabella, A. 2009. Highly efficient diamond Raman laser. *Optics Letters*, 34, 2811–2813.

Mildren, R. P., Convery, M., Pask, H. M., Piper, J. A. & McKay, T. 2004. Efficient, all-solid-state, Raman laser in the yellow, orange and red. *Optics Express*, 12, 785–790.

Mildren, R. P., Pask, H. M., Ogilvy, H. & Piper, J. A. 2005. Discretely tunable, all-solid-state laser in the green, yellow, and red. *Optics Letters*, 30, 1500–1502.

Mildren, R. P., Ogilvy, H. & Piper, J. A. 2007. Solid-state Raman laser generating discretely tunable ultraviolet between 266 and 320 nm. *Optics Letters*, 32, 814–816.

Mildren, R. P., Butler, J. E. & Rabeau, J. R. 2008. CVD-diamond external cavity Raman laser at 573 nm. *Optics Express*, 16, 18950–18955.

© Woodhead Publishing Limited, 2013

Mildren, R. P., Coutts, D. W. & Spence, D. J. 2009. All-solid-state parametric Raman anti-Stokes laser at 508 nm. *Optics Express*, 17, 810–818.

Mochalov, I. V. 1997. Laser and nonlinear properties of the potassium gadolinium tungstate laser crystal $KGd(WO_4)_2:Nd^{3+}$–(KGW:Nd). *Optical Engineering*, 36, 1660–1669.

Murray, J. T., Powell, R. C., Peyghambarian, N., Smith, D., Austin, W. & Stolzenberger, R. A. 1995. Generation of 1.5 μm radiation through intracavity solid-state Raman shifting in $Ba(NO_3)_2$ nonlinear crystals. *Optics Letters*, 20, 1017–1019.

Murray, J. T., Austin, W. L. & Powell, R. C. 1998. End-pumped intracavity solid-state raman lasers. *OSA Trends in Optics and Photonics Series*, 19, 129–135.

Murray, J. T., Austin, W. L. & Powell, R. C. 1999. Intracavity Raman conversion and raman beam cleanup. *Optical Materials*, 11, 353–371.

Omatsu, T., Lee, A., Pask, H. M. & Piper, J. 2009. Passively Q-switched yellow laser formed by a self-Raman composite $Nd:YVO_4/YVO_4$ crystal. *Applied Physics B: Lasers and Optics*, 97, 799–804.

Palero, J. A., Boer, V. O., Vijverberg, J. C., Gerritsen, H. C. & Sterenborg, H. J. C. M. 2005. Short-wavelength two-photon excitation fluorescence microscopy of tryptophan with a photonic crystal fiber based light source. *Optics Express*, 13, 5363–5368.

Parrotta, D. C., Lubeigt, W., Kemp, A. J., Burns, D., Dawson, M. D. & Hastie, J. E. 2011. Continuous-wave Raman laser pumped within a semiconductor disk laser cavity. *Optics Letters*, 36, 1083–1085.

Pask, H. M. 2003. The design and operation of solid-state Raman lasers. *Progress in quantum electronics*, 27, 3–56.

Pask, H. M. 2005. Continuous-wave, all-solid-state, intracavity Raman laser. *Optics Letters*, 30, 2454–2456.

Pask, H. M. & Piper, J. A. 1999. Efficient all-solid-state yellow laser source producing 1.2-W average power. *Optics Letters*, 24, 1490–1492.

Pask, H. M. & Piper, J. A. 2000. Diode-pumped $LiIO_3$ intracavity Raman lasers. *IEEE Journal of Quantum Electronics*, 36, 949–955.

Pask, H. M., Myers, S., Piper, J. A., Richards, J. & McKay, T. 2003. High average power, all-solid-state external resonator Raman laser. *Optics Letters*, 28, 435–437.

Pask, H. M., Dekker, P., Mildren, R. P., Spence, D. J. & Piper, J. A. 2008a. Wavelength-versatile visible and UV sources based on crystalline raman lasers. *Progress in Quantum Electronics*, 32, 121–158.

Pask, H. M., Mildren, R. P. & Piper, J. A. 2008b. Optical field dynamics in a wavelength-versatile, all-solid-state intracavity cascaded pulsed Raman laser. *Applied Physics B: Lasers and Optics*, 93, 507–513.

Penzkofer, A., Laubereau, A. & Kaiser, W. 1979. High intensity Raman interactions. *Progress in Quantum Electronics*, 6, 55–140.

Piper, J. A. & Pask, H. M. 2007. Crystalline Raman lasers. *IEEE Journal on Selected Topics in Quantum Electronics*, 13, 692–704.

Rini, M., Cristiani, I. & Degiorgio, V. 2000. Numerical modeling and optimization of cascaded CW Raman fiber lasers. *IEEE Journal of Quantum Electronics*, 36, 1117–1122.

Rong, H., Jones, R., Liu, A., Cohen, O., Hak, D., Fang, A. & Paniccia, M. 2005. A continuous-wave Raman silicon laser. *Nature*, 433, 725–728.

Rong, H., Kuo, Y. H., Xu, S., Liu, A., Jones, R., Paniccia, M., Cohen, O. & Raday, O. 2006. Monolithic integrated Raman silicon laser. *Optics Express*, 14, 6705–6712.

Sabella, A., Piper, J. A. & Mildren, R. P. 2010. 1240 nm diamond Raman laser operating near the quantum limit. *Optics Letters*, 35, 3874–3876.

© Woodhead Publishing Limited, 2013

Sato, Y., Taira, T., Pavel, N. & Lupei, V. 2003. Laser operation with near quantum-defect slope efficiency in Nd:YVO_4 under direct pumping into the emitting level. *Applied Physics Letters*, 82, 844–846.

Savitski, V. G., Burns, D. & Kemp, A. J. 2011a. Low-loss synthetic single-crystal diamond: Raman gain measurement and high power Raman laser at 1240 nm. *CLEO/Europe*, Munich, 22 May 2011.

Savitski, V. G., Hastie, J. E., Dawson, M. D., Burns, D. & Kemp, A. J. 2011b. Multi-watt continuous-wave diamond Raman laser at 1217 nm. *CLEO/Europe*, Munich, 22 May 2011.

Simons, J., Pask, H., Dekker, P. & Piper, J. 2004. Small-scale, all-solid-state, frequency-doubled intracavity Raman laser producing 5 mW yellow-orange output at 598 nm. *Optics Communications*, 229, 305–310.

Spence, D. J., Dekker, P. & Pask, H. M. 2007. Modeling of continuous wave intracavity Raman lasers. *IEEE Journal on Selected Topics in Quantum Electronics*, 13, 756–763.

Spence, D. J., Granados, E. & Mildren, R. P. 2010. Mode-locked picosecond diamond Raman laser. *Optics Letters*, 35, 556–558.

Takei, N., Suzuki, S. & Kannari, F. 2002. 20-Hz operation of an eye-safe cascade Raman laser with a $Ba(NO_3)_2$ crystal. *Applied Physics B: Lasers and Optics*, 74, 521–527.

Taylor, L. R., Feng, Y. & Calia, D. B. 2010. 50 W CW visible laser source at 589 nm obtained via frequency doubling of three coherently combined narrow-band Raman fibre amplifiers. *Optics Express*, 18, 8540–8555.

Wang, J., Zhang, H., Wang, Z., Ge, W., Zhang, J. & Jiang, M. 2006. Growth, properties and Raman shift laser in tungstate crystals. *Journal of Crystal Growth*, 292, 377–380.

Wang, Z. P., Hu, D. W., Fang, X., Zhang, H. J., Xu, X. G., Wang, J. Y. & Shao, Z. S. 2008. Eye-safe Raman laser at 1.5 μm based on $BaWO_4$ crystal. *Chinese Physics Letters*, 25, 122–124.

Xia, J., Lü, Y. F., Zhang, X. H., Cheng, W. B., Xiong, Z., Lu, J., Xu, L. J., Sun, G. C., Zhao, Z. M. & Tan, Y. 2011. All-solid-state CW Nd:$KGd(WO_4)_2$ self-Raman laser at 561 nm by intracavity sum-frequency mixing of fundamental and first-stokes wavelengths. *Laser Physics Letters*, 8, 21–23.

Yu, H., Li, Z., Lee, A. J., Li, J., Zhang, H., Wang, J., Pask, H. M., Piper, J. A. & Jiang, M. 2011. A continuous wave $SrMoO_4$ Raman laser. *Optics Letters*, 36, 579–581.

Zong, N., Cui, Q. J., Ma, Q. L., Zhang, X. F., Lu, Y. F., Li, C. M., Cui, D. F., Xu, Z. Y., Zhang, H. J. & Wang, J. Y. 2009. High average power 1.5 μm eye-safe Raman shifting in $BaWO_4$ crystals. *Applied Optics*, 48, 7–10.

Zverev, P. G., Murray, J. T., Powell, R. C., Reeves, R. J. & Basiev, T. T. 1993. Stimulated Raman scattering of picosecond pulses in barium nitrate crystals. *Optics Communications*, 97, 59–64.

© Woodhead Publishing Limited, 2013

18
Cryogenic lasers

D. RAND, J. HYBL and T. Y. FAN, Massachusetts Institute of Technology, USA

DOI: 10.1533/9780857097507.2.525

Abstract: Cryogenic lasers are an attractive technology for average power scaling in a single aperture with a near-diffraction-limited beam. This chapter begins with a history of cryogenic lasers, followed by a discussion on how thermo-optic material properties play a critical role in the design of such systems. We present a survey of thermo-optic property measurements for various host materials at both room and cryogenic temperatures, including thermal conductivity, coefficient of thermal expansion and temperature. Finally, we review recent progress in lasers using cryogenically cooled Yb^{3+}-doped gain media, with an emphasis on high average power.

Key words: cryogenic lasers, ytterbium-doped gain media, diode-pumped solid-state lasers, thermo-optic material properties.

18.1 Introduction

The properties of solid-state laser materials at cryogenic temperature are attractive for high-power operation and efficiency. This enables high-average-power scaling and high efficiency within simple laser architectures. Operation of lasers at cryogenic temperatures has been used since the very earliest days of laser development. Generally, operation at low temperatures has been viewed as an undesirable and impractical means to improve laser performance. However, many of the fundamental laser materials properties (thermal conductivity, thermal expansion, change in refractive index with temperature (d*n*/d*T*), saturation intensity and fluence) improve significantly as the temperature decreases. Additionally, many rare-earth-ion-doped solid-state lasers that are quasi-three-level lasers at 300 K become four-level lasers at cryogenic temperature, leading to more efficient laser operation. The overhead associated with cryogenic cooling has been mitigated over time as cryogenics have become increasingly ubiquitous. This chapter provides an overview of the use of cryogenically cooling to ~77 K for the purpose of power scaling pulsed and continuous wave (cw) lasers. The chapter is organized into four sections: (1) History of cryogenically cooled lasers, (2) Laser material properties at cryogenic temperatures, (3) Recent cryogenic laser achievements, and (4) Conclusion and future trends.

© Woodhead Publishing Limited, 2013

18.2 History of cryogenically cooled lasers

The history of cryogenic solid-state lasers is nearly as long as the history of lasers. For example, the second solid-state laser material ever demonstrated was U-doped CaF_2 operated with liquid He cooling (Sorokin and Stevenson 1960), just a few months after the demonstration of the ruby laser. Operation at low temperature depleted the thermally induced lower-laser-level population, leading to the first demonstration of a four-level laser. Throughout the subsequent decades, cryogenic cooling was used extensively in demonstrations of solid-state lasers. For example, 50 W of power at 2-μm wavelength from a Ho-doped yttrium aluminum garnet (YAG) laser was demonstrated by the mid-1970s (Beck and Gürs 1975). However, over much of this time, the use of cryogenic cooling was viewed negatively because of the difficulties in cryogenic engineering and the operational difficulties in supporting cryogenic systems. With the rise in performance of cryogenic lasers that has occurred over the past decade, this perception has changed to a more balanced view. As a result, the significant benefits of cryogenic cooling – power scalability with good beam quality – must be weighed against its costs. For some applications, cryogenic cooling will never be viable, but for others the enhancements in performance are easily worth the costs.

There are four primary reasons why solid-state lasers have been cooled significantly below room temperature. First, many common solid-state laser transitions in rare-earth ions have lower laser levels in the ground-state manifold, just a few hundred wavenumbers above the ground state (quasi-three-level laser). Consequently, these lasers are quasi-three-level lasers at room temperature, but the lower laser levels become thermally depopulated at lower temperatures, resulting in a much lower laser threshold. Second, for some lasers, particularly transition-metal-ion based, e.g., Co:MgF_2 (Johnson *et al.* 1966, Moulton and Mooradian 1979) and Fe:ZnSe (Adams *et al.* 1999), the metastable-level lifetimes greatly increase at lower temperature because the nonradiative relaxation rates are reduced, which reduces laser threshold. Third, in the early development of diode-pumped solid-state lasers, the pump lasers themselves needed cryogenic cooling to operate. For example, in the first diode-pumped laser, U-doped CaF_2, both the pumps and the gain element were cooled to liquid He temperature (Keyes and Quist 1964). Finally, cryogenic cooling enables thermo-optic effects to be greatly reduced at a given power level because of improved thermo-optic properties of the gain medium at lower temperatures. Fundamentally, crystalline dielectrics generally have significantly higher thermal conductivity at liquid nitrogen temperature than at room temperature and the thermal expansion coefficient decreases significantly too, leading to smaller dn/dT. All of these differences in materials properties lead to improved power scalability at cryogenic temperature as compared with room temperature.

© Woodhead Publishing Limited, 2013

The need for cryogenic cooling for the first three reasons has been mitigated by various technical advancements. The development of high-brightness diode pumps has enabled quasi-three-level lasers to be operated efficiently at room temperature with reasonable threshold (although cryogenic operation fundamentally retains the advantage of lower threshold and higher optical-to-optical efficiency). Improvements in nonlinear optical materials have enabled optical parametric oscillators to operate in spectral regions where transition-metal-ion lasers might otherwise operate. Finally, highly efficient room-temperature diode pumps are available, eliminating the need for their cryogenic cooling.

On the other hand, progress has been slow in the area of average-power scaling of solid-state lasers at room temperature. The host materials used for high-power lasers are basically no different from those used two decades ago although there have been improvements, primarily limited to YAG, in available size with the advent of ceramic materials. Much of the progress over the past two decades has been primarily driven by the large improvements in diode pumps. There have also been continuing efforts in refinement of gain-element geometries (slabs, disks, etc.) and improvements in compensation of thermo-optic effects. Because of the fundamental improvement in materials properties, cryogenic cooling offers the possibility of ~100 times average-power based on the improved properties alone.

An early experiment that attributed improved laser performance to better thermal properties at cryogenic temperature was in Ti:Al_2O_3 lasers. Moulton (1986) at MIT Lincoln Laboratory found that the output power in a Ti:Al_2O_3 laser improved from 150 mW at 300 K to 450 mW at 77 K, which he ascribed to improved thermal conductivity. Continued efforts with improved quality Ti:Al_2O_3 materials later raised this power to 9 W with near-diffraction-limited beam quality pumped by 45 W from two argon ion lasers (Bass *et al.* 1989). Schulz and Henion (1989, 1991), also at MIT Lincoln Laboratory, recognized that the combination of higher thermal conductivity and lower dn/dT should enable >100 times average power at 77 K compared with 300-K operation in Ti:Al_2O_3 and demonstrated 350-W quasi-cw output power (60 mJ in 170 μs) in a near-diffraction-limited beam. Given the high thermal diffusivity and the beam size, the laser was operating nominally in thermal steady state at these pulse lengths. This work set the stage for continuing efforts in average-power scaling in cryogenic Ti:Al_2O_3 lasers (Zavelani-Rossi *et al.* 2000, Backus *et al.* 2001, Pittman *et al.* 2002, Matsushima *et al.* 2006), and at this point up to 40-W average power with good beam quality has been demonstrated from a cryogenic Ti:Al_2O_3 amplifier (Matsushima *et al.* 2006).

At about the same time as the Schulz and Henion work, significant progress was being made in the development of diode-pumped bulk Yb:YAG lasers. Yb:YAG (and other bulk Yb-doped lasers) had previously been operated

© Woodhead Publishing Limited, 2013

as cryogenic lasers because the laser threshold was too high to be reached with flashlamp pumps (Johnson *et al.* 1965, Bagdasarov *et al.* 1974) and low-brightness semiconductor light-emitting diodes (Reinberg *et al.* 1971). Lacovara *et al.* (1991) demonstrated the first room-temperature Yb:YAG laser enabled by the development of high-power InGaAs diode lasers at 940 nm, but part of their experiments compared 77-K operation with 300-K operation and found more efficient operation at the lower temperature, as anticipated. The lower laser level for the 1030-nm transition in Yb:YAG is at 612 cm^{-1} above the ground state, which is approximately 3 kT at 300 K and 9 kT at 100 K. Consequently, room-temperature Yb:YAG operates as a quasi-three-level laser with higher threshold and lower slope efficiency (Fan and Byer 1987) than at cryogenic temperatures at which the laser is essentially a four-level laser. Developments in room-temperature Yb:YAG have led to lasers with hundreds of watts of output power (Bruesselbach *et al.* 1997, Giesen 2004, and Patel *et al.* 2006), but managing thermo-optic effects to obtain good beam quality and achieving high efficiency remain significant engineering challenges. Recent work (Fan *et al.* 1998, Brown 1997) has recognized that going back to cryogenic cooling of Yb-doped lasers could help solve some of these power-scaling issues while enhancing the efficiency at the same time.

18.3 Laser material properties at cryogenic temperatures

The key thermal and thermo-optic material attributes often associated with laser performance — thermal conductivity κ, coefficient of thermal expansion (CTE) α, and dn/dT – usually improve as the temperature is lowered in crystalline dielectrics. This improvement is illustrated in Fig. 18.1, which plots these three parameters as a function of temperature for the canonical case of undoped YAG, a popular laser host material. As can be seen, each thermo-optic property scales favorably as the temperature is decreased from room temperature to 100 K; that is, the thermal conductivity increases, while dn/dT and CTE both decrease. These general temperature dependencies under cryogenic cooling play a vital role in improving the performance of a solid-state laser.

The thermal conductivity is deduced from the values of material density ρ, thermal diffusivity β, and specific heat at constant pressure C_p, using the following equation:

$$k(T) = \rho(T)\beta(T)C_p(T) \qquad [18.1]$$

The thermal diffusivity β determines the rate at which thermal equilibrium is reached. It is given by

$$\beta(T) = \Lambda(T)u(T) \qquad [18.2]$$

© Woodhead Publishing Limited, 2013

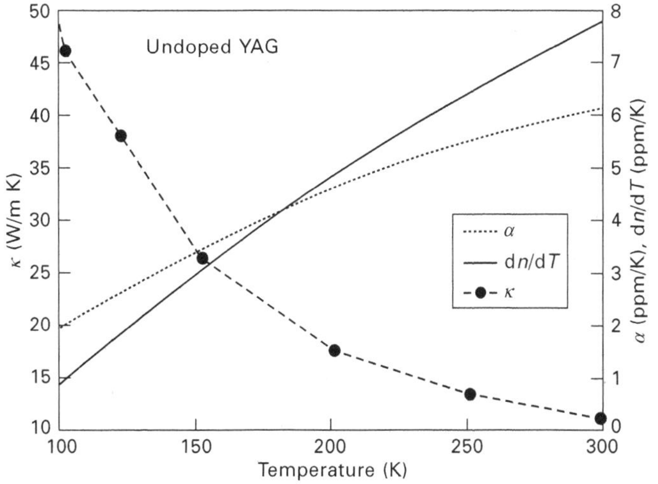

18.1 Plots of thermal conductivity κ, coefficient of thermal expansion α, and d*n*/d*T* as a function of temperature for undoped YAG.

where Λ and u are the mean free path and speed of the phonons, respectively. Upon cooling, the density ρ increases slightly while C_p decreases. The thermal diffusivity is expected to increase upon cooling from room temperature to cryogenic temperatures in the temperature regime where phonon–phonon scattering is dominant, because the mean free path generally increases upon cooling. As a result, thermal conductivity is expected to increase upon cooling from room temperature.

The CTE α is expected to decrease upon cooling, approaching zero as temperature approaches absolute zero. For isotropic materials, α is proportional to the specific heat according to the Grüneisen relation

$$\alpha(T) = 1/3K(T)\gamma_G\rho(T)C_v(T) \qquad [18.3]$$

where K is the compressibility, γ_G is the Grüneisen constant of the order of 2, and C_v is the specific heat at constant volume. Since the thermal coefficients of K and ρ are relatively small, the variation of α with temperature is essentially the same as that of C_v. At very low temperatures (T << Debye temperature), C_v is proportional to T^3. Hence, α will be nearly proportional to T^3 at very low temperatures.

The dn/dT is a function of both the wavelength of light and the temperature of the material. The wavelength region of interest for lasers lies between the electronic absorption edge on the short wavelength side and the phonon absorption bands on the long wavelength side. In this wavelength region, dn/dT is determined by the thermal coefficients of the absorption edge and the phonon bands, which decrease upon cooling. It turns out that in this

© Woodhead Publishing Limited, 2013

wavelength region, the sign of dn/dT is generally positive and its magnitude decreases upon cooling, as desired. However, there is no simple relationship for predicting the variation of dn/dT with temperature.

18.3.1 Laser material measurements

While general temperature dependence is known, measurements are required in order to provide accurate, quantitative values. Clearly, these data are needed in order to make critical assessments as to the desirability of various host materials and to provide engineering inputs for thermo-optic performance. Until recently, few data points below room temperature for laser hosts have been available. In this section, we compile data from several publications (Aggarwal *et al.* 2005, Fan *et al.* 2007, Rand *et al.* 2011b) that have begun to fill this void.

Of the materials measured, the hosts $Y_3Al_5O_{12}$ (YAG), $LiYF_4$ (YLF) and $YAlO_3$ (YALO) belong to the cubic, tetragonal and orthorhombic crystal classes, respectively. In the case of the cubic crystals, the sides a, b and c of the unit cell are equal to one another. In the case of the tetragonal crystals, two of the sides are equal but different from the third side, that is, $a \neq b \neq c$. In the case of the orthorhombic crystals, all three sides are different, that is, $a \neq b \neq c$. The axes of YALO are labeled according to the standardized space group *Pnma* (Diehl and Brandt 1975), where a = 5.3286, b = 7.3706, and c = 5.1796. Optically, YAG, YLF and YALO crystals are isotropic, uniaxial and biaxial, respectively. The principal refractive indices n_1, n_2 and n_3 are all equal to one another for isotropic crystals with $n_1 = n_2 = n_3 = n$. In the case of uniaxial crystals, two of the principal indices are equal to each other, but different from the third one. That is, $n_1 = n_2 \neq n_3$ with $n_1 = n_2 = n_o$, where n_o is called the ordinary refractive index for the light polarization perpendicular to the c-axis, and $n_3 = n_E$, where n_E is called the extraordinary refractive index for the light polarization parallel to the c-axis. In the case of biaxial crystals, all three principal refractive indices have different values with $n_1 = n_a$, $n_2 = n_b$ and $n_3 = n_c$. $Lu_3Al_5O_{12}$ (LuAG) and $LiLuF_4$ (LuLF) are isotropic and uniaxial, respectively. BaY_2F_8 (BYF), $KGd(WO_4)_2$ (KGW) and $KY(WO_4)_2$ (KYW) have monoclinic crystal structure and are optically biaxial. Thermo-optic properties of BYF, KGW and KYW were measured with samples oriented along the two-fold b-axis, which is the only axis of symmetry in these crystals.

The sesquioxides (Y_2O_3, Lu_2O_3, and Sc_2O_3) are cubic crystals and are known to have relatively good thermal conductivity. They have been made in both ceramic and single-crystal form. Here, the Yb:Y_2O_3, Yb:Lu_2O_3, Sc_2O_3, and Yb:Sc_2O_3 are ceramic samples and thus should have somewhat lower thermal conductivity than a single-crystal sample, although essentially the same CTE and dn/dT. Ytterbium-doped Y_2SiO_5 (YSO) and YLF are of

© Woodhead Publishing Limited, 2013

particular interest as materials for sub-picosecond laser operation because of their large gain bandwidths. Although YSO has monoclinic symmetry, we report here properties only along the direction of the two-fold symmetry axis. Also of interest is $Gd_3Sc_2Al_3O_{12}$ (GSAG), which is a YAG isomorph and so is expected to have similar thermal expansion and d*n*/d*T* to YAG. Cryogenic Yb:GSAG in combination with Yb:YAG has been used to provide a composite gain bandwidth that is larger than that of Yb:YAG alone (Rand *et al.* 2011a).

The measurement methods are the same as described in previous work (Aggarwal *et al.* 2005, Fan *et al.* 2007, Rand *et al.* 2011b). These methods have given consistent results; in particular, we have shown that the CTE and d*n*/d*T* measurements are self-consistent (Aggarwal *et al.* 2005), which is not possible without good accuracy. The laser-flash method is used to measure thermal diffusivity and specific heat, from which thermal conductivity can be calculated (Parker *et al.* 1961, Smith and Campbell 2002, ASTM). At cryogenic temperature, the thermal diffusivity can change rapidly with temperature in crystalline dielectrics, so to minimize errors associated with this variation, the temperature rise is kept to ~1 K or less for measurements at low temperature. Thermal expansion is measured using a double Michelson laser interferometer (Wolff and Savedra 1985) at Precision Measurements and Instruments Corporation on ~25-mm-long samples. The change in the optical path length with temperature is also measured using interferometry, and d*n*/d*T* is computed from those two measurements.

Table 18.1 compiles the thermal conductivities of the measured samples, as derived from measurements of thermal diffusivity and the specific heat (Aggarwal *et al.* 2005, Fan *et al.* 2007, Rand *et al.* 2011b). As expected, the thermal conductivity rises significantly. Also as anticipated, doping reduces the thermal conductivity, caused by the mass difference of the dopant ion (Gaume *et al.* 2003).

The CTE measurements are fitted to a second-order polynomial of the form

$$\alpha = M_0 + M_1 T + M_2 T^2, \qquad [18.4]$$

and the fits are valid in the range of approximately 90–320 K. The fitted coefficients are listed in Table 18.2, along with a computation from the polynomial fit at 300 K and 100 K. Comparison to data from other CTE measurements of these materials can be found (Aggarwal *et al.* 2005, Fan *et al.* 2007, Rand *et al.* 2011b), particularly at 300 K. As expected, the CTE, in general, drops significantly as the temperature decreases in these materials.

The d*n*/d*T* data are fitted to a third-order polynomial of the same form as Eq. [18.4]. The fitted coefficients and a computation of *dn/dT* at 300 K and 100 K from the fit are shown in Table 18.3. Comparison to this data from

© Woodhead Publishing Limited, 2013

Table 18.1 Thermal conductivity (W/mK)

Material	~100 K	120–150 K	~200 K	~250 K	~300 K
YAG	46.1 (102 K)	26.4 (152 K)	17.6 (201 K)	13.4 (251 K)	11.2 (299 K)
Ceramic YAG	38 (102 K)	30.3 (120 K)	16.7 (200 K)		10.6 (298 K)
Yb(2%):YAG	33.8 (98 K)	18.7 (149 K)	13.0 (199 K)	10.2 (250 K)	8.6 (298 K)
Yb(4%):YAG	25.0 (101 K)	15.5 (151 K)	11.8 (201 K)	9.6 (251 K)	8.2 (298 K)
Yb(15%):YAG	16.4 (101 K)	11.3 (152 K)	9.0 (202 K)	7.4 (251 K)	6.7 (298 K)
GSAG	11.6 (83 K)	8.50 (120 K)			4.76 (298 K)
LuAG	25.4 (101 K)	16.5 (151 K)	12.2 (201 K)	9.6 (250 K)	8.3 (298 K)
GGG	33 (91 K)	22 (121 K)	12.5 (201 K)		7.5 (298 K)
Ceramic Y_2O_3	52 (92 K)	36 (121 K)	21 (200 K)		13.0 (298 K)
Ceramic Yb(3%): Y_2O_3	52 (87 K)	35 (121 K)	22 (201 K)		13.4 (298 K)
Ceramic Yb(10%):Y_2O_3	11.0 (88 K)	9.89 (117 K)			6.12 (298 K)
Ceramic Yb(10%):Lu_2O_3	32 (92 K)	27 (121 K)	17 (200 K)		11.1 (298 K)
Sc_2O_3	19.8 (88 K)	18.9 (117 K)			12.4 (298 K)
Ceramic Yb(9%):Sc_2O_3	7.00 (88 K)	6.35 (117 K)			4.57 (298 K)
YALO (ll a)	64.9 (101 K)	32.7 (152 K)	20.3 (202 K)	14.6 (251 K)	11.7 (298 K)
Yb(5%):YALO (ll a)	21.1 (100 K)	13.5 (150 K)	9.9 (200 K)	8.1 (250)	7.1 (298 K)
YALO (ll b)	54.4 (101 K)	25.6 (152 K)	16.9 (202 K)	12.4 (251 K)	10.0 (298 K)
Yb(5%):YALO (ll b)	25.0 (100 K)	15.3 (150 K)	11.6 (199 K)	9.6 (250 K)	8.3 (298 K)
YALO (ll c)	77.6 (101 K)	35.4 (152 K)	23.3 (202 K)	17.0 (251 K)	13.3 (298 K)
Yb(5%):YALO (ll c)	21.2 (100 K)	13.5 (150 K)	10.7 (199 K)	8.7 (250 K)	7.6 (298 K)
$GdVO_4$ (ll c)	44 (91 K)	29 (121 K)	17 (200 K)		10.7 (298 K)
$GdVO_4$ (ll a)	38 (92 K)	25 (121 K)	14 (201 K)		8.6 (298 K)

© Woodhead Publishing Limited, 2013

Table 18.1 Continued

Material	~100 K	120–150 K	~200 K	~250 K	~300 K
YSO (II b)	16.0 (88 K)	10.5 (117 K)			4.66 (298 K)
YLF (II a)	24.2 (101 K)	12.4 (152 K)	8.3 (202 K)	6.3 (251 K)	5.3 (298 K)
Yb(5%):YLF (II a)	11.3 (100 K)	7.6 (150 K)	5.6 (200 K)	4.7 (250 K)	4.1 (298 K)
Yb(25%):YLF (IIa)	7.78 (83 K)	5.59 (120 K)			2.68 (298 K)
YLF (II c)	33.7 (101 K)	17.1 (152 K)	11.7 (202 K)	8.8 (251 K)	7.2 (298 K)
Yb(5%):YLF (II c)	13.7 (101 K)	9.6 (150 K)	7.3 (200 K)	6.0 (250 K)	5.2 (298 K)
Yb(25%):YLF (IIc)	9.35 (83 K)	6.96 (120 K)			3.41 (298 K)
LuLF (II a)	23.6 (101 K)	11.6 (152 K)	8.1 (202 K)	6.2 (251 K)	5.0 (298 K)
LuLF (II c)	31.3 (101 K)	15.1 (152 K)	10.6 (202 K)	7.8 (251 K)	6.3 (298 K)
BYF (II b)	9.7 (101 K)	5.8 (152 K)	4.9 (202 K)	4.1 (251 K)	3.5 (298 K)
KGW (II b)	6.8 (101 K)	4.7 (152 K)	3.6 (202 K)	3.0 (251 K)	2.6 (298 K)
KYW (II b)	7.4 (101 K)	4.8 (152 K)	3.6 (202 K)	3.0 (251 K)	2.7 (298 K)

other d*n*/d*T* measurements can be found (Aggarwal *et al.* 2005, Fan *et al.* 2007, Rand *et al.* 2011b).

18.4 Recent cryogenic laser achievements

In this section, we review recent results in the field of cryogenically cooled Yb^{3+}-doped solid-state lasers. This is divided into three parts: cw results, nanosecond-class pulsed results, and ultrashort laser results.

18.4.1 Continuous wave operation

Due to their small quantum defect and low saturation intensity, cryogenically cooled Yb-doped lasers are attractive for average power scaling cw waveforms and are competitive with fiber lasers in terms of efficiency and brightness. At reasonable intracavity intensities, >70% optical efficiency at kilowatt-class power levels and near-ideal beam quality can be achieved using simple

© Woodhead Publishing Limited, 2013

Table 18.2 Coeffiicient of thermal expansion (ppm/K)

Material	M_0	M_1	M_2 (× 10^{-3})	300 K	100 K
YAG	−1.8496	0.04 368	−0.056 844	6.14	1.95
GSAG	−1.156	0.040 551	−0.0478	6.7	2.4
LuAG	−1.4132	0.045 552	−0.06 798	6.13	2.46
GGG	−1.0278	0.043 952	−0.054 602	7.24	2.82
Y_2O_3	−4.2499	0.085 464	−0.23 333	0.39	1.96
Sc_2O_3	−2.785	0.049 076	−0.06 184	6.4	1.5
YALO (II a)	0.80 841	0.049 318	0.35 367	2.32	−1.16
YALO (II b)	−2.171	0.0 068 212	−0.15 492	8.08	3.24
YALO (II c)	−3.7894	0.088 694	−0.23 308	8.7	3.0
$GdVO_4$ (II c)	−4.3876	0.10 833	−0.31 269	−0.03	3.32
$GdVO_4$ (II a)	0.54 663	−0.006 105	0.063 961	4.47	0.58
YVO_4 (II a)	−0.254	0.008 297	−0.003 871	1.9	0.54
YVO_4 (II c)	−3.335	0.071 231	−0.1058	8.5	3.7
YSO (II b)	−1.231	0.042 648	−0.05 909	6.3	2.4
YLF (II a)	−7.9972	0.1181	−0.14 579	14.31	2.36
YLF (II c)	−4.7454	0.094 241	−0.14 974	10.05	3.18
LuLF (II a)	−8.6582	0.13 791	−0.21 263	13.6	3.0
LuLF (II c)	−4.1475	0.096 003	−0.15 365	10.8	3.9
BYF (II b)	−0.78 163	0.12 477	−0.21 644	17.2	9.5
KGW (II b)	1.2115	−0.0 073 998	0.032 473	1.9	0.8
KYW (II b)	0.01 338	0.010 215	−0.0 010 818	3.0	1.0

end-pumped rod geometries. For example, building from of previous work (Ripin *et al.* 2004, 2005), Fan *et al.* (2007) demonstrated 455-W average power with 71% optical efficiency and an M^2 of 1.4 using a simple oscillator with an end-pumped rod. Utilizing a similar laser design, Feve *et al.* (2011) demonstrated near-diffraction-limited performance, $M^2 = 1.1$, up to 450-W output power. In this experiment, the beam began to show minor degradation at power levels >500 W, $M^2 = 1.1 \times 1.2$, due to the onset of astigmatic thermal lensing. These demonstrations were performed using simple resonator designs and end-pumped laser rods (or slabs).

For further power scaling, gain element geometries and more aggressive cooling schemes must be developed to manage the larger heat loads and avoid beam quality degradation and maintain efficiency. Brown *et al.* (2010) demonstrated power scaling to 963-W cw output power at 44% optical efficiency using a single-pass amplifier with thin laser crystals sandwiched between sapphire heat sinks, which were cooled with flowing LN_2. Relative to the end-pumped rods, this geometry reduces the path length for heat removal, and the two-sided cooling symmetrizes the thermal profile, as well as stress, in the laser crystal. Furuse *et al.* (2009) take a somewhat different approach with the total-reflection active-mirror (TRAM) laser. The TRAM is constructed using a composite ceramic laser crystal. The top of the crystal is a trapezoidal piece of undoped YAG and is used to couple pump and laser light into the thin bottom layer of cryogenically-cooled Yb:YAG. In

© Woodhead Publishing Limited, 2013

Table 18.3 Change in refractive index with temperature d*n*/d*T* at 1.06-μm wavelength (in ppm/K)

Material	M_0	M_1	M_2 ($\times 10^{-3}$)	M_3	300 K	100 K
YAG	–3.946	0.05 294	–0.045 605		7.8	0.9
LuAG	–2.3926	0.029 863	0.018 715		8.3	0.7
GGG	–5.8321	0.13 021	–0.031 604	3.12×10^{-7}	13.2	4.3
Y_2O_3	4.0719	–0.038 973	0.025 755	-3.51×10^{-7}	6.1	2.4
Sc_2O_3	–2.67	0.06 313	–0.171	2.68×10^{-7}	10.8	2.2
YALO (II a)	–3.761	0.052 286	–0.047 418		7.7	1.0
YALO (II b)	–3.12	0.089 199	–0.13 228		11.7	4.5
YALO (II c)	–3.3004	0.048 643	–0.033 055		8.3	1.2
$GdVO_4$ (II c)	–1.7052	0.053 633	–0.017 304	2.89×10^{-7}	6.6	2.2
$GdVO_4$ (II a)	–2.356	0.059 418	–0.00 662	6.15×10^{-8}	11.2	3.0
YVO_4 (II a)	2.86	–0.03 646	0.5623	-1.16×10^{-6}	11.2	3.7
YVO_4 (II c)	4.40	–0.0610	0.4453	-7.80×10^{-7}	5.1	2.0
YSO (II b)	14.52	–0.16 426	0.74	-9.27×10^{-7}	6.8	4.6
YLF (II a)	–0.10 364	0.0 010 536	–0.053 254		–4.6	–0.5
YLF (II c)	–0.038 934	–0.015 482	–0.021 742		–6.6	–1.8
LuLF (II a)	28.459	–0.54 623	2.6191	-3.8485×10^{-6}	–3.6	–7.2 (140 K)
LuLF (II c)	33.941	–0.62 295	2.9489	-4.388×10^{-6}	–6.0	–7.5 (140 K)

© Woodhead Publishing Limited, 2013

this geometry, the heat is generated only in the thin active layer, where it can be efficiently removed. However, since the pump and laser beams enter the doped region at a shallow angle, reasonable gain and absorption lengths can be achieved with very thin (hundreds of microns) layers of Yb:YAG. Furuse *et al.* (2009) have demonstrated 273 W of output power with 63% optical efficiency using this approach.

Yb:YLF is also an attractive candidate due to the potential to use the 960 nm pump/995 nm lasing transitions to realize a <4% quantum defect. Zapata *et al.* (2010) demonstrated 224-W average power with 68% slope efficiency (relative to absorbed power) using an end-pumped rod of cryogenically cooled Yb:YLF. These published results confirm that cryogenic solid-state lasers are competitive with other candidate technologies, e.g. fiber lasers, for efficient high-average power scaling in cw waveforms with excellent beam quality.

18.4.2 Nanosecond-class results

Solid-state lasers are very well suited to the generation of Q-switched pulses, allowing for high pulse energies in a nanosecond-class waveform. Typically, the radiative lifetime of the upper laser level in Yb-doped materials ranges from a few hundred microseconds to a few milliseconds at cryogenic temperature, enabling a high energy storage capability for Q-switched operation. As mentioned previously, cryogenic cooling of Yb-doped gain media also offers substantial improvements to laser kinetics by increasing efficiency, increasing emission cross-sections (and consequently reducing the saturation fluence or intensity), and thermally depopulating the terminal state to achieve four-level operation. As an example, the peak stimulated emission cross-section of Yb:YAG increases by a factor of about five as temperature is reduced from room temperature to 77 K. Likewise, the product of emission cross-section and upper-state lifetime increases by a similar factor (Fan *et al.* 2007, Dong *et al.* 2003). Saturation fluence and intensity are reduced accordingly, which in turn enables efficient operation of Q-switched oscillators at lower intracavity fluence.

Tokita *et al.* (2005) reported acousto-optically (AO) Q-switched Yb:YAG lasers that generate 30-mJ, 32-ns pulses at 1.5-kHz pulse repetition frequency (PRF) and as much as 70-W average power at higher pulse rates (5-kHz PRF with 75-ns pulsewidth). More recently, Feve *et al.* (2011) demonstrated AO Q-switching of cryo-Yb:YAG with 340-W output power for repetition rates in the range of 40–100 kHz. At 40 kHz and 340 W, the pulse duration was 75 ± 5 ns with 8.5-mJ pulse energy (Feve *et al.* 2011). An alternative to achieving high pulse energy in a nanosecond waveform is by using a master oscillator/power amplifier (MOPA) architecture. Kawanaka *et al.* (2010) demonstrated a MOPA system with a cryogenically cooled ceramic

© Woodhead Publishing Limited, 2013

Yb:YAG regenerative amplifier and a four-pass power amplifier consisting of a double-end-pumped Yb:YAG rod, also at cryogenic temperature. Pulse energies exiting the regenerative amplifier were 6.5 mJ and 1.5 mJ at repetition rates of 200 Hz and 1 kHz, respectively, in a 10-ns waveform. Beam quality was near diffraction limited, with $M^2 < 1.1$. The pulse energy at the output of the power amplifier was 140 mJ at 200-Hz PRF, with no reported beam quality.

Another way of achieving shorter duration Q-switched pulses of <20 ns full width at half-maximum (FWHM) in a high-power, high-gain cryo-Yb oscillator is by using an electro-optic Q-switch (Pockels cell) to maximize extracted pulse energy and minimize Q-switched pulse duration. Recently, we have demonstrated an electro-optically Q-switched, end-pumped, cryogenic Yb:YAG laser that simultaneously achieves 114-W average power at 5-kHz PRF, a 16-ns pulse duration, and near-diffraction-limited beam quality (Manni *et al.* 2010).

The laser configuration of our electro-optically Q-switched laser is shown in Fig. 18.2. A 1%-doped Yb:YAG crystal having an undoped endcap is single-end pumped with a fiber-coupled laser diode array at 940 nm. The crystal is 23 mm long (including a 1 mm thick endcap), with a 5 mm × 5 mm cross-section. Pump light is then focused into the Yb:YAG crystal through the high-reflector (HR) mirror of the laser resonator using a 150-mm focal length (FL) spherical lens. The diameter of the gain region in the Yb:YAG crystal is about 1.5 mm.

The HR mirror of the laser resonator is highly reflecting (>99.5%) at 1030 nm and highly transmitting (>98%) at the 940-nm pump wavelength. The mirror has a 6-meter convex radius of curvature. The resonator axis makes a single pass through the Yb:YAG crystal and then reflects off two thin-film dielectric polarizers. The BBO Pockels cell is a transverse-field, two-crystal device having a 5-mm clear aperture and a quarter-wave voltage of 4.4 kV at 1030 nm. The output coupler is a 10% reflecting flat mirror, and overall physical resonator length is 43 cm. Both the output coupler and resonator length were chosen to achieve a pulse duration of 16 ns at 100-W output power.

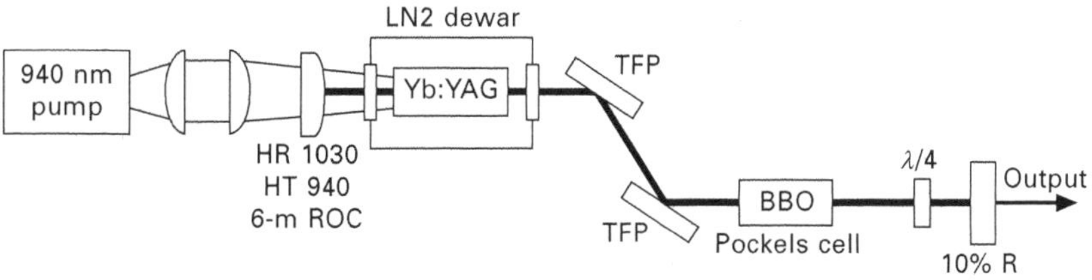

18.2 Layout of electro-optic Q-switched oscillator. HR/HT: highly reflecting (1030 nm)/highly transmitting (940 nm) with 6-m radius of curvature (ROC); TFP: thin-film dielectric polarizer.

© Woodhead Publishing Limited, 2013

Laser performance data for the resonator, operating in both cw mode and Q-switched at 5-kHz PRF, is shown in Fig. 18.3. At 244-W maximum pump power, cw power is 123 W and Q-switched average power is 114 W. Laser threshold is approximately 60 W. Average slope efficiency is 68% in cw mode and 63% in Q-switched mode. Absolute efficiency at 244-W pump power is 50% in cw mode and 47% in Q-switched mode.

The Q-switched laser pulse at 114-W average output power is shown in Fig. 18.4(a). The measured pulse width is 16 ns FWHM. The temporal profile was recorded using the full 500-MHz bandwidth of the oscilloscope and represents an average of 16 pulses. Shown in Fig. 18.4(b) is the near-field beam profile at 114-W output power and 5-kHz PRF. The profile is a 1:1 image of the beam at the laser's output coupler; at 114 W the beam diameter ($1/e^2$) is 1.2 mm. The far-field beam profile at the focus of a 25-cm FL lens is shown in Fig. 18.4(c). Far-field beam diameter at 114 W is 260 μm. Beam quality is near diffraction limited.

18.4.3 Ultrashort pulse lasers

Over the last decade, ultrashort pulse solid-state lasers with cryogenically cooled Yb-doped gain media have undergone a remarkable progression. Figure 18.5 plots a collection of ultrashort pulse laser demonstrations as a function of peak and average power. Early cryo-Yb demonstrations were focused on operation in one of two operating regimes: high peak power or high average power (Kawanaka *et al.* 2003, Tokita *et al.* 2007, Akahane *et al.* 2007, Ogawa *et al.* 2007). More recent results aim to demonstrate

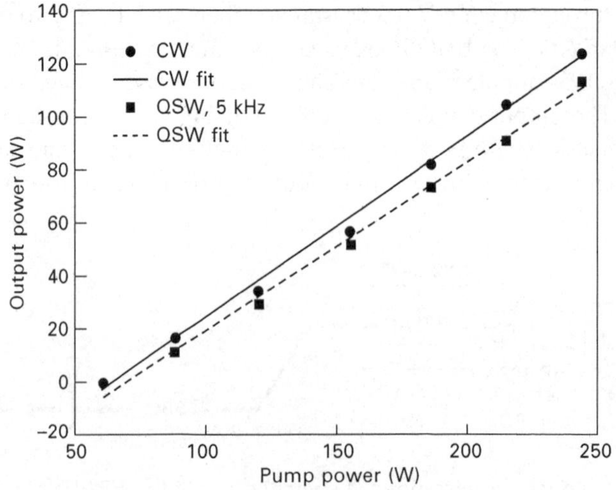

18.3 Continuous-wave (CW) and Q-switched (QSW) input–output data. Slope efficiencies are 68% CW and 63% QSW.

© Woodhead Publishing Limited, 2013

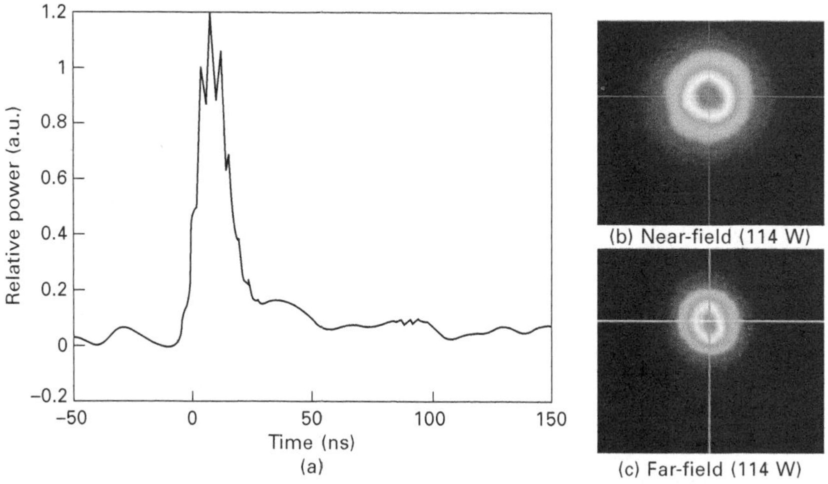

18.4 (a) Q-switched pulse shape (at 114 W, 5-kHz PRF). The FWHM of the pulse determined by a polynomial fit is 16 ns. The structure in the pulse is due to longitudinal mode beating that is only partially resolved by the oscilloscope (500-MHz bandwidth). (b) 1:1 image of the beam at the output coupler (Q-switched, 114 W). Beam diameter is 1.2 mm ($1/e^2$). (c) Far-field beam profile at focus of 25-cm FL lens. Beam diameter ($1/e^2$) at focus is 260 μm at 114-W output power.

average power scalability (Hong *et al.* 2008, Brown *et al.* 2010), as well as simultaneously scale both peak and average power (Rand *et al.* 2011a, Pugžlys *et al.* 2009, Furch *et al.* 2009, Hong *et al.* 2010, Curtis *et al.* 2011). Also shown, for comparison, are ultrashort pulse laser demonstrations using different gain media, including room-temperature Yb:YAG (Tümmler *et al.* 2009, Metzger *et al.* 2009, Russbueldt *et al.* 2010, Klingebiel *et al.* 2011, Schulz *et al.* 2011), Yb-doped fiber (Röser *et al.* 2007, Eidam *et al.* 2010), and cryogenically cooled Ti:sapphire (Backus *et al.* 2001, Matsushima *et al.* 2006).

Many ultrashort laser applications require high average and high peak power in a near-diffraction-limited beam. We define a figure of merit that is the product of the peak and average power. Loci of equal figures of merit are depicted in Fig. 18.5, e.g., by the line representing 10^{10} W^2. From this standpoint, one trend emerges for the cryo-Yb demonstrations plotted in Fig. 18.5, the rapidity of which is noteworthy. An arrow showing the general direction towards high peak and high average power is depicted, wherein the early demonstrations with high peak power are encircled on the left of Fig. 18.5 and the more recent demonstrations are encircled on the right.

Ultrashort-pulse lasers must confront additional challenges to those that afflict high-average-power lasers in the cw and nanosecond-pulse regime. High peak power may result in undesired nonlinear effects as well as damage

© Woodhead Publishing Limited, 2013

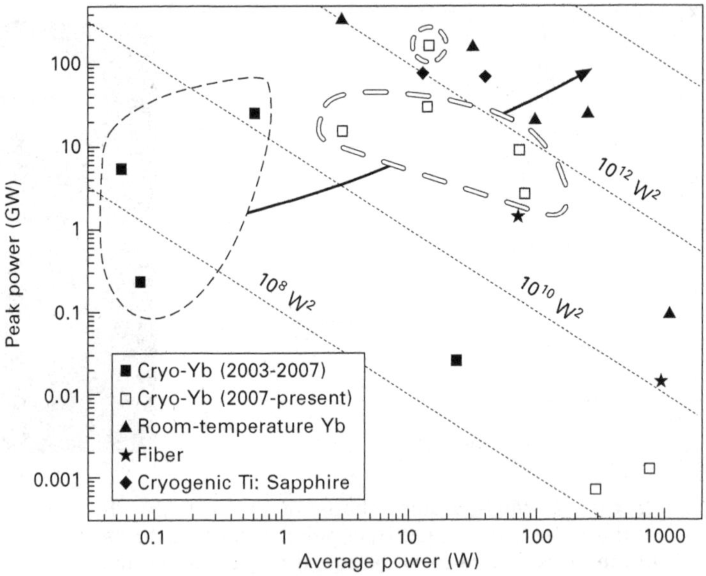

18.5 A compilation of recent cryogenically-cooled Yb-doped ultrashort pulse lasers, plotted as a function of average and peak power. Also included are representative demonstrations of different gain media, including Yb:YAG at room temperature, Yb-doped fiber, and cryogenically-cooled Ti:sapphire. For consistency, peak power is defined with respect to the FWHM pulse duration, and average power is defined prior to pulse compression.

of optical components, and is usually mitigated by temporally broadening the pulses during amplification (Strickland and Mourou 1985, Backus *et al.* 1998). Even with this technique, the saturation fluence can be significantly higher than the damage threshold, limiting extraction efficiency. Cryogenic cooling typically enhances the emission cross-section, to the benefit of energy extraction. However, this generally comes at the expense of a reduction in the gain bandwidth, and therefore an increase in the minimum pulse duration achievable.

For example, in Yb:YAG, the 5.3-nm FWHM bandwidth at room temperature reduces to 1.5 nm at liquid nitrogen temperature. In this section, we present the performance of a 100-W, kHz-class cryo-Yb:YAG amplifier. We then discuss two approaches to improve upon the minimum pulse duration: (1) the use of multiple laser host materials to exploit a composite gain bandwidth; and (2) the use of laser host materials with greater bandwidth at cryogenic temperature. With the former approach, Rand *et al.* (2011a) demonstrated 12-mJ pulse energy (post-compression) at 5-kHz PRF and 1.6-ps pulse duration using a combination of cryogenically cooled Yb:YAG and Yb:GSAG in the power amplifier. Taking the latter approach, Rand *et*

© Woodhead Publishing Limited, 2013

al. (2011b) demonstrated 1-mJ pulse energy at 10-kHz PRF from a cryo-Yb:YLF regenerative amplifier. Near-diffraction-limited performance was achieved with a spectral bandwidth of 2.2 nm FWHM (corresponding to a transform-limited pulse duration of 500 fs).

A cryo-Yb:YAG ultrashort power amplifier

We present here the results of a laser system (Rand *et al.* 2011b) which consists of a mode-locked Yb:KYW laser, a room-temperature Yb:YAG regenerative amplifier, and a Yb:YAG power amplifier at cryogenic temperature. The Yb:KYW laser operates with 450-mW average power at 30.5-MHz PRF with 450-fs-long output pulses. The output is fed to a grating-based stretcher and then to a regenerative amplifier, where a Pockels cell pulse picks at 5-kHz PRF. The pulse length of the stretched pulse downstream from the regenerative amplifier is about 150 ps. A schematic of the four-pass power amplifier is shown in Fig. 18.6. The gain crystals are 2.5-cm-long, 1%-doped Yb:YAG (including a 3.5-mm-long, undoped Yb:YAG endcap), which are placed in series in a liquid nitrogen cryostat. Each crystal is pumped from a single end by a fiber-coupled (0.4-mm diameter, 0.22 NA) diode laser system, with the pumps imaged to 4.5-mm-diameter spots in the gain crystals. The four passes are implemented using standard polarization multiplexing. The laser beam is re-imaged from the first to the second pass (and from third to fourth).

The average output power of the stretched pulse beam as a function of incident pump power is shown in Fig. 18.7, along with an inset showing the measured near-field at full power. Operation up to 115 W was achieved in a near-diffraction-limited beam.

Measurements of the input and output optical spectra are shown in Fig. 18.8. The spectrum from the regenerative amplifier is centered at 1029.8 nm with a bandwidth of approximately 1 nm FWHM. This center wavelength is slightly longer than the gain peak of cryo-Yb:YAG but with sufficient overlap to provide efficient extraction (*cf.* Fig. 18.10). After amplification, the peak has blue-shifted to 1029.5 nm and the width has narrowed, although a spectral tail to the red is still present. The recompression of the amplified pulse was not attempted, but a calculation of the transform-limited pulse duration demonstrates potential for 2–3 ps FWHM.

A cryo-Yb:YAG/Yb:GSAG ultrashort power amplifier

In a recent demonstration, we employed the technique of using multiple gain media to provide a larger bandwidth (Rand *et al.* 2011a). This approach is attractive because it directly leverages existing high-power cryogenic Yb:YAG technology. In this system, the second gain medium was Yb:GSAG, which was chosen because its properties are similar to those of YAG, both being

© Woodhead Publishing Limited, 2013

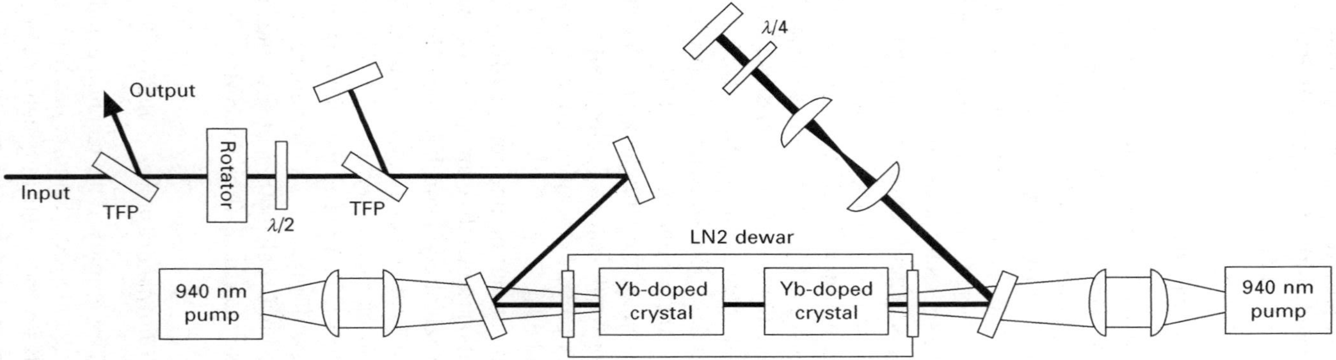

18.6 Schematic layout of four-pass power amplifier; TFP: thin-film polarizer.

© Woodhead Publishing Limited, 2013

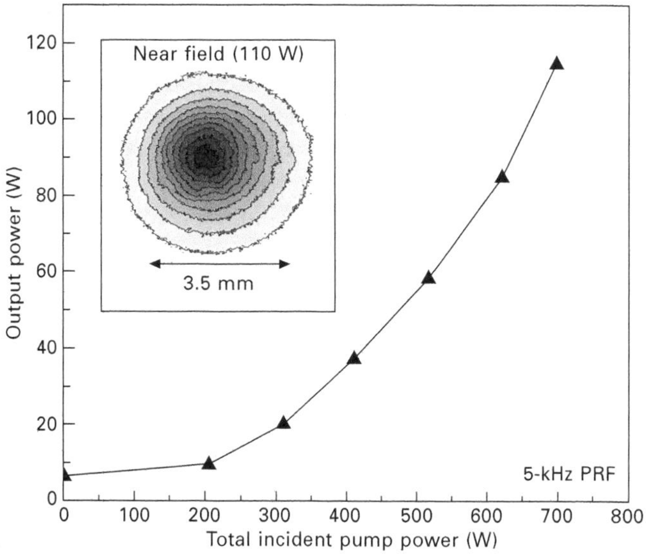

18.7 Average output power as a function of total incident pump power for the Yb:YAG power amplifier at 5-kHz PRF. The inset shows the near-field intensity profile at 110 W.

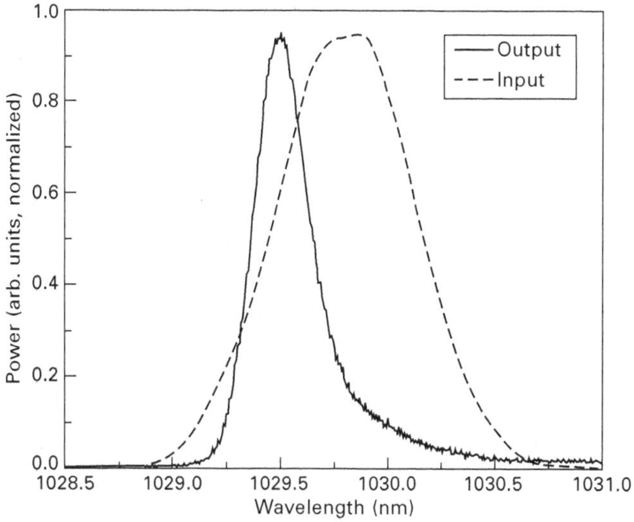

18.8 Spectral performance of Yb:YAG power amplifier at full power (115 W). The dashed curve and solid curve show the input and output spectrum of the power amplifier, respectively.

oxide garnets, and its gain peak at cryogenic temperature is offset from, but overlaps with, the gain spectrum in YAG.

For this demonstration, several modifications were made to the power

© Woodhead Publishing Limited, 2013

amplifier discussed in the previous section (see Fig. 18.6). One of the gain elements was replaced with a 1.5-cm-long, 2%-doped Yb:GSAG crystal. The pump beam diameter was decreased from 4.5 mm to 3 mm, and the spectrum of the regenerative amplifier was wavelength-tuned to the red to better match the cryo-Yb:GSAG gain peak. Finally, the quarter-wave plate is replaced with a Faraday rotator, which provides birefringence compensation (Sherman 1998, Lü *et al.* 1996). We estimate that uncompensated birefringence would cause a 3% depolarization loss for two passes, owing primarily to residual stress in the as-grown Yb:GSAG. Thermally induced stress is mitigated by cryogenic cooling.

The amplification results are shown in Fig. 18.9 (average-power performance) and Fig. 18.10 (spectral performance) operating at 5-kHz PRF. The power, measured at the output of the amplifier but before recompression, is shown as a function of the total pump power. The inset of Fig. 18.9 shows the measured near-field output. Operation up to 73-W average power was achieved.

The spectrum from the regenerative amplifier is centered at 1030.3 nm with a bandwidth of 1.1 nm FWHM, as shown in Fig. 18.10. This center wavelength is between the gain peaks of Yb:YAG and Yb:GSAG, closer to that of Yb:GSAG. After amplification, the peak has blue-shifted to 1029.7 nm and narrowed to 0.7 nm FWHM, although there is a significant tail to

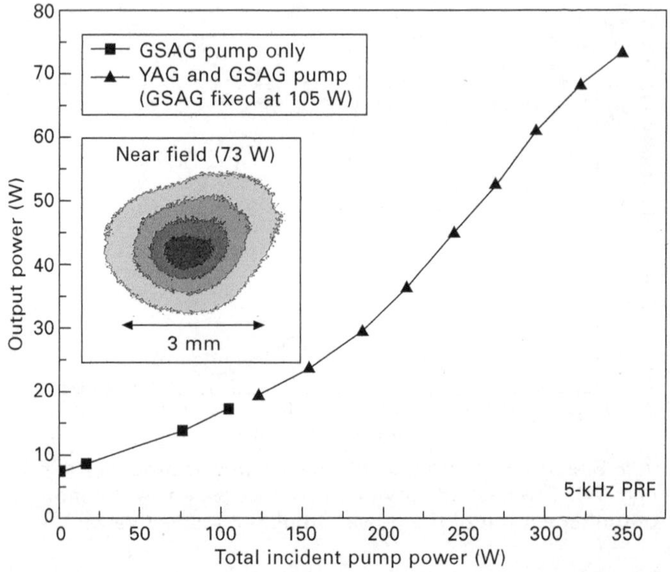

18.9 Average output power as a function of total incident pump power for the Yb:YAG/Yb:GSAG power amplifier at 5-kHz PRF. The inset shows the near-field intensity profile at 73 W.

© Woodhead Publishing Limited, 2013

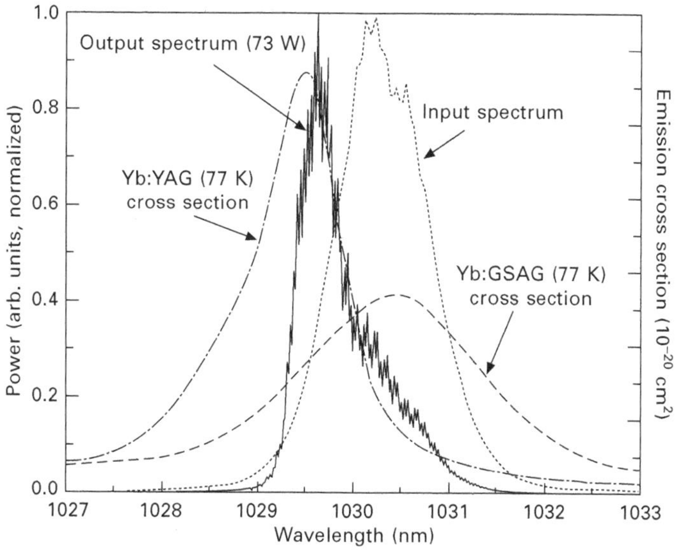

18.10 Spectral performance of Yb:YAG/Yb:GSAG power amplifier at full power (73 W). The dotted curve shows the input spectrum to the power amplifier, and the solid curve shows the output spectrum (at 73 W). The emission cross sections for both Yb:YAG and Yb:GSAG at 77 K are shown for reference (dash-dot and dash curves, respectively).

the red. The compressibility of the amplified pulse using a Treacy pulse compressor consisting of multilayer dielectric gratings with a transmission of 84% was tested. The input pulse was compressed to 1.4 ps (FWHM), and the output was compressed to 1.6 ps (FWHM) at full power, yielding a compressed output power of 60 W.

18.5 Conclusion and future trends

Cryogenically cooling solid-state laser materials to liquid nitrogen temperature significantly improves thermo-optic properties, which enables significant average-power scaling within simple laser architectures. For Yb^{3+}-doped materials, which transition from three-level to four-level laser operation, cryogenic cooling provides efficient power scaling in both cw and pulsed waveforms. We believe that the full potential for power-scaled cryogenically cooled lasers has not yet been realized. In the future, cryogenic cooling will continue to be combined with traditional methods, such as thin-disk and slab geometries and advanced cooling techniques, for managing thermo-optics at high power. Additionally, cryogenic cooling will be applied to laser materials such as Er:YAG (Ter-Gabrielyan *et al.* 2009, Setzler *et*

© Woodhead Publishing Limited, 2013

al. 2010) to power-scale at eyesafe wavelengths, and to materials such as Yb:YLF (Zapata *et al.* 2010) and Yb:CaF_2 (Ricaud *et al.* 2010) to generate femtosecond-class pulses. These advances will enable efficient, multi-kilowatt lasers with near-diffraction-limited beam quality producing waveforms that range from cw to 100-fs-class pulses. Although cryogenic cooling creates some logistical burden, in many applications where the lasers are used in facilities such as accelerators or in applications that require a limited-duty cycle, the fundamental advantages of cryogenic laser materials will enable performance that would otherwise be unattainable.

18.6 Acknowledgment

This work was sponsored by the Department of the Army under Air Force Contract #F19628-95C-0002. Opinions, interpretations, conclusions and recommendations are those of the author and are not necessarily endorsed by the United States Government.

18.7 References

Adams J J, Bibeau C, Page R H, Krol D M, Furu L H and Payne S A (1999), '4.0–4.5-μm lasing of FeZnSe below 180 K, a new mid-infrared laser material', *Opt Lett*, 24, 1720–1722.

Aggarwal R L, Ripin D J, Ochoa J R and Fan T Y (2005), 'Measurement of thermo-optic properties of $Y_3Al_5O_{12}$ (YAG), $Lu_3Al_5O_{12}$ (LuAG), $YAlO_3$ (YALO), $LiYF_4$ (YLF), $LiLuF_4$ (LuLF), BaY_2F_8 (BYF), $KGd(WO_4)_2$ (KGW), and $KY(WO_4)_2$ (KYW) laser crystals in the 80–300 K temperature range', *J Appl Phys*, 98, 103514.

Akahane Y, Aoyama M, Ogawa K, Tsuji K, Tokita S, Kawanaka J, Nishioka H and Yamakawa K (2007), 'High-energy, diode-pumped, picosecond Yb:YAG chirped-pulse regenerative amplifier for pumping optical parametric chirped-pulse amplification,' *Opt Lett*, 32, 1899–1901.

ASTM, *Annual Book of ASTM Standards*, Vol 14.02, 'Thermal Measurements' (ASTM International, West Conshohocken, PA).

Backus S, Durfee C G, Murnane M M and Kapteyn H C (1998), 'High power ultrafast lasers', *Rev Sci Instrum*, 69, 1207–1223.

Backus S, Bartels R, Thompson S, Dollinger R, Kapteyn H C and Murnane M M (2001), 'High-efficiency, single-stage 7-kHz high-average-power ultrafast laser system', *Opt Lett*, 26, 465–467.

Bagdasarov Kh S, Bogomolova G A, Vylegzhanin D N, Kaminskii A A, Kevorkov A M, Petrosyan A G and Prohkorov A M (1974), 'Luminescence and stimulated emission of Yb^{3+} ions in aluminum garnets', *Sov Phys-Dokl*, 19, 358–359.

Bass I L, Bonnano R E, Braun D, Delos-Santos K, Erbert G V, Hackel R P, Haynman C A and Puisner J A (1989), 'Ti:sapphire laser performance above 15 w cw with line narrowing', in Shand M L and Jenssen H P, eds, *Tunable Solid-state Lasers*, Vol 5 of *OSA Proc Ser*, Washington, DC, Optical Society of America, 33–35.

Beck R and Gürs K (1975), 'Ho laser with 50-W output and 6.5% slope efficiency', *J Appl Phys*, 46, 5224–5225.

© Woodhead Publishing Limited, 2013

Brown D C (1997), 'Ultrahigh-average-power diode-pumped Nd:YAG and Yb:YAG lasers', *IEEE J Quantum Electron*, 33, 861–873.

Brown D C, Singley J M, Kowalewski K, Guelzow J and Vitali V (2010), 'High sustained average power cw and ultrafast Yb:YAG near-diffraction-limited cryogenic solid-state laser', *Opt Express*, 18, 24770–24792.

Bruesselbach H W, Sumida D S, Reeder R A and Byren R W (1997), 'Low-heat high-power scaling using InGaAs-diode-pumped Yb:YAG lasers', *IEEE J Sel Top Quantum Electron*, 3, 105–116.

Curtis A H, Reagan B A, Wernsing K A, Furch F J, Luther B M and Rocca J J (2011), 'Demonstration of a compact 100 Hz, 0.1 J, diode-pumped picosecond laser', *Opt Lett*, 36, 2164–2166.

Diehl R and Brandt G (1975), 'Crystal structure refinement of $YAlO_3$, a promising laser material', *Mater Res Bull*, 10, 85–90.

Dong J, Bass M, Mao Y, Deng P and Gan F (2003), 'Dependence of the Yb^{3+} emission cross section and lifetime on temperature and concentration in yttrium aluminum garnet', *J Opt Soc Am B*, 20, 1975–1979.

Eidam T, Hanf S, Seise E, Andersen T V, Gabler T, Wirth C, Schreiber T, Limpert J and Tünnermann A (2010), 'Femtosecond fiber CPA system emitting 830 W average output power', *Opt Lett*, 35, 94–96.

Fan T Y and Byer R L (1987), 'Modeling and cw operation of a quasi-three-level 946 nm Nd:YAG laser', *IEEE J Quantum Electron*, QE-23, 605–612.

Fan T Y, Crow T and Hoden B (1998), 'Cooled Yb:YAG for high power solid-state lasers', in *Airborne Laser Advanced Technology, Proc SPIE*, 3381, 200–205.

Fan T Y, Ripin D J, Aggarwal R L, Ochoa J R, Chann B, Tilleman M and Spitzberg J (2007), 'Cryogenic Yb^{3+}-doped solid-state lasers', *IEEE J Sel Top Quantum Electron*, 13, 448–459.

Feve J-P M, Shortoff K E, Bohn M J and Brasseur J K (2011), 'High average power diamond Raman laser', *Opt Express*, 19, 913–922.

Furch F J, Reagan B A, Luther B M, Curtis A H, Meehan S P and Rocca J J (2009), 'Demonstration of an all-diode-pumped soft X-ray laser', *Opt Lett*, 34, 3352–3354.

Furuse H, Kawanaka J, Takeshita K, Miyanaga N, Saiki T, Imasaki K, Fujita M and Ishii S (2009), 'Total-reflection active-mirror laser with cryogenic Yb:YAG ceramics', *Opt Lett*, 34, 3439–3441.

Gaume R, Viana B, Vivien D, Roger J-P and Fournier D (2003), 'A simple model for the prediction of thermal conductivity in pure and doped insulating crystals', *Appl Phys Lett*, 83, 1355–1357.

Giesen A (2004), 'Results and scaling laws of thin disk lasers', in *Proc Solid-State Lasers XIII: Technology and Devices, LASE 2004*, paper 5332–42.

Hong K-H, Siddiqui A, Moses J, Gopinath J, Hybl J, Ilday F Ö, Fan T Y and Kärtner F X (2008), 'Generation of 287 W, 5.5 ps pulses at 78 MHz repetition rate from a cryogenically cooled Yb:YAG amplifier seeded by a fiber chirped-pulse amplification system', *Opt Lett*, 33, 2473–2475.

Hong K-H, Gopinath J T, Rand D, Siddiqui A M, Huang S-W, Li E, Eggleton B J, Hybl J D, Fan T Y and Kärtner F X (2010), 'High-energy, kHz-repetition-rate, ps cryogenic Yb:YAG chirped-pulse amplifier', *Opt Lett*, 35, 1752–1754.

Johnson L F, Geusic J E and Van Uitert L G (1965), 'Coherent oscillations from Tm^{3+}, Ho^{3+}, Yb^{3+}, and Er^{3+} ions in yttrium aluminum garnet', *Appl Phys Lett*, 7, 127–128.

Johnson L F, Guggenheim H J and Thomas R A (1966), 'Phonon-terminated optical masers', *Phys Rev*, 149, 179–185.

© Woodhead Publishing Limited, 2013

Kawanaka J, Yamakawa K, Nishioka H and Ueda K-I (2003), '30-mJ, diode-pumped, chirped-pulse Yb:YLF regenerative amplifier', *Opt Lett*, 28, 2121–2123.

Kawanaka J, Takeuchi Y, Yoshida A, Pearce S J, Yasuhara R, Kawashima T and Kan H (2010), 'Highly efficient cryogenically-cooled Yb:YAG laser', *Laser Phys*, 20, 1079–1084.

Keyes R J and Quist T M (1964), 'Injection luminescent pumping of CaF_2: U^{3+} with GaAs diode lasers', *Appl Phys Lett*, 4, 50–52.

Klingebiel S, Wandt C, Skrobol C, Ahmad I, Trushin S A, Major Z, Krausz F and Karsch S (2011), 'High energy picosecond Yb:YAG CPA system at 10 Hz repetition rate for pumping optical parametric amplifiers', *Opt Express*, 19, 5357–5363.

Lacovara P, Choi H K, Wang C A, Aggarwal R L and Fan T Y (1991), 'Room-temperature diode-pumped Yb:YAG laser', *Opt Lett*, 16, 1089–1091.

Lü Q, Kugler N, Weber H, Dong S, Müller N and Wittrock U (1996), 'A novel approach for compensation of birefringence in cylindrical Nd: YAG rods', *Opt Quantum Electron*, 28, 57–69.

Manni J G, Hybl J D, Rand D, Ripin D J, Ochoa J R and Fan T Y (2010), '100-W Q-switched cryogenically cooled Yb:YAG laser', *IEEE J Quantum Electron*, 46, 95–98.

Matsushima I, Yashiro H and Tomie T (2006), '10 kHz 40 W Ti:sapphire regenerative ring amplifier', *Opt Lett*, 31, 2066–2068.

Metzger T, Schwarz A, Teisset C Y, Sutter D, Killi A, Kienberger R and Krausz F (2009), 'High-repetition-rate picosecond pump laser based on a Yb:YAG disk amplifier for optical parametric amplification', *Opt Lett*, 34, 2123–2125.

Moulton P F (1986), 'Spectroscopic and laser characteristics of Ti:Al_2O_3', *J Opt Soc Am B*, 3, 125–133.

Moulton P F and Mooradian A (1979), 'Broadly tunable cw operation of Ni:MgF_2 and Co:MgF_2 lasers', *Appl Phys Lett*, 35, 838–840.

Ogawa K, Akahane Y, Aoyama M, Tsuji K, Tokita S, Kawanaka J, Nishioka H and Yamakawa K (2007), 'Multi-millijoule, diode-pumped, cryogenically-cooled Yb:$KY(WO_4)_2$ chirped-pulse regenerative amplifier', *Opt Express*, 15, 8598–8602.

Parker W J, Jenkins R J, Butler C P and Abbott G L (1961), 'Flash method of determining thermal diffusivity, heat capacity, and thermal conductivity', *J Appl Phys*, 32, 1679–1684.

Patel F D, Harris D G and Turner C E (2006), 'Improving the beam quality of a high power Yb:YAG rod laser', in *Solid-State Lasers XV: Technology and Devices*, *Proc SPIE*, 6100, 610018.

Pittman M, Ferre S, Rousseau J P, Notebaert L, Chambaret J P and Cheriaux G (2002), 'Design and characterization of a near-diffraction-limited femtosecond 100-TW 10-Hz high-intensity laser system', *Appl Phys B*, 74, 529–535.

Pugžlys A, Andriukaitis G, Baltuška A, Su L, Xu J, Li H, Li R, Lai W J, Phua P B, Marcinkevičius A, Fermann M E, Giniūnas L, Danielius R and Ališauskas S (2009), 'Multi-mJ, 200-fs, cw-pumped, cryogenically cooled, Yb,Na:CaF_2 amplifier', *Opt Lett*, 34, 2075–2077.

Rand D A, Shaw S E J, Ochoa J R, Ripin D J, Taylor A, Fan T Y, Martin H, Hawes S, Zhang J, Sarkisyan S, Wilson E and Lundquist P (2011a), 'Picosecond pulses from a cryogenically cooled, composite amplifier using Yb:YAG and Yb:GSAG', *Opt Lett*, 36, 340–342.

Rand D, Miller D, Ripin D J and Fan T Y (2011b), 'Cryogenic Yb^{3+}-doped materials for pulsed solid-state laser applications,' *Opt Mater Express*, 1, 434–450.

© Woodhead Publishing Limited, 2013

Reinberg A R, Riseberg L A, Brown R M, Wacker R W and Holton W C (1971), 'GaAs:Si LED pumped Yb-doped YAG laser', *Appl Phys Lett*, 19, 11–13.

Ricaud S, Papadopoulos D N, Camy P, Doualan J L, Moncorgé R, Courjaud A, Mottay E, Georges P and Druon F (2010), 'Highly efficient, high-power, broadly tunable, cryogenically cooled and diode-pumped Yb:CaF_2', *Opt Lett*, 35, 3757–3759.

Ripin D J, Ochoa J R, Aggarwal R L and Fan T Y (2004), '165-W cryogenically cooled Yb:YAG laser', *Opt Lett*, 29, 2154–2156.

Ripin D J, Ochoa J R, Aggarwal R L and Fan T Y (2005), '300-W cryogenically cooled Yb:YAG laser', *IEEE J Quantum Electron*, 41, 1274–1277.

Röser F, Eidam T, Rothhardt J, Schmidt O, Schimpf D N, Limpert J and Tünnermann A (2007), 'Millijoule pulse energy high repetition rate femtosecond fiber chirped-pulse amplification system', *Opt Lett*, 32, 3495–3497.

Russbueldt P, Mans T, Weitenberg J, Hoffmann H D and Poprawe R (2010), 'Compact diode-pumped 1.1 kW Yb:YAG Innoslab femtosecond amplifier', *Opt Lett*, 35, 4169–4171.

Schulz M, Riedel R, Willner A, Mans T, Schnitzler C, Russbueldt P, Dolkemeyer J, Seise E, Gottschall T, Hädrich S, Duesterer S, Schlarb H, Feldhaus J, Limpert J, Faatz B, Tünnermann A, Rossbach J, Drescher M and Tavella F (2011), 'Yb:YAG Innoslab amplifier: efficient high repetition rate subpicosecond pumping system for optical parametric chirped pulse amplification', *Opt Lett*, 36, 2456–2458.

Schulz P A and Henion S R (1989), 'Efficient, high-average-power, liquid-nitrogen-cooled Ti:Al_2O_3 laser', in Shand M L and Jenssen H P, eds, *Tunable Solid-state Lasers*, Vol 5 of *OSA Proc Ser*, Washington, DC, Optical Society of America, 36–38.

Schulz P A and Henion S R (1991), 'Liquid-nitrogen-cooled Ti:Al_2O_3 laser', *IEEE J Quantum Electron*, 27, 1039–1047.

Setzler S D, Shaw M J, Kukla M J, Unternahrer J R, Dinndorf K M, Beattie J A and Chicklis E P (2010), 'A 400W cryogenic Er:YAG laser at 1645 nm', *Proc SPIE*, 7686, 76860C.

Sherman J (1998), 'Thermal compensation of a cw-pumped Nd:YAG laser', *Appl Opt*, 37, 7789–7796.

Smith S E and Campbell R C (2002), 'Flash diffusivity method: A survey of capabilities', *ElectronicsCooling* (May 2002), ITEM Publications, available from http://www.electronics-cooling.com/2002/05/flash-diffusivity-method-a-survey-of-capabilities/

Sorokin P P and Stevenson M J (1960), 'Stimulated infrared emission from trivalent uranium', *Phys Rev Lett*, 5, 557–559.

Strickland D and Mourou G (1985), 'Compression of amplified chirped optical pulses', *Opt Commun*, 56, 219–221.

Ter-Gabrielyan N, Dubinskii M, Newburgh A, Arockiasamy M and Merkle L (2009), 'Temperature dependence of a diode-pumped cryogenic Er:YAG laser', *Opt Express*, 17, 7159–7169.

Tokita S, Kawanaka J, Fujita M, Kawashima T and Izawa Y (2005), 'Efficient high-average-power operation of Q-switched cryogenic Yb:YAG laser oscillator', *Jpn J Appl Phys*, 44, L1529–L1531.

Tokita S, Kawanaka J, Izawa Y, Fujita M and Kawashima T (2007), '23.7-W picosecond cryogenic-Yb:YAG multipass amplifier', *Opt Express*, 15, 3955–3961.

Tümmler J, Jung R, Stiel H, Nickles P V and Sandner W (2009), 'High-repetition-rate chirped-pulse-amplification thin-disk laser system with joule-level pulse energy', *Opt Lett*, 34, 1378–1380.

Wolff E G and Savedra R C (1985), 'Precision interferometric dilatometer', *Rev Sci Instrum*, 56, 1313–1319.

© Woodhead Publishing Limited, 2013

Zapata L, Ripin D and Fan T Y (2010), 'Power scaling of cryogenic Yb:$LiYF_4$', *Opt Lett*, 35, 1854–1856.

Zavelani-Rossi M, Lindner F, Le Blanc C, Chériaux G and Chambaret J P (2000), 'Control of thermal effects for high-intensity Ti:sapphire laser chains', *Appl Phys B*, 70, S193–S196.

© Woodhead Publishing Limited, 2013

19
Laser induced breakdown spectroscopy (LIBS)

C. PASQUINI, Universidade Estadual de Campinas – UNICAMP, Brazil

DOI: 10.1533/9780857097507.2.551

Abstract: Laser induced breakdown spectroscopy (LIBS) is a relatively new analytical technique based on pulsed laser sources. In this chapter the fundamentals of the LIBS technique are introduced, followed by a description of its experimental set-up. The relevant aspects of modern lasers to LIBS are discussed in the context of their analytical applications, including the effect of the main laser characteristics and beam quality on the analytical performance of the technique. Finally, the future of LIBS technique is discussed considering the impact of modern lasers.

Key words: laser induced breakdown spectroscopy, LIBS, analytical spectroscopy, laser produced plasma.

19.1 Introduction to laser induced breakdown spectroscopy (LIBS)

One interesting and still in development application of solid-state lasers is that of the analytical technique named laser induced breakdown spectroscopy (LIBS). The technique has emerged among other spectroscopic counterparts thanks to several outstanding characteristics, such as rapid chemical multielemental qualitative and quantitative analysis, almost non-destructive features, direct use on samples without pretreatment, independent of whether they are non-conductive solids, liquids or gases, and presenting facility to perform micro, stand-off and in-field analyses.

In brief, the technique is based on the interaction of a high-power short pulse (nanosecond to femtosecond) of a laser beam focused in or on a sample. The fluence (J m^{-2}) of the laser pulse at the focal point must exceed the breakdown threshold of the material. The mechanism of laser–matter interaction is a well-studied area, although some related issues of the interaction still require research. Certainly, the interaction depends on the type of sample material and laser characteristics.

However, some relevant effects will always occur as consequences of the laser pulse–matter interaction. In solid samples this interaction causes the ablation of a tiny amount of material from the sample surface and the production of a high temperature and high electron density plasma, where

© Woodhead Publishing Limited, 2013

the ablated species are excited and decay to emit electromagnetic radiation of characteristic wavelengths. Liquid samples experience a similar effect. Gaseous samples, including aerosols, break down under the effect of the laser radiation on the molecules and/or suspended particles, also generating the plasma. In all cases, the emitted radiation is sampled and analyzed by optical instrumentation to extract qualitative and quantitative information about the probed sample.

The interest in LIBS can be accessed by looking at the exponential growth of the number of publications, shown in Fig. 19.1, during the last two decades that have included the term 'laser induced breakdown spectroscopy' in their title and/or abstract. Although this figure is conclusive on the current tendency, it is worth noting that the numbers presented refer only to the results obtained after searching for the term 'Laser induced breakdown spectroscopy' found in the database WebofScience (2011).

19.1.1 Brief history and fundamentals

The history of LIBS can be traced back to the origins of the laser. In 1960 the first laser was demonstrated (Maiman, 1960) and two years later, Brech and Cross (1962) showed the first useful plasma laser-generated from a surface. A few years later Brech (1967) reviewed the first works on laser-excited spectroscopy. However, the lack of suitable robust pulsed lasers or

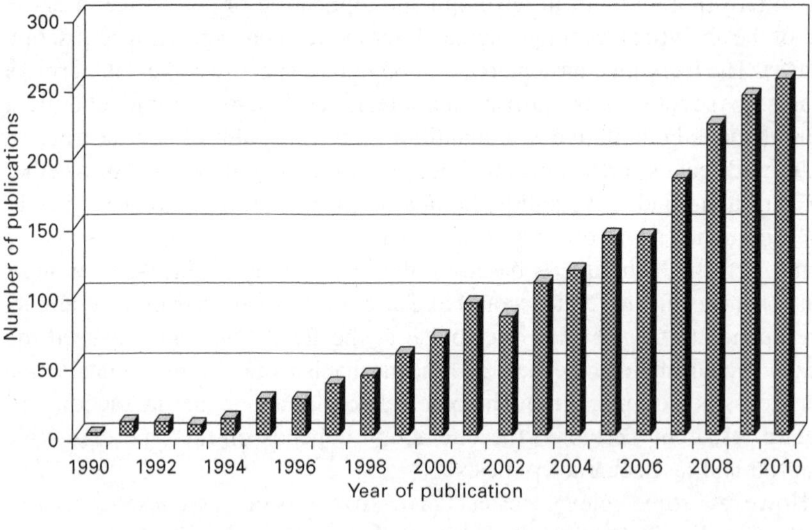

19.1 Evolution of the number of publications using the term 'laser induced breakdown spectroscopy' since 1990. Source: Webofscience, 25 June 2011.

© Woodhead Publishing Limited, 2013

optical instrumentation, and detectors, as well as their high cost, delayed the evolution of the technique.

The technique of LIBS has found a notable revival of interest since the beginning of the 1980s. The robustness of the laser systems, mainly the solid-state Q-switched pulsed devices, had been improved, and analytical spectroscopic instrumentation based on gated solid-state detector arrays became available, both at affordable cost. The pioneering works of this new age are represented by the development of LIBS-based methods for analysis of dangerous aerosols made by Radziemski *et al.* (1983) and Cremers and Radziemski (1983). Since then the application of LIBS has been extended to several types of samples and fields of applications, such as metallurgy, analysis of liquids, gases and aerosols, stand-off analysis, cultural and heritage objects, and microanalysis. The progress of the technique and its application to solve analytical problems can be inferred from Fig. 19.1. Presently, commercial instruments are available from a number of suppliers.

The fundamentals of the LIBS technique are in the interaction of the pulsed laser beam with matter. The reasons are obvious if one considers that this interaction will cause the breakdown and ablation of material from solid and liquid samples, and provide the energy to multiply and energize the electrons removed from the sample and sustain a short-lifetime plasma where the ablated species will be excited to finally decay, emitting electromagnetic radiation. This radiation is the source of analytical information. For the usual laser pulse durations in the range of nanoseconds, the whole event lasts for few hundred microseconds. The optical breakdown is an essential phenomenon and the laser pulse must supply a power higher than the breakdown threshold of the sample material. There are many factors affecting the interaction of a laser beam pulse with the sample material. The most relevant are the laser beam quality, pulse width and energy, the wavelength of the laser radiation, the composition and pressure of the atmosphere surrounding the sample and, of course, the nature of the sample.

The ideal scenario for LIBS would be where the reproducibility of the analytical signal (emitted radiation at characteristic wavelengths) from laser pulse to laser pulse could be ensured by a reproducible non-fractionated ablation (the vaporized material would maintain the elemental composition of the original sample from where it was ablated, and the total ablated mass remains constant for each laser pulse) and by the characteristics of the resulting plasma, such as electron density, plasma temperature and the spatial distribution around the focusing area. Furthermore, the ablation and evolved plasma should be independent of the sample matrix. Of course, these requirements are not completely fulfilled and analysts are always attempting to overcome the limitations found in practical applications of LIBS.

Whatever the way the material ablation, plasma formation and plasma evolution occur, being dependent on the interaction between the laser radiation

© Woodhead Publishing Limited, 2013

and matter and on the instrument employed for a given LIBS system, there is, apparently, agreement on the behavior of the plasma and the spectral analytical signal characteristics with time.

Figure 19.2 depicts a typical behavior of the luminous plasma obtained after a laser pulse of 5 ns has been fired through a focusing lens on a solid sample, generating a typical fluence (J m^{-2}) on its surface capable of breaking down the material and producing the plasma. An initial highly luminous, high temperature and high electron density plasma produces a continuous signal following the end of the laser pulse. This continuous signal is mainly caused by emission of decelerating electrons of high kinetic energy (a phenomenon known as *bremsstrahlung*), as well as recombination of free electrons with ions. The analytical information is hidden by this intense continuous signal and only the most intense emission lines can be observed. The elements initially ablated during the first few picoseconds of the laser pulse are predominantly ions. Shortly after (some tens of nanoseconds), the intensity of the continuous signal starts to decrease and a fraction of the elements are in their neutral atomic form. Both ions and neutral species produce characteristic emission lines whose wavelengths and intensities are used to identify and quantify the elements, respectively. For some time (1 to 10 μs), these persistent lines can be observed on a low continuous background, ensuring correct access to their intensity and wavelength. Finally, after expansion against the atmosphere, the plasma cools down, allowing for elements to recombine into molecules that continue emitting radiation at longer wavelengths (lower energy).

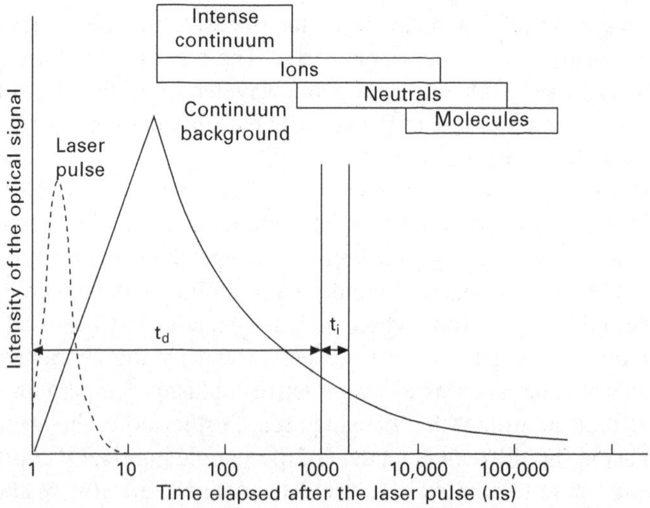

19.2 Relevant events and changes in the species population during plasma lifetime. t_d = time delay since laser has been fired; t_i = integration time interval.

Based on the plasma evolution in time, most of the LIBS systems employing nanosecond pulsed lasers also employ some type of temporal discrimination in the attempt to optically probe the plasma only during a time window where the high initial continuous background has decayed to a tolerable level. Figure 19.2 shows the two relevant time intervals employed by the so-called time resolved LIBS (TRELIBS) technique. In TRELIBS, an initial delay time (t_d) has elapsed from the moment the laser has been Q-switched until a gated detector is triggered to initiate the integration of the emitted radiation. The radiation is integrated for a time window t_i. Typical values for t_d and t_i are in the range 0.1–2 and 2–10 μs, respectively. Figure 19.3 shows the typical effect of the time delay (t_d) on the profile of the LIBS spectra of a copper foil probed by a 5 ns laser pulse of 1,064 nm with an irradiance of 16.7×10^{10} W cm^{-2}. The integration time (t_i) was 1 μs for both spectra.

19.1.2 LIBS apparatus

Figure 19.4 depicts a typical laboratory LIBS apparatus. The heart of the system is the pulsed laser. Usually, the laser provides an external signal to synchronize the time sequence (t_d and t_i) necessary for data acquisition. The pulse duration and energy, the radiation wavelength and the beam quality are all of importance concerning the generation of the LIBS signal.

The collection of the radiation emitted by the plasma is usually made by using an optical fiber coupled to a focusing lens. In this case, commonly the fiber diameter determines the slit width, whose usual dimension is 50 μm. Alternative arrangements are based on focusing the emitted radiation directly onto the entrance slit of the spectrograph. Spectrographs are employed to disperse the emitted radiation into its wavelengths with a high resolution (typical resolution power: $\lambda/\Delta\lambda$ = 5,000–10,000; i.e., 0.04–0.02 nm at 200 nm). They can be based on common Czerny-Turner arrangements or can be of the echelle type (Bauer *et al.*, 1998; Haisch *et al.*, 1998), capable of acquiring a broadband spectrum, typically from 200 to 900 nm.

After dispersion, the radiation is impinged on a gated detector array. The number of detectors in the array can vary from 2,096 for linear arrays coupled to Czerny–Turner dispersing optics to more than a million for bidimensional arrays coupled to echelle spectrographs.

The Q-switch synchronizing signal generated by the laser electronics is passed to a pulse generator that delays the detector gate signal by the selected time interval (t_d) and generates the integration pulse according to the programmed time interval (t_i). Usually the pulse generator is incorporated into the detector electronics and can be programmed from the software accompanying the detector package.

Many types of detectors have been used in LIBS. As mentioned before, at least for LIBS systems based on nanosecond pulsed lasers, the detectors

© Woodhead Publishing Limited, 2013

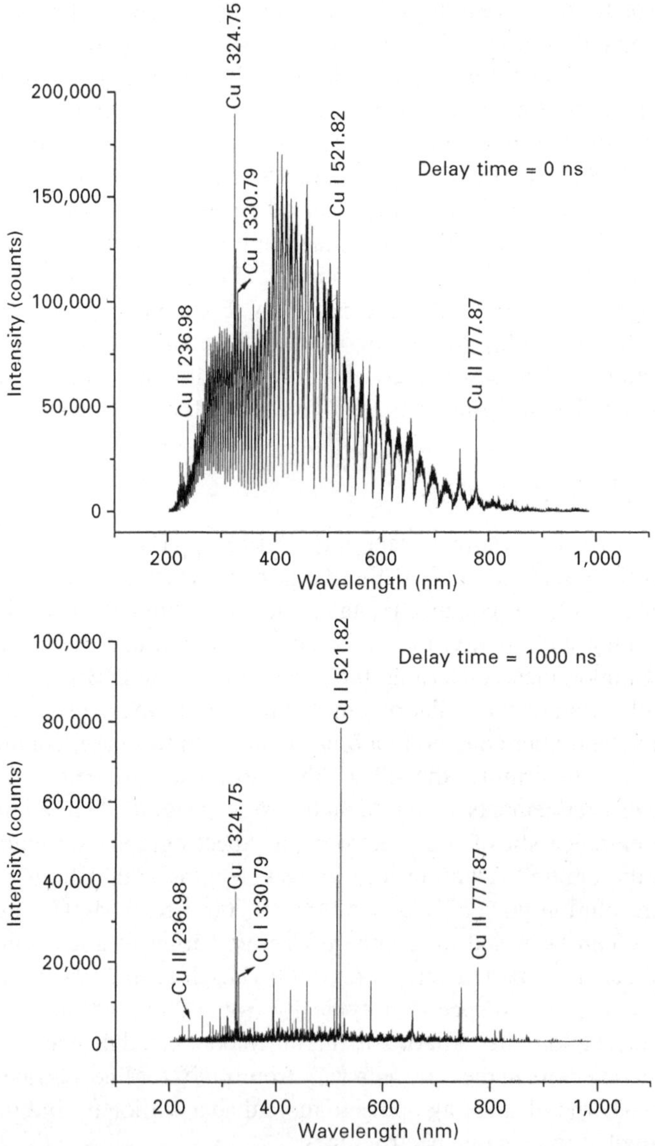

19.3 Effect of the time delay since the pulse delivery on the LIBS emission spectrum of a copper foil. Pulse energy = 110 mJ; pulse width = 5 ns; lens focal length = 25 cm. Integration time = 1 μs for both spectra.

need to feature a gated operation in order to achieve the high precision time resolution necessary to acquire the emission spectrum after the intense continuum signal has faded, through a short integration time interval. Arrays

© Woodhead Publishing Limited, 2013

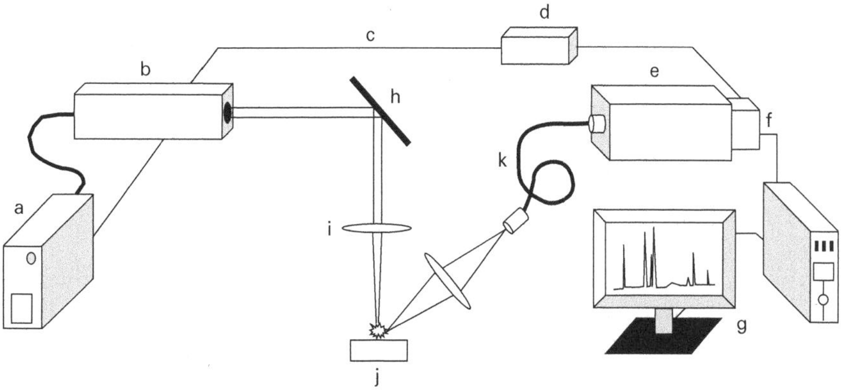

19.4 Typical laboratory LIBS apparatus: a, laser power supply; b, laser head; c, Q-switch synchronizing signal; d, programmable time delay and pulse length generator; e, spectrograph; f, detector array; g, microcomputer and monitor; h, dichroic mirror; i, focusing lens; j, sample; k, optical fiber.

of charge coupled devices (CCD), intensified charge coupled devices (ICCD), and more recently CMOS (Gonzaga and Pasquini, 2008) devices have been used with success in the detection of LIBS.

The bidimensional arrays of ICCD are perhaps the most frequently used detector system due to their high time resolution (nanoseconds) and outstanding sensitivity provided by the opto-electronic amplifying device placed between the optical spectrograph output and the ICCD array (Bauer *et al.*, 1998; Sabsabi *et al.*, 2003). The amplification stage resembles that of a photomultiplier tube where a high voltage is used to accelerate and multiply the number of photo-electrons initially produced by the photons of the dispersed radiation. The photo-electrons are further reconverted to photons of fixed wavelength by hitting a screen containing a photo-sensitive material and directed to the CCD array for optimized detection. The high voltage can be switched on or off at very fast speed, ensuring, as mentioned before, the time resolution of the detection.

19.1.3 Analytical information from laser induced plasmas

The analytical information about a sample probed by LIBS is both qualitative and quantitative. The first type is employed for pattern recognition and classification of samples. In this case, the source of information is in the wavelength of the characteristic emission lines of the elements present in the sample and in their relative intensities. The presence of an element in a sample can be attested to by identification of at least three characteristic lines of that element in the LIBS spectrum. A given organic compound such an

© Woodhead Publishing Limited, 2013

explosive can be identified by the relative intensities of the emission lines of carbon, hydrogen, nitrogen and oxygen that characterize a fingerprint of that compound (DeLucia *et al.*, 2003; Gottfried *et al.*, 2009). Oil-contaminated soils can be classified according to their elemental fingerprint of emission lines showing specific patterns for carbon and hydrogen. Heavy metal contamination in toys and soil can be easily confirmed by looking at the many characteristic emission lines of the contaminating elements (Fortes *et al.*, 2010).

Multivariate techniques for pattern recognition can help in the qualitative use of information provided by LIBS. SIMCA, discriminant partial least squares (PLS-DA), linear discriminant analysis (LDA) and hierarchical cluster analysis can be used to fully exploit the information present in the spectra in order to identify plastics (Anzano *et al.*, 2011), classify toys regarding the presence of toxic elements (Godoi *et al.*, 2011), classify soils (Pontes *et al.*, 2009) and identify historic building materials (Colao *et al.*, 2010), just to mention a few recent examples.

Quantitative information relies on the line emission intensities and on their proportionality to the content of a given element in the sample. In this case, there are two approaches presently in use. One is based on the use of reference materials whose matrix (the majority composition of the sample) resembles that of the samples to be analyzed. Analytical calibration curves are constructed by taking the intensities of an emission line characteristic of an element plotted against the known content of that element in the reference material. A linear curve is expected, although some phenomena such as self-absorption occurring in the plasma can degrade the linearity at higher contents of the element. The intensity of characteristic elemental emission lines observed in a LIBS system is strongly dependent on the sample matrix. Therefore, analytical curves constructed with reference materials can be used only for analysis of samples with very similar matrices.

Multivariate regression techniques can be employed to exploit in full the very large amounts of information present in a LIBS spectrum. The usual multivariate regression methods, such as principal component regression (PCR) and partial least squares (PLS) regression, can be used to model even the entire spectrum as a function of the content of an element present in the sample. The advantages and disadvantages of univariate and multivariate approaches to quantitative and qualitative use of LIBS information have been discussed (Stipe *et al.*, 2010; Andrade *et al.*, 2010). However, it appears that both approaches are finding their use in LIBS applications.

Another approach to the quantitative use of LIBS is that known as 'calibration-free' technique (Tognoni *et al.*, 2010). In this approach, the plasma experimental parameters are estimated from spectroscopic data generated by the plasma itself (Ciucci *et al.*, 1999). The most relevant parameters are the plasma temperature and the electron density and they are calculated assuming

© Woodhead Publishing Limited, 2013

the local thermodynamic equilibrium (LTE) condition. Boltzmann and Saha equations are employed to find out the plasma temperature while the electron density of the plasma can be determined by measuring the widths of some emission lines caused by the Stark effect. Concentrations of the elements present in the sample are determined computationally by finding the relative concentration of the elements detected and predicting the theoretical spectrum by using the plasma experimental parameters (i.e., temperature and electron density, previously determined) until it matches as closely as possible the experimental spectrum. The elemental composition producing the best match is taken as the sample original composition. Up to now, this approach has produced good results only for major elements present in a sample (Praher *et al.*, 2010).

Improvements in plasma modeling are being continuously pursued to obtain better comprehension of the LIBS phenomena and to achieve better analytical performance of the calibration-free method (Gornushkin *et al.*, 2011).

19.1.4 Overview of the LIBS literature

The LIBS technique has been the subject of a series of textbooks (Cremers and Radziemski, 2006; Miziolek *et al.*, 2006; Radziemski and Cremers, 1989; Singh and Thakur, 2007) and reviews (Pasquini *et al.*, 2007; Burakov *et al.*, 2010; Gadiuso *et al.*, 2010; Michel, 2010; Gottfried *et al.*, 2009; Fantoni *et al.*, 2008, Winefordner *et al.*, 2004; Singh *et al.*, 2011; Russo *et al.*, 2011) published in the major scientific journals devoted to analytical spectroscopy. The information found in this literature is complete and will help anyone interested to obtain knowledge in depth on LIBS.

19.2 Types of laser induced breakdown spectroscopy (LIBS) systems and applications

Historically most of the LIBS systems are very similar to that described in Fig. 19.4 and are designed to work on the laboratory bench and employed mostly for analysis of solid samples, mainly metallic alloys. However, since the different goals of use of the technique were identified, new proposals for LIBS instrumentation have appeared in the literature.

As an example of a recent LIBS application, some works have shown the feasibility of using LIBS to determine isotopic ratios, a very difficult task for which a portable field instrument is needed. A high resolution spectrograph has been employed to distinguish between the ^{235}U and ^{238}U emission lines at 424.437 nm as well as H and Li isotopes (Cremers *et al.*, 2011).

© Woodhead Publishing Limited, 2013

Alternatively, Doucet *et al.* (2011a) describe a LIBS system employing a low resolution spectrograph and partial least squares (PLS) regression aiming to determine elemental isotopic ratios ($^{235}U/^{238}U$ and hydrogen/tritium). Russo (2011) proposed the use of rotational-vibrational emissions occurring in the near-infrared portion of the spectrum for the same purpose of quantifying the isotopic ratios. The separation between the emission lines attributed to each isotope is increased substantially in this case as a consequence of the effect of the reduced mass on the oscillating frequency of molecular species detected in the near-infrared spectral range.

One of the principal characteristics of LIBS is its *in-situ* analysis feasibility. Therefore, portable, stand-alone equipment has been developed to permit data acquisition in the field (Fortes and Laserna, 2010; Agresti *et al.*, 2009; Yamamoto *et al.*, 1996). Compact and robust solid-state lasers were accommodated in a backpack or in a handled case, while fiber optics conduct the laser pulse to the target and collect the emitted radiation back to a compact spectrograph to be analyzed for qualitative and quantitative information. The high power supply necessary for the laser used in this type of equipment still restrains overall size as well as time of stand-alone operation and the portability of the system.

Portable LIBS systems have been employed in many applications for sampling materials and objects that cannot be moved away from their original location. Recently, Cuñat *et al.* (2008) and Fortes *et al.* (2007) have described a portable LIBS instrument and applied it to *in-situ* analysis of karstic formations inside a cave and to the production of a chemical image of the Malaga cathedral façade in Spain, respectively. Depth profile analyses of Sr, Mg and Ca found in the cave formations were used for Sr/Ca and Mg/Ca calculations, useful for paleoclimate applications. The analysis of the façade of Malaga cathedral revealed the distribution of different materials (sandstone, limestone, marble, and cement mortar) which were discriminated on the basis of different ratios of intensities of emission between Si/Ca and Ca/Mg. The results were considered satisfactory to help in the restoration works. Goujon *et al.* (2008) described a portable LIBS system capable of operating in the single and double pulse mode and applied it to the *in-situ* identification of several materials, such as pigments in paintings and icons, metals and ceramics.

Certainly, the use of portable LIBS instruments will increase in the future because it competes favorably with other analytical techniques based on X-ray fluorescence spectroscopy (XRF), Raman spectroscopy (Raman), laser induced fluorescence (LIF) and ion mobility spectrometry (IMS) due to its ability to analyze both molecular compounds (through the ratio of elements such as C, H, O and N) and light elements such as sodium, beryllium and lithium. Also, the combination of laser based techniques such as LIBS and Raman spectroscopy will add versatility to portable instruments, ensuring

© Woodhead Publishing Limited, 2013

their application in molecular and atomic analysis (Fortes and Laserna, 2010; Moros and Laserna, 2011).

Stand-off (remote) LIBS systems have been developed to exploit a unique characteristic of the technique that allows probing a sample object at long distances of up to 120 m (Fortes and Laserna, 2010; Gonzales *et al.*, 2009). Powerful lasers are used with appropriate optics to deliver the light pulse to a distant target and collect and analyze the radiation emitted from the generated plasma. Despite the problems associated with several atmospheric effects, such as temperature, humidity, pressure and wind, on the propagation of the laser beam to the sample and of the emitted radiation back to the LIBS spectrograph, this approach has been used to analyze samples whose temperature forbids the operator to be close to the sample and to screen mobile objects for traces of explosives.

A miniaturized LIBS system can be assembled in a small vehicle to be transported close to the sample and to perform analysis by exploiting the unique stand-off characteristic of the technique. This approach has been adopted to construct LIBS systems aimed at space exploration in the near future (Rauschenbach *et al.*, 2010; Gadiuso *et al.*, 2010; Salle *et al.*, 2005).

Systems based on dual laser pulses can also be used in the field (Goujon *et al.*, 2008) but their performance has been demonstrated mainly in the laboratory. In these types of LIBS systems, two successive pulses are delivered to the same sample spot with an inter-pulse delay optimized to achieve higher analytical detectivity. Apparently high detectivity (sometimes a 10-fold or higher increase in the emission line intensity of an element is found when compared with the single pulse technique) is produced due to the extra ablation provided by the second pulse arriving in an environment that was previously modified by the first pulse (Stratis *et al.*, 2000; Scaffidi *et al.*, 2006). Dual-pulse LIBS systems require very good synchronization of two pulsed lasers to produce repeatable signals. Typical inter-pulse delays are in the range 1 to 100 μs when the usual Nd:YAG pulsed lasers pumped by flashlamps are employed.

The dual-pulse approach to LIBS has recently been proposed as an alternative to overcome the limitations of the low energy lasers required to compact LIBS systems (Doucet *et al.*, 2011b). It can be demonstrated that the combination of a short and a long pulse on the nanosecond and microjoule scale can help to maintain the gain in detectivity achieved in the millijoule range of laser pulse energy. The authors report that the best results in terms of detectivity were found when the long pulse is delivered first to the sample.

The use and construction of LIBS systems aiming at microanalysis have been described in the literature (Bette *et al.*, 2005; Nicolas *et al.*, 2007, Zorba *et al.*, 2011). This is another outstanding characteristic of the technique.

© Woodhead Publishing Limited, 2013

The capability of performing chemical analysis with 3D resolution results from the small area on which a good quality laser beam can be focused. This area can be typically about 10 μm or smaller. Also, the ablated layer of the surface material can be managed to perform a depth profile analysis with a few microns' resolution. Fast pulse rate diode-pumped or microchip lasers, and more recently fiber lasers, can be used in LIBS systems for microanalysis.

Zorba *et al.* (2011) have recently demonstrated the viability of LIBS microanalysis to reach below-micrometer (420 nm) spatial resolution by using femtosecond laser pulses and detecting a few hundreds of atogram (ag) of ablated material.

Nowadays, complete commercial LIBS instruments can be purchased from several manufacturers. The list includes™ Applied Photonics (ST-LIBS™, FO-LIBS™ and LIBSCAN™ 50/100), Applied Spectra (RT-100 series LIBS system), BAe System (Tracer 2100), Foster and Freeman (ECCO™), Laser Analysis Technologies (Spectrolaser system), Ocean Optics (LIBS2500), Pharmalaser (PharmaLIBS™250), Photon Machines (Insight™ and Crossfire™) and StellarNet, Inc. (Porta-LIBS-2000). However, most of these instruments are designed to be operated in the laboratory and are similar to the one shown in Fig. 19.4. There is still a lack of commercial portable, stand-alone and microanalytical equipment.

19.3 Solid-state lasers for laser induced breakdown spectroscopy (LIBS)

As mentioned before, the heart of a LIBS system is the laser source. Several types of lasers have been employed with success in LIBS systems. In common, the lasers employed in a LIBS system must be pulsed and provide a synchronization signal, usually an active Q-switch triggering pulse, to permit the proper timing employed for signal acquisition as described in Fig. 19.3 and Section 19.1.1. The pulse energy and length should produce a fluence or irradiance high enough to cause sample matrix breakdown, to ablate a tiny portion of material, and to initiate and sustain the plasma until the analytical signal (emission spectrum) is acquired.

The most common type of laser, and the one responsible for LIBS popularization, is the active Q-switched Nd:YAG pulsed laser pumped by flashlamps. This type of laser produces a fundamental beam at 1064 nm. The pulse energy can be in the range from 20 mJ to 1 J and the pulse length is in the range of few nanoseconds (typically 5–50 ns). The fundamental wavelength can be doubled or tripled by employing ready-to-use harmonic generators based on non-linear optical devices. Therefore, the radiation can be produced at 532 and 355 nm, extending the use up to the ultraviolet region.

© Woodhead Publishing Limited, 2013

19.3.1 Effect of beam quality and laser radiation wavelength

The beam quality expressed by the parameter M^2 influences the LIBS signal due to its effect in determining the illuminated area at the lens focal distance. This area determines the maximum irradiance (or the fluence) which can be attained by a given laser system.

The equation relating the beam waist of a collimated laser beam at the focus of an aberration-free lens of the optical arrangement to the beam quality (M^2) is given by:

$$w = \frac{2f\lambda M^2}{\pi D} \qquad [19.1]$$

where w is the minimum beam waist radius at the focal distance f of the focusing lens, D is the laser beam diameter, λ is the wavelength of the laser radiation.

The maximum fluence, F (J m^{-2}), and irradiance, I (W cm^{-2}) found at the focal distance of the lens are given, respectively, by:

$$F = \frac{E_P}{\pi w^2} \qquad [19.2]$$

$$I = \frac{E_P}{\pi w^2 P_{BL}} \qquad [19.3]$$

where E_P is the laser pulse energy (J) and $P_{BL}$is the pulse band length in seconds.

Usually the distance from the lens to the sample is kept lower than the lens focal distance in order to control the fluence and increase the probed area of the sample surface, and also to avoid, in some cases, breakdown of the surrounding atmosphere (commonly, air).

The wavelength of the laser pulse affects the fluence and irradiance on the probed surface of the sample, as shown in equation 19.1. The diffraction limit determines the minimum beam waist radius (w) possible, which is smaller for shorter wavelengths, as shown by equation 19.1. This fact is also relevant to achieving better lateral resolution for microanalysis LIBS techniques where the sample surface is scanned by successive pulses in the x–y directions in order to map its chemical composition (Taschunk *et al.*, 2007).

Another aspect important for LIBS related to the laser radiation wavelength is the interaction with the sample material during the very initial stage of the ablation process and with the ablated material during the plasma lifetime. For shorter wavelengths, two-photons absorption by the sample matrix is possible. The effect of plasma shielding, discussed in detail below, can cause

© Woodhead Publishing Limited, 2013

significant modifications in the analytical information obtained from a LIBS system. This phenomenon depends on the laser wavelength because the collective behavior of the electrons in the plasma defines a plasma-resonant frequency depending on the electron density. When the electron density reaches a critical value the plasma frequency equals the laser frequency and the radiation is strongly absorbed. If the laser frequency is lower than the plasma frequency the radiation is reflected by the plasma.

19.3.2 Effect of pulse length (from nanosecond to femtosecond pulsed lasers)

Nanosecond (typically in the 5 to 50 ns band range) solid-state pulsed lasers are widespread, compact, of low cost and have high pulse energy (from 0.1 to 1 J). Most of the work done on LIBS found in the literature was based on this type of laser. However, lasers with shorter pulse duration (down to a few tens of femtoseconds) have recently become available at reducing costs. The effects of the pulse duration on the LIBS spectral signal are beyond the obvious increase in the irradiance at the focal point that a shorter pulse of the same energy can produce in relation to a longer pulse, as predicted by equation 19.3. The pulse duration is also related to the ablation stage and possible interactions with the generated plasma. It can be demonstrated that the increase in the irradiance caused by an increase in the pulse energy for a fixed pulse duration of the order of 5 ns will reach a value above which the mass ablated from the sample surface will remain practically constant (Russo *et al.*, 2004). Therefore, the analytical signal also stops increasing.

Considering pulse durations in the range of nanoseconds, the interaction of the radiation with the sample has enough time to occur with a significant contribution of thermal effects. Thermal effects will prevail for long-duration pulses of a few nanoseconds because the phonon relaxation after absorption of the optical energy is in the order of tenths of a picosecond. Therefore, some sample characteristics such as the different boiling and fusion temperatures of the metals present in a metallic alloy, for instance, can impart a significant effect on the composition of the removed material (fractionation). In this example, the composition of the removed material can depart from the real composition of the sample because it will be enriched in the metal with the higher vapor pressure under the thermal effect.

The effect of longer laser pulses on LIBS extends beyond the pulse–sample interaction to the pulse-plasma interaction. After a few picoseconds the plasma is generated by the front of a nanosecond-timescale laser pulse, but there is still enough energy arriving for a significant time interval to promote an optical-plasma interaction. This interaction is referred to as a type of plasma shielding which prevents the remainder of the pulse from reaching the sample surface; in this way, the ablation of the sample material ceases

© Woodhead Publishing Limited, 2013

before the laser pulse has finished. Plasma shielding can occur by absorption of the optical energy by free electrons (the inverse of *bremsstrahlung*) or by scattering when the remaining energy of the laser pulse is reflected by the plasma. The threshold for absorption and scattering of the laser radiation depends on the electron density of the plasma. High wavelengths require high electron densities to be absorbed or reflected. Therefore, the laser wavelength can also play a role in the behavior of a LIBS system, affecting the extension with which the laser pulse interacts with the sample (increasing the ablation) and/or how it interacts with the generated plasma (by absorption or reflection).

On the other hand, pulses with lengths between pico- and femtoseconds have durations of the same magnitude or even shorter than the phonon relaxation time (~100 fs). The thermal response of the targeted material may play a secondary role in this situation. The optical energy is transferred to the sample material before any thermal effect can be manifested. In addition, fast pulses do not last long enough to interact with the formed plasma. One advantage of this is the absence of plasma shielding and its consequences, such as the high background continuous signal found at the beginning of the plasma lifetime.

The use of short-laser pulses in LIBS can reduce the energy and fluence threshold necessary to achieve ablation. Better depth resolution in LIBS microanalysis can be found as a consequence of lower ablation depths. The reduction of the thermal effects can reduce fractionation of the ablated material, which becomes more representative of the original sample composition. Furthermore, the background continuous radiation decays faster in plasmas produced by laser pulses in the femtosecond scale than in the nanosecond scale (Angel *et al.*, 2001). This can reduce the necessity for the LIBS system to require the gated, more expensive, detectors usually employed with nanosecond pulses, as shown in Fig. 19.4. Integration of the emitted radiation over many successive laser pulses is also possible and can overcome part of the loss in detectivity found in LIBS systems based on short pulses. The most employed short pulse laser in LIBS is the self mode-locked Ti:sapphire laser which employs chirped pulse amplification.

Despite the advantages pointed out above, no significant gains in the analytical performance have been observed when LIBS systems based on short (fs) or long (ns) laser pulse length were compared (Sabsabi, 2007). However, the use of short laser pulses in LIBS is still new and deserves more profound studies before a definitive conclusion can emerge.

19.3.3 Pulse repetition rate

The pulse repetition rate of a laser is not a critical factor in LIBS. Usually, for bulk analysis of a material, the time elapsed between pulses is shorter

© Woodhead Publishing Limited, 2013

than the time interval required for the detector reading and data transfer to a mass storage device in a computer. High pulse rates in the nanosecond pulse length regime are not recommended because they can cause fractionation by thermal effects and modify the ablated composition from that of the sample. However, there are some special cases where the use of fast repetition rates becomes valuable.

Solid-state lasers pumped by flashlamps present pulse repetition rates typically in the range 10–20 Hz. This repetition rate is suitable for use in LIBS systems aimed at the determination of bulk concentrations of the elements in a sample. However, the repetition rate becomes critical when the LIBS technique is aimed at microanalysis with lateral resolution on the order of 20 μm because an excessively long time would be necessary to scan a 1 cm^2 surface, for example. Diode-pumped lasers can achieve pulse rates of 1 kHz, although the pulse energy is limited to few millijoules. This pulse rate can significantly reduce the scanning time in microanalysis applications of LIBS.

The pulse rate is also important in short pulse lengths on the femtosecond scale considering the low luminosity plasma obtained under these conditions and the necessity of integration of many successive pulses in order to find the required analytical detectivity.

19.3.4 Microchip and diode pumped lasers

Diode pumped lasers (Hoehse *et al.*, 2011; Loebe *et al.*, 2003) and microchip lasers (Gornushkin *et al.*, 2004; Godwal *et al.*, 2008) have become available recently for LIBS and are described in detail in Chapter 14. In this approach, the broad spectral flashlamps are replaced by narrower-band-emitting laser diodes to pump a variety of lasing crystals. The pump efficiency increases while the high dissipation found by using broadband flashlamps decreases. The pulse repetition rate is above 1 kHz with these devices.

Microchip lasers are compact devices employing passive Q-switch technology with the active medium being also pumped by a coupled diode laser and capable of producing laser pulses with durations shorter than 1 nanosecond with peak energies between 2 and 50 μJ. The pulse repetition rate is above 1 kHz in these devices.

The relevant achievements of diode pumped lasers and microchip lasers regarding their use in LIBS are related to the higher pulse rate these devices can reach (1–30 kHz), the high quality of the laser beam they generate, and their short pulse length. Equation 19.1 shows that the beam quality directly affects the focusing quality and the pulse fluence on the sample surface. Active Q-switch-operated diode-pumped lasers easily reach 1 kHz pulse rates and their beam quality allows for sharp focusing with spots smaller than 20 μm. The pumped diode lasers can have lifetimes as long as 10,000

© Woodhead Publishing Limited, 2013

hours or 10^{10} pulses which are substantially longer than the lifetimes of a flashlamp.

The pulse energy of a diode-pump laser can reach a few millijoules. Microchip devices provide lower pulse energies (20 to 50 μJ per pulse), although they can be used in LIBS systems since their high beam quality can be associated with short focal distance optics (microscope objectives) to produce a fluence above the threshold of the target material (Gonzaga and Pasquini, 2008). These low energy lasers can reach a pulse length of a tenth of a nanosecond (typically 100 ps). Therefore, additional advantages related to the short pulse length regime (see discussion above) can be found when these lasers are employed in a LIBS system. An additional advantage comes from the small size of the microchip lasers and of their power supplies which allow for construction of compact, low weight, battery-operated portable LIBS systems.

19.3.5 Fiber lasers

Fiber lasers comprise a new variety of lasers and their principles and construction are described in Chapter 15. They have recently been evaluated for use in LIBS systems and have shown very good performance, considering they are still in their initial stages of development (Gravel *et al.*, 2011; Schill *et al.*, 2007).

Again the benefits of employing a fiber laser in a LIBS system come from the high beam quality ($M^2 \sim 1$), the pulse energy (in the millijoule range) and the pulse length (10 to 100 ns). The pulse repetition rate can also be very high, contributing to the use of these devices in fast LIBS instruments dedicated to microanalysis. In addition, the use of fiber as the lasing medium imparts a high flexibility to the laser and no critical alignment is necessary to achieve very good performance of the laser source. This characteristic is in agreement with the actual demand for portable and robust LIBS instruments.

19.4 Future trends

The most important applications of LIBS are not, at least in the view of the present results, in the determination of the bulk contents of elements in a sample, as this can be done with almost unbeatable quality by other, well-established techniques, such as ICP OES, X-ray fluorescence or even electric arc and spark excitation for metallic/conductive samples. The real contribution of LIBS to analytical science will be that of attending to the requirement for fast (screening), in-field and/or stand-off analysis, to supply a fast microanalysis with 3D resolution on the order of microns and to provide analysis in real time in hazardous environments, inaccessible to human beings.

© Woodhead Publishing Limited, 2013

The future of LIBS as an analytical technique is certainly linked to the future of solid-state lasers as they are being improved to reduce the cost while achieving better beam quality, higher pulse energy and lower pulse duration, down to the scale of femtoseconds. High compactness and robustness are also a requirement for LIBS when one wants to exploit its unique characteristics for stand-off and in-field applications.

The role of the laser source on the phenomena associated with sample ablation, plasma generation, sustainability and species excitation in LIBS is being enlightened as a consequence of the large amount of research being carried out in the field. At the same time, the technique is finding its preferred applications. Both of these facts could lead laser suppliers to consider tailoring their lasers to specifically attend to the requirements of LIBS, accelerating, in this way, its development in the near future.

19.5 References

Agresti J, Mencaglia A A and Siano S (2009), 'Development and application of a portable LIPS system for characterizing copper alloy artifacts', *Anal. Bioanal. Chem.*, 395, 2255–2262.

Andrade J M, Cristoforetti G, Legnaioli S, Lorenzetti G, Palleschi V and Shaltout A A (2010), 'Classical univariate calibration and partial least-squares for quantitative analysis of brass samples by laser-induced breakdown spectroscopy', *Spectrochim. Acta*, Part B, 65, 658–663.

Angel S M, Startis D N, Eland K L, Lai T S, Berg M A and Gold D M (2001), 'LIBS using dual-and ultra-short laser pulses', *Fresenius*, 369, 320–327.

Anzano J, Bonilla B, Montull-Ibor B and Casas-Gonzalez J (2011), 'Plastic identification and comparison by multivariate techniques with laser-induced breakdown spectroscopy', *J. Appl. Pol. Sci.*, 121, 2710–2716.

Bauer H E, Leis F and Niemax K (1998), 'Laser induced breakdown spectrometry with an echelle spectrometer and intensified charge coupled device detection', *Spectrochim. Acta, Part B*, 53, 1815–1825.

Bette H, Noll R, Muller G, Jansen H W, Nazikkol C and Mittelstadt H (2005), 'High-speed scanning laser-induced breakdown spectroscopy at 1000 Hz with single pulse evaluation for detection of inclusions in steel', *J. Laser Appl.* 17, 183–190.

Brech F (1967), 'A review of achievements in laser-excited spectrochemistry', *Appl. Spectrosc.*, 21, 376.

Brech F and Cross L (1962), 'Optical micromission stimulated by ruby maser', *Appl. Spectrosc.*, 16, 59.

Burakov V S, Raikov S N, Tarasenko N V, Belkov M V and Kiris V V (2010), 'Development of laser-induced spectroscopy method for soil and ecological analysis (review)', *J. Appl. Spectrosc.*, 77, 595–608.

Ciucci A, Corsi M, Palleschi V, Rastelli S, Salvetti A and Tognoni E (1999), 'New procedure for quantitative elemental analysis by laser-induced plasma spectroscopy', *Appl. Spectrosc.*, 53, 960–964.

Colao F, Fantoni R, Ortiz P, Vazquez M A, Martin J M, Ortiz R and Idris N (2010), 'Quarry identification of historic building materials by means of laser induced

© Woodhead Publishing Limited, 2013

breakdown spectroscopy, X-ray fluorescence and chemometric analysis', *Spectrochim. Acta, Part B*, 65, 688–694.

Cremers D A and Radziemski L J (1983), 'Detection of chlorine and fluorine in air by laser-induced breakdown spectrometry', *Anal. Chem.*, 55, 1252– 1256.

Cremers D A and Radziemski L J (2006), *Handbook of Laser-induced Breakdown Spectroscopy*, Chichester, John Wiley & Sons.

Cremers D A, Beddingfield A, Smithwick R, Chinni R C, Bostian M, Smith G, Multari R, Jones C R, Beardsley B and Karch L (2011), 'Detection of RNE threats and isotopes of U, H and Li in air using a LIBS instrument', *Abstract Book*, 6th EuroMediterranean Symposium on Laser Induced Breakdown Spectroscopy, 32.

Cuñat J, Fortes F J, Cabalín L M, Carrasco, F, Simón M D and Laserna J J (2008), 'Man-portable laser-induced breakdown spectroscopy system for in situ characterization of karstic formations', *Appl. Spectrosc.*, 62, 1250–1255.

DeLucia F C, Harmon R S, McNesby K L, Winkel R J and Miziolek A W (2003), 'Laser-induced breakdown spectroscopy of energetic materials', *Appl. Opt.*, 42, 6148–6152.

Doucet F R, Lithgow G, Kosierb R, Bouchard P and Sabsabi M (2011a), 'Determination of isotope ratios using laser-induced breakdown spectroscopy in ambient air at atmospheric pressure for nuclear forensics', *J. Anal. At. Spectrosc.*, 26, 536–541.

Doucet F R, Gravel, J Y, Elnasharty I, Bouchard P, Harith M A and Sabsabi M (2011b), 'Combining short and long nanosecond pulses in the microjoule range for use in double pulse LIBS', *Abstract Book*, 6th EuroMediterranean Symposium on Laser Induced Breakdown Spectroscopy, 58.

Fantoni R, Caneve L, Colao F, Fornarini L, Lazic V and Spizzichino V (2008), 'Methodologies for laboratory laser induced breakdown spectroscopy semi-quantitative and quantitative analysis – A review', *Spectrochim. Acta, Part B*, 63, 1097–1108.

Fortes F J and Laserna J J (2010), 'The development of fieldable laser-induced breakdown spectrometer: No limits on the horizon', *Spectrochim. Acta – Part B*, 65, 975–990.

Fortes F J, Cuñat J, Cabalín L M and Laserna J J (2007), 'In situ analytical assessment and chemical imaging of historical buildings using a man-portable laser system', *Appl. Spectrosc.,* 61, 558–564.

Fortes F J, Ctvrtnickova T, Mateo M P, Cabalin L M, Nicolas G and Laserna J J (2010), 'Spectrochemical study for the in-situ detection of oil spill residues using laser-induced breakdown spectroscopy', *Anal. Chim. Acta*, 683, 52–57.

Gadiuso R, Dell'Aglio M, De Pascale O, Senesi G S and De Giacomo A (2010), 'Laser induced breakdown spectroscopy for elemental analysis in environmental, cultural heritage and space applications: A review of methods and results', *Sensors*, 10, 7434–7468.

Godoi Q, Leme F O, Trevizan L C, Pereira E R, Rufini I A Santos D and Krug F J (2011), 'Laser-induced breakdown spectroscopy and chemometrics for classification of toys relying on toxic elements', *Spectrochim Acta* – Part B, 66, 138–143.

Godwal Y, Kaigala G, Hoang V, Lui S L, Backhouse C, Tsui Y and Fedosejevs R (2008), 'Elemental analysis using micro laser-induced breakdown spectroscopy (μLIBS) in microfluidic platform', *Opt. Expr.*, 16, 12435–12445.

Gonzaga F B and Pasquini C (2008), 'A complementary metal oxide semiconductor sensor array based detection system for laser induced breakdown spectroscopy: Evaluation of calibration strategies and application for manganese determination in steel', *Spectrochim. Acta, Part B*, 63, 56–63.

Gonzales R, Lucena P, Tobaria L M and Laserna J J (2009), 'Standoff LIBS detection of

© Woodhead Publishing Limited, 2013

explosive residues behind a barrier', *J. Anal. At. Spectrom.*, 24, 1123–1126.

Gornushkin I B, Amponsah-Manager K, Smith B W, Omenetto N and Winefordner J D (2004), 'Microchip laser-induced breakdown spectroscopy: A preliminary feasibility investigation', *Appl. Spectrosc.*, 58, 762–769.

Gornushikin I B, Shabanov S V and Panne U (2011), 'Abel inversion applied to a transient lasr induced plasma: implications from plasma modeling', *J. Anal. At. Spectrom.*, 26, 1457–1465.

Gottfried J L, De Lucia F C, Munson C A and Miziolek A W (2009), 'Laser-induced breakdown spectroscopy for detection of explosives residues: a review of recent advances, challenges, and future prospects', *Anal. Bioanal. Chem.*, 395, 283–300.

Goujon J, Giakoumaki A, Piñon V, Musset O, Anglos D, Georgiou E and Boquillon J P (2008), 'A compact and portable laser-induced breakdown spectroscopy instrument for single and double pulse applications', *Spectrochim. Acta Part B*, 63, 1091–1096.

Gravel J F Y, Doucet F R, Bouchard P and Sabsabi M (2011), 'Evaluation of a compact high power pulsed fiber laser source for laser-induced breakdown spectroscopy', *J. Anal. At. Spectrom.*, 26, 1354–1361.

Haisch C, Panne U and Niessner R (1998), 'Combination of an intensified charge coupled device with an echelle spectrograph for analysis of colloidal material by laser-induced plasma spectroscopy', *Spectrochim. Acta, Part B*, 53, 1657–1667.

Hoehse M, Gornushkin I, Merk S and Panne U (2011), 'Assessment of suitable diode pumped solid state lasers for laser induced breakdown and Raman spectroscopy', *J. Anal. At. Spectrom.*, 26, 414–424.

Loebe K, Uhl A and Lucht H (2003), 'Microanalysis of tool steel and glass with laser-induced breakdown spectroscopy', *Appl. Opt.*, 42, 6166–6173.

Maiman T H (1960), 'Stimulated optical radiation in ruby', *Nature*, 187, 493–494.

Michel A P M (2010), 'Review: Applications of single-shot laser-induced breakdown spectroscopy', *Spectrochim. Acta, Part B*, 65, 185–191.

Miziolek A W, Palleschi V and Schechter I (2006), *Laser-induced Breakdown Spectroscopy (LIBS)*, Cambridge, Cambridge University Press.

Moros J and Laserna J J (2011), 'New Raman–laser-induced breakdown spectroscopy identity of explosives using parametric data fusion on an integrated sensing platform', *Anal. Chem.*, 83, 6275–6285.

Nicolas G, Mateo M P and Pinon V (2007), '3D chemical maps of non-flat surfaces by laser-induced breakdown spectroscopy', *J. Anal. At. Spectrom.*, 22, 1244–1249.

Pasquini C, Cortez J, Silva L M C and Gonzaga F B (2007), 'Laser induced breakdown spectroscopy', *J. Braz. Chem. Soc.*, 18, 463–512.

Pontes M J C, Cortez J, Galvão R K H, Pasquini C, Araújo M C U, Coelho R M, Chiba M K, de Abreu M F and Madari B E (2009), 'Classification of Brazilian soils by using LIBS and variable selection in the wavelet domain', *Anal. Chim. Acta*, 642, 12–18.

Praher B, Palleschi V, Viskup R, Heitz J and Pedarnig J D (2010), 'Calibration free laser-induced breakdown spectroscopy of oxide materials', *Spectrochim. Acta, Part B*, 65, 671–679.

Radziemski L J and Cramers D A (1989), *Laser-induced Plasmas and Applications*, New York, Marcel Decker.

Radziemski L J, Loree T R, Cremers D A and Hoffman N M (1983), 'Time-resolved laser-induced breakdown spectrometry of aerosols', *Anal. Chem.*, 55, 1246–1252.

Rauschenbach I, Jessberger E K, Pavlov S G and Hubers H W (2010), 'Miniaturized laser-induced breakdown spectroscopy for in-situ analysis of Martian surface: Calibration and quantification', *Spectrochim. Acta, Part B*, 65, 758–768.

© Woodhead Publishing Limited, 2013

Russo R E (2011) 'Laser ablation molecular isotopic spectroscopy (LAMIS): LIBS for isotopes at atmosphere pressure', Abstract Book, 6th EuroMediterranean Symposium on Laser Induced Breakdown Spectroscopy, 41.

Russo R E, Mao X L, Liu C and Gonzalez J (2004), 'Laser assisted plasma spectrochemistry: laser ablation', *J. Anal. At. Spectrom.*, 18, 2084–1089.

Russo R E, Suen T W, Bol'shakov A A, Yoo J, Sorkhabi O, Mao X L, Gonzalez J, Oropeza D and Zorba V (2011), 'Laser plasma spectrochemistry', *J. Anal. At. Spectrom.*, 26, 1596–1603.

Sabsabi M (2007), 'Femtosecond LIBS' in Singh J P and Thakur S N, *Laser-induced Breakdown Spectroscopy*, Amsterdam, Elsevier, 151–171.

Sabsabi M, Detalle V, Harith M A, Tawfik W and Imam H (2003), 'Comparative study of two new commercial echelle spectrometers equipped with intensified CCD for analysis of laser-induced breakdown spectroscopy', *Appl. Optics*, 42, 6094–9098.

Salle B, Cremers D A, Maurice S and Wiens R C (2005), 'Laser-induced breakdown spectroscopy for space exploration: Influence of the ambient pressure on the calibration curves prepared from soil and clay samples', *Spectrochim. Acta, Part B*, 60, 479–490.

Scaffidi J, Angel S M and Cremers D A (2006), 'Emission enhancement mechanism in dual-pulse LIBS', *Anal. Chem.*, 78, 24–32.

Schill A W, Heaps D A, Stratis-Cullum D N, Arnold B R and Pellegrino M P (2007), 'Characterization of near-infrared low energy ultra-short laser pulses for portable applications of laser induced breakdown spectroscopy', *Opt. Express*, 15, 14045–14056.

Singh J P and Thakur S N (2007), *Laser-induced Breakdown Spectroscopy*, Amsterdam, Elsevier.

Singh J P, Almiral, J R, Sabsabi M and Miziolek A W (2011), 'Laser-induced breakdown spectroscopy (LIBS)', *Anal. Bioanal. Chem.*, 400, 3191–3192.

Stipe C B, Hensley B D, Boersema J L and Buckley S G (2010), 'Laser-induced breakdown spectroscopy of steel: A comparison of univariate and multivariate calibration methods', *Appl. Spectrosc.*, 64, 154–160.

Stratis D N, Eland K L and Angel S M (2000), 'Dual-pulse LIBS using a pre-ablation spark enhanced ablation and emission', *Appl. Spectrosc.*, 54, 1270–1274.

Taschunk M T, Cravetchi I V, Tsui Y Y and Fedosejevs R (2007), in Singh J P and Thakur S N (2007), *Laser-induced Breakdown Spectroscopy*, Amsterdam, Elsevier, 173–196.

Tognoni E, Cristoforetti G, Legnaioli S and Palleschi V (2010), 'Calibration-free laser-induced breakdown spectroscopy: State of the art', *Spectrochim. Acta, Part B*, 65, 1–14.

Webofscience (2011) search from http://apps.webofknowledge.com/ (accessed 25 June 2011).

Winefordner J D, Gornushikin I B, Correl T, Gibb E, Simith B W and Omenetto N (2004), 'Comparing several atomic spectrometric methods to the super stars: special issue on laser induced breakdown spectrometry, LIBS, a future super star', *J. Anal. Atm. Spectrom.*, 19, 106–108.

Yamamoto K Y, Cremers D A, Ferris M J and Foster L E (1996), 'Detection of metals in the environment using a portable laser-induced breakdown spectroscopy instrument', *Appl. Spectrosc.*, 50, 222–233.

Zorba V, Mao X L and Russo R E (2011), 'Ultrafast laser induced breakdown spectroscopy for high spatial chemical analysis', *Spectrochim. Acta, Part B*, 66, 189–192.

© Woodhead Publishing Limited, 2013

20
Surgical solid-state lasers and their clinical applications

D. G. KOCHIEV, A. M. Prokhorov General Physics Institute, Russian Academy of Sciences, Russia,
A. V. LUKASHEV, Stemedica Cell Technologies, USA,
I. A. SHCHERBAKOV and S. K. VARTAPETOV,
A. M. Prokhorov General Physics Institute,
Russian Academy of Sciences, Russia

DOI: 10.1533/9780857097507.2.572

Abstract: The chapter describes the basics of laser–tissue interaction based on the analysis of radiation transport equations. Interaction mechanisms of laser radiation with biological tissue are considered in detail. Clinical applications of lasers with emphasis on ophthalmology and urology are presented.

Key words: Solid-state lasers, laser–tissue interaction, medical applications of lasers, urology, ophthalmology.

20.1 Introduction

In solid-state lasers the active medium is a paramagnetic ion doped dielectric crystal and glass or a dielectric crystal with intrinsic defects of the crystalline lattice. Laser action has been achieved in hundreds of different crystals and glasses (Weber, 1999); however, there are only a few laser crystals that have found real practical applications (Shcherbakov, 1988; Kaminskii, 1996).

Investigation of the interaction of laser radiation with biological tissue started right after laser generation had been achieved in ruby crystal (Maiman, 1960). The first experimental results on laser coagulation of the retina had been published within two to three years after the initial work (Koester *et al.*, 1962; Zaret *et al.*, 1963).

At the early stage of the development of laser medicine the Nd:YAG laser was the most common surgical laser among all other solid-state lasers. Huber *et al.* (2010) has pointed out that in the mid-1980s, a renaissance began in solid-state laser research. The advances of semiconductor lasers enabled efficient diode laser pumping of laser materials and the realization of all solid-state devices in a compact and robust manner. In parallel new laser materials have been discovered and developed. Besides Nd^{3+} various efficient diode pumped lasers have been investigated on the basis of the

© Woodhead Publishing Limited, 2013

rare-earth ions Er^{3+}, Tm^{3+}, Ho^{3+} and Yb^{3+} in combination with various host crystals. Transition metal ion lasers based on Cr^{3+}-doped and Ti^{3+}-doped crystals have contributed substantially to the solid-state laser renaissance, as well as new laser classes obtained with tetravalent Cr^{4+} and divalent Cr^{2+} as active ions.

A specific feature of solid-state lasers is an ability to obtain a wide variety of operation modes and output parameters of laser radiation. That is the basis of the huge diversity of different uses of solid-state lasers in medicine. For example, the wavelength range of solid-state lasers covers the spectrum from the ultraviolet up to the mid-infrared. At the same time the range of energy fluence can cover three orders of magnitude (from 1 J/cm^2 to 10^3 J/cm^2); the range of power density covers 18 orders (from 10^{-3} W/cm^2 to 10^{15} W/cm^2); and the laser pulse duration covers 16 orders, from continuous wave (CW, ~10 s) to femtosecond pulses (10^{-15} s). Such a wide range of output parameters of the laser light gives a unique possibility of inducing different mechanisms of laser–tissue interaction and providing a variety of clinical applications (Shcherbakov, 2010).

20.2 Laser–tissue interaction

20.2.1 Optical properties of biological tissues: absorption and scattering of laser radiation

The spatial distribution of volumetric energy density while the laser light propagates in tissue defines the processes of laser–tissue interaction. Those processes depend on the intensity of the laser radiation and the optical properties of the tissue. The amount of reflected, absorbed and scattered light is defined, in general, by the type of tissue and the wavelength of laser radiation through the refraction index, absorption and scattering coefficients of the tissue.

Light propagation in turbid biological tissue has certain features. The energy of light absorbed in tissue could be transformed into heat, re-emitted in the form of fluorescence or phosphorescence, or dissipated due to photochemical reactions. The type of losses of the laser light energy (reflection, absorption or scattering) that are dominant while it propagates in tissue primarily depend on the type of tissue and the wavelength of the laser radiation. In laser surgery information on absorption and scattering properties of selected tissue is necessary to estimate the distribution of the laser radiation in the tissue, to choose the most appropriate dose of exposure and for planning the results of laser action on the tissue.

The attenuation of the intensity of the laser beam in biological tissue due to absorption and scattering is governed by Lambert–Beer's law:

$$I(z) = I_0(1 - r_s) \exp(-\mu_t z) \qquad [20.1]$$

© Woodhead Publishing Limited, 2013

where $r_s = [(n_2 - n_1)/(n_2 + n_1)]^2$ is a coefficient of Fresnel reflection from the interface of the media at the normal incidence; n_1 and n_2 are refraction indices of the medium and biotissue; $I(z)$(W/cm^2) is the intensity of laser radiation at distance z (cm) from the surface of tissue; I_0 (W/cm^2) is the intensity of the incident radiation at the surface of tissue; $\mu_t = \mu_a + \mu_s$ is an extinction coefficient (cm^{-1}); μ_a is a coefficient of absorption; and μ_s is a coefficient of scattering (Tuchin, 1997; Welch and Gardner, 2001).

By disregarding Fresnel losses, it follows from [20.1] that

$$z = \frac{1}{\mu_t} \ln(I_0 / I(z)) \qquad [20.2]$$

and a mean free path of a single time scattered photon in biotissue is defined as $l_{ph} = 1/(\mu_a + \mu_s)$.

The impact of pulsed laser radiation on a biological tissue is defined by the spatial distribution of the volumetric power density. The optical properties of tissue, primarily its ability to absorb and scatter light, are the main factors which determine the distribution of the laser radiation in the volume of tissue. A typical value of the scattering coefficient of most biological tissue in the visible and near-infrared regions lies in the range 100–500 cm^{-1} and monotonically decreases with an increase of wavelength (Jacques and Patterson, 2004; Cheong *et al.*, 1990). The coefficient of scattering of biological tissue is 10–100 times higher than the coefficient of absorption for most wavelengths, with the exception of the ultraviolet and far-infrared bands.

The interaction of laser radiation with biological tissue can be considered as the propagation of photons in a medium of homogeneously distributed centers of scattering and absorption. In the framework of this model, the propagation of laser radiation in a scattering medium can be described in terms of the radiation transport theory. The incidence of a flat wavefront laser radiation onto a turbid medium is described by this basic non-stationary transport equation for monochromatic light (Ishimaru, 1978; Wang and Wu, 2007):

$$\begin{aligned} \frac{1}{c}\frac{\partial L(\bar{r}, \bar{s}, t)}{\partial t} &= -\bar{s} \cdot \nabla L(\bar{r}, \bar{s}, t) - \mu_t L(\bar{r}, \bar{s}, t) \\ &\quad + \mu_s \int_{4\pi} L(\bar{r}, \bar{s}', t)\, p(\bar{s}', \bar{s})\, \mathrm{d}\Omega' + S(\bar{r}, \bar{s}, t) \end{aligned} \qquad [20.3]$$

where $L(\bar{r}, \bar{s}, t)$ an angular spectrum of laser intensity, i.e. the intensity at the point $\bar{r}$ in the direction of $\bar{s}$; $p(\bar{s}', \bar{s})$ is a phase function, such that $p(\bar{s}', \bar{s})\mathrm{d}\Omega$ is a probability for the radiation propagating in the direction of $\bar{s}'$ to be scattered into solid angle around the direction $\bar{s}$; $\mu_t = \mu_a + \mu_s$ is the total attenuation coefficient in turbid medium; $\bar{s}'$ and $\bar{s}$ are unit vectors in the direction of incident and scattered photons; $\mathrm{d}\Omega'$ is a unit solid angle in

© Woodhead Publishing Limited, 2013

the direction of $\bar{s}'$; and $S(\bar{r}, \bar{s}, t)$ is a function of the source for the incident light. The phase function $p(\bar{s}', \bar{s})$ describes the scattering properties of the medium and is the function of probability density for scattering in the direction s of a photon, propagating in the direction s'. Equation [20.3] represents the rate of change of energy in unit volume of the medium, in unit solid angle in the direction $\bar{s}$.

For most practical applications, equation [20.3] does not have analytical solutions. Thus, when one chooses a model to describe radiation propagation in a turbid tissue, the appropriate statistical approximations should be defined depending upon what kind of absorption and scattering processes are most dominant. These models include a method of first approximation, the Kubelka–Munk model, and diffusion approximation. The Monte-Carlo method is used for a number of different numerical calculations of the radiation transport equation; this method was also developed and applied for tissue optics. The Monte-Carlo method is based on a numeric simulation of photon transport in a turbid medium. Using this method, photon random walk in a turbid sample of biotissue can be traced from the point of incidence to the sample all the way up until the act of absorption in the tissue or exiting from the sample.

Scattering dominates over absorption in visible and near-infrared spectral bands for most biotissues. In such conditions, some reliable estimates could be obtained by using diffusion approximation to analyze light propagation in a turbid medium. (Ishimaru, 1978; Wang and Wu, 2007).

One of the basic tasks in the optics of biological tissue is to predict the distribution of radiation in the medium. Some aspects of the calculation of spatial distribution of the radiation intensity, in the case of multiple scattering and the possibility of direct measurement of the volume distribution of the light intensity of the pulsed laser radiation in biological tissue, are discussed in the work of Pelivanov *et al.* (2006).

20.2.2 Mechanisms of laser–tissue interactions

The basic method of traditional surgical treatment is an intervention in a pathological process or a disease state. If laser radiation is used as a surgery tool its main purpose would be to induce changes in biological tissue: from resection to initiation of chemical reactions, as in photodynamic therapy.

As mentioned above, the output parameter of laser radiation (wavelength, intensity, pulse duration) can be widely varied, which gives an opportunity to perform different types of interaction of the laser radiation with biological objects: linear and non-linear interaction, single and multi-photon processes, coherent and non-coherent processes, thermal and non-thermal, etc. Thus, one can induce different types of effects of laser radiation in biological

© Woodhead Publishing Limited, 2013

tissues: photo-chemical changes, thermal destruction, explosive ablation, optical breakdown, shock wave generation, photo-destruction, etc. (Letokhov, 2003).

Photochemical mechanism

In laser medicine the photochemical mechanism of laser–tissue interaction plays a major role in photodynamic therapy (PDT). During PDT some specially selected chromophores (photosensitizers) are injected into the body. Monochromatic laser radiation induces selective photochemical reactions, which are followed by biological conversions in living tissues. After resonance excitation by laser radiation, a molecule of photosensitizer undergoes several synchronous or sequential decays, which induce intramolecular transfer reactions. As a result of a chain of such reactions a cytotoxic reagent is released, which induces non-reversible oxidative damage to cellular structures. The main idea is to use a chromophore receptor, which acts as a catalyst. Usually laser exposure takes place at a relatively low power density (~1 W/cm^2) and long exposure time, from seconds to continuous irradiation. Visible laser light with deep penetration into biotissue, which is important to irradiate deeply located tissue structures, is used for this method.

Thermal interaction

If photochemical processes occur as a result of a chain of specific photochemical reactions, thermal effects of laser–tissue interaction are usually non-specific (Parrish and Deutsch, 1984). At the microscopic level, thermal effects originate as a result of the volumetric absorption of laser radiation that takes place in molecular vibration-rotational bands and is followed by non-radiative relaxation. A rise of temperature in the tissue occurs very efficiently due to a large number of vibrational-rotational levels in biological molecules and numerous pathways of collisional relaxation. Typical photon energy for different lasers is 0.35 eV for Er: YAG, 1.2 eV for Nd:YAG, and 6.4 eV for ArF. That energy is substantially larger than the kinetic energy of a molecule at room temperature, which is only 0.025 eV (Niemz, 2004).

The loss of material at the tissue surface during thermal interaction is a result of three major thermal processes – evaporation, boiling and explosive boiling, which take place in tissue due to the absorption of laser radiation (Miotello and Kelly, 1995). The significance of each of these processes depends on both the laser pulse duration and the temperature which is achieved in the irradiated material (Miotello and Kelly, 1999).

Thermal effects play a predominant role for continuous wave (CW) lasers and lasers that operate in normal spiking or free-running mode. Tissue removal starts when the temperature of the surface layer becomes higher

© Woodhead Publishing Limited, 2013

than 100°C and is accompanied by a pressure rise in the tissue. Histological analysis of tissue at this stage reveals the presence of ruptures and vacuoles (sap cavities). Continuing laser exposure leads to a temperature rise in the tissue of up to 350–450°C, when burning out and carbonization of tissue takes place. A thin layer of carbonized tissue (~0.02 mm) and a layer of vacuoles (~0.03 mm) support a high gradient of pressure along the tissue removal interface, while the rate of tissue removal remains constant and depends on the type of tissue (LeCarpentier, 1993).

A similar mechanism of tissue removal is applicable to pulsed Er:YAG/YSGG and Nd:YAG lasers with a pulse duration of $\tau_L \geq 100$ µs. Venugopalan *et al.* (1996) showed that tissue removal at those wavelengths and pulse durations takes place during the laser pulse and is not due to explosive processes but is due to the superficial effects of thermal melting followed by the release and vaporization of tissue. Heating and evaporation at the surface of the tissue are accompanied by heating and evaporation of water under the surface without any possibility of leaving the tissue, which is not yet damaged. This process leads to further temperature and pressure build-up underneath the tissue surface until the tissue ruptures and material blows out.

The presence of the extracellular matrix (ECM) influences the development of the phase transition processes during pulsed laser irradiation of tissue. Boiling of water inside the tissue volume occurs when the difference in the chemical potential of vapor and liquid phases, needed for bubble growth, exceeds not only surface tension at the phase interface but also the energy of elastic tension of ECM, needed for matrix deformation of the surrounding tissue. Bubble growth in tissue requires higher internal pressure compared to a pure liquid. The larger pressure results in an increase of the boiling temperature. Pressure build-up during boiling continues until it exceeds the ultimate stress of ECM and results in subsequent tissue ejection. Such a process of bulk boiling without a free surface is often called ‘confined boiling’, taking into account the presence of the ECM (Vogel and Venugopalan, 2003). The process of bulk boiling in tissue initially develops as isobar heating until it reaches binodal and continues along it until the saturated vapor pressure exceeds the stretch limit of the ECM.

Depending on the laser power density and exposure time, thermal damage of tissue can vary from carbonization and melting at the tissue surface, to hyperthermia several millimeters in depth from the surface. As a rule, the goal of laser surgery is one of those results; thus accurate estimation of laser parameters plays a critical role.

Spatial distribution of heat transfer could be described by a time-dependent parameter, called the thermal penetration depth, defined as:

$$z_{therm} = \sqrt{4\chi t} \qquad [20.4]$$

© Woodhead Publishing Limited, 2013

where χ is the thermal diffusivity (temperature conductivity) and z_{therm} is a distance at which the temperature drops by e times from its peak value. A spatially limited surgical effect (selective phototermolysis) (Anderson and Parrish, 1983) occurs when the pulse duration is less than the thermal diffusion time of the particular volume. The volume of tissue subject to heating is defined in most cases by the penetration depth of the laser light into the tissue, where the laser intensity is attenuated e times:

$$\delta = 1/\mu_{eff} = [3\mu_a(\mu_a + \mu'_s)]^{1/2}$$

and thermal relaxation time t_d is defined as (Furzikov, 1987):

$$t_d = \frac{\delta^2}{4\chi} \quad [20.5]$$

Heat confinement in the interaction area is possible if the ratio of the laser pulse duration to the thermal relaxation time meets the condition $\tau_L/t_d \leq 1$. If the laser pulse duration $\tau_L \leq t_d$, heat does not spread out evenly at the distance of the optical penetration depth and as a result thermal damage of the adjacent tissue is negligible. If we define dimensionless parameter τ_d^* as a metric of laser pulse duration relative to the time of thermal relaxation, $\tau_d^* = (\tau_L/\tau_d)$, then using (20.5) the condition of absence of thermal interaction can be expressed as (Vogel and Venugopalan, 2003):

$$\tau_d^* = \frac{4\chi\tau_L}{\delta^2} \leq 1 \quad [20.6]$$

Continuous wave lasers and lasers with a 'long' pulse duration ($\tau_L \geq 100$ μs) produce quite a large zone of thermal damage in the adjacent tissues. The size of the zone of unwanted changes in the tissue depends on the intensity of the incident laser radiation, its wavelength and exposure time. To control collateral thermal damage one should limit the time of laser exposure using a short laser pulse duration in order to minimize heat transfer out from the zone of interaction according to condition (20.6).

Ablation

In the technical literature the word ablation (from Latin *ablatio*) is interpreted as a combination of physical and chemical processes which leads to the removal of material from the surface or volume of an object. There are several attributes of laser ablation: (1) ablation is intrinsically related to the absorbed laser energy; (2) ablation may take place in vacuum or an inert environment; (3) a result of the laser ablation is formation of a vapor-gas (or plasma-gas) cloud of the products of ablation (Anisimov and Luk'yanchuk, 2002).

The reduction of the laser pulse duration changes the pattern and dynamics

© Woodhead Publishing Limited, 2013

of thermal processes during laser tissue interaction. An increase of the energy deposition rate in a tissue due to a shorter pulse duration as well as its spatial distribution results in the generation of substantial thermal and mechanical transient processes, which in turn are drivers of pulsed laser ablation.

Photomechanical processes, which develop in the tissue volume during laser-tissue interaction, can lower the threshold of ablation. During the absorption of the laser energy and heating of the tissue, the tissue material expands in order to maintain equilibrium with respect to its thermodynamic parameters and the state of the surrounding tissue. If a non-homogeneous distribution of temperature appears after the absorption of laser energy in the tissue, it invokes the development of thermoelastic deformations and the generation of propagating compression waves in the tissue material.

However, it takes a characteristic time t_m for the material of a tissue to expand or to establish mechanical equilibrium. This time is equal to the time needed for a transverse acoustic wave to travel across the characteristic length of the system. Thus, $t_m = \delta/c_a$, where c_a is the speed of sound in the tissue and δ is the characteristic length of the heated volume. The characteristic length is usually the lesser of two parameters: the laser beam radius at the surface of the tissue, r, or the effective penetration depth of the radiation in the tissue, which for simplicity may be considered equal to μ_{eff}^{-1}.

When the heating time or laser pulse duration t_L is greater than t_m, or $\tau_m^* = (\tau_L/t_m) >> 1$, tissue material expansion occurs during the laser pulse and the value of the induced pressure changes together with the intensity of the laser pulse. If $\tau_m^* = (\tau_L/t_m) << 1$, then the energy deposition into the system occurs faster than the system can mechanically react. In this case, the speed of expansion is determined by the inertia of the heated layer and does not depend on the laser intensity, the pressure in the tissue being proportional to the bulk energy density absorbed in the tissue at every moment of time. Upon the absorption of a laser pulse with a pulse duration τ_L, which is far less than the traveling time of an acoustic wave across the area of heat deposition t_m, so-called 'inertial confinement' of the tissue takes place, i.e. the tissue does not have time to expand and its heating occurs at a constant volume (Itzkan *et al.*, 1995).

If the condition of inertial confinement is expressed in terms of laser pulse duration relative to the time of acoustic wave travel, then as a metric of inertia, dimensionless parameter $\tau_m^* = (\tau_L/t_m)$ can be expressed as (Venugopalan, 1995):

$$\tau_m^* = \frac{c_a \tau_L}{\delta} < 1 \quad [20.7]$$

where c_a is the speed of sound in the tissue.

Albagli *et al.* (1994) showed that tissue ablation by laser pulses with $\tau_m^* \leq 1$ is achieved at a much lower level of the radiation energy density compared

© Woodhead Publishing Limited, 2013

to laser pulses with a greater value of τ_m^*. However, such a transition to a greater ablation threshold will only take place when $\tau_m^* \geq 10$.

If the rate of energy deposition in the tissue volume during laser irradiation is much faster than the speed of internal energy loss due to water evaporation and boiling, then the water in the tissue transfers to a superheated metastable state. When the thermodynamic state of the water approaches spinodal a fluctuation mechanism of nucleation takes effect (homogeneous nucleation), enabling the fast decay of the metastable phase. The most pronounced homogeneous nucleation is observed during fast pulsed heating of a liquid phase; it leads to a spontaneous explosive boiling of a highly overheated liquid (phase explosion) (Martynyuk, 1977; Miotello and Kelly, 1995).

The process of phase separation spontaneously fills up the whole volume of superheated liquid. During this process superheated, unstable liquid at atmospheric pressure spontaneously transfers into an equilibrium state of mixed phase composed of saturated liquid and vapour at pressures that can approach the saturation pressure corresponding to the spinodal temperature (p = 9.2 MPa). If the pressure surge leads to tissue rupture and destruction, then there is an ejection of a mixture of vapor-droplets of saturated vapor and liquid. If the pressure surge does not lead to the destruction of the ECM then the tissue transfers to an intermediate state at the binodal curve. In this case, the temperature and pressure in the tissue are determined by its elasticity and an additional energy deposition is needed to destroy and remove tissue (Vogel and Venugopalan, 2003).

Photoablation

The dissociation energy of chemical bonds in organic molecules is close to or less than the energy of photons of ultraviolet lasers (4.0–6.4 eV). During tissue exposure the photons are absorbed by molecules that can lead to direct breaking of chemical bonds, producing so-called ablative 'photodecomposition' of tissue (Srinivasan and Braren, 2003).

Transformation of the absorbed energy of laser photons through vibrational relaxation of molecules is a basis for the photothermal mechanism of tissue ablation. It had been thought that specific features of photochemical decomposition of tissue using UV lasers can be easily distinguished from photothermal processes and that only photochemical decomposition provides accurate tissue removal. However, it was shown that the quantum yield of photochemical processes is low enough to explain ablation at the energy density levels observed in experiments (Nikogosyan and Gorner, 1999). The ablation of tissue by UV lasers is not only due to photochemical dissociation but always comes together with thermal processes (Schmidt *et al.*, 1998).

© Woodhead Publishing Limited, 2013

Plasma-induced laser ablation and photodestruction

'Laser plasma-induced ablation' (Niemz, 2004), or 'laser-induced plasma-mediated ablation' (Vogel and Venugopalan, 2003) is characterized by plasma formation during the interaction of laser radiation with tissue. For laser pulses with duration from 10 ps to 10 ns the mechanism of interaction is classified as electromechanical (Boulnois, 1986). This means the generation of plasma in a very intense electrical field and the subsequent tissue removal due to shock wave propagation, cavitation and jet formation.

Niemz (2004) interprets the process of tissue removal as two separate processes: tissue removal due to ionization and tissue removal due to mechanical processes, which develop after plasma generation (photodestruction). There are still some difficulties in identifying specific results of ablation in each of these two case orders to separate those two processes. Thus, very often, those two processes are considered as one process and all is interpreted as photodestruction of tissue (Ogura *et al.*, 2002; Vogel *et al.*, 1994).

For laser radiation with intensity in the range of about 10^{10}–10^{12} W/cm^2 at the surface of the tissue and with a pulse duration of less than nanoseconds, plasma formation is typical. This corresponds to the local electrical field strength of ~10^6–10^7 V/cm (Boulnois, 1986). In materials in which the temperature has substantially risen due to the high absorption of laser radiation, plasma formation is induced and maintained by thermal emission of free electrons. In media with low optical absorption, plasma is formed at much higher laser intensities because of electrons released due to multiphoton absorption of laser radiation and avalanche ionization of tissue molecules (optical breakdown). This feature of optical breakdown means that plasma can form not only in strongly absorbing tissues, for example tissue containing pigment, but also in transparent, low absorbing tissues.

The thresholds of optical breakdown in water and most transparent biological media (cornea, vitreous body) are close to each other (Vogel *et al.*, 1994). The threshold of plasma formation for tissues with high absorption is lower due to the thermal emission of free electrons.

Dynamics of ablation processes

Tissue removal by pulsed laser radiation requires the destruction of the ECM and cannot be considered as a process of dehydration during laser heating. Laser action on the ECM of tissue during pulsed laser exposure results in the weakening of the intermolecular bonds in the tissue. Pressure waves generated during the phase of explosion and confined boiling destroy the ECM of tissue. As a result, an explosive ejection of tissue material takes place without its full evaporation. The threshold of ablation of such a process turns out to be lower than the specific enthalpy of evaporation of water

© Woodhead Publishing Limited, 2013

$(c_p\Delta T + h_{fg})$ = 2580 J/cm^{-3} (h_{fg} = 2255 J/cm^{-3} is the specific heat of vaporization of water). Tissues with a high value of tensile strength require higher temperature to destroy the ECM. For those tissues, the threshold of volumetric energy density comparable to or exceeding the enthalpy of evaporation is required to destroy the ECM (Kaufmann and Hibst, 1996).

The relation between the threshold energy density H'''_{th} and the dimensionless parameter τ^*_m was revealed by Albagli *et al.* (1994). For the tissue of the aorta H'''_{th} = 300 J/cm^3 at $\tau^*_m \leq 1$ and H'''_{th} = 2200 J/cm^3 at $\tau^*_m \geq 20$, which is close to the value of the specific enthalpy of vaporization of water.

Phase transitions occurring in tissue lead to the formation of an ablation plume, which consists of water vapor, organic products in gaseous form, water droplets and/or particles of tissue fragments. Expansion of this plume into outer space is accompanied by an acoustic transient process, which at high energy density leads to shock wave formation. A flow of removed materials perpendicular to the tissue surface produces a recoil pressure at the surface, which provides an additional material ejection and collateral effects in the tissue volume. Flow components parallel to the tissue surface, which are developed at the later stages, affect tissue removal and induce material relocation in the area of interaction. Scattering and absorption of laser radiation in the plume becomes a limitation of ablation efficiency at high intensities of the incident laser pulses (Vogel and Venugopalan, 2003).

20.3 Clinical applications of solid-state lasers

The complexity of biological tissue as the object for scientific investigation, and the range of required scientific disciplines needed for its study (from non-equilibrium thermodynamics to photochemistry; from plasma physics to biomechanics) are distinctive features which define the difficulties of investigating the interaction of high intensity pulsed laser radiation with tissue (Vogel and Venugopalan, 2003). The chemical composition, optical properties, molecular structure, mechanical and elastic properties of tissue must be taken into account to investigate and understand the mechanism of laser–tissue interaction. Further studies of tissue properties, the development of fiber delivery systems for laser radiation, progress in *in situ* visualization, as well as the development of new laser systems are the basis of continuing growth and evolution of using laser technology in surgery and medicine. The variety of clinical applications of solid-state lasers in different areas of surgery explains the increasing interest in laser surgery.

The Nd:YAG laser is one of the most common surgical lasers, which is used in free-running mode with endoscopic delivery in pulmonology, gastro-enterology and urology (Peng *et al.*, 2008). In aesthetic medicine it is used for hair removal and spider vein treatment (Littler, 1999); in oncology for interstitial laser coagulation of tumors (Pech *et al.*, 2007). In Q-switch

© Woodhead Publishing Limited, 2013

mode with a laser pulse duration about 10 ns the Nd:YAG laser is used in ophthalmology, for example for the treatment of glaucoma (Barnes *et al.*, 2004).

The absorption of most tissues is low at the main line (λ = 1064 nm) of the Nd:YAG laser. The effective penetration depth into tissue at this wavelength could be up to several millimeters and thus it provides good homeostasis and bulk coagulation. However, the volume of ablated tissue is relatively small. Tissue resection and ablation are accompanied by substantial thermal damage of adjacent tissues, edemas and inflammatory processes.

An important advantage of the Nd:YAG laser is that its radiation can be delivered at the impact zone through an optical fiber. Lasers at 2790 nm (Er:YSSG) and 2940 nm (Er:YAG) are more efficient for laser ablation; however, high absorption of optical fiber materials at these wavelengths limits their use for surgical applications. The delivery of laser radiation using commercially available optical fibers is possible for 2010 nm of Cr:Tm:YAG and 2120 nm of Cr:Tm:Ho:YAG lasers. At these wavelengths, the penetration depth is relatively small and can provide efficient ablation and minimize unwanted thermal collateral effects. The penetration depth is ~170 microns at 2010 nm (thulium laser) and ~350 microns at 2120 nm (holmium laser) (Vogel and Venugopalan, 2003).

Laser dermatology exploits almost all types of solid-state lasers from visible (ruby at 693 nm, alexandrite at 755 nm, KTP laser at 532 nm) to near-infrared (Nd:YAG at 1064 nm, 1320 nm and 1440 nm, Er:glass at 1540 nm) as well as mid-infrared lasers like Ho:YAG at 2120 nm, Er:Cr:YSGG at 2790 nm and Er:YAG at 2940 nm. Selective photothermolysis is the main effect which is used in laser dermatology for treating skin tissue. Lasers in dermatology are used for the treatment of different vascular lesions of the skin, benign and malignant tumors, pigmentation, tattoo removal and various cosmetic procedures (Wheeland, 1995). Recently Nd:YAG, Er:glass, Er:Cr:YSGG and Er:YAG lasers have been used for so-called fractional photothermolysis. In order to achieve a therapeutic effect on the skin, but decrease the time of skin recovery after laser treatment, laser radiation is delivered on the skin in the form of small dots (fractions). Each dot produces the thermal effects described above, for example a microablation channel, but the density of dots on the skin is chosen to produce faster recovery after ablative damage while maintaining the long-term skin modification effect (Manstein *et al.*, 2004).

Er:Cr:YSGG at λ = 2780 nm and Er:YAG at λ = 2940 nm lasers are used in dentistry for the ablation of hard dental tissue to treat caries and prepare for a root canal. These laser procedures are painless, they do not bring discomfort to the patient, they do not have unwanted collateral thermal effects and they do not damage adjacent tooth structures (Iaria, 2008; Kang *et al.*, 2008). KTP, Nd:YAG, Er:Cr:YSGG and Er:YAG are used for soft tissue surgery in the oral cavity (Deppe and Horch, 2007).

© Woodhead Publishing Limited, 2013

The endoscopic technique and optical fiber catheters enable the delivery of laser radiation almost non-invasively to the upper and lower sections of the gastrointestinal tract. The most common laser in gastroenterology is the Nd:YAG laser, which provides homeostasis of soft tissues. Laser re-canalization reduces the complications of swallowing in patients with esophagus and forestomach cancer; however, repeated procedures or installation of stents might be required (Nishioka, 1995).

20.3.1 Lasers in ophthalmology

An eye, being an optical system, is the only human organ which is naturally accessed by a laser in its both anterior and posterior segments. Ophthalmology was the first field of medicine where lasers were used as an instrument. The first studies on the applications of lasers in ophthalmology were performed in the late 1960s and were targeted to the treament of a detached retina. 'Laser ophthalmology' has become a general usage term and nowadays it is impossible to imagine a present-day ophthalmologic clinic without a laser. Laser treatment of a detached retina has been discussed for many years; however, it was the discovery of laser photocoagulation of the retina that made it a routine procedure in ophthalmologic clinical practice (L'Esperance, 1968).

Efforts to study the use of pulsed Nd:YAG lasers for destroying the lens capsule in the case of secondary cataract development started in the late 1970s to early 1980s (Aron-Posa *et al.*, 1980). Today, capsulotomy performed with a Q-switch neodymium laser is a standard surgical procedure to treat secondary cataracts. During this procedure a membrane, which grows near the artificial lens after a primary cataract procedure, is removed by focusing intensive laser radiation on that membrane and inducing an optical ablation at the surface and breakdown in the volume of the membrane.

The discovery of the possibility of using short wavelength UV laser radiation to perform non-thermal cornea modification to change its curvature and thus correct vision acuity became quite a revolution in ophthalmology (Trokel *et al.*, 1983). Today laser surgery to correct vision with the help of short wavelength UV lasers (PRK – photorefractive keratotomy; LASIK–laser-assisted in situ karetomileusis) is commonly performed in many clinics.

Significant progress in refractive surgery and some other microsurgical interventions has been achieved by using lasers with short and ultrashort pulse durations. The use of femtosecond lasers (Juhasz *et al.*, 1999) enabled the development of minimally invasive microsurgical intervention for refractive surgery, cornea transplantation, and the intrastromal channel.

At the present time the number of laser procedures in ophthalmology is quite large and continues to grow. Today lasers are routinely used for almost

© Woodhead Publishing Limited, 2013

all areas of vitro-retinal and corneal surgical procedures: for glaucoma treatment, for cataract treatment, for retina treatment; for transplantation of the cornea with a femtosecond laser, and for correction of refraction – myopia, hypermetropia, astigmatism and presbyopia.

Practically all types of lasers – gas (argon ion and excimer), solid-state (CW, pulsed, with harmonics generation), fiber lasers, and diode lasers, including optically pumped diode lasers – are used in ophthalmology. Solid-state lasers have a special position among these – laser ophthalmology started with solid-state lasers and they still occupy a niche in the ophthalmologic laser system marketplace.

Nowadays Nd:YAG and Nd:YLF lasers (CW, pulsed Q-switched with pulse duration of several nanoseconds as well as femtosecond lasers) are the most commonly used in ophthalmology practice. To a lesser extent, other lasers such as free-running Nd:YAG at 1440 nm, Ho:YAG at 2100 nm and Er:YAG at 2940 nm are also used in ophthalmology.

Different segments of the eye have a different composition and absorption coefficient of the laser radiation at the same wavelength. Thus, the preferred choice of laser (wavelength, energy and pulse duration) depends on the location where the procedure is to be performed, as well as the type of interaction and the expected results in the impact zone. Based on the spectral properties of the anterior segment and cornea, lasers with wavelengths ranging from 180 nm to 315 nm are the most suitable for surgical applications on these tissues. It is possible to achieve deeper penetration, up to the eye lens, by using lasers in the spectral range 315–400 nm. All other segments of the eye behind the lens can be treated by laser radiation with wavelengths from 400 nm all the way up to 1400 nm. At wavelengths longer than 1400 nm water absorption in the tissue is a dominant factor that restricts and defines the use of those lasers.

The effect of the laser radiation on the tissue depends on the laser pulse duration and its repetition rate. For CW lasers and lasers with long pulse duration the effect of laser action is mostly thermal: the shorter the pulse duration, the lesser the thermal load in the tissue. Laser pulse duration and pulse peak intensity are the most important characteristics to consider. For very short pulses and high peak intensities, multi-photon ionization can be achieved and a direct transition from the solid-state to the gas-plasma phase is observed. A similar mechanism takes place when short wavelength radiation is used; a transition from the solid-state to the ionized gas phase (material ablation) due to direct tissue ionization is possible in this case.

The sections below give a short review of solid-state lasers that are currently in use or have been thoroughly investigated for applications in ophthalmology.

© Woodhead Publishing Limited, 2013

Refractive surgery: keratoplasty

Several techniques are used for laser refractive surgery. Two of them are direct cornea reprofiling due to laser evaporation of the outer layers of the cornea, and keratoplasty. In keratoplasty, a laser produces local coagulation zones at the periphery of the cornea. These coagulation zones induce stretching of the cornea and increase its curvature. In the case of direct reprofiling, the change of the cornea curvature could be positive or negative, thus a correction of both myopia and hypermetropia (hyperopia) is possible. Laser keratoplasty enables the correction of only hypermetropia, i.e. one can only increase the curvature of the cornea.

Short wavelength laser radiation in the UV band near 200 nm is the most preferable for direct cornea reprofiling. The absorption coefficient of the cornea is very high at this wavelength ($>2 \times 10^3$ cm^{-1}) and in this case photoablation has minimal unwanted thermal effects.

The standard in this area is the use of excimer ArF lasers at 193 nm wavelength (Munnerlyn, 2003). However, with the development of solid-state laser technology, the use of solid-state lasers with harmonics generation into the UV spectral band might become a real alternative for gas excimer lasers (Dair *et al.*, 1999). Solid-state lasers considered for refractive surgery are Nd:YAG and Nd:YLF lasers with flash lamp or diode pumping and nonlinear conversion of the fundamental wavelength into the fifth harmonics (210–213 nm).

The Er:YAG laser at λ = 2940 nm was tried for refraction correction (Lasser *et al.*, 1992). Due to its high water content, the cornea strongly absorbs 3-micron laser radiation and so efficient ablation of corneal material can be achieved. Later it was shown that despite the much higher efficiency of material removal by 2940 nm laser radiation, the quality of the walls and the bottom of the ablated cavity is not as smooth and collateral thermal damage of adjacent tissue is greater compared to short wavelength excimer laser radiation (Mrochen *et al.*, 2000).

Recently solid-state lasers with femtosecond pulse durations have found use in refractive surgery. The use of ultra-short laser pulses creates super-high electrical field strength in the focus and thus induces multi-photon ionization and as a result material removal. Due to their high precision, femtosecond laser pulses were tried for the removal of material inside a cornea to change its refractive properties. They were also used to resect the epithelium layer on the eye as a first step of a LASIK procedure, with the goal of replacing mechanical keratome with precision laser technology (Kurtz *et al.*, 1998; Juhasz *et al.*, 1999). Besides those applications, femtosecond laser radiation was used for intrastromal interactions that lead to the precision cutting of corneal layers. In keratoplastic surgery, another mechanical surgical instrument, the trepan, could be replaced by a laser, to provide high-precision cutting free of pain.

© Woodhead Publishing Limited, 2013

Typical output parameters of solid-state lasers for refractive surgery and keratoplasty are the following: wavelength of 1000 nm falling in the transparency range of the cornea; laser pulse duration less than 800 fs; repetition rate that defines the time of procedure, 0.1–1.0 MHz; energy of the pulse 1–3 μJ (Binder, 2008). A remarkable success of using femtosecond lasers in ophthalmology is mainly related to progress in the development of femtosecond solid-state lasers. During the upcoming years, all femtosecond fiber lasers are expected to substitute solid-state bulk lasers in medical systems for refractive surgery. However, the obtainable single pulse energy in fiber lasers has a fundamental limit, thus solid-state lasers will retain their significance and a perspective for development in applications where the laser pulse should be delivered to posterior eye segments and where higher pulse energy is required. In the near term, femtosecond solid-state lasers with single pulse energy above 10 μJ and a repetition rate above 100 kHz will be used for precision treatment in the posterior segments of the eye.

In laser keratoplasty, laser–tissue interaction should take place in the volume of the cornea but without penetration into the endothelium to avoid damaging it. Solid-state lasers with wavelengths having relatively low absorption in the cornea are used for laser keratoplasty. Typical lasers for this purpose are an erbium-doped glass laser at λ = 1540 nm or a Ho:YAG laser at λ = 2090 nm (Seiler *et al.*, 1990). Both lasers are flashlamp pumped, operated in free-running mode with laser pulse duration of about 100–300 μs and a repetition rate of several pulses per second (1–3 Hz); single pulse energy is several hundreds of millijoules. Laser radiation at 1540 nm penetrates relatively deep into corneal tissue (~500 μm), it absorbs in the bulk tissue, the coagulation area is in inside the cornea, and the refractive effect due to stretch tissue is more stable.

Cataract: glaucoma

The method of phacoemulsification is currently a standard procedure to remove cataracts. A special system – the facoemulsifier – is used to mechanically remove the cataract by a needle knife, which vibrates at a very high frequency. On the other hand, laser removal of a matured cataract has the advantage over the mechanical method of phacoemulsification.

A pulsed Nd:YAG laser operated at an uncommon emission line of 1440 nm is used for this procedure. This laser wavelength is strongly absorbed by water (absorption coefficient ~30 cm^{-1}) and can be easily transmitted through standard quartz optical fibers. Typical energy and temporal output parameters of a laser cataract extractor are pulse duration 100 microseconds, single pulse energy up to 500 μJ, and repetition rate 10–30 Hz (Fyodorov *et al.*, 1998).

© Woodhead Publishing Limited, 2013

Pulsed Nd:YAG lasers with flashlamp or diode pump with short nanosecond pulses (Q-switch mode) are used for secondary cataract treatment and for some cases of anti-glaucoma therapy. The laser pulse duration is 3–10 ns, the repetition rate is several pulses per second, and the single pulse energy is up to 10 mJ. In some cases, a similar laser with second-harmonic generation at 532 nm is used to achieve better selectivity of laser action.

A new approach to laser cataract extraction with the help of a femtosecond laser was recently reported (Nagy *et al.*, 2009). It is quite likely that the collateral mechanical damage of the eye elements and tissues will be minimal during the eye lens destruction with femtosecond pulses and this method of photodestruction can be used on all types of cataracts.

Retina: vitreous body

Laser–tissue coagulation is the most successful method for treating various complications of the retina and vitreous body. Coagulation is achieved through direct thermal interaction or in combination with selective interaction with an appropriately chosen wavelength of laser radiation.

The most common lasers for laser coagulation are solid-state lasers with diode pumping with intracavity conversion to the second harmonic (527–540 nm), diode lasers at 810 nm and optically pumped diode lasers (OPDL) at 577 nm. Quasi CW lasers, in which a variable pulse duration in the range 10–100 ms is regulated by modulation of pumping power and variable repetition rate with an interpulse interval of 1 ms and longer, are used for traditional laser photocoagulation. Despite the high efficiency of laser photocoagulation, the use of laser pulses with a pulse duration greater than 50–100 μs may induce undesirable thermal damage of adjacent tissues. In order to minimize thermal damage it is possible to use a train of individual short pulses (<10 μs) with low energy (a fraction of a microjoule or around 1 μJ) at a high repetition rate (1 kHz and higher); this technique is referred to as micropulse technology (Roider *et al.*, 1993). The micropulse regime in combination with the correct choice of laser radiation wavelength minimizes thermal damage to surrounding tissues. Because of the ability to tune up the lasing wavelength, OPDL are very attractive for micropulse technology; however, the single pulse duration in these lasers cannot be shorter than 20–50 μs and the average power is not more than 5 W, thus OPDL is not the perfect choice for a 'micropulse' approach because of the long pulse duration. On the contrary, Nd-doped solid-state lasers (YLF, YAG, YAP) with second-harmonics generation can be used to obtain laser pulses at 527 nm, 532 nm and 540 nm respectively with a pulse duration of 1–2 μs, energy up to 1 mJ and a repetition rate of 100–300 Hz. These lasers do not perfectly match to the retina absorption peak at 577 nm, but in terms of pulse duration they fit better to the concept of micropulses.

© Woodhead Publishing Limited, 2013

Another area of application of solid-state lasers in ophthalmology is vitrectomy (partial or full removal of vitreous body). In a number of papers (Peterson *et al.*, 2000; Binder *et al.*, 2000) it was demonstrated that the Er:YAG laser at λ = 2940 nm with pulse energy 20–40 mJ and a repetition rate of 30 Hz is an efficient instrument for vitrectomy. Despite positive outcomes of the preliminary studies, which showed higher efficiency of this method in comparison to mechanical instruments (faster speed of cutting, smaller increase of intraocular pressure), this method is not yet used in everyday clinical practice. The main problem remains the absence of a reliable and commercially available laser light delivery system.

20.3.2 Lasers in urology

Urology is one of the surgery fields where all the feasible parameters of laser radiation produced by solid-state lasers are widely used. Similarly to other areas of medicine, the use of lasers in urology and the character of laser–tissue interaction is determined both by the output parameters of the laser radiation, such as pulse duration τ_L, wavelength λ, and the energy density of the incident laser radiation, and by the parameters of the biological tissue: its optical and mechanical properties, composition and morphological structure.

Nowadays in urology Ho:YAG, Nd:YAG, Tm:YAG and KTP solid-state lasers are the most widely used laser systems. They are used for the whole range of urological tasks, including the interaction with both solid and soft tissues. The safety and efficacy of solid-state lasers has been demonstrated for treatment of urolithiasis, urethral stricture, benign prostatic hyperplasia (BPH), superficial bladder cancer, kidney cancer, and ureterocele resection (Marks and Teichman, 2007).

Laser lithotripsy and laser treatment of benign prostatic hyperplasia are the most frequently used areas of laser applications in urology. Laser lithotripsy as a practical method appeared in the middle of the 1980s by using pulsed flashlamp pumped dye lasers in clinical practice. The high cost and complexity of using those systems have led to their replacement by Ho:YAG lasers. In contrast to the pulsed-dye laser whose therapeutic efficacy is based on photoacoustic properties, the Ho:YAG laser's effectiveness as a lithotripter is due to the high absorption of 2120-nm radiation by stone and water.

By selecting the wavelength and laser pulse duration one can change both the mechanism and the result of laser action on tissue. By increasing the pulse duration of a Nd:YAG laser in the Q-switch mode to 200–800 ns, it is possible to use an optical fiber with a core diameter of 200–400 μm for stone fragmentation (Helfmann *et al.*, 1992). In an experiment, it was shown that the efficiency of stone fragmentation at those pulse durations (the

© Woodhead Publishing Limited, 2013

submicrosecond range) increases if dual-wavelength laser radiation is used. The ratio of energy of short and long wavelength radiation was determined in order to obtain better clinical results. The results of those studies brought new technology to clinical urology – the lithotripter on all solid-state KTP lasers with microsecond pulse duration (Muller *et al.*, 1993).

Prostate diseases are socially significant throughout the world. BPH in the USA is the fifth most prevalent non-cancer-related disorder among men aged 50 years and older, and it is estimated that 6.5 million US white men between 50 and 79 years of age meet the criteria for treatment (Hollingsworth and Wei, 2006). Transurethral resection of the prostate (TUR) is nowadays the 'gold standard' for BHP treatment, but it has a number of drawbacks; complications occur in more than 20% of patients. Numerous patients face concomitant inconveniences, such as prolonged disuria, hemoturia, and retrograde ejaculation (Bouchier-Hayes *et al.*, 2006). New laser technologies are considered the most perspective and promising alternative methods of treatment.

One of the first laser techniques of treatment, known as visual laser ablation of prostate (VLAP), was performed by Nd:YAG lasers (Costello *et al.*, 1992). Laser radiation at λ = 1064 nm has low absorption in prostate tissue and deep penetration depth. The result of interaction is coagulation of a large volume of tissue with delayed necrosis, a low extent of direct ablation, an extended period of cauterization and a long period of necrotic tissue rejection.

Later on, the technique of using Ho:YAG lasers became widespread. New procedures were developed: holmium laser ablation of prostate (HoLAP), holmium laser resection of prostate (HoLRP), and holmium laser enucleation of prostate (HoLEP) (Wilson and Gilling, 2006). The range of laser power used is 40–100 W and varies depending on the size of prostate hyperplasia and the specific technique of its removal.

In 2002 an alternative method of photoselective vaporization of prostate (PVP) was suggested by using the second harmonic of a Nd:YAG laser operated in quasi-CW mode. Laser photoselective vaporization of prostate competes with the current 'gold standard' of treatment in urology – transurethral resection of prostate (Bouchier-Hayes *et al.*, 2006). Laser radiation at λ = 532 nm is not absorbed by water but very well absorbed by hemoglobin of blood in prostate tissue. This laser surgery method shows very good post-operative results, has fewer complications compared to other laser techniques and has become widely used (Malek, 2008). The cost of the procedure and its duration limit its use to prostates of small volume.

The laser techniques developed so far suffer from low speed of prostate tissue removal and long duration of the procedure. In all the techniques described above, a slow heating of tissue to a temperature above 100°C takes place at a constant pressure. The laser parameters used to perform the

© Woodhead Publishing Limited, 2013

procedure and the properties of the prostate tissue explain the root cause of the inadequate ablation rate.

Prostate consists of parenchyma, presented by granular tissue, which in turn consists of prostatic glands and stroma with fibromuscular structure. The proportion of stromal component in the whole volume of prostate hyperplasia is 50%–75% (Aoki *et al.*, 2001; Deering *et al.*, 1994) and corresponds to a complex multilevel system of stromal microenvironment of granular epithelium.

The glandular component of prostate tissue exposed to laser radiation evaporates faster than the stromal component (Eure *et al.*, 2009). Prostate treated with Nd:YAG laser vaporization had a layer of granular collagenous tissue separating the regenerated urothelium from the residual prostatic glands (Kuntzman *et al.*, 1996). Fibrous tissue has much higher ultimate tensile strength and requires the application of much higher pressure inside the tissue for rupture of the ECM and ablation onset (Vogel and Venugopalan, 2003).

The remaining coagulated tissue and stroma fragments, which are formed at the walls of the ablated cavity, act like a screen that shields the incident laser radiation and precludes further removal of tissue. In cases of large hyperplastic prostate a combination technique was used to enlarge the cavity in the gland. The first step was laser photoselective vaporization of prostate (PVP), then transurethral resection (TUR) was used to remove white-gray coral-type stromal remains (Verger-Kuhnke *et al.*, 2006).

A common trend in the laser treatment of BPH is to increase the output power of laser radiation and to choose a laser wavelength with higher absoption in tissue to make the tissue removal rate less dependent on its properties and structure.

One of the promising areas of using solid-state lasers is their application for cancer treatment and, particularly, for renal cancer. A goal of improving quality of life of cancer patients stimulated investigation of novel methods of renal cancer surgery. In part, this increase was due to better diagnostic techniques, which enabled the detection of tumors at earlier stages. With advances in imaging technology, many patients are diagnosed before any symptoms occur (Mabjeesh *et al.*, 2004).

For a long time the method of choice for surgical treatment of renal cancer was radical nephrectomy. Kidney resection was the only way in the case of a double side tumor, or a tumor of a single or the only functional kidney (Aliaev and Krapivin, 2001). Despite continuing improvement of laparoscopic partial technique, surgeons are facing two common challenges: adequate tissue cutting and homeostasis. Lasers are at the forefront of the investigation because of their dual ability to cut and coagulate tissue.

The main reason for doubts among urologists regarding the effectiveness of laparoscopic partial nephrectomy (LPN) is technical difficulties in implementing

© Woodhead Publishing Limited, 2013

reliable homeostasis during surgery and lack of confidence that the surgery is adequately radical. Using lasers for laparoscopic resection of kidney has an advantage in simultaneous dissection of tissue and implementation of reliable homeostasis while delivering laser energy via thin flexible fibers.

The first clinical cases of LPN using a Ho:YAG laser were described in Lotan *et al.* (2002). LPN was performed on three patients with minimal blood loss and without renal vessels clamping. According to the authors, a laser pulse energy of 0.2 J and a repetition rate of 60 Hz, or a laser pulse energy of 0.8 J and a repetition rate of 40 Hz, of a Ho:YAG laser can produce efficient homeostasis during LPN. The authors believed that the main advantages of the Ho:YAG laser were the simultaneous resection and coagulation of tissue with minimal damage to surrounding renal parenchyma.

A quasi-CW Nd:YAG laser with second-harmonic generation (KTP-laser) was tested in the series of LPN surgeries on calves (Moinzadeh *et al.*, 2005). Eleven out of 12 surgeries were performed without renal vessels clamping. The histology of the dissected tissue revealed minimal collateral damage of the surrounding tissue. Less blood sputtering compared to a Ho:YAG laser was pointed out.

Gruschwitz *et al.* (2008) reported about five successful cases of LPN without renal vessels clamping on human patients. Regular follow-up ultrasound tests did not reveal post-surgical bleeding. The authors believed that LPN was a safe alternative to the classical surgical resection if patients were carefully selected.

The results of using a thulium laser (λ = 2013 nm) for LPN on nine patients were reported by Mattioli *et al.* (2008). One laparoscopic and eight open resections were performed with laser powers of 12 W and 15 W. In all the cases except three open surgeries, clamping was used. The authors point to good homeostasis, the ability to accurately dissect the cortical layer of kidney, as well as the absence of small bubbles and minimal gas/smoke formation.

It is important to point out that all published examples of successful clinical use of laser resection of the kidney were performed using Ho:YAG, Tm:YAG and KTP lasers whose radiation is strongly absorbed by tissue. As a result the coagulation ability of these lasers is limited due to the low effective penetration depth of the laser radiation into the tissue. The main advantages of using these lasers in urology is their ability to resect, ablate and vaporize tissue due to high energy deposition into the tissue and small collateral damage to the surrounding tissue. Diode lasers at λ = 980 nm and λ = 810 nm have a greater penetration depth but were used at a medium to low output power level of laser light (23 W).

An Nd:YAG laser was used in experiments for kidney resection with maximum depth of kidney parenchyma coagulation 2.75 mm at the laser output of 60 W (Stein, 1986); this should provide reliable homeostasis

© Woodhead Publishing Limited, 2013

during kidney resection. The results of successful use of a Nd:YAG laser (λ = 1064 nm) for LPN on 27 patients without renal ischemia were reported by Teodorovich *et al.* (2011). Due to tissue coagulation during kidney resection by the Nd:YAG laser, the hemorrhage of renal parenchyma was substantially reduced and additional homeostasis was not required. There were no cases of intra-operative complications. None of the treated patients required a hemotransfusion.

20.4 Current and future trends in laser surgery

The research and development of laser techniques for surgery covers such areas as gynecology, urology and neurosurgery as well as treatment of different types of tumors by photodynamic therapy (PDT) and interstitial laser coagulation (ILC). The progress in instrumentation of minimally invasive surgery, implementation of miniature catheters and endoscopes promotes laser application in endosurgery.

The development of new laser systems and the number of clinical studies of new laser techniques for medical treatments continue to grow and laser technologies are becoming an integral part of modern medicine.

The analysis of the applications of solid-state lasers in medicine presented in this chapter has outlined the main trends in the development of laser technologies for medicine. Firstly, the expansion of the spectral range of laser radiation makes it possible to selectively affect different tissues and pigments. Secondly, control of the temporal parameters of laser radiation (variation of pulse duration and repetition rate) provides better spatial selectivity of the thermal effect. Thirdly, the use of lasers with a short pulse duration enables nonthermal interaction with tissue due to nonlinear multiphoton processes occurring in small localized volumes.

20.5 References

Albagli D, Perelman L T, Janes G S, von Rosenberg C, Itzkan I, Feld M (1994), 'Inertially confined ablation of biological tissue', *Lasers Life Sci*, 6, 55–68.

Aliaev Yu G, Krapivin A A (2001), *Kidney Resection at Renal Cancer*, Moscow, Medicine.

Anderson R R, Parrish J A (1983), 'Selective photothermolysis: precise microsurgery by selective absorption of pulsed radiation', *Science*, 220: 4596, 524–527.

Anisimov S I, Luk'yanchuk B S (2002), 'Selected problems of laser ablation theory', *Phys Usp*, 45, 293–324.

Aoki Y, Arai Y, Maeda H (2001), 'Racial differences in cellular composition of benign prostatic hyperplasia', *Prostate*, 49, 243–250.

Aron-Posa D, Aron J, Griesemann M, Thyzel R (1980), 'Use of the neodymium-YAG laser to open the posterior capsule after lens implant surgery: a preliminary report', *J Am Intraocul Implant Soc*, 6:4, 352–354.

© Woodhead Publishing Limited, 2013

Barnes E A, Murdoch I E, Subramaniam S, Cahill A, Kehoe B, Behrend M (2004), 'Neodymium:yttrium-aluminum-garnet capsulotomy and intraocular pressure in pseudophakic patients with glaucoma', *Ophthalmology*, 111:7, 1393–1397.

Binder S (2008), 'Femtosecond lasers. How do you select a system?', *Cataract & Refractive Surgery Today*, 10, 53–56.

Binder S, Stolba U, Kellner L, Krebs I (2000), 'Erbium:YAG laser vitrectomy: clinical results', *Am J Ophthalmol*, 130:1, 82–86.

Bouchier-Hayes D M, Anderson P, Appledorn S V, Bugeja P, Costello A J (2006), 'KTP laser versus transurethral resection: early results of a randomized trial', *J Endourol*, 20:8, 580–585.

Boulnois J L (1986), 'Photophysical processes in recent medical laser developments: a review', *Laser Med Sci*, 1, 47–66.

Cheong W F, Prahl S A, Welch A J (1990), 'A review of the optical properties of biological tissues', *IEEE J Quantum Electron*, 26, 2166–2185.

Costello A J, Bowsher W G, Bolton D M, Braslis K G, Burt J (1992), 'Laser ablation of the prostate in patients with benign prostatic hypertrophy', *Br J Urol*, 69, 603–608.

Dair G T, Pelouch W S, van Saarloos P P, Lloyd D J, Paz Linares S M, Reinholz F (1999), 'Investigation of corneal ablation efficiency using ultraviolet 213-nm solid state laser pulses', *Invest Ophthalmol Vis Sci*, 40:11, 2752–2756.

Deering R E, Bigler S A, King J, Choongkittaworn M, Aramburu E, Brawer M K (1994), 'Morphometric quantitation of stromain human benign prostatic hyperplasia', *Urology*, 44, 64–70.

Deppe H, Horch H H (2007), 'Laser applications in oral surgery and implant dentistry', *Lasers Med Sci*, 22, 217–221.

Eure G, Gonzalez R, Alivizatos G, Malloy T, Sandhu J (2009), 'Creating a new standard of care for BPH: A standardized approach to photoselective vaporization of the prostate using GreenLight HPS™ laser therapy (Clinical Report)', *Urology Times*, 15 April 2009.

Furzikov N (1987), 'Different lasers for angioplasty: thermooptical comparison', *IEEE J Quantum Electron*, 23:10, 1751–1755.

Fyodorov S N, Kopaeva V G, Andreev Yu V, Erofeev A V, Belikov A V, Bogdalova E G, Skripnik A V, Frolova O A (1998), 'Laser extraction of cataract', *Ophthalmosurgery*, 3, 3–10.

Gruschwitz T, Stein R, Schubert J, Wunderlich H (2008), 'Laser-supported partial nephrectomy for renal cell carcinoma', *Urology*, 71:2, 334–336.

Helfmann J, Mihailov V A, Konov V I, Muller G, Nikolaev D A, Pak S K, Shcherbakov I A, Silenok A S (1992), 'Efficiency of stone fragmentation by long pulses of a Q-switched Nd:YAG laser', *Proc SPIE*, 1643, 78–85.

Hollingsworth J M, Wei J T (2006), 'Economic impact of surgical intervention in the treatment of benign prostatic hyperplasia', *Reviews in Urology*, 8:3, s9-s15.

Huber G, Kränkel C, Petermann K (2010), 'Solid-state lasers: status and future', *J Opt Soc Am B*, 27:11, B93-B105.

Iaria G (2008), 'Clinical, morphological, and ultrastructural aspects with the use of Er:YAG and Er,Cr:YSGG lasers in restorative dentistry', *Gen Dent*, 56:7, 636–639.

Ishimaru A (1978), *Wave Propagation and Scattering in Random Media*, New York, Academic Press.

Itzkan I, Albagli D, Dark M L, Perelman L T, von Rosenberg C, Feld M (1995), 'The thermoelastic basis of short pulsed laser ablation of biological tissue', *Proc Natl Acad Sci USA*, 92, 1960–1964.

© Woodhead Publishing Limited, 2013

Jacques S, Patterson M (2004), 'Light-tissue interactions', in Webb C E and Jones J D, *Handbook of Laser Technology and Applications. Volume III: Applications*, IOP Publishing.

Juhasz T, Loesel F H, Kurtz R M, Horvath C, Bille J F, Mourou G (1999), 'Corneal refractive surgery with femtosecond lasers', *IEEE J Sel Top Quantum Electron*, 5:4, 902–910.

Kaminskii A A (1996), *Crystalline Lasers: Physical Processes and Operating Schemes*, CRC Press.

Kang H W, Oh J, Welch A J (2008), 'Investigations on laser hard tissue ablation under various environments', *Phys Med Biol*, 53:12, 3381–3390.

Kaufmann R, Hibst R (1996), 'Pulsed erbium:YAG laser ablation in cutaneous surgery', *Lasers Surg Med*, 19:3, 324–330.

Koester C J, Snitzer E, Campbell C J, Rittler M C (1962), 'Experimental laser retina photocoagulation', *J Opt Soc Am*, 52, 607.

Kuntzman R S, Malek R S, Barrett D M, Bostwick D G (1996), 'Potassium-titanyl-phosphate laser vaporization of the prostate: A comparative functional and pathologic study in canines', *Urology*, 48:4, 575–583.

Kurtz R M, Horvath C, Liu H H, Krueger R R, Juhasz T (1998), 'Lamellar refractive surgery with scanned intrastromal picosecond and femtosecond laser pulses in animal eyes', *J Refractive Surgery*, 14, 541–548.

Lasser T, Ludwig K, Lukashev A, Heymann (1992), 'Photoablation on bovine cornea with a Q-switch Er:YAG laser', *Proc SPIE*, 1644, 301–308

LeCarpentier G L, Motamedi M, McMath L P, Rastegar S, Welch A J (1993), 'Continuous wave laser ablation of tissue: analysis of thermal and mechanical events', *IEEE Trans Biomed Eng*, 40, 188–200.

L'Esperance F A (1968), 'An opthalmic argon laser photocoagulation system: design, construction, and laboratory investigations', *J Trans Am Ophthalmol Soc*, 66, 827–904.

Letokhov V S (2003), 'Laser light in biomedicine and the life sciences: from the present to the future', in Tuan Vo-Dinh, *Biomedical Photonics Handbook*, CRC Press, 5.4-5.16.

Littler C M (1999), 'Hair removal using an Nd:YAG laser system', *Dermatol Clin*, 17:2, 401–430.

Lotan Y, Gettman M T, Ogan K, Baker L A, Cadeddu J A (2002), 'Clinical use of the holmium:YAG laser in laparoscopic partial nephrectomy', *J Endourol*, 16:5, 289–292.

Mabjeesh N J, Avidor Y, Matzkin H (2004), 'Emerging nephron sparing treatments for kidney tumors: a continuum of modalities from energy ablation to laparoscopic partial nephrectomy', *J Urol*, 171, 553–560.

Maiman T (1960), 'Simulated optical radiation in ruby', *Nature*, 187:6, 493–494.

Malek R S (2008), 'GreenLight HPS Laser therapy for BPH: Clinical outcomes and surgical recommendations from the international GreenLight user group (IGLU)', *Eur Urol Suppl*, 7, 361–362.

Manstein D, Herron G S, Sink R K, Tanner H, Anderson R R (2004), 'Fractional photothermolysis: A new concept for cutaneous remodeling using microscopic pattern of thermal injury', *Lasers Surg Med*, 34, 426–438

Marks A J, Teichman J M (2007), 'Lasers in clinical urology: state of the art and new horizons', *World J Urol*, 25, 227–233

Martynyuk M M (1977), 'Phase explosion of a metastable fluid', *Combust Explos Shock Waves* 13, 178–184.

© Woodhead Publishing Limited, 2013

Mattioli S, Munoz R, Recasens R, Berbegal C, Teichmann H (2008), 'What does Revolix laser contribute to partial nephrectomy?', *Arch Esp Urol*, 61:9, 1126–1129.

Miotello A, Kelly R (1995), 'Critical assessment of thermal models for laser sputtering at high fluences', *Appl Phys Lett*, 67, 3535–3537.

Miotello A, Kelly R (1999), 'Laser-induced phase explosion: new physical problems when a condensed phase approaches the thermodynamic critical temperature', *Appl Phys A*, 69, S67–S73.

Moinzadeh A, Gill I S, Rubenstein M, Ukimura O, Aron M, Spaliviero M (2005), 'Potassium-titanyl-phosphate laser laparoscopic partial nephrectomy without hilar clamping in the survival calf model', *J Urol*, 174, 1110–1114.

Mrochen M, Semshichen V, Funk R H, Seiler T (2000), 'Limitations of erbium:YAG laser photorefractive keratectomy', *J Refract Surg*, 16:1, 51–59.

Muller G, Hetfmann J, Pashinin V P, Pashinin P P, Konov V I, Tumorin V V, Shklovsky E J (1993), 'New alternative for laser lithotripsy, long pulse passively Q-switched solid-state laser with fibre-based resonator', *Proc SPIE*, 2086, 103–110.

Munnerlyn C R (2003), 'Lasers in ophthalmology: past, present and future', *J Mod Optics*, 50, 2351–2360.

Nagy Z, Takacs A, Filkorn T, Sarayba M (2009), 'Initial clinical evaluation of an intraocular femtosecond laser in cataract surgery', *J Refract Surg*, 25:12, 1053–1060.

Niemz M H (2004), *Laser–tissue Interactions: Fundamentals and Applications*, Springer, Berlin.

Nikogosyan D, Gorner H (1999), 'Laser-induced photodecomposition of amino acids and peptides: extrapolation to corneal collagen', *IEEE J Sel Top Quantum Electron*, 5:4, 1107.

Nishioka N S (1995), 'Applications of lasers in gastroenterology', *Lasers Surg Med*, 16:3, 205–214.

Ogura M, Sato S, Ishihara M, Kawauchi S, Arai T, Matsui T, Kurita A, Kikuchi M, Ashida H, Obara M (2002), 'Myocardium tissue ablation with high-peak-power nanosecond 1,064- and 532-nm pulsed lasers: Influence of laser-induced ilasma', *Lasers Surg Med*, 31, 136–141.

Parrish J, Deutsch T (1984), 'Laser photomedicine', *IEEE J Quantum Electron*, 20, 1386–1396.

Pech M, Wieners G, Freund T, Dudeck O, Fischbach F, Ricke J, Seemann M D (2007), 'MR-guided interstitial laser thermotherapy of colorectal liver metastases: efficiency, safety and patient survival', *Eur J Med Res*, 12:4, 161–168.

Pelivanov I M, Belov S A, Solomatin V S, Khokhlova T D, Karabutov A A (2006), 'Direct opto-acoustic in vitro measurement of the spatial distribution of laser radiation in biological media', *Quantum Electron*, 36:12, 1089–1096.

Peng Q, Juzeniene A, Chen J, Svaasand L O, Warloe T, Giercksky K, Moan J (2008), 'Lasers in medicine', *Rep Prog Phys*, 71, 056701.

Peterson H, Mrochen M, Seiler T (2000), 'Comparison of erbium:yttrium-aluminum-garnet-laser vitrectomy and mechanical vitrectomy: a clinical study', *Ophthalmology*, 107:7, 1389–1392.

Roider J, Hillenkamp F, Flotte T, Birngruber R (1993), 'Microphotocoagulation: Selective effects of repetitive short laser pulses', *Proc Natl Acad Sci USA*, 90, 8643–8647.

Schmidt H, Ihlemann J, Wolff-Rottke B, Luther K, Troe J (1998), 'Ultraviolet laser ablation of polymers: spot size, pulse duration, and plume attenuation effects explained', *J Appl Phys*, 83, 5458–5468.

Seiler T, Matallana M, Bende T (1990), 'Laser thermokeratoplasty by means of a

© Woodhead Publishing Limited, 2013

pulsed holmium:YAG laser for hyperopic correction', *Refract Corneal Surg*, 6:5, 335–339.

Shcherbakov I A (1988), 'Solid-state laser', in Prokhorov A, *Physical Encyclopedia*, Moscow, Sovetskaya Entsiklopediya, 4, 49–50.

Shcherbakov I A (2010), 'Laser physics in medicine', *UFN*, 180:6, 661–665.

Srinivasan R, Braren B (2003), 'Ablative photodecomposition of polymer films by pulsed far-ultraviolet (193 nm) laser radiation: dependence of etch depth on experimental conditions', *J Polym Sci Tech*, 22:10, 2601–2609.

Stein B S (1986), 'Urologic dosimetry studies with the Nd:YAG and CO_2 lasers: bladder and kidney', *Lasers Surg Med*, 6:3, 353–363.

Teodorovich O, Zabrodina N, Galljamov E, Yankovskaya I, Kochiev I, Lukashev A (2011), 'Laser laparoscopic partial nephrectomy in clinical cases (N = 17)', *Proc SPIE*, 7883, 78831G1–8.

Trokel S L, Srinivasan R, Braren B (1983), 'Excimer laser surgery of the cornea', *Am J Ophthalmol*, 96, 710–715.

Tuchin V V (1997), 'Light scattering study of tissues', *Phys Usp*, 40, 495.

Venugopalan V (1995), 'Pulsed laser ablation of tissue: surface vaporization or termal explosion?', *Proc SPIE*, 2391, 184–189.

Venugopalan V, Nishioka N S, Mikic B B (1996), 'Thermodynamic response of soft biological tissues to pulsed infrared-laser irradiation', *Biophys J*, 70, 2981–2993.

Verger-Kuhnke A B, Reuter M, Epple W, Ungemach G, Beccaría M L (2006), 'Combined treatment of prostate adenoma with the 80 watt KTP-laser and low-pressure transurethral resection', *Actas Urol Esp*, 30:4, 394–401.

Vogel A, Venugopalan V (2003), 'Mechanisms of pulsed laser ablation of biological tissues', *Chem Rev*, 103:2, 577–644.

Vogel A, Busch S, Jungnickel K, Birngruber R (1994), 'Mechanisms of intraocular photodisruption with picosecond and nanosecond laser pulses', *Lasers Surg Med*, 15, 32–43.

Wang L V, Wu H I (2007), *Biomedical Optics: Principles and Imaging*, Wiley-Interscience.

Weber M J (1999), *Handbook of Laser Wavelengths*, CRC Press.

Welch A. J, Gardner C (2001), 'Optical and thermal response of tissue to laser radiation', in Waynant R W, *Lasers in Medicine*, CRC Press.

Wheeland R G (1995), 'Clinical uses of lasers in dermatology', *Lasers Surg Med*, 16:1, 2–23.

Wilson L C, Gilling P J (2006), 'Lasers for prostate surgery – an update. Benign prostate hyperplasia', *Business Briefing: European Kidney & Urological Disease*, 52–53.

Zaret M M, Ripps H, Siegel I M, Breinin G V (1963), 'Laser photocoagulation of the eye', *Arch Ophthalmol*, 69:1, 97–104.

© Woodhead Publishing Limited, 2013

21
Solid-state lasers (SSL) in defense programs

Y. KALISKY, Nuclear Research Center, Israel

DOI: 10.1533/9780857097507.2.598

Abstract: This chapter reviews the updated status and applications of a defensive weapon based on high power lasers, in the battlefield. The laser weapon is a novel concept which utilizes a high power laser beam to traverse the distance to an incoming object at the speed of light, and then destroy or disable it. Various types of solid-state and fiber lasers and configurations will be discussed in this review. For reference we include in this review also gas lasers and free-electron lasers since they play a major role in any present or future high power laser applications. Although the emphasis of this chapter is on solid-state lasers, it includes discussions on various configurations such as airborne lasers (ABL), diode pumped crystals and disk lasers as well as heat-capacity lasers. Recent applications of ultrafast solid-state lasers for non-lethal or low collateral damage applications will be presented.

Key words: laser weapon, chemical laser, solid-state laser, ABL, fiber laser, disk laser.

21.1 Introduction

Laser weapons based on high power lasers (HPL) are currently considered as tactical as well as strategic beam weapons, as part of a general layered defense system against ballistic missiles and short-range rockets. This chapter is an updated review of two recent papers (Kalisky, 2009; Kalisky and Kalisky, 2010). The laser weapon is a kind of weapon that can disable or destroy military targets or incoming objects by approaching the target at the speed of light, and this is attractive against short-range rockets and mortars. Laser weapons, unlike kinetic weapons, are effective in principle, at long or short distances, owing to the laser beam's unique characteristics such as narrow bandwidth, high brightness, and coherence in both time and space. *The objective of a laser weapon is to disable or destroy efficiently military targets or incoming objects by intercepting the target at the speed of light.* The effect of the laser beam on an incoming object is a thermal one. The laser beam, aimed at a small area on the rocket's skin, is converted into a large amount of heat, followed by a temperature increase and finally catastrophic failure by material ablation or melt.

The usefulness of laser light as a weapon has been studied for decades but only in recent years has it become feasible owing to advances in solid-state laser materials and thermal management technologies. There are two types

© Woodhead Publishing Limited, 2013

of lasers based on the active lasing media that are being used: gas lasers and solid-state lasers, including fiber lasers, which belong to the rare-earth-doped glass laser family. All these types of lasers will be discussed below.

21.2 Background

Modern warfare is completely different from wars or battles we know from the past. Rockets, either guided or unguided, as well as missiles are currently used by terror organizations, and are considered as a global threat to the free world. There are several types of rockets and missiles, for both short range (5–100 km) and long range (several hundreds to thousands of kilometers). The rocket/missile weapon is effective in both short and long range theaters, particularly in regional conflicts, and mainly in low-intensity conflict (LIC), due to its obvious advantages outlined below:

- It is able to strike and destroy targets.
- It is relatively fast, easy to operate, inexpensive, mobile and flexible to use.
- It can be fired onto targets without any warning.
- The design, operation and maintenance of this kind of weapon do not require skilled manpower.
- Missiles and rockets can be fired in large quantities to compensate for inherent inaccuracies. (A guided missile, which is much more accurate, can be used instead.)

Terrorists and potential enemies consider rockets, missiles and any other unmanned weapons as an integral part of airpower. Accordingly, they consider the massive use of such weapons as a legitimate means of obtaining air superiority and victory and of subduing countries or organizations.

21.3 Properties of laser weapons

Laser beams are considered as directed energy weapons (DEW), compared with kinetic weapons such as anti-rocket missiles. Laser weapons are categorized into two main classes. The first one is the tactical type of laser, which is used mainly against short range rockets, artillery or mortar (RAM) or against unmanned aerial vehicles (UAV). The second category of weapon is the strategic laser, used against medium or long range missiles or intercontinental ballistic missiles (ICBM). In the long range, it can be considered as a weapon that can operate effectively 'beyond the horizon'.

Laser weapons have several specific advantages compared with kinetic weapons, through both types of weapons complement each other. The laser beam has unique characteristics such as high brightness, B, which is defined as:

© Woodhead Publishing Limited, 2013

$$B = P/(A \cdot \Omega) \qquad [21.1]$$

where P is the laser power, A is the laser beam area and Ω is the beam solid angle (in steradians, sr). From Eq. [21.1] it is observed that the brightness is proportional to the laser power and inversely proportional to the square of the beam quality (defined as the ratio between the laser beam divergence and the diffraction-limited beam divergence). When the distance to the incoming target increases, the laser beam area increases and the beam quality deteriorates sharply due to scattering by atmospheric particles, turbulence, or thermal blooming effects. The laser power is also attenuated as a result of laser radiation absorption, mainly by water molecules, and hence the power density will be insufficient for any significant thermal damage.

A laser beam is also characterized by a specific spectral bandwidth, and by its coherence in both time and space. Above all, a laser beam travels at the speed of light and therefore approaches targets practically instantaneously, with multiple target engagement, and this unique property allows for retargeting easily in a few seconds to another target. Lasers can follow fast targets easily and maintain their beam on maneuvering air targets precisely. Precision engagement means reduced collateral damage. Another advantage of lasers is their functional flexibility, since target materials have a gradual response towards lasers depending on their power level.

Lasers have several significant disadvantages which limit their operation and limit the susceptibility of the target's skin to laser damage. These limitations are listed below:

- Weather conditions, dust and clouds
- Atmospheric absorption
- Scattering and air turbulence
- Thermal blooming and laser beam defocusing
- Saturation attacks due to multiple weapons that attack a target simultaneously
- Countermeasures against lasers based on target design features such as special coating, highly reflective surfaces, or ceramic shielding
- Beam jittering on the target – problems of point stability.

21.3.1 Laser–rocket interaction

The interaction between the laser beam and an incoming missile or rocket is based on thermal effects. In other words, the energy stored in the laser beam is converted into heat. The ability of a laser beam to destroy a target depends on several parameters such as engagement geometry, dwell time, and laser wavelength. Usually, the laser beam is focused at specific sectors in the missile or rocket, such as the *liquid fuel,* the *storage tank,* the *warhead case,* the *navigation system* or the *engine area.* These specific sectors are

© Woodhead Publishing Limited, 2013

vulnerable, mainly due to the internal high pressure at the boost phase. The method of attacking incoming missiles or rockets at the boost phase is termed *boost phase intercept* or *BPI*. At the boost phase, the internal pressure inside the liquid fuel tank can reach values of 20 tons/m^2. Another contribution to the huge stresses on the missile skin is the axial pressure due to aerodynamic forces and atmospheric resistance at the supersonic speed of the fast accelerating missile, usually about 7 m/s^2 for the Scud B missile. Also, attacking the missile at the BPI will increase the potential damage around the launching site, far away from the target. For a detailed discussion on this subject, see Forden (1998).

The failure temperature of Al is 180°C, and that of steel is 460°C. Melting of these metals occurs at power densities of 10 kW/cm^2 for Al and 75 kW/cm^2 for steel during an exposure time of ~20–70 s. When the missile skin is warmed in an arc around its circumference to such power densities, this can lead to catastrophic failure of the missile due to material fracture, ablation or melt. However, irreversible damage to the target occurs even at power densities that are an order of magnitude lower than the values indicated above.

The catastrophic failure depends on the laser fluence, on the skin material and on its thickness. The skin material and thickness of missiles like the Scud B, Taepo-Dong 2, the American intercontinental ballistic missile (ICBM) such as Minuteman, or the Russian SS-18 are either steel or Al, with thickness in the range of 1–3 mm. It is estimated that laser fluence levels ranging from 1 to 10 kJ/cm^2 are necessary for catastrophic failure and that laser fluence levels of 20–25 kJ/cm^2 are needed for metal vaporization during 20–70 s illumination time. Also in this case irreversible damage to the target occurs even at fluence levels which are an order of magnitude lower than the values indicated. For example, the amount of energy per unit area needed to increase the temperature of 3 mm thickness of steel to failure levels (specific heat 0.5 J/g°C, density 7.8 g/cm^3) is estimated to be 600 J/cm^2.

21.4 Gas lasers

The subject of gas lasers is beyond the scope of this book. However, no description of the application and performance of solid-state lasers in the battlefield will be complete without a brief description of the performance and operation of high power gas lasers.

Chemical lasers are a type of gas laser where the population inversion is produced directly by the heat generated by the exothermic chemical reaction. The heat of reaction of a typical $H_2 + F_2$ reaction mixture is very large, 2000 J/liter, while 1000 J/liter are necessary for vibrational excitation. The exothermic chemical reactions can be either associative or dissociative, such as $A + B \rightarrow AB$ or $ABC \rightarrow A + BC$, respectively. A well-known

© Woodhead Publishing Limited, 2013

example of an associative exothermic reaction is the reaction between H_2 or D_2 and F_2 to form HF/DF products: $H_2 + F_2 \rightarrow 2HF^*$ or $D_2 + F_2 \rightarrow 2DF^*$, where HF* or DF* indicates excited states of the products. When the excited molecules decay to their ground state they emit photons in the NIR spectral range, at discrete wavelengths of 2.6–3.3 μm for HF (≈10% atmospheric transmission) and 3.5–4.2 μm for DF (≈100% atmospheric transmission). This type of laser belongs to the family of vibrational–rotational lasers, since the laser action is between several rotational–vibrational excited levels of HF (or DF). In order to initiate the reaction, it is necessary to dissociate the fluorine–donor molecules such as F_2, NF_3 or SF_6 by pulsed power electrical discharge devices.

There are two types of military gas lasers which will be described below.

21.4.1 Tactical high energy laser (THEL)

This type of laser is based on an HF/DF (2.6–3.3/3.5–4.2 μm) chemical laser, which is based on a technological demonstrator termed also as MIRACL – Mid Infrared Advanced Chemical Laser. The active product is a DF laser (3.8 μm), with output power in the range of about 1 MW. The MIRACL was the precursor of the Nautilus DF-based laser system, developed jointly by the US and Israel. An advanced version of the ground-based THEL, the Skyguard (see Fig. 21.1), was introduced in 2006 for a tactical range of 5–10 km.

21.1 Ground-based THEL (Skyguard) DF chemical laser during field test in the US Army's White Sands missile range in New Mexico, USA (courtesy of Northrop-Grumman, available from http://www.military.com/soldiertech/0,14632, Soldiertech_MTHEL,,0.html, accessed 29 May 2007).

© Woodhead Publishing Limited, 2013

The basic reaction of the HF/DF chemical laser is a 'chain reaction', between atomic fluorine (generated by electrical discharge from F_2, NF_3, SF_6, etc.) and molecular hydrogen (or deuterium), or other sources such as H_2O_2 or C_2H_4, namely:

$$F + H_2 \rightarrow HF^* + H \text{ (cold reaction, exothermic, 31.6 kcal/mole)} \quad [21.2]$$

$$H + F_2 \rightarrow HF^* + F \text{ (hot reaction, exothermic, 98 kcal/mole)} \quad [21.3]$$

The vibrational population inversion in the excited HF^* molecule is achieved by the heat generated in the chemical reaction, and the emission wavelength is in the range of 2.7–4.2 μm.

The reaction [21.2] is repeated by the atomic fluorine. The vibrational laser transition in the cold reaction is between $v = 2$ to $v = 1$ vibrational states, while in the hot reaction, the laser transition is $v = 5 \rightarrow v = 4$. The reaction is initiated by electrical discharge which dissociates the molecular fluorine into reactive fluorine atoms. The excess of HF gas has to be evacuated from the pumping chamber to avoid reabsorption of the laser emission by the HF/DF molecules that are in the ground state. A novel design of a plasma-cathode-preionized HF laser yielded an efficient room temperature, atmospheric pressure chemical laser (Kalisky, *et al.*, 1997).

The mobile version of this laser (MTHEL–Mobile Tactical High Energy Laser) can be ground based with 100 kW (or more) output power needed to destroy short range rockets or artillery shells, or airborne by C-130 C-17 or C-5 airplanes. The airborne version of this DF chemical laser was a megawatt-class laser, with beam dimensions of 14 × 14 cm, and it operated successfully for 70 s, over 150 km range. All these lasers are equipped with subsystems such as an advanced target acquisition system as well as the Pointer Tracker Subsystem (PTS).

21.4.2 Airborne laser (ABL)

The Chemical Oxygen-Iodine Laser (COIL) is a megawatt-class laser, emitting within the atmospheric window with ≈95% transmission at 1.315 μm, and is considered as a strategic weapon. The population inversion is achieved by a chemical reaction between excited singlet molecular oxygen and molecular iodine, leading to energy transfer between singlet oxygen and atomic iodine and to population inversion in the excited state. The iodine molecules are both dissociated and excited by a singlet oxygen, denoted by $O_2^{1\Delta^*}$, which is generated by reaction between molecular chlorine and hydrogen peroxide according to reaction [21.4]. The excited singlet oxygen is a long-lived state (~45 min), and it reacts with molecular iodine. The result of this reaction is dissociation of molecular iodine into iodine atoms and a triplet ground state oxygen molecule. Finally, the excited state singlet oxygen transfers its extra energy to atomic iodine which decays to the ground state and emits the radiation via $^2P_{1/2} \rightarrow {}^2P_{3/2}$ electronic transition according to reactions [21.5] and [21.6]:

$$H_2O_2 + Cl_2 \rightarrow 2HCl + O_2{}^{1\Delta*} \qquad [21.4]$$

$$O_2{}^{1\Delta*} + I_2 \rightarrow O_2{}^{3\Sigma} + 2I \qquad [21.5]$$

$$2I + O_2{}^{1\Delta*} \rightarrow O_2{}^{3\Sigma} + 2I^* \rightarrow \hbar\nu\ ({}^2P_{1/2} \rightarrow {}^2P_{3/2}, \lambda = 1.315\ \mu m) \qquad [21.6]$$

The laser operates, similarly to other chemical lasers, at low pressure and fast flow conditions, so that heat removal from the lasing medium is easier than for high power solid state lasers. The last reaction [21.6] is enhanced at lower temperatures towards the production of more excited iodine atoms, therefore supersonic expansion nozzles (Mach ≈ 3–4) are used to cool the reactants. It can produce several megawatts of output power, in 20–40 shots, during 300–400 s. The laser is carried on a Boeing 747–400 F freighter aircraft at 40,000 feet, and is also named YAL-1A (see Fig. 21.2).

The laser weapon system includes also the BMC4I – battle management, command, control, communications, computers, and intelligence. This laser is intended to destroy the missile before separation of warheads, at the boost phase, at a distance of about 400 km, when the missile is relatively slow and vulnerable to attack. In addition to the main laser, the HEL requires a

21.2 Boeing 747-400F with the COIL Airborne Laser that is capable of destroying ballistic missiles at their boost phase; the nose-mounted turret is the laser beam exit (courtesy of Northrop-Grumman, available from http://www.airforce-technology.com/projects/abl, accessed July 2009).

© Woodhead Publishing Limited, 2013

target acquisition and tracking system. They are the Active Ranger System (ARS), which is a CW CO_2 laser at 11.15 μm for primary tracking and prioritizing; the Track Illuminating Laser (TILL), a kilowatt-class Yb:YAG laser emitting at 1030 nm to track and stabilize the laser; and the Beacon Illuminating Laser (BILL), a kilowatt-class Nd:YAG laser to compensate for atmospheric distortion along the path of the laser beam to the target. In a demonstration in February 2010, an ABL system destroyed ballistic missiles at their boost phase, at an estimated (undisclosed) range of 400–700 km.

An Advanced Tactical Laser (ATL) version of this laser, mounted on a C-130 aircraft, was recently fired during flight and hit a target on the ground, for the first time. The ATL is a smaller version of ABL and is designed to destroy and disable ground based targets at a long firing range (about 20 km effective range). It generates output power in the range of 100–300 kW and is capable of firing 100 shots accurately, with reduced collateral damage and with minimal non-combat fatalities, therefore it is suitable for combat activities in urban environments and congested areas, as well as covert operations.

21.5 Solid-state lasers

Solid-state lasers (SSL) are presently considered as a future tactical anti-missile, anti-rocket, and anti-mortar laser weapon. Tactical laser weapons based on SSL are used for ultra-precision missions with reduced collateral damage, suitable as tactical directed energy weapons both in the battlefield and in urban environments. They are based on solids (crystals such as YAG, or crystalline YAG-ceramics) doped with rare-earth ions such as Nd^{3+} or Yb^{3+} and diode-pumped at 808 nm or 940/970 nm, respectively. An energy level scheme of Nd^{3+} and Yb^{3+}-doped YAG crystal is presented in Fig. 21.3. (See Chapter 1 for an energy level scheme of Nd^{3+} and Yb^{3+}-doped YAG crystals (Kalisky, 2006).) Solid-state lasers have some significant advantages over other candidates in their simplicity of operation, their size, the possibility of generating multi-wavelength sources by non-linear optical devices, and the absence of toxic gases which provides simple logistics and handling. Owing to their relatively compact size, solid-state lasers are highly mobile, and can be integrated on land or air platforms.

On the other hand, SSL is a short-range tactical laser weapon (5–10 km), and unlike the COIL, it is limited by atmospheric transmission, which is ~50–60%. Weather conditions can also be a significant factor. It should be emphasized that the wavelength-reflective properties of SSL from the target are higher compared with longer-wavelength gas lasers. Solid-state lasers are in the development phase with the following challenges: reducing the size, increasing the pumping efficiency and output power, and removing heat from the crystal and diodes. Accordingly, there are two approaches towards achieving the goal of high power solid-state lasers:

© Woodhead Publishing Limited, 2013

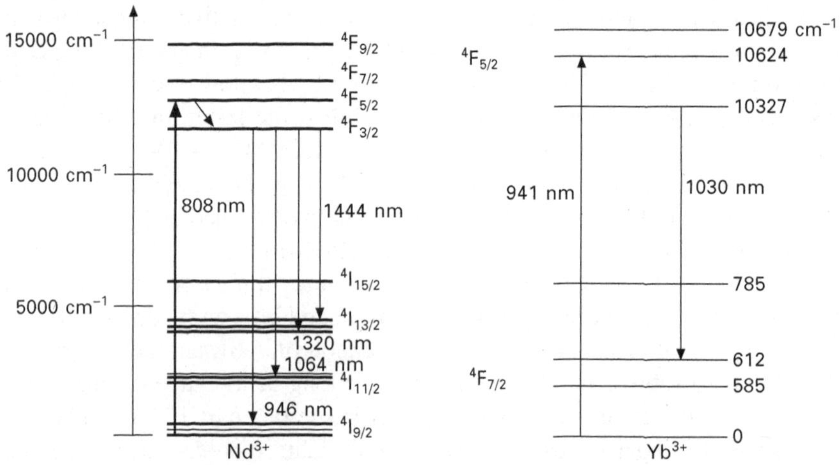

21.3 Energy level scheme and relevant optical transitions of Nd:YAG and Yb:YAG laser crystals.

1. Operating independent laser chains in parallel, and then combining the laser beams into a single coherent beam.
2. Serial mode operation – using a single cavity, but increasing the size of the active lasing centers or adding more crystals and pumping units inside the cavity, in order to upgrade the laser performance.

To summarize, the two approaches provide a way of scaling the laser performance: in the first one, parallel operation, scaling is achieved by adding chains. In the second one, serial operation, scaling is done by increasing the size and number of the active lasing elements.

21.5.1 Diode-pumped solid state crystalline lasers

A laser has been developed under Joint High Power Solid State Lasers (JHPSSL) by Northrop Grumman which produced 105 kW output power at ~2 times diffraction-limited beam quality. The ≈100 kW laser is a combined slab MOPA, where seven of 15 kW laser chain slabs ('building blocks') are phase locked. Each chain is actually a combination of a 200-W Yb-doped fiber amplifier (YDFA), and a series of four end-pumped, gain module slabs, with a total amplified output power of 15 kW. The 15-kW module can be operated as a compact stand-alone unit and is called Firestrike. The amplifying chains are arranged in parallel, and are driven by the same oscillator. Figure 21.4 presents a view of one of four such end-pumped gain module slabs of the amplifying chain. A maritime laser prototype with additional 'building blocks' aimed at addressing and defeating threats relevant to naval warfare is currently being developed.

© Woodhead Publishing Limited, 2013

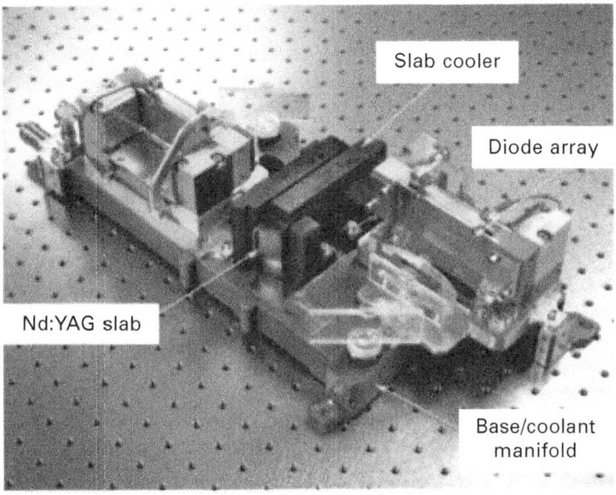

21.4 A gain module slab which is a part of the amplifying chain developed by Northrop-Grumman, with a total amplified output power of 15 kW per module (courtesy of Northrop-Grumman, available from http://www.laserfocusworld.com/display article/292398/12/none/none/Feat/Photonic-Frontiers:-laser-weapons-pumping, accessed July 2009).

Each amplified beam is wavefront-corrected, and all seven amplified beams are combined to form a single-aperture coherent beam. A two-chain laser produced 30-kW output power with 2.1 times diffraction limited beam quality, 20% optical efficiency, and several minutes of operation.

Another approach, that of serial operation, namely increasing the size of the lasing centers, has been developed by Textron and is called the 'ThinZag' configuration. It utilizes Nd:YAG ceramic slabs placed between pieces of quartz, and an index-matched coolant, tilted at such an angle that allows total internal reflection of the laser beam passing through the slabs. The 'ThinZag' laser that utilizes index-matched coolant is also called HELLADS (High Energy Liquid Laser Area Defense System), and its final goal is to develop an efficient, lightweight, compact, kilowatt-class solid state laser.

By 2007 this laser produced 15 kW, and by February 2010 100 kW by using two ceramic slabs in series. It is expected to yield 300–400 kW output power by 2018. The zigzag motion compensates for variations in the off-axial refractive index, significantly improving the laser beam quality. It also eliminates the need for AR coating which is very sensitive to optical damage at high peak power levels. The ceramic material has several unique advantages (Taira, 2007) and this allows for scaling of the output power by increasing the gain medium size instead of adding lasers, thus circumventing

© Woodhead Publishing Limited, 2013

challenges of homogeneous and distortion-free beam combiners. Ceramic crystals have added value in their capability of doping the active lasing ions (such as Nd^{3+}, Yb^{3+} and others) to high ion densities and uniformly. They also possess higher shock parameters and consequently a higher damage threshold compared with crystals (Taira, 2007).

21.5.2 Heat capacity lasers (HCL)

A solid-state pulsed laser is being developed at the Lawrence Livermore National Laboratory (LLNL). It is a large aperture (100 cm^2, 2 cm thick) Yb or Nd:YAG ceramic slab, unstable resonator, wavefront-corrected laser beam. This laser operates on the basis of serial mode operation as described above. It is therefore a robust and reliable solid-state laser, with simple laser architecture and simple maintenance.

The laser medium is not actively cooled during the pulsed burst operation, which means that thermal gradients in the laser rod are minimized by keeping thermal load *at a constant level* on the laser during operation (Lafortune *et al.*, 2004). After each burst the laser is turned off to cool. Therefore, thermally induced refraction index gradients as well as mechanical and thermo-optical stresses inside the crystal are significantly reduced (Hecht, 2007). In 2007 this laser produced 67 kW of average output power at 20% duty cycle, for a short time, and with 2–3 times diffraction limited beam quality and 10% total efficiency using five Nd:YAG ceramic slabs in series. The output energy was 335 J/pulse at a 200-Hz pulse repetition rate. An initial demonstration of material (steel) removal by applying 25-kW HCL with a beam size of 2.5 × 2.5 cm and a repetition rate of 200 Hz for 10 s of continuous laser operation showed a significant amount of heat removal and ablation. Interaction of this laser beam with aluminum sheet resulted in softening and metal rupture before melting.

Scaling of the output power to 400 kW or towards higher levels will be achieved by increasing both the size and the number of slabs and the duty cycle of the high power diode arrays. A cross-sectional view of one end of the diode-pumped Nd:YAG ceramics SSL developed at LLNL is presented in Fig. 21.5.

21.5.3 Disk laser

The disk laser is a face-cooled laser medium with a high ratio of surface to volume, and a multi-pass pump configuration via several HR parabolic mirrors. The crystal is Yb:YAG thin disk, 100–200 μm thick, pumped at 940 nm or 970 nm and lasing at 1030 nm. The axial heat flow coincides with the laser beam direction and therefore eliminates radial temperature gradients, refraction index gradients, and consequently thermal lensing

© Woodhead Publishing Limited, 2013

21.5 A cross-sectional view of the 10 × 10 cm^2 diode-pumped, heat capacity, Nd:YAG ceramic slab laser developed at LLNL (courtesy of Lawrence Livermore National Laboratory, available from http://www.laserfocusworld.com/display article/292398/12/none/none/Feat/Photonic-Frontiers:-laser-weapons-pumping, accessed July 2009).

effects. Scaling of a single-disk output power of approximately 5 kW can be achieved currently by increasing the number of pump modules or by using a multiple-disk configuration in series.

Boeing Directed Energy Systems have produced a thin-disk diode-pumped Yb:YAG laser operated at power levels of >25 kW at 70% efficiency, with suppressed radial ASE, with a nearly diffraction limited beam quality and operated for several seconds (Avizonis *et al.*, 2009). An increase in the output power to 6.5 kW was obtained by using a Yb:YAG ceramic thin-disk laser. The laser system is a series of Trumpf, Inc., commercial lasers used in the automotive industry – see Fig. 21.6(a) for more technological details (Deile *et al.*, 2009). A 16-kW commercial product of Trumpf (TruDisk disk laser) is presented in Fig. 21.6(b). This laser is scalable to 100 kW of output power, using the same technology. This is supported by predictions that 30 kW of output power can be extracted from a single disk, and therefore a target of 100 kW with good beam quality is feasible.

21.5.4 Fiber lasers

High power fiber lasers (HPFL), which are devices based on rare-earth-doped glassy fibers, are also potential candidates for laser weapons. The advantages of rare-earth-doped glasses are clear: high absorption and emission cross-sections owing to the asymmetric nature of the glass structure, high doping levels, hosts with low phonon energies and consequently, low non-radiative losses, and

© Woodhead Publishing Limited, 2013

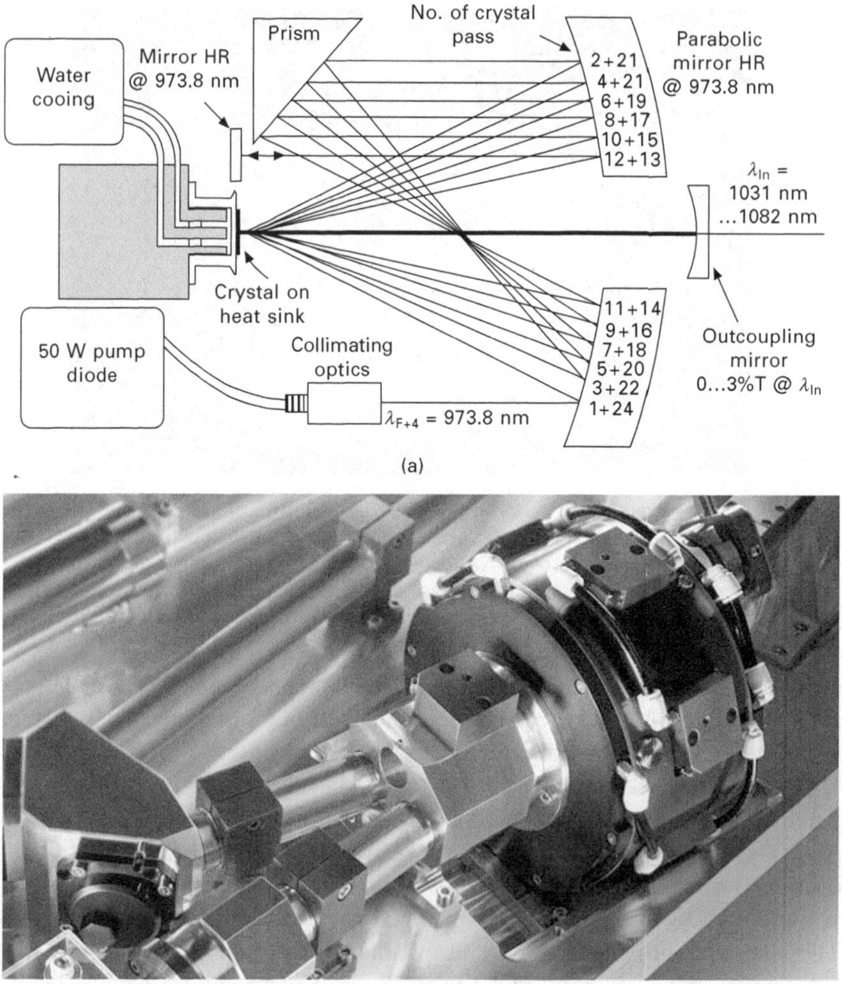

(a)

(b)

21.6 (a) A schematic layout of a commercial disk laser produced by Trumpf, Inc. (courtesy of Trumpf, Inc.); (b) a commercial 16-kW TruDisk disk laser produced by Trumpf, Inc. (available from http://www.trumpf-laser.com/en/products/solid-state-lasers/disk-lasers.html, accessed 20 July 2011).

large spectral bandwidth (inhomogeneous broadening). High power fiber lasers are currently used in industry for materials processing (cutting, welding) and consequently can be used for military applications (Kalisky, 2006).

Fiber lasers are efficient sources, with reduced thermally induced effects owing to their relatively large area per unit volume. The laser is a master oscillator power amplifier (MOPA) configuration. Companies like IPG (Shiner, 2009) have reported the operation of a 10-kW single-mode output power

© Woodhead Publishing Limited, 2013

Yb^{3+} doped silicate fiber laser at ≈1 μm, nearly diffraction limited beam ($M^2 < 1.1$), 500 μm^2 effective mode area, and 80–90% optical efficiency with inner core diameter of 30 μm, for a 10-kW unit. The use of chirally coupled-core fiber design with a side core that decouples higher modes is another option with potential technology towards military applications to obtain single mode high power fiber lasers (Carter, 2008).

The fiber laser is an efficient source, compact, modular, expandable into a wide range of wavelengths with reduced thermo-optical effects, and no thermal degradation. The reduction in the thermal load is achieved by pumping at 1018 nm, close to the 1070 nm peak emission wavelength, with quantum defect of 5%. This robust laser can deliver through a beam combiner up to ≈50 kW multimode output power through 100 or 200 μm fiber with about 4 times diffraction limited beam and wall-plug efficiency higher than 30%. Field propagation experiments by NRL demonstrated high power fiber laser performance in a kilometer-range by incoherent beam combining. About 3 kW could be delivered to a target at about 1.2 km by NRL, USA (Sprangle *et al.*, 2007, 2008, 2009). The Navy fiber laser prototype demonstrator is also called the Laser Weapon System (LaWS). It is a multi-mission laser, intended to be used against UAV, or alternatively for jamming electro-optical (EO) sensors, or EO guided missiles. In June 2009, a 33 kW fiber laser (5.5 kW/fiber laser × 6) with beam quality of $M^2 \approx 7–10$, successfully engaged several UAVs in combat-representative scenarios over the desert, and a year later over the ocean. In these field tests the fiber laser disabled small boat materials and jammed EO sensors (O'Rourke, 2011). The goal of the project is to obtain ~100 kW output power combined from multiple single-mode, linearly polarized fiber lasers by 2014, with potential to be a megawatt-class laser.

Disadvantages of fiber lasers such as optical nonlinearities (mainly above 5 kW), optical damage due to thermal load at ~200 W/m inside the fiber inner core, losses due to ground-state absorption of Yb^{3+} at 1070 nm, and dopant non-uniformities are still major issues to be considered in future upgrading of the fiber laser performance up to 30 kW single-mode output power.

Scaling of laser performance can be obtained by using various types of beam combiners such as spectral, coherent and incoherent beam-combining technologies (Loftus *et al.*, 2007; Sprangle *et al.*, 2008). The coherent beam combination requires lasers which are spectrally narrowband. In this context a 10-GHz, 1-kW Yb fiber laser has been demonstrated, and it is anticipated that coherent combining of a couple of such lasers will produce several hundred kilowatts in the future.

Finally, the shortfalls of both beam quality and laser efficiency can be reduced significantly by using cryogenically cooled high power diode-pumped Yb:YAG lasers. The intrinsic, well-known advantages of Yb:YAG suggest that Yb-doped crystals are excellent candidates for high power lasers, with

© Woodhead Publishing Limited, 2013

reduced thermal effects. Output power of 963 W at 91.9% slope efficiency and beam quality of $M^2 < 1.3$ has been reported recently (Brown and Tornegard, 2010). Table 21.1 summarizes the main characteristics in terms of laser weapons of solid-state and chemical lasers.

21.6 Alternative lasers

It is worth mentioning other laser candidates currently under development for high power military applications. One candidate is the Free Electron Laser (FEL) for the Navy, currently being developed under contract from the Office of Naval Research by Boeing and Raytheon. The FEL is an all-electric *tunable* source, spanning a large bandwidth spectral range. The Navy proposal is aimed at the development of a tunable 100-kW source and further to a megawatt-class laser, that can intercept incoming missiles in a range of 2 km. Depending on the environmental conditions, the laser can be tuned on a day-to-day basis (O'Rourke, 2011). FEL provides various wavelengths for various atmospheric transmissions and types of target. It is also possible to easily adjust the laser output power for disabling or destroying a target. This laser provides high resolution imaging capabilities, thus efficient use and time-saving procedures are expected (Francoeur, 2009).

Other efforts are under way to develop efficient (60%), low divergence and reliable pumping sources for solid-state lasers such as multi-kilowatt conductively cooled laser diode stacks that are compatible with solid-state laser technology, and ultrafast solid-state lasers (high peak power in the femtosecond time regime) to produce induced plasma, local heating and ultrasonic shock waves on the target – an effect known as dynamic pulse detonation (DPD). A mobile version of 'flash bang' technology using an HF

Table 21.1 A comparison between the main weapon characteristics of solid-state and chemical lasers

Solid-state lasers	Chemical lasers
Tactical	Strategic (except ATL)
Compact	Large footprint
Kilowatt-class	Megawatt-class
Deployable (air ground), portable	Airborne, ground-based
Reduced logistics	Logistics required
Inert components	Toxic, corrosive materials
Short range operation	Long range operation
Ultra-precision, reduced collateral damage	Aimed mainly at missiles
Limited by atmosphere, weather conditions, clouds	Operates at high altitude (DF) Suitable for low and high altitude (COIL)
Technology not yet mature	Mature technology

© Woodhead Publishing Limited, 2013

chemical laser is called Pulsed Energy Projectile (PEP). This is a non-lethal alternative in various scenarios and applications such as border control, protection of sensitive facilities, law enforcement operations, protection of security checkpoints, etc. The very high peak power is accompanied by a very intense electromagnetic pulse as well as filamentation effects in the air towards the target, with the ability to stun, disorientate or paralyze target enemies.

Mine standoff detection and clearing is another application of high power heat capacity lasers. The high peak power laser vaporizes the moisture in the soil, and consequently generates a series of micro-explosions, which displace the soil and allow further laser beam propagation through the soil. The high energy pulse enables the laser beam to penetrate and raise the temperature of the explosive material contained in the buried landmine up to its deflagration. The high peak power of a solid state laser is also utilized for a system termed the Airborne Laser Mine Detection System (ALMDS). This system has been designed for mine detection in the sea – a system that detects and locates floating or submerged mines. It is mounted on an MH-60 helicopter.

Last but not least is the space-borne laser (SBL) designed in the 1980s for extreme range targets, which is a part of the 'planetary defense system', based on high power HF chemical lasers. Further and more detailed information about high power lasers and their various applications (including military ones) is found in Injeyan and Goodno (2011).

There are other military applications of lasers which require low power or energy levels. This subject is beyond the scope of this review, which is focused on high power lasers. One type of low power/energy application is laser rangefinders, and laser designators, directed infrared countermeasures (DIRCAM) or laser detection and ranging (LADAR) (Zafrani *et al.*, 2010).

Another application of low power/energy lasers is standoff detection of explosives and chemicals by laser induced breakdown spectroscopy (LIBS). In this method a UV laser such as frequency–quadrupled Q-switched Nd:YAG (20 ns pulse width, with energies in the range 10–50 mJ, at 266 nm) or an excimer laser is used. The detection sensitivity is at levels of parts per billion. The system detects various substances including chemical and biological ones and explosives like TNT, RDX, TATP, DNT, etc., at distances exceeding 50 m (Lazic *et al.*, 2010).

21.7 Conclusions and future trends

Lasers are going to revolutionize the battlefield by providing *ultrafast* and *ultra-precision* capabilities, with reduced collateral damage. They are divided into two types: gas lasers (chemical lasers) and solid-state lasers (SSL). Lasers

© Woodhead Publishing Limited, 2013

are considered as both strategic weapons and tactical weapons, according to the laser type. Solid-state lasers are compact, deployable systems with reduced logistics that can be used as tactical weapons on the ground, on bombers or on strike fighters, mainly against short range rockets or missiles. They can also be used for ultra-precision missions with reduced collateral damage, both in the battlefield and in urban environments. It should be noted here that wavelengths in the eye-safe region (1.5–2 μm) are needed especially in urban areas or heavily populated regions. The operation of SSL is limited by atmospheric distortion, high laser beam reflectivity from the target, clouds and weather conditions. Scaling the output power requires optimizing laser materials, advanced cooling architectures, novel laser materials, efficient, low-divergence multi kilowatt laser diode systems and diode stacks (60% efficiency, 10-kW output power), reducing non-linearities in HPHL and effective coherent beam-combining optics.

Ground-based gas lasers such as the HF/DF chemical lasers are megawatt-class lasers. Chemical lasers operating around 100 kW are considered as tactical weapons. These lasers are efficient (30%) and technologically mature, with successful performance as both tactical and strategic weapons. Also, heat removal from the lasing medium is easier than for high power solid-state lasers. The airborne version of this laser (ABL) has the capability of intercepting and destroying missiles before the separation of warheads, at the boost phase, well above clouds. A proof of the concept was demonstrated in February 2010 by Boeing Defense Space and Security, which is the main contractor of the ABL Test bed (ABLT), where its airborne laser destroyed, for the first time, a ballistic missile in the boost phase. Although no other details were released, hitting and destroying an in-flight ballistic missile at the boost phase marks a significant success for the project. The main drawbacks of gas chemical lasers are their large size, their logistics, and the potentially toxic and corrosive gases involved in the lasing process.

Despite the limitations of each type, the laser weapon has unique characteristics which make it a weapon of the 'space age'.

21.8 References

Avizonis P V, Bossert D J, Curtin M S and Killi A (2009), 'Physics of high performance Yb:YAG thin disk lasers', in *The Conference on Lasers and Electro-Optics (CLEO)/ The International Quantum Electronics Conference (IQEC),* optical Society of America, Washington, DC, 2009, paper number CThA2.

Brown D and Tornegard S (2010), 'High-average-power cryogenic lasers for new applications', *Photonics Spectra*, 44(11), 36–39.

Carter A (2008), 'Increased output and efficiency of fiber lasers', *Photonics Spectra*, August, 60–63.

Deile J, Brockmann R and Havrilla D (2009), 'Current status and most recent developments of industrial high power disk lasers', in *The Conference on Lasers and Electro-Optics*

© Woodhead Publishing Limited, 2013

(CLEO)/The International Quantum Electronics Conference (IQEC), Optical Society of America, Washington, DC, 2009, paper number CThA4.

Forden G E (1998), 'The airborne laser (ABL), the American view of BPI', in *Ballistic Missiles: the Threat and the Response*, A Stav, ed., B Koroth, associate ed., 227–246, Tel Aviv, Yedioth Ahronoth Chemed Books.

Francoeur A (2009), 'Megawatt laser power at sea', *Photonics Spectra*, 43(11), 34–35.

Hecht J (2007)' 'Pumping up the power', *Laser Focus World*, 43, 72–76.

Injeyan H and Goodno G D (2011), *High Power Laser Handbook*, New York, McGraw-Hill.

Kalisky Y (2006), *The Physics and Engineering of Solid State Lasers*, Bellingham, WA, SPIE Press USA.

Kalisky Y (2009), 'Applications of high power lasers in the battlefield', in *Technologies for Optical Countermeasures VI*, D H Titterton and M A Richardson, eds, *Proc. SPIE* 7483, 748305-1–748305-6.

Kalisky Y and Kalisky O (2010), 'The status of high power lasers', *Optical Engineering*, Special Section Commemorating the Anniversary of the Laser, 49(9), 091003-1–09100-3-4.

Kalisky Y, Waichman K, Kamin S and Chuchem D (1997), 'Plasma cathode preionized atmospheric pressure HF chemical laser', *Opt. Commun.*, 137, 59–63.

Lafortune K N, Hurd R L, Joansson E M, Dane C B, Fochs S N and Brase J M (2004), 'Intracavity adaptive correction of a 10 kW, solid state, heat capacity laser', in *Laser Resonators and Beam Control VII*, A V Kudryashovand and A H Paxton, eds, *Proc. SPIE Meeting*, Bellingham, WA, 5333, 53–61.

Lazic V, Palluci A, Jovicevic S, Carapanese M, Poggi C and Buono E (2010), 'Detection of explosives at trace levels by laser induced breakdown spectroscopy (LIBS)', *Proc SPIE (Chemical, Biological, Radiological, Nuclear, and Explosives (CBRNE) Sensing XI)*, 7665, 76650V/1–76650V/9.

Loftus T H, Thomas A M, Hoffman P R, Norsen M, Royse R, Liu A and Honea E C (2007), Spectrally beam-combined fiber lasers for high-average-power applications', *IEEE J. Quantum Electron.,* 13(3), 487–497.

O'Rourke R (2011), 'Navy Shipboard lasers for surface, air and missile defense: background and issue for Congress', Congressional Research Service, R41526. Available from http://www.fas.org/sgp/crs/weapons/R41526.pdf (accessed 12 July 2011).

Shiner B (2009), 'Recent progress in high power fiber lasers', presented at *Laser Applications Workshop, ALAW 2009*, Plymouth MI, 12–14 May.

Sprangle P, Penano J, Ting A and Hafizi B (2007), 'Incoherent combining of high-power fiber lasers for long range directed energy applications', NRL Report, NRL/MR/6790-06-8963, June 2007.

Sprangle P, Ting A, Penano J, Fischer R and Hafizi B (2008), 'Beam combining: High-power fiber-laser beams are combined incoherently', available from http://www.laserfocusworld.com/displayarticle/331428, December 2008 (accessed 14 July 2011).

Sprangle P, Ting A, Penano J, Fischer R and Hafizi B (2009), 'High-power fiber-laser beams are combined incoherently', Available from: www.accessmylibrary.com/laser-focus-world/february-2009.html, February 2009 (accessed 17 July 2011).

Taira T (2007), 'RE^{3+}-ion-doped YAG ceramic lasers', *IEEE J. Quantum Electron.,* 13(3), 798–809.

Zafrani N, Sacks Z, Greenstein S, Peer I, Tal E, Luria E, Ravnitzki G, David D, Zajman A and Izhaky N (2010), 'Forty years of lasers at Elop-Elba systems', *Optical Engineering*, Special Section Commemorating the Anniversary of the Laser, 49(9), 091004-1–09100-3-12.

© Woodhead Publishing Limited, 2013

22
Environmental applications of solid-state lasers

A. CZITROVSZKY, Institute for Solid State Physics and Optics, Hungary

DOI: 10.1533/9780857097507.2.616

Abstract: This chapter is aimed at scientists and students but should also be useful to environmental experts dealing with measurement and monitoring of atmospheric pollution by dusts, fibres, fumes, mists, smokes, vapours, gases or biological agents. It summarizes the main airborne atmospheric pollutants, measurement principles and instrumentation based on optical methods using lasers (particle counters and sizers, aerosol analysers, gas monitors, LIDARs, etc.) and also important aspects of the identification and control of atmospheric contaminants. The possibilities of the application of different instruments are also presented.

Key words: remote sensing, LIDAR, aerosols, light scattering and absorption, atmospheric pollution.

22.1 Introduction

After the development of lasers by Charles Townes, Alexander Prochorov and Nikolay Basov (winners of the Nobel Prize in 1964), their application to optical measurement techniques developed very quickly. The high power and high spectral density, low divergence, high coherence and intensity of laser light give the possibility of performing different measurements based on light scattering, absorption, extinction spectroscopy, Doppler velocimetry, interferometry, etc. These measurement methods are now widespread and are involved in the study of the contamination of the atmosphere and industrial environments, health promotion, toxicology, pharmacology, environmental protection, colloid chemistry, and the study of aerosols and hydrosols.

The investigation of particulate and gaseous substances using lasers has been especially intensively developed in the last decades, with optical particle counters and sizers, aerosol analysers, PIV (particle imaging velocimeter) devices, micro-Raman spectrometers and powerful remote sensing methods – LIDAR (light detection and ranging) and LDV (laser doppler velocimeters).

The common feature and main benefit of these methods is the possibility of performing *non-contact*, *real-time*, mainly *in situ* measurements with a *short sampling time and high accuracy*. As the medium holding the information

© Woodhead Publishing Limited, 2013

on measured values is the light itself, in most cases the sampling can be done at a distance from the sample, which is beneficial especially in the case of atmospheric measurements. A further advantage of these optical methods is that of combining different optical measurement methods in the same instrument, thus being able to determine simultaneously a number of parameters, concentration and size distribution of aerosol particles, complex refractive index (real and imaginary parts), density and electrical charge, distance to the scattering media and velocity of their movement, etc. Knowing the specific absorption spectra of different gaseous components, we can determine the concentration of toxic gases, their spread and attenuation.

The aim of this chapter is to give a short overview of existing methods of optical measurements of atmospheric pollution by aerosols and gases using laser-based optical methods and to demonstrate their advantages. At the same time we will describe the measuring principle and development of the new instrumentation, emphasizing measurement principles and instruments developed and patented by us in recent years. Comparing the features and possibilities of different instruments, we present a short list of their possible applications. The operational principles and novelties are briefly described, as well as their advantages and limitations. Further on, we compare the technical parameters of several commercially available instruments with newly developed ones.

We hope that this chapter will help to give a better understanding of laser-based optical methods applied to the characterization of atmospheric pollution, its spread and development.

22.2 Classification of atmospheric contaminants

According to the classification of the National Occupational Health and Safety Commission [1], environmental atmospheric contaminants consist of dust, mist, fume, smoke, vapour gases and biological agents (Table 22.1).

- *Dust* contains airborne solid particles generated usually during grinding, crushing or chipping of hard materials or from the mechanical dispersion of fine powders.
- *Mist* particles are airborne droplets of substances that are normally liquid at ambient temperatures. Mist may form through condensation of vapour or through spraying of liquids.
- *Fume* contains airborne solid particles condensed from the vaporous state, usually through volatilization of molten metals.
- *Smoke* particles are generated from the incomplete combustion of fuel. Smoke usually contains gases and vapours in addition to solid particles.

These particulate contaminants form the aerosols which are in permanent

© Woodhead Publishing Limited, 2013

Table 22.1 The main types of environmental contaminants and their sources

Type	Examples	Industry/process
Dusts including fibres	Silica dust, coal dust, grain dust, asbestos fibres	Construction and mining industries, agriculture
Mists	Acid/alkali mists, chrome plating mist, pesticide mist	Metal pre-treatment, electroplating, aerial spraying
Fumes	Metal fumes, welding fumes	Smelters, welding, foundries
Smokes	Emission from coke ovens	Steelworks
Vapours	Paint solvent vapours, chlorinated hydrocarbons	Spray painting, solvent degreasing, dry-cleaning
Gases	SO_2, NO_x, CO, PAH, ozone, carbon monoxide, chlorine, hydrogen sulphide, radon	Combustion products, steelworks, caustic soda manufacture, sewage works
Biological agents	Pollen, bacteria, viruses, the Q-fever organism	Healthcare workers, meatworks, agriculture

motion and interconversion in the atmosphere (the average mass of these particles is ~1 kg above each square metre of continental area). Depending on the topological features, meteorological conditions and local emission/immission, the concentration and size distribution of these contaminants can vary in a wide range covering several orders of magnitude.

Gaseous contaminants consist of gases and vapours:

- *Gas* is a molecular dispersion of material in the air.
- *Vapour* is also a molecular dispersion of material in the air, but in this case the material is normally liquid at ambient temperature.

For the physical characterization of aerosols (measurement of their concentration, size distribution, shape, density, optical parameters, etc.) the most conventional and appropriate method is light scattering, and for definition of gaseous contaminants measurement of specific absorption spectra. The size ranges of the most relevant particulate environmental pollutants are shown in Fig. 22.1.

22.3 Light scattering as a powerful method for the measurement of atmospheric contamination by aerosols

The measurement of the contamination of the atmosphere by aerosols and gases is based on light scattering and specific spectrum-dependent light absorption (Table 22.2).

Light scattering can be described as an interaction of light with matter, which leads to a change of the direction, intensity, sometimes the polarization,

© Woodhead Publishing Limited, 2013

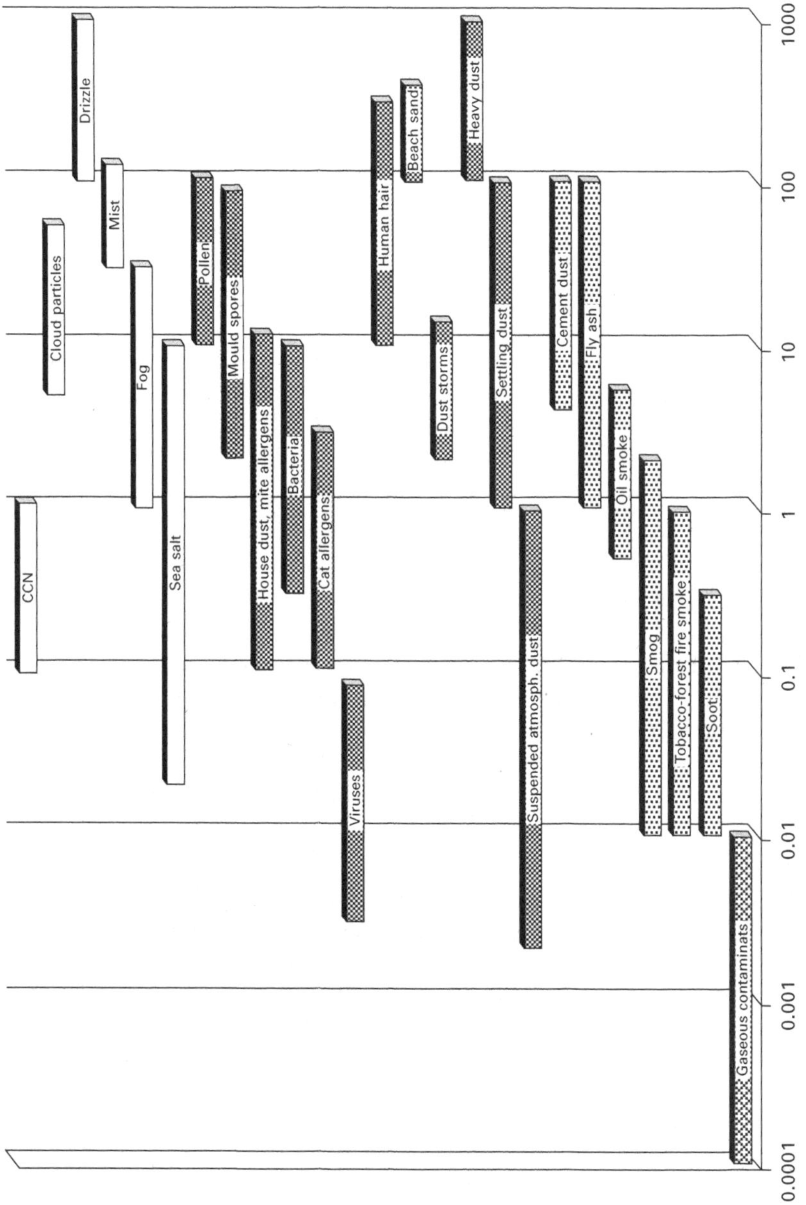

22.1 Typical size range of atmospheric pollutants (in micrometers).

© Woodhead Publishing Limited, 2013

Table 22.2 Optical methods for studying aerosols

Method	Measuring method	Measured parameters
Optical microscopy	Optical microscopes	Size, size distribution, shape, form factor, colour
Elastic light scattering	Optical particle counters and sizers, aerosol analysers	Concentration. size distribution, optical parameters (complex refractive index), density
Non-elastic light scattering	Spectroscopic instruments, micro-Raman spectrometers	Characterization of absorption spectra, identification of chemical composition
Doppler velocimetry	Laser Doppler velocimeters	Velocity, velocity distribution, size distribution using phase detection methods
Photon correlation spectroscopy and dynamic light scattering	Photon correlators, multiparticle light scattering instruments – LIDAR	Optical density, spatial distribution of particles, concentration

and in the case of non-elastic scattering the length, of the incident light wave. The incident light interacts with the optical heterogeneity centre of the medium (local changing of refractive index due to the particles) and is deflected from its incident direction, so scattering occurs as an effect of heterogeneity. Scattering is usually accompanied by absorption, which leads to the attenuation of the light energy propagating in the initial direction and transformation to some other form of energy. Different gases have different specific absorption spectra, so their concentration can be defined by comparison of the spectrum of the illuminating light and light transmitted through the absorbing media.

Many authors have studied theoretically and experimentally the phenomena of light scattering and absorption during interaction of light with various media. In this chapter we present two basic approaches: for single particles (spheres) much smaller than the wavelength of the incident light, and for particles comparable to or bigger than the wavelength of the incident light. As the relations between the scattering intensity and the properties of the particles in these two cases are different, the scattering coefficient β_p can be described as the sum [2]:

$$\beta_p = \beta_M + \beta_R \qquad [22.1]$$

where β_M corresponds to the particles comparable to or bigger than the wavelength of the incident light (Mie scattering) and β_R corresponds to the particles smaller than the wavelength of the incident light (Rayleigh scattering).

© Woodhead Publishing Limited, 2013

Both components depend on the optical and geometrical parameters of the particles, the scattering angle, the concentration and the wavelength of the incident light.

For small spherical particles (radius $a << \lambda$), when the incident light is linearly polarized and the concentration of the particles is not too high (multiple scattering can be neglected), in the case of Rayleigh scattering, the scattering coefficient will be [2]:

$$\beta_{\mathrm{R}} = \frac{8\ {}^4 a^6 N}{\lambda^4} \frac{n^2 - 1}{n^2 + 1} (1 + \cos^2 \Theta) \qquad [22.2]$$

where λ is the wavelength of the light, a is the radius of the particle, N is the concentration, n is the refractive index of the particle, and Θ is the scattering angle. As we can see, in this case the scattering intensity is proportional to the sixth power of the radius of the particle and inversely proportional to the fourth power of the wavelength. The calculations show that, depending on the wavelength, β_{R} in the range $0.6 < \lambda < 1.0$ micrometres varies by one order of magnitude and in the range $1.0 < 1 < 10$ micrometres by nearly four orders of magnitude.

When the size of the particles is neither very large nor very small compared to the wavelength of light, Mie theory must be used to obtain an accurate prediction of the scattering [3]. This theory represents an exact solution of Maxwell's equation for scattering spheres with complex refractive index. We present here only the main consideration and final result of this theory; the complete derivation can be found in the monography of Van de Hulst and Kerker [4–6]. In case of Mie scattering the scattered light is polarization dependent and we denote by $S_1(\theta)$ and $S_2(\theta)$ the normalized amplitudes of the flux scattered through angle Θ; the subscripts 1 and 2 refer to flux polarized normal to the scattering plane and parallel to it respectively. The normalization is made relative to the amplitude incident on the particle (sphere) cross-section. According to Mie theory,

$$\left.\begin{aligned} S_1(\theta) &= \sum_{m=1}^{\infty} \frac{2m+1}{m(m+1)} [a_m \pi_m(\cos\theta) + b_m \tau_m(\cos\theta)] \\ S_2(\theta) &= \sum_{m=1}^{\infty} \frac{2m+1}{m(m+1)} [b_m \pi_m(\cos\theta) + a_m \tau_m(\cos\theta)] \\ &\text{where} \\ \pi_m(\cos\theta) &= \frac{1}{\sin\theta} P_m^1(\cos\theta) \\ \tau_m(\cos\theta) &= \frac{\mathrm{d}}{\mathrm{d}\theta} P_m^1(\cos\theta) \end{aligned}\right\} \qquad [22.3]$$

P_m^1 is an associated Legendre polynomial, and the coefficients a_m and b_m are given by

© Woodhead Publishing Limited, 2013

$$\left.\begin{aligned}
a_m &= \frac{\psi'_m(\hat{n}x)\psi_m(x) - \hat{n}\psi_m(\hat{n}x)\psi'_m(x)}{\psi'_m(\hat{n}x)\xi_m(x) - \hat{n}\psi_m(\hat{n}x)\xi'_m(x)} \\
b_m &= \frac{\hat{n}\psi'_m(\hat{n}x)\psi_m(x) - \psi_m(\hat{n}x)\psi'_m(x)}{\hat{n}\psi'_m(\hat{n}x)\xi_m(x) - \psi_m(\hat{n}x)\xi'_m(x)} \\
&\text{where} \\
\psi_m(x) &= \sqrt{x/2}\,J_{m+1/2}(x) \\
\xi_m(x) &= \sqrt{x/2}\,[J_{m+1/2}(x) + (-1)^m i J_{-m-1/2}(x)] \\
x &= ka = \frac{2\ a}{\lambda}
\end{aligned}\right\} \qquad [22.4]$$

In these equations:

- a_m, b_m are coefficients characterizing the geometrical and optical parameters of the scattering particle,
- π_m, τ_m describe the angular dependence of the scattered light,
- ξ_m, ψ_m are Ricati–Bessel functions,
- J_m are Bessel functions,
- a is the radius of the sphere (particle),
- x is the size parameter – the ratio of the sphere's circumference to the wavelength.

As we can see, in this case the scattering coefficient consists of two components – one is connected with the geometrical and optical parameters of the particles, and the other describes the angular distribution of the scattered light.

In the case of a polydisperse mixture of particles the scattering on bigger particles (Mie scattering) is dominant. In this case the dependence on the wavelength is not as strong as in the case of smaller particles (Fig. 22.2).

As can be seen from the Mie scattering theory, the scattered intensity depends also on the complex refractive index. This dependence is especially important in optical measurement methods of aerosols like optical particle sizing, because usually we try to determine the size distribution of the mixture containing particles with different refractive indices.

In both previous theories we suppose that the particles are frozen in the medium where they are located. If the incident light is coherent and monochromatic (e.g. a laser beam), with an appropriate detector working in a photon counting regime, it is possible to observe time-dependent fluctuations in the scattered intensity (dynamic light scattering).

Particle size distributions can be calculated either assuming some standard form such as log-normal or without any such assumption. In the latter case, it becomes possible, within certain limitations, to characterize multimodal or skewed distributions. The size range for which dynamic light scattering is appropriate is typically submicron with some capability of dealing with

© Woodhead Publishing Limited, 2013

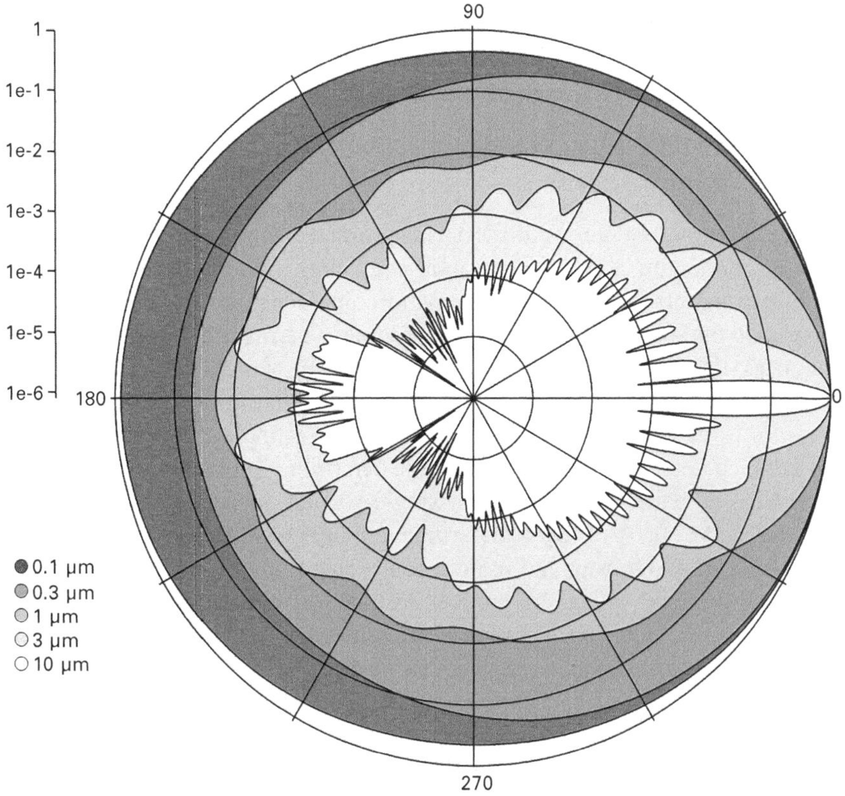

22.2 Angular distribution of the scattered intensity for different sizes calculated by us using Mie scattering theory. The curves are normalized to the maximal value for forward scattering. We note that the maximal scattered intensity for a 10 micron particle is more than six orders of magnitude larger than for a 0.1 micron particle.

particles up to a few microns in diameter. The lower limit of particle size depends on the scattering properties of the particles concerned (relative refractive index of particle and medium), incident light intensity (laser power and wavelength) and detector/optics configuration.

Dynamic light scattering (also known as quasi elastic light scattering, QELS, and photon correlation spectroscopy, PCS) is particularly suited to determining small changes in mean diameter such as those due to adsorbed layers on the particle surface or slight variations in manufacturing processes [7–9].

© Woodhead Publishing Limited, 2013

22.4 Instrumentation based on laser light scattering and absorption for the measurement of aerosols

22.4.1 Optical particle counters and sizers for environmental monitoring

Optical particle counters (OPC) or spectrometers employ a small sensing volume, either by a focused incandescent lamp or by a laser source. In such instruments it is important to avoid coincidence errors resulting from more than one particle in the sensing volume. The instrument manufacturer specifies the maximum number concentration which can be handled. Generally, commercial instruments handle concentrations of up to $\sim 10^6$ particles/litre. Beyond this concentration limit, sample dilution is usually used, which decreases the accuracy of the determination of the concentration. The range of particle diameter that single-particle instruments are capable of handling is ~0.1 μm. Optical particle counters have found wide use, first in cleanroom monitoring and more recently in community air pollution and industrial hygiene studies. A number of laboratory instruments employing single-particle scattering have been constructed. A critical review of such instruments is given by Chigier and Stewart [10].

Based on previous light scattering experiments [11–13], the development of airborne particle counters and sizers started in the middle of the twentieth century and accelerated in the 1960s after the invention of lasers.

The particle size distribution in single-particle counters is determined by comparing the detected pulse heights of the optical signals that correspond to the single particle flow through a small illuminated zone with a standard calibration curve, obtained from a set of uniform particles of known diameter. For determination of the concentration, the flow rate (the volume to be tested) is also measured simultaneously. The different light detection geometries applied in different particle counters yield a wide range of instrument designs and constructions. As the dependence of the scattered intensity on the refractive index of the particles is less pronounced around 90° of scattering angle, in most airborne particle counters perpendicular scattering geometry is implemented. A typical light scattering geometry is shown in Fig. 22.3. The aerosol to be tested passes through the illuminated zone and scatters the light of the laser beam in all directions. Part of this scattered light at a certain acceptance angle is collected by the optical system of the detector. The amplitude of the photoelectric signal generated on the detector by the particle crossing the illumination zone is compared with the standard calibration curve obtained from the set of known particles.

Calibration curves calculated by us using Mie theory for different scattering and integration angles are shown in Fig. 22.4. As we can see, especially in

© Woodhead Publishing Limited, 2013

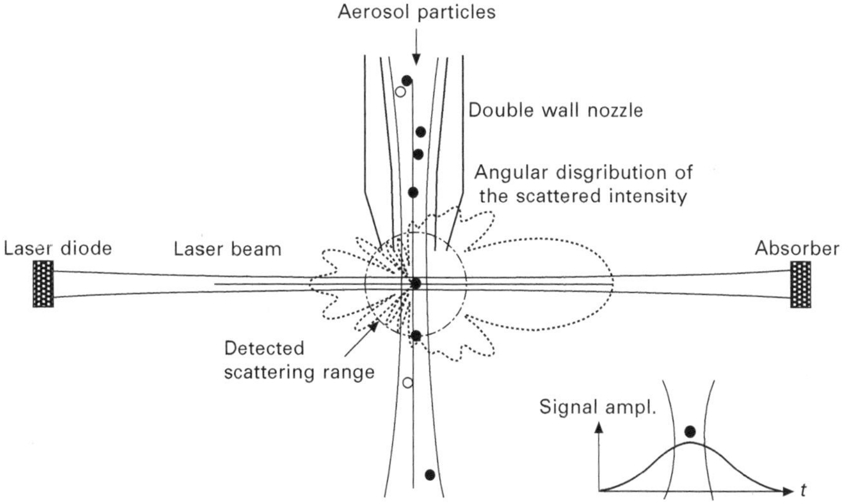

22.3 The principle of airborne particle counters. The particle crossing the illumination zone produces a photoelectric impulse, which is connected with the size of the particle to be measured.

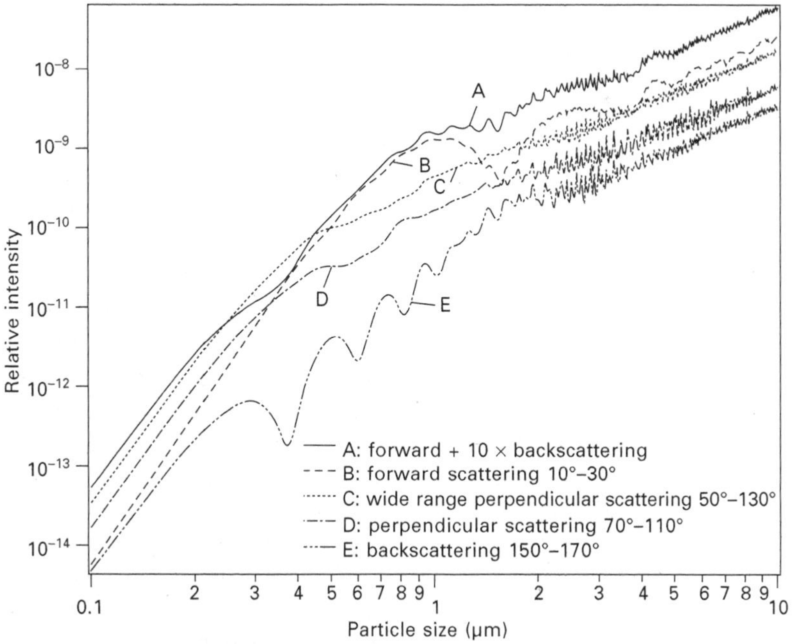

22.4 Calculated calibration curves for different scattering angles and integration ranges for polystyrene latex.

© Woodhead Publishing Limited, 2013

the case of forward and backward scattering, the relation between the size and the scattered intensity in certain size ranges is not uniform, the long-period oscillations for back scattering being pronounced in the 0.4–1 micrometre range and for forward scattering in the 1–3 micrometre range. In case of perpendicular scattering shorter oscillations are pronounced in the 2–5 micron range, but if we measure the size distribution, this geometry is even better. In this case the dependence on the refractive index of the particles is also less pronounced.

For further consideration we plotted also the calculations of the combination of forward and backward scattering, which we applied on our new aerosol analyser, described later. This device is capable of measuring not only the size distribution and concentration but also the complex refractive index of aerosol particles.

From the 1980s devices based on 90° scattering geometry were manufactured by Technoorg-Linda Ltd (Fig. 22.5). They were applied to cleanroom monitoring measurements [14–18], tests of various types of air filtering equipment (laminar boxes, climatic and air cleaning systems in operating theatres, hi-tech laboratories), checking the particle size distribution and concentration in the inhalation chambers for toxicological experiments (studying the toxicity of agricultural chemicals), monitoring the aerosols and air in the various stages of the pharmacological processing of medicines, and measuring the urban aerosols within the city of Budapest. Such devices were applied also to monitoring the air quality in some industrial works – the Forte Photochemical Works (Vac), the Chinoin Pharmaceutical Factory – (Budapest), the Parma Pharmaceutical Factory, the Viscosa Works (Nyergesújfalu) and in the operation theatres of several hospitals [19–27].

Previous studies of the measurement of particle size distribution in aerodynamic particle sizers [28–30] showed that devices operating by measuring the velocity of each individual particle in the incoming aerosol stream accelerated through an orifice nozzle underestimate the aerodynamic diameter by an average of 25% [31, 32], even in the case of regularly shaped non-spherical particles. The reason for such an underestimation is that these devices work under ultra-Stokesian flow conditions (within the measurement zone, particles having an aerodynamic diameter greater than 1 μm have a Reynolds number greater than 1 [33]). Similar effects are also obtained when measuring liquid droplets larger than a few microns, since liquid spheres are distorted when passing through the region of strong acceleration in the tapered nozzle [34, 35]. The measurement results depend on particle density even in the case of spherical particles [36, 37]. The same devices overestimate the size of particles that have a density appreciably greater than unity. As a result, several manufacturers have introduced an option in the operating software to correct for deviations in particle density [38]. An additional problem is that calibration curves obtained at reduced ambient pressure are

© Woodhead Publishing Limited, 2013

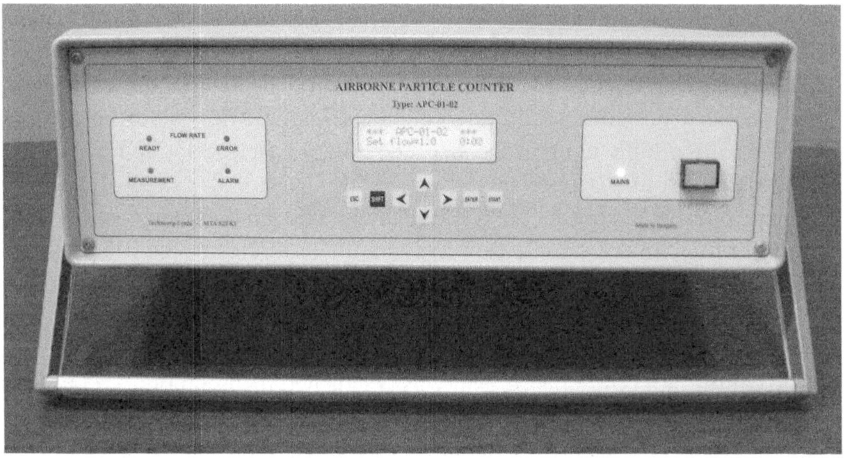

22.5 Airborne particle counters APC-03-2C and APC-01-02 produced by Technoorg Ltd (Hungary).

different from the manufacturer's data, indicating that recalibration of the device is required if other than standard operating conditions occur [39].

The density of the particles can be calculated from the motion equations in the case of accelerating particles. The electric charge can be measured by analysing the trajectory of the particles moving in the electric field. The refractive index can be determined from forward- and backward-scattering at different wavelengths. The velocity of the particles can be measured using laser Doppler methods. These methods are described in the following paragraphs.

In several particle counters (e.g. in the Dual Wavelength Particle Spectrometer – DWOPS), simultaneous measurement of some other parameters is also possible (in DWOPS, the complex refractive index) [40]. A special sampling and detection geometry providing the measurement of very high concentrations is used in particle counters of Topas GmbH. Several particle counters have some additional modular units to measure specific aerosols, e.g. the particle counters of Grimm have a modular unit for measurement of PAHs or nanoparticles, which can be attached to different Grimm instruments.

22.4.2 Aerosol analysers for the measurement of the complex refractive index of aerosol particles and identification of the contamination sources

Recently a new method was introduced for the real-time measurement of size and the complex refractive index of aerosol particles [40]. The method is based on a dual wavelength illumination system where the scattered light is

© Woodhead Publishing Limited, 2013

collected over four angular ranges, forward and backward scattering directions compared to the two illumination beams of Nd:YAG and frequency doubled Nd:YAG lasers (Fig. 22.6). The measured and digitized quartet is then compared to a pre-computed table calculated using Mie scattering theory. The rows of the table contain the size, the complex refractive index and the corresponding four scattered signals from the four angular ranges.

The resolution and the accuracy of the method depend on many parameters, e.g. instabilities in the sample flow system, uncertainties in the illumination and detection electronics, the applied analogue-to-digital converter and also the dimensions of the evaluation table. A comprehensive numerical and experimental study showed the feasibility of the method [41].

The calibration of the instrument was performed using laboratory particle standards (PSL). Initially, scale factors were determined for each detector to link the scattered intensities calculated using Mie theory with the detector responses.

To evaluate the performance of the method, particles were generated from a suspension using pneumatic atomization, and electrostatic classification (DMA) was used to obtain monodisperse distributions. These particles were then introduced to the instrument's measuring volume after aerodynamic focusing. The four signals from the four detectors were caught by a four-channel peak detection system which utilizes certain logic to minimize the effects of possible crosstalk between channels. The detected signals were

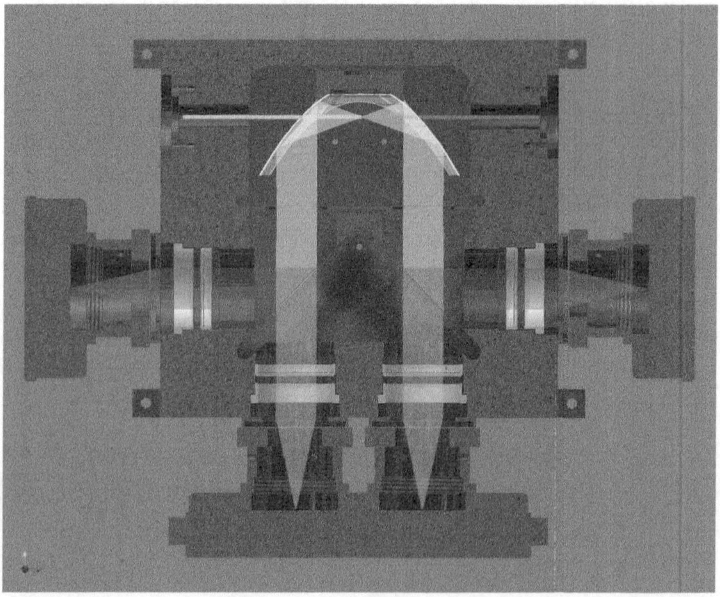

22.6 Scheme of the dual wavelength optical spectrometer (DWOPS) and its sampling chamber.

© Woodhead Publishing Limited, 2013

digitized and using the scale factors a search was performed in the evaluation table to obtain the particle size and complex refractive index.

PSL particles with different sizes from aqueous suspension, DEHS particles generated from an isopropyl alcohol–DEHS solution, paraffin oil and carbon-like absorbing particles generated from black ink diluted in pure distilled water were used for the test measurements. The Dual Wavelength Optical Particle Analyzer developed by us recently togeher with Technoorg-Linda Ltd is shown in Fig. 22.7.

22.4.3 New methods used to characterize the electrical charge and density of particles

After the development of the different airborne and liquid-borne particle counters the possibility of extending the laser method of aerosol sizing to obtain particle size and charge distribution was proposed [42, 43]. This possibility can be realized by using an electrical field perpendicular to both the aerosol jet and the laser beam. Deflection of charged particles in the electric field results in a change of intensities of the scattered light pulses because of the inhomogeneous transverse intensity distribution of the laser beam. Aerosol particle distribution with respect to the scattering amplitudes is measured at

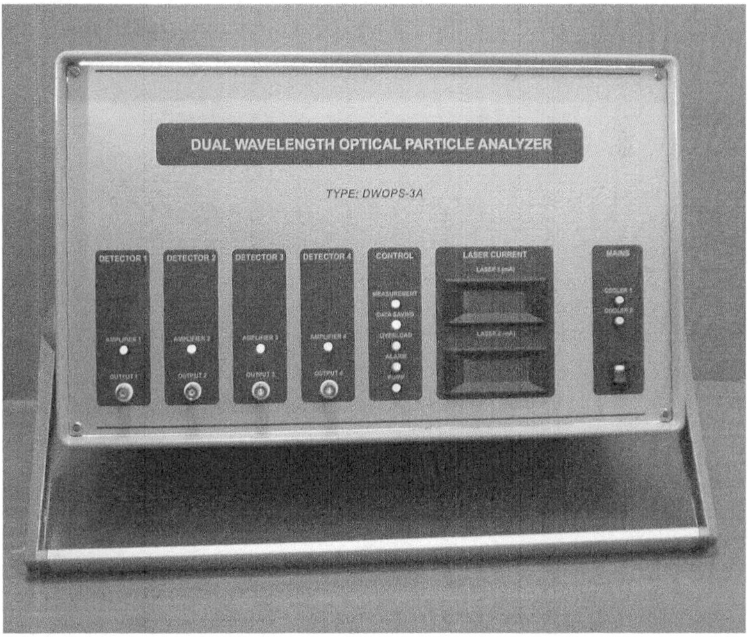

22.7 The Dual Wavelength Optical Analyzer developed by Technoorg Ltd (Hungary).

© Woodhead Publishing Limited, 2013

several values of the field strength. The set of distributions thus obtained can be transformed to a two-dimensional size and charge distribution. Suppose a focused air jet containing aerosol particles is crossing a laser beam. If a voltage is applied to the capacitor plates, the charged aerosol particles will be deflected from their original trajectory (Fig. 22.8).

The amplitude of the scattered light pulse detected in the experiment produced by particles crossing the laser beam depends on the value of the deflection. For simplicity we assume that in the absence of the electric field the aerosol jet crosses the laser beam axis. A particle deflected from its original trajectory by distance *L(Eqa)* gives rise to a light pulse with intensity depending on the electrical field strength inside the capacitor E, the radius of the particle a, and its charge q. Analysing aerosol samples at different values of E_i, $i = 1, 2, 3, \ldots, i_{\max}$, one can determine the number of particles giving rise to scattered light intensity Y at a fixed value of E_i. In other words, one can determine $NE_i(Y)$, the distribution function of the aerosol particles with respect to Y. If a certain range of values of DE_i from $E_i - \mathrm{d}E_i/2 + \mathrm{d}E_i/2$ can be put in correspondence to each dependence N_{E_i} (Y), then the function N_{E_i} (Y) can be transformed to the normalized two-dimensional distribution of aerosol particles with respect to E_i and Y_i. The problem is to derive aerosol particle charge and size distribution from the dependence $N^{(2)}$ (E_i, Y). This derivation can be done by using one-dimensional distribution functions [43]. The accuracy of calculation of the distribution functions is determined by the ratio of the width l of the air jet containing aerosols to the characteristic laser beam radius (ideally, one should use an air jet as thin as possible).

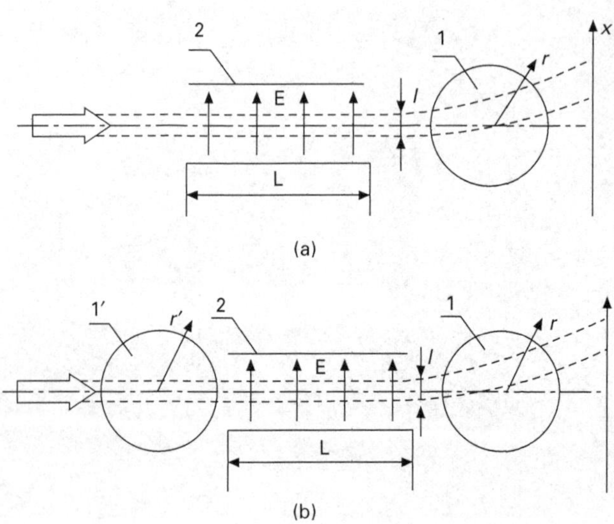

22.8 Scheme for the measurement of electrical charge on aerosol particles.

© Woodhead Publishing Limited, 2013

At l/a <0.2 the accuracy of the method proposed is better than that of the traditional ones, based on the light scattering amplitude [43]. However, this possibility can be realized only for stationary (during the time of measurement) states of the aerosol. One can estimate this drawback, provided the range of charge distribution is not too broad, so that it is sufficient to measure the distribution of the aerosol particles with respect to scattering amplitudes at a fixed value of the field. In this case it is possible to make two measurements simultaneously by illuminating the aerosol jet at two different points [43]. In the first laser spot, processing the scattered signal traditional for laser sizers is used to determine the particle size distribution. Thus, after passing through the electric field of the capacitor, particles arrive at the second laser spot, where the second measurement is made simultaneously, which allows determination of the charge distribution.

Thus, the use of an electric field to deflect the charged particles in the transverse direction with respect to the laser beam in the aerosol particle sizer allows one to determine the two-dimensional size and charge distribution function of small-sized aerosol particles. Preliminary experiments show the applicability of this method.

The determination of the density of the aerosol particles is based on the measurement of the time of flight of the particle between two illuminated spots in the acceleration region after aerodynamic focusing by an orifice nozzle (Fig. 22.9).

22.4.4 Overview of other optical instruments for the measurement of atmospheric pollution by aerosols

Optical particle measuring instruments may be further divided according to whether the sensing zone contains one or numerous particles at a given

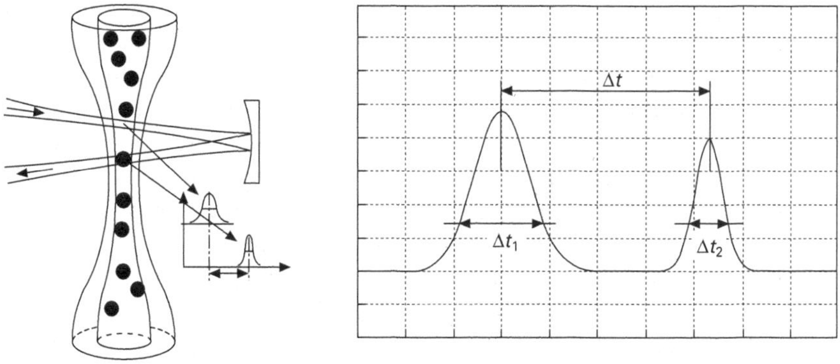

22.9 The principle of the estimation of the density of aerosol particles.

© Woodhead Publishing Limited, 2013

time. We will consider multiparticle instruments first, than single-particle light scattering direct-readout instruments.

Light-scattering photometers are multiparticle sensing zone instruments, in which light scattered from particles in the sensing zone falls on the detector off the optical axis. As the number of particles increases, the light reaching the detector increases. The angular pattern of scattering from a spherical particle is a complicated function of particle diameter, refractive index, and wavelength. Forward-scattering photometers, which employ a laser light source and optics similar to dark field microscopy, have been commercially produced. A narrow cone of light converges on the aerosol cloud but is prevented from falling directly on the photodetector by a dark stop, only light scattered in the near-forward direction falling on the detector. The readout of these instruments is in mass or number concentration, but the calibration may change with composition and size distribution of the particles to be measured [44].

In *integrating nephelometers* the particles are illuminated in a long sensing volume and scattered light reaches the detector at an angle from about 8° to 170° off axis. This simplifies the complex angular scattering relationship by summing the scattering over nearly the entire range of angles. In some cases, the scattering was shown to be well correlated with the atmospheric mass concentration. Some caution must be exercised when using nephelometers in an environment with sooty particles since the scattering will be attenuated because of light absorption. In this case the particle concentration will be lower than expected [44, 45].

A multiparticle, light-scattering instrument which employs a long path back-scattering light collection is *LIDAR* [46]. It is a powerful remote-sensing instrument that can measure the distance, and several optical parameters of the target. A powerful pulsed laser is used (in wavelengths ranging from about 10 micrometres to the UV), and the temporal analysis of back-scattered light indicates the spatial distribution of the target. Different types of scattering are used for different LIDAR applications, the most common being Rayleigh scattering, Mie scattering, Raman scattering and Fluorescence. Suitable combinations of wavelengths can allow for remote mapping of atmospheric contents by looking for wavelength-dependent changes in the intensity of the back-scattered signal. In LIDAR there are two kinds of detection methods – incoherent or direct energy detection (amplitude measurement) and coherent detection (phase sensitive or Doppler measurements). Coherent (Doppler) systems use optical heterodyne detection which, being more sensitive than direct detection, allows operation at a much lower power. In both coherent and incoherent LIDAR, there are two types of pulse models: *micropulse* systems and *high energy* systems. Micropulse systems need increasing amounts of computer power available combined with advances in laser technology and use less energy in the laser, typically on the order

© Woodhead Publishing Limited, 2013

of one microjoule. High-power systems are more applicable to atmospheric research, where they are widely used for measuring the height, layering and densities of clouds, cloud particle properties, extinction coefficient, backscatter coefficient, depolarization, temperature, pressure, wind, humidity, and trace gas concentration, e.g. of ozone, methane, nitrous oxide, etc. This type of instrument has been used to map smoke plume opacity in the vicinity of a stack. Unless the size distribution and composition of the particles are known, only a qualitative comparison of aerosol concentration at different locations can be made. The application of LIDARs is very wide, from environmental monitoring, agriculture, biology, meteorology and archeology through geology, forestry, transportation and astronomy up to military applications [47–53].

Another method which is similar to scattering is the *laser diffraction method*. This is a multiparticle method which can be used only for size distribution measurement of aerosol or hydrosol particles; the concentration of the particles cannot be determined [54]. The principle of the measurement is based on the observation of diffraction parameters of laser light on an ensemble of polydisperse particles.

Condensation nuclei counters or *condensation particle counters* (CPC) are used to measure the total number concentration of airborne particles much smaller than 0.5 μm which cannot be detected directly by light scattering. These instruments use the principle of adiabatic expansion or cooling in a vapour-saturated chamber. Vapour condenses upon nuclei and the particles grow to a detectable final diameter. They are then counted by light scattering. The instrument measures total nuclei number concentration since the final particle size is relatively independent of the number of nuclei present. Three types of instruments are currently in use. The first is a manual type in which a single expansion is performed in a water vapour-saturated chamber. In the second type, the expansions are performed cyclically, two per second, in a smaller water vapour-saturated chamber. A third type of CNC has been developed in which the particles are passed continuously through an alcohol chamber and are grown by cooling [55].

In the above instruments, the light interaction with a particle (or particles) is measured to obtain information directly. The particle motion is not important as long as it remains for an appropriate time in the sensing volume. Several instruments have been developed in which optical detection is used to infer particle motion.

In *aerodynamic particle counters* the aerodynamic diameter of a particle can be measured by determining particle velocity in an accelerating flow by measuring the time of flight of the particle between two spots in the acceleration region after aerodynamic focusing by an orifice nozzle. Commercially available instruments have been designed for aerosols in the range of 0.5–10 μm. The advantage of this method is that it is independent

© Woodhead Publishing Limited, 2013

of the optical properties of the particles as long as the scattered light can be detected [29–31, 33].

Particle counters and sizers were initially developed for cleanroom monitoring, but now their other branch of application is the field of high particle concentration, e.g. industrial air quality control, environmental monitoring, toxicology, aerosol testing, health control, etc. For these applications a series of instruments have been developed by Amherst, TSI, Topas, Grimm, Technoorg, etc.

Below we will describe some of the airborne particle counters developed by the authors.

Portable particle counters have an entire, fully functioning air monitoring system built into one compact unit. Extremely versatile, mobile and with a high airflow sample rate, they can be used cost-effectively for contamination control via cleanroom certification or monitoring. These counters generally have built-in printers, displays, alarms, and input–output ports such as serial, RJ-45 or USB for computer connectivity. Some portable particle counters also have other features such as for temperature, relative humidity, pressure differential and air velocity. Data from these sensors can be downloaded into a computer as well. Most portable particle counters from different manufacturers have been made to do the same job, but come in a variety of packages.

Remote particle counters are small particle counters that are used to monitor a fixed location, typically inside a cleanroom or mini-environment to continuously monitor particle levels 24 hours a day, seven days a week. These smaller counters typically do not have a local display and are connected to a network of other particle counters and other types of sensors to monitor the overall cleanroom performance. This network of sensors is typically connected to a facility monitoring system (FMS), a data acquisition system or a programmable logic controller. This computer-based system can be integrated into a database, notifying facility or process personnel by means of alarms (and perhaps also emails) when conditions inside the cleanroom exceed predetermined environmental limits. Remote particle counters are available in several different configurations, from single channel to models that detect up to eight channels simultaneously. Remote particle counters can have a particle size detection range from 0.1 to 100 micrometres and may feature one of a variety of output options including 4-20 mA, RS-485 Modbus, Ethernet, USB and pulse output.

Modified aerosol portable particle counters are attached to a sequencing sampling system. The sequencing sampling system allows for one particle counter to sample multiple locations, via a series of tubes drawing air from up to 32 locations inside a cleanroom. Although they are typically less expensive then utilizing remote particle counters, each tube is monitored in sequence.

© Woodhead Publishing Limited, 2013

Handheld particle counters are small self-contained particle counters that are easily transported and used. Although they have lower flow rates of 0.1 ft^3/min (0.2 m^3/h) than larger portables with 1 ft^3/m (2 m^3/h), handhelds are useful for most of the same applications. However, longer sample times may be required when doing cleanroom certification and testing. Most handheld particle counters have direct mount isokinetic sampling probes.

Detailed descriptions of the above instruments can be found in references 56–89. Further development of such instruments can be achieved by widening the size ranges and combining particle counting and sizing with other facilities, such as determination of the electrical charge, refractive index, density of the particles, etc.

22.5 Gas monitors based on optical measurement methods using lasers

Gas pollution monitoring of the atmosphere is based on the principle of different absorption of radiation by different gases at different wavelengths, e.g. for measurement of ozone and SO_2 the radiation between λ = 290 and 340 nm is convenient. The instruments presently used are double monochromators which compare UV intensity at a few fixed pairs of wavelengths. One wavelength is chosen to absorb ozone or SO_2 strongly, and the other of the pair somewhat less. The spectrophotometer is controlled by an external PC and provides practically continuous data without an observer present.

Other instruments based on cavity ring-down technology using *a pulsed quantum cascade laser spectrometer* are capable of field-deployable, real-time, ambient gas measurement of atmospheric levels of carbon dioxide, methane, hydrogen sulphide, ammonia and NO_X with parts-per-billion (ppbv) sensitivity and of water vapour with parts-per-million (ppmv) sensitivity (Table 22.3) [90–97]. The instruments capable of parts-per-billion sensitivity and fast 1 Hz or 10 Hz measurement rates may be used in even the most complex gas streams. The high accuracy and excellent precision of such commercially available monitors make them ideally suited to addressing the demanding requirements of many atmospheric air and emissions continuous monitoring applications. A high precision wavelength monitor makes certain that only the spectral absorption feature of interest is being monitored, greatly reducing the analyser's sensitivity to interfering gas species. As a result, the analyser maintains high linearity, precision and accuracy over changing environmental conditions with minimal calibration required.

The main manufacturers (Table 22.4), among many others, of atmospheric gas monitors are

- Picardo (http://www.picarro.com/gas_analyzers)
- Endee Engineers Pvt-Ltd (http://www.endee-engineers.com/products/portable-gas-monitors/model-400.php)

© Woodhead Publishing Limited, 2013

Table 22.3 Typical composition of the atmosphere in ppm (1 vol% = 10,000 ppm; for moist air 68% v.h. is assumed at 20°C)

Gas	Dry	Humid
Main gases		
N_2 – Nitrogen	780 840	768 543
O_2 – Oxygen	209 450	206 152
H_2O – Water vapour	0	15 748
Ar – Argon	9340	9193
CO_2 – Carbon dioxide	340	335
Trace gases		
Ne – Neon	18	18
He – Helium	5	5
CH_4 – Methane	1.8	1.8
Kr – Krypton	1.1	1.1
H_2 – Hydrogen	0.5	0.5
N_2O – Nitrous oxide	0.3	0.3
CO – Carbon monoxide	0.09	0.09
Xe – Xenon	0.09	0.09
O_3 – Ozone	0.07	0.07
Additional trace gases	3.05	3.0
Total	1000 000	1000 000

Table 22.4 List of different types of instruments to measure atmospheric gases

Manufacturer	Model	Type of gas to measure	Range/resolution
Endee Engineers Pvt. Ltd	Mod. 400	NO	2–2000 ppbv (2 ppmv)
Endee Engineers Pvt. Ltd	Mod. 205	O_3	1 ppbv to 100 ppmv
AFC International, Inc., Toxic gas detector with different sensors	Ammonia sensor	NH_3	0–100 ppm/1 ppm
	Carbon monoxide sensor	CO	0–500 ppm/1 ppm
	Chlorine sensor	Cl	0–50 ppm/0.1 ppm
	Nitric oxide sensor	NO	0–250 ppm/0.5 ppm
	Nitrogen dioxide sensor	NO_2	0–20 ppm/0.1 ppm
	Sulfur dioxide sensor	SO_2	0–20 ppm/0.1 ppm
Ferret	MultiRAE	VOC – volatile organic compounds	
Ferret	MeshGuard	CO, SO_2, H_2S, flammables	ppb sensitivity
Picarro	XRDS Analyzer	CO, CO_2, CH_4	ppb sensitivity

© Woodhead Publishing Limited, 2013

- AFC International, Inc. (http://www.afcintl.com/product/tabid/93/productid/172/sename/g460-multi-gas-detector-with-pid-from-gfg/default.aspx)
- Ferret (http://www.ferret.com.au/c/Active-Environmental-Solutions/Portable-Single-and-Multi-Gas-Detectors-from-Active-Environmental-Solutions-p16778).

22.6 Remote sensing using lasers and ground-based and airborne light detection and ranging (LIDAR)

In environmental science active remote sensing (where the detector measures the radiation that is illuminated and reflected or back-scattered from the target) is widely used for the study of vegetation rates, forestry, erosion, air and earth pollution, monitoring of natural environments, terrestrial surveillance, etc. The process of remote sensing is also helpful for city planning, archaeological investigations, military observation and geomorphological surveying.

LIDAR is an example of active remote sensing where the time delay between emission and detection, the attenuation at different wavelengths and spectra is measured, establishing the location, height, speed and composition of an object. Remote sensing makes it possible to collect data in dangerous or inaccessible areas, e.g. monitoring deforestation in areas such as the Amazon Basin, or glacial features in Arctic and Antarctic regions. Reliable predictions using laser scan data can be retrieved for different forest heights, for parameters related to forest density, such as stem number, stem diameter, biomass, timber volume or basal area, and for herbaceous vegetation height. It also replaces costly and slow data collection on the ground, ensuring in the process that areas or objects are not disturbed. Orbital platforms collect and transmit data from different parts of the spectrum, which in conjunction with larger-scale aerial or ground-based sensing and analysis, provides monitoring of El Niño and other natural long- and short-term phenomena. Other applications include different areas of the earth sciences such as natural resource management, agricultural fields such as land usage and conservation, and national security, ground-based and stand-off collection in border areas.

Monitoring of three-dimensional vegetation structure is essential for ecological studies, as well as for hydrodynamic modelling of rivers. Height and density of submerged vegetation and density of emergent vegetation are the key characteristics from which roughness parameters in hydraulic models are derived. Airborne laser scanning is a technique with broad applications in vegetation structure mapping, which therefore may be a promising tool in monitoring floodplain vegetation for river management applications. Based on laser scanning techniques, the extraction of vegetation height can

© Woodhead Publishing Limited, 2013

be determined. Laser data from meadows and unvegetated areas show too much noise to predict vegetation structure correctly. These data clearly show the large structural differences both between and within vegetation units that currently are the basis of floodplain vegetation and roughness mapping. The results indicate that airborne laser scanning is a promising technique for extracting the 3D structure of floodplain vegetation in winter, except for meadows and unvegetated areas.

Space and airborne sensors and LIDARs have been used to map burned areas, assess characteristics of active fires and characterize post-fire ecological effects. The fire severity burn severity and related terms can also be estimated by remote sensing practitioners who require unambiguous remote sensing products for fire management [98–102].

22.6.1 Spaceborne laser scanning systems

Spaceborne laser scanning systems are designed for surveying and monitoring natural environments, terrestrial surveillance and the study of geodynamic processes. In this system usually a two-colour picosecond-pulse laser ranging system is used operating at a pulse rate of about 20–40 Hz from about a 1500–2000 km circular orbit. In the case of high precision study of geological processes, for better resolution a ground-based passive array of retroreflector targets is used which are deployed in tectonically highly active zones. In appropriate atmospheric conditions a precision within ±1 cm can be achieved. Since the atmospheric correction of the laser ranging data requires the atmospheric surface pressure at each target site at the satellite's pass to be known with a precision to within a few millibars, the precise determination of the surface pressures will be accomplished by measuring the difference in the roundtrip transit time of laser pulses emitted at two widely separated wavelengths. It is planned to employ a frequency-doubled pulsed alexandrite laser source emitting a 380 nm and a 760 nm pulse pair. The projected ranging receiver will operate with the help of a circular-scan streak tube, which has a scan diameter of 30 mm, swept at a scan rate of 300 MHz, and is expected to provide a limiting temporal resolution of 6 ps [103, 104].

22.6.2 The laser vegetation imaging sensor (LVIS)

The laser vegetation imaging sensor (LVIS) is an airborne, scanning laser altimeter, designed and developed at NASA's Goddard Space Flight Center (GSFC). LVIS operates at altitudes up to 10 km above ground, and is capable of producing a data swath up to 1000 m wide nominally with 25-m wide footprints. The entire time history of the outgoing and return pulses is digitized, allowing unambiguous determination of range and return pulse

© Woodhead Publishing Limited, 2013

structure. Combined with aircraft position and attitude knowledge, this instrument produces topographic maps with decimetre accuracy and vertical height and structure measurements of vegetation. The laser transmitter is a diode-pumped Nd:YAG oscillator producing 1064 nm, 10 ns, 5 mJ pulses at repetition rates up to 500 Hz. LVIS has recently demonstrated its ability to determine topography (including sub-canopy) and vegetation height and structure on flight missions to various forested regions in the US and Central America. The LVIS system is the airborne simulator for the Vegetation Canopy Lidar (VCL) mission, providing simulated data sets and a platform for instrument proof-of-concept studies. The topography maps and return waveforms produced by LVIS provide Earth scientists with a unique data set allowing studies of topography, hydrology and vegetation with unmatched accuracy and coverage [100, 105].

22.6.3 Differential absorption LIDAR (DIAL) to measure tropospheric water vapour (H_2O) profiles from airborne and spaceborne platforms

To measure tropospheric water vapour profiles a pulsed laser beam is sent out into the atmosphere and a small proportion of the light is back-scattered by particles along the beam path to a sensitive detector. In this sense dust particles and aerosols are being used as reflectors. The laser light is in short pulses and the time resolution of the back-scattered light gives range resolution as in a simple LIDAR. For concentration measurement the DIAL system relies on a differential return from two closely spaced wavelengths, only one of which is absorbed strongly by the target gas. The size of the differential return signal at different distances along the laser beam path indicates concentration. The concentrations can be converted into mass emissions by making a series of scans with the DIAL along different lines within a plume and combining these with meteorological data. These measurements are then used to produce a mass emission profile for a whole site. Such a system is convenient for measurement of water vapour profiles from high altitude aircraft. The analysis shows that a $<10\%$ H_2O profile measurement accuracy is possible with a vertical and horizontal resolution of 200 m and 10 km, respectively, at night and 300 m and 20 km during the day. Global measurements of H_2O profiles from spaceborne DIAL systems can be made to a similar accuracy with a vertical resolution of 500 m and a horizontal resolution of 100 km [106, 107].

22.7 Conclusion

Different kinds of contaminations of the atmosphere affect the quality of our life in many different ways. First of all, they influence the Earth's

© Woodhead Publishing Limited, 2013

radiation balance directly by scattering and absorbing solar radiation, and indirectly by acting as cloud condensation nuclei. The interaction between atmospheric aerosols and climate systems is the dominant uncertainty in predicting radiative forcing and the future climate. Secondly, aerosol particles and gases adversely affect human health in urban areas. The interactions between air quality and climate are largely unknown, although some links have been identified. Thirdly, aerosol particles and several gases modify the intensity and distribution of radiation that reaches the Earth's surface, having direct influences on the terrestrial carbon sink. Better understanding and quantifying of the above effects in the atmosphere requires detailed information on the main contamination parameters, definition of different sources and study of their changes.

Based on light scattering, absorption, fluorescence and spectroscopy, a number of different measurement methods have been developed which, accelerated by the application of lasers, widen the possibilities of simultaneous measurement of the main parameters of atmospheric contamination. The newly developed instruments described in this chapter, with competitive technical specifications, make possible the determination of atmospheric contamination, to study the kinetic stability, aggregation and coagulation kinetics of airborne particles, the dynamics of aerosols, the spreading of gases, etc. They have been successfully applied in environmental air pollution monitoring, in medicine, in toxicology, in pharmacology, in chemistry, in filter testing, in industrial environment monitoring, in nuclear air ingress experiments, etc.

Besides demonstrating the principles of measurement and the capability of the described instruments, the newly obtained results are also pointing the way to possible future applications.

Worldwide, about 30 large companies are involved in laser light scattering and the above-mentioned applications. Most of these companies are in the USA.

22.8 References

1. National Occupational Health and Safety Commission, Atmospheric Contaminants, December 1989, Australian Government Publishing Service Canberra, WAP 89/031 GS 008-1989.
2. L. Levi, *Applied Optics*, Wiley, New York, 1968.
3. G.A. Mie, A contribution to the optics of turbid media, *Ann. Physik*, vol. 25, no. 4, pp. 377–445, 1908.
4. H.C. Van de Hulst, *Light Scattering by Small Particles*, Wiley, New York, 1957.
5. M. Kerker, *The Scattering of Light and Other Electromagnetic Radiation*, Academic Press, New York, 1969.
6. M. Kerker *et al.*, Color effects in the scattering of white light by micron ard submicron spheres, *J. Opt. Soc. Am.*, vol. 56, pp. 1248–1258, 1966.

© Woodhead Publishing Limited, 2013

7. W.G. Driscoll, W. Vaughan (eds), *Handbook of Optics*, McGraw-Hill, New York, 1978.
8. C.F. Boren, D.R. Huffman, *Absorbing and Scattering of Light by Small Particles*, Wiley, New York, 1983.
9. B.E. Dahneke (ed.), *Measurement of Suspended Particles by Quasi-elastic Light Scattering*, Wiley, New York, 1983.
10. N. Chigier, G. Stewart, Particle sizing and spray analysis, *Opt. Eng.*, vol. 23, p. 554, 1984.
11. V.P. Zaitsev, S.V. Krivokhizha, I.L. Fabelinski, A. Czitrovszky, L.L. Chaikov, P. Jani, Correlation radius near the critical points of guoiacol–glycerin solution, *JETP Lett.*, vol. 43, pp. 112–116, 1986.
12. S.V. Krivokhizha, O. Lugovaia, I.L. Fabelinski, L.L. Chaikov, A. Czitrovszky, P. Jani, Temperature dependence of the correlation radius of the fluctuation of the concentration of guoiacol–glycerin solution, *JETP*, vol. 103, pp. 115–124, 1993.
13. L.L. Chaikov, I.L. Fabelinski, S.V. Krivikhizha, O. Lugovaia, A. Czitrovszky, P. Jani, Light scattering spectrum in the region of the upper, lower and double critical point in guoiacol–glycerin solution, *J. Raman Spectrosc.*, vol. 25, no. 7–8, pp. 463–468, 1994.
14. *Recommended particle for testing clean rooms*, Institute of Environmental Science, Chicago, IES-CC-006-84T, 1984.
15. A. Czitrovszky, P. Jani, New design for a light scattering airborne particle counter and its application, *Opt. Eng.*, vol. 32, no. 10, pp. 2557–2562, 1993.
16. A. Czitrovszky, P. Jani, Design and application of light scattering airborne particle counter developed in KFKI, *J. Aerosol Sci.*, vol. 24s, pp. 227–228, 1993.
17. A. Czitrovszky, P. Jani, Performance of the new type of general purpose airborne particle counter, *SPIE*, vol. 1983, pp. 998–999, 1993.
18. A. Czitrovszky, P. Jani, P.P. Poluektov, G. Yu. Kolomeitsev, V.G. Yefankin, A new method for laser analysis of aerosol particles to determine the distribution of electric charge, size, density and concentration, *SPIE*, vol. 1983, pp. 969–970, 1993.
19. A. Czitrovszky, P. Jani, Efficiency of sampling by the APC-03-2 single particle counter, *J. Aerosol Sci.*, vol. 25s, pp. 465–466, 1994.
20. A. Czitrovszky, J. Frecska, L. Matus, P. Jani, Investigation of aerosol released from heated LWR fuel rod and its properties, *Proc. of IV Int. Aerosol Conf.*, Los Angeles, CA, pp. 784–785, 1994.
21. A. Czitrovszky, P. Jani, Application examples of APC-03-2 and APC-03-2A Airborne Particle Counters in a highly contaminated environment, *J. Aerosol Sci.*, vol. 26s, pp. 793–794, 1995.
22. A. Czitrovszky, J. Frecska, P. Jani, L. Matus, A. Nagy, Size distribution of aerosols released fom heated LWR fuel rods, *J. Aerosol Sci.*, vol. 27s, pp. 467–468, 1996.
23. A. Czitrovszky, P.L. Csonka, P. Jani, Á. Ringelhann, J. Bobvos, Experimental investigation of altitude dependence of size distribution and concentration of dust particles within the city of Budapest, *J. Aerosol Sci.*, vol. 27s, pp. 117–118, 1996.
24. A. Czitrovszky, P.L. Csonka, P. Jani, Á. Ringelhann, J. Bobvos, Comparison of different measurement methods of airborne dust pollution within the city of Budapest, *J. Aerosol Sci.*, vol. 27s, pp. 19–20, 1996.
25. P. Jani, A. Nagy, A. Czitrovszky, Aerosol particle size determination using a photon

© Woodhead Publishing Limited, 2013

correlation laser Doppler anemometer, *J. Aerosol Sci.*, vol. 27s, pp. 531–532, 1996.
26. A. Czitrovszky, P.L. Csonka, P. Jani, Á. Ringelhann, J. Bobvos, Measurement of airborne dust pollution within the city of Budapest, *15th Annual Conference of the American Association for Aerosol Research – 96*, Orlando, FL, p. 210, 1996.
27. A. Czitrovszky, P.L. Csonka, P. Jani, Á. Ringelhann, J. Bobvos, Complex measurement of airborne dust pollution within the city of Budapest, *SIP*, vol. 2965, pp. 103–112, 1996.
28. R.J. Remiraz, J.K. Agarval, F.R. Quant, G.J. Sem, Real time aerodynamic particle size analyzer, in *Aerosols in the Mining and Industrial Work Environments, Vol. 3, Instrumentation*, V.A. Marple and Y.H. Liu (eds), pp. 879–895, Ann Arbor Science Publishers, Collingwood, MI, 1983.
29. P.A. Baron, Sampler evaluation with aerodynamic particle sizer, in *Aerosols in the Mining and Industial Work Environments, Vol. 3, Instrumentation*, V.A. Marple and Y.H. Liu (eds), pp. 861–877, Ann Arbor Science Publishers, Collingwood, MI, 1983.
30. W.D. Griffiths, P.J. Iles, N.P. Vaughan, Calibration of the APS 33 aerodynamic particle sizer and its usage, *TSI J. Particle Instrum.*, vol. 1, no. 1, pp. 3–9, 1986.
31. I.A. Marshal, J.P. Mitchell, W.D. Griffiths, The behavior of regular-shaped non-spherical particles in a TSI aerodynamic particle sizer, *J. Aerosol Sci.*, vol. 22(I), 73–89.
32. I.A. Marshal, J.P. Mitchell, The behavior of non-spherical particles in a Malvern aerosizer, API, Progress report, March–August; AEA Technology Report AEA RS 5167, 1991.
33. P.A. Baron, Calibration and use of the aerodynamic particle sizer APS 3300, *Aerosol Sci. Technol.*, vol. 5, pp. 55–67, 1986.
34. I.A. Marshal, J.P. Mitchell, The behavior of spheroidal particles in the time-of-flight aerodynamic particle sizer, *J. Aerosol Sci.*, vol. 23, suppl. L, pp. S227–S300, 1992.
35. W.D. Griffiths, P.J. Iles, N.-P. Vaughan, The behavior of liquid droplet aerosol in an APC 3300, *J. Aerosol Sci.*, vol. 17, no. 6, pp. 921–930, 1986.
36. H.C. Wang, W. John, Particle density correction for aerodynamic particle sizer, *Aerosol Sci. Technol.*, vol. 6, pp. 191–198, 1987.
37. G. Ananth, J.C. Wilson, Theoretical analysis of the performance of the TSI aerodynamic particle sizer, *Aerosol Sci. Technol.*, vol. 9, pp. 189–199, 1988.
38. H.C. Wang, W. John, A simple iteration procedure to correct for the density effect in aerodynamic particle sizer, *Aerosol Sci. Technol.*, vol. 10, pp. 501–505, 1989.
39. B.T. Chen, Y.S. Cheng, H.C. Yeh, Performance of a TSI aerodynamic particle sizer, *Aerosol Sci. Technol.*, vol. 4, pp. 89–97, 1985.
40. W.W. Szymanski, A. Nagy, A. Czitrovszky, P. Jani, A new method for the simultaneous measurement of aerosol particle size, complex refractive index and particle density, *Measurement Sci. Technol.*, vol. 13, pp. 303–307, 2002.
41. A. Nagy, W.W. Szymanski, A. Golczewski, P. Gál, A. Czitrovszky, Numerical and experimental study of the performance of the Dual Wavelength Optical Particle Spectrometer (DWOPS), *J. Aerosol Sci.*, vol. 38, pp. 467–478, 2007.
42. A. Czitrovszky, A. Nagy, P. Jani, Development of a new particle counter for simultaneous measurement of the size distribution, concentration and estimation of the shape-factor of liquid-borne particles, *SPIE*, vol. 3749 pp. 574–575, 1999.

© Woodhead Publishing Limited, 2013

43. V.A. Kashcheev, P.P. Poluektov, A.N. Semikin, A. Czitrovszky, P. Jani, Measurement of aerosol particle charge and size distribution by a laser particle analyser, *Laser Phys.*, vol. 2, no. 4, pp. 613–616, 1992.
44. A.P. Waggoner, R.J. Charlson, Measurement of aerosol optical parameters, in *Fine Particles*, B.Y.H. Liu (ed.), Academic Press, New York, 1976.
45. P. Lillenfeld, Current mine dust monitoring instrument development, in *Aerosols in the Mining and Industrial Work Environments*, V.A. Marple and Y.H. Liu (eds), pp. 879–895, Ann Arbor Science Publishers, Collingwood, MI, 1982.
46. C.S. Cook *et al.*, Remote measurement of smokeplume transmittance using LIDAR, *Appl. Opt.*, 11, no. 8, pp. 1742–1748, 1972.
47. A.P. Cracknell, L. Hayes, *Introduction to Remote Sensing*, 2nd edn, Taylor and Francis, London, 2007.
48. S. Koukoulas, G.A. Blackburn, Mapping individual tree location, height and species in broadleaved deciduous forest using airborne LIDAR and multi-spectral remotely sensed data, *Int. J. Remote Sensing*, vol. 26, no. 3, pp. 431–455.
49. N.C. Coops, M.A. Wulder *et al.*, Comparison of forest attributes extracted from fine spatial resolution multispectral and LIDAR data, *Can. J. Remote Sensing*, vol. 30, no. 6, pp. 855–866.
50. S. Hese, W. Lucht *et al.*, Global biomass mapping for an improved understanding of the CO_2 balance – the Earth observation mission Carbon-3D, *Remote Sensing of Environment*, vol. 94, no. 1, pp. 94–104.
51. E. Naesset, T. Gobakken *et al.*, Laser scanning of forest resources: the Nordic experience, *Scand. J. Forest Res.* vol. 19, no. 6, pp. 482–499.
52. R. Nelson, A. Short *et al.*, Measuring biomass and carbon in Delaware using an airborne profiling LIDAR, *Scand. J. Forest Res.*, vol. 19, no. 6, pp. 500–511.
53. T. Eid, T. Gobakken, E. Naesset, Comparing stand inventories for large areas based on photo-interpretation and laser scanning by means of cost-plus-loss analyses, *Scand. J. Forest Res.*, vol. 19, no. 6, pp. 512–523.
54. H.G. Barth (ed.), *Modern Methods of Particle Size Analysis*, Wiley, New York, pp. 137–157, 1984.
55. J. Bricard *et al.*, Detection of ultra fine particles by means of a continuous flux condensation nuclei counter, in *Fine particles*, B.Y.H. Liu (ed.), Academic Press, New York, 1976.
56. W.W. Szymanski, A. Majerowicz, Direct optical aerosol concentration detemination in multiply scattering media, *J. Aerosol Sci.*, vol. 23, pp. S353–S356, 1992.
57. API Mach 2 Aerosizer, Instruction manual, Amherst Process Instruments Inc., Mountain Farms Technology Park, Hadley, MA 01035-9547.
58. Par-Tec 200/300, Instruction manual, Lasentec LSI, Inc., 523 North Belt East, Suite 100, Houston, TX 77066.
59. TSI Mod. 3775 Laser Particle Counter, Instruction manual, TSI Inc., 500 Cardigan Road, PO Box 64394, St Paul, MN 55164.
60. Malvern Mod 3600 Particle Sizer, Instruction manual, Malvern Instruments, Inc., 187 Oaks Road, Framingham, MA 01701.
61. Hiac-Royco Mod. 5100 Aerosol Particle Counter, Instruction manual, Hiac-Royco, Inc., 141 Jefferson Drive, Melno Park, CA 94025.
62. Climet Mod. CI 6400 Airborne Particle Counter, Instruction manual, Climet Instruments Co., PO Box 151, Redlands, CA 92373.
63. PMS Mod. LPC-525 Laser Particle Counter, Particle Measuring System, Inc., 1855 South 57th Court, Boulder, CO 80301.

© Woodhead Publishing Limited, 2013

64. APC-032A Airborne Particle Counter, Instruction manual, Technoorg-Linda Ltd, Budapest 1077, Rozsa u. 24.
65. LQB-1-200 Liquidborne Particle Counter, Instruction manual, Technoorg-Linda Ltd, Budapest 1077, Rozsa u. 24.
66. Topas, SeriesLAP Instruction manual, Topas GmbH, Dresden D-01277, Hofmannstrasse 37.
67. Dantec, Instruction manual, Measurement Technology Division, Skovlunde, Denmark.
68. W.W. Szymanski, A. Nagy, A. Czitrovszky, P. Jani, A new method for the simultaneous measurement of aerosol particle size, complex refractive index and particle density, *Measurement Sci. Technol.*, vol. 13, pp. 303–307, 2002.
69. A. Nagy, W.W. Szymanski, A. Golczewski, P. Gál, A. Czitrovszky, Numerical and experimental study of the performance of the Dual Wavelength Optical Particle Spectrometer (DWOPS), *J. Aerosol Sci.*, vol. 38, pp. 467–478, 2007.
70. A. Czitrovszky, A. Nagy, P. Jani, Development of a new particle counter for simultaneous measurement of the size distribution, concentration and estimation of the shape-factor of liquid-borne particles, *SPIE*, vol. 3749, pp. 574–575, 1999.
71. P. Jani, A. Nagy, A. Czitrovszky, Nano-particle size distribution measurement in photon correlation experiments, *SPIE*, vol. 3749, pp. 458–459, 1999.
72. P. Jani, A. Czitrovszky, A. Nagy, R. Hummel, Investigation of size distribution and concentration of aerosols released at CODEX AIT experiments, *J. Aerosol Sci.*, vol. 30s, pp. 101–102, 1999.
73. A. Czitrovszky, P. Jani, A. Nagy, C. Schindler, W.W. Szymanski, An approach to a simultaneous assessment of aerosol particle size and refractive index by multiple angle detection, *J. Aerosol Sci.*, vol. 31, pp. 761–762, 2000.
74. P. Jani, A. Nagy, Z. Lipp, A. Czitrovszky, Velosizer – photon correlation LDA system, *J. Aerosol Sci.*, vol. 31, pp. 390–391, 2000.
75. A. Nagy, W.W. Szymanski, A. Czitrovszky, P. Jani, Modeling of a new optical aerosol particle analyzer for the simultaneous measurement of size, complex refractive index and density, *J. Aerosol Sci.*, vol. 32, pp. 83–84, 2001.
76. P. Jani, Z. Lipp, A. Nagy, A. Czitrovszky, Propagation delay statistics of scattered intensities, *J. Aerosol Sci.*, vol. 32, pp. 87–88, 2001.
77. P. Jani, M. Koniorczyk, A. Nagy, A. Czitrovszky, Probability distribution of scattered intensities, *J. Aerosol Sci.*, vol. 32, pp. 563–564, 2001.
78. P. Jani, A. Nagy, Z. Lipp, A. Czitrovszky, Simultaneous velocity and size measurement of nanoparticles in photon correlation experiments: *SPIE*, vol. 4416, pp. 236–240, 2001.
79. W.W. Szymanski, A. Nagy, A. Czitrovszky, P. Jani, A new method for the simultaneous measurement of aerosol particle size, complex refractive index and particle density, *Measurement Sci. Technol.*, vol. 13, pp. 303–307, 2002.
80. P. Jani, M. Koniorczyk, A. Nagy, Z. Lipp, B. Bartal, A. László, A. Czitrovszky, Probability distribution of scattered intensities, *J. Aerosol Sci.*, vol. 33, no. 5, pp. 697–704, 2002.
81. A. Czitrovszky, A. Nagy, P. Jani, Environmental monitoring of the atmospheric pollution by aerosols, *J. Aerosol Sci.*, vol. 34, pp. S953–954, 2003.
82. P. Jani, A. Nagy, A. Czitrovszky, Field experiments for the measurement of time of flight statistics of scattered intensities on ensemble of aerosol particles, *J. Aerosol Sci.*, vol. 34, pp. S1209–1210, 2003.
83. L. Wind, L. Hofer, A. Vrtala, A. Nagy, W.W. Szymanski, Optical effects caused by

© Woodhead Publishing Limited, 2013

non-soluble inclusions in microdroplets and implications for aerosol measurement, *J. Aerosol Sci.*, vol. 34, pp. S1245–1246, 2003.
84. A. Golczewski, A. Nagy, P. Gál, A. Czitrovszky, W.W. Szymanski, Performance modelling and response of the Dual-Wavelength Optical Particle Spectrometer (DWOPS), *J. Aerosol Sci.*, vol. 35, pp. S839–840, 2004.
85. L. Wind, L. Hofer, A. Nagy, P. Winkler, A. Vrtala, W.W. Szymanski, Light scattering from droplets with inclusions and the impact on optical measurement of aerosols, *J. Aerosol Sci.*, vol. 35, pp. 1173–1188, 2004.
86. Á. Farkas, I. Balasházy, A. Czitrovszky, A. Nagy, Simulation of therapeutic and radioaerosol deposition, *J. Aerosol Med.*, vol. 18, no. 1, pp. 102–104, 2005.
87. A. Czitrovszky, Gy. Farkas, G. Bánó, A. Nagy, D. Oszetzky, P. Jani, P. Gál, Z. Donkó, Á. Kiss, K. Rózsa, M. Koós, P. Varga, L. Csillag, Development of lasers and their application in KFKI, *Hungarian Science (Magyar Tudomány)*, no. 9, pp. 1499–1510, 2005.
88. Z. Hózer, L. Maróti, P. Windberg, L. Matus, I. Nagy, Gy. Gyenes, M. Horváth, A. Pintér, M. Balaskó, A. Czitrovszky, P. Jani, A. Nagy, O. Prokopiev, B. Tóth, The behavior of VVER fuel rods tested under severe accident conditions in the CODEX facility, *Nucl. Technol.*, vol. 154, no. 3, pp. 302–317, 2006.
89. A. Czitrovszky, Application of optical methods for micron and sub-micron particle measurements, Chapter 7 in *Aerosols – Science and Technology*, I. Agranovski, (ed.), pp. 203–241, Wiley, New York, 2010.
90. A. Nagy, W.W. Szymanski, A. Golczewski, P. Gál, A. Czitrovszky, Numerical and experimental study of the performance of the Dual Wavelength Optical Particle Spectrometer (DWOPS), *J. Aerosol Sci.*, vol. 38, no. 4, pp. 467–478, 2007.
91. W.W. Szymanski, A. Nagy, A. Czitrovszky, Optical particle spectrometry problems and perspectives, *Journal of Quantitative Spectroscopy and Radiative Transfer*, vol. 110, no. 11, pp. 918–929, 2009.
92. B.H. Lee, E.C. Wood, M.S. Zahniser, J.B. McManus, D.D. Nelson, S.C. Herndon, G.W. Santoni, S.C. Wofsy, J.W. Munger, Simultaneous measurements of atmospheric HONO and NO_2 via absorption spectroscopy using tunable mid-infrared continuous-wave quantum cascade lasers, *Appl. Phys. B.*, vol. 102, pp. 417–423, 2011.
93. J.B. McManus, M.S. Zahniser, D.D. Nelson Jr., J.H. Shorter, S. Herndon, E. Wood, F. Wehr, Application of quantum cascade lasers to high-precision atmospheric trace gas measurements, *Opt. Eng.*, vol. 49, paper 111124, 2010.
94. K. Stimler, D. Nelson, D. Yakir, High precision measurements of atmospheric concentrations and plant exchange rates of carbonyl sulfide using mid-IR quantum cascade laser, *Glob. Change Biol.*, vol. 16, pp. 2496–2503, 2010.
95. S.C. Herndon, M.S. Zahniser, D.D. Nelson Jr., J. Shorter, J.B. McManus, R. Jiménez, C. Warneke, J.A. DeGouw, Airborne measurements of HCHO and HCOOH during the New England Air Quality Study 2004 using a pulsed quantum cascade laser spectrometer, *J. Geophys. Res.*, vol. 112, paper D10S03, 2007.
96. R. Jimenez, S. Herndon, J.H. Shorter, D.D. Nelson, J.B. McManus, M.S. Zahniser, Atmospheric trace gas measurements using a dual quantum-cascade laser mid-infrared absorption spectrometer, *Proc. SPIE*, vol. 5738, pp. 318–330; *Novel In-Plane Semiconductor Lasers IV*, C. Mermelstein, D.P. Bour (eds), 2005.
97. J. Wormhoudt, J.H. Shorter, C.C. Cook J.J. Zayhowski, Diode-pumped 214.8-nm Nd:YAG/Cr^{4+}:YAG microchip laser system for the detection of NO, *Appl. Opt.*, vol. 39, pp. 4418–4424, 2000.
98. J.B. Blair, D.B. Coyle, J.L. Bufton, D.J. Harding, Optimisation of an airborne laser

© Woodhead Publishing Limited, 2013

altimeter for remote sensing of vegetation and tree canopies, *Proc. Int. Geosci. Remote Sens. Symp.*, vol. II, pp. 938–941, 1994.

99. J.L. Bufton, J.B. Garvin, J.F. Cavanaugh, L. Ramos-Izquierdo, T.D. Clem, W.B. Krabill, Airborne lidar for profiling of surface topography, *Opt. Eng.*, vol. 30, pp. 72–78, 1991.
100. R. Dubayah, J.B. Blair, J.L. Bufton, D.B. Clark, J. JáJá, R. Knox, S.B. Luthcke, S. Prince, J. Weishampel, The vegetation canopy lidar mission, *Proc. Conf. Land Satellite Information in the Next Decade II*, American Society for Photogrammetry and Remote Sensing, Bethesda, M.D. pp. 100–112, 1997.
101. J.B. Garvin, J. Bufton, J.B. Blair, D. Harding, S. Luthcke, J. Frawley, D. Rowlands, Observations of the Earth's topography from the Shuttle Laser Altimeter (SLA): laser-pulse echo-recovery measurements of terrestrial surfaces, *Phys. Chem. Earth*, vol. 23, no. 9–10, pp. 1053–1068, 1998.
102. W.B. Krabill, R. Thomas, K. Jezek, K. Kuivinen, S. Manizade, Greenland ice sheet thickness changes measured by laser altimetry, *Geophys. Res. Lett.*, vol. 22, no. 17, pp. 2341–2344, 1995.
103. D.E. Smith, M.T. Zuber, H.V. Frey, J.B. Garvin, J.W. Head, D.O. Muhleman, G.H. Pettengill, R.J. Phillips, S.C. Solomon, H.J. Zwally, W.B. Banerdt, T.C. Duxbury, Topography of the northern hemisphere of Mars from the Mars Orbiter Laser Altimeter, *Science*, vol. 279, no. 5357, pp. 1686–1692, 1998.
104. C.W. Wright, R.N. Swift, Application of new GPS aircraft control/display system to topographic mapping of the Greenland ice cap, *Proc. 2nd Int. Airborne Remote Sens.* Conf., San Francisco, II, pp. 210–212, 1996.
105. J.B. Blair, M.A. Hofton, Modeling laser altimeter return waveforms over complex vegetation using high-resolution elevation data, *Geophys. Res. Lett.*, vol. 26 pp. 2509–2512, 1999.
106. S. Ismail, E.V. Browell, Airborne and spaceborne lidar measurements of water vapor profiles, *Appl. Opt.*, vol. 28, no. 17, pp. 3603–3615, 1989.
107. H.P. Lutz, W. Krause, G. Barthel, International Astronautical Federation, *Proc. 33rd International Astronautical Congress*, Paris, 27 September to 2 October, p. 27, 1982.

© Woodhead Publishing Limited, 2013

Index

© Woodhead Publishing Limited, 2013

© Woodhead Publishing Limited, 2013

© Woodhead Publishing Limited, 2013

© Woodhead Publishing Limited, 2013

© Woodhead Publishing Limited, 2013

© Woodhead Publishing Limited, 2013

© Woodhead Publishing Limited, 2013

© Woodhead Publishing Limited, 2013

© Woodhead Publishing Limited, 2013

© Woodhead Publishing Limited, 2013

© Woodhead Publishing Limited, 2013

© Woodhead Publishing Limited, 2013

© Woodhead Publishing Limited, 2013

© Woodhead Publishing Limited, 2013